Metallic, Magnetic and Carbon-Based Nanomaterials

THE ELECTROCHEMICAL SOCIETY SERIES

The Electrochemical Society
Advancing solid state & electrochemical science & technology

ECS - The Electrochemical Society
65 South Main Street
Pennington, NJ 08534-2839
http://www.electrochem.org

A complete list of the titles in this series appears at the end of this volume.

Metallic, Magnetic and Carbon-Based Nanomaterials

Synthesis and Biomedical Applications

Ajit Khosla
School of Advanced Materials and Nanotechnology, Xidian University, Xian
Shaanxi, China

Irshad A. Wani
Department of Chemistry, Govt. Degree College Anantnag, Anantnag, University of Kashmir
Jammu and Kashmir, India

Mohammad N. Lone
Department of Chemistry, Central University of Kashmir, Ganderbal
Jammu and Kashmir, India

WILEY

Published by John Wiley & Sons, Inc., Hoboken, New Jersey.
Published simultaneously in Canada.

Library of Congress Cataloging-in-Publication Data Applied for:

Hardback ISBN: 9781119870654

Cover Design: Wiley
Cover Image: © ARTUR PLAWGO/SCIENCE PHOTO LIBRARY/Getty Images

Set in 9.5/12.5pt STIXTwoText by Straive, Chennai, India

Contents

About the Authors

Ajit Khosla, PhD, is a distinguished professor; he is a fellow of the Royal Society of Chemistry (UK); fellow of the Electrochemical Society (USA), School of Advanced Materials and Nanotechnology, Xidian University, China (https://amn.xidian.edu.cn/); founding editor-in-chief in ECS Sensors Plus; editor in *The Electrochemical Society* journals (JES and JSS). He is one of top 2% scientists listed by Elsevier and Stanford University.

Irshad A. Wani, PhD, is currently serving as an Assistant Professor in Chemistry at Postgraduate Department of Chemistry, Govt. Postgraduate Degree College Anantnag, University of Kashmir, India. He has earlier served as Lecturer in Chemistry from 2010–2017 in J&K School Education Department. Since 2017, Dr. Wani has been teaching Chemistry at Undergraduate & Postgraduate level in various college of J&K Higher Education Department. Dr. Wani has more than fourteen years of teaching experience. Dr. Irshad A. Wani did his masters in chemistry in the year 2007 & Ph.D. in Nanochemistry in the year 2012 from Jamia Millia Islamia, New Delhi, India. He has received various prestigious awards notably ICTSGS Services Award (2021) conferred by SPAST foundation & B N Kailoo Memorial ISCAS Medal (2009) conferred by ISCAS, India etc. He has a good track record of publications & has published research papers in various esteemed international journals. He is a co-editor of some edited books & has also authored several book chapters. Dr. Wani has served as a Topical Editor of S-Cube: Sustainable Solutions and Society (SPAST) Journal. He is serving as a peer reviewer of various international journals. Dr. Wani is also a life member of various prestigious scientific societies. His current research interests include Green Synthesis of Inorganic nanoparticles, Inorganic-Organic Hybrid Nanoparticles, Conjugation of nanoparticles with various significant molecules for biomedical, therapeutic & sensing applications.

Dr. Mohammad N. Lone, PhD, is working as an Assistant Professor of Chemistry in the School of Life Sciences, Central University of Kashmir. Dr. Nadeem received his M.Sc. in Chemistry from Jamia Millia Islamia (Central University) New Delhi in 2012. For his Ph.D. he worked with Prof. Imran Ali in the same university with focus on development of anticancer heterocycles and chiral drugs. During his Ph.D., Dr. Nadeem was awarded Basic Scientific Research Meritorious Juniour and Senior Research Fellowships by University Grants Commission, New Delhi. He has research interests in heterocyclic synthesis, nano formulations, metallodrugs and anticancer studies. Dr. Nadeem has authored and co-authored more than nineteen research publications (original research papers, reviews, book chapters, books and conference papers) of national and international repute.

Preface

Nanotechnology, situated at the convergence of physics, chemistry, biology, and engineering, represents an unparalleled domain in scientific exploration, offering unprecedented avenues for precise manipulation of matter at atomic and molecular scales. This intricate manipulation has led to the development of materials with unique properties, notably metallic, magnetic, and carbon-based nanomaterials, which have found myriad applications, particularly within biomedicine.

The book *Metallic, Magnetic, and Carbon-Based Nanomaterials: Synthesis and Biomedical Applications* serves as a comprehensive compendium of cutting-edge research and collaborative interdisciplinary efforts aimed at understanding the synthesis, characterization, and biomedical utility of these extraordinary materials. Authored by experts in the field, the book aims to present the forefront of nanomaterial science while offering practical insights into their deployment in biomedicine.

Chapter 1 serves as a foundational exploration of nanomaterial synthesis methods and characterization techniques. From the intricacies of top-down and bottom-up approaches to the synthesis and characterization of gold, silver, magnetic ferrite, and carbon-based nanoparticles, this chapter lays the groundwork for understanding the structural and physicochemical properties that underpin their biomedical functionalities.

In Chapter 2, delves into the multifaceted biomedical applications of gold and silver nanoparticles. These noble metal nanoparticles exhibit extraordinary optical and physicochemical properties that render them indispensable in bioimaging, biosensing, drug delivery, and antimicrobial therapies. Yet, a comprehensive understanding of their toxicity profiles is essential to mitigate potential adverse effects and ensure their safe integration into clinical practice.

Chapter 3 delves into the biomedical applications of magnetic nanoparticles, particularly ferrites. From their role as MRI contrast agents to enabling targeted drug delivery and magnetic hyperthermia, magnetic nanomaterials offer unprecedented opportunities for diagnostic and therapeutic innovation. However, unraveling the intricacies of their toxicity is paramount for their successful translation from the bench to the bedside.

Chapter 4 explores the biomedical applications of quantum dots, semiconductor nanocrystals endowed with tunable optical properties. These quantum wonders hold immense potential for bioimaging, cell tracking, pathogen detection, and tumor biology investigations. Nonetheless, comprehensive studies on their toxicity are imperative to ensure their safe and effective utilization in clinical diagnostics and therapeutics.

In Chapter 5, delves into the biomedical applications of carbon-based nanomaterials, including carbon nanotubes and graphene. With their exceptional mechanical, electrical, and biocompatible properties, carbon-based nanomaterials offer unprecedented opportunities for controlled drug delivery, medical nanorobotics, tissue engineering, and biosensing. However, rigorous assessment of their toxicity profiles is essential for their seamless integration into biomedical applications.

Chapter 6 explores the biomedical applications of silica-based and polymeric nanoparticles, elucidating their roles as versatile drug carriers, imaging agents, antibacterial treatments, and vaccine adjuvants. Understanding the intricacies of their toxicity profiles is paramount for harnessing their full potential in addressing pressing biomedical challenges and improving patient outcomes.

Finally, Chapter 7 paints a visionary picture of nanotechnology-based treatments, devices, and applications across various medical domains, including cardiovascular health, diabetes management, wound healing, antibacterial treatments, and cancer therapy. Nanotechnology offers a transformative paradigm for personalized medicine and precision healthcare, ushering in an era of unprecedented therapeutic efficacy and patient-centric care.

In conclusion, *Metallic, Magnetic, and Carbon-Based Nanomaterials: Synthesis and Biomedical Applications* stands as a testament to the collective ingenuity and interdisciplinary collaboration driving the field of nanotechnology forward. May this book inspire future generations of researchers, innovators, and clinicians to continue pushing the boundaries of scientific discovery and biomedical advancement.

January 2025

Ajit Khosla, Ph.D
School of Advanced Materials and Nanotechnology
Xidian University, Xian, Shaanxi, China

Irshad A. Wani, Ph.D
Department of Chemistry, Govt. Degree College Anantnag,
Anantnag, University of Kashmir, Jammu and Kashmir, India

Mohammad N. Lone, Ph.D
Department of Chemistry, Central University of Kashmir,
Ganderbal, Jammu and Kashmir, India

1

Nanomaterials

Synthesis Methods and Characterization Techniques

Ajit Khosla[1], Irshad A. Wani[2], and Mohammad N. Lone[3]

[1] *School of Advanced Materials and Nanotechnology, Xidian University, Xian, Shaanxi, China*
[2] *Department of Chemistry, Govt. Degree College Anantnag, Anantnag, University of Kashmir, Jammu and Kashmir, India*
[3] *Department of Chemistry, Central University of Kashmir, Ganderbal, Jammu and Kashmir, India*

1.1 Nanoscience and Nanotechnology: An Outline of Terms and Concepts

The definition of nanotechnology remains a central concern in the scientific community, prompting questions about its precise scope and boundaries. Nanotechnology generally pertains to the study and manipulation of materials and phenomena at the nanoscale, typically below 100 nm, where size-dependent quantum mechanical effects become prominent [1]. While this range captures the essence of nanotechnology and its unique properties, some experts caution against strictly adhering to the 100 nm boundary, as certain devices and materials in the pharmaceutical domain may be excluded [2].

An additional crucial aspect in defining nanotechnology is to consider whether the materials involved are natural, synthetic, or manufactured [3]. Naturally occurring biomolecules in living cells often exist at the nanoscale, which could redefine disciplines like biochemistry and molecular biology as integral parts of nanotechnology.

The following definition of nanotechnology is provided by the U.S. National Nanotechnology Initiative (NNI) [3]: "The understanding & control of matter at dimensions between approximately 1 & 100 nanometers, where unique phenomena enable novel nanotechnology applications." Other noteworthy definitions of nanotechnology encompass its broad scope and applications [2]. For example, it can be described as "The design, characterization, production, and utilization of structures, devices, and systems attained through deliberate manipulation of size and shape at the nanometer scale (atomic, molecular, and macromolecular scale) that demonstrate at least one unique or enhanced characteristic or property." Nanotechnology can also be viewed as "Developing objects and surfaces with unique functionalities stemming directly from nanoscale dimensions and/or arrangement." These unique properties may manifest in mechanical, electrical, or photochemical attributes not observed in bulk materials.

The term "nanotechnology" originates from the ancient Greek word "νᾶνος" and the Latin word "nanus," both meaning "dwarf" or "very small." Within the International System of Units (SI), "nano" represents a reduction factor of 10–9 times, indicating manipulation of matter at a scale one billionth of a meter (10–9 m) [4].

Nanotechnology represents more of an engineering approach than a standalone science, drawing extensively from fields such as biology, physics, chemistry, and materials science [5]. Its potential

Metallic, Magnetic and Carbon-Based Nanomaterials: Synthesis and Biomedical Applications, First Edition.
Ajit Khosla, Irshad A. Wani, and Mohammad N. Lone.

impact on these sciences is expected to be transformative. The term "nanotechnology" was initially coined by Eric Drexler in his publication "Engines of Creation" (1986). It refers to the manipulation of individual atoms and molecules to create structures characterized by exact atomic specifications [6]. It is observed that the concept of nanotechnology has its roots in physicist Richard Feynman's seminal lecture from 1959, titled "There's Plenty of Room at the Bottom" [7, 8]. In 1974, Norio Taniguichi of Tokyo Science University coined the term "nanotechnology" in reference to semiconductor processes involving control on the order of nanometers, which continues to serve as the foundational statement [9]: "Nanotechnology primarily involves the manipulation of materials through processes such as separation, consolidation, and deformation at the level of individual atoms or molecules." Nanoscience encompasses the examination of phenomena and the control of materials at atomic, molecular, and macromolecular levels, where properties exhibit notable deviations from those observed at larger scales. Nanotechnologies, on the other hand, entail the design, characterization, manufacturing, and utilization of structures, devices, and systems through precise control of shape and size at the nanometer scale [10].

Nanotechnology pertains to the manipulation and study of materials and phenomena at the nanoscale, corresponding to one-billionth of a meter or one-millionth of a millimeter [1, 4]. The nanoscale, defined by the order of magnitude 10^{-9} in the metric system, encompasses volumes, weights, and units of time. To illustrate, a nanometer bears a similar relation to a meter as the diameter of a hazelnut does to the diameter of the Earth. For instance, a water molecule spans approximately 0.3 nm, a single gold atom's diameter is about a third of a nanometer, a red blood cell is roughly 7000 nm wide, and a sheet of paper is about 100,000 nm thick.

At the nanoscale, materials exhibit unique physical, chemical, and biological properties at the nanoscale often exhibit substantial disparities from their bulk counterparts and individual atoms or molecules [11]. These distinctive properties have captured scientific interest due to the emergence of unusual phenomena and behaviors in nanomaterials. The term "nanoparticle" derives from "nanos" (Greek for dwarf) and "particulum" (Latin for a particle) [12]. In the context of nanotechnology, "nano" predominantly refers to particle length. Nanoparticles encompass objects ranging from 1 to several hundred nanometers in size [12–14]. On the other hand, nanocrystals represent crystalline clusters consisting of a few hundred to a few thousand atoms typically possess dimensions within the nanoscale range. Their properties are influenced primarily by their surfaces rather than their bulk volumes due to their small size. Nanoparticles can be composed of materials, including metals, metal oxides, micelles, biomolecules, and synthetic polymers [15]. The quality of colloidal inorganic nanocrystals depends on factors such as their crystalline nature, narrow size distribution, unique shape, and dispersion stability. "Nanocluster" refers to small aggregates of atoms and molecules with diameters typically in the nanoscale, ranging from a few units to several thousand [16]. Although there is not a strict demarcation between clusters and nanoparticles, the behavior of nanoparticles is significantly influenced by their nanometer dimensions. As such, characterizing nanoparticles involves investigating size, shape, surface charge, and porosity to comprehend and predict their behavior.

Nanoscience and nanotechnology are not entirely novel concepts. For instance, polymers composed of nanoscale monomers have been studied by chemists for over two decades [17]. However, recent advancements in fabrication tools have allowed direct probing and examination of atoms and molecules with great precision, further advancing nanoscience and nanotechnology [18]. Incorporating nanomaterials into bulk materials can drastically alter their properties, resulting in materials with enhanced strength and unique characteristics due to quantum confinement effects and increased surface area-to-volume ratios at the nanoscale.

In summary, nanotechnology explores and manipulates materials and phenomena at the nanoscale, where distinctive properties emerge. Nanoparticles and nanoclusters play vital roles in this domain, offering unique attributes with potential applications across various fields, promising transformative advancements in science and technology.

1.1.1 Nanomaterials: Properties and Classification

Nanomaterials are materials wherein 50% or more of the constituent particles possess external dimensions falling within the size range of 1–100 nm [19]. Determining whether a material qualifies as a nanomaterial can be done through specific conditions, one of which is the volume-specific surface area (VSSA) or surface area-to-volume ratio. A VSSA greater than 60 m^2/cm^3 serves as a reliable indicator of nanomaterials; however, certain nanomaterials, such as metal nanoparticles, may have a VSSA below this threshold. Thus, the VSSA criterion confirms nanomaterial classification but is not applicable in the reverse direction [20].

Nanoparticles boast significantly larger surface areas per unit mass compared to larger particles [21, 22]. To illustrate the impact of particle size on surface area, let us consider an American Silver Eagle coin weighing 31 g and possessing a total surface area of approximately 3000 mm^2. If this coin were to be subdivided into spherical nanoparticles of, for instance, 10 nm size, the total surface area of all the resulting nanoparticles would amount to around 7000 m^2. This remarkable increase in surface area is equivalent to nearly the size of a soccer field. In essence, the total surface area of all nanoparticles with a size of 10 nm is over two million times greater than the surface area of the silver dollar coin.

As materials are reduced to the nanoscale, quantum effects and surface phenomena begin to exert significant dominance over their properties [23]. When approaching the lower end of the nanoscale, various characteristics of the material, such as optical, electrical, and magnetic behavior, undergo alterations. These effects are particularly pronounced in materials like QDs and quantum well lasers, which find applications in optoelectronics. For crystalline solids and other materials, diminishing their size to the nano range results in a substantial increase in interfacial area. This phenomenon profoundly impacts both the mechanical and electrical properties of these solids [24]. For instance, in materials like metals, the boundaries between grains play a vital role in slowing down or arresting the propagation of defects under stress, thereby providing strength to the material. When the grain size is reduced to the nanoscale, the interfacial area within the material significantly rises, leading to a remarkable enhancement in strength. This is precisely why nanocrystalline materials exhibit much greater strength compared to their bulk counterparts. An exemplary illustration of this phenomenon is nanocrystalline nickel, which demonstrates a strength comparable to that of hardened steel.

Two key factors that markedly differentiate nanomaterials from conventional materials are the enhanced relative surface area and the manifestation of quantum effects [25, 26]. These factors have the potential to modify or enhance properties like strength, reactivity, and electrical characteristics. A key characteristic that makes nanoparticles intriguing from a technical perspective is their surface-to-volume ratio, which grows as their particle diameter decreases. Consequently, nanoparticles consist of a few to several thousand atoms, with a considerable portion of these atoms located on the particle surface. As the particle size diminishes, a higher proportion of atoms occupies the surface in comparison to the interior. For instance, in a 30 nm particle, 5% of its atoms are on the surface, whereas a 10 nm particle has 20% of approximately 30,000 atoms positioned on its surface. This trend continues, with a 5 nm particle containing 40% of approximately 4000 atoms on its surface, and a 1 nm diameter particle having almost all 30 atoms on the surface. The surface atoms, distinct from those inside the material, possess unsaturated bonds, which contribute to the heightened reactivity of the particle surface.

Nanomaterials showcase heightened reactivity, serving as a cornerstone for their diverse range of applications [27]. Precise manipulation of particle size enables the creation of a new generation of exceptionally selective catalysts, capable of expediting specific chemical processes and generating targeted products from raw materials [28]. The increased reactivity of nanomaterials also accounts for their lower melting point, facilitating the reduction of firing temperatures in ceramics through the incorporation of nanoparticulate raw materials. Moreover, these composite materials exhibit minimized shrinkage during the hardening process, making them highly valuable for dental prosthetics. Nanoporous materials, with their expansive pore-specific surface area, find utility in efficient substance filtration, offering an ideal platform for the deposition of filtered substances [29]. These materials also function as potent catalytic agents and promoters of adsorption due to their heightened reactivity [30]. Additionally, nanomaterials possess remarkable optical attributes, such as transparency, absorption, luminescence, and scattering, opening up new possibilities for their utilization.

Given that nanoparticles possess sizes well below the visible light wavelength range (380–780 nm), they have a propensity to absorb light of specific wavelengths [14, 31]. Understanding this light absorption by nanoparticles requires a quantum mechanics-level comprehension. Semiconductor nanoparticles, commonly known as QDs, offer a means to tailor fluorescence wavelength [32]. These optical properties make nanoparticles particularly intriguing candidates for applications in optoelectronics, medical diagnostics, and cosmetics [33, 34]. Furthermore, magnetic materials in the nanometer range can be magnetized as permanent magnets in one direction only [35]. As a result, magnetic nanoparticles (MNPs) contribute to increased storage capacity in magnetic data storage devices. Additionally, the magnetic characteristics of nanomaterials display a relatively low sensitivity to temperature fluctuations.

Nanomaterials are categorized based on the number of dimensions that fall within the nanoscale range (<100 nm) [36, 37]. These classifications are as follows:

1. ***Zero-dimensional (0D) nanomaterials***. These materials have all three dimensions measured within the nanoscale, and none of the dimensions exceed 100 nm. 0D nanomaterials are commonly known as nanoparticles and include examples such as metal nanoparticles and QDs.
2. ***One-dimensional (1D) nanomaterials***. This category comprises materials in which one of the three dimensions extends beyond the nanoscale, while the other two dimensions are confined within the nanorange. Examples of 1D nanomaterials include nanotubes, nanorods, and nanowires.
3. ***Two-dimensional (2D) nanomaterials***. In this type, two dimensions are outside the nanoscale range, while the material is limited to the nanorange in the third dimension. Such nanomaterials exhibit plate-like shapes, and notable examples include graphene, nanofilms, nanolayers, and nanocoatings.
4. ***Three-dimensional (3D) nanomaterials***. These materials are not confined to the nanoscale in any dimension. The 3D nanomaterials category encompasses bulk powders, bundles of nanowires and nanotubes, as well as multi nanolayers.

1.1.2 Forms of Nanoparticles

Nanoparticles can exhibit diverse chemical compositions, encompassing inorganic materials like metals, metal oxides, and semiconductor materials, as well as organic materials such as carbon-based compounds, including fullerenes and carbon nanotubes [27]. Engineered or synthetic nanoparticles can be categorized into specific types based on their distinct chemical and physical attributes.

Carbon-based nanoparticles can be synthesized in various forms, such as spherical nanoparticles like fullerenes or cylindrical nanotubes like carbon nanotubes [38]. Carbon black, an industrially produced soot, is a precisely defined material, purposefully synthesized under controlled conditions with a carbon content exceeding 96%. On the other hand, soot resulting from chimney and diesel combustion contains organic and inorganic contaminations to varying degrees, making them less well-defined combustion by-products of hydrocarbons or coal.

Metal nanoparticles are submicron-scale entities composed of pure metals like gold, platinum, silver, titanium, zinc, cerium, iron, and thallium. They find extensive applications, particularly in catalysis and the biomedical field [39].

Metal oxide-based nanoparticles are formed from metal oxides like zinc oxide (ZnO) or titanium dioxide (TiO_2) [40]. Certain oxide nanoparticles are widely used in consumer goods, such as paints, cosmetics, and varnish [41, 42]. Others, such as aluminum oxide (Al_2O_3) and Zircon (ZrO_2) nanopowders, are employed in technical ceramics components to enhance hardness and breaking strength [43, 44].

Semiconductor nanoparticles originate from diverse compounds [45]. Depending on the elements they are derived from and their placement in the periodic table, they are classified as II–VI, III–V, or IV–VI semiconductors [46]. For instance, silicon and germanium fall under group IV, while GaN, GaP, GaAs, InP, and InAs are categorized as III–V semiconductors. II–VI semiconductors include ZnO, ZnS, CdS, CdSe, and CdTe. These semiconductor nanocrystals, with their distinctive optical properties, find applications in laboratory and medical diagnostics.

Composites, wherein the matrix is often composed of polymers, offer a remarkable approach to combine the unique characteristics of nanoparticles with those of the composite matrix.

1.2 Special Effects of Nanosystems

Nanosystems, often involving structures and materials at the nanometer scale, exhibit an array of exceptional and unique effects that transcend the boundaries of conventional science and engineering. These special effects manifest in various ways, influencing the behavior of materials, biological systems, and devices.

1.2.1 Quantum Confinement Effect

Inorganic nanosystems refer to chemical entities composed solely of inorganic materials. They demonstrate novel behaviors stemming from quantum size effects and the presence of significant surface areas and interfaces at the nanoscale, leading to unique phenomena. Indeed, individual molecules showcase characteristics governed by quantum mechanical principles, whereas the chemical and physical attributes of bulk materials adhere to quantum mechanics laws. Nanosystems exhibit electronic, electrochemical, photochemical, magnetic, optical, mechanical, or catalytic properties that deviate notably not only from individual molecular units but also from macroscopic systems.

Quantum-size effects manifest in nanosized entities due to their overall dimensions being on the same scale as the characteristic wavelength for fundamental excitations in materials [47, 48]. These excitations, which encompass the wavelengths of electrons, photons, and similar particles, transport energy quanta within materials, thereby dictating the dynamics of their transmission and transformation from one state to another. Yet, when the size of structures aligns with the order of magnitude of these characteristic wave functions, the transmission and behavior of quanta undergo significant perturbation. Consequently, quantum mechanical selection rules, typically

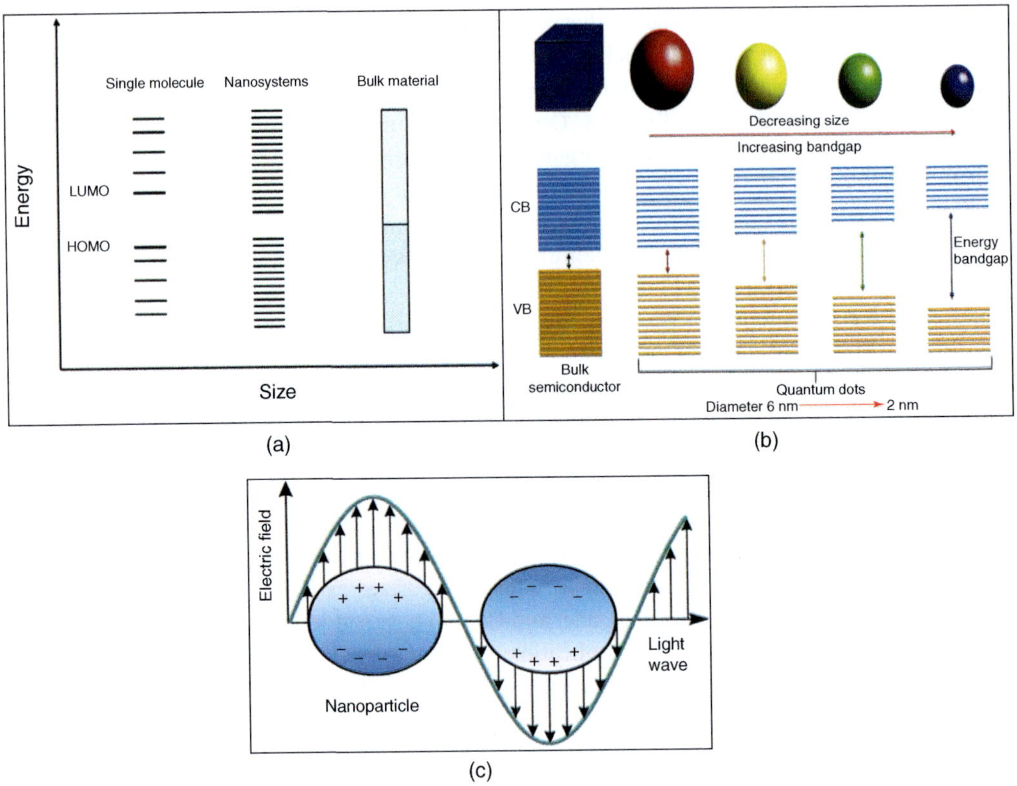

Figure 1.1 Variation of bandgap energy with size of (a) metals and (b) semiconductors. (c) Illustrations of plasmon oscillations in a spherical object, demonstrating the movement of the conduction electron charge distribution concerning the atomic nuclei. Source: (a, b) Reproduced with permission from Sumanth Kumar et al. [49]/Elsevier, (c) Unser et al. [50]/MDPI/CC BY 4.0.

imperceptible on a larger scale, become apparent. Take metals as an example; their conventional "metallic" attributes, such as conductivity, diminish as their size decreases as the number of atoms comprising the sample significantly reduces. In fact, the electronic conduction band of a metal transitions from continuous levels in bulk infinite materials to discrete states as size diminishes, leading to a rise in bandgap energy (Figure 1.1a).

Likewise, QDs possess electronic characteristics that fall between those of bulk semiconductors and isolated molecules [31]. Their optoelectronic attributes are influenced by their dimensions and shapes, and these attributes change depending on these factors. For instance, when QDs are stimulated by a photon with energy represented as hν (where ν signifies the frequency of the incoming photon), larger QDs, typically around 5–6 nm in size, emit light in the orange or red wavelength range [49]. Smaller QDs emit shorter wavelengths, falling within the blue or green spectrum. Consequently, one can intentionally adjust these features by modifying the size and shape of the QDs to attain the desired results. Figure 1.1b provides a visual representation of how the bandgap of QDs changes with varying sizes. These QDs can be produced from single-element materials like silicon or germanium, or from compound semiconductors such as CdSe, PbSe, CdTe, and PbS [51–56]. Sometimes, QDs are referred to as "artificial atoms" since they exhibit discrete electronic states, akin to what is observed in atoms and molecules.

1.2.2 Concept of Surface Plasmon Resonance (SPR)

"How light interacts with matter" has been a subject of inquiry, notably pursued by Michael Faraday in his investigation of metal colloids. Faraday's pioneering work revealed that even when metal particles are substantially smaller than the wavelength of light, they exhibit substantial light scattering and absorption, resulting in vivid colors, even in dilute solutions [57, 58]. Notably, Faraday's exploration led to the understanding that the collective oscillation of conduction electrons within metal particles is accountable for the scattering and absorption of light at specific frequencies, giving rise to distinct colors, particularly in the cases of silver and gold. This occurrence, now known as SPR, derives its name from the polarization of surface charges generated by collective electron oscillations (surface) and the analogy to electron oscillations in gaseous plasma (plasmon) [59].

Localized surface plasmons (LSPs) refer to charge density oscillations constrained within metallic nanoparticles, often referred to as metal clusters, and metallic nanostructures (Figure 1.1c) [59]. The excitation of LSPs by an incident electric field, at wavelengths corresponding to resonance, results in robust light scattering, the emergence of intense surface plasmon (SP) absorption bands, and an augmentation of local electromagnetic fields. The frequency (absorption maxima or color) and intensity of these SP absorption bands are intrinsic to the material type, typically gold, silver, or platinum, and are remarkably sensitive to factors like size, size distribution, shape of nanostructures, and the surrounding environment [50]. These specific characteristics have spurred considerable interest in LSPs, leading to the development of a wide array of LSP-based sensors and devices.

The striking red color observed in aqueous colloidal gold dispersions is a direct manifestation of localized SPR. Throughout history, colloidal gold has intrigued many, having been used to provide color in medieval cathedral windows and believed, until the eighteenth century, to possess life-prolonging and rejuvenating properties when ingested as *aurum potabilis* [60, 61]. In contemporary times, colloidal gold has witnessed exponential growth in its utilization as a biological label, marker, and stain in various microscopic applications [62]. More recently, metallic nanoparticles and nanostructures have found successful use as molecular recognition elements and amplifiers in sensors and biosensors, as well as integral components in nanoscale optical devices [63–65]. Notably, surface plasmons are distributed unevenly around nonspherical metallic nanoparticles and nanostructures, resulting in shape-dependent LSPR absorption spectra [66]. For instance, in metallic nanorods, the plasmon resonance splits into low and high-energy absorption bands, with the high-energy band corresponding to electron oscillations perpendicular to the major axis and the low-energy band arising from electron oscillations along the major axis. This separation becomes more pronounced with increasing aspect ratio of the nanorods [67–69]. Triangular particles exhibit multiple plasmon resonances, including a longitudinal (bulk) plasmon mode and a highly localized enhancement at their sharp tips [70, 71].

Mie theory has been extended to cylindrical and needle-like nanoparticles, and optical properties, including absorption, extinction, and scattering efficiencies, have been calculated using discrete dipole approximations for metallic nanoparticles with arbitrary shapes [72–74]. The analysis of scattering spectra from differently shaped individual nanoparticles (e.g., spherical, rod-like, triangular, pentagonal, and tetrahedral) of silver, gold, and nickel through total internal reflection spectroscopy has obviated the necessity for producing highly monodisperse nanoparticle populations in specific shapes and provided valuable insights into the correlations between shape, environment, and spectra.

It is worth noting that Mie theory is only applicable to noninteracting nanoparticles that are well separated in their solid state or present at low concentrations in dispersions. For interacting particles, the plasmon resonance red-shifts, followed by the emergence of a lower-energy absorption band, similar to the longitudinal absorption band observed in nanorods. Two main types of interactions occur between arrayed metallic nanoparticles: near-field coupling, which involves particles in close proximity, and far-field dipolar interactions [75]. Dipolar interactions are electrodynamic in nature, with dipole fields from a plasmon oscillation in one particle inducing surface plasmon oscillation in adjacent particles [76].

1.3 Aspects of Synthesis of Nanomaterials

Atoms and molecules serve as the fundamental constituents of all matter, shaping its properties and interactions. Understanding how these basic units assemble is crucial for comprehending material behavior. Precise control over synthetic pathways becomes imperative in crafting nanoscale building blocks of varied sizes and shapes, which in turn can facilitate the development of advanced devices and technologies with enhanced capabilities. To achieve this, two complementary approaches are pursued: a top-down strategy focusing on miniaturizing existing components and materials, and a bottom-up approach involving the construction of increasingly intricate molecular structures at the atomic or molecular level [77]. These divergent methods underscore the organizational complexity of nanosystems, situated at the intersection between molecular entities and bulk materials (Figure 1.2). Richard Feynman famously advocated for the top-down approach in his seminal 1959 lecture, emphasizing the ample opportunities for exploration at the nanoscale [79]. This method excels in achieving structures with long-range order and bridging connections with the macroscopic world. Conversely, Jean-Marie Lehn championed the bottom-up approach, recognizing the vast potential at the nanoscale and the ability to establish short-range order through assembly [80]. Integrating these two techniques promises to offer the most comprehensive toolkit for nanofabrication, theoretically enabling a wide array of possibilities.

1.3.1 Top-Down Approach

The top-down approach utilizes techniques such as machining, templating, or lithography to achieve miniaturization. Typically, this method starts with patterns created at larger scales, often in the microscale, which are then scaled down to the nanoscale [81]. An important advantage of this strategy is that components are both patterned and constructed in situ, obviating the need for additional assembly steps.

Various nanostructures can be crafted through methods like electronic, ionic, or X-ray lithography [27, 82]. Initially, a quantum well, characterized by two finite dimensions, is fashioned, followed by the creation of a quantum wire with one finite dimension. Ultimately, a quantum dot, a zero-dimensional structure with all dimensions in the nanoscale, is produced. While current short-wavelength optical lithography techniques can achieve dimensions of at least 100 nm, which traditionally mark the nanoscale threshold, efforts are ongoing to develop extreme ultraviolet and X-ray sources to extend lithographic printing capabilities to dimensions ranging from 10–100 nm. However, challenges persist, particularly related to beam focalization. Furthermore, scanning beam techniques like electron-beam lithography can generate patterns down to approximately 20 nm, with even smaller features attainable using scanning probes for depositing or removing thin layers.

Mechanical printing techniques usually start with the creation of a high-resolution master "stamp" using lithographic methods, as previously described [83]. This stamp, or its copies, is then

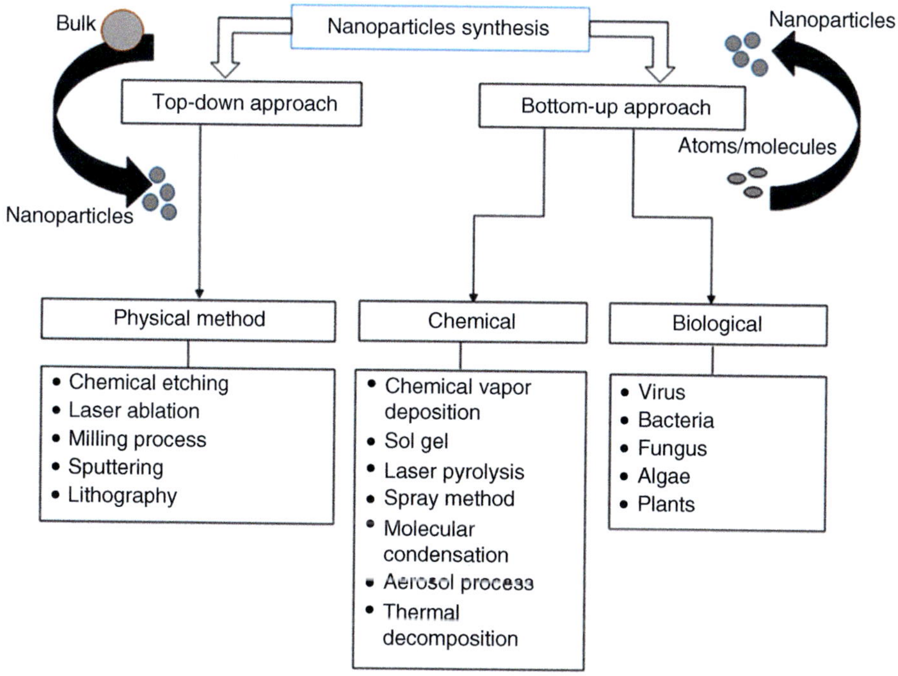

Figure 1.2 Various methods used for the synthesis of nanoparticles. Source: Bloch et al. [78]/Frontiers Media S.A./CC BY 4.0.

used to transfer the pattern onto a surface. Finally, the thin layer of masking material under the stamped regions is removed. These nanoscale printing methods provide several benefits, including the capability to work with various materials on curved surfaces.

These techniques are generally expensive and involve complex manufacturing processes. Despite their success at the microscale, top-down methods encounter difficulties when working at nanoscale dimensions. Moreover, they often yield two-dimensional structures, limiting the production of arbitrary three-dimensional objects due to their reliance on layer-by-layer patterning. Notably, the progress of top-down methodology has been mainly propelled by disciplines like materials engineering and physics, with limited input from inorganic chemists in leveraging these approaches.

1.3.2 Bottom-Up Approach

Bottom-up techniques in nanofabrication involve the step-by-step accumulation of atoms or clusters, utilizing either chemical or physical forces at the nanoscale to guide the assembly of basic units into more extensive structures [77]. Typically, this method entails the chemical synthesis of nanoscale materials within colloidal or supramolecular setups, often undergoing phase changes like sonochemical surface deposition or solid-phase precipitation from solutions. These strategies draw inspiration from biological systems, where chemical interactions drive the formation of diverse vital structures. Researchers seek to emulate nature's ability to produce precise atom clusters, enabling their spontaneous arrangement into sophisticated configurations. Obtaining the desired shape and size requires meticulous control over the material's nucleation and growth. This synthesis method is particularly captivating for chemists. Bottom-up approaches, which initiate from individual atoms and molecules, resonate deeply with the principles of chemistry and molecular biology. Given that a significant portion of chemistry already revolves around manipulating nanoscale entities or coordinating the self-assembly of molecules into more extensive structures, this methodology offers a natural progression for exploration and innovation. Bottom-up techniques present numerous advantages. They offer a wide range of preparation methods, facilitating meticulous control over scale dimensions, including at the atomic or molecular level. Moreover, they typically incur lower costs compared to top-down approaches [13, 14, 84–86]. Nonetheless, despite these benefits, realizing preprogrammed self-assembly of arbitrarily large systems with complexities similar to those observed in natural systems remains a formidable hurdle.

Preparation of nanomaterials can be classified into physical and chemical methods. The physical methods are based on subdivision of bulk metals, including pulverization or mechanical crushing of bulk metal, arc discharge between metal electrodes, etc. Metal nanoparticles thus produced are usually large in size and have a wide size distribution [46]. Several physical methods have been reported for the synthesis of nanosized particles. These include vapor condensation methods, spray pyrolysis, mechanical deformation, thermochemical decomposition of metal–organic precursors in flame reactor, and other aerosol processes named after the energy sources applied to provide the high temperature during gas-particle conversion [47, 48]. The chemical methods are grounded on the concept of reduction of metal ions or decomposition of precursors to form atoms, followed by aggregation of the atoms. Nanoparticles prepared by chemical methods usually have a narrow size distribution [46]. Increasing interest in chemical synthesis of nanoparticles is clearly indicated by the number of reports and reviews on this subject [49, 51–55]. However, it is notable that some methods can be considered as either chemical or physical routes depending on the media, precursors, and operating conditions such as milling.

The formation of metal nanoparticles by chemical methods can be carried out by reduction of metal ions with chemical reductants or decomposition of metal precursors with extra energy. The chemical reductants involve molecular hydrogen, alcohol, hydrazine (HA), $NaBH_4$, $LiAlH_4$, citrate, etc. Energy provided from the outside involves photoenergy (ultraviolet and visible light), γ-ray, electricity, thermal energy (heat), nonchemical energy, etc. In order to produce metal nanoparticles with a narrow size distribution, agents stabilizing colloidal dispersion of metal nanoparticles are of vital importance [46]. Various methods of the nanoparticle's synthesis can be classified based on the process media, including vapor, liquid, and solid-state processing routes, and combined method, such as vapor–solid–liquid approach [53].

Many of the properties associated with nanoparticles are directly related to the relatively higher energetic state of atoms and molecules at a surface when compared with those in the bulk. In many cases, the production of nanoparticles involves techniques to hinder the natural course of thermodynamics through the manipulation of kinetics. In other cases, it is possible to hinder the natural growth of phases through the use of dilution or via protection of surfaces using surface-active agents or by coating and encapsulation of nanoparticles in a glassy media such as those used for instance in the case of polymers [56].

1.4 Synthesis and Characterization of Gold Nanoparticles

The synthesis and characterization of Au NPs stand as an intriguing and essential domain within the realm of nanotechnology and materials science. Gold nanoparticles, celebrated for their distinct size-dependent properties, have been widely applied across diverse fields such as electronics, medicine, catalysis, and environmental science. This research endeavor involves the controlled creation of gold nanoparticles, often on a nanometer scale, and the subsequent detailed examination of their physical, chemical, and optical properties. Understanding how to synthesize these nanoparticles and precisely characterize their attributes is fundamental to harnessing their potential in various technological advancements and innovations. In this paper, we delve into the intricacies of synthesizing and characterizing gold nanoparticles, shedding light on their diverse applications and the remarkable insights they offer into the nanoscale world of materials. Numerous methodologies have been employed for the fabrication of gold nanoparticles, encompassing techniques such as chemical, electrochemical, thermal, and sonochemical pathways. Synthesis of Au NPs by employing these methods is discussed below.

1.4.1 Chemical Methods of Synthesis

Synthesis of Au NPs by chemical reduction technique involves two steps. The primary step involves reducing the gold salt through the utilization of reducing agents like borohydrides, HA, amino boranes, hydroxyl amine, formaldehyde, and hydrogen peroxide and organic reducing molecules such as citric acid, oxalic acid, polyols, sugars, and acetylene. Electron-rich molecules such as metal sandwich complexes have also been used for the reduction of gold salts to produce Au NPs. The second step involves the stabilization of the synthesized Au NPs by various stabilizers such as trisodium citrate, sulfur-containing ligands such as thiolates, oxygen-based ligands, phosphorus-based ligands, nitrogen-based ligands including heterocyclic compounds, polymers, surfactants, dendrimers, etc. One of the well-known techniques for fabricating Au NPs was designed by Turkevisch in 1951 [57]. Turkevisch method involves the conventional chemical route by reducing gold salt ($HAuCl_4$) in aqueous sodium citrate solution. It involves adding sodium citrate to the boiling $HAuCl_4$ solution under vigorous stirring. Wine red Au NP solution is produced within few minutes of stirring with Au NP size of around 20 nm. Polte et al. [87] studied the mechanistic details of the Au NP formation using classical Turkevisch method. The study analyzed various aspects of Au NP formation, including their size, shape, and growth mechanisms, employing techniques such as SEM, HRTEM, UV-vis spectra, XANES, and SAXS. The findings and their interpretation are as follows. The nanoparticles produced in the study closely matched values reported in the literature for similar syntheses. The particles were predominantly spherical and nearly uniform in size, with a polydispersity of around 12% and a mean radius of 7.6 nm. The UV-vis spectra showed an absorbance maximum in the range of 520–550 nm, which is typical for the plasmon resonance of gold nanoparticles. During the synthesis, the peak maximum shifted slightly from about 540 nm to approximately 523 nm, indicating changes in particle size. The curve depicting the intensity at each peak position versus time demonstrated the typical time-dependent behavior, with a slow initial intensity increase followed by more rapid growth. However, the growth mechanism based on UV data remained controversial. Evolution of XANES spectra revealed an initially slow reduction process, followed by a significant decrease in Au(III) content after approximately 50 minutes, coinciding with the rapid escalation in the intensity of the Au NPs plasmon resonance detected via UV-Vis studies. The data suggested that a considerable portion of Au(III) species was being converted into metallic Au(0). SAXS analysis provided information on particle growth during the synthesis. The data indicated different phases of GNP formation. Initially, there was rapid nucleation, resulting in small particles with a mean radius of 2 nm and high polydispersity. Subsequently, particles grew by coalescence, followed by further growth through a diffusion-limited process. The final phase showed a rapid growth in particle size, accompanied by the near-complete consumption of Au(III) species. The study varied reaction temperature and the concentration of the gold precursor. Higher temperatures and increased precursor concentrations led to significantly faster Au NP formation, suggesting that the reduction of Au(III) is an activated chemical process with a reaction order greater than one. The experiments proposed a sequential process of Au NP formation that did not fully align with previously proposed mechanisms. It revealed rapid nucleation, followed by coalescence and further precursor reduction contributing to particle growth. These findings challenged the classical theory of self-nucleation and highlighted the importance of coalescence processes in obtaining monodisperse GNPs. Varying the concentration of sodium borohydride produced the Au NPs of various morphologies such as nanospheres, nanowires, nanorods, and nanotriangles [88]. Frens in 1973 modified the Turkevisch method to produce Au NPs in size from 15 to 150 nm by changing the ratio of reducing agent and gold salt [89]. Several research teams have modified the Turkevich and Frens method to synthesize citrate-stabilized Au NPs with diverse sizes and diameters. For example, it was demonstrated by Kimling et al. that smaller-sized Au NPs are produced at higher concentration of citrate whereas lower concentration of citrate produced large-sized Au NPs by aggregation of small Au NPs [90]. Kumar et al. [91] investigated the mechanism behind the Turkevich–Frens method, illustrating the formation process of dicarboxy acetone (DCA) through the citrate oxidation, followed by the reduction of gold salt to Au atoms, which then assemble to form gold nanoparticles (Au NPs). The dicarboxyl acetone produced serves as a stabilizing agent, distinct from the role of citrate itself. Turkevisch method was further modified by various research groups to study the effect of varying the pH [92–95], reaction temperature, citrate concentration [96], and gold salt concentration [97] on Au NP shape, size distribution, and other characteristic properties. Utilizing ascorbic acid as a reducing agent in chemical reduction and an aqueous surfactant, cetyltrimethylammonium bromide (CTAB) as a shape director also resulted in the production of anisotropic Au NPs. The size and characteristics of gold nanoparticles (Au NPs) were found to be significantly affected by the alteration in the sequence of reagent addition. In a study by Jiménez et al. [98] the order of reagent addition was modified keeping factors such as the SC : Au ratio (13.6), temperature (100 °C), final volume (150 mL), and total reaction time (5 minutes) constant. The two methods were employed: (i) direct synthesis (SC was added to a hot $HAuCl_4$ solution) and (ii) inverse synthesis ($HAuCl_4$ solution was added to a boiling sodium citrate solution). A blue shift in the SPR peak, from 527.5 nm (direct synthesis) to 518 nm (inverse synthesis), indicating smaller Au NPs in the inverse method, was revealed by UV-vis spectra (Figure 1.3A). This shift was corroborated by TEM analysis (Figure 1.3B), which demonstrated a decrease in mean particle size and an improved size distribution, from 36.6 ± 6.8 nm (SD ~ 18%) for direct synthesis to 9.0 ± 1.2 nm (SD ~ 13%) for inverse synthesis. The same trend was observed for a lower SC : $HAuCl_4$ ratio (6.8), where a blue shift in the SPR peak from 522 nm (direct) to 519 nm (inverse) was observed, accompanied by a decrease in the absorbance maximum. Once again, TEM analysis confirmed a reduction in Au NP size and polydispersity, from 17.8 ± 2.5 nm (SD ~ 14%) for direct synthesis to 14.9 ± 1.5 nm (SD ~ 10%) for inverse synthesis. The addition of DCA to the reducing mixture had a notable impact on Au NP size and reaction rate. With a fixed SC : $HAuCl_4$ ratio (13.6), changing the SC : DCA ratios in the reducing mixture influenced reaction kinetics. A decrease in SPR peak from 523 to 519 nm was observed when increasing DCA content from 100 : 0 to 90 : 10 (SC : DCA), but it returned to 523 nm at a 50 : 50 ratio (Figure 1.3C). TEM analysis confirmed a correlation between SC : DCA ratio, final Au NP size, and monodispersity (Figure 1.3C). When using a 0 : 100 SC : DCA ratio, no particle formation was observed, likely due to the solution's pH of 3.7. This suggests that heating SC before adding Au^{3+} could induce SC oxidation. The role of pH in Au NP synthesis was also explored. A decrease in pH from 6.2 to 5.9 was observed when reducing SC content in the reducing mixture from a 100 : 0 to 90 : 10 SC : DCA ratio. However, the influence of SC : DCA ratio on reaction kinetics, SPR peak, and final Au NP size was more significant than pH changes. Adjusting the pH value from 5.1 to 6.2 resulted in a blue shift in the SPR peak and smaller Au NPs. The presence of DCA in the reaction mixture induced a decrease in NP size and an improvement in monodispersity, which was not attributed to pH changes. In conclusion, both the order of reagent addition and the addition of DCA influenced the size, shape, and reaction kinetics of Au NPs, indicating the complex interplay

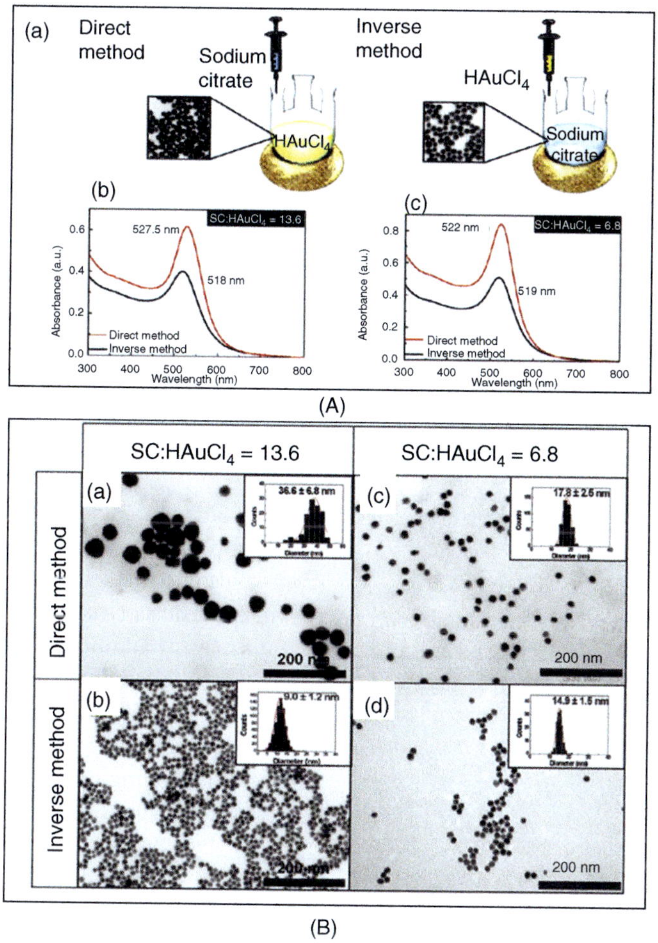

Figure 1.3 (A) Illustration of a comparison between the (a) direct and inverse sequences of reagent addition and the characterization of gold nanoparticles (Au NPs) through optical analysis (UV-vis spectroscopy). The synthesis was carried out at 100 °C for five minutes, with two sets of conditions: (b) 0.165 mM $HAuCl_4$ and an SC: $HAuCl_4$ ratio of 13.6 and (c) following the Turkevich conditions with 0.25 mM $HAuCl_4$ and an SC: $HAuCl_4$ ratio of 6.8. (B) The morphology of the synthesized gold nanoparticles (Au NPs) examined using transmission electron microscopy (TEM) at a temperature of 100 °C for five minutes. This characterization was performed under two different sets of conditions: (a and b) with 0.165 mM $HAuCl_4$ and an SC: $HAuCl_4$ ratio of 13.6, and (c and d) following the Turkevich conditions, which involved 0.25 mM $HAuCl_4$ and an SC: $HAuCl_4$ ratio of 6.8. The TEM images for both the direct and reverse sequences of reagent addition are presented. Additionally, the size distribution of the Au NPs is illustrated in the inset graphs within the TEM images. (C) Analysis of the gold nanoparticles (Au NPs) involved both optical assessment through UV-vis spectroscopy and morphological examination using transmission electron microscopy (TEM). These Au NPs were synthesized under conditions that included 0.165 mM of $HAuCl_4$ and an SC: $HAuCl_4$ ratio of 13.6, with the reaction temperature set at 100 °C. The study examined different ratios of the reducing mixture SC: DCA, specifically 100 : 0 (a and d), 90 : 10 (b and e), and 50 : 50 (c and f). The TEM images feature inset graphs that provide measurements of the size distribution of the Au NPs. Source: Reproduced with permission from Ojea-Jimenez et al. [98]/American Chemical Society.

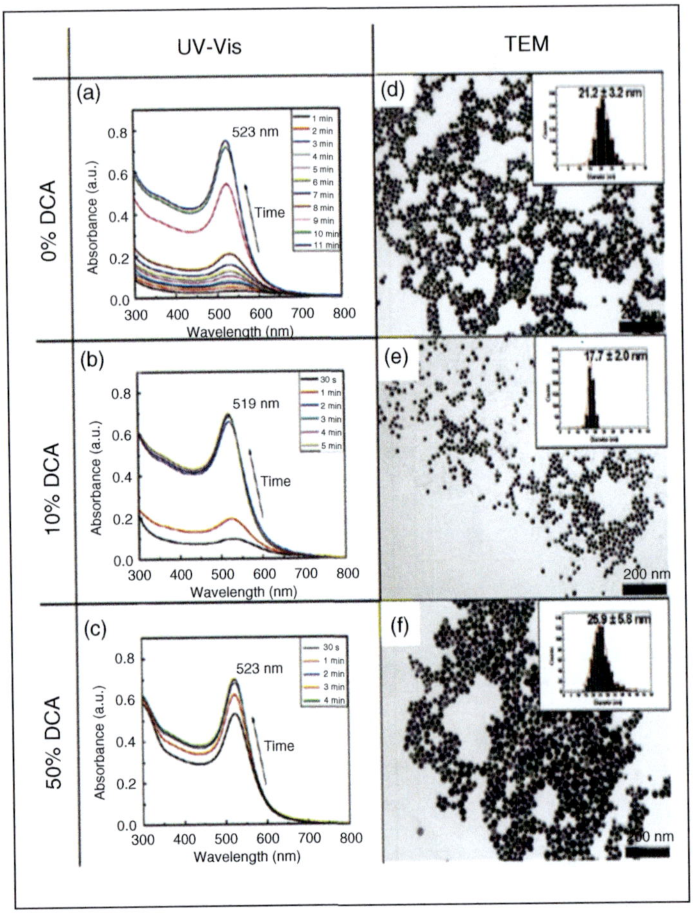

(C)

Figure 1.3 (*Continued*)

of these factors in the synthesis process. Due to the mild reducing ability of sodium citrate, Au NPs produced by Turkevisch method resulted in the production of Au NPs of bigger sizes, with diameters greater than 10 nm. The reducing ability of citrate ions was found to vary by changing the reaction medium, hence size of Au NPs can be tailored by changing the solvent polarity. In another study, Ojea-Jiménez et al. [99] reported that using D_2O as a solvent instead of water enhanced the reducing ability of citrate ions which resulted in the production of smaller-sized Au NPs of diameter of 5 nm. The synthesis of Au NPs was conducted at 100 °C under the same reaction conditions, but with varying amounts of D_2O in the solvent (% D_2O = 0, 50, and 100%). The growth of the particles was tracked through UV-Vis studies of the solutions, and the results discovered a blue-shift in the SPR band, shifting from 519 to 512.5 nm as % D_2O increased. This shift in the SPR band corresponds to changes in particle size, supported by a reduction in the hydrodynamic diameter from 15.4 to 9.7 nm and a decrease in negative surface charge from −43.0 to −33.9 mV. Transmission electron microscopy (TEM) and image analysis further confirmed that higher % D_2O led to a reduction in particle diameter. Using pure D_2O as the solvent resulted in the production of smaller Au NPs, measuring 5.3 ± 1.1 nm, in comparison to those produced in pure H_2O, which had a diameter of 9.0 ± 1.2 nm. It is noteworthy that this method could generate a small fraction of even smaller Au NPs, down to 2–3 nm, which could be isolated using standard separation techniques (detailed in the

Supporting Information), potentially yielding higher Au NP yields. Temperature and the presence of D_2O in the reaction were found to influence Au NP size. At lower temperatures (90 °C), reactions in D_2O proceeded more slowly, resulting in larger Au NPs compared to syntheses in H_2O at 90 °C or D_2O at 100 °C. This is consistent with observations that reducing the reaction temperatures resulted in the production of larger NPs. The study postulates a reaction mechanism for the reduction of Au(III) by sodium citrate, based on experimental evidence and existing literature. The mechanism involves two steps: a rapid ligand exchange with citrate to form an intermediate complex, followed by a slow rate-determining step involving a concerted decarboxylation and reduction of Au(III) species. The involvement of deuterium in D_2O affects the rate of this reaction, enhancing the reducing strength of the citrate molecule and influencing the fabrication of Au NPs. In conclusion, this research demonstrates that particle size can be regulated by altering the solvent system and provides insights into the intricate mechanism affected by the presence of deuterium in the solvent. It emphasizes the sensitivity of water-based systems, where hydrogen bonding interactions play a dominant role in all reaction steps. The citrate/aqueous synthesis method for Au NPs is widely used in the field of nanotechnology.

Another novel approach discovered by Brust and Schiffrin used tetraethylammonium bromide (TOAB) as a phase transfer agent to transfer $AuCl^{4-}$ from aqueous phase to the organic phase to synthesize thermally stable Au NPs employing $NaBH_4$ as a reductant in presence of dodecanthiol [100]. The reduction of the gold salt caused a transition in the color of the organic phase, shifting from orange to a rich brown hue, signifying the formation of both cuboctahedral and icosahedral gold nanoparticles (Au NPs). This approach demonstrated an easy and straightforward method for producing thiolate-stabilized gold nanoparticles (Au NPs) with good air stability and repeated dissolution and isolation without agglomeration. Further, this method produced small-sized (<5 nm) Au NPs with narrow size distribution and easy substitution and functionalization by different ligands. It was further shown that larger molar ratios of dodecane-thiol/$AuCl^{4-}$ produced smaller-sized Au NPs [101]. Further fast addition of $NaBH_4$ to cooled solutions produced smaller and mono-dispersed Au NPs. The reaction between thiol and Au NPs led to the oxidative addition of the S—H bond onto two Au atoms on the surface of the Au NP, leading to the creation of an S—Au bond and the elimination of an H_2 molecule through reductive elimination. Li et al. [102] employed Raman spectroscopy to demonstrate the formation of S—Au bonds on the surface of Au NPs, revealing that these bonds formed only after the addition of $NaBH_4$, thus elucidating the mechanism of Au NP formation. The study presented detailed spectroscopic investigations into the two-phase synthesis of metal nanoparticles (NPs), with a focus on the addition of thiol or ligands and $NaBH_4$ in the process. Raman spectroscopy revealed that no Au—S bonds were formed before the addition of $NaBH_4$, confirming that the metal precursor was not the commonly believed [M(I)SR]n-like polymer but rather the [TOA][MX_2] complex. 1H NMR spectra suggested the formation of inverse micelle structures involving TOAB (TOA), encapsulating water, and encapsulated metal ions. The proposed NP formation mechanism highlighted the role of inverse micelles, where Au NPs formed within micelles after $NaBH_4$ addition, and ligands later diffused through the TOA shell to form Au—S bonds at the interface. A reversed synthetic route successfully produced stable, ligand-protected ultrasmall Au NPs with better size control, and increasing stirring time led to larger NP formation. Additionally, the use of disulfide and organo-chalcogen ligands in the synthesis demonstrated the same principles, with the reversed route producing smaller, more monodisperse Au NPs. This study provides invaluable insights into the two-phase synthesis of metal nanoparticles, with implications for control and optimization in NP size and distribution.

Burst–Schiffrin method was further improved in 1995 by using methanol solution as a medium for Au NP synthesis in absence of phase transfer agent TAOB [103]. Since both gold salt and p-mercaptophenol are soluble in methanol, hence it acted as an excellent solvent for a single-phase synthesis. This modified procedure proved to be advantageous since it avoided the incorporation of TAOB impurities in the synthesized Au NPs. Later all such solvents have been exploited for single-phase synthesis of Au NPs in which both $HAuCl_4$ and thiols are soluble such as methanol, ethanol, and water. Thus, a number of publications emerged on the single-phase synthesis of thiol-stabilized Au NPs [101, 104]. Brust et al. synthesized water soluble and biocompatible Au NPs stabilized with (1-mercaptoundec-11-yl) tetraethylene glycol of size less than 10 nm with hydrophilic tetra ethylene glycol (TEG) shell and hydrophobic alkanethiol core [105]. A variety of thiolate ligands such as alkane thiolates, arene thiolates, and other functionalized thiolates have been used for the synthesized of functionalized Au NPs [106–110]. Au NPs have also been synthesized by changing both gold salt precursors and reducing agents. Moores et al. documented the synthesis of phosphine-stabilized Au NPs using $[AuCl(SMe_2)]$ complex in place of $HAuCl_4$ and sodium naphthalene ($C_{10}H_8Na$) as a reductant [111]. The formation of Au NPs was confirmed through UV-Visible spectroscopy which revealed the emergence of SPR band around 580 nm. The presence of phosphine on the Au NP surface was confirmed by FTIR and TGA analysis. IR spectra gave the peaks corresponding to in plane and out-of-plane bending vibrations of aromatic C—H bonds (700 and 750 cm^{-1}), P—C bond vibrations (760 cm^{-1}) and characteristic aromatic C–L–C vibrations (1450 cm^{-1}). TGA showed the weight loss which revealed the presence of phosphine on surface of Au NPs. Both the techniques revealed the existence of phosphine ligands on Au NP surfaces even after several washings which indicated a strong coordination the phosphine ligands and Au NPs. TEM images showed the small average particle size of Au NPs of 5.7 nm $\pm$ 1.3 and 12.7 nm $\pm$ 3.7 nm. A single-step synthesis of phosphine-stabilized Au NPs has also been achieved by reducing Et_3PAuCl with 9-BBN employed as a reducing agent [112]. The coordination between the phosphine ligand and Au NPs was studied by NMR spectroscopic technique. The particle size was determined to be 1.4 ± 0.2 nm and 1.2 ± 0.2 nm for reaction carried out at 40 and 50 °C, respectively. Further the results illustrated that the core size of Au NPs decrease with increase in reaction temperature. Other examples of synthesis of Au NPs by phosphine derivatives include triphenyl phosphine [113], 1,1-binaphthyl-2,2-dithiol [114], calixarene phosphine [115, 116], and tetrakis-(hydroxymethyl)phosphonium chloride (THPC) [117] for applications such as proactive etching agents, optics, catalysis, bimetallic nano shell formation, respectively.

A variety of alkyl thiolate-functionalized gold nanoparticles have been synthesized using a modified single-phase method, where 9-borabicyclo[3.3.1]nonane (9-BBN) served as a mild reducing agent [118]. Nitrogen and oxygen ligands containing electronegative groups such as amine, carboxyl, phenolic, and carbonyl groups have also been used for stabilization of Au NPs [119–123]. Amines have been predominantly used for this purpose. For example, octadecyl amine-stabilized Au NPs have been synthesized by Crook et al. [120]. Ligands like 4-(*N*,*N*-dimethylamino) pyridine (DMAP) have been frequently used for stabilizing Au NPs for ligand substitution reactions and in heterogenous catalysis [124, 125]. DMAP-stabilized Au NPs are synthesized via a phase transfer process by adding aqueous DMAP solution to TOAB-stabilized Au NPs [126–129]. Among oxygen-containing ligands, phenolic and carboxyl derivatives are frequently employed stabilizers which include folic acid [130], *N*-2-hydroxyethylpiperazine-*N*-2-ethanesulphonic acid (HEPES) [131], and *N*-vinylpyrrolidone [132].

Au NPs have been functionalized by either direct Au NP synthesis or by ligand substitution reactions such as halide nucleophilic substitution [133], coupling reactions such as amidic coupling by employing amino termini [134], polymerization [135], nucleophilic addition reactions [136], and click functionalization. The weakness of Au-citrate bond enables it to be substituted or functionalized with stronger ligands such as thiolates. Such thiolate functionalized Au NPs have been synthesized by Mulvaney and Giersig from citrate-stabilized Au NPs by substituting citrate ligands with thiol ligands [137]. This strategy was later used by Mirkin et al. to synthesize thiolate DNA functionalized Au NPs [138]. Functionalization of citrate-stabilized Au NPs with thiolated PEG was also carried out for use in in vivo computer tomography as a contrast agent [139]. Nitrile and amine-terminated dendrimer functionalized Au NPs have been synthesized by reaction between poly(propyleneimine) dendrionic thiols and citrate-stabilized Au NPs (19.5 nm) for drug delivery application [140]. Alkylthiolate ligands on Au NP surface have also been substituted by functionalized alkyl thiolate ligands by ligand substitution reaction by employing the corresponding functional thiols [141]. There are various reports where some other labile ligands on the Au NP surface have been substituted by thiol ligands [142–144]. However, ligand exchange reaction suffers some major drawbacks such as chances of incomplete ligand substitution and irreversible aggregation of the Au NPs. There is also a chance of incompatibility between new ligand and the initial medium in which Au NPs are already dispersed [145].

In summary, this section outlines several methods for synthesizing gold nanoparticles (Au NPs) using chemical reduction techniques. The synthesis process typically comprises two key steps: the reduction of gold salt and the stabilization of the resulting Au NPs. Various reducing agents, including borohydrides, HA, amino boranes, and organic molecules like citric acid, are utilized in the reduction step. The subsequent stabilization of Au NPs involves employing a diverse array of stabilizers such as trisodium citrate, sulfur-containing ligands, oxygen-based ligands, phosphorus-based ligands, nitrogen-based ligands, polymers, surfactants, and dendrimers. The Turkevich method, established in 1951, stands as a prominent technique for Au NP synthesis, involving the reduction of gold salt in an aqueous sodium citrate solution to yield wine-red Au NP solutions. Various modifications of this method by different research groups have been investigated to explore the impacts of factors like pH, reaction temperature, citrate concentration, and gold salt concentration on the size, shape, and distribution of Au NPs.

The section further explores the utilization of ascorbic acid as a reducing agent and CTAB as a shape-directing agent for the synthesis of anisotropic Au NPs. It highlights how the order of reagent addition and the introduction of DCA can impact the size, shape, and reaction kinetics of Au NPs. Another method, pioneered by Brust and Schiffrin, employs TOAB as a phase transfer agent to transfer $AuCl^{4-}$ ions from the aqueous phase to the organic phase, resulting in the production of thermally stable Au NPs. This Brust–Schiffrin approach has since been enhanced by using methanol as a medium for Au NP synthesis without the need for a phase transfer agent. Various ligands, such as alkyl thiolates, nitrogen and oxygen ligands, and phosphine derivatives, have been employed to stabilize and functionalize Au NPs. Ligand substitution reactions, including halide nucleophilic substitution, coupling reactions, polymerization, nucleophilic addition reactions, and click functionalization, have been utilized to functionalize Au NPs. The section underscores the importance of ligand exchange reactions while recognizing their limitations, such as incomplete ligand substitution and the potential for irreversible aggregation of Au NPs.

1.4.2 Electrochemical Method

The electrochemical method for the fabrication of transition metal NPs was first used by Reetz et al. [146, 147] employing tetra alkyl ammonium salts as stabilizer metal nanoclusters in nonaqueous medium. Electrochemical synthesis of Au NPs has been achieved using various quantities of surfactant, tetradodecylammonium bromide TTAB as stabilizer [148]. The particle size of the Au NPs varied with the amount of the TTAB surfactant, temperature, and current density used. The particle diameter decreased from 58.3 ± 12.6 to 10.6 ± 6.5 nm with increase in amount of the TTAB from 1 to 90 mg (Figure 1.4A). Thus, the particle size was found to be inversely proportional to the amount of the surfactant used. The Au particle diameter was found to vary directly with the current density. The particle diameter increased from 53.7 ± 17.2 to 12.2 ± 7.3 nm with an increase in the current

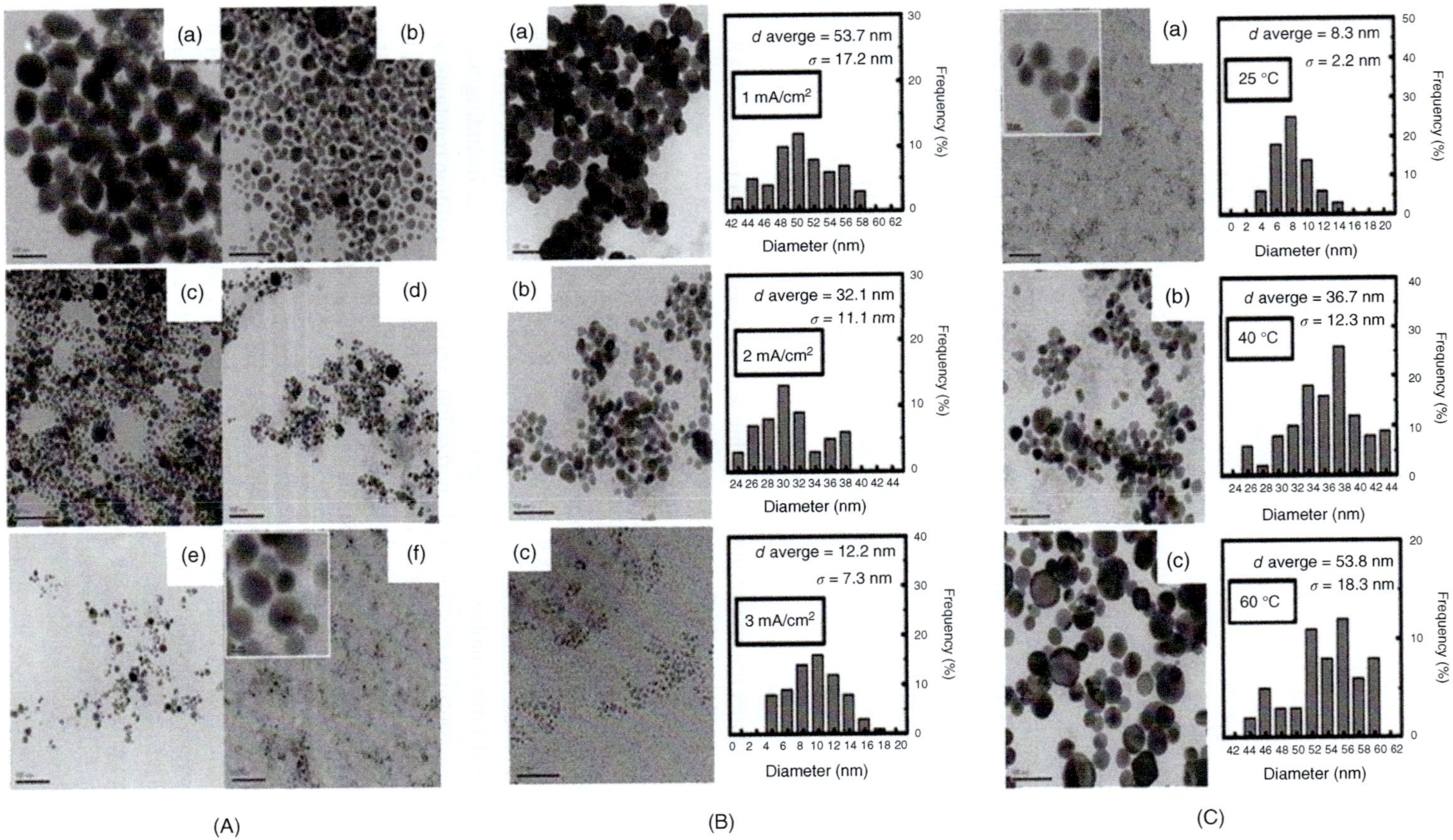

Figure 1.4 (A) Transmission electron microscope images of gold nanoparticles produced with varying quantities of TTAB surfactant, depicted as 1 mg (a), 10 mg (b), 30 mg (c), 50 mg (d), 70 mg (e), and 90 mg (f); the scale bar denotes a length of 100 nm. (B) Transmission electron microscopy images of gold nanoparticles generated under different current densities, specifically at 1 mA/cm^2 (a), 2 mA/cm^2 (b), and 3 mA/cm^2 (c), the scale bar indicates a size of 100 nm. (C) Transmission electron microscope pictures of gold nanoparticles produced under distinct growth temperatures, namely at 25 °C (a), 40 °C (b), and 60 °C (c); the scale bar denotes a size of 100 nm. Source: Reproduced with permission from Huang et al. [148]/IOP Publishing.

density from 1 to 3 mA/cm^2 (Figure 1.4B). The cathode was found to be strongly gold colored at the current density of 3 mA/cm^2. This decrease in particle size was also confirmed by blue shift in the SPR peak from 538 to 523 nm as the particle size decreased. Effect of varying growth temperature on the Au particle size was also studied for a constant current density and surfactant amount (30 mg, 2 mA/cm^2). Higher temperature resulted in heavy gold plating of the electrode. The particle size of Au NPs increased from 8.3 ± 2.2 to 53.8 ± 18.3 nm as the temperature raised from 25 to 60 °C (Figure 1.4C). Further UV-Visible spectroscopy revealed a red-shifting of the SPR peak from 521 to 538 nm with increase in particle diameter from 8.3 to 53.8 nm. Au NPs were synthesized by electrochemical deposition technique employing ITO film coated glass electrode using 0.1–1 mM $HAuCl_4$ in 0.5 M H_2SO_4 solution. Au NPs with size between 70 and 300 nm were produced under variable $HAuCl_4$ concentrations and deposition times [149]. Au NPs were found dispersed on the electrode surface with higher deposition time resulted in bigger Au NPs, whereas higher $HAuCl_4$ concentration resulted in larger concentration of Au NPs on the electrode surface. ITO-coated glass electrodes based Au NP formation by electrochemical method has been reported by various research groups [150–152]. High density and large surface area gold NP arrays were synthesized on a silicon substrate by electrochemical deposition technique using high overpotential. It was observed that high overpotential led to the fast nucleation and small particle size. Electrochemical deposition of Au NPs on *n*-doped silicon substrate by employing various electrochemical techniques with variable reaction parameters such as pH, temperatures, reagent concentration, deposition time, and reaction kinetics have been studied [153]. Of all the electrochemical methods, cyclic voltammetry was found to produce large-sized Au NPs due to electrodeposition taking place at the slower rate. Further for the similar reaction conditions such as same temperature and pH, the particle size was observed to be less in diffusion regime as compared to electrochemical deposition process. Various gold nanostructures were synthesized by one-step electrochemical deposition on GCE employing 5 mM $HAuCl_4$ solution [154]. The morphologies of Au NPs were controlled by changing the pH and temperature. For example, hierarchical waxberry-like Au NPs with high catalytic performance for oxygen reduction were produced at pH of 4 which also exhibited remarkable super hydrophobicity. The contact angle was found to increase with rise in temperature and deposition time. Electrochemical deposition of Au NPs was carried out in a solution containing thiosulfate and sulfite to study the nucleation and crystal growth of Au NPs on the GCE [155]. It was observed that electrodeposition for one second resulted in the production of uniform-sized Au NPs with diameters of 80 ± 5 nm. Coalescence of Au NPs was observed for 10 s, whereas large aggregates of Au NPs with diameters of 100–300 nm were obtained for 100 s of electrodeposition. Various research groups have reported the fabrication of Au NPs by the electrochemical method on a GCE [156–160]. A method known as substrate-enhanced electrolytic deposition has been used for the conversion of Au ions into Au NPs on CNTs supported on metal substrate which is having lower reduction potential than Au ions itself [161]. Here CNTs act as templates or cathode for deposition other than the reducing agents. Here size distribution of electrodeposited Au NP size on side walls of CNTs can be controlled by various parameters gold precursor concentration, nucleation potential, and deposition time. Since the particle size of the electrodeposited Au NPs on CNTs has bigger particle size usually between 50 and 500 nm, hence obtaining smaller Au NP with good dispersion is a challenge. This method produces Au NPs on CNTs, which act as a sensitive platform for biosensing applications, with signal intensity three times higher than nonmodified carbon paste electrodes [162]. Au NPs with cauliflower morphology with size of 200 nm have been electrodeposited on CNT-modified electrode, which exhibited good electrocatalytic activity for oxidation of dopamine, ascorbic acid, and uric acid [163]. Au NPs of different morphologies have been produced on a graphitic surface which in turn were produced by micromechanical cleavage of organic graphitic flakes on silicon wafer with SiO_2 insulating layer of 300 nm [164]. Room temperature resulted in production of 4 nm thick dense cluster of Au on graphene surface [165]. Au polygons were obtained at 100 °C and irregular Au islands at 80 °C [166]. Gold dendrites were obtained as the substrate was cooled from 80 °C to room temperature [166]. As the temperature of the substrate reached above 200 °C, coalescence and growth resulted in the production of three dimensional and geometrical metal islands [166]. Au NPs with positive surface charge with a particle diameter of 2–6 nm have been self-assembled on graphene sheets functionalized with 1-pyrene butyric acid [63]. The synthesis was achieved by simple mixing of the aqueous dispersion of various components where electrostatic force acted as a driving force for Au NP formation. An amount of Au NPs assembled on functionalized graphene sheets could be easily modulated by changing the weight ratio of the various components. Further the functionalized graphene sheets exhibited the loading capacity on Au NPs up to 300 times more than its own weight. Glassy carbon electrode covered by this Au NP-G composite exhibited electrochemical stability and electrocatalytic ability. Furthermore, the Au NP-G composite-modified GC electrode acted as an efficient biosensor for electrochemical detection of uric acid with good sensitivity. Porous Au nanofoams have been prepared by electrochemical deposition of 0.1 M $HAuCl_4$ and 0–3 M NH_4Cl [167]. The deposition of Au took place on a Pt/Ti/Si electrode within the range of E = 0 to −8 V over the period of 15–120 seconds. The gold nanofoams had a pore size of 0.01–10 μm and was ascertained to be impurity free. It was observed that Au foam deposition was possible only when enough NH_4Cl was incorporated to the electrolytic solution as a hydrogen source. The Au nanofoam had a well-known nanoporous structure with promising sensing and electrocatalytic applications. Au nanorods with two different aspect rations have been prepared by electrochemical reduction technique using cationic surfactant as a soft template for shape-directed synthesis [168]. The synthesized Au nanorods had a mean diameter of 10 nm. The electrochemical method is found to be superior to other methods due to the use of modest equipment, low processing temperatures, low cost, and high purity.

In summary, the section discusses the electrochemical method for synthesizing transition metal nanoparticles (NPs), focusing on gold nanoparticles (Au NPs). The electrochemical method was initially utilized by Reetz et al., using tetraalkyl ammonium salts as stabilizers in a nonaqueous medium for metal nanoclusters. The synthesis of Au NPs through electrochemical methods involves varying parameters such as surfactant quantity, temperature, and current density. The particle size of Au NPs is found to be inversely proportional to the amount of the surfactant (TTAB) used, with smaller particles obtained at higher surfactant concentrations. Additionally, the current density and growth temperature are identified as factors influencing particle size. Higher current density and lower growth temperature lead to smaller Au NPs. The synthesis of Au NPs through electrochemical deposition on ITO film-coated glass electrodes is discussed, with variations in $HAuCl_4$ concentrations and deposition times resulting in Au NPs of different sizes.

Different research groups have reported the electrochemical deposition of Au NPs on various substrates, including silicon, silicon with CNTs, and graphitic surfaces. The morphologies of the Au NPs are controlled by adjusting pH, temperature, and deposition time. The text highlights the use of cyclic voltammetry, which produces larger-sized Au NPs due to slower electrodeposition rates. A method called substrate-enhanced electrolytic deposition is introduced, where Au ions are converted into Au NPs on CNTs supported on a metal substrate. The control over the size distribution of gold nanoparticles (Au NPs) electrodeposited onto carbon nanotubes (CNTs) can be achieved by manipulating factors like the concentration of the gold precursor, nucleation potential, and deposition time. This method is particularly useful for biosensing applications. The electrochemical deposition technique is praised for its simplicity, low processing temperatures, cost-effectiveness, and high purity in synthesizing Au NPs. Various morphologies, including cauliflower-like structures, dendrites, and nanofoams, are achieved through this method, demonstrating its versatility for different applications such as electrocatalysis and sensing. Overall, the electrochemical method is presented as a superior technique due to its use of modest equipment and its potential for low-cost, high-purity synthesis of metal nanoparticles.

1.4.3 Polymer and Dendrimer-Mediated Synthesis

Various promising potential applications of gold nanoparticles such as photocatalysis, chemisorption, sensing, therapeutic, and magnetic device fabrication are dependent upon its size and morphology [26]. Controlling such features of the gold nanoparticles is a challenging and attractive with tremendous progress made in this direction. Several novel strategies have been developed for this purpose. To control the size and morphology of the Au NPs, various synthetic methods involve several chemical or biological entities have been employed for this purpose. It has been shown by various research studies, synthesis of Au NPs in the presence of surfactants can greatly affect the size shape, stability, and particle size distribution [18]. Employing polymeric stabilizers greatly enhances the long-term stability, improves the solubility, and provides amphiphilicity of Au NPs. Polymers also provide surface density, improve compatibility, and tailor the properties of Au NPs. Mention of polymer-stabilized Au NPs dates to 1718 in Helcher's treatise which report water soluble starch-stabilized Au NPs. However, nanotechnology has provided a rapid momentum for synthesizing polymer-stabilized Au NPs for use in various applications such as optics, catalysis, and biology. Two major techniques used for the fabrication of polymer-stabilized Au NPs are "grafting to" and "grafting from" techniques. Grafting from technique involves the attachment of polymer molecules on the Au NP surfaces in presence of radical initiators. This approach can be considered as a technique of functionalization of Au NPs. In contrast, the grafting-to method utilizes polymers containing sulfur, phosphorus, or nitrogen groups either at the terminus or within the polymer chain to stabilize gold nanoparticles (Au NPs). This route is synonymous with the Brust–Schiffrin process or ligand substitution method. Here the gold salt precursor is first mixed with functionalized polymer and latter reducing agent is added to obtain polymer-stabilized Au NPs. Various ligand polymers have been used for synthesized Au NPs such as thiolated PEG, polystyrene (PS), poly(*N*-isopropylacrylamide) (PNIPAM), poly(vinyl pyridine) (PVP), five-arm PEG-b-PCL star block copolymers and amine groups containing ligands such as poly(ethylenimine) (PEI), amine, and arsenic group containing poly(acryloylaminophenylarsonic acid) (PAAPHA), ionic polymers, poly(*N*-vinyl caprolactam) (PVCL) functionalized with oethyl-S-(1-methoxycarbonyl)ethyl dithiocarbonate, poly(*N*-vinyl caprolactam) (PVCL), and polymer ligands (DDT-PVAc and PTMP-PVAc) functionalized with thioether groups. Size of Au NPs can be regulated by varying the polymer/Au ratio, and the change in particle size can be easily detected by a change in intensity of SPR peak. Certain polymers serve dual roles as both stabilizers and reducing agents such as PS and PAAPHA. However, the problem with such reducing agents is that due to their weak reducing ability, Au NPs formed attain bigger particle sizes in comparison to those formed with strong reducing agents such as $NaBH_4$. Gold nanoparticles have also been fabricated using functionalized co-polymers. For example, thermoresponsive hybrid core–shell Au NP have been synthesized in situ using thiol and amino functionalized poly (nIsopropylacrylamide) (NIPAM)-b-(glycidyl methacrylate) (GMA) (poly NIPAM-b-GMA) as template for stabilizing Au NPs inside polymer shells. Synthesis of Au NPs by ligand substitution reaction has a significant advantage in that the Au nano seeds synthesized by Turkevich or Brust method are nearly monodisperse which after ligand substitution reaction leads to the formation of monodispersed polymer-stabilized Au NPs. PEO-b-PS-b-P4VP-stabilized AuNPs were obtained by Azzam et al. by adding PEO-b-PS-b-P4VP to TOAB-stabilized Au NP solution in toluene with subsequent stirring for 24 hours. Various research groups have synthesized polymer-stabilized Au NPs by this technique. "Grafting to" technique employs polymer as templates for the stabilization of Au NPs as core–shell structures. A simple solution phase synthesis was employed to reduce gold ions into Au NPs of controlled size and shape in presence of poly-(methylphenylphosphazene) (PMPP) of various concentrations which prevents the unlimited growth of the gold crystals under ambient temperature [169]. The stability of the polymer-mediated Au NPs can be further enhanced by encapsulation with block co polymers. For example, Xinjiao et al. documented the fabrication of Au NPs dimer systems using citric acid, which were stabilized in this state by further encapsulation within 154 block copolymers of polystyrene and poly (acrylic acid) [170]. Among various polymers, DNA is considered to be a good candidate for the stabilizing gold nanoparticles at various sizes and shapes due to its advanced features such as specificity, photochemical stability, and mechanical rigidity [171, 172]. Polymers have proved to be one of the best choices for the shape-directing synthesis and stability of gold nanoparticles [173, 174].

Dendrimers are synthetic macromolecules which are cauliflower shaped, highly branched with well-defined structure and composition [175, 176]. Dendrimers are used to stabilize Au NPs by various ways such as entrapping them into the macromolecular structure by the N, S, O, and P containing functional groups and by steric effects [177–181]. Such Au NPs are called as dendrimer-encapsulated Au NPs (Au-DENS). Au NPs are also stabilized by encapsulation by small dendrimer molecules at their surface boundaries which are known as dendrimer stabilized Au NPs. Au NPs can function as dendrimer cores, with coordinating ligands situated at the focal points of the dendrons to stabilize them. Such Au NP systems contain functional groups at the periphery of the dendrons, hence they are found to be useful for application as redox sensors [182, 183]. Dendrimer-encapsulated Au nanoparticles were first synthesized using polyamidoamine (PAMAM) as stabilizers by Crooks and Tomalia's research groups [184–186]. Since then, several dendrimers stabilized Au NPs of various types have been synthesized. Various factors which govern the synthesis of dendrimer-stabilized Au NPs such as dendrimer to Au Salt ratio [187–189] and type of functional groups in the dendrimer [190–193]. PEG dendrimer-stabilized Au NPs have been synthesized and stabilized by 1,2,3-triazole coordination by Click method [194]. In this method, Au NPs are either encapsulated by large dendrimer macromolecules or surrounded by small dendrimer molecules which are known as zeroth generation dendrimers. Au NPs have been synthesized using phosphorylcholine-functionalized dendrimers as stabilizers for use in biosensing applications due to good stability and controlled surface properties [195]. The inner structure of dendrimer-stabilized Au NPs has been investigated by Voelker's group using NICISS technique [191]. Cho et al. [196] reported the synthesis of four types of Newkome-type Dendron-stabilized Au NP (Figure 1.5A). The study encapsulates a comprehensive exploration of gold nanoparticles (AuNPs) conjugated with a variety of dendrons. It commences with the synthesis phase, which introduced four distinct dendron types, encompassing both two-directional dendrons with cystamine cores (G1—COOH and G2—COOH), a one-directional monothiol dendron (SH—G1—COOH), and a unique "di-anchored" dendron (TA—G1—COOH). The dendron conjugates, denoted as Au—G1—COOH, —G2—COOH, —SH—G1—COOH, and —TA—G1—COOH, were achieved through the amalgamation of dendron solutions with citrate-stabilized gold colloids. The characterization of these conjugates is delineated, featuring a multitude of analytical methods such as DLS (Figure 1.5B), UV-vis spectroscopy, XPS, and AFM (Figure 1.5C). These analyses unveiled crucial attributes, including optimal z-average sizes, minor redshifts in the SPR band, and highly uniform size distributions. The application of XPS further validated the presence of essential chemical functional groups on the AuNPs, while it became evident that steric effects wielded significant influence over the surface coverage of dendrons on the AuNPs. The stability studies constituted a pivotal aspect of the study, elucidating that the conjugates retained stability under conditions pertinent to biological contexts. These studies demonstrated the remarkable resilience of the conjugates across various media, including high-saline environments and a spectrum of pH values. Notably, the conjugates exhibited robustness even under highly acidic

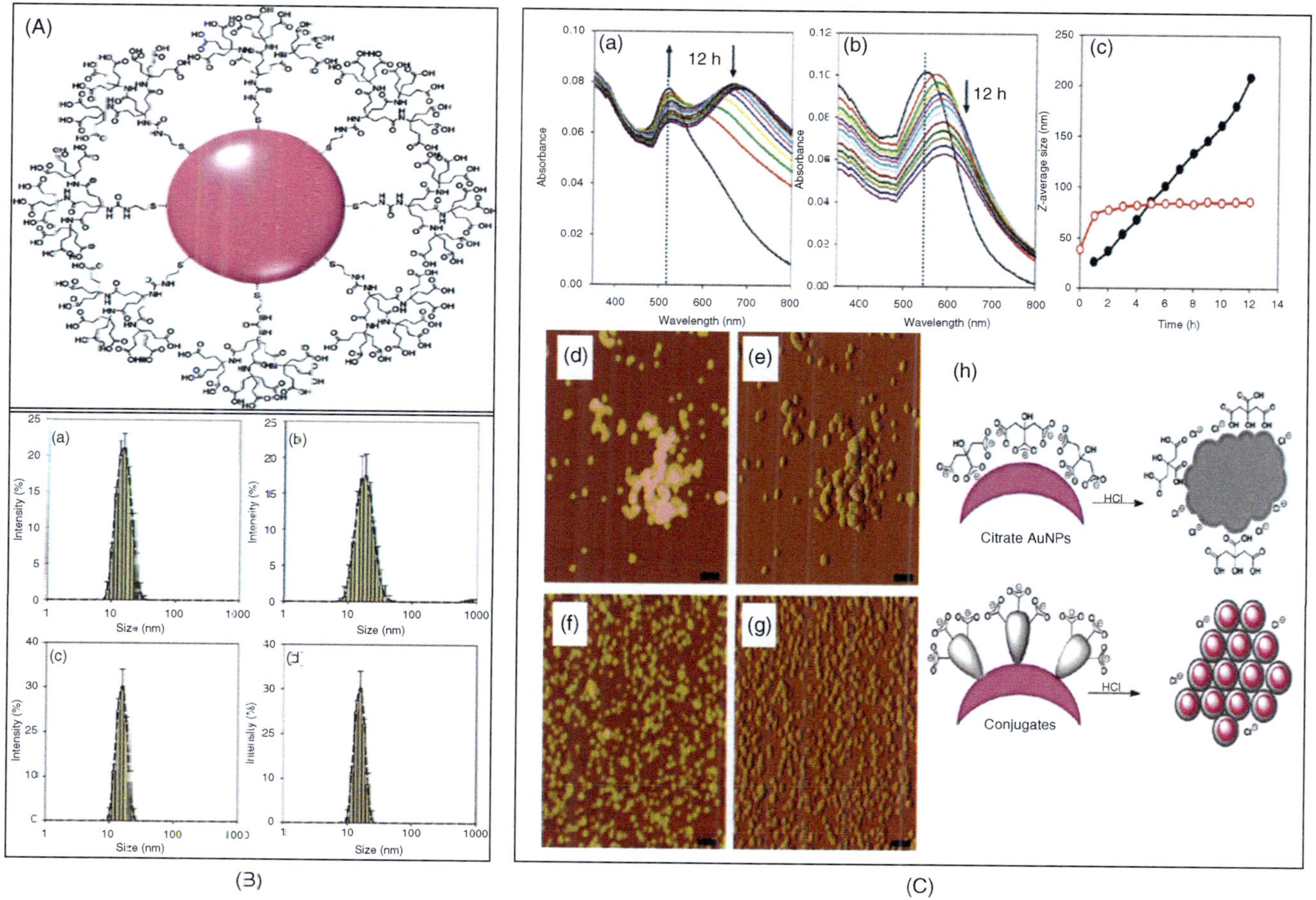

Figure 1.5 (A) Schematic representation of Newkome-type stabilized Au NPs. (B) DLS intensity-weighted size distributions of Au conjugates using NNLS: (a) Size profiles of Au-G1-COOH, (b) size profiles of Au-G2-COOH, (c) size profiles of Au-SH-G1-COOH, (d) size profiles of Au-TA-G1-COOH. (C) Time-dependent exploration of AuNPs in acidic environment (pH 3): (a) UV-vis absorbance patterns for citrate-stabilized AuNPs, (b) UV-vis absorbance patterns for Au-TA-G1-COOH with a notable 520 nm marker, (c) evolution of Z-average sizes over time for citrate-stabilized AuNPs and Au-TA-G1-COOH. Atomic force microscopy (AFM) imaging outcomes: (d) AFM height images of citrate-stabilized AuNPs, (e) AFM amplitude images of citrate-stabilized AuNPs, (f) AFM height images of Au-TA-G1-COOH, (g) AFM amplitude images of Au-TA-G1-COOH. Illustrating the mechanism of cluster formation: (h) Comparative schematic of cluster formation mechanism in citrate-capped AuNPs and AuNP-dendron conjugates under acidic conditions. Source: (a, b, c, h) Reproduced with permission from Cho et al. [196]/American Chemical Society.

conditions, with pH 3 causing minimal disruption. The revelation that chemical stability was governed by both dendron size (branch effect) and the S/Au ratio (Sulfur coverage effect) was underlined, with the latter emerging as the more prominent factor. Lastly, the paper addressed the issue of shelf life and explores the viability of lyophilization for semidry storage. Impressively, the conjugates proved to maintain stability over a minimum of six months under standard laboratory conditions. Furthermore, the preliminary foray into lyophilization hinted at the potential for extending their shelf life while retaining colloidal stability after resuspension.

In short, the section discusses the synthesis of Au NPs using polymer and dendrimer-mediated methods, exploring their potential applications across diverse fields such as optics, catalysis, and biology. The size and morphology of Au NPs are crucial for their functionalities, and the section delves into strategies for controlling these features. Polymeric stabilizers are highlighted for enhancing the long-term stability, solubility, and amphiphilicity of Au NPs. Two major synthesis techniques, "grafting to" and "grafting from," are discussed. The former involves using polymers with specific end groups to stabilize Au NPs, while the latter attaches polymer molecules to Au NP surfaces in the presence of radical initiators. Various ligand polymers, including thiolated PEG, polystyrene (PS), and poly(*N*-isopropylacrylamide) (PNIPAM), are employed to synthesize polymer-stabilized Au NPs. The size of Au NPs can be controlled by adjusting the polymer-to-gold ratio, and some polymers act as both stabilizers and reducing agents. Functionalized co-polymers, such as thermoresponsive hybrid core–shell Au NPs, are synthesized in situ using templates for stabilizing Au NPs inside polymer shells. Ligand substitution reactions are highlighted for their advantage in producing nearly monodisperse polymer-stabilized Au NPs.

The section also explores the use of block copolymers for enhancing the stability of polymer-mediated gold nanoparticles. DNA is recognized as a suitable candidate for stabilizing Au NPs due to its specificity, photochemical stability, and mechanical rigidity. Dendrimers, synthetic macromolecules with a highly branched and well-defined structure, are discussed as stabilizers for Au NPs. Different types of dendrimer-stabilized Au NPs, such as dendrimer-encapsulated Au NPs (Au-DENS) and dendrimer-stabilized Au NPs, are synthesized. Factors affecting the synthesis, including dendrimer-to-Au salt ratio and the type of functional groups in the dendrimer, are explored. The section concludes with a study by Cho et al., which investigates gold nanoparticles conjugated with various dendrons. The study introduces four distinct dendron types and discusses their synthesis, characterization, stability, and potential applications. The conjugates demonstrate resilience in various conditions, including high saline environments and different pH values, and exhibit stability over an extended period, even under lyophilization for semidry storage.

1.4.4 Photoreduction Method

The photoreduction process has demonstrated to be one of the most effective methods for synthesizing single-crystalline Au NPs [197]. Ultraviolet light spans between visible light and ionizing radiations and have the wavelength from 100 to 400 nm [198, 199]. UV light can be categorized into three groups, viz., UV-A having wavelength from 315 to 400 nm, UV-B with wavelength from 290 to 315 nm, and UV-C with wavelength between 100 and 290 nm [197]. In UV photochemical synthesis of Au NPs, the reducing agent may or may not facilitate the reduction of metal ions [200, 201]. In some cases, UV radiation is absorbed by the photosensitizer such as ketones, which produce free radicals by photolysis. These free radicals facilitate the reduction of metal ions into metal atoms. The metal atoms form metal cluster which subsequently grow to form metal nanoparticles [202, 203].

Shape-controlled synthesis of Au NPs by UV-irradiation technique was successfully achieved by Shengchun Yang et al. [204]. Size and shape-controlled synthesis of Au NPs by photochemical technique was found to be more effective in presence of various macromolecules such as dendrimers and surfactants. UV radiation encourages the photochemical reduction of aqueous Au (111) solution, whereas macromolecules act as soft templates for the production of Au NPs, thereby providing the steric hindrance and preventing the agglomeration of the particles [205]. In most cases, the particle size decrease with increase in the degree of polymerization. Heparin-assisted synthesis of negatively charged functionalized gold nanoparticles was obtained using photochemical synthesis method and were used for the SERS [206]. Au NPs of the size of 80 nm were obtained by irradiation of chloroauric acid with highly energetic γ-radiations. These highly energetic radiations result in the fast nucleation of the metal ions thereby giving no or less time to control shape of the resulting nanoparticles. However, UV radiations being less energetic results in the formation of radicals with lifetime of the order of milliseconds, which results in reduction of the metal ions over longer time periods, hence, provides the opportunity to exert control over the particle morphology [207]. During photochemical synthesis, morphology of Au NPs can be regulated by various reaction parameters such as temperature, pH, precursor concentration, stabilizers, and reaction time [208, 209]. The influence of pH and reagents on the fabrication of Au NPs was studied by Rodriguez et al. using UV lamp (256 nm). The results demonstrated that basic solution (pH 9) resulted predominantly in the formation of nanospheres [209]. The UV-Vis studies revealed only one SPR band at 515 nm. Yet, in acidic solutions with pH levels of 3 or 5, nanorods were generated, exhibiting two SPR bands. Cheng et al. [210] similarly explored the impact of pH on Au NP morphology and reported findings consistent with those described earlier. The effect of pH on the size and morphology of Au NPs was also studied by Unal et al. using red cabbage extract as a reducing and a stabilizing agent under the influence of UV light source [211]. Anthocyanins, which are natural pigments and secondary metabolites found in RCE, are detailed as key components responsible for the pH-dependent color variations in the synthesized AuNPs. The synthesis process, depicted in Scheme (Figure 1.6A), is described as a step-by-step transformation involving anthocyanin-Au^{3+} complex formation, charge transfer, and the ultimate production of stable zero-valent AuNPs. Anthocyanins are highlighted as efficient photo-reductants. The results of AuNP characterization at pH 7.4 are then presented, demonstrating time-dependent Au formation, the transition to colloidal AuNPs, and their stable, spherical shape. The impact of RCE extract concentrations on AuNP formation is also discussed, with higher concentrations favoring stable AuNP synthesis, while lower concentrations lead to inadequate reduction. Furthermore, the influence of pH on AuNP formation is examined, showing that more stable and colloidal AuNPs are obtained at pH 11 compared to pH 2.5 (Figure 1.6B). The impact of using single or multiple reducing or stabilizing agents on AuNP formation is addressed, with anthocyanins from RCE acting efficiently as both reducing and stabilizing agents (Figure 1.6C). The colloidal stability of RCE AuNPs in different NaCl concentrations is highlighted, emphasizing their enhanced stability compared to commercial AuNPs. Finally, the RCE AuNPs are discussed as effective catalysts for the conversion of 4-nitrophenol to 4-aminophenol (4-NP to 4-AP) based on their catalytic performance. The findings suggest that RCE AuNPs could have potential applications in various scientific and biomedical fields due to their stability and catalytic capabilities. The morphology of Au NPs is also influenced by the concentration of the precursor [212]. In most of the Au NPs synthesis, $HAuCl_4$ is the most commonly employed gold precursor [205, 209, 213, 214]. Silver nitrate has also been used to produce anisotropic Au NPs, as silver ions directly contribute to the emergence of the longitudinal SPR peak in Au NP solutions. The impact of varying concentrations of $HAuCl_4$ and $AgNO_3$ on the morphology of Au NPs was investigated by Sanabria-Cala et al. [215] using UV light (256 nm) in presence of CTAB, a cationic surfactant. Their findings indicated that an increase in the concentration of $HAuCl_4$ led to the formation of spherical Au NPs, attributed to the constrained growth of CTAB micelles. Higher concentrations of silver nitrate favor the

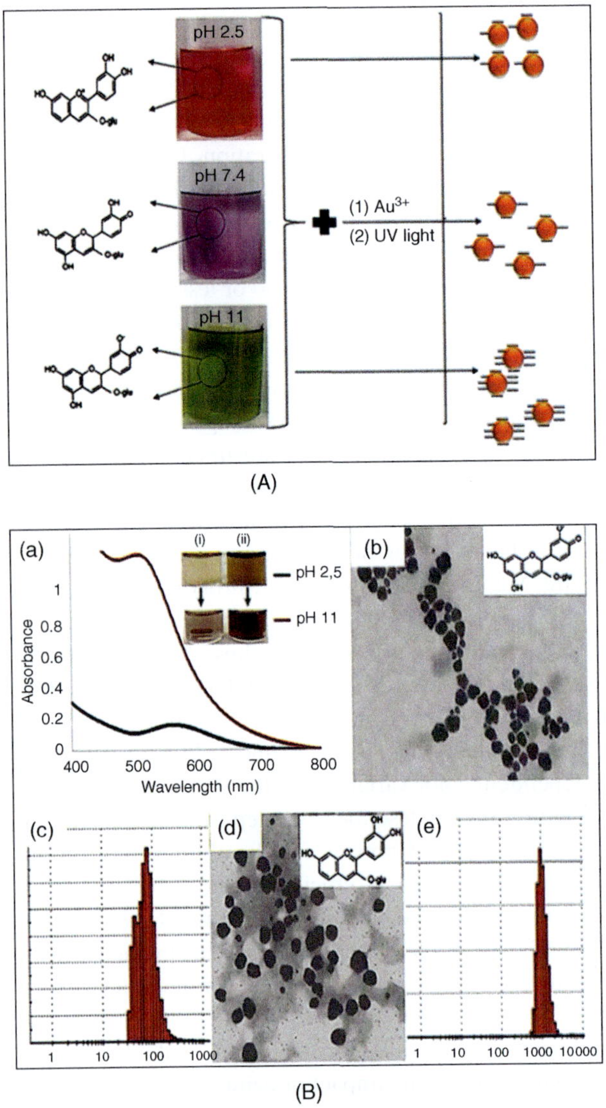

Figure 1.6 (A) Depiction of pH-dependent RCE Au NPs when exposed to UV irradiation. The color of RCE solutions, the electronic structure of anthocyanin, and the charge density on the surface of RCE Au NPs directly correlate with the reaction pHs. (B) Analysis of pH-dependent RCE Au NPs. (a) UV–Vis spectra of the Au NPs synthesized at pH 2.5 (depicted by the black line) and pH 11 (indicated by the red line). Inset in (a) Visual representation of the color of Au NP solutions at pH 2.5 (i) and pH 11 (ii). (b) Scanning electron microscopy (SEM) image of the Au NP formed at pH 11. Inset in (b) the molecular structure of anthocyanin carrying a negative charge. (c) Size distribution of the Au NP synthesized at pH 11. Similar characterization is performed for Au NPs formed at pH 2.5. (d) SEM image of the Au NPs. Inset in (d) the molecular structure of anthocyanin with a positive charge. (e) Size distribution of the Au NPs. (C) Analysis of the size and shape of Au NPs formed by varying reducing agents/capping agents (a) UV–Vis spectra showing Au NP absorption peaks. RCE alone: 537 nm (blue). Adding ascorbic acid to Au^{3+} shifted LSPR to 559 nm (red) and 547 nm (black) before and after RCE addition. (b) TEM image showing 27 nm Au NPs with RCE alone. (c) DLS spectra showing 70 nm effective diameter with RCE alone. (d) TEM image showing Au NPs size range (18–53 nm) with RCE added to Au^{3+} and ascorbic acid. (e) TEM image showing Au NPs size range (21–35 nm) with ascorbic acid added to Au^{3+} and RCE. (f) DLS spectra showing 185 nm effective diameter with RCE added to Au^{3+} and ascorbic acid. (g) DLS spectra showing 90 nm effective diameter with ascorbic acid added to Au^{3+} and RCE. Source: (a–c) Reproduced with permission from Unal et al. [211]/Elsevier.

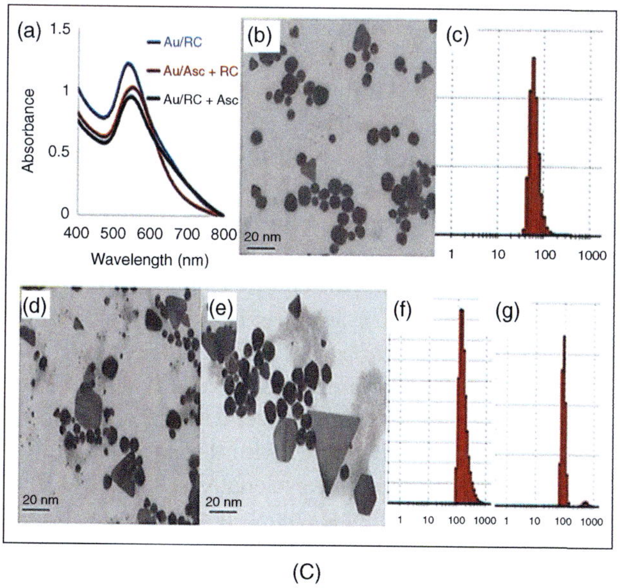

Figure 1.6 (*Continued*)

formation of anisotropic Au NPs. Numerous photochemical techniques for synthesizing Au NPs have utilized stabilizing agents including trisodium citrate, PVP, Triton X-100, citric acid, chitosan, and glycine [205, 213, 216–218]. Shiraishi et al. [219] investigated the influence of citric acid concentration on the morphology of Au NPs under UV radiation (254 nm). Higher concentration of citric acid produced smaller-sized Au NPs in higher yields which is indicated by the amplified intensity of the SPR peak around ~530 nm with increasing concentration of citric acid. Higher concentration of citric acid resulted in fast nucleation which subsequently led to the generation of tiny Au NPs. In some cases, formation of Au NPs was induced by UV radiations alone without the use of any chemical or biological reducing agent. For instance, Teixeira et al. fabricated Au NPs using UV light and polyethyleneimine (PEI) as the stabilizing agent without the use of any external reducing agent [217]. The influence of adjusting the concentration of $HAuCl_4$ (precursor) and PEI ([PEI] : [$HAuCl_4$] ratios of 5 : 1, 10 : 1, and 20 : 1) on the morphology of Au NPs was investigated. It was observed that Au NP formation did not occur in absence of PEI which indicated that it played a key role in the nucleation of Au atoms for the subsequent formation of Au NPs. Further increase in the ratio of PEI to precursor concentration led to the production of spherical Au NPs. Surfactants serve as commonly employed stabilizing agents, playing a pivotal role in the generation of Au NPs with diverse morphologies. Surfactants act as templates which regulate the growth the Au NPs and direct their shape and size [220]. Various types of surfactants have been used for this purpose such as anionic surfactant, e.g., SDS, SDBS, cationic surfactants such as CTAB, DTAB, and nonionic surfactants such as Tween X-80, Triton X-100, and hydrogenated castor oil (HCO). The nature of the surfactant and their structure also affects the NP size. For instance, the study by Shang et al. [221] systematically explored the influence of spacer length and hydrophobic chain length within gemini surfactants on the properties of synthesized gold nanoparticles. Gemini surfactants, specifically 16-s-16, were the focus of the investigation, varying the spacer length (s) through values of 3, 4, 6, 10, and 12. The research uncovered a fascinating interplay between these molecular parameters and the resulting nanoparticle properties. In the context of spacer length, initial observations from UV-Visible spectra of the $HAuCl_4$ solution revealed a charge transfer band at 222 nm and a moderate D–d transition band at 290 nm. However, after exposure to irradiation, a significant shift occurred, with the D–d transition band moving to 320 nm. When Gemini surfactants were introduced to the $HAuCl_4$ solution, the quaternary amine groups within these molecules interacted with the negatively charged $AuCl_4$ ions, leading to the formation of turbid suspensions. Crucially, after 12 hours of irradiation, these suspensions transformed into transparent solutions with a distinct purple color, implying the successful generation of Au NPs. The UV-Visible bands of these NPs displayed a notable absorption peak at around 540 nm, characteristic of the gold plasmon band. Importantly, the spacer length played a pivotal role in determining the absorption peak's position. Longer spacer lengths led to a blue-shift in the wavelength. This trend was further corroborated through TEM and AFM characterization, which confirmed that particle size exhibited an inverse relationship with spacer length, decreasing from about 22 to 7 nm as spacer length increased. This intricate relationship between spacer length and nanoparticle properties unveiled the potential for fine-tuning gold nanoparticles according to specific design requirements. The study delved into the effect of hydrophobic chain length while maintaining a consistent spacer length. This analysis demonstrated that the size of synthesized gold nanoparticles increased with the extension of the hydrophobic chains, with a significant influence on the absorption wavelengths. When the chain length was increased from 16 to 18, the impact of the hydrophobic chain on nanoparticle properties became less pronounced. In particular, it led to a red-shift in the absorption wavelengths. This experimental exploration showcased the pivotal role that both spacer length and hydrophobic chain length play in modulating the properties of gold nanoparticles, offering a glimpse into the tailored design of these nanoparticles for various applications. The research extended to investigate the synthesis of gold nanoparticles using traditional surfactants, including anionic (SDBS, SDS), cationic (DTAB, CTAB), and nonionic (Triton X-100) surfactants. The choice of surfactant was revealed to significantly impact the size and properties of the generated Au NPs. In particular, cationic surfactants led to the production of smaller nanoparticles. While anionic and nonionic surfactants produced nanoparticles with different characteristics. The fundamental role of surfactants in mediating the production of Au NPs was evident, with their presence leading to transparent solutions with a distinct purple hue, in comparison to the formation of macro-scale gold particles in the absence of surfactants. The study also proposed a mechanism for the formation of Au NPs in the presence of cationic gemini surfactants. This mechanism hinged on the photolysis and reduction of ($Au^{III}Cl_4$) ions dissociated from $HAuCl_4$. These ions were reduced to form Au metal, a process facilitated by the micelles formed by surfactants and their interaction with ions. Importantly, both spacer length and hydrophobic chain length influenced the aggregation of these micelles and the concentration of ($Au^{II}Cl_3$) ions, resulting in varying nanoparticle sizes and absorption wavelengths.

In short, the section discusses the photoreduction method as an effective route for synthesizing single-crystalline gold nanoparticles (Au NPs). Ultraviolet (UV) light, spanning between visible light and ionizing radiations, is utilized for the UV photochemical synthesis of Au NPs. This method involves the reduction of metal ions, facilitated by UV radiation or photosensitizers like ketones that generate free radicals through photolysis, subsequently reducing metal ions to form metal clusters and nanoparticles. Shape-controlled fabrication of Au NPs through UV irradiation is demonstrated, with size and shape being effectively controlled in the presence of macromolecules such as dendrimers and surfactants. UV radiation encourages the photochemical reduction of aqueous Au solutions, while macromolecules act as soft templates, providing steric hindrance and preventing particle agglomeration. The effect of various parameters, like pH, temperature, precursor concentration, stabilizers, and reaction time, on the morphology of Au NPs during photochemical synthesis is explored. pH, in particular, is shown to affect the formation of nanospheres or nanorods, providing insights into the control of Au NP morphology.

The section presents a case study using red cabbage extract (RCE) as a reducing and stabilizing agent under UV light, resulting in the fabrication of stable, spherical Au NPs. The study investigates the impact of RCE extract concentrations, pH, and the use of single or multiple reducing or stabilizing agents on Au NP formation, highlighting the stability and catalytic capabilities of RCE Au NPs. Furthermore, the fabrication of Au NPs using different precursors such as tetra chloroauric acid ($HAuCl_4$) and silver nitrate is discussed. Surfactants, including anionic (SDS, SDBS), cationic (CTAB, DTAB), and nonionic (Triton X-100) surfactants, are commonly used stabilizing agents that play a fundamental role in controlling Au NP growth, shape, and size. The study also investigates the influence of spacer length and hydrophobic chain length in gemini surfactants on the properties of synthesized Au NPs. Overall, the photoreduction method is presented as a versatile and controllable approach for synthesizing gold nanoparticles with tunable properties for various applications.

1.4.5 Using Reverse Micelles and Liposomes

Surfactants have been utilized to synthesize Au NPs within colloidal aggregates known as reverse micelles or water-in-oil (W/O) microemulsions [84, 222–225]. Reverse micelles are thermodynamically stable, transparent, and isotropic water in oil systems which are composed of water core or droplet dispersed in oil phase and stabilized by the surfactant molecules present at the interface of water and oil [226–230]. Water is confined in the inside polar core of the reverse micelles, which act as nanoreactors for the fabrication and growth of the Au NPs. The surfactant molecules present at the interface form a rigid or semirigid barriers that prevent the agglomeration of the Au NPs [231]. Further, the size of Au NPs is governed by the diameter of the reverse micellar water core [232]. Reverse micelles can be defined as nanoreactors that offer an optimal environment for the controlled nucleation and growth of nanostructures. Furthermore, the surfactant molecules provide the steric stabilization that prevents the Au NPs from aggregation. Different low molecular weight surfactants, including cetyl trimethyl ammonium bromide (CTAB), sodium dodecyl sulfonate (SDS), sodium bis(2-ethylhexyl) sulfosuccinate (AOT), Triton X-100, and others, can form reverse micelles to facilitate the creation of Au NPs with diverse sizes and shapes. Nanoparticles are prepared in reverse micelles using either a single microemulsion method or two microemulsion systems [233]. Single microemulsion method further includes energy triggering and microemulsion plus reactant technique [232]. In the energy-triggering approach, the introduction of a triggering agent into the microemulsion medium initiates the reaction. This activation of the medium sets off a cascade, resulting in the formation of nanoparticles. Pulse radiolysis and photolysis have been used as triggering agents for the formation of NPs [234]. In one microemulsion plus reactant method, one of the reactants such as gold salt is dispersed inside the aqueous cores of the microemulsion system, and the second reactant which usually acts as a precipitating agent (reducing agent such as $NaBH_4$) in the form of aqueous/nonaqueous or gas phase is introduced into the microemulsion medium. This microemulsion system can be prepared in an inverse manner with precipitating agent taken inside the aqueous cores of the microemulsions and the second gold salt fed from outside [232]. The reducing agent induces reduction of metal ions inside the aqueous cores of the microemulsion resulting in the formation of small nuclei which further aggregate to from NPs of specific size and shapes, which is governed by the size and shape of the reverse micelles. The whole process is diffusion driven since the second reactant diffusions into the water droplets of the microemulsions containing second reactant. In two microemulsion method, two separate microemulsion systems are prepared designated as ME-1 and ME-II. The aqueous phase of one of the microemulsions contain dissolved gold salt whereas second one contains reducing agent. The two microemulsions are then mixed under constant stirring during due to which Brownian motion of the reverse micelles leads to the intermicellar collusion and mixing of the micellar contents. Fusion and fission events occur during the micellar collusion leading to the formation of temporal intermittent reverse micellar structures. The intermittent reverse micellar systems allow exchange of the reaction precursors inside the aqueous cores resulting in the occurrence of the chemical reduction of metal ions. After the chemical reduction, the intermittent reverse micellar structures then break to from reverse micelles containing metal atoms which aggregate to form metal nanoclusters. Metal nanoclusters then aggregate to form nuclei which grow further to form NPs. Employing reverse micelles for the synthesis of NPs suffers from a major drawback. During Au NP synthesis, surfactant is employed in large quantities which gets incorporated in Au NPs during separation step [145]. Since most of the surfactant molecules are biologically toxic in nature, which limits their use in biological applications. The single-step purification of microemulsion synthesized Au NPs has been done by Hollamby et al by inducing phase transition and subsequent separation [235]. The study explored the synthesis and purification of gold nanoparticles (Au-NPs) using a microemulsion system. The color change from yellow to ruby red observed upon mixing the microemulsion with $AuCl^{4-}$ was indicative of successful Au-NP formation. UV spectroscopy confirmed this transformation, showing the disappearance of the peak around 400 nm and the emergence of a strong absorbance band cantered at 525 nm. This change happened rapidly, with UV-vis stopped flow data revealing an initial decrease in absorbance followed by a rapid increase, reaching a maximum within 35 ± 3 ms. Phase separation was induced by adding equal volumes of water to the Au-NP-containing microemulsion, resulting in two coexisting phases: an octane-rich upper phase and a water-rich lower phase. Surprisingly, the Au-NPs partitioned strongly into the upper phase, as evidenced by the distinct color difference between the two phases (dark ruby red for the upper phase and light pink for the lower phase). UV-vis spectroscopy further quantified this concentration difference, revealing that sample contained twice the concentration of Au-NPs as the initial homogeneous sample, and 18 times that of the lower aqueous phase. The concentration calculations showed that approximately 82% of the mass of Au-NPs from the original microemulsion was partitioned into the top phase after separation. This strong partitioning and purification process allowed for the efficient separation and purification of the Au-NPs from any excess CTAB, the surfactant used. These purified Au-NPs could then be dried and redispersed in fresh octane. This preference of Au-NPs for the nonaqueous phase is somewhat unexpected, considering the typical behavior of CTAB to form stable bilayers on negatively charged surfaces. The mixed octane and butanol organic solvent in the upper phase appeared to disfavor CTAB bilayer formation, facilitating the hydrophobization of the Au-NP surface. Any residual Au-NPs in the lower phase might still be stabilized by a CTAB bilayer, but further investigation is required. TEM images confirmed that the purified Au-NPs retained their size and shape, consisting of a polymorphic collection of spheres, triangles, and short rods. These nanoparticles exhibited a wide length-scale range from approximately 100 to 400 Å. There are numerous studies on the Microemulsion synthesis of Au NPs for various applications [84, 222, 223, 236–239].

However, using co-polymer-based micelles offers several advantages over classical surfactant-based micelles. The critical micellar concentration (CMC) of co-polymers is small in comparison to surfactants. The also exhibit higher kinetic stability as compared to surfactants. Further size and shape of co-polymer-based micelles can be easily altered by varying the composition, architecture, and length of constituent blocks of a co-polymer. The stability of Au NPs formed inside co-polymer-based reverse micelles can be improved by raising the length of the coronal blocks [240, 241]. Gold NPs have been synthesized using various types of co-polymers such as poly(ethylene oxide)-poly(propylene oxide)-poly(ethylene oxide), (PEO-PPOPEO) [242–246], Poly(styrene) blockpoly(vinylpyridine) (PS-b-PVP) [247, 248], polymer micelle stabilized poly(styrene) block-poly(4-vinylpyridine)((PS-b-P4VP)4) [249], polystyrene-block-poly(2 vinylpyridine) [250], and poly(ethylene oxide)-poly(*n*-butyl acrylate) (PEO-PnBA) [251]. Gold NPs synthesized in block co-polymers have been used for various applications such as biosensing [246] and catalysis [248, 252]. Gold NPs were synthesized by reducing the gold salt ($HAuCl_4$) in the polar cores of poly(styrene)-block-poly(2-vinylpyridine) (PS-*b*-P2VP) copolymer micelles by Moller et al. [253]. Three different reducing agents were used to reduce gold salt inside the microemulsions of different strengths such as HA, triethylsilane (TES), and potassium triethylborohydride (PTB). The size of the Au NPs was found to vary with the strength of the reducing agent used. The average diameter of the Au NPs formed was calculated to be 1.7, 2.6, and 8 nm using HA, TES, and PTB reductants, respectively. Rapid reduction of Au (III) ions occurred with HA and PTB, whereas TES proved to be milder reducing agent. DLS measurements demonstrated that incorporation of $HAuCl_4$ and reducing agents into PS-*b*-P2VP micelles caused the swelling of the resulting micelles. Gold dodecahedra and triangular nanoplates have been synthesized via one-step process reducing $HAuCl_4$ in presence of citric acid and Tetronic T904, an amphiphilic copolymer comprised of polyethylene oxide and propylene oxide units. Tetronic T904 co-polymer acted both reducing and stabilizing agent. TEM images showed the formation of dodecahedral Au NPs (85%) when using 0.5 mM $HAuCl_4$, 30 mM citric acid and 0.8 mM T904 and 1 mM HCl at 80 °C. The Au dodecahedra exhibited pentagonal symmetry the lateral size of 30 ± 4 nm and apex sizes of 19 ± 3 nm. The remaining 15% of the Au NP sample is composed of Au nameplates (Figure 1.7A). However, as the concentration of co-polymer was increased from 0.5 to 6 mM, the concentration of small triangular Au nanoplates increased in the sample. The product was found to contain 90% of Au nanoplates with an edge length of 40 nm. The remaining 10% included other shapes such as decahedra and octahedra (Figure 1.7B). The shape of the Au nanotriangles was found to evolve with time. After one hour of reaction, the sample mainly contained small Au NPs of undefined shapes. Spherical and polyhedral Au NPs were observed after six hours of reaction. Small triangular nanoplates with size of 15 nm were observed after 15 hours of reaction. As the reaction prolonged to 24 hours and then to 36 hours, a gradual improvement of size uniformity was observed with Au nanoplates of certain smaller-sized started disappearing whereas relatively larger Au nanotriangles became bigger (Figure 1.7C). The whole development suggested occurrence of Ostwald ripening and growth process. The synthesis of Au NPs inside co-polymers takes place in three steps [255]. First step involves the reduction of Au ions inside the crown ether like regions of the block co-polymers in solution. Second step involves the absorption of block co-polymers on the surface of Au NPs. Third step involves the growth of the smaller Au NPs stabilized by the polymer molecules. Various factors such as temperature, size, and shape of the reverse micelles and micellar environment are the main factors which govern the overall growth kinetic which lead to the Au NPs of specific morphologies [245, 256].

Liposomes are the supramolecular assemblies of amphiphilic liquids, which can be utilized as reverse micelles for the fabrication of Au NPs. A two-step approach for the synthesis of Au NPs inside liposomes has been performed by incorporating a reductant, $NaBH_4$ to a mixture of liposomes and $HAuCl_4$ [257]. In the first step, Au nanoparticles of 1 nm diameter were produced by employing an aqueous solution of 1,2-dipalmitoyl-sn-glycero-3-phosphothio-ethanol-phospholipid (PL). Subsequently, the size of Au NPs was increased from 1 to 5 nm through the interaction of PL with the Au NPs. Furthermore, it was determined that PL forms a double layer on the Au NP surfaces, and this interaction was facilitated by the thiol functionality of PL. Gold NPs have been synthesized inside glycerol incorporated nanosized liposomes of 1, 2-dioleoyl-sn-glycero-3-[phosphor-rac-(1-glycerol)] (DOPG) and 1-palmitoyl-2-hydroxy-sn-glycero-3-phosphocholine (lyso-PPC) in presence of a short-chain alkanethiol, 6-mercapto-1-hexanol (MCH) as the capping agent [258]. Effect of varying concentration of glycerol, reducing agent and capping agent on the size and morphology of Au NPs have been examined. Nanoliposomes were prepared using a combination of DOPG and lyso-PPC in the molar ratio of 88 : 12 with varying glycerol concentration (3-15% [v/v]). The authors initially explored the synthesis of Au NPs in glycerol. They observed the

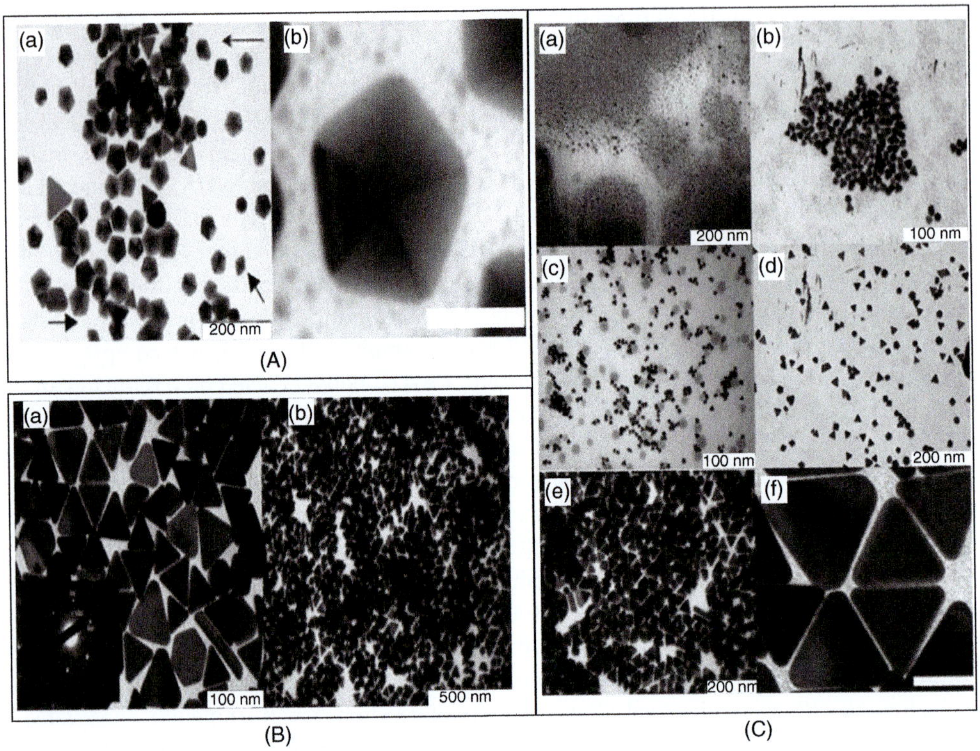

Figure 1.7 (A) Displays (a) transmission electron microscopy (TEM) images of Au decahedra that were synthesized at 80 °C in a solution containing 0.8 mM T904, 30 mM citric acid, 0.5 mM $HAuCl_4$, and 1 mM HCl. (b) The images reveal the fivefold twinned symmetry characteristic of decahedra. Notably, when the orientation of the decahedra to the electron beam changes, they exhibit a bipyramid geometry, as indicated by the arrows. The scale bar represents a length of 50 nm. (B) TEM images depict the ultrasmall Au nanoplates that were synthesized under the conditions of 80 °C, 6 mM T904, 12 mM citric acid, 0.5 mM $HAuCl_4$, and 3 mM HCl. Additionally, the inset in (a) shows the selected area electron diffraction (SAED) pattern, confirming the crystalline nature of the triangular nanoplates. (C) Provides a sequence of TEM images that illustrate the progressive formation of ultrasmall triangular nanoplates using the same conditions as in (A), captured at different time points: (a) 1 hour, (b) 6 hours, (c) 15 hours, (d) 24 hours, and (e) 36 hours. Furthermore, (f) presents a magnified view of the triangular nanoparticles. The scale bar in these images represents a length of 20 nm. Source: (a–c) Reproduced with permission from Goy-Lopez et al. [254]/Royal Society of Chemistry.

formation of relatively large and heterogeneous particles (10–50 nm) when chloroauric acid was mixed with glycerol in PBS solution at 25 °C in the absence of liposomes. However, when glycerol was incorporated into liposomes and a pH jump was applied, smaller and more homogeneous particles (5–10 nm) were obtained. This suggests the reduction of Au(III) to Au(0) was more effective when liposomes were used. Next, liposomes were introduced as nanoreactors for Au NP synthesis. The incorporation of glycerol within the liposome structure was expected to provide a semisolid reaction environment, favoring controlled nanoparticle formation. TEM images displayed the stages of the reduction reaction in the presence of glycerol-incorporated liposomal nanoreactors (Figure 1.8). Particles were observed both within the nanoreactors and in the surrounding solution after the elimination of the lipidic membrane. The authors then assessed the effect of glycerol concentration and the presence of a capping agent (6-mercapto-1-hexanol, MCH) on the particle size. They found that higher glycerol concentrations led to larger particles (7.7–6.4 nm), which indicated oxidation by-products of glycerol reacting with Au(III). However, the inclusion of MCH

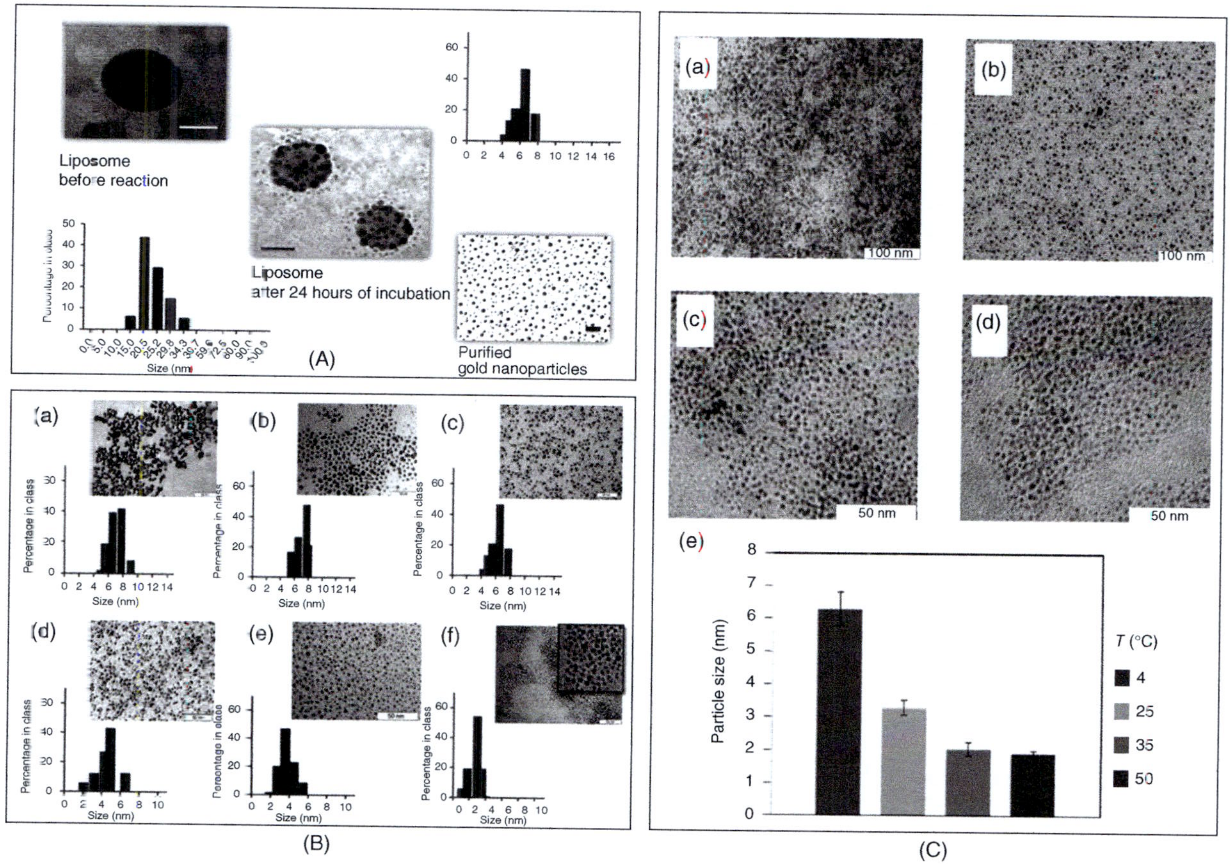

Figure 1.8 (A) Illustrates the progression of gold nanoparticle formation within liposomes containing 15% v/v glycerol. The images depict liposomes before the reaction (upper left), liposomes during the reaction (middle) (scale bars = 10 nm), and purified gold nanoparticles synthesized inside the nanoreactor (lower right, scale bar 20 nm). The graph in the lower left corner shows the size distribution of liposomes before the reaction (PI = 0.232), while the graph in the upper right corner displays the calculated nanoparticle size distribution using TEM. (B) The impact of capping agent and glycerol concentration is presented. The calculated particle size distribution and corresponding TEM images of particles are shown for two scenarios: (a–c) in liposomes without a capping agent (MCH) and (d–f) in the presence of MCH, with varying concentrations of glycerol (a, d) at 3%, (b, e) 10%, and (c, f) at 15%, after 24 hours of incubation at 25 °C. Scale bars in the images are 50 nm. (C) Demonstrates the influence of temperature on particle size and shape in glycerol (15% v/v)-incorporated liposomes with the capping agent MCH at different temperatures: (a) 4 °C, (b) 25 °C, (c) 35 °C, and (d) 50 °C after 24 hours of incubation. Additionally, graph (e) presents the calculated mean diameter of the particles ($n \geq 3$, scale bars = 20 nm). Source: (a–c) Reproduced with permission from Genc et al. [258]/American Chemical Society.

in liposomes resulted in a reduction in particle size (2.9–4.9 nm) due to the reducing effect of MCH (Figure 1.8B). The influence of temperature on particle properties was also investigated. Higher temperatures led to smaller particle sizes (1.9 nm at 50 °C compared to 6.3 nm at 4 °C) because of the increased reaction rate and rapid nucleation at elevated temperatures, leading to more efficient particle formation (Figure 1.8C). A quick one-step environmentally friendly synthesis of Au NPs has been performed by employing giant uni-lamellar vesicles (GUV) and ascorbic acid as a reductant [259]. Spherical Au NPs were formed at lower precursor concentrations, whereas nonspherical Au NPs were observed at higher precursor concentrations. Further GUVs were doped with cationic and anionic liquids to study the effect on Au NP size. The presence of anionic dopants led to the formation of larger Au NPs compared to cationic dopants.

In summary, this section explores the use of surfactants in synthesizing gold nanoparticles (Au NPs) within colloidal aggregates known as reverse micelles. Reverse micelles represent thermodynamically stable systems consisting of a water core dispersed within an oil phase, stabilized by surfactant molecules at the water/oil interface. The water core acts as nanoreactors for Au NP formation, with surfactants preventing NP aggregation by forming rigid or semi-rigid barriers. Various low molecular weight surfactants, such as CTAB, SDS, AOT, and Triton X-100, are employed to create reverse micelles for Au NP synthesis. Two methods, single microemulsion and two microemulsion systems, are discussed. Single microemulsion methods include energy-triggering and microemulsion-plus-reactant methods, utilizing triggering agents like pulse radiolysis or photolysis. Two microemulsion methods involve mixing separate microemulsion systems (ME-I and ME-II) to induce chemical reduction and NP formation.

A major drawback discussed is the incorporation of a large quantity of surfactant during Au NP synthesis, limiting biological applications due to the potential toxicity of surfactant molecules. A study by Hollamby et al. is mentioned, where a single-step purification method using phase transition and subsequent separation is employed to address this issue. The section then delves into the advantages of using co-polymer-based micelles over classical surfactant-based micelles. Co-polymers exhibit small CMCs, higher kinetic stability, and easy alteration of micelle size and shape by varying composition, architecture, and length. Various co-polymers, such as PEO-PPOPEO, PS-b-PVP, PS-b-P2VP, PEO-PnBA, have been used for synthesizing gold nanoparticles with applications in biosensing and catalysis. The use of liposomes, supramolecular assemblies of amphiphilic lipids, as reverse micelles for Au NP synthesis is also discussed. A two-step approach involves incorporating a reductant, $NaBH_4$, to a mixture of liposomes and $HAuCl_4$. The size and morphology of Au NPs can be manipulated by adjusting glycerol concentration, the choice of reducing agent, and the capping agent present within liposomes.

Overall, the section provides insights into diverse methods and techniques for synthesizing gold nanoparticles using surfactants, co-polymers, and liposomes, each with its advantages and applications.

1.4.6 Green Chemistry Methods

Though chemical methods of synthesis of NPs are most used due to good yield, but the toxic nature of the organic solvents and various organic/inorganic reducing agents limit their applications in biomedical areas [14, 84–86, 225]. Further chemical methods of synthesis make use of expensive stabilizing agents and other reagents. These and various other limitations of chemical methods of synthesis forced the researchers to work for the alternative green methods of synthesis which are eco-friendly as well as cost effective in nature [258, 260–262]. This endeavor for greener methods of synthesis gave rise to use of biological reducing agents and capping agents obtained from plants and animal sources. Hence, biological methods of synthesis became the focal point for synthesizing nanoparticles. Biological methods use plant or plant extracts, plant-derived macromolecules such as polysaccharides, microbes, and enzymes for the synthesis of NPs [263–265].

Plants have been gaining major attention for the synthesis of Au NPs due to their wide availability and low cost. Plants contain phytochemicals with hydroxyl, carboxy, carbonyl groups, and various other macromolecules such as polysaccharides in natural abundance which easily reduce Au^{3+} to Au and finally stabilize the Au NPs as well [266, 267]. Further, the plant-mediated synthesis involves simple procedure of mixing the plant extract with $HAuCl_4$ and wait till color of the resulting mixture turns red/purple. Synthesis of Au NPs has been carried out by using various plants such as *Aloe vera* [268], *Azadirachta indica* [269], *Medicago sativa* [270], *Coriandrum sativum* [271], *Lemon grass* [272], *Terminalia catappa* [273], and *Pelargonium graveolens* [274]. Spherical Au NPs of 20 nm size have been synthesized by using *Croton caudatus Geisel* extract. The resulting Au NPs have been studied for antibacterial, antifungal, and anticancer activity [262]. *Simarouba glauca* plant extract has been used by N. Thangamania and N. Bhuvaneshwari for synthesizing Au NPs [275]. The size and morphology of Au NPs have been affected by varying leaf broth concentration. HRTEM images reflect the Au NPs of various sizes and morphologies. However, the Au NP size decreased with increase in the leaf broth concentration. Promising antimicrobial activity against a variety of microbial strains was observed in the resulting Au NPs. Gold nanoparticles were synthesized utilizing aqueous leaf extract from *Mentha longifolia* [260]. Characterization involved XRD, UV-visible studies, SEM, TEM, EDX, and FTIR spectroscopy. XRD analysis revealed the presence of pure phasic Au with high crystallinity. TEM images exhibited the formation of spherical Au NPs with an average particle size of 36.4 nm. These Au NPs demonstrated efficacy against human breast cancer cell lines. Najlaa S. Al-Radadi synthesized Au NPs with a particle size of 43.17 nm using *licorice* root extract [261]. Gold NPs have been synthesized by using gum Arabic obtained from plants [276–278]. Due to good water solubility and biocompatibility, chitosan which is a second most abundant natural polymer has been used to synthesize Au NPs [218, 279]. The resulting Au NPs have been stabilized by the amine groups as well as by steric effects.

The environmentally friendly synthesis of Au NPs using microbes has garnered significant attention in recent years. However, bacterial-mediated synthesis of NPs has its own advantages as well as disadvantages. The main advantage lies in the easy handling and genetic manipulation of the bacterial systems for NP synthesis. The lack of proper control over size and shape of NPs and inability of large-scale synthesis is the main challenge for the bacterial-mediated synthesis. Control over size, shape, and crystallinity can be accomplished by properly understanding the underlying molecular and genetic mechanism of the synthesis of NPs which is yet to be properly understood. Gold NPs have been fabricated utilizing four different microbes, viz., bacteria, fungi, actinomycetes, and yeasts. However, only a few species of bacteria are enabled to reduce metal ions to metal atoms. These bacteria reduce metal ions extracellularly and intracellularly. For example, Au NPs in the range of 5–25 nm have been prepared intracellularly using *Bacillus subtilis* by Beveridge and Murray. These bacteria reduce Au^{III} to Au atoms and finally to Au NPs inside the cell wall. Gold NPs of different morphologies such as spherical, cubical, and hexagonal with sizes in the range of 5–200 nm have been produced intracellularly by various sulfate-reducing bacteria [280, 281]. Numerous reports also exist regarding the extracellular synthesis of Au NPs [282–284]. For example, the synthesis of Au NPs has been accomplished using cell-free supernatant of *Paracoccus haeundaensis* BC74171T [282]. TEM micrographs and DLS revealed that the synthesized Au NPs were sphere shaped, with an average grain size of 20.93 ± 3.46 nm. The surface of Au NPs was found to be negatively charged as revealed by the zeta potential measurements which gave a surface potential of −4.26 mV. EDX spectrum displayed a strong peak at 2.15 keV corresponding to the elemental gold. The reaction rate enhanced with an increase in temperature as is evident by the emergence of the SPR peak at 534 nm at the temperature 60 °C. Below 60 °C, no SPR peak was observed. The maximum peak intensity was recorded at 70 °C. Further, the peak intensity of the SPR peak was enhanced with an increase in incubation time which indicated the increase in number of Au NPs with time.

Since fungal mycelia can withstand higher pressures and agitation in bedchambers, hence they have been advantageously used for the fabrication of Au NPs as compared to microbial systems [285]. In recent years, a wide range of fungi has been recognized as potential candidates [286–289]. They also secrete numerous enzymes in large quantities, which act as bioreductants for the synthesis of Au NPs. Further, high metal tolerance and bioaccumulation capability of fungi makes them advantageous in Au NP synthesis as compared to bacteria and fungi. Further, large-scale synthesis of Au NPs is possible with fungi due to easy handling of biomass, downstream processing, and the ability to secrete large quantities of extracellular enzymes. Further, the Au NP size and morphology are controlled by various active biomolecules involved in energy metabolism of the cells like proteins such as ATPase, reducing sugars, glyceraldehyde-3-phosphate dehydrogenase, and binding proteins like 3-glucan. Like in bacteria, Au NPs can be fabricated both intracellularly and extracellularly using fungi. In intracellular synthesis, Au^{3+} ions are absorbed by the fungal cells and subsequently reduced by the biomolecules intracellularly, which result in the fabrication of Au NPs [290, 291]. The fungus *Verticillium* sp. has been utilized for the intracellular production of Au NPs. Gold nanoparticles displaying well-defined morphology and dimensions, measuring 20 nm in size, were detected on the cytoplasmic membrane of the fungal mycelium [292]. Ultrathin sections of the fungal mycelium when viewed TEM illustrated the presence of Au NPs exhibiting hexagonal, spherical, and triangular shapes on the cell wall. However, the Au NPs with quasi- hexagonal morphologies were observed on the cytoplasmic membrane. Spherical Au NPs with size of less than 10 nm were synthesized when Au (III) ions were added to the *Verticillium luteoalbum* and incubated at the pH of 3.0. However, the incubation at the higher pH of 5.0 led in the creation of Au NPs of triangular and hexagonal morphologies [293, 294]. In some cases, extracellular synthesis of Au NPs took place due to the significant extracellular production of bioenzymes, which acted as both reducing agents and capping agents simultaneously [294]. Shankar et al. employed an endophytic fungus, *Colletotrichum* sp., isolated from geranium plant leaves, for the rapid production of Au NPs [295]. Similarly, Au NPs have been produced extracellularly by using the fungus *Trichothecium* sp. to reduce Au (III) ions [296]. TEM images displayed the formation of highly polydisperse Au Nps which have a triangular, hexagonal, and rod like morphologies with size between 5 and 200 nm. It has been postulated that loosely bound proteins and enzymes secreted by the fungus is involved in the synthesis of Au NPs of various morphologies [296].

Numerous reports document the synthesis of Au NPs by yeasts. For example, bakers' yeast (*Saccharomyces cerevisiae*) has been employed to synthesize Au NPs. It was reported that aldehyde groups from the reducing sugars on the cell wall of the yeast reduce Au^{3+} ions to Au atoms and finally to Au NPs [55]. Gold NPs with spherical, triangular, and hexagonal morphologies with size less than 100 nm have been synthesized using *Yarrowia lipolytica* NCIM 3589, a tropical marine yeast. The Au NPs were found to be distributed mainly inside the cytoplasm of the yeast. Further, the reduction of Au^{3+} ions was observed to be pH dependent. Hexagonal and triangular Au nanocrystals were formed when the yeast cells were incubated at pH 2.0. At higher pH of 7.0 and 9.0, smaller Au NPs of the average grain size of 15 nm were produced [56]. Gold NPs of various morphologies have been synthesized using the yeast, *Magnusiomyces ingens* LH-F1. UV-Visible studies displayed a time-evolved SPR peak at 540 nm with little on no peak shifting with time. This unshifted SPR peak indicates the no significant agglomeration of Au NPs with time due to efficient capping by biomolecules of the yeast. Gold NPs were synthesized by employing the cell-free extract of *M. ingens* LH-F1 yeast [297]. The confirmation of Au NP formation was additionally validated through UV-Visible spectroscopy, where a distinct SPR band was observed at around 540 nm. TEM images displayed the spherical and pseudospherical and anisotropic Au NPs with size less than 28 nm. The lower concentration ((25–50 mg/L) of cell-free extract led to the formation of Au NPs. of various morphologies such as triangular, hexagonal, and anisotropic structures. Whereas higher concentration (100–200 mg/L) produced mostly spherical Au NPs of smaller sizes.

In short, this section discusses the limitations of chemical methods for synthesizing nanoparticles (NPs), for instance, the utilization of hazardous organic solvents, expensive stabilizing agents, and reagents. Due to these drawbacks, researchers have explored alternative "green" methods of synthesis, particularly biological methods using reducing and capping agents derived from plants, microbes, and enzymes.

Plants, with their abundant phytochemicals and macromolecules like polysaccharides, are highlighted for their role in synthesizing gold nanoparticles (Au NPs). Various plants, such as *A. vera*, *Azadirachta indica*, *Medicago sativa*, and others, have been utilized for the eco-friendly and cost-effective synthesis of spherical Au NPs with different sizes. The resulting NPs have shown activities against bacteria, fungi, and cancer cells.

Microbes, including bacteria, fungi, actinomycetes, and yeasts, are investigated for their capacity to synthesize Au NPs. Bacterial synthesis is advantageous for easy handling and genetic manipulation, but challenges include controlling NP size and shape. Bacteria like *B. subtilis* and sulfate-reducing bacteria are discussed for their intracellular and extracellular synthesis of Au NPs.

Fungi are highlighted as potential players for Au NP biosynthesis because of their high tolerance to metals, bioaccumulation capability, and the ability to secrete large quantities of extracellular enzymes. Both intracellular and extracellular synthesis methods using fungi such as *Verticillium* sp. and *Colletotrichum* sp. are discussed. Fungal mycelia's ability to withstand higher pressures and agitation is noted as an advantage. Yeasts, including *Saccharomyces cerevisiae* and *Y. lipolytica*, are mentioned for their role in synthesizing Au NPs through reducing sugars on their cell walls. The pH dependence of Au NP synthesis by *Y. lipolytica* is highlighted.

Overall, the section emphasizes the shift toward green synthesis methods using biological entities for the production of Au NPs, offering eco-friendly and cost-effective alternatives to traditional chemical approaches. The specific examples provided showcase the potential applications of these synthesized NPs in various fields, including medicine and nanotechnology.

1.5 Synthesis and Characterization of Silver Nanoparticle

The synthesis and characterization of silver nanoparticles (AgNPs) represent a vibrant and essential area of nanotechnology research, offering numerous applications in diverse fields. Silver NPs are particularly intriguing because of their distinctive electronic, optical, and antibacterial characteristics. The process of synthesizing these nanoparticles involves reducing silver ions to nanoscale silver structures, and various methods, like chemical, physical, and biological approaches, are employed. Characterization techniques such as UV-Visible spectroscopy, TEM, SEM, XRD, and FTIR are crucial for understanding the size, shape, crystallinity, and surface characteristics of the synthesized AgNPs. The ability to precisely control the synthesis parameters and characterize the resulting NPs is pivotal for tailoring their properties to meet specific application requirements, ranging from medical and environmental uses to catalysis and electronics. This interdisciplinary field continues to advance, driven by the need for efficient and sustainable nanomaterials in various technological domains. Different methods for synthesizing silver nanoparticles can be categorized under the following subheadings.

1.5.1 Physical Approaches

Silver nanoparticles have been produced using physical methods such as the vaporization-condensation method and laser ablation process [298]. The evaporation–condensation method is conducted within a furnace tube under constant atmospheric pressure. The material is vaporized inside the furnace in presence of a carrier gas stream. However, this approach of synthesis of AgNPs has certain limitations such as large space occupied by the furnace, large power consumption, and long preheating time to achieve a stable operational temperature. However, the method has been modified by Jung et al. [299] to synthesize AgNPs by employing a tiny ceramic heater to generate a localized heating effect. The Ag metal source was heated on a heater, and the Ag vapors were conveyed by the carrier gas. Due to the rapid cooling of the aerosol generator in comparison to the conventional furnace tube furnace, the Ag vapors were transformed into Ag nanoparticles. The geometric mean diameter of the Ag particles, as well as the particle concentration, was observed to be directly proportional to the temperature of the heater surface and increased from 6.2 to 21.5 nm. Silver NPs were fabricated by Raffi et al. [300] by inert gas condensation method. The influence of diverse experimental parameters, including evaporation temperature and pressure of the inert gas on the size and morphology of AgNPs, was studied. Low evaporation temperature and inert gas pressure led to in the production of smaller AgNPs. With the evaporation temperature of 1123 K, the size of the Ag nanoparticles rose from 9 to 24 nm with an elevation in inert gas pressure from 0.5 to 100 Torr. Similarly, for the same inert gas pressure (e.g., 5 Torr), the particle size increased from 14 ± 2 to 20 ± 2 nm. Spherical AgNPs with particle size of 50, 90, and 130 nm have been synthesized by Harra et al. by evaporation–condensation technique [301]. Silver NPs have been produced via laser ablation of the bulk materials. Laser ablation synthesis of AgNPs can be traced to the year 2000 when Mafune et al. [302] synthesized AgNPs by laser ablation in aqueous solution of dodecyl sulfate (SDS). The effect of the SDS concentration and laser power on the average particle size of AgNPs has been studied. The average particle diameter decreased from 16.2 ± 4.0 to 11.7 ± 5.3 nm with increased in SDS concentration from 0.003 to 0.05 M, respectively. However, the average particle diameter increased from 7.9 ± 3.3 nm to 12.8 ± 4.1 nm with rise in the laser power from 40 to 70 mJ/pulse, respectively. Recently AgNPs have been fabricated using Nd : YAG laser by laser ablation method [303]. PLA has been focused on 99.9% pure Ag plate in water medium. The laser shots resulted in the production of AgNPs. The production of AgNPs was verified through UV-Visible spectroscopy, showing an SPR peak around 450 nm. Silver NPs have been created by Rafique et al. [304] in a single-step green method using two different lasers in de ionized water medium. Here the effect of laser type on the size and morphology of AgNPs has been studied. TEM images displayed the production of AgNPs with spherical shapes with size in between 10 and 30 nm using CW diode laser, whereas pulsed Nd : YAG laser resulted in the formation of monodispersed AgNPs with size between 4 and 14 nm. Further, the bandgap was found to be vary inversely to the Ag NP size. The bandgap values obtained for extrapolation of UV-Visible spectroscopy data were observed to be 2.47 eV for AgNPs synthesized by CW laser and 2.67 eV for AgNPs synthesized by pulsed laser. Although the physical methods produce AgNPs with high yield and high purity without using environmentally hazardous chemicals and production of toxic by-products, but the control over nanoparticle size and agglomeration is a great challenge. Further, the physical methods use complex equipment which have high-power consumption and requires long durations for nanoparticle synthesis.

1.5.2 Chemical Approach

Chemical methods of synthesis are found to be more convenient, efficient, and easy to handle. Chemical methods of synthesis involve the reduction of silver salts by suitable reducing agents to synthesize Ag nanostructures with high productivity and at a very minimal cost. The chemical fabrication of AgNPs involves three components, viz., silver precursor, reducing agent, and capping agent. The size and morphology of AgNPs are governed by the dual stage, nucleation, and growth. Even the formation of primary nuclei is regulated by different reaction parameters like pH, precursor concentration, temperature, nature of reducing, and capping agent. [305–307]. The reduction of silver salts can be accomplished using a variety of inorganic and organic reducing agents which include metal complexes, block co-polymers, and biological macromolecules [85, 86, 266, 306–313]. Silver salts get reduced by receiving electron from reducing agents with the result Ag is obtained in a zerovalent state. The Ag atoms then agglomerate and grow by undergoing nucleation and growth to form colloidal Ag of variable sizes and shapes. The strength of the reducing agent greatly affects the size and morphology of AgNPs. For example, strong reductants like $NaBH_4$ yield small and monodisperse AgNPs with low stability [86]. However, weak reductants result in the production of bigger AgNPs due to slower reduction rate. Further weaker reducing agents provide enough time to control shape and size distribution of AgNPs [314]. The most critical aspect of Ag NP synthesis is the stabilization of the dispersive AgNPs [315, 316]. In order to achieve stabilization of AgNPs, various capping agents have been used such as polymers like PEG, PVP, PMMA, PMAA, biopolymers like cellulose, chitosan, and various other macromolecules like oleylamine gluconic acid [316–320]. Stabilization of AgNPs by stabilizing agents can be accomplished through either electrostatic or steric repulsions. Some capping agents provide steric stabilization to AgNPs by imparting negative charge to the NP surface such as halides, citrate, carboxylates, and polyoxoanions. Whereas few capping agents impart positive charge to the Ag NP surface such as branched PEI, etc. There are some macromolecules which impart steric stabilization to Ag NP solutions such an organic polymers and alkyl ammonium cations. Such bulky groups on the Ag NP surface undergo steric repulsion and impart stabilization to the Ag NP solution. For instance dodecanthiol capped AgNPs have been fabricated by Oliveira et al. which exhibit good stability in aqueous solutions [321]. Numerous reports have been documented on the synthesis of colloidal Ag using various stabilizing agents. Luo et al. [322] investigated the impact of PEG on the growth of Ag nanoparticles in the absence of other chemical agents. Here PEG acted both as a reducing agent and a stabilizing agent. The chain length of the glycol significantly influenced the formation of Ag nanoparticles. Ethylene glycol proved to be inactive in reducing Ag^+ ions even at high temperatures of 80 °C. However, AgNPs were effectively synthesized by using PEG 2000 under similar experimental conditions. With increase in polymer chain length of PEG, the rate at which Ag^+ ions are reduced to Ag nanoparticles increased significantly. The particle size enhanced from 10 to 80 nm with rise in reaction temperature. Silver nanoparticles have been fabricated using formaldehyde as a reductant and PVP as a stabilizer employing two different organic bases, triethyl amine and pyridine [323]. Triethyl amine being the more basic in nature resulted in the formation of AgNPs of larger sizes (20–30 nm). Whereas pyridine being less basic in nature as compared to triethyl amine produced smaller-sized AgNPs (10–20 nm). Monodisperse AgNPs have been produced using oleyl amine-paraffin system by Chen et al. [324]. It was reported that Ag NP synthesis occurred in three stages, viz., growth, incubation, and Ostwald ripening. This method used liquid paraffin as a solvent and Oleyl amine as a reductant and a capping agent. Colloidal silver nanoparticles with stability have been successfully synthesized by utilizing 2-(dimethylamino) ethanol (DMAE) as a reductant and PVP as a stabilizer [325]. The influence of the reducing agent and the dispersant on the particle size and morphology has been studied. Room temperature synthesis of AgNPs was achieved by this method. The AgNPs obtained were small and displayed uniform size distribution, with a diameter of less than 10 nm. Further, the Ag NP sample was found to be redispersible in water with good electroconductivity. Silver nanoparticles have also been conveniently prepared using low molecular weight compounds such as aldonic acid like sodium gluconate as a dispersing agent instead of high molecular weight compounds such as PVP [326]. This method produced AgNPs with sizes less than 10 nm with good electrical conductivity. Aldonic acid, a low molecular weight compound, demonstrated efficacy as a stabilizing agent for the fabrication of dispersible Ag nanoparticles compared to high molecular weight compounds like PVP [326]. Further, this dispersing agent reduced the sintering temperature (150 °C) of AgNPs as compared to those prepared using other polymers such as PVP (200 °C). Ag nanoparticles have also been synthesized using the polyol method. In this method, silver salt is mixed with ethylene glycol in various rations and then heated. The stability of the AgNPs can be enhanced by using other polymeric stabilizers such as PVP. For instance, AgNPs have been produced by Kim et al. using polyol method [327]. Two distinct synthetic methods have been employed to produce Ag nanoparticles, aiming to investigate the influence of reaction parameters on particle size and morphology. In the first method called precursor solution method, silver nitrate was dissolved in ethylene glycol along with PVP (Mw = 10,000). The mixture was then refluxed at the temperature of 100–150 °C with heating rate of 1–7.5 °C/minute. Once the reaction was complete, AgNPs were extracted from the solution by centrifugation. In the second method called as injecting precursor method, first PVP was dissolved in ethylene glycol and heated to achieve the desired reaction temperature. Next aqueous solution of silver nitrate (40%) was injected into the hot mixture of PVP-ethylene glycol solution maintained at reaction temperature. Later, the AgNPs were recovered by centrifugation. Silver nanoplates and nanobelts have been prepared by refluxing suspension of silver nano seeds of average diameter of 3.5 nm [328]. The silver nano seeds were in turn synthesized by reducing $AgNO_3$ with $NaBH_4$ in presence of sodium citrate and PVP. The Ag nano seeds exhibited as sharp SPR peak at 400 nm. After refluxing the silver nanosuspension, four distinct SPR peaks emerged which were located at 330, 525, 855, and 1190 nm. The emergence of these SPR peaks reflected the transformation of Ag nano seeds into different morphologies. It was observed that Ag nano seeds transformed into triangular nanoplates (95%) and nanobelts or nanowires (5%). The silver nanoplates self-assembled into monolayered structures on TEM grid. The ED revealed each nanoplate to be a single crystal and the hexagonal symmetry of the patterned spots indicated two triangular faces of each nanoplate were bound by {111} faces. The nanowires were separated from the nanoplates by ultracentrifugation. Such Ag nanowires were found to have a diameter of 9 nm with wire length of 25 μm. Single crystal Ag nanocubes have been obtained by Im et al. by a typical polyol method by reducing $AgNO_3$ with ethylene glycol (EG) in presence of PVP [329]. The change in morphology of the twinned Ag nanocrystals was induced by addition of hydrochloric acid in minute quantities along with nitric acid formed *in situ* of the reaction. At lower concentration of nitric acid, a mixture of silver nanocubes, tetrahedrons, and nanowires were obtained after 25 hours of the reaction when 0.125 mM of HCl was used. This mixture of morphologies resulted due to incomplete itching of Ag-twinned seeds. Monodisperse Ag nanocubes with an edge length of 130 nm were formed after 26 hours exclusively when concentration of HNO_3 was increased to 0.25 mM. However, a totally different morphology of Ag nanostructures comprising of Ag nanowires and irregular particles was obtained with further increase in concentration of HCl to 0.37 mM (Figure 1.9A). The Ag nanocubes synthesis was scaled up by increasing the volume of the solutions by five times. The SEM images of the Ag NP sample obtained from the scale-up synthesis displayed the formation of Ag nanocubes exclusively with average edge length of 125 nm (Figure 1.9B). XRD pattern displayed the occurrence of (200) peak with abnormal intensity. The formation of single crystal Ag nanocubes with sharp edges and corners is observed in TEM image. The inset of the figure provides the EDS pattern of the AgNPs which also confirm the formation of single Ag nanocrystals. The polyol method was also used by Tao et al. to synthesize polyhedral Ag nanocrystals by reducing silver salt in diol solvent in presence of PVP as a stabilizer and

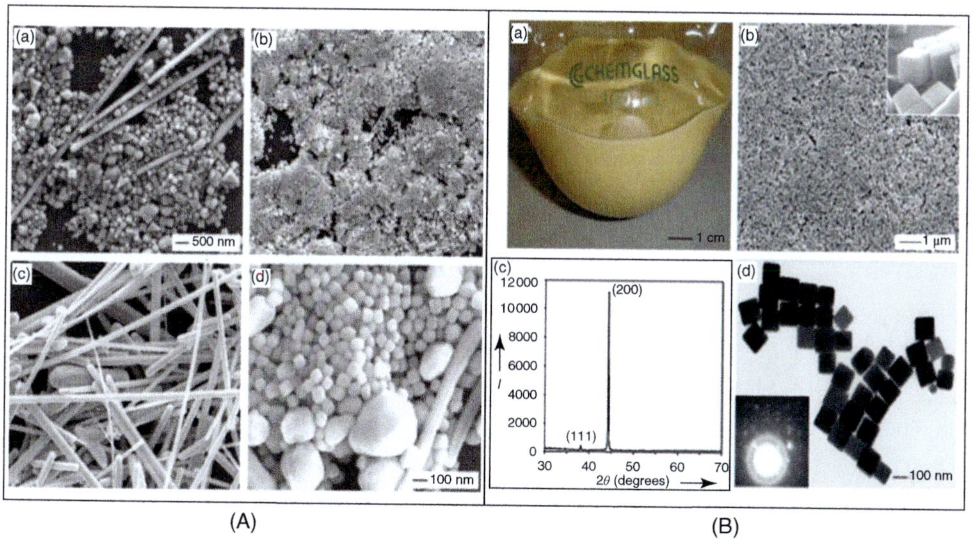

Figure 1.9 (A) Scanning electron microscopy (SEM) images depict silver nanoparticles synthesized under various concentrations of HCl, with a scale bar of 500 nm for each: (a) at 0.125 mm, (b) at 0.25 mm, and (c) at 0.375 mm. The reaction mixture consisted of $AgNO_3$ (23.5 mm) and PVP (36.7 mm, calculated in terms of the repeating unit), and the synthesis process involved heating in an oil bath maintained at 1408C. In (d), silver nanoparticles were produced using the same conditions as before, except that HNO_3 (0.25 mm) was used instead of HCl (0.25 mm). (B) (a) Image of the reaction mixture in a scaled-up (five times larger) synthesis. (b) Representative SEM pictures of the silver nanocubes produced in their as-synthesized state. The inset provides a closer view of the SEM image, highlighting the well-defined sharp corners and edges of these nanocubes. (c) XRD pattern obtained from the same batch of silver nanocubes. (d) TEM image displaying the silver nanocubes, with the inset showing an electron diffraction pattern recorded by aiming the electron beam perpendicularly at the (100) facet of a silver nanocube. Source: (a, b) Reproduced with permission from Im et al. [329]/John Wiley & Sons.

shape-controlling agent [330]. A variety of Ag nanocrystal shapes have been obtained such as cubes (80 nm) and truncated cubes (120 nm), cuboctahedra (150–200 nm), octahedra (250–300 nm), and truncated octahedra (200–250 nm), etc. (Figure 1.10A). Silver nano bars (aspect ratio; 2.7) and nano rice have been synthesized by Wiley et al. by polyol method by varying the concentration of bromide ions [331]. The nano bars were anisotropic with length of 146 nm and width of 56 nm. Silver nano bars have sharp edges and corners and are single crystals bounded by (100) facets. The edges and corners were latter rounded and transformed into single crystal nano rice by storing the Ag nano bars in 5% aqueous solution of PVP at room temperature for one week (Figure 1.10B). Both the types of Ag nanostructures exhibited two SPR peaks in visible and near IR regions. In case of both Ag nanobars and nano rice, the position of longitudinal SPR peaks red-shifted with increase in the aspect ratio, whereas not much change in position of transverse SPR was noticed (Figure 1.10C). Silver nanocubes, Ag bi pyramids, and Ag nanowires have been synthesized by the same group, i.e., Wiley et al. by a polyol process with PVP and Ag NO_3 in the molar ratio of 1.5. Silver nanocubes, right bipyramids, and nanowires were produced by the subsequent addition of Cl^-, Br^-, polyethyleneimine and Fe^{3+} ions to the PVP solution [332]. Single crystal Ag was grown into truncated cubes or sharp cubes by the addition of chloride ions which depended upon the kinetics of the reaction. Addition of NaCl as a Cl^- ions source produced truncated Ag nanocubes after 26 hours of the reaction, however, replacement of NaCl with HCl produced sharp Ag nanocubes. The 75 nm truncated Ag nanocubes exhibited dipolar SPR at 440 nm with two small shoulders associated with the main SPR suggested the occurrence of small amount of anisotropy. However, the 90 nm sized

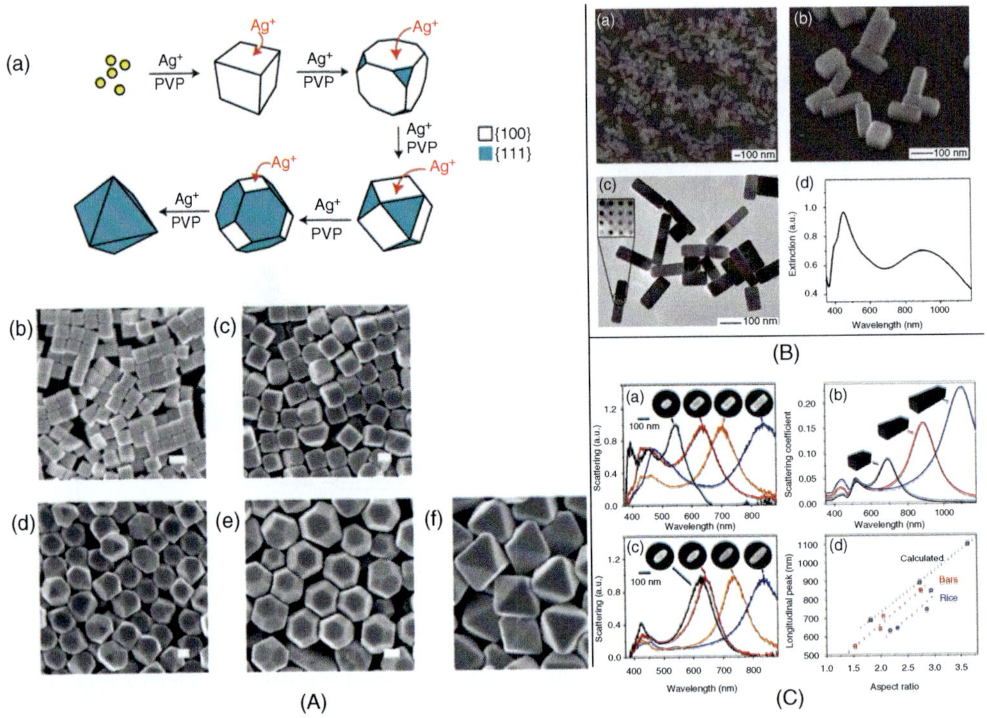

Figure 1.10 (A) By prolonging the polyol reaction for a specific duration, it is possible to achieve a high yield of various polyhedral shapes that are enclosed by {100} and {111} facets. (a) An illustration of the nucleation and growth process, where silver consistently accumulates on the {100} facet, eventually forming a fully {111}-enclosed octahedron. (b–f) Scanning electron microscope (SEM) pictures showing cubes, truncated cubes, cuboctahedra, truncated octahedra, and octahedra, in that order (scale bar: 100 nm). Source: Adapted from Ref. [330] with permission. Copyright (2006) John Wiley & Sons. (B) (a) Scanning electron microscope (SEM) image of silver nanobars. (b) SEM image of the same sample tilted at a 45° angle. (c) Transmission electron microscope (TEM) image of the silver nanobars, with the inset displaying a convergent beam electron diffraction pattern, confirming that the nanobars are single crystals enclosed by (100) facets. (d) The visible and near-infrared (vis-NIR) extinction spectrum of the nanobars exhibits a transverse plasmon resonance in the visible range and a longitudinal resonance in the near-infrared region. (C) (a) SEM images of individual nanobars along with their corresponding normalized scattering spectra. The longitudinal plasmon peak of the nanobars shifts toward longer wavelengths as the aspect ratio increases. (b) Scattering spectra of nanobars with lengths of 100, 150, and 200 nm, while maintaining a width of 55 nm and a height of 50 nm, calculated using discrete dipole approximation (DDA). (c) SEM images of individual nanorice and their corresponding normalized scattering spectra. (d) A plot illustrating the relationship between the longitudinal plasmon peak position and the aspect ratio. Both nanobars and nanorice exhibit redshifts in their peaks as their lengths increase, but, on average, the nanobar peaks are shifted 80 nm more towards longer wavelengths compared to nanorice. Source: (a) Reproduced with permission from Wiley et al. [331]/John Wiley & Sons, (b, c) Reproduced with permission from Wiley et al. [332]/American Chemical Society.

sharp Ag nanocubes exhibited a sharp dipolar SPR at 500 nm. Using NaBr in place of NaCl resulted in the formation of single twin crystals. These single twin seeds latter grew to form right bi pyramids with two different sizes of 75 and 150 nm. The most intense dipole SPR of 75 nm bipyramids exhibited a red-shifting of 100 nm from sharp Ag nanocubes. With increase in size of Ag nano bi pyramids, there was a shift in peak intensity from 530 to 742 nm due to enhanced charge separation and energy loss in 150 nm sized bi pyramids as compared to 75 nm sized bi pyramids. Silver nanowires have been produced from multiple twinned Ag seeds by preventing the etching either by running reaction under argon atmosphere or in presence of Fe (II) or Fe (III) in the molar ratio of Ag : Fe = 11,000. Multiple twinned Ag nano seeds produced initially by the addition of both iron (II) acetylacetonate ($Fe(acac)_2$) and Cl^- grew into nanorods of 20 nm diameter and 150 nm length which latter rapidly grew into Ag nanowires of micrometer length after 60 minutes of the reaction. UV-Visible spectrum showed the occurrence of SPR peak at 400 nm for multiple twinned dodecahedral Ag nano seeds which shifted to 470 nm as the nano seeds transformed into Ag nanorods. But as the nanorods transformed into nanowires, the extinction band broadened further, but the most intense SPR peak blue shifted to 382 nm. Silver nano prisms have been synthesized by Metraux and Martin by chemical reduction by employing sodium borohydride and sodium citrate as reducing agents in presence of PVP and H_2O_2 [333]. The production of nano prisms was confirmed spectroscopically by occurrence of in plane dipole surface plasmon band. A large red-shifting in bands has been observed with increase $NaBH_4$ concentration. This red-shifting in the SPR bands has been ascribed to the increase in edge lengths and tip sharpness of the Ag nano prisms. The edge length of the nano prisms enhanced from 31 ± 7 to 39 ± 8 nm with rise in $NaBH_4$ concentration from 0.3 to 0.8 mM. Multibranched Ag nanoflowers have been synthesized by Zaheer and Rafiuddin employing citric acid as a reducing and shape-directing agent [334]. Cong et al. [335] reported the formation of Y and K-shaped Ag nanostructures using a polyol method with PVP as a capping agent.

In brief, the section discusses the synthesis and characterization of AgNPs through chemical methods, emphasizing their convenience, efficiency, and cost-effectiveness. Chemical synthesis involves reducing silver salts using various agents, such as metal complexes, block co-polymers, and biological macromolecules. The size and morphology of AgNPs are controlled by nucleation and growth stages, influenced by parameters like precursor concentration, temperature, pH, and types of reducing and capping agents. Strong reducing agents like sodium borohydride produce small, monodisperse AgNPs, while weak reducing agents result in larger NPs with controlled size distribution. Stabilizing agents, including polymers and biopolymers, are crucial for achieving stability, and they impart either electrostatic or steric repulsions. The synthesis methods vary, employing chemicals like PEG, PVP, PMMA, and low-molecular-weight compounds such as aldonic acid or using polyol methods. The diverse morphologies obtained include nanocubes, nanoplates, nanowires, nanorods, nanoflowers, and nanoprisms, showcasing the versatility of chemical synthesis in tailoring Ag NP properties for specific applications.

The section discusses the synthesis and characterization of silver nanoparticles (AgNPs) through chemical methods, emphasizing their convenience, efficiency, and cost-effectiveness. Chemical synthesis involves reducing silver salts using various agents, such as metal complexes, block co-polymers, and biological macromolecules. The size and morphology of AgNPs are controlled by nucleation and growth stages, influenced by parameters like precursor concentration, temperature, pH, and types of reducing and capping agents. Strong reducing agents like sodium borohydride produce small, monodisperse AgNPs, while weak reducing agents result in larger NPs with controlled size distribution. Stabilizing agents, including polymers and biopolymers, are crucial for achieving stability, and they impart either electrostatic or steric repulsions. The synthesis methods vary, employing chemicals like PEG, PVP, PMMA, and low-molecular-weight compounds such as aldonic acid or using polyol methods. The diverse morphologies obtained include nanocubes, nanoplates, nanowires, nanorods, nanoflowers, and nanoprisms, showcasing the versatility of chemical synthesis in tailoring Ag NP properties for specific applications.

1.5.3 Photoinduced Reduction Synthesis of AgNPs

Due to the good versatility, high spatial resolution, and convenient handling, photoinduced reduction have been employed for the synthesis of silver nanoparticles with diverse sizes and morphologies. This method enables the use of various mediums to synthesize AgNPs using this approach like polymers, emulsions, glasses, and surfactant micelles. Utilizing a photoinduced reduction method, Ag nanoparticles with a size of 8 nm were synthesized through the utilization of micro reactor capsules composed of poly(styrene sulfonate)/poly(allylamine hydrochloride) [336]. Moreover, conversion of spherical AgNPs to triangular Ag nanocrystals with edge length of 30–12 m has been achieved by photoinduced method [337]. Dual-beam illumination was employed to achieve shape conversion of AgNPs, utilizing citrate and poly styrene sulfonate as stabilizing agents.

In another study, monodispersed silver nanoparticles (AgNPs) with ligand coatings were rapidly synthesized by swiftly reducing Ag^+ ions through α-aminoalkyl radicals. These radicals originated from thioxanthone (TX) and its derivatives, including the 2-substituted thioxanthone series, TX—O—CH_2—COO— (TO) and TX—S—CH_2—COO— (TS), in their excited triplet state via hydrogen abstraction toward an aliphatic amine. The reduction of Ag^+ ions to the Ag^0 state was kinetically controlled by the substituent effect on thioxanthones. This control also led to the tuning of the quantum yield in the reaction [338]. The application of irradiation to the solution led to a change in the absorption spectra of the SPR in the range of 320–66 nm, which suggests an increase in the rate of Ag NP formation. Whereas the absorption spectra of the chromophore remained constant which indicates the regeneration process of thioxanthones. The formation of NPs was established by TEM, which displayed a marked difference in morphologies of AgNPs by reduction in solution of TX and those of TO and TS. Multimodal AgNPs have been obtained by TO, whereas monodispersed AgNPs of an average size of 4.8 ± 1 nm have been achieved with TS. Silver NP aggregates have been obtained using TX with smaller-sized AgNPs seen on the periphery of the Ag NP aggregates. All the structural variation has been corroborated with the development of SPR bands during irradiation. Silver nanoparticles (AgNPs) were successfully synthesized through the photoreduction of $AgNO_3$ using sodium citrate. Irradiation with a variety of light sources, including UV, white, blue, green, cyan, and orange, was used to accomplish photoreduction. The location and intensity of optical absorption bands changed as a result of the light source shift, and this was attributed to changes in the AgNPs' sizes and morphologies [339]. A simple UV photo activation method has been employed by Ghosh et al. to fabricate stabilized AgNPs in Trition X-100 solution [340]. The surfactant Triton X-100 not only acted as reducing and stabilizing agents but also growth controlling agent. Through a reduction in the system's diffusion/mass transfer coefficient, the surfactant managed the growth of Ag NP through a diffusion-controlled mechanism. The fabrication of AgNPs was ascribed to the autocatalytic growth mechanism. Ligands of the salt precursors profoundly affected the photochemical reduction efficiency owing to electron scavenging property, solubility, and stability. Synthesis of AgNPs was realized by UV light irradiation of an alkaline solution of silver nitrate ($AgNO_3$) and water-soluble chitosan derivative, carboxymethylated chitosan (CMCTS) at variable pH and $AgNO_3$ concentrations [341]. CMCTS served simultaneously both reducing and a stabilizing agent for the Ag NP synthesis. The particle size and particle density increase with increase in pH from 0.7 to 2.2. Further, the Ag NP size and particle size distribution increased remarkably with increase in pH up to 9.6. But solution with pH of 12.4 resulted in the creation of large-sized AgNPs with relatively limited size distribution (2–8 nm) which led to the strongest SPR peak at pH of 12.4. Likewise, the decrease in concentration of CMCTS from 0.5 to 0.1% caused the SPR peak's intensity to drop with corresponding red-shifting. Further decrease in concentration of CMCTS from 0.1 to 0.033% resulted in increase in SPR peak intensity with corresponding blue shifting. Change in concentration of CMC resulted in change in particle size of the AgNPs. Thus, it was speculated that higher concentration of CMCTS resulted in the formation of nucleation templates by coiling of the CMCTS chains which restrict the activity of Ag^+ ions and CMCTS. Fabrication of Ag nano colloids was realized by photo-induced reduction of gelatin-stabilized aqueous silver bromide using ascorbic acid as a reducing agent [342]. The effect of the molar ratios of AgBr to ascorbic acid has been studied. The absorption peaks shifted to shorted wave lengths from 429 to 389 nm with rise in molar ratio from 1 : 1.5 to 1 : 9.0. This SPR absorption shift has been attributed to AgNPs' declining size. Further, the reaction time shortened from 90 to 5 minutes with increase in the reductant concentration. The effect of the radiation type on the efficiency of the reaction was also studied using various radiation sources viz., xenon flash, high-pressure mercury vapor and halogen lamps. The spectroscopic measurements have demonstrated that photoreduction of AgBr to AgNPs was most efficient while using radiation from the halogen lamp. Silver nanoparticles have been grown on the surface of cellulose fibers by surface activation of cellulose fiber surface and reduction of the silver nitrate [343]. The effect of the intensity of the UV radiation on the growth and concentration of AgNPs was investigated. It was discovered that the AgNPs' particle size increased from 56 to 85 nm with increase in the UV radiation intensity from 40 to 48 mW/cm^2. However, it has been perceived that size of the AgNPs is not much influenced by the change in $AgNO_3$ concentration. UV light source has been used to induce Ag NP formation and cross-linking of the PVP polymer from a solution of $AgNO_3$ and PVP [344]. The emergence of an SPR signal at 420 nm indicated the formation of AgNPs. With an increase in irradiation time, the SPR band strength increased, and its peak somewhat shifted to longer wavelengths. When the irradiation period was prolonged from 30 to 480 minutes, the particle size went from 2.5 to 3.45 nm. Highly reflective ultrathin films of Ag NP have been generated by on a glass slide by the photoinduced approach using silver precursor and free radical photo generator without adding any extra reductant [345]. The resulting Ag NP film was characterized by AFM and TEM. AFM displays the surface roughness of the films ranging from 7 to 100 nm. The production of spherical AgNPs with a diameter ranging from 2 to 10 nm was visible in TEM pictures. Additionally, green-reducing agents sourced from plant sources and microbial species have been used to facilitate photoinduced green synthesis of AgNPs [346–350].

In brief, this section explores the use of photoinduced reduction methods for synthesizing silver nanoparticles (AgNPs) with various sizes and morphologies. The versatility of this method allows for synthesis in different mediums, including polymers, emulsions, glasses, and surfactant micelles. Examples include the use of micro-reactor capsules, ligand-coated monodispersed AgNPs, and UV photoactivation methods. The photoinduced approach enables the rapid reduction of Ag^+ ions to Ag^0 state, leading to the formation of AgNPs with controlled morphologies. The influence of various parameters such as pH, concentration, and types of stabilizing agents is discussed, showcasing the tunability of this method for specific applications. The section also covers studies on the efficiency of different light sources and the use of green reducing agents derived from plant and microbial sources for environmentally friendly Ag NP synthesis.

1.5.4 Electrochemical Synthesis of Silver Nanoparticles

Electrochemical techniques have been utilized for the synthesis of silver nanoparticles (AgNPs), achieving excellent uniformity through the careful control of electrolysis parameters and manipulation of the electrolytic solution composition. In the study conducted by Johnas et al., AgNPs coated with polyphenyl pyrrole were synthesized using the electrochemical method at the liquid–liquid interface. *N*-phenyl pyrrole played a crucial role in facilitating the transfer of silver ions (Ag^+) from the aqueous phase to the organic phase, followed by a gradual transfer of electrons from the metal electrode to the Ag^+ ions, resulting in the formation of metal silver clusters. The confirmation of AgNPs formation was established through the observation of a SPR peak at 390 nm [351]. Monodisperse silver nanoparticles (AgNPs) within the size range of 1–18 nm were successfully synthesized using the electrochemical reduction method both inside and outside zeolite crystals. This was accomplished by using electrodes modified with ultrathin faujasite zeolite sheet

(CZFMEs-FAU) [352]. The zeolite film prevents the AgNPs from agglomeration during electrochemical reduction of the silver ions. The control of AgNPs' size was achieved by varying the degree of ion exchange in CZFMEs. A low degree of silver exchange yielded abundant AgNPs with a uniform diameter of 18 nm on the zeolite film surface through electrochemical reduction. A high exchange rate, involving Na^+ ions replacing Ag^+ ions, resulted in the formation of smaller AgNPs ranging from 1 to 2 nm. Intermediate exchange rates produced AgNPs with intermediate sizes, around 5 nm. Additionally, Ma et al. successfully synthesized spherical AgNPs with an average diameter of 20.6 nm through electrochemical reduction of Ag ions in the presence of PVP [353]. PVP has been employed to increase the rate of Ag NP formation and to reduce the rate of silver film deposition. Low molar ratios of PVP/Ag^+ did not produced any absorption peak in the UV-Visible region. As the molar ratio was increased to 50 : 1 a broad SPR absorption band could be observed around 420 nm. However, as the molar ratio was increased from 50 : 1 to 1000 : 1, a highly symmetrical SPR band was noticed. The occurrence of symmetrical SPR band confirmed the fabrication of AgNPs of small size and uniforms size distribution. The particle size was found to be influenced by the rotational speed of the working electrode. Low rotational speeds (500 and750 rpms) produce large-sized AgNPs with high degree of polydispersity. Smaller-sized AgNPs with narrow particle size distribution were produced as the rotational speed was increased (1000 and 1500 rpms). The absorption spectra exhibited a reduction in the intensity of the SPR absorption bands as the rotation speed of the electrode increased. This decline in band intensity was attributed to a decrease in particle size. The introduction of SDS into the solution led to the formation of even smaller silver nanoparticles (AgNPs) with a narrower distribution of the particle sizes [353]. Silver nanoparticles of approximately 20 nm in size were produced by electrochemical approach using potentiostatic and galvanostatic methods in ethanol solution [354]. A simple electrochemical set up has been employed by khaydarov et al. to synthesized AgNPs without the use of any stabilizing agent using two silver plates as cathode and anode [355]. The electrochemical process involved the oxidation dissolution of silver from the anode, resulting in the formation of Ag^+ ions. Subsequently, Ag^+ ions migrated from the anode to the cathode, where they underwent reduction to form zerovalent Ag atoms. These atoms then nucleated and grew to develop into silver nanoparticles (AgNPs). The stability and dispersibility of the AgNPs were improved by incorporating PVP as a protective agent. TEM images showed that spherical AgNPs with an average particle size of 7 nm were created, with a size range of 2–20 nm. The electrochemical synthesis of AgNPs was accomplished through potentiostatic electrolysis, utilizing an Ag anode [356]. The silver ions are reduced to the metallic silver and deposited on the electrode. DLS data show the formation of nanosized AgNPs with particle size between 14 and 79 nm. Electrochemical synthesis of AgNPs was also carried out using sodium polyacrylate (NaPA) as a stabilizer [357]. Silver nanoparticles up to the size of 10 nm have been obtained by this method. Silver nanoparticles have been obtained by electrochemical reduction method using sacrificial Ag silver anode as a silver ions source and rhamnolipid, RL as a stabilizer [358]. Investigations was conducted into the impact of temperature and stabilizer on the rate of Ag dissolution and cathodic reductive production of Ag-RL complex ions. It was found that the rate at which AgNPs formed increased in response to an increase in RL concentration. Also, the rate of anodic and cathodic currents increased up to 1.5 times with increase in temperature from 40 to 60 °C. UV–Visible spectra demonstrated an increase in optical density of the SPR peak with peak shifting toward shorter wavelengths with increase in the number of CVA cycles. TEM images showed the fabrication of spherical AgNPs with size in the range of 2–20 nm when the RL concentration was within 0.5–2 g/L in the temperature range of 40–60 °C. A decrease in Ag NP size was observed with increase in RL concentration inside the solution. However, increase in Ag NP size and particle size dispersion has been observed with increase in temperature of the solution.

To summarize, this section explores the utilization of electrochemical techniques to produce silver nanoparticles (AgNPs). The electrochemical approach offers precise control over electrolysis parameters and electrolytic solution composition, leading to homogeneous and controlled Ag NP formation. Various techniques, including liquid–liquid interfaces and zeolite film-modified electrodes, are discussed for achieving monodisperse AgNPs with controlled sizes. The role of stabilizing agents such as polyphenyl pyrrole, PVP, and rhamnolipid in the electrochemical synthesis is highlighted. The influence of factors like rotational speed and temperature on the grain size and distribution is investigated. Additionally, comparisons between potentiostatic and galvanostatic methods, as well as the impact of different stabilizers on the rate of Ag NP formation, are discussed. The findings indicate the versatility of electrochemical methods for tailoring the size and properties of AgNPs for various applications.

1.5.5 Microwave-Assisted Synthesis of Silver Nanoparticles

The microwave heating technique proves to be a convenient and promising approach for producing highly crystalline silver nanostructures characterized by variable sizes and a narrow size distribution [359]. This method features shorter reaction times and reduced energy consumption, leading to the formation of silver nanoparticles (AgNPs) with high yield and minimal agglomeration [359]. The synthesis of silver nanoparticles has been accomplished through a microwave-assisted approach, utilizing *Cyanthillium cinereum* (*C. cinereum*) both as a reductant and a capping agent [360]. This method can also reduce chemical waste by using environmental friendly reaction media [361]. The AgNPs obtained were crystalline in nature and spherical in shape with the average particle size of 19.25 ± 0.44 nm as revealed by RTEM images. The highly crystalline nature of AgNPs was evident from SAED pattern by displaying the bright spots in ordered concentric fashion. The creation of NPs at an $AgNO_3$ concentration of 1×10^{-3} M was indicated by the existence of an SPR signal at 418 nm. Microwave-assisted fabrication of AgNPs has been accomplished inside hybrid hydrogel (CGH) of carboxy methyl cellulose (CMC) and gelatine. CMC acted as a reducing agent and stabilizing agent in the CGH. The formation of AgNPs was confirmed by the appearance of SPR peak around 414 nm. Silver NPs were obtained at different pH and different concentrations of CMC as reductant and silver nitrate. Stable SPR peak at 414 nm was observed at pH of 12.0 and $AgNO_3$ concentration of 0.8 mM. TEM images revealed the production of uniform and spherical CGH-AgNPs with the mean particle diameter of 12 nm. EDX analysis displayed a spectral peak at 3 keV indicating the formation of pure silver [362]. Microwave-assisted fabrication of anisotropic AgNPs has been realized from the decomposition of silver oxalate using PVP as a reducing agent [363]. Time evolution of SPR spectra shows the appearance of a peak at 367 nm, which exhibits red shifting with subsequent increase in peak intensity with time. The peak shifting occurs from 367 after 60 seconds to 422 nm after 75 seconds and then to 462 nm after 90 seconds of irradiation (Figure 1.11A). The fabrication of spherical AgNPs after 60 seconds of irradiation was confirmed by the HRTEM mages which show spherical AgNPs between 3 and 9 nm in size (Figure 1.11B). After 75 seconds of irradiation, HRTEM images display the fabrication of anisotropic AgNPs like rods, triangles, and other morphologies with the dimensions of 15–45 nm diameter and 20–70 nm length (Figure 1.11C). After 90 seconds of microwave irradiation, the formation of anisotropic AgNPs with size between 50 and 120 nm occurred which has been ascribed to the emergence of a broad peak around 569 nm due to longitudinal interaction. Microwave-assisted procedure has been used to create AgNPs using starch as a reductant and a stabilizing agent. The AgNPs obtained were pseudo-spherical in morphologies with a 12 nm average particle size [364]. Fabrication of AgNPs was carried out by polyol process under microwave irradiation conditions using silver nitrate as a metal salt precursor. The metal salt precursor was combined with ethylene glycol serving as a

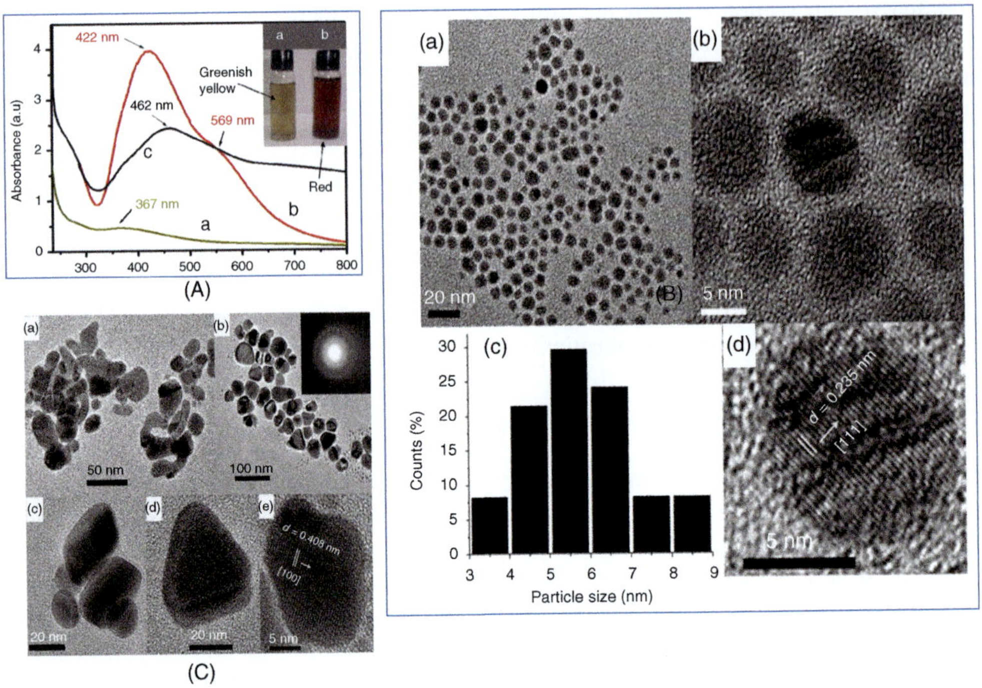

Figure 1.11 (A) UV-visible spectra of Ag nanoparticle colloids generated through microwave irradiation in ethylene glycol for different durations: (a) 60 seconds, (b) 75 seconds, and (c) 90 seconds. (B) High-resolution transmission electron microscopy (HRTEM) pictures of silver nanoparticles produced by subjecting silver oxalate to microwave exposure in ethylene glycol for 60 seconds, shown in (a) and (b). In (c), a size distribution histogram of these Ag nanoparticles is presented. (d) Depicts a detailed image of an Ag nanoparticle in the center of the (b) image, showing its lattice structure. (C) High-resolution transmission electron microscopy (HRTEM) pictures of silver nanoparticles created by exposing silver oxalate to microwave radiation in ethylene glycol for 75 seconds. Inset in (b) displays the selected area electron diffraction (SAED) pattern of one of the silver nanoparticles. Source: (a–c) Reproduced with permission from Navaladian et al. [363]/IOP Publishing.

reductant and PVP functioning as a capping agent under for 15 minutes at 150° in a microwave digestive system. The reaction yielded silver nanoparticles (AgNPs) of diverse shapes with discreet Ag nanorods of the size of 50 nm [365]. Similarly, AgNPs were obtained by microwave irradiation of the solution of $AgNO_3$ with sodium citrate as capping agent and formaldehyde acting as a reductant [366]. The range particle size of AgNPs varied with the concentration of all the three reactions components. As the concentration of silver nitrate was reduced, the particle size shrank, and the particle size distribution narrowed. As the concentration of sodium citrate increased, there was a corresponding decrease in the average particle size. Silver NPs of variable shapes Ag nanorods, nanowires, cubes, triangles and triangular bi pyramids have been fabricated by microwave polyol process utilizing ethylene glycol as a reducing agent in presence of $H_2[PtCl_6]$ and PVP [367]. PVP acted as a shape-directing agent and gets adsorbed on the [100] facet resulting in the production of Ag nanorods, nanowires, nanocubes, and triangular bipyramids. Ag nanostructures were obtained without PVP, but in the presence of Cl^- ions, the yields were lower. This indicates that chloride ions also play a role as shape directors, influencing the synthesis of Ag nanostructures with varying shapes. TEM pictures of the supernatant revealed the fabrication of nanorods/nanowires with length of 1 μm (yield 21%), Ag nanocubes (6%), Ag triangular bipyramids (4%), Ag nanoflowers (0.5%), and spherical Ag nanoparticles of average diameter of 60 nm (68%) (Figure 1.12A). However, a significant amount of Ag NP precipitate has been observed in the

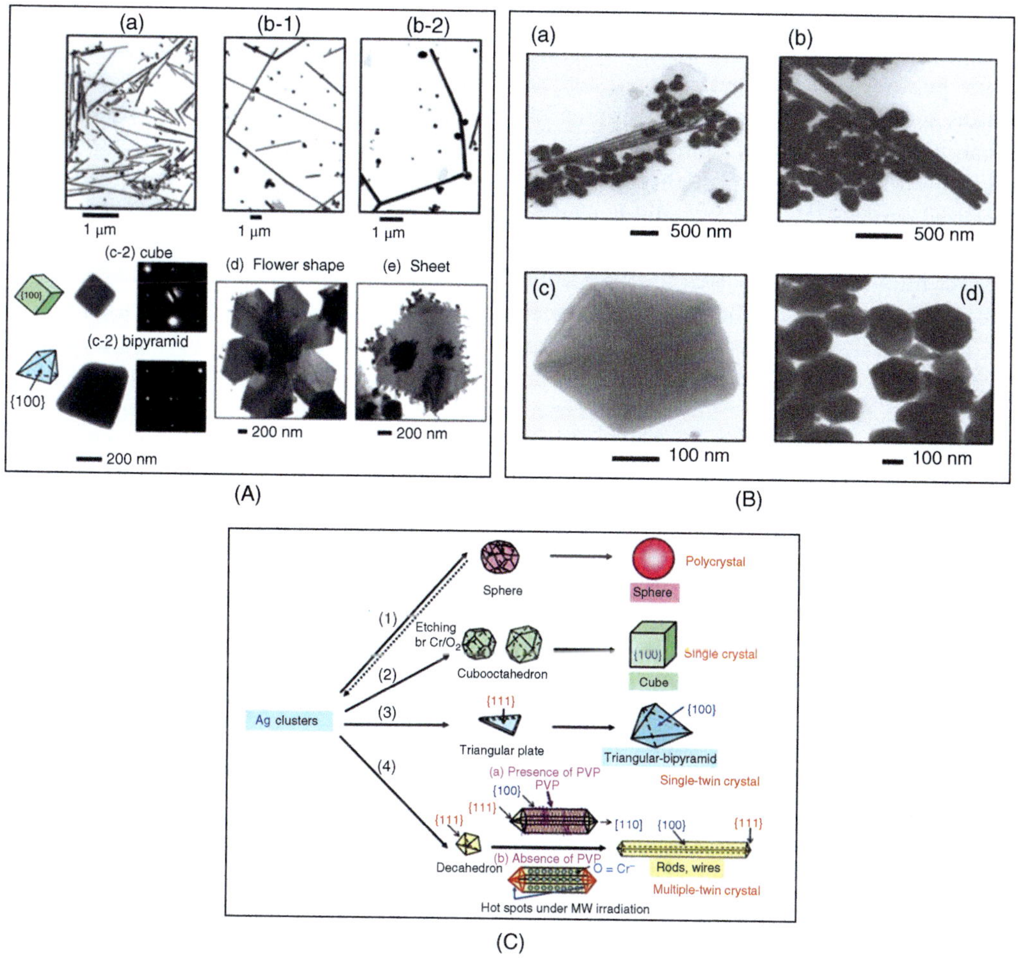

Figure 1.12 (A) Transmission electron microscopy images of silver nanostructures synthesized from surfactants using (a) $AgNO_3$ (46.5 mM)/$H_2[PtCl_6].6H_2O$ (115 mM)/PVP (264 mM)/EG and (b–e) $AgNO_3$ (46.5 mM)/$H_2[PtCl_6].6H_2O$ (115 mM)/EG Solutions. Illustration of selected area electron diffraction patterns for cube and bipyramid shapes in (c-1) and (c-2). (B) Transmission electron microscopy images of silver nanostructures derived from precipitates formed through the utilization of $AgNO_3$ (46.5 mM)/$H_2[PtCl_6].6H_2O$ (115 mM)/EG solutions. (C) Mechanism of silver nanostructure formation during microwave-assisted heating. Source: (a–c) Reproduced with permission from Tsuji et al. [367]/Oxford University Press.

absence of PVP. TEM pictures of the precipitate revealed the production of large-sized spherical Ag particles, 1-D wires and do decahedral silver structures (Figure 1.12B). The mechanistic steps of the growth of various Ag nanostructures are presented schematically in Figure 1.12C. A green chemistry approach utilizing microwave radiations was employed for the synthesis of silver nanoparticles (AgNPs). Amino acids like L-lysine and L-arginine served as reducing agents, while water-soluble starch acted as a capping agent. Starch-capped AgNPs self-assembled to form multilayered mirror-like films on the glass surface [368]. Highly monodispersed AgNPs have been fabricated by microwave irradiation of ethanolic solution of $AgNO_3$ using PVP as a stabilizer [369]. The confirmation of silver nanoparticles (AgNPs) formation was established through the emergence of a SPR peak at 416 nm. Stable and polycrystalline AgNPs with particle size of 10 ± 5 nm of were obtained after 5 seconds of microwave irradiation. Sodium alginate was used as a reducer and stabilizer in the microwave irradiation process to create silver nanoparticles [370]. The impact of reaction parameters, including the concentration of sodium alginate, sodium nitrate, pH, and irradiation time, was examined. XRD studies revealed the presence of face-centered cubic silver in pure phase. UV visible spectroscopic studies revealed the SPR peak at 417 nm and the peak intensity increased with rise in concentration of sodium alginate from 0.1 to 0.5%. However, as the concentration of sodium alginate increased from 0.5 to 1.5%, the peak intensity reduced and shifted toward a longer wavelength, indicating the agglomeration of AgNPs and the creation of large-sized particles. Using 0.5% sodium alginate, and varying the concentration of silver nitrate, it was found that SPR peak intensity centered at 418 nm augmented with rise in concentration of the silver nitrate which indicated the fabrication of AgNPs in higher amounts at higher sodium nitrate concentrations. Further increase in irradiation times also resulted in an increase in SPR peak intensity at 418 nm with slight red-shifting to 427 nm which implied the increased rate of formation of AgNPs. At low pH, no reduction occurs as evidenced by the absence of SPR peak, whereas a band at 270 nm has been seen at pH 5 which corresponds to the presence of Ag^+ clusters. An elevation in pH led to the appearance of a new peak at 418 nm, indicating an augmented reduction rate of Ag^+ ions by sodium alginate due to more and more exposed carboxyl groups. TEM images revealed the development of fewer AgNPs at lower silver nitrate concentrations (0.8 mm/L). Nonetheless, the density and dispersibility of AgNPs increased as the $AgNO_3$ concentration was raised from 4 to 8 mm/L, while maintaining an average grain size of 12 nm at the higher $AgNO_3$ concentration. Average particles size also increased from 10 to 15 nm with increase in pH from 5 to 11. The average grain size of spherical AgNPs exhibited a decrease from 20 to 10 nm as the sodium alginate concentration increased from 0.1 to 0.5%. Numerous reports have been published on the microwave-assisted fabrication of AgNPs utilizing various reductants and stabilizers [371–376].

In brief, microwave heating methods provide a convenient and promising approach for the fabrication of highly crystalline AgNPs with variable sizes and narrow size distribution. This method, renowned for its abbreviated reaction times, diminished energy consumption, and minimized chemical waste, has been employed for the microwave-assisted synthesis of silver nanoparticles (AgNPs). Various reducing agents and stabilizers, such as *Cyanthillium cinereum*, carboxy methyl cellulose, gelatine, PVP, starch, ethylene glycol, amino acids (L-lysine and L-arginine), and sodium alginate, have been employed to achieve controlled Ag NP formation with specific morphologies. The microwave-assisted method proves to be versatile, offering a green and efficient synthesis route for tailored silver nanostructures with diverse applications.

1.5.6 Bio Fabrication of Silver Nanoparticles

In recent times, eco-friendly synthetic methods utilizing reductants sourced from biological entities like microbes and plant extracts have emerged as a practical substitute for traditional chemical synthesis routes in the creation of silver nanoparticles (AgNPs) [377–379]. The eco-friendly synthesis of AgNPs has become a practical substitute for physicochemical synthesis methods, primarily due to the elimination of toxic chemicals and solvents, as well as the nonproduction of hazardous by-products [380, 381].

Microorganisms like bacteria and fungi have been used for remediation of the toxic metal ion pollutants from the toxic environments [382, 383]. These microorganisms facilitate the reduction of toxic metal ions to metal nanoparticles. Synthesis of AgNPs by microbes involve both intracellular as well as extracellular syntheses. Intracellular synthesis involves use of intracellular components as reducers as well as stabilizers [384].

1.5.6.1 AgNPs Synthesized Using Bacteria

Bacteria have demonstrated significant efficacy as bio-factories for the synthesis of silver nanoparticles (AgNPs), occurring both intracellularly and extracellularly. Certain bacterial strains with silver resistance have been found to accumulate silver on their cell wall, reaching up to 25% of their dry weight [385]. The bacterial strain resistant to silver, specifically *Pseudomonas stutzeri AG259*, was first utilized for the synthesis of silver nanoparticles (AgNPs) through the cultivation of the bacteria in a concentrated silver nitrate solution. The resultant AgNPs, averaging 200 nm in size, were noted for their accumulation inside bacterial cell wall [386]. Spherical silver nanoparticles (AgNPs) with sizes ranging from 5 to 15 nm were produced through the utilization of silver-resistant *Bacillus* species [387]. The silver nanoparticles (AgNPs) accumulated within the periplasmic space of the bacterial cells. Furthermore, successful extracellular synthesis of AgNPs was achieved by employing the *actinomycete* strain *Streptomyces glaucus* 71 MD [388]. TEM images unveiled the creation of spherical silver nanoparticles (AgNPs) with an average size measuring 13 nm. Large-scale production of AgNPs has been achieved by using *Proteus mirabilis PTCC 1710*. Extracellular synthesis of silver nanoparticles ranging in size from 5 to 25 nm was achieved using cultural supernatants derived from *Escherichia coli* ATCC 8739, *B. subtilis* ATCC 6633, and *Streptococcus thermophilus* ESh1 [389]. The rapid synthesis of silver nanoparticles (AgNPs) has been accomplished using supernatants of *Klebsiella pneumonia, E. coli, Enterobacter cloacae* [390], and *Bacillus megaterium* [391]. The rapid biosynthesis of silver nanoparticles (AgNPs) has been accomplished using the S-27 strain, *Bacillus flexus* family [392]. Upon adding the cultural supernatant to the silver nitrate solution, rapid color change yellow to brown was observed. The brown colored solution showed the SPR peak at 420 nm. AFM and FE-SEM images displayed anisotropic AgNPs (spherical and triangular) with average size of 12 and 61 nm. Both intracellular and extracellular syntheses of AgNPs have been achieved the bacterial strain *Proteus mirabilis* PTCC 1710 [393]. The AgNPs displayed SPR peak at 400–410 nm. TEM images revealed the generation of spherical silver nanoparticles (AgNPs) within the size range of 10–20 nm. Extracellular biosynthesis of AgNPs has been reported by Das et al. using CS11 strain of *Bacillus* sp. obtained from heavy the soil contaminated with heavy metals [394]. The silver nanoparticles were seen after 24 hours of interaction between $AgNO_3$ and bacteria at room temperature. The presence of silver nanoparticles (AgNPs) was spectroscopically confirmed through the appearance of a surface plasmon resonance (SPR) peak at 450 nm. The nanoparticles exhibited a spherical shape of the sizes of 42 and 92 nm. Silver nanoparticles (AgNPs) were synthesized using *Bacillus cereus* isolated from contaminated soil [395]. Optimization of the experimental conditions was done using central composite design. Optimized values of temperature (48.5 °C), inoculum size (8.7 mL), silver nitrate concentration (1 mM), and pH (9) were employed for fabrication of AgNPs. The synthesis yielded the spherical silver nanoparticles (AgNPs) with a size in the range of 5–7 nm. Spherical AgNPs with diameter of 40–50 nm have been synthesized using cultural supernatant from *Lactobacillus plantarum*. While numerous studies have been conducted on the synthesis of silver nanoparticles (AgNPs) utilizing various bacteria, a notable limitation of this approach is the relatively slow synthesis rate, and a restricted range of sizes and shapes has been achieved.

1.5.6.2 AgNPs Synthesized Using Fungi

Good metal tolerance and bioaccumulation ability of fungi have made them a good candidate for the biosynthesis of metal nanoparticles [396]. Also, fungi are much more convenient to handle inside the laboratories and secrete enzymes as reductant in larger quantities as compared to bacteria. These features enable fungi to produce metal nanoparticles at higher rates and in large quantities [397]. Further simplicity in handling of fungi has made them a good eco-friendly candidate for the green synthesis of NPs. The extracellular synthesis of AgNPs was first reported by Mukherjee et al. by exposing fungus *verticillium* to silver nitrate [398]. The Ag NP formation took place on mycelial surface due the reduction of the Ag^+ ions by the carboxylate ions of the enzymes secreted by the fungal cell. The silver atoms then underwent nucleation and growth to form AgNPs.

Silver NPs with the size of 25 ± 12 nm have been produced. Fast extracellular biogenesis of silver nanoparticles (AgNPs) has been documented employing fungus *Aspergillus fumigatus*, and NPs were formed within minutes of the reaction. The generated AgNPs had a spherical shape and ranged in size from 5 to 25 nm [399]. Naqvi et al. employed the fungus *Aspergillus flavus* to produce AgNPs by extracellular mycosynthesis, combining silver nitrate with the fungus's culture filtrate. Silver NPs in the size range of 5–30 nm have been obtained with the SPR peak around 400 nm [400]. The fungus y *Fusarium solani* (USM-3799) have been used by Ignle et al. for synthesizing AgNPs by mixing the 1 mM silver nitrate with fungal cell filtrate [401]. The development of an SPR peak at 420 nm optically identified the production of AgNPs. The presence of biomolecules like proteins as stabilizing agents on the Ag NP surface was ascertained by FTIR analysis. The formation of spherical AgNPs with an average grain size of 16.23 nm in the size range of 5–35 nm was shown by the TEM pictures. Synthesis of AgNPs has been achieved by Syed et al. by employing the thermophilic fungus *Humicola* sp. [402]. The synthesis was conducted by combining the suspension of fungal mycelia with the 100 mL of silver nitrate at the temperature of at 50 °C by maintaining the reaction pH at 9. The fabrication of AgNPs was monitored spectroscopically by measuring the absorption of the reaction mixture. SPR absorption peak around 415 nm has been identified. Ag NP formation was identified by a yellow to brown color shift in the reaction mixture. TEM pictures displayed spherical AgNPs with fcc symmetry and sizes ranging from 5 to 25 nm. The presence of protein molecules on the surface of silver nanoparticles (AgNPs), acting as stabilizing agents, has been verified through FTIR spectroscopy. The FTIR analysis revealed two bands at around 1644 and 1523 cm^{-1}, known as amide I and II bands. These bands are caused by the stretching vibrations of the amide linkages at —C=O (amide I) and —N—H (amide II). Rapid biosynthesis of AgNPs has been accomplished by the bio reduction of 1 mM $AgNO_3$ with different concentrations of hot water extract of basidiocarps, *Pleurotus cornucopiae var. citrinopileatus. P. cornucopiae* [403]. After the reaction mixture was incubated for 72 hours, the color of the mixture changed from yellow to yellowish brown, indicating the synthesis of AgNPs. AgNPs were identified by an SPR peak at about 450 nm in UV-visible spectroscopy. The presence of polysaccharides and proteins as reducing and capping agents has been confirmed by FTIR spectroscopically, and the FTIR spectrum of the Ag NP sample displays various peaks at 3304, 2200, 2066, 1969, 1636, 1261, 1094, and 611 cm^{-1}, which represent the absorptions corresponding to stretching and bending vibrations of the various functional groups of various biomolecules responsible form reduction and subsequent stabilization of AgNPs. The generation of spherical silver nanoparticles (AgNPs) has been corroborated through HRTEM, revealing sizes ranging between 20 and 30 nm. Silver NPs have been synthesizing by *Pleurotus ostreatus*, a tree oyster mushroom [404]. The color of the reaction mixture changed from pale yellow to dark brown after the mushroom aqueous extract and 1 mM silver nitrate were incubated, confirming the production of AgNPs. TEM images revealed the creation of spherical AgNPs with an average size of 28 nm. Numerous reports exist on the synthesis of AgNPs using various fungi, both intra and extracellularly [405–411]. Various yeast strains have also been explored for the synthesis of silver nanoparticles (AgNPs) with promising outcomes [397, 412, 413].

1.5.6.3 AgNPs Synthesized Using Actinomycetes, Algae, and Viruses

Like fungi biosynthesis of AgNPs has been achieved using actimycetes both intracellularly as well as extracellularly [414–420]. Synthesis of AgNPs has been carried out by the cell wall enzymes which reduced Ag^+ ions to Ag seeds which finally grow to produce using marine actinobacteria, *Streptomyces* sp *LK3b* [415]. Silver NPs exhibited SPR peak 420 nm, and XRD studies showed intense Bragg reflections corresponding to the fcc structure of nano silver. TEM and AFM micrographs displayed the formation of spherical silver nanoparticles (AgNPs) with an average grain size of 5 nm. The mechanistic examination disclosed the involvement of enzymes, such as

NADH-dependent nitrate reductase, in the reduction of Ag^+ ions to metallic Ag via an electron transfer mechanism. Biological synthesis of silver nanoparticles (AgNPs) has been carried out employing *Streptomyces antimycoticus L*, a strain of endophytes isolated from the leaves of *Mentha longifolia L* [421]. Confirmation of successful biosynthesis of silver (Ag) was ascertained through XRD, TEM, DLS, and UV-Visible studies. Utilizing these techniques, it was verified that spherical silver nanoparticles (AgNPs) were formed with sizes ranging from 13 to 40 nm. The stability of AgNPs was confirmed by carrying out the optical absorption measurement using UV-Visible spectroscopy. The UV-Visible spectrum proved the remarkable stability of silver nanoparticles (AgNPs) as evidenced by the presence of the SPR peak at 415 nm, which remained intact for nearly one month. Additionally, silver nanoparticles (AgNPs) were determined to possess a negative surface charge of −19.2 mV, as indicated by the zeta potential analysis. Green synthesis of AgNPs was achieved using cultural filtrate extracts of two endosymbiotic actinomycete strains, 192ANMG (F2) and 17ANMG (F7) isolated from *Crella cyathophora*, a marine sponge [422]. The colloidal and surface stability of AgNPs was checked by DLS and zeta potential measurements. Dynamic light scattering (DLS) demonstrated an increase in the hydrodynamic diameter of the particles upon the addition of the cultural filtrate to the silver nitrate solution, indicating the reduction of silver nitrate by the extract to form silver nanoparticles (AgNPs). The zeta potential measurements of the Ag NP samples indicated values of −24.6 and −25 mV for *Streptomyces* sp. 192ANMG (F2) and *Streptomyces* sp. 17ANMG (F7), respectively. TEM measurements revealed the formation of spherical AgNPs with average particle size of ~8.66 ± 2 nm in case of *Streptomyces* sp. 192ANMG (F2) and nonspherical AgNPs with average particle size of 35 ± 2 nm in case of *Streptomyces* sp. 17ANMG (F7). The hexagonal structure of silver (Ag) in silver nanoparticles (AgNPs) was revealed through SAED analysis. Numerous reports have been published on the synthesis of AgNPs employing actinomycetes [423–427].

Certain investigations have indicated that the synthesis of silver nanoparticles (AgNPs) can be achieved through some of the aquatic microorganisms like algae. Such aquatic microorganisms have shown the tendency not only to accumulate Ag ions but also to synthesize the AgNPs. For example, Ag nanoplates have been fabricated at room temperature by the extract of *Chlorella vulgaris*, a unicellular green alga [428]. Initially, the algal extract, appearing as a light-yellow liquid, underwent a color change to pinkish red upon contact with a 1 mM $AgNO_3$ solution for 12 hours at room temperature. This color transformation was attributed to the SPR of Ag nanocrystals in the visible region. UV–vis spectroscopy revealed three distinct plasmon bands at approximately 465, 392, and 347 nm for the synthesized Ag nanocrystals (Figure 1.13a). TEM analysis unveiled the presence of circular, rod-like, and triangular Ag nanocrystals in the synthesized sample. Triangular particles, along with some circular particles, displayed uniform contrast, suggesting they were single-crystalline nanoplates (Figure 1.13b). FESEM imaging confirmed the platelike morphology, with rod-like particles identified as Ag nanoplates standing perpendicularly on the TEM grid. The Ag nanoplates exhibited an average diameter of 44 nm, with a standard deviation of 6 nm. The average thickness, estimated from the width of rod-like particles, was 20 ± 4 nm. High-resolution TEM images demonstrated the single crystallinity of the nanoplates, supported by electron diffraction patterns showing a sixfold symmetry indicative of {111} faces (Figure 1.13c). The diffraction spots' *d*-spacings corresponded to the {422}, {220}, and {422} planes, consistent with previous observations on Ag or Au nanocrystals (Figure 1.13d). The synthesis mechanism was attributed to active species in the algal extract, primarily identified as algal proteins (aP). Modifications made to algal proteins indicated that tyrosine (Tyr) residues played a role in reducing silver ions, while carboxyl groups in aspartic acid (Asp) and/or glutamic acid (Glu) residues contributed to the anisotropic growth of silver nanocrystals, resulting in the formation of nanoplates. Further experiments separated the algal proteins into low- and high-molecular-weight fractions. The high-molecular-weight fraction

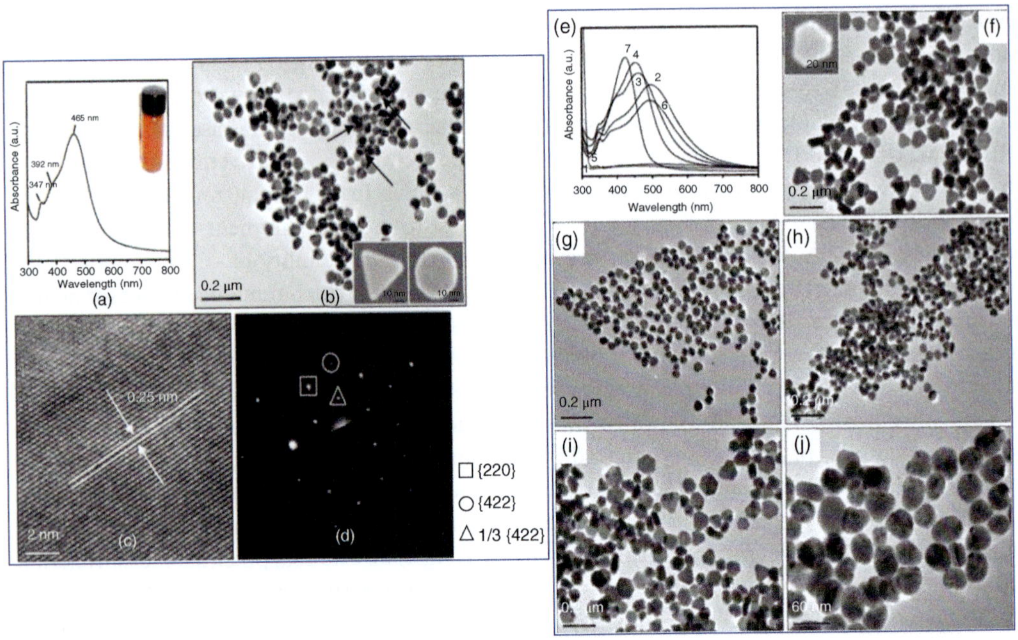

Figure 1.13 (a) UV–visible spectrum of Ag nanocrystals produced through the reduction of an aqueous Ag ion solution using the algal extract. The inset displays a digital image of the resultant Ag colloidal solution. (b) Representative transmission electron microscopy (TEM) image of the synthesized Ag nanocrystals, with arrows indicating areas where flat particles have overlapped. The inset features characteristic field emission scanning electron microscopy (FESEM) images portraying circular and triangular Ag nanocrystals exhibiting a planar structure. (c) High-resolution TEM (HRTEM) image captured from the vertex of an isolated Ag nanoplate. (d) Selected area electron diffraction (SAED) pattern of an Ag nanoplate, with spots enclosed in a box, a circle, and a triangle corresponding to {220}, {422}, and 1/3{422} reflections, respectively. (e) UV–visible spectra of Ag nanocrystals obtained under varied reaction conditions. Curves 1–7 were generated by adding 1 mL of 10 mM $AgNO_3$ to 9 mL of (curve 1) the low-molecular-weight fraction of the algal extract; (curve 2) the high-molecular-weight fraction of the algal extract (aP); (curve 3) heat-denatured aP; (curve 4) urea-denatured aP; (curve 5) NAI-modified aP; (curve 6) deacetylated NAI-modified aP; and (curve 7) amine-modified aP, respectively. The pH of all solutions was adjusted to 7.0. (f–j) Exemplary transmission electron microscopy (TEM) images of Ag nanocrystals prepared under conditions described in curves 2, 3, 4, 6, and 7, respectively. The inset in (b) displays a typical field emission scanning electron microscopy (FESEM) image of the truncated triangular plate, confirming the planar structure. Source: Reproduced with permission from Xie et al. [428]/American Chemical Society.

(aP) demonstrated significant absorption in the UV–vis spectrum, indicative of Ag nanoplate formation (Figure 1.13e). TEM analysis of Ag nanocrystals formed in the aP solution revealed mainly truncated triangular plates (Figure 1.13f–j). Heat-denatured and urea-denatured aP solutions exhibited similar UV–vis spectra and TEM images, suggesting that protein conformation did not significantly affect Ag nanoplate formation. The red- or blue-shifting of SPR peaks in denatured protein solutions correlated with the size of the formed nanoplates, indicating a direct relationship between protein conformation and nanoplate dimensions. The research findings led to the conclusion that algal proteins, particularly those containing tyrosine (Tyr) and carboxyl groups, played a vital role in both the reduction of silver ions (Ag ions) and the regulated growth of silver nanoplates. The mechanism involved a specific interaction between protein molecules and Ag ions, resulting in anisotropic growth and the fabrication of planar Ag nanostructures. The work opens avenues for understanding the active amino acid residues in protein molecules accountable for Ag ion reduction and nanoplate growth.

As the silver nitrate solution was brought in contact with algal extract, pinkish red colloidal suspension was obtained due to the SPR phenomenon exhibited by the A marine alga, *Caulerpa racemose* have been employed by Kathriaven et al. for the synthesis of AgNPs [429]. XRD studies revealed the fabrication of pure and crystalline face-centered cubic silver. The morphology of silver nanoparticles was observed to be both spherical and triangular, and the average particle size was determined to be 10 nm. Similarly, *Sargassum wightii*, a marine algae, has been discovered to extracellularly produce silver nanoparticles (AgNPs) [430]. Various other reports are there for the fabrication of AgNPs by various algal species [431–438].

In summary, this section delves into the potential of various microorganisms, including bacteria, fungi, actinomycetes, algae, and even viruses, to serve as efficient bio-factories for the environmentally friendly synthesis of silver nanoparticles (AgNPs). Bacteria, such as *Pseudomonas stutzeri* and *Proteus mirabilis*, demonstrate both intracellular and extracellular synthesis capabilities, producing AgNPs with varying sizes and distributions. Fungi, like *Verticillium*, *Aspergillus*, and *Fusarium solani*, exhibit metal tolerance and bioaccumulation abilities, offering a convenient and eco-friendly approach for Ag NP synthesis. Actinomycetes, algae, and even viruses also contribute to the green synthesis, providing a diverse range of potential biofabrication sources for tailored silver nanostructures. The section emphasizes the advantages of these biological methods, such as higher synthesis rates, large-scale production, and eco-friendly processes.

1.5.6.4 Mechanisms of the Fungal-Mediated Synthesis of AgNPs

Components present in the cell wall, such as proteins, enzymes, and other biomolecules, have been identified to play a crucial role in the synthesis of silver nanoparticles (AgNPs). These biomolecules facilitate the reduction of Ag^+ ions into Ag atoms, initiating nucleation and subsequent growth to fabricate silver nanoparticles (AgNPs). The AgNPs formed are then stabilized by proteins, peptides, and other biomolecular components. A number of reports have discussed the mechanistic details of the fungal-mediated synthesis of AgNPs. As an illustration, Duran et al. have connected cell wall polymers and other biomolecules like quinones, anthraquinones, and naphthoquinones to the production of silver nanoparticles (AgNPs) using *Fusarium oxysporum*. These biomolecules have been demonstrated to function as electron shuttles, transferring electrons to Ag^+ ions and reducing them to Ag atoms [439]. Mukherjee et al. have put forward a reasonable mechanistic detail for the synthesis of silver nanoparticles (AgNPs) by *Trichoderma asperellum* analyzing the FTIR spectroscopic peaks of the AgNPs and fungal cell-free extrate. As per the findings, it has been suggested that fungal extract contain protein molecules containing S-H bonds like cysteine, which plays a pivotal role for the reduction of silver ions. Cysteine has been found to undergo dehydrogenation and de-electronation in presence of Ag^+ ions which itself get reduced to Ag ions [440]. However, difference in the mechanistic details has been reported by several groups for extracellular and intracellular synthesis of AgNPs. In the process of extracellular synthesis, proteins, and enzymes released by fungal cells have been documented to play a pivotal role in the reduction of Ag^+ ions to Ag atoms, ultimately leading to the formation of silver nanoparticles (AgNPs) [441]. Whereas, in intracellular synthesis, synthesis of AgNPs occurs inside the cytoplasm and vesicles of the fungal cell which are then expelled out of the fungal cell by exocytosis [442].

1.5.7 Plant-Mediated Synthesis of Silver Nanoparticles

The realm of nanotechnology has experienced a significant shift in nanoparticle synthesis, with an increasing focus on green and sustainable methods. Notably, the plant-mediated fabrication of AgNPs emerges as an environmentally friendly and biocompatible approach. Plants, equipped with numerous bioactive compounds, serve as potent reservoirs for both reducing and stabilizing agents in the metallic nanoparticle synthesis. This section delves into the captivating domain of plant-mediated synthesis, investigating the underlying mechanisms, benefits, and the varied applications of AgNPs generated through this eco-friendly chemistry approach. The green synthesis of AgNPs using whole plant or plan components is discussed below.

Plant extracts or various parts of plants have been employed in the synthesis of silver nanoparticles (AgNPs). The synthesis process involves utilizing plant extracts, serving the dual function of both reducing and stabilizing agents. Various plant species ranging from angiosperms to algae have been used for this purpose [443–447]. The first article on the synthesis of AgNPs has been documented by Gardea-Torresdey et al. using sprouts of *Alfa alfa* plant [448]. The roots of the plant absorbed Ag from the agar medium and transported to shoots where silver atoms agglomerated to form silver nanostructures. The AgNPs were arranged in one-dimensional array with size ranging from 2 to 20 nm. The fabrication of silver nanoparticles (AgNPs) has been effectively accomplished using Geranium (*P. graveolens*) leaf extract [449]. Upon adding the extract to aqueous $AgNO_3$ solution, spherical and crystalline AgNPs have been produced, with an average particle size between 16 and 40 nm. Silver nanoparticles were synthesized using *Capsicum annuum L.* extract [450]. The effect of reaction time and $AgNO_3$ concentration on the Ag NP size and morphology has been investigated. The silver nanoparticles (AgNPs) exhibited a spherical morphology, with the particle size increasing from 10 ± 2 nm at 5 hours to 25 ± 3 nm at 9 hours and further to 40 ± 5 nm at 13 hours [450]. Silver NPs were synthesized using *Apiin* extract obtained from henna plant [451]. In just two minutes of combining *Apiin* extract with silver nitrate, there was a noticeable increase in the intensity of the SPR peak, which was observed at 456 nm. TEM images reveal the development of quasi-spherical silver nanoparticles (AgNPs) with an average particle size of 39 nm. Silver nanoparticles (AgNPs) with a quasi-spherical morphology and a size range of 10–30 nm were effectively synthesized by employing the aqueous leaf extract of the *Chenopodium album* plant [452]. Silver NPs have been synthesized using the leaf extract of *Aesculus hippocastanum* plant [453]. The nanoparticles (NPs) exhibited a spherical shape with a size of 50 ± 5 nm. Furthermore, these NPs were characterized by a negative zeta potential value of −29.1 mV. Block and Rod like Ag nanostructures have been synthesized with sizes of 40, 50, and 57 nm have been synthesized by Vinodhini et al. by employing the leaf extracts of *Tabernaemontana divaricate, Basella alba*, and *Allium fistulosum*, respectively [454]. The obtained silver nanoparticles (AgNPs) demonstrated remarkable antimicrobial, antidiabetic, and antioxidant activities. Numerous studies have been published on the green synthesis of AgNPs utilizing plant leaf extracts, resulting in various sizes and shapes [455–464].

Ethanolic extracts of some medicinal plants have been used for synthesizing AgNPs [465]. The morphological attributes of silver nanoparticles (AgNPs) were assessed using TEM, DLS, and AFM. The AFM images displayed the development of spherical silver nanoparticles (AgNPs) with smooth surfaces. DLS studies revealed the mean diameter of AgNPs as 90.87 nm and PDI = 0.119 (*Gelsemium sempervirens*), 112.2 nm and PDI = 0.131 (*Thuja occidentalis*), 111.3 nm and PDI = 0.221 (*Phytolacca decandra*), and 122.8 nm and PDI = 0.281 (*Hydrastis canadensis*). The zeta potential values of −15.6, −16.8, −18.2, and −18.9 mV have been reported. Fruit extract of *Malus domestica* has been used by Roy et al. for synthesizing AgNPs of average diameter of 20 nm [466]. Rhizome extract of *Acorus calamus* has been exploited for the synthesis of spherical AgNPs with average diameter of 31.86 nm. Velmurugan et al. employed peanut shell extract for the production of silver nanoparticles (AgNPs) that exhibited spherical and oval shapes, with a size range between 10 and 50 nm. *Nyctanthes arbortristis* flower extract (ethanol) has been utilized to synthesize spherical and oval AgNPs that range in size from 5 to 20 nm. The extract's reducing capability was confirmed by the emergence of a brown color in the reaction mixture, along with the detection of the SPR peak at 440 nm [467]. The existence of lattice fringes observed in the

HRTEM images indicated the crystalline nature of the synthesized silver nanoparticles (AgNPs). Furthermore, the protective capping effect of phytochemicals present in the ethanolic extract was confirmed by the appearance of absorption peaks in the infrared (IR) spectrum of the Ag NP sample at 3430 cm^{-1} (–OH)), 2917 cm^{-1}, and 2848 cm^{-1} (alkanes), 1696 cm^{-1} (carbonyl), 1435 cm^{-1} (geminal methyl), and 1044 cm^{-1} (aliphatic amines). However, the FTIR spectrum of the extract only displays the characteristic IR bands of hydroxyl (3220–3250 cm), alkanes (715, 2895, and 2960 cm^{-1}), C=C of benzene (1627 cm^{-1}), aromatic amines (1374 cm^{-1}) and aliphatic amine (1040 and 1053 cm^{-1}). The existence of biomolecules on the surface of silver nanoparticles (AgNPs) was validated through thermogravimetric analysis/differential thermal analysis (TGA/DTA) analysis of the Ag NP sample which revealed the significant weight loss of about 70% corresponding to the degradation of bioactive compounds from the NP surfaces at high temperatures. Bark extract of *Butea monosperma* has been used to fabricate AgNPs [468]. Upon mixing the extract with aqueous $AgNO_3$, the mixture turned reddish brown in color which demonstrated the fabrication of AgNPs. The brown colored solution also displayed SPR at 424 nm. TEM images revealed the presence of spherical silver nanoparticles (AgNPs) with an average size of 35 nm. High crystallinity of AgNPs was evident from the occurrence of lattice fringes in HRTEM images. The phase purity of AgNPs was ascertained by EDX-X-ray analysis which displayed a sharp absorption peak at 3 keV corresponding to pure Ag. The IR peak at 1758 cm^{-1} corresponding to C=O absorption was found to be present in IR spectrum of the bark extract but absent in the Ag NP sample. Thus, it has been attributed that carboxylic acid group plays a pivotal role in the conversion of Ag^{+} to Ag^{0}. The rapid one-step synthesis of silver nanoparticles (AgNPs) was accomplished by reducing $AgNO_3$ with the hot water extract obtained from *Citrus sinensis* (sweet orange) peels [469]. The color of the reaction mixture turned from yellow to brown with SPR peak observed a round 420 nm. The intensity of the SPR peak increased with temperature and at maximum intensity was reported at 40 °C. Further increase in temperature up to 60 °C resulted in decrease in intensity of the SPR band. Further pH changes were also found to affect the intensity of SPR peaks. Low pH ranges (pH 3–5) of the reaction medium resulted in low-intensity SPR peaks which indicated the acidic medium hinders the formation of AgNPs. The absorption intensity of SPR peak increased with increase in pH till pH 9 at which maximum intensity has been reported [469]. SPR intensity decreased with further increase in pH above 9. Tea extract (*Camellia sinensis*) resulted in rapid synthesis of AgNPs in a short time of one minute [470]. Particle shape was found to be spherical in AFM images with particle size of 3.9 ± 1.6 nm. The TGA of the pure green extract and AgNPs was carried out between 25 °C and 900 °C under nitrogen atmosphere. The TGA plots of the produced silver nanoparticles (AgNPs) displayed weight losses, with the initial loss (3.4–4.9%) observed at 130 °C, corresponding to the removal of adsorbed water molecules from the surfaces of the AgNPs. The subsequent weight loss (11.4–16.9%) detected in the temperature range of 130–150 °C was attributed to the thermal degradation of adsorbed phytochemicals from the surfaces of the silver nanoparticles (AgNPs). The weight loss (~8.9–10%) observed at 900 °C was a result of the thermal destruction of resistant aromatic compounds present in the green tea extract, which became adsorbed on the surfaces of the silver nanoparticles (AgNPs). Numerous reports have been published for the fabrication of AgNPs using various plant parts [471–477]. Besides the plant extracts, Pant-based polymers have been used to synthesize AgNPs and their derivatives like starch, polysaccharide, guargum, cellulose, hemicellulose D-glucuronic acid, arabinoxylan, and lignin [266, 478–485].

To summarize, the section discusses the transformative shift in nanotechnology toward green and sustainable approaches, focusing on the plant-mediated synthesis of silver nanoparticles (AgNPs). Harnessing the diverse bioactive compounds present in plants, this environmentally friendly method explores the mechanisms, advantages, and versatile applications of AgNPs. The synthesis of AgNPs employing plant extracts as both reducing and stabilizing agents is explored, encompassing a wide array of plant species ranging from angiosperms to algae. Key examples include the use of *Alfalfa* sprouts, *Geranium* leaves, *Capsicum annuum* L., and various other plants, each resulting in unique nanoparticle characteristics. The section highlights the role of different plant parts, such as leaves, fruits, rhizomes, and even bark, in Ag NP synthesis. Noteworthy examples include the use of *Malus domestica* fruit extract, *Acorus calamus* rhizome extract, and *Butea monosperma* bark extract. The diverse sources contribute to the broad spectrum of sizes and shapes of AgNPs, expanding the possibilities for applications. Additionally, the utilization of ethanolic extracts from medicinal plants showcases the synthesis of AgNPs with specific morphological features. The use of extracts from *G. sempervirens*, *T. occidentalis*, *P. decandra*, and *H. canadensis* results in well-defined spherical AgNPs with smooth surfaces. Furthermore, the section explores the rapid synthesis of AgNPs using hot water extracts, such as those from *Citrus sinensis* peels and *Camellia sinensis* tea. The impact of factors such as temperature and pH on the synthesis process is discussed, emphasizing the one-minute synthesis achieved with tea extract. The synthesis of AgNPs is not limited to plant extracts alone, as the section notes the successful use of plant-based polymers and derivatives like starch, polysaccharide, guargum, cellulose, hemicellulose D-glucuronic acid, arabinoxylan, and lignin. These materials contribute to the diverse and eco-friendly arsenal of methods for producing silver nanoparticles.

In brief, the plant-mediated fabrication of AgNPs emerges as a promising avenue, demonstrating not only the feasibility of green and sustainable practices but also providing a rich variety of nanoparticle characteristics for a multitude of applications across different fields.

1.6 Synthesis of Magnetic Nanoparticles

The fabrication of MNPs with diverse characteristics has garnered significant attention, leading to the exploration of various synthesis methods [486–493]. This section will discuss the different approaches employed to generate shaped, controlled, and stable MNPs with various dimensions and morphologies. Among the numerous techniques, the co-precipitation method stands out as the most widely adopted for MNP synthesis. This method has proven effective in producing MNPs with controlled morphologies and magnetic properties. The co-precipitation process involves adding a base to an aqueous salt solution under an inert atmosphere, either at room temperature or at elevated synthesis temperatures [494]. The ensuing steps in the co-precipitation synthesis of MNPs are outlined below.

$$Fe^{2+} + 2Fe^{3+} + 8OH^- \rightarrow Fe(OH)_2$$

$$+2Fe(OH)_3 \rightarrow Fe_3O_4 + 4\,H_2O$$

In this approach, the size and shape of MNPs are influenced by several factors, including the type of metal salts employed, such as sulfates, chlorides, nitrates, perchlorates, the ratio of ferrous to ferric ions, reaction media temperature, pH, ionic strength, stirring rate of the reaction mixture, and the rate of dropwise base addition [495–497]. As an example, the fabrication of Fe_3O_4 NPs involved the hydrolysis of ferrous and ferric salts (sulfates and chlorides) with varying molar ratios, utilizing 1,6 hexane diamine as a base [498]. The choice of ferrous to ferric salts ratio was identified as a critical factor influencing the mechanism of Fe_3O_4 formation. Specifically, the use of ferrous salts alone led to the production of larger-sized MNPs, while employing both ferrous and ferric salts resulted in finer-sized MNPs. Moreover, the mean particle diameter increased from 9 to 37 nm with an elevated molar ratio of ferrous ions concerning the total iron ions in solution. The observed variation in particle size was attributed to distinct reaction mechanisms associated with metal salts of different valences [499, 500]. For instance, the formation of MNPs using ferrous ions alone occurs through the dehydration reaction of ferrous hydroxide and ferric oxyhydroxide [499, 500]. Fe_3O_4 synthesis was also achieved via a co-precipitation reaction of ferrous and ferric chlorides without the use of surfactants [501]. The resulting magnetite nanoparticles exhibited an average diameter of 10 nm with a limited size distribution. Subsequently, these magnetite nanoparticles were converted into γ-Fe_3O_4 through direct aerial oxidation.

The microemulsion method has been utilized for the fabrication of MNPs, presenting advantages over alternative techniques by allowing superior control over particle size and morphology. Particle size and shape are also influenced by the type and structure of the surfactant utilized. Pilant et al. utilized microemulsions with ferrous dodecyl sulfate and a micellar solution to synthesize nanoscale MNPs, adjusting the concentration and temperature (25–50 °C) to control particle size. MNPs with an average particle size in the range of 3.7–11.6 nm were achieved, with saturation magnetization decreasing as particle size increased due to nonlinear structure changes at the interface [502]. Lopez et al. utilized the microemulsion method with Brij-97 surfactant to synthesize magnetic iron oxide nanoparticles (MIONPs) with a size range of 2–8 nm [503]. Results indicate that the microemulsion method, employing inverse micelles, facilitates the formation of water nanodroplets within an organic compound, functioning as nanoreactors to regulate particle size. These nanodroplets are surrounded by a surfactant coat that manages their size and separates them from the organic compound. By using a nonionic surfactant (Brij-97) and cyclohexane as the organic phase, microemulsions are created in the cyclohexane/Brij-97/water ternary system at low water concentrations across a broad range of Brij-97 concentrations. The synthesis process involves blending two microemulsions, M(I) and M(II), at 65 °C. M(I) consists of cyclohexane/Brij-97/aqueous iron salt solution, while M(II) is similar but lacks salts in the aqueous solution and contains a 20% larger volume of cyclohexylamine. An acidic medium is used to prevent Fe^{2+} oxidation, and an alkaline medium is employed for iron salt precipitation. After aging and cooling, the resulting precipitate is washed and dried, yielding different samples (A, B, C, D, E), each associated with varying x values in the microemulsions. X-ray diffraction patterns reveal a gradual broadening of peaks from sample A to E, indicating a reduction in particle size. The diffractograms are in line with γ-Fe_2O_3. TEM micrographs display particle aggregates with a noticeable size reduction from sample A to E. Magnetic measurements demonstrate alterations in the zero-field-cooled (ZFC) and field-cooled (FC) data, suggesting variations in the average blocking temperature ($\langle TB \rangle$) and broadening of the size distribution with increasing particle size. Magnetization versus field curves at room temperature show a decline in maximum magnetization values from A to E, aligning with the observed reduction in magnetic particle volumes from TEM and X-ray fits. The use of stable microemulsions at elevated temperatures is revealed as an advancement in the synthesis of IONPs compared to the conventional room temperature microemulsion approach. The study emphasizes the capability to regulate the final particle size through simple adjustments in the synthesis procedure.

Magnetic iron oxide NPs were synthesized in bicontinuous microemulsions composed of dodecyltrimethylammonium bromide and didodecyldimethylammonium bromide (3,2 weight ratio) [487]. The results indicate that, consistent with a prior study, a system composed of DTAB/DDAB (3/2, w/w) mixture, MMA as the oleic phase, and an aqueous solution of $FeCl_3{\cdot}6H_2O$ and $FeCl_2{\cdot}4H_2O$ in a 3/2 molar ratio at an overall concentration of 0.75 M can form a microemulsion region at 80 °C. This region suggests a transition from reverse microemulsions to bicontinuous microemulsions as the system moves from the MMA-(DTAB/DDAB) side to the central zone of the diagram. Electrical conductivity measurements indicate a transition from low conductivity (up to ~15% aqueous solution) corresponding to reverse microemulsions to higher conductivities (from ~20 wt% aqueous solution) associated with bicontinuous microemulsions. The transition region from reverse to bicontinuous microemulsions is estimated to occur between 15 and 20 wt% aqueous solutions. Selected microemulsion compositions were used for precipitation reactions, resulting in black powder products, indicative of magnetite formation. The yield of these reactions, calculated as 1.4–1.6 g of dried solids per 100 g of the total mixture, significantly exceeds the theoretical yield calculated from reported reverse microemulsion recipes. STEM micrographs reveal small-diameter particles (6.9–7.9 nm) with a relatively narrow size distribution when precipitation reactions are conducted in bicontinuous microemulsions. The diameters of these particles align with the channels in bicontinuous microemulsions, acting as nanoreactors to limit particle growth. Increasing surfactant concentration correlates with a reduction in particle size. X-ray diffraction patterns indicate the presence of magnetite or a mixture of magnetite–maghemite in the particles. Room temperature magnetization curves exhibit superparamagnetic behavior, consistent with essentially single domains in magnetic particles below 10 nm. The magnetization values decrease slightly with decreasing surfactant concentration, revealing a direct relationship between magnetization and particle size. Overall, the study demonstrates the advantage of bicontinuous over reverse microemulsions in terms of yield and control over particle size during the synthesis of MNPs. The magnetization behavior is consistent with previous findings and supports the notion that crystallinity influences the magnetic properties of the nanoparticles.

Microemulsions composed of CTAB surfactant were employed for the synthesis of MIONPs [504]. The size and morphology of ME-MIONs were characterized using TEM, highlighting their nearly spherical shape and a size range of 7–10 nm. Some agglomeration was observed, likely owing to the high surface energy of the nanoparticles. The oxidation of iron oxide is found to depend strongly on experimental factors such as temperature, reaction environment, and surface conditions. XRD results indicate that the obtained particles are likely composed of magnetite (Fe_3O_4), maghemite (γ-Fe_2O_3), or a combination of both, reflecting variations in crystal structure based on different experimental conditions. TEM-derived particle size values align with those obtained from XRD analysis, confirming the consistency of the results. The XRD patterns of ME 1-MION and ME 2-MION appear quite similar after calcination at 450 °C, suggesting a common spinel crystal structure. However, differences emerge after drying at 70 °C, making it challenging to definitively determine whether the nanoparticles consist of magnetite or maghemite. The broad diffraction peaks indicate a very small original particle size, in line with similar findings reported in the literature. The particle size, as determined from XRD analysis of samples dried at 70 °C, is measured at 9.2 nm for ME 1-MION and 7.8 nm for ME 2-MION. These findings highlight the influence of drying conditions on the ultimate particle size and crystalline structure of ME-MIONs.

Lakshman et al. also utilized a CTAB-based microemulsion to synthesize MIONs with sizes of 7–10 nm and a resulting surface area of 142 $m^2\,g^{-1}$ [505]. Vidal et al. synthesized maghemite NPs through a one-pot microemulsion route using oleyl amine as a precipitating agent [506]. DLS measurements were conducted to investigate the size of microemulsion nanodroplets. The results indicate that stable microemulsions are formed above 50 °C, with instability issues arising below this temperature due to nanodroplet coalescence. The nanodroplet size increases as the temperature decreases, attributed to better hydration of the surfactant's hydrophilic part, leading to phase inversion. TEM and DLS were employed to examine the size distribution of particles obtained with oleylamine. The nanoparticles exhibited a size range of 8.2 ± 0.6 and 7.8 ± 0.5 nm for oleylamine-coated and oleic acid-coated nanoparticles, respectively. Upon washing and redispersion in heptane, the average sizes increased to 10.6 ± 0.7 and 9.6 ± 0.7 nm for oleylamine and oleic acid-coated particles, respectively. TEM micrographs of washed particles synthesized with cyclohexylamine revealed irregularly shaped particles with a broad size distribution, prone to aggregation. XRD patterns indicated that uncoated particles correspond to crystalline maghemite, and coated particles exhibited similar core structures. FTIR spectroscopy ascertained the presence of oleic acid and oleylamine on coated particles. Low wavenumber regions in mid-infrared (MIR) and far-infrared (FIR) spectra suggested absorption bands consistent with maghemite nanoparticles. Thermogravimetric analysis (TGA) revealed mass losses for water, organic shell, and solvent residues. Coated particles showed similar mass-loss profiles, with oleic acid-coated particles having 4.2% and 45.0% mass loss for water and the organic shell, respectively. Oleylamine-coated particles displayed 4.4% mass loss for water and 45.4% for the organic shell. Magnetization studies indicated that size reduction and surfactant adsorption contributed to the reduction in magnetization. Oleylamine-coated particles exhibited a slightly lower magnetization than oleic acid-coated particles, suggesting differences in charge donation at the particle surface. Overall, the study provides detailed insights into the synthesis and characterization of magnetic IONPs, considering various surfactants and solvents, offering valuable information for potential applications. Numerous reports are available on the microemulsion synthesis of MNPs and their alloys [507–512].

The polyol process was utilized to produce MNPs. In this procedure, the liquid polyol served as the solvent for the metallic precursor, acting as both the reducing agent and, in certain instances, a complexing agent for metallic cations. The metal precursor's solubility in the polyol could vary. The solution underwent stirring and heating to a specific temperature, reaching the polyol's boiling point for less reducible metals. For metals with low reducibility, precise control of precipitation kinetics resulted in nonagglomerated metal particles with well-defined shapes and sizes. Improved regulation of the average size of metal particles was attained by incorporating foreign particles through seeding the reactive medium (heterogeneous nucleation). This separation of nucleation and growth steps led to the formation of uniform particles. A one-spot polyol process, as employed by Abbas et al., was utilized for synthesizing high magnetization magnetite nanoparticles [513]. The size of MNPs varied with changes in reaction parameters like temperature, time, and the ratio of polyol to metal precursor, yielding sizes of 32.3 and 9.2 nm. Exceptionally high magnetization values of 85.87 emu/g at 300K and 91.7 emu/g at 5K, with low coercivity, were observed. Superparamagnetic iron oxide MNPs were synthesized through a polyol process involving stepwise phase transformation using sodium acetate as a hydrolyzing agent [514]. The iron salt precursor underwent a sol–gel process, transforming into amorphous ferrihydrite, which, upon heating, transitioned to crystalline FeO(OH) with lepidocrocite (γ-FeO(OH)) phase, further reduced to Fe_3O_4. Varying the concentration of sodium acetate influenced the crystallinity and cluster size of iron oxide NPs. The cluster size increased from 143 to 338 nm with a rise in sodium acetate concentration from 0.9 to 1.8 mol% and then decreased with further increases (3.6%, 7%, and 10%) (Figure 1.14A). Particle size also varied with changes in the molar ratios of $FeCl_3$ and H_2O, decreasing from 104 to 11 nm with a drop in the molar ratio from 0.10 : 5 to 0.02 : 5. A rise in the molar ratio to 0.20 : 5 resulted in a particle size of 98 nm (Figure 1.14B). XRD studies indicated a decrease in diffraction peak intensity with decreasing molar ratios of $FeCl_3$ and H_2O. XRD investigations revealed a reduction in the intensity of diffraction peaks as the molar ratio of $FeCl_3$ to H_2O decreased from 0.10 : 5 to 0.02 : 5 (Figure 1.14C). This reduction in peak intensity corresponded to a decrease in crystallinity associated with the diminishing particle size of Fe_3O_4 NPs. Iacovita et al. employed a high-temperature polyol process for the fabrication of IOMNPs [515]. This synthesis was conducted under elevated temperature and pressure, resulting in ferromagnetic IOMNPs with a remarkable saturation magnetization of 83 emu/g at room temperature. The polyol method has been widely utilized for synthesizing MNPs with diverse compositions [516–522]. In addition to the discussed methods, MNPs have been produced through various other techniques, such as sonochemical [523–529], sol gel [530–532] and solvothermal method [533–537].

In short, this section discusses the production of MNPs and various synthesis methods to achieve controlled characteristics. The co-precipitation method is highlighted as widely adopted, involving the addition of a base to an aqueous salt solution under an inert atmosphere. Factors influencing the size and shape of MNPs include metal salts, their ratios, reaction media temperature, pH,

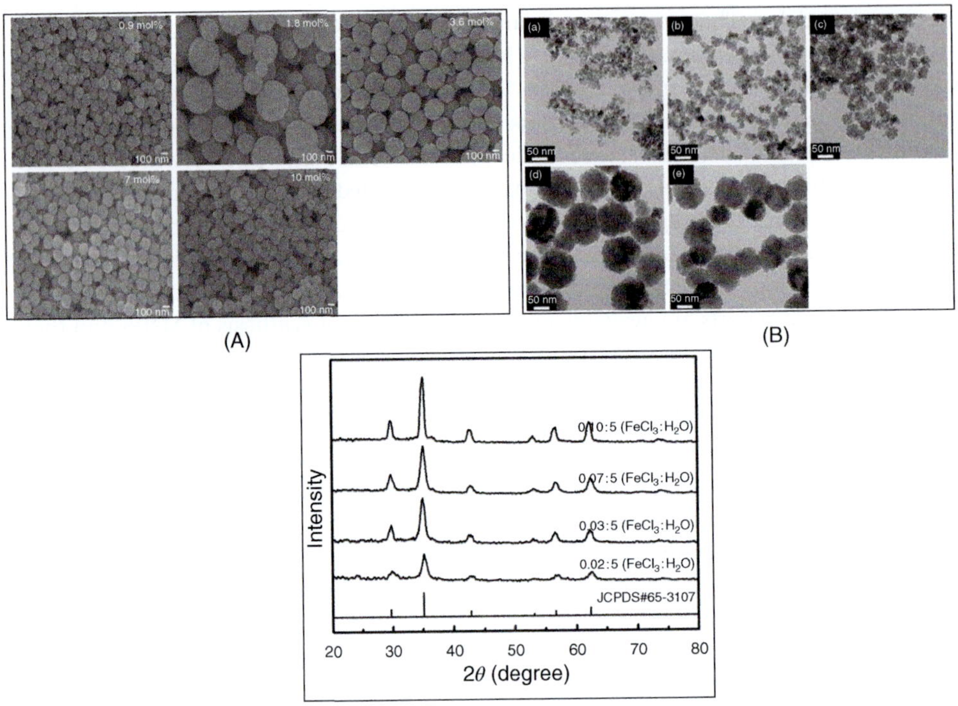

Figure 1.14 (A) Displays scanning electron microscopy (SEM) visuals of the Fe_3O_4 nanoparticles created using varying concentrations of NaOAc. In (B), transmission electron microscopy (TEM) images exhibit the Fe_3O_4 particles synthesized in accordance with different molar ratios of $FeCl_3$ and H_2O: (a) 0.02 : 5, (b) 0.03 : 5, (c) 0.07 : 5, (d) 0.10 : 5, (e) 0.20 : 5. Furthermore, (C) showcases the X-ray diffraction patterns of the synthesized Fe_3O_4 nanoparticles with distinct molar ratios of $FeCl_3$ and H_2O. Source: (a–c) Reproduced with permission from Oh et al. [514]/Elsevier.

ionic strength, stirring rate, and base addition rate. The microemulsion method is explored for MNP synthesis, offering advantages in size and morphology control. Different surfactants lead to varied results, and stable microemulsions at elevated temperatures are deemed advantageous. Bicontinuous microemulsions are presented as superior in yield and particle size control compared to reverse microemulsions. The polyol process is employed, utilizing liquid polyol as a solvent and reducing agent for metallic precursors. Precise control over parameters yields uniform particles. The one-spot polyol process produces high-magnetization magnetite nanoparticles, while step-wise phase transformation using sodium acetate results in superparamagnetic iron oxide MNPs with varying cluster and particle sizes. The high-temperature polyol process is utilized for ferro-magnetic iron oxide MNPs with significant saturation magnetization. The section concludes by mentioning other synthesis techniques, such as sonochemical, sol–gel, and solvothermal methods, highlighting the diverse approaches to MNP production.

1.7 Synthesis of Classical Quantum Dots

Classical QDs refer to nanoscale particles made of semiconducting material, typically characterized by diameters in the range of a few nanometers [31]. The quantum confinement effects in these QDs lead to a widening of the band gap [32]. As a result, the state of unbound charge carriers within

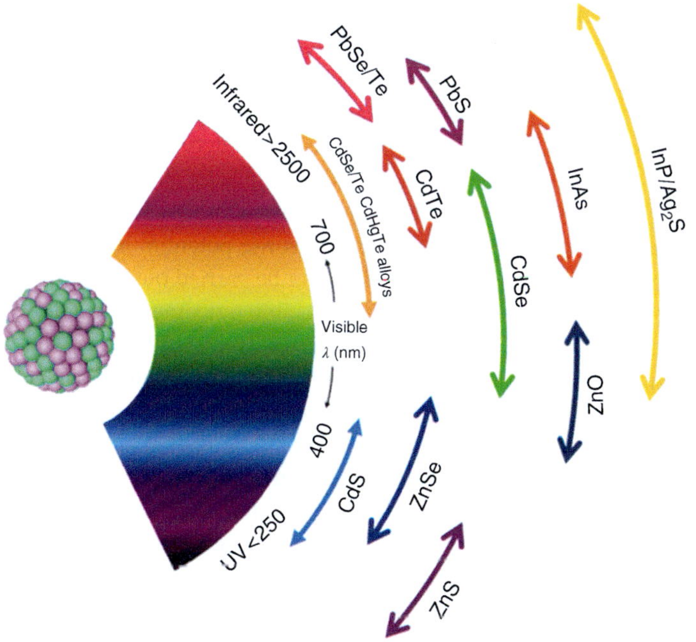

Figure 1.15 Illustrates core materials of representative quantum dots (QDs), scaled based on their emission wavelength and superimposed onto the spectrum. Source: Reproduced with permission from Pu et al. [539]/American Chemical Society.

the quantum dot (QD) is quantized, and the arrangement of distinct energy states is linked to the dimensions of the QD. This feature enables the adjustment of the emission color of QDs by managing their size and composition [538]. By choosing an appropriate core material with a desired bandgap and fine-tuning the size of QDs, the emission wavelength can cover the visible, ultraviolet (UV), and near-infrared (NIR) regions [539] (Figure 1.15). In comparison to organic dyes, QDs display similar quantum yields, yet they boast extinction coefficients 10–50 times greater, markedly reduced photobleaching rates, and fluorescence that is 10–20 times brighter. Additionally, they exhibit 100–200 times enhanced photostability [540]. Currently, CdSe and CdTe QDs are widely favored, showcasing attractive optical properties within the visible spectrum ranging from 400 to 700 nm [541–543]. Furthermore, QDs with NIR emission characteristics, including CuInSe, Ag_2S, and PbS QDs, have garnered interest for their benefits in *in vivo* animal imaging studies where NIR light is advantageous [544–546] Recently, there has also been interest in perovskite QDs because of their possible use in extremely effective photoelectric devices [547]. Indeed, recent literature has used the terms "carbon QDs" or "graphene QDs" to describe various carbon-based nanoparticles [548]. Upon further investigation, it becomes apparent that several of these particles display features more closely associated with fluorescent organic dots. In this article, the focus is on QDs composed of semiconductor nanocrystals. Several colloidal methods have been used to synthesize QDs, which are discussed briefly below.

1.7.1 Organometallic Methods of Synthesis of Quantum Dots

Conventional techniques for producing QDs typically entail injecting semiconductor precursors into appropriate organic solvents through a hot injection method. As an illustration, Murray et al. conducted the synthesis of CdSe QDs by injecting a liquid mixture of Cd precursor ($Cd(CH_3)_2$)

and Se metal in tri-*n*-octylphosphine (TOP) into heated trioctylphosphine oxide (TOPO) [51]. The TOPO solvent functions as a stabilizing agent, enabling the reaction to occur at elevated temperatures around 320 °C. Through precise control of the reaction temperature, a range of QDs with sizes spanning from 1.5 to 11.5 nm was successfully produced. However, the cadmium precursor, $Cd(CH_3)_2$, employed in this process proved to be highly toxic, rendering it impractical for large-scale synthetic applications. To address this issue, Peng et al. later substituted this toxic cadmium precursor with a less harmful reagent, CdO, aiming for a more environmentally friendly synthesis of CdTe, CdSe, and CdS QDs [55]. This method has the potential to generate additional QDs with reasonably high yields. Subsequent syntheses can be carried out at comparatively lower temperatures (250–300 °C) with consistent and reproducible outcomes. Deng et al. further streamlined the quantum dot synthesis process by developing Zinc-blend CdSe. They achieved this by substituting coordinating solvents such as TOP/TOPE with more economical and environmentally friendly options like paraffin liquid and oleic acid [549]. Nanocrystals of CdSe, with sizes that can be adjusted within the range of 2–5 nm and distinct UV-Visible peaks, were successfully produced. The hot injection method, with precise selection of synthetic parameters, has been applied to synthesize various QDs, including lead sulfide (PbS) [54, 550], lead selenide, PbSe [54], Indian phosphide InP [551], silver sulfide Ag_2S [552], and silver selenide Ag_2Se [553]. Size-controlled synthesis of fluorescent perovskite QDs has also been achieved by hot injection method [554, 555].

While injection-based reaction techniques yield QDs with desirable physical and chemical attributes, challenges in regulating post-injection reaction temperatures have led to reproducibility issues. In response to this challenge, various noninjection methods have been introduced for large-scale QD production. For instance, Efrima et al. developed a straightforward and adaptable low-temperature approach for synthesizing high-quality metal sulfide QDs by heating metal xanthate in hexadecylamine solvent [556]. The process was conducted at a moderate temperature range of 50–150 °C, generating metal sulfide QDs even at a low reaction temperature of 70 °C. Nevertheless, the QD size was manipulated by adjusting both the reaction temperature (ranging from 50 °C to 150 °C) and the concentration of metal xanthate. Elevated metal xanthate concentration led to the fabrication of smaller particles, attributed to rapid nucleation following the principles of La Mer behavior [557]. Liu et al. reported the low-temperature synthesis of PbS QDs using non injection approach [558]. The impact of various synthetic reaction parameters on the development of PbS QDs has been examined. The reactions were conducted within the temperature range of 30–120 °C, utilizing octadecane as the solvent. Lead oxide (PbO) and lead acetate served as the lead sources, while bis(trimethylsilyl)sulfide ((TMS)2S), thioacetamide (TAA), and elemental sulfur (S) were employed as sulfur sources. Different acids, such as oleic acid (OA), were explored as stabilizing agents. Optimal quality PbS QDs were reported when using (TMS)2S and OA. The reactions yielded QDs with a narrow bandwidth as small as 100 nm. These QDs displayed emissions spanning from 700 to 900 nm, dependent on factors such as reaction temperature, growth periods, acid-to-lead ratio, and the feed molar ratio of lead to sulfur. Du et al. synthesized monodisperse and single-crystalline Ag_2S QDs using a strategy involving a single-source precursor (Ag(DDTC)), resulting in single-crystalline Ag_2S QDs with a size of approximately 10 nm [559]. The reaction scheme for the synthesis of Ag_2S QDs by this method is shown schematically in Figure 1.16A. The synthetic procedure involved the heating of a mixture containing Ag (DDTC), oleic acid (OA), octadecylamine (ODA), and 1-octadecane (ODE), followed by precipitation, washing, and drying. The selection of capping ligands and solvents played a pivotal role in achieving uniform Ag_2S QDs. Adjusting the solution composition allowed for the tuning of QD size and size distribution. The Ag_2S QDs in the monoclinic phase exhibited near-infrared (NIR) emission at 1058 nm, indicating their potential for mitigating autofluorescence in bioimaging. X-ray diffraction (XRD)

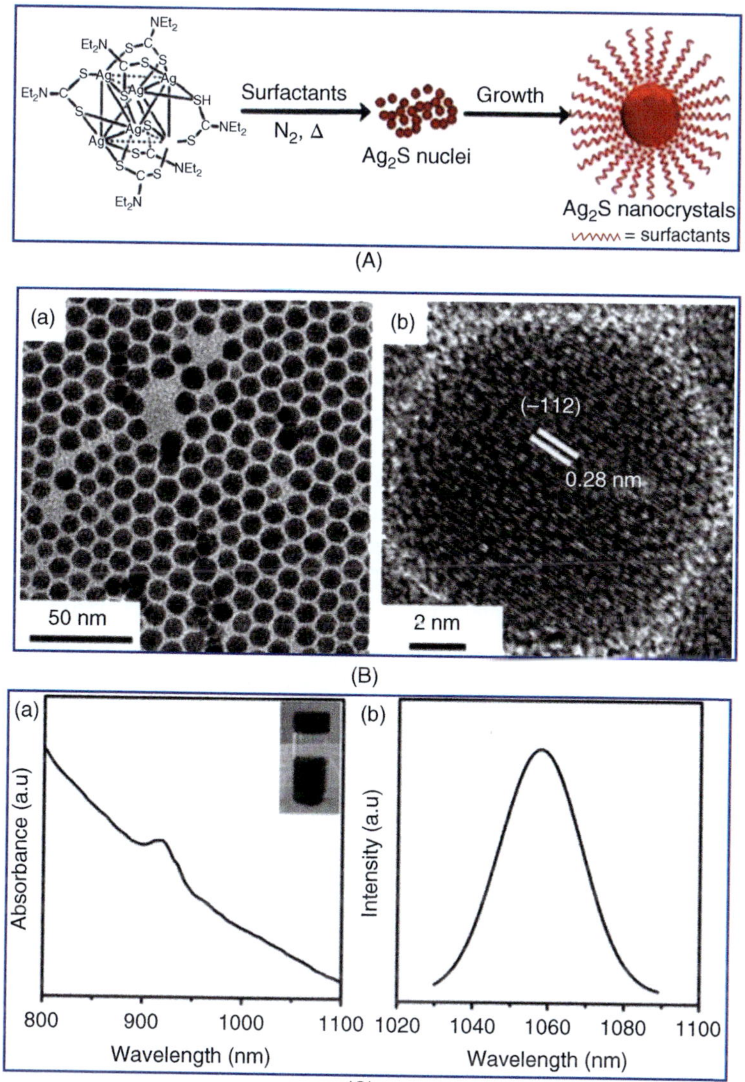

Figure 1.16 (A) Illustrates the process of synthesizing near-infrared (NIR) Ag_2S quantum dots (QDs) from a single-source precursor, Ag (DDTC). (B) Displays transmission electron microscopy (TEM) images, including (a) an overall view of Ag_2S QDs and (b) a high-resolution TEM (HRTEM) image focusing on a single Ag_2S QD. (C) Presents (a) the NIR absorption spectrum of the obtained Ag_2S QDs, with an inset showing the dispersion of Ag_2S in cyclohexane and (b) the NIR fluorescence emission spectrum of Ag_2S QDs at room temperature, excited at λex = 785 nm. Source: (a–c) Reproduced with permission from Du et al. [559]/American Chemical Society.

patterns validated the monoclinic nature of the Ag_2S nanocrystals, and TEM images portrayed well-dispersed QDs with a two-dimensional ordered arrangement. The TEM images illustrated the Ag_2S QDs to be uniformly sized, of an average size of 10.2 ± 0.4 nm. (Figure 1.16B). The single crystallinity of the QDS was confirmed by HRTEM images which interplanar distance of ~0.28 nm which corresponds to (112) facets of monoclinic Ag_2S (Figure 1.16B). Whereas NIR-UV-Visible spectrum showed a discreet and lowest energy absorption peak at 920 nm (Figure 1.16C). The NIR photoluminescence spectrum of the Ag_2S QDs, excited with a 785 nm laser diode, exhibited a symmetric emission peak at 1058 nm with a small full-width at half-maximum (FWHM), indicating a narrow size distribution (Figure 1.16C). The quantum yield of the QDs was not examined in this study. The results suggest that the as-prepared Ag_2S QDs have potential as nontoxic alternatives for in vivo bioimaging, offering a unique optical window for tissue penetration. Further investigations may explore the incorporation of a core–shell structure to enhance quantum yield and photostability.

1.7.2 Aqueous Synthesis of QDS Using Chemical and Biological Methods

While organometallic methods for synthesizing QDs provide advantages such as high crystallinity, a narrow particle size distribution, and a high fluorescence quantum yield, they have limitations in biomedical applications. Organometallic QDs exhibit low water solubility, necessitating additional phase transfer steps to enhance their stability and dispersibility in water. Moreover, the organometallic synthesis process involves the use of various harmful solvents, posing environmental concerns. To address these limitations, water, a more environmentally friendly solvent, has been employed for QD synthesis. Aqueous-phase synthesis of QDs offers several benefits, including the use of water-soluble ligands as stabilizers for biomedical applications and lower reaction temperatures typically below 100 °C. The fabrication of QDs in an aqueous phase was initially documented by Henglein [560] in 1982 and Brus et al. [561] in 1983, where CdS QDs were synthesized. Vossemeyer et al. successfully produced monodisperse CdS QDs utilizing 1-thioglycerol as a stabilizer, and $Cd(C1O_4)$ $6H_2O$ and H_2S as reactant precursors. The size of the CdS QDs ranged from 13–19 Angstroms. Rogach et al. [562] described the aqueous phase synthesis of thiol-stabilized CdTe QDs with commendable luminescence properties using 2-mercaptoethanol and 1-thioglycerol as stabilizers. The synthesis involved adding NaHTe solutions to $Cd(C1O_4)2\text{-}6H_2O$ solutions in a nitrogen atmosphere at pH 9.2 in the presence of stabilizers. The size of the QDs was adjusted by regulating the initial concentrations of NaHTe precursor, temperature, and the type of stabilizer. The CdTe QDs exhibited fluorescence properties attributed to the 1s–1s electronic transition. High-quality CdZnS and Cu-doped CdZnS QDs were synthesized by Yakoubi et al. [563] using aqueous phase synthesis with 3-mercaptopropionic acid (MPA) or *N*-acetylcysteine (NAC) as capping agents. Initially, nearly monodisperse Cu:CdZnS QDs were produced in an aqueous solution, utilizing (MPA as a capping ligand. Subsequently, Cu:CdZnS/ZnS core/shell QDs were prepared by carefully depositing ZnS onto preformed Cu:CdZnS nanocrystals through the addition of Zn^{2+} and S^{2-} precursors and the ligand at 100 °C. The synthesis of host CdZnS QDs involved injecting an aqueous solution of Na_2S into a mixture of cationic Cd^{2+}, Zn^{2+}, and the ligand at basic pH, followed by heating at 100 °C. Optimal conditions for obtaining high-quality CdZnS QDs included a $Cd^{2+} + Zn^{2+}$ concentration of 5 mM and a $(Cd^{2+} + Zn^{2+})/S^{2-}$ molar ratio of 1/1. The CdZnS QDs displayed photoluminescence quantum yields (PL QYs) up to 7%, with variations in precursor concentrations or molar ratios affecting the PL efficiency. Different Cd and Zn precursors were explored, and $CdBr_2$ and $ZnBr_2$ resulted in the best PL QY (5%) for CdZnS QDs. The pH of the Cd^{2+}-Zn^{2+}-MPA precursor solution influenced the QDs' properties, with pH 8 yielding the highest PL QY (5%). The formation of alloyed $Cd_xZn_{1-x}S$ QDs was evidenced by continuous shifts in absorption and PL spectra, indicating the incorporation of wider bandgap ZnS with narrower bandgap CdS (Figure 1.17A). However, attempts at certain compositions like $Cd_{0.1}Zn_{0.9}S$ and $Cd_{0.8}Zn_{0.2}S$ failed to produce fluorescent nanocrystals. Cu-doped CdZnS QDs were synthesized by adding $CuCl_2$ to the reaction mixture. The PL emission redshifted upon Cu^{2+} doping, and the optimal doping concentration was found to be 0.5%. The Cd/Zn ratio influenced the emission wavelength, with higher Cd concentrations leading to a red-shift. The synthesis was versatile,

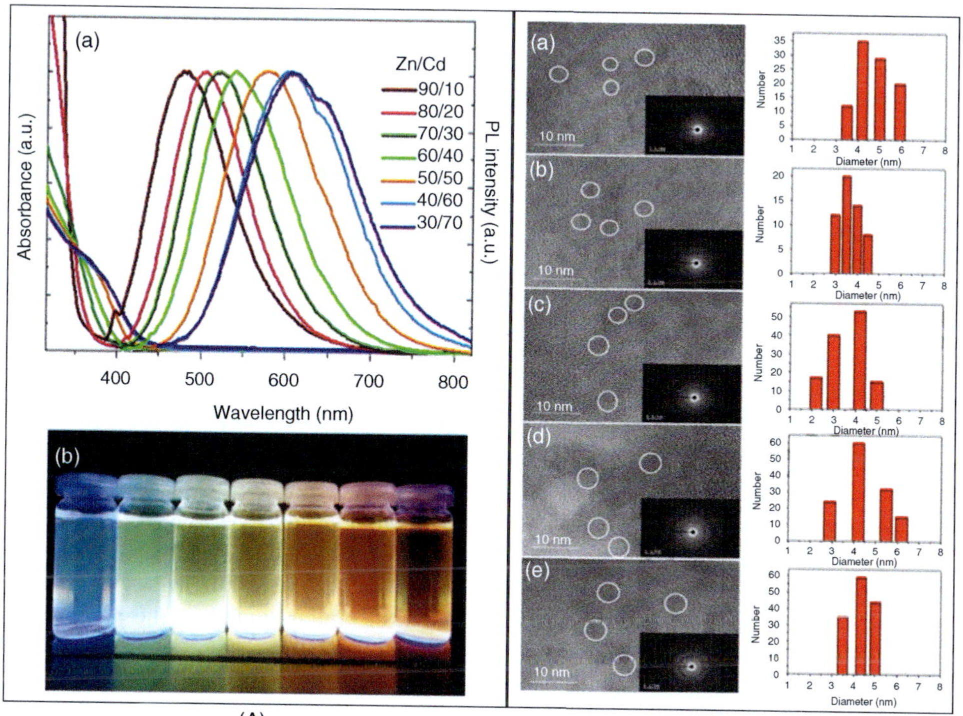

Figure 1.17 (A) Shows (a) UV-visible absorption and photoluminescence (PL) spectra of Cu-doped $Cd_xZn_{1-x}S$ quantum dots (QDs). Additionally, (b) presents a photographic representation of aqueous dispersions of Cu: $Cd_xZn_{1-x}S$ QDs under UV light irradiation, with variations in *x* from 0.1 to 0.7 and an excitation wavelength of 350 nm. Moving to (B), it includes high-resolution transmission electron microscopy (HR-TEM) images, selected area electron diffraction (SAED) patterns, and size distributions of Cu-doped QDs for different compositions: (a) $Cd_{0.2}Zn_{0.8}S$, (b) $Cd_{0.3}Zn_{0.7}S$, (c) $Cd_{0.4}Zn_{0.6}S$, (d) $Cd_{0.5}Zn_{0.5}S$, and (e) $Cd_{0.6}Zn_{0.4}S$. Source: (a, b) Reproduced with permission from Yakoubi et al. [563]/Elsevier.

allowing the use of different water-soluble thiol ligands. Among those tested, *N*-acetylcysteine (NAC), MPA, and glutathione (GSH) provided the highest PL QYs (27, 9, and 3%, respectively). To enhance stability and optical performance, Cu:CdZnS/ZnS core/shell QDs were prepared. The ZnS shell was grown using the successive ion layer adsorption and reaction (SILAR) technique, resulting in a significant increase in PL QY (from 9 to 42%). Further coating beyond five monolayers of ZnS led to aggregation and decreased PL intensity. Excited state lifetime decays of Cu-doped CdZnS QDs indicated triexponential decays, with lifetimes ranging from 0.46 to 1.55 μs. The addition of a ZnS shell increased the average PL lifetime, indicating efficient surface passivation. The hydrodynamic diameters of Cu:CdZnS and Cu:CdZnS/ZnS QDs in a phosphate buffer solution (pH 7.4) were 7.3 ± 1.2 and 9.2 ± 1.5 nm, respectively (Figure 1.17B). Both QDs exhibited stability over ten days, with a slight increase in diameter and a reduction in PL intensity. The dots were stable over a range of pH (7–12), with a slight decrease in PL QY at pH 6 and rapid decay below pH 4. Photostability testing demonstrated that the QDs retained their PL QY even after prolonged UV-light exposure. In summary, a synthetic approach was developed that allowed for the fabrication of stable and highly luminescent Cu:CdZnS and Cu:CdZnS/ZnS QDs, holding promise for various applications, including biological labeling and tumor detection. Further tuning the particle size with different stoichiometric ratios of Cd/Zn precursors in CdZnS QDs allowed for the adjustment of photoluminescence properties across the entire visible spectrum. Aqueous phase synthesis of QDs has been further extend to microwave [564], hydrothermal [565], ultrasonic [56, 566, 567], and microemulsion methods [568–570] of synthesis. For instance microwave-aided synthesis of water-dispersable CdTe/CdS/ZnS CSS QDs was reported by He et al. [571]. As synthesized QDS exhibited high photoluminescence quantum yield (40–80%) and excellent photostability. The size of these core shell shell (CSS) QDs was found to be around 4.5 nm. A one-pot aqueous phase microwave synthesis of CdTe QDs has been reported by employing Na_2TeO_3 as tellurium (Te) source and MPA as a stabilizing agent [572]. The QDs were produced in a brief reaction time of 10–14 minutes and achieved 40–60% emission in 550–640 nm wavelength range. These CdTe QDs were spherical in size with a mean diameter of 3 nm. By controlling the reaction temperature, various sizes of CdTe QDs with tuneable UV-Visible/PL emissions were produced. Several reports are available for the aqueous phase synthesis of QDs by employing various chemical methods of synthesis [573–580]. However, due to toxicity issues with heavy metal QDs, efforts have been on to synthesize greener QDs based on nontoxic elements like ZnO [581], ZnS [582], ZnSe [583] in aqueous phase which emit in the blue range. In the past decade, there has been a growing interest in synthesized near-infrared fluorescent QDs for in vivo imaging applications. Example of such QDs is Ag_2S QDs, which has been synthesized by several research groups [584–588].

Apart from the aqueous phase methods, QDs of various types have also been synthesized by utilizing the bio machinery or biocomponents derived from the various living things especially microorganisms such as bacteria and fungi [589]. In the intracellular synthesis, the precursors such as metal and sulfide ions enter the cell through the transmembrane and the synthesis is carried out by the intracellular enzymes present inside the cytoplasm. Whereas in extracellular synthesis, formation of QDs takes place due to the enzymes secreted by the cell to the outside medium and hence the synthesis occurs either on the cell membrane or inside the medium. Various QDs have been synthesized by this approach such as ZnS QDs [590, 591], CdS QDs [592–596], PbS QDs [597, 598], CdSe QDS [599, 600], Ag_2S QDs [601, 602], Ag_2Se QDs [603], and SnO_2 QDs [604].

In brief, this section discusses the synthesis of classical QDs, which are nanoscale particles of semiconducting material. These QDs exhibit quantum confinement effects, resulting in a quantized state of free charge carriers and tunable emission colors based on size and composition control. The advantages of QDs over organic dyes include larger extinction coefficients, reduced photobleaching rates, and brighter fluorescence. CdSe and CdTe QDs are popular in the visible spectrum, while CuInSe, PbS, and Ag2S QDs with near-infrared (NIR) emission properties have gained attention for in vivo imaging. Perovskite QDs are also explored for efficient photoelectric devices.

The synthesis methods are discussed, starting with organometallic approaches involving hot injection methods. Various examples, including CdSe QDs, are presented, and efforts to substitute toxic precursors with environmentally friendly alternatives are highlighted. Aqueous synthesis methods are then introduced to address limitations in organometallic approaches, focusing on water-soluble ligands for biomedical applications. Examples include CdS, CdTe, and CdZnS QDs synthesized in aqueous phases. The section also touches on alternative synthesis methods such as microwave, hydrothermal, ultrasonic, and microemulsion techniques.

The versatility of aqueous phase synthesis is demonstrated in the synthesis of stable and luminescent Cu:CdZnS and Cu:CdZnS/ZnS QDs, showing promise for applications like biological labeling and tumor detection. The discussion extends to near-infrared fluorescent QDs for in vivo imaging, particularly Ag_2S QDs. Additionally, the section explores the use of biological components for QD synthesis, involving microorganisms like bacteria and fungi. Both intracellular and extracellular synthesis approaches are discussed, highlighting the synthesis of various QDs, including ZnS, CdS, PbS, CdSe, Ag_2S, Ag_2Se, and SnO_2 QDs.

1.8 Synthesis of Carbon-Based Nanomaterials

Intensive research has focused on exploring the remarkable attributes of carbon nanomaterials, including fullerenes, carbon nanotubes, and graphene. These materials exhibit exceptional properties, such as outstanding electrical and thermal characteristics, high mechanical strength, making them suitable for diverse applications ranging from water treatment, hydrogen storage, optoelectronics, sensing, to controlled drug delivery. Additionally, their utilization as protective agents in various engineering products underscores their significance. In the biomedical field, these carbon nanomaterials have demonstrated promising results. This section presents a brief summary of the synthetic methods used to fabricate various types of carbon-based nanomaterials, such as carbon nanotubes, fullerenes, and graphene.

1.8.1 Synthesis of Carbon Nanotubes (CNTs)

Carbon nanotubes (CNTs) is a versatile carbon allotrope comprised of rolled up sheets of graphene [605]. Carbon nanotubes (CNTs) are found in two primary forms: single-walled carbon nanotubes (SWCNTs) and multiwalled carbon nanotubes (MWCNTs) [606]. SWCNTs are composed of a single layer of carbon atoms and have a diameter of around 3 nm [607], while MWCNTs are composed of two or more layers and have a diameter in the range of 7–100 nm [608]. Both types of CNTs can reach lengths up to 1 mm. Their extensive applications in the fields of sensors, electronics, mechanics, chemistry, and biology are attributed to their remarkable tensile strength, Young's modulus, and twisting properties compared to metals like steel and iron [609]. Various synthesis methods, including laser ablation, arc discharge, chemical vapor deposition (CVD), and plasma-enhanced CVD, have been employed to produce CNTs.

1.8.1.1 Plasma-Based Synthesis

Plasma-based synthesis stands out as a cutting-edge and versatile approach for the production of CNTs, offering unique advantages in terms of precision and control over the synthesis process [610]. Leveraging the fundamental principles of plasma physics, this method harnesses the energy of a high-temperature plasma to induce the growth of CNTs from carbon-containing precursors. The intricate interplay of reactive species in the plasma environment allows for the tailoring of CNT characteristics, including size, structure, and purity.

The techniques used in plasma-based synthesis of CNTs are discussed briefly below.

Arc Discharge Method Arc discharge technique was first used by Iijima to synthesize CNTS who observed MWCNTs in the sooth of graphite electrodes [611]. This technique involves the production of plasma between the two graphite electrodes, which finally results in the production of good-quality CNTs [612]. The anode of the setup is usually filled with carbon precursor and a catalyst (Fe, Co, etc.), whereas cathode is composed of pure graphite. The two electrodes are then mounted at a pot difference of 20 V and are kept at 1–2 mm distance. The electrodes are enclosed in a setup, which is filled with inert gas at very low pressures. As the arc is triggered between the electrodes, plasma is produced which is usually composed of carbon, catalyst vapors, and inert gas. During the process, CNTs get deposited on the graphite cathode and at the same time anode gets consumed. This method produces fine CNTs with fewer defects [613, 614]. The yield and quality of CNTs can be optimized by varying the reaction parameters such as type of the catalyst, inert gas type, inert gas pressure, electrode distance, and arc current density. For instance, using hydrogen in the growth region results in the production of highly crystalline MWCNTs [615–618]. Single-walled

CNTS are usually synthesized in the presence of catalytic material such as Gd, Ni, Fe, Co, Pt, Pd, and Ag or mixtures of metal catalysts such as Ni, Co, and Fe with different components such as Co-Ru, Co-Pt, Ni-Y, and Co-Ni [619, 620]. Multiwalled CNTs have usually been synthesized without the presence of a catalyst [621–623]. The mass production of SWCNTs has been achieved by employing Ni-Y bimetallic catalyst in helium ambient gas [624]. Various reports are available for the synthesis of CNTs using arc discharge method [624–626].

Laser Ablation Method Laser ablation method is the second method, which has been used to synthesize CNTs. For instance, synthesis of SWCNTs in high yields (70%) was achieved by Thess et al. in 1996 using Ni and Co as a catalyst at 1200 °C [627]. These CNTs were uniform in diameter and self-organized in the shape of ropes with each rope composed of 100–500 NTs. In this method, the reaction usually takes place inside quartz tube furnace. Graphitic material embedded with a catalyst is placed inside the quartz tube and heated by the pulsed laser due to which the material converts into vapor form. The pressure is maintained constant during the heating by employing the inert gases. Metal catalyst NPs formed in the graphitic vapor during pulsed heating catalyze the growth of CNTs [628]. This method is employed to produce mostly SWCNTs [629].

Chemical Vapor Deposition (CVD) The main problems associated with the plasma methods is low yield, high reaction temperature, and several purification process for obtaining pure CNTs. To overcome these limitations, chemical vapor deposition (CVD) method has been devised in 1996 for the large-scale production of CNTs [630]. Infect, arc method, and laser ablation methods are superior to CVD to produce more crystalline CNTs, but CVD has a special advantage of producing CNTs in bulk quantities and in high yields. In this method, the carbon substrate materials such as fullerene are placed inside an oven and heated to the temperature of 700–900 °C [631]. During the heating process, carbon-bearing gases such as methane, acetylene, and ethylene are added to the reaction chamber. During heating, the carbon-based materials are converted into atomic form, which interacts with the substrate and forms CNTs. The heating can be accomplished by either using energy from heater coil or plasma. CVD results in the production of CNTs in yields greater than 90% and high percentage purity. Based on catalyst location, CVD can be further classified into supported CVD [632] and floating catalyst CVD (FC-CVD) [633, 634]. In the former method, catalyst is first prepared and reduced before the synthesis of SWCNTs. This method is complicated due to interaction between catalyst and support. However, in latter method, catalyst and SWCTs synthesis occurs in the gaseous environment, thereby making it more scalable and a single-step process. Further, FC-CVD results in the production of SWCNTs with high purity. The morphology and chirality of the NCTs can be tuned by proper selection of the catalyst [635]. For instance, using bimetallic catalyst material results in the production of SWCNTs with narrow chirality [636]. The carbon source in CVD can be either in liquid phase or in solid phase. In CVD process using liquid hydrocarbon as a carbon source such as benzene and alcohol, the liquid carbon source is heated in a flask and purged by an inert gas, which carries the vapors hydrocarbon vapors into the reaction zone where decomposition and further reaction result in the production of CNTs [38]. In CVD using solid carbon source, the hydrocarbon material that may be a low volatile material such as camphor, naphthalene, and ferrocene is placed at a low-temperature zone of a reaction chamber, where it is vaporized and carried toward the high-temperature reaction zone loaded with the catalyst. The catalyst used in CVD may also be used either in solid, liquid, or gas form. The catalyst can be either placed inside the reactor or fed from outside. The catalyst can either be placed in the hot zone to catalyze the growth of CNTs or catalyst is subjected to pyrolysis at suitable temperatures which liberates the metal NPs in situ. The MNPs then catalyze the growth of the CNTs. This method is flexible compared to other methods like arc discharge or laser ablation method in the sense that plenty of hydrocarbon sources in any of the three states can be used as substrates. It also permits the growth of CNTs in a variety of forms like powders and films owing to the lower growth temperatures (550–1000 °C), and CVD offers better control over the growth parameters. This approach has been utilized to cultivate carbon nanotubes (CNTs) tailored for specific applications [607, 637–643].

Plasma-Enhanced Chemical Vapor Deposition (PECVD) The thermal CVD method described above has a disadvantage of operating at moderately higher temperature, which are quite unbearable for certain substances to withstand without decomposition. Therefor an alternative method known as plasma-enhanced chemical vapor deposition (PECVD) method has been designed which operates at lower temperatures, room temperature in some cases for the deposition of CNTs [644]. In this method, a high voltage is applied between the two electrodes in a reaction chamber, which results in the production of the electric discharge. The substrate is put on the electrodes and the reactant gas is flushed through the chamber to deposit a uniform film [645]. The substrate is itself loaded with transition metal catalysts such as Fe, Ni, and Co by thermal CVD or sputtering technique. The electric discharge results in the production of fine metal particles which then catalyze the growth of CNTs from the carbon gas source such as CH4, C_2H_2, C_2H_4, C_2H_6, and CO, which is fed to the reaction chamber during the discharge [646]. The type of the catalyst affects the morphological features of the CNTS such as diameter, wall thickness by controlling the growth rate. For instance, aligned MWCNTs of the diameter of 15 nm have been produced by using Ni as a catalyst. This method operates in different modes like radio frequency (RF) mode known as RF-PECVD, diffusion current mode, DC-PECVD, microwave, MW-PECVD, and hot filament-PECVD. Hot filament-PECVD uses thermal energy for the generation of plasma and has successfully produced CNTs. DC-PECVD resulted in the production of mostly vertically aligned CNTs. For instance, Teo et al. also produced vertically aligned CNTs by DC-PECVD using n+ doped Si (100) substrate [647]. Here the substrate was biased at −600 V which initiated a DC glow plasma discharge. The CNTs were grown vertically by electric field in the plasma sheath. The transmission electron micrograph in Figure 1.18A illustrates a typical carbon nanotube (CN) grown via direct current plasma-enhanced chemical vapor deposition (dc PECVD). The presence of Ni catalysts at the tip of all CNs suggests favorable conditions for the tip-growth diffusion mechanism during deposition. The CNs exhibited well-crystallized graphene walls, numbering between 20 and 40, indicative of excellent conductivity along their axis. Electrical conductivity measurements revealed a room temperature bulk resistivity comparable to multiwall carbon nanotubes produced by arc discharge, emphasizing their potential utility. Graphitic bamboo-like fringes were observed along the length of the CN, consistent with findings from other PECVD studies. Upon the termination of growth and cooling of the catalyst, disordered carbon was expelled, resulting in mostly hollow, tubular structures, even for CNs with large diameters (300 nm). Examining the nucleation of multiple and single isolated CNs, it was observed that a 7 nm thick Ni film patterned into a 100 nm wide line yielded multiple CNs, while an array of 100 nm wide square dots produced more uniform, single CNs. The critical size for deterministic single CN nucleation was found to be 300 nm. A statistical study on the effect of catalyst dot size revealed that only at 100 nm were single CNs deterministically obtained. Beyond 300 nm, the average number of CNs per catalyst dot increased linearly with dot size, indicating some degree of control over CN number by adjusting dot dimensions (Figure 1.18B). To achieve good structural uniformity, single isolated CNs were grown, exhibiting tight distributions in diameter and height. The observed CN diameter was slightly smaller than the catalyst dot size, and a statistical approach revealed good uniformity. Field emission measurements on individual, isolated CNs showed similar Fowler–Nordheim characteristics and emission currents, confirming the uniformity of emission properties. The statistical predictions for field enhancement factor uniformity were validated by electrical measurements, demonstrating

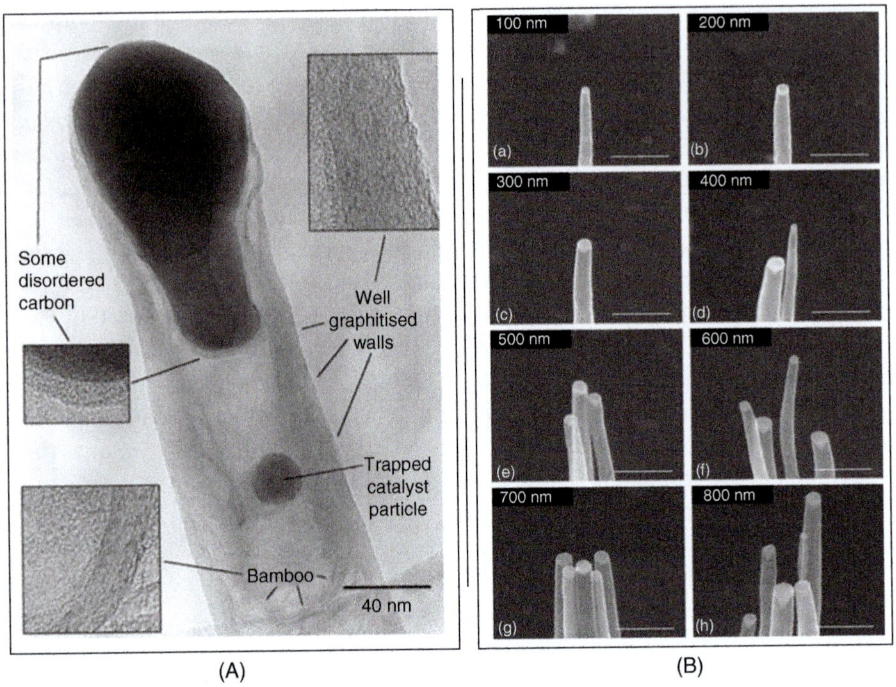

Figure 1.18 (A) Presents a transmission electron micrograph of a carbon nanotube (CN) grown via dc plasma-enhanced chemical vapor deposition (PECVD). The accompanying insets are magnified by 2.5×. In (A), high-resolution images of the CN heads nucleated from Ni dots, ranging in width from (a) 100 to (h) 800 nm, are displayed. The scale bar, 400 nm in length, and a sample tilt of 40° provide context for the images. Source: (a, b) Reproduced with permission from Teo et al. [647]/IOP Publishing.

the reproducibility and reliability of the deposition process. Boskovic et al. synthesized CN fibers (CNFs) at three different temperatures including room temperature using RF-PECVD using Ni powder as a catalyst and methane or methane/hydrogen as a carbon source [648]. TEM images revealed the spaghetti morphology of CNFs with CNF growth and diameter quite independent on substrate temperature. HRTEM reveals the Ni catalyst located at the tip of the CNF. Vertically aligned CNTs were synthesized by Hussain et al on TiN/SiO_2/Si substrate and Ni as a catalyst using RF-PECVD [649]. SEM/XPS studies revealed that the surface of CNTs covered with ultra-thin layer of PANI. PECVD is widely used for the growth of CNTs for various applications. PECVD has been used to grow CNTs for various applications [650–659].

To summarize, this section discusses the synthesis of CNTs, a versatile carbon allotrope with applications across diverse fields such as sensors, electronics, mechanics, chemistry, and biology. Two main forms of CNTs are SWCNTs and MWCNTs, differing in the number of layers and diameter. The focus is on plasma-based synthesis methods for CNTs, offering precision and control over the synthesis process. The following techniques are explored:

Arc discharge method. This method involves the production of plasma between two graphite electrodes, leading to the deposition of high-quality CNTs. It utilizes carbon precursors and catalysts, and the parameters like catalyst type, inert gas, pressure, and electrode distance influence the yield and quality of CNTs.

Laser ablation method. In this method, a pulsed laser is used to heat graphitic material embedded with a catalyst inside a quartz tube furnace. The metal catalyst nanoparticles formed catalyze the growth of CNTs. This method is primarily employed for the synthesis of SWCNTs.

Chemical vapor deposition (CVD). CVD is introduced as a method to overcome limitations associated with plasma methods, offering large-scale production of CNTs. Carbon substrate materials are heated in an oven, and carbon-bearing gases are added, leading to the formation of CNTs. The catalyst's location classifies CVD into supported CVD and floating catalyst CVD, with the latter offering scalable and single-step processes.

Plasma enhanced chemical vapor deposition (PECVD). PECVD operates at lower temperatures compared to traditional CVD, making it suitable for substances that cannot withstand high temperatures. The method utilizes electric discharge and transition metal catalysts to deposit uniform CNT films. PECVD operates in different modes like RF, diffusion current, DC, microwave, and hot filament, each with specific advantages and applications. The discussion includes specific examples, such as the use of Ni-Y bimetallic catalyst for mass production of SWCNTs, the production of vertically aligned CNTs by DC-PECVD, and the use of PECVD for various applications.

In short, plasma-based synthesis methods offer a range of techniques to produce CNTs with controlled characteristics, enabling their application in diverse fields.

1.8.2 Synthesis of Graphene

The synthesis of graphene can be broadly categorized into three approaches that are discussed below.

1.8.2.1 Exfoliation: Mechanical and Liquid-Phase Exfoliation

Exfoliation methods involve the separation of the individual graphene sheets from the high-quality graphitic carbon using mechanical or chemical energy. The graphene sheets in graphitic carbon of which it is made of are bound together by weak van der Waals forces of attraction. Exfoliation involves the breaking of these weak bonds by chemical or mechanical energy to separate the graphene sheets. The first report of the synthesis of graphene by this method involved intercalation of the pure graphitic sheets with potassium metal and then exfoliation in ethanol to from dispersion of graphene sheets [660]. The sonication of the graphene suspension resulted in the production of graphene scrolls of the thickness of 40 ± 15 layers. Graphene scrolls tough several graphene layers thick provided the idea that it is possible practically to separate graphene sheets of few layers of thickness known as FLG from graphitic carbon. The production of few layer graphene (FLG) and single-layer graphene (SLG) was first reported by Novoselov et al. [661] who produced stable monocrystalline graphitic films with a few atom thickness by mechanical exfoliation of pyrolytic graphite. This process produced graphene proved to be reliable and fast, hence garnered the attention of the research community [662, 663]. Single and few layers of graphene were produced by Shukla et al. [664] using the exfoliation method by bonding the bulk graphite to the borosilicate glass and then subjected to exfoliation to get single or multilayers of graphene on the substrate. Mechanical exfoliation has resulted in the production of graphene sheets of variable thickness using various carbon sources such as natural graphite [665], highly ordered pyrolytic graphite (HOPG) [666, 667], and monocrystal graphite [668]. Although mechanical exfoliation produces high-quality graphene, but this method suffers from the main disadvantage of giving low yield and insufficient control over the number of graphene layers needed for a particular application [669]. A little different approach was used by exfoliation in the liquid phase. Liquid-phase exfoliation can produce high-quality graphene directly from graphite [670–672]. This method increases the interplanar spacing between graphitic layers by either intercalation of metal atoms or large-sized molecules or by chemical oxidation [673]. Mostly sonication of the graphite in liquid solvent exfoliates it into graphene of monolayer to few layers thickness [673–676]. The size of graphene flakes as well as the number of graphene sheets can be altered by changing the sonication parameters such as sonication time, frequency, and sonication power [674, 676]. The yield and stability of the product graphene is affected by the type of the liquid solvent [677]. The proficiency of the solvent for graphite exfoliation in turn depends upon the surface tension of the solvent. Solvents demonstrating superior performance are those that effectively overcome the van der Waals interactions between graphene layers. For this reason, solvents with surface tension values from 40 to 50 mJ/m^2 are considered as better exfoliating solvents. Water is one of the good solvents used for exfoliation [678]. Some additional chemical species have been used which when added to the solvent such as water enhance the exfoliation by reducing the surface energy. Various solvents such as cyclohexanone, benzylamine, dimethyl sulfoxide (DMSO), dimethyl formamide (DMF), and tetramethyl urea have been employed for the exfoliation of the graphite exfoliation [679–683]. For instance, graphene dispersion with concentration up to 0.01 mg/mL was produced by Hernandz et al. [677] by exfoliation of graphite in *N*-methyl-pyrrolidone solvent. The yield was initially 1% which was further improved up to 12% by further processing. Liquid-phase exfoliation method has the advantage of large-scale production of graphene. Single to few layers graphene was produced by Lotya et al. [684] by dispersing graphite in sodium dodecylbenzene sulfonate (SDBS) followed by sonication. A comprehensive study on the dispersion, exfoliation, and characterization of graphene flakes has been carried out. The optimization of dispersion conditions was a crucial step in the process. The absorption coefficient (R) was determined, providing insights into the concentration of dispersed graphite. The study revealed empirical relationships, such as $CG = 0.01 \times CG,i$, shedding light on the concentration dynamics during centrifugation. The surfactant concentration (CSDBS) and graphite concentration (CG,i) were varied to optimize the dispersion conditions, leading to concentrations comparable to those achieved in surfactant-stabilized nanotube dispersions. The evidence of exfoliation was thoroughly examined through TEM analysis, revealing various types of graphene flakes, including monolayers, few-layer graphene, and larger disordered flakes (Figure 1.19A). TEM images reflected the presence of flakes with 40% flakes containing less than five layers and 3% flakes composed of monolayers of graphene. These graphene layers were found to be free of defects. TEM and HRTEM images of graphene flakes are shown in Figure 1.19B. The ability to recycle sediment for improved exfoliation was a notable finding, indicating the potential for enhancing the overall yield of graphene. Dispersion stability, characterized by the zeta potential, suggested effective electrostatic repulsion between surfactant-coated graphene flakes. The temporal stability of dispersions was investigated, revealing a biexponential sedimentation behavior. The proposed stabilization mechanism, based on DLVO theory, indicated a potential energy barrier opposing aggregation, contributing to the stability of surfactant-coated graphene sheets. The study extended to the fabrication of graphene films through vacuum filtration. SEM and optical images demonstrated the morphology of the films, emphasizing the presence of large and small graphene flakes. The characterization of flake quality through FTIR, Raman spectroscopy, and X-ray photoelectron spectroscopy indicated low levels of defects and oxidation, showcasing the high quality of the produced graphene flakes. The optical and electrical properties of the films were explored, with transparency and sheet resistance measurements indicating the potential application of these films. The deposition of individual graphene flakes onto surfaces, such as mica, was achieved, offering opportunities for further characterization and applications. The study provides valuable insights into the dispersion, exfoliation, and characterization of graphene flakes, paving the way for potential applications in various fields, including electronics and materials science. The comprehensive approach to understanding the underlying mechanisms and optimizing conditions contributes to the advancement of graphene research. Both single layer and FLG has been produced by exfoliation method using various modifications for a variety of applications [685–689].

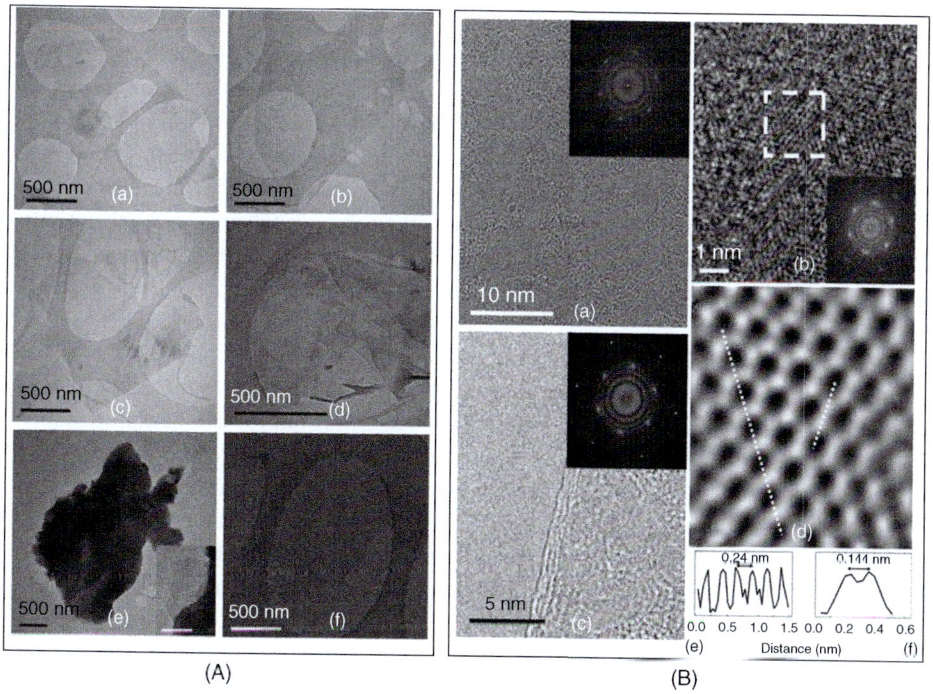

Figure 1.19 (A) Displays chosen TEM illustrations of flakes produced through surfactant processing. (a) Illustrates a monolayer, albeit with a small square debris piece near its left-hand edge. (b) Represents a bilayer, (c) a trilayer, and (d) a disordered multilayer. (e) Depicts an exceptionally large flake, with an inset offering a close-up of an edge displaying a small multilayer graphene flake protrusion. Additionally, (f) exhibits a monolayer from a sample prepared through sediment recycling. Moving to (B); it showcases high-resolution TEM images of surfactant-exfoliated graphene flakes. (a) Presents a HRTEM image of a segment of a graphene monolayer, accompanied by an inset of a fast Fourier transform equivalent to an electron diffraction pattern of the image. (b) Features a HRTEM image of a segment of a trilayer, also with an inset of a Fast Fourier transform. (c) Displays a HRTEM image of part of a graphene monolayer, along with an inset of a fast Fourier transform of the region enclosed by the white square. The scale bar in all images is 1 nm. Additionally, (d) provides a filtered image of part of the region in the white square. Lastly, (e) presents intensity analysis along the left white dashed line, revealing a hexagon width of 2.4 Å, while (f) shows intensity analysis along the right white dashed line, indicating a C—C bond length of 1.44 Å. Source: Reproduced with permission from Lotya et al. [684]/American Chemical Society.

In short, this section, the focus is on exfoliation methods for the synthesis of graphene, which involve separating individual graphene sheets from high-quality graphitic carbon using mechanical or chemical energy. Graphene sheets within graphitic carbon are held together by weak van der Waals forces, and exfoliation entails breaking these weak bonds through chemical or mechanical means to isolate the graphene sheets.

The initial report on graphene synthesis through exfoliation involved intercalating pure graphitic sheets with potassium metal, followed by exfoliation in ethanol to form a dispersion of graphene sheets. Subsequent sonication of the graphene suspension resulted in the production of graphene scrolls of approximately 40 ± 15 layers. This method provided the insight that separating graphene sheets of few layers, known as FLG, from graphitic carbon is practically feasible. The production of FLG and single-layer graphene (SLG) was pioneered by Novoselov et al., who mechanically exfoliated pyrolytic graphite to produce stable monocrystalline graphitic films with a few atom thickness. This method proved to be reliable and fast, gaining attention from the research community.

Mechanical exfoliation has been achieved using various carbon sources, such as natural graphite, HOPG, and monocrystal graphite. However, while this method produces high-quality graphene, it suffers from the main disadvantage of low yield and insufficient control over the number of graphene layers needed for particular applications. An alternative approach is liquid-phase exfoliation, which can produce high-quality graphene directly from graphite. This method increases the interplanar spacing between graphitic layers through intercalation or chemical oxidation. Sonication of graphite in liquid solvents, such as water or other organic solvents, exfoliates it into graphene sheets of monolayer to few layers in thickness. The size and number of graphene flakes can be controlled by adjusting sonication parameters, and the proficiency of the solvent depends on its surface tension.

Liquid-phase exfoliation offers the advantage of large-scale production of graphene. Various solvents, including *N*-methyl-pyrrolidone, cyclohexanone, benzylamine, dimethyl sulfoxide (DMSO), dimethyl formamide (DMF), and tetramethyl urea, have been employed for exfoliation. Notably, water has been identified as a good solvent for exfoliation. A comprehensive study on the dispersion, exfoliation, and characterization of graphene flakes has been conducted. The optimization of dispersion conditions was crucial, and the study revealed empirical relationships and stabilization mechanisms contributing to the stability of surfactant-coated graphene sheets. The ability to recycle sediment for improved exfoliation and the fabrication of graphene films through vacuum filtration were notable findings.

In summary, exfoliation methods, whether mechanical or liquid phase, offer pathways for the large-scale fabrication of graphene with controlled characteristics. These methods have found applications across various fields and the comprehensive studies provide valuable insights for advancing graphene research and applications in fields such as electronics and materials science. Both single-layer and few-layer graphene have been successfully produced through exfoliation methods with modifications tailored for specific applications.

1.8.2.2 Chemical Vapor Deposition (CVD)

This is a bottom-up approach in which organic molecules in either solid (camphor, polymers, and sucrose), liquid (benzene, hexane, ethanol, and turpentine oil), or gas (methane, ethane, and ethyne) phase are decomposed at high temperatures in without the presence of oxygen and the carbon atoms are self-assembled to yield graphene structures. This method produces graphene with fewer layers with large surface areas and long-range order. CVD offers flexibility over the choice of the carbon precursors. The formation of graphene layers is catalyzed by the presence of mostly transitional metals such as Ni, Cu, Co, Au, Pt, Pd, Ru, Ir, Re, and Rh. Depending upon the substrate material used, growth mechanism is a multistep process which involves such as decomposition, adsorption, dehydrogenation, diffusion, nucleation, segregation, and precipitation. The use of copper (Cu) as a catalyst is prevalent in the synthesis of graphene due to its low solubility for carbon, making it a widely adopted choice and offering effective control over the number of layers in the resulting graphene. For instance, synthesis of graphene has been achieved by Zhang et al. [690] by employing water less cleaning method of Cu foils. Here the contaminated Cu foils were first annealed at high temperatures, and the oxidized surface layer was removed during the cooling step. Next the diluted methane, Ar, and H_2 were then introduced into the chamber for the growth of graphene layers. It was revealed that annealing of Cu foils in the absence of He resulted in the production of defect-free multilayer graphene. The large-scale fabrication of 3D graphene with high electrical conductivity and a high specific area has been achieved by Mint et al. [691] using 3D Ni nanowire foam as a catalyst at lower temperatures (670 °C). Different types of catalysts and carbon precursors have been used for the growth of single to multilayer graphene for

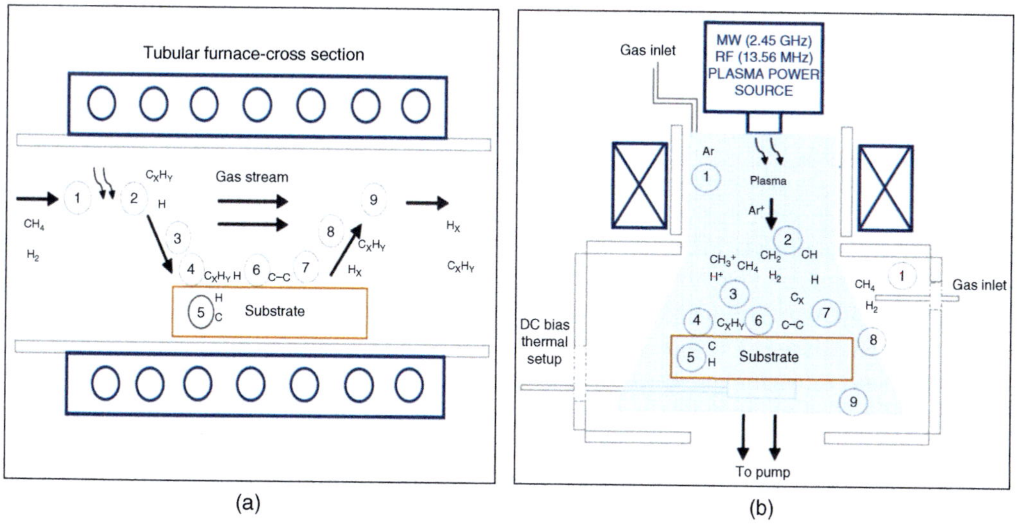

Figure 1.20 Illustrates schematic diagrams of (a) thermal CVD and (b) plasma-assisted CVD processes, specifically applied to graphene synthesis from CH_4/H_2 mixtures. Source: Reproduced with permission from Muñoz et al. [697]/John Wiley & Sons.

various applications. Researchers have also shown interest in plasma-assisted synthesis (PACVD) of graphene. The first report of the synthesis of single to multilayer graphene by PACVD dates to 2004 using radiofrequency PACVD [692, 693]. Graphene was synthesized on a variety of substrates without any catalyst deposition or surface preparation. The subnanometer thickness graphene sheets were produced from a mixture of CH_4 and H_2 at a total pressure of 12 Pa and a power of 900 W with a substrate temperature of around 680 °C. Due to simplicity of this process, it drew the interest of the scientific community worldwide to produce graphene with some modifications [694–696]. Schematic representation of the production of graphene by thermal CVD and plasma CVD is shown in Figure 1.20. The process involves several key steps: (i) Transportation of reactants through forced convection, (ii) thermal activation (in the case of (a)) or plasma activation (in the case of (b)), (iii) prevention of homogeneous gas reaction with particles and powder production in graphene synthesis, achieved by controlling kinetic parameters (P, T, n), (iv) transportation of reactants through gas diffusion originating from the main gas stream and traversing the boundary layer, (v) adsorption of reactants on the surface of the substrate, (vi) dissolution and bulk diffusion of species based on substrate solubility and physical properties. This step involves thermal activation-mediated-surface processes, encompassing chemical decomposition (catalytic), reaction, surface migration to attachment sites (such as atomic-level steps), incorporation, and other heterogeneous surface reactions, ultimately leading to film growth, (vii) desorption of by-products from the surface, (viii) transportation of by-products through diffusion in the boundary layer and back to the main gas stream, and (ix) forced convection-driven transportation of by-products away from the deposition region. CVD method has been used to prepare single to multilayer graphene for a variety of applications [698–701].

In short, this section discusses the CVD method, a bottom-up approach for synthesizing graphene. In CVD, organic molecules in solid, liquid, or gas phases, such as camphor, polymers, sucrose, benzene, hexane, ethanol, methane, ethane, and ethyne, are decomposed at high temperatures in the absence of oxygen. The carbon atoms then self-assemble to form graphene structures. This approach allows to produce graphene with fewer layers, large surface areas, and long-range order.

CVD offers flexibility in choosing carbon precursors, and the formation of graphene layers is catalyzed by transitional metals like Ni, Cu, Co, Au, Pt, Pd, Ru, Ir, Re, and Rh. Copper (Cu) is widely used as a catalyst due to its low solubility of carbon, enabling precise control over the number of graphene layers. For example, Zhang et al. achieved the synthesis of defect-free multilayer graphene by annealing Cu foils in the absence of helium. Large-scale production of 3D graphene with high electrical conductivity and specific area was achieved using 3D Ni nanowire foam as a catalyst at lower temperatures. Various catalysts and carbon precursors have been explored for growing single to multilayer graphene for diverse applications. Researchers have also explored plasma-assisted synthesis (PACVD) of graphene, with the first report dating back to 2004 using radiofrequency PACVD. Graphene was synthesized on various substrates without catalyst deposition or surface preparation. The process involves transporting reactants through forced convection, thermal or plasma activation, preventing homogeneous gas reactions, adsorption on the substrate surface, dissolution and bulk diffusion of species, desorption of by-products, and transportation of by-products away from the deposition region.CVD has been extensively used for preparing single to multilayer graphene for various applications, showcasing its versatility and potential in graphene synthesis.

1.8.2.3 Laser-Induced Synthesis

Laser-based synthesis has a special advantage of simple and efficient localized energy delivery [702]. The desired number of graphene sheets can be obtained by changing the parameters such as laser power, pulse width, and scanning rate [703]. Laser-induced processing of graphene has been achieved by techniques such as pulsed laser deposition, laser exfoliation, laser-induced CVD, and laser writing [704–707]. This method involves creating localized heating of a substrate or catalyst, either in a vacuum or a gas atmosphere, to facilitate the direct deposition or patterning of graphene. The regions of laser irradiation achieve the desired temperature instantly. The rapid temperature rise due to rapid heating rate decomposes the carbon source at a fast rate [708]. The forward movement of laser spots on the substrate lets local temperatures to drop which allows the diffusion of additional carbon atoms on the catalyst surface which results in the formation of graphene layers. Graphene layers have been fabricated on Ni foils using laser-induced CVD by irradiating the Ni foils with laser radiations in CH_4/H_2 atmosphere [709]. The control of the laser scan rate has been employed to manage the number of graphene layers. Jiang et al. successfully achieved the formation of few-layer graphene on Ni plated using laser-induced CVD in a CH_4/H_2 atmosphere under ambient conditions [710]. The quantity of graphene layers has been controlled by tuning the laser power and scan speed. There are various reports of the formation of high-quality FLG by laser-induced CVD. The fabrication of FLG on Ni films has been achieved by Wang et al. [711] using pulsed laser ablation of the graphite target at 650 °C. The thickness of the graphene films, correlating with the number of graphene layers, was regulated by adjusting the ablation time. Several metal films have been explored for the generation of few-layer graphene through the pulsed laser ablation technique. The number of graphene layers was influenced by the cooling rate and energy of the pulsed laser [709, 710]. The formation of high-quality graphene has been achieved by direct laser writing on various types of substrates, which has proved to be a low cost and simple method [712]. For instance, direct laser wring of graphene patterns on various insulating substrates such as glass, Si/SiO_2 has been achieved by Xiong et al. [712] using femtosecond laser under ambient conditions. Graphene patterns such as spirals, texts, electric circuits, and line arrays have been drawn using this technique. Laser synthesis has been used to synthesize graphene for various applications [713–717].

In short, this section explores laser-based synthesis as a method for graphene production, emphasizing its advantage of providing simple and efficient localized energy delivery. Laser parameters such as power, pulse width, and scanning rate can be adjusted to control the number of graphene sheets obtained. Various techniques, including pulsed laser deposition, laser-induced CVD, laser exfoliation, and laser writing, have been employed for laser-induced processing of graphene. Laser-induced CVD involves the localized heating of a substrate or catalyst, either in a gas atmosphere or vacuum, for direct deposition or patterning of graphene. The rapid temperature rise, facilitated by the laser, decomposes the carbon source quickly, and the local temperature drop during laser movement allows the diffusion of additional carbon atoms on the catalyst surface, resulting in the fabrication of graphene layers. Control over the laser scan rate allows for the manipulation of the number of graphene layers.

Research has demonstrated the production of graphene layers on Ni foils through laser-induced CVD in a CH_4/H_2 atmosphere. The number of layers was regulated by manipulating the laser scan rate. Additionally, few-layer graphene formation on Ni-plated substrates was accomplished using laser-induced CVD under ambient conditions, with control over the number of layers achieved through adjustments in laser power and scan speed. Pulsed laser ablation methods were also employed for generating high-quality few-layer graphene on diverse metal films. The thickness of the graphene films, corresponding to the number of layers, is controlled by changing the ablation time, cooling rate, and laser energy. Direct laser writing on different substrates, such as glass and Si/SiO_2, has emerged as a low-cost and simple method for producing graphene patterns. For example, femtosecond lasers have been employed for ambient condition direct laser writing, creating graphene patterns like spirals, texts, electric circuits, and line arrays. Laser synthesis of graphene has found applications in various fields, highlighting its versatility and potential for practical use.

1.9 Conclusion and Future Perspective

The synthesis methods discussed, spanning from gold nanoparticles to graphene, showcase the versatility and innovation within nanomaterial fabrication. The synthesis of gold nanoparticles employs diverse methods, including chemical reduction, citrate reduction, and green synthesis using plant extracts. These methods yield nanoparticles with unique properties suitable for various applications such as medicine, catalysis, and sensing. Future perspectives may involve refining green synthesis approaches, exploring new reducing agents, and expanding applications in targeted drug delivery and diagnostics.

Like gold nanoparticles, silver nanoparticles are synthesized through chemical and green methods. The antibacterial and antiviral properties of silver nanoparticles make them valuable in medical applications. Future directions could focus on optimizing synthesis for specific medical uses, addressing toxicity concerns, and exploring broader applications in water purification and consumer products.

MNPs find applications in medicine, particularly in targeted drug delivery and imaging. Synthesis methods include co-precipitation, thermal decomposition, and microemulsion. Future perspectives involve enhancing biocompatibility, improving magnetic properties, and advancing functionalization for precise biomedical applications.

QDs offer unique optical properties with applications in imaging and sensing. Synthesis methods include colloidal synthesis and organometallic synthesis. Future research may emphasize improving quantum dot stability, reducing toxicity, and expanding their applications in fields like solar cells and bioimaging.

Carbon nanotubes exhibit remarkable properties, leading to applications in various fields. Synthesis methods include arc discharge, laser ablation, CVD, and plasma-enhanced CVD.

Future perspectives may involve scalability improvements, optimization for specific applications, and exploration of novel catalysts and precursors.

Graphene synthesis employs exfoliation methods, CVD, and laser-based approaches. Techniques like liquid-phase exfoliation and CVD offer scalable production. Future directions may include further optimization of CVD for large-scale applications, exploring new exfoliation methods, and advancing applications in electronics, materials science, and energy storage.

In conclusion, the future of nanomaterial synthesis lies in refining existing methods, exploring novel approaches, and expanding applications in diverse fields. The continuous advancement in nanomaterial synthesis contributes to the development of innovative technologies with far-reaching implications in medicine, electronics, and environmental science.

References

1 Bhushan B 2017 Introduction to Nanotechnology *Springer Handbook of Nanotechnology* Springer Handbooks ed B Bhushan (Berlin, Heidelberg: Springer) pp 1–19.

2 Nasrollahzadeh M, Sajadi S M, Sajjadi M and Issaabadi Z 2019 An Introduction to Nanotechnology Interface Science and Technology *An Introduction to Green Nanotechnology*, vol 28, ed M Nasrollahzadeh, S M Sajadi, M Sajjadi, Z Issaabadi and M Atarod (Elsevier) pp 1–27.

3 Roco M C 2011 The long view of nanotechnology development: the National Nanotechnology Initiative at 10 years *J. Nanopart. Res.* 13 427–45.

4 Litter M I and Ahmad A 2023 The World of Nanotechnology *Industrial Applications of Nanoparticles* (CRC Press) pp 1–15.

5 Nikulainen T and Palmberg C 2010 Transferring science-based technologies to industry – does nanotechnology make a difference? *Technovation* 30 3–11.

6 Toumey C 2010 Tracing and Disputing the Story of Nanotechnology *International Handbook on Regulating Nanotechnologies* (Edward Elgar Publishing).

7 Feynman R P 1992 There's plenty of room at the bottom [data storage] *J. Microelectromech. Syst.* 1 60–6.

8 Junk A and Riess F 2006 From an idea to a vision: there's plenty of room at the bottom *Am. J. Phys.* 74 825–30.

9 Singh N B, Susan M A B H and Chaudhary R G 2023 *Emerging Nanomaterials and Their Impact on Society in the 21st Century* (Materials Research Forum LLC).

10 Dvorsky R, Svoboda L and Bednář J 2022 *Nanoparticles' Preparation, Properties, Interactions and Self-Organization* (Cham: Springer International Publishing).

11 Vaseashta A and Dimova-Malinovska D 2005 Nanostructured and nanoscale devices, sensors and detectors *Sci. Technol. Adv. Mater.* 6 312–8.

12 Chansoriya T, Khilwani B and Ansari A S 2023 Nanoparticle-Based Drug Delivery System for Beginners *Interaction of Nanomaterials With Living Cells* ed F A Sheikh, S Majeed and M A Beigh (Singapore: Springer Nature) pp 557–80.

13 Wani I A 2021 Recent advances in biogenic silver nanoparticles & nanocomposite based plasmonic-colorimetric and electrochemical sensors *ECS J. Solid State Sci. Technol.* 10 047003.

14 Wani I A and Ahmad T 2013 Size and shape dependant antifungal activity of gold nanoparticles: a case study of Candida *Colloids Surf. B Biointerfaces* 101 162–70.

15 Moeinzadeh S and Jabbari E 2017 Nanoparticles and Their Applications *Springer Handbook of Nanotechnology* Springer Handbooks ed B Bhushan (Berlin, Heidelberg: Springer) pp 335–61.

16 Sousa A A, Schuck P and Hassan S A 2021 Biomolecular interactions of ultrasmall metallic nanoparticles and nanoclusters *Nanoscale Adv.* 3 2995–3027.

17 Moore J S 1997 Shape-persistent molecular architectures of nanoscale dimension *Acc. Chem. Res.* 30 402–13.

18 Loos J 2005 The art of SPM: scanning probe microscopy in materials science *Adv. Mater.* 17 1821–33.

19 Singh V, Yadav P and Mishra V 2020 Recent Advances on Classification, Properties, Synthesis, and Characterization of Nanomaterials *Green Synthesis of Nanomaterials for Bioenergy Applications* ed M Srivastava, V K Gupta, P K Mishra and N Srivastava (John Wiley & Sons, Ltd) pp 83–97.

20 Saleh T A 2020 Nanomaterials: classification, properties, and environmental toxicities *Environ. Technol. Innov.* 20 101067.

21 Sudha P N, Sangeetha K, Vijayalakshmi K and Barhoum A 2018 Nanomaterials History, Classification, Unique Properties, Production and Market *Emerging Applications of Nanoparticles and Architecture Nanostructures* Micro and Nano Technologies ed A Barhoum and A S H Makhlouf (Elsevier) pp 341–84.

22 Dazon C, Fierro V, Celzard A and Witschger O 2020 Identification of nanomaterials by the volume specific surface area (VSSA) criterion: application to powder mixes *Nanoscale Adv.* 2 4908–17.

23 Ghosh C K 2015 Quantum Effect on Properties of Nanomaterials *Introduction to Nano: Basics to Nanoscience and Nanotechnology* Engineering Materials ed A Sengupta and C K Sarkar (Berlin, Heidelberg: Springer) pp 73–111.

24 Roduner E 2006 Size matters: why nanomaterials are different *Chem. Soc. Rev.* 35 583–92.

25 Rotello V 2004 *Nanoparticles: Building Blocks for Nanotechnology* (Springer Science & Business Media).

26 Ghosh Chaudhuri R and Paria S 2012 Core/shell nanoparticles: classes, properties, synthesis mechanisms, characterization, and applications *Chem. Rev.* 112 2373–433.

27 Baig N, Kammakakam I and Falath W 2021 Nanomaterials: a review of synthesis methods, properties, recent progress, and challenges *Mater. Adv.* 2 1821–71.

28 Kumar Sharma R, Bandichhor R, Mishra V, Sharma S, Yadav S, Mehta S, Arora B, Rana P, Dutta S and Solanki K 2023 Advanced metal oxide-based nanocatalysts for the oxidative synthesis of fine chemicals *Mater. Adv.* 4 1795–830.

29 Polarz S and Smarsly B 2002 Nanoporous materials *J. Nanosci. Nanotechnol.* 2 581–612.

30 Notario B, Pinto J and Rodriguez-Perez M A 2016 Nanoporous polymeric materials: a new class of materials with enhanced properties *Prog. Mater. Sci.* 78–79 93–139.

31 Alivisatos A P 1996 Semiconductor clusters, nanocrystals, and quantum dots *Science* 271 933–7.

32 Baskoutas S and Terzis A F 2006 Size-dependent band gap of colloidal quantum dots *J. Appl. Phys.* 99 013708.

33 Aavani F, Khorshidi S and Karkhaneh A 2019 A concise review on drug-loaded electrospun nanofibres as promising wound dressings *J. Med. Eng. Technol.* 43 38–47.

34 Delehanty J B, Medintz I L, Pons T, Brunel F M, Dawson P E and Mattoussi H 2006 Self-assembled quantum dot– peptide bioconjugates for selective intracellular delivery *Bioconjug. Chem.* 17 920–7.

35 Akbarzadeh A, Samiei M and Davaran S 2012 Magnetic nanoparticles: preparation, physical properties, and applications in biomedicine *Nanoscale Res. Lett.* 7 1–13.

36 Rizwan M, Shoukat A, Ayub A, Razzaq B and Tahir M B 2021 Types and Classification of Nanomaterials *Synthesis, Characterization, Hazards and Safety* Micro and Nano Technologies ed M B Tahir, M Sagir and A M Asiri (Elsevier) pp 31–54.

37 Sannino D 2021 Types and Classification of Nanomaterials *Nanotechnology: Trends and Future Applications* ed M B Tahir, M Rafique and M Sagir (Singapore: Springer) pp 15–38.

38 Rao N, Singh R and Bashambu L 2021 Carbon-based nanomaterials: synthesis and prospective applications *Mater. Today Proc.* 44 608–14.

39 Jamkhande P G, Ghule N W, Bamer A H and Kalaskar M G 2019 Metal nanoparticles synthesis: an overview on methods of preparation, advantages and disadvantages, and applications *J. Drug Deliv. Sci. Technol.* 53 101174.

40 Chavali M S and Nikolova M P 2019 Metal oxide nanoparticles and their applications in nanotechnology *SN Appl. Sci.* 1 607.

41 Chauhan R, Kumar A, Tripathi R and Kumar A 2021 Advancing of Zinc Oxide Nanoparticles for Cosmetic Applications *Handbook of Consumer Nanoproducts* ed S Mallakpour and C M Hussain (Singapore: Springer) pp 1–16.

42 Haas K-H 2021 Application of Metal Oxide Nanoparticles and Their Economic Impact *Metal Oxide Nanoparticles* ed O Diwald and T Berger (John Wiley & Sons, Ltd) pp 29–65.

43 Al-Amin M, Sarker S, Mumu H and Gafur M 2021 Fabrication methods and fracture behaviour of zirconia toughened alumina (ZTA) nano composite ceramic: a review *Mater. Today Proc.*

44 Duntu S H, Agyapong J, Ahmad I, Islam M and Boakye-Yiadom S 2022 Combined effect of low ZrO_2 content and carbon nanostructures on mechanical and wear properties of hot-pressed Al_2O_3/ZrO_2 hybrid nanocomposites *Ceram. Int.* 48 34803–16.

45 Terna A D, Elemike E E, Mbonu J I, Osafile O E and Ezeani R O 2021 The future of semiconductors nanoparticles: synthesis, properties and applications *Mater. Sci. Eng. B* 272 115363.

46 Yacobi B G 2003 Types of Semiconductors *Semiconductor Materials: An Introduction to Basic Principles* Microdevices (Boston, MA: Springer) pp 135–70.

47 Thambidurai M, Muthukumarasamy N, Agilan S, Murugan N, Vasantha S, Balasundaraprabhu R and Senthil T S 2010 Strong quantum confinement effect in nanocrystalline CdS *J. Mater. Sci.* 45 3254–8.

48 Liu J, Nie Y, Xue W, Wu L, Jin H, Jin G, Zhai Z and Fu C 2020 Size effects on structural and optical properties of tin oxide quantum dots with enhanced quantum confinement *J. Mater. Res. Technol.* 9 8020–8.

49 Sumanth Kumar D, Jai Kumar B and Mahesh H M 2018 Quantum Nanostructures (QDs): An Overview *Synthesis of Inorganic Nanomaterials* Micro and Nano Technologies ed S M Bhagyaraj, O S Oluwafemi, N Kalarikkal and S Thomas (Woodhead Publishing) pp 59–88.

50 Unser S, Bruzas I, He J and Sagle L 2015 Localized surface plasmon resonance biosensing: current challenges and approaches *Sensors* 15 15684–716.

51 Cb M, Norris D J and Bawendi M G 1993 Synthesis and characterization of nearly monodisperse CdE (E = sulfur, selenium, tellurium) semiconductor nanocrystallites *J. Am. Chem. Soc.* 115 8706–15.

52 Pattantyus-Abraham A G, Kramer I J, Barkhouse A R, Wang X, Konstantatos G, Debnath R, Levina L, Raabe I, Nazeeruddin M K, Grätzel M and Sargent E H 2010 Depleted-heterojunction colloidal quantum dot solar cells *ACS Nano* 4 3374–80.

53 Kumar D S, Kumar B J and Mahesh H M 2016 Optical properties of colloidal aqueous synthesized 3 mercaptopropionic acid stabilized CdS quantum dots *AIP Conf. Proc.* 1728 020162.

54 Li H, Chen D, Li L, Tang F, Zhang L and Ren J 2010 Size-and shape-controlled synthesis of PbSe and PbS nanocrystals via a facile method *CrystEngComm* 12 1127–33.

55 Peng Z A and Peng X 2001 Formation of high-quality CdTe, CdSe, and CdS nanocrystals using CdO as precursor *J. Am. Chem. Soc.* 123 183–4.

56 Rosiles-Perez C, Serrano-Estrada M A, Sidhik S, Alatorre-Ordaz A, Torres-Castro A, Vallejo M A, Jiménez-González A E and López-Luke T 2020 Synthesis of high quality PbS colloidal quantum dots by ultrasonic bath as photosensitizers in a TiO_2 solar cell *J. Solid State Chem.* 292 121720.

57 Turkevich J, Stevenson P C and Hillier J 1951 A study of the nucleation and growth processes in the synthesis of colloidal gold *Discuss. Faraday Soc.* 11 55–75.

58 Amendola V, Pilot R, Frasconi M, Maragò O M and Iatì M A 2017 Surface plasmon resonance in gold nanoparticles: a review *J. Phys. Condens. Matter* 29 203002.

59 Hutter E and Fendler J H 2004 Exploitation of localized surface plasmon resonance *Adv. Mater.* 16 1685–706.

60 Louis C and Pluchery O 2012 *Gold Nanoparticles for Physics, Chemistry and Biology* (World Scientific).

61 Lohse S E and Murphy C J 2013 The quest for shape control: a history of gold nanorod synthesis *Chem. Mater.* 25 1250–61.

62 Giasuddin A, Jhuma K and Haq A M 2013 Use of gold nanoparticles in diagnostics, surgery and medicine: a review *Bangladesh J. Med. Biochem.* 5 56–60.

63 Hong W, Bai H, Xu Y, Yao Z, Gu Z and Shi G 2010 Preparation of gold nanoparticle/graphene composites with controlled weight contents and their application in biosensors *J. Phys. Chem. C* 114 1822–6.

64 Akbari Nakhjavani S, Afsharan H, Khalilzadeh B, Ghahremani M H, Carrara S and Omidi Y 2019 Gold and silver bio/nano-hybrids-based electrochemical immunosensor for ultrasensitive detection of carcinoembryonic antigen *Biosens. Bioelectron.* 141 111439.

65 Anker J N, Hall W P, Lyandres O, Shah N C, Zhao J and Van Duyne R P 2010 Biosensing with plasmonic nanosensors *Nanosci. Technol. Collect. Rev. Nat. J.* 308–19.

66 Tamiya E 2016 Nano-materials for LSPR and electrochemical biosensors *Nanomed. Nanotechnol. Biol. Med.* 2 450.

67 McMillan B G, Berlouis L E A, Cruickshank F R, Pugh D and Brevet P-F 2005 Transverse and longitudinal surface plasmon resonances of a hexagonal array of gold nanorods embedded in an alumina matrix *Appl. Phys. Lett.* 86 211912.

68 Ghosh S K and Pal T 2007 Interparticle coupling effect on the surface plasmon resonance of gold nanoparticles: from theory to applications *Chem. Rev.* 107 4797–862.

69 El-Brolossy T A, Abdallah T, Mohamed M B, Abdallah S, Easawi K, Negm S and Talaat H 2008 Shape and size dependence of the surface plasmon resonance of gold nanoparticles studied by photoacoustic technique *Eur. Phys. J. Spec. Top.* 153 361–4.

70 Eustis S and El-Sayed M A 2006 Why gold nanoparticles are more precious than pretty gold: noble metal surface plasmon resonance and its enhancement of the radiative and nonradiative properties of nanocrystals of different shapes *Chem. Soc. Rev.* 35 209–17.

71 Ren Y, Qi H, Chen Q, Wang S and Ruan L 2017 Localized surface plasmon resonance of nanotriangle dimers at different relative positions *J. Quant. Spectrosc. Radiat. Transf.* 199 45–51.

72 Klar T, Perner M, Grosse S, von Plessen G, Spirkl W and Feldmann J 1998 Surface plasmon resonances in single metallic nanoparticles *Phys. Rev. Lett.* 80 4249–52.

73 Pathak N K and Sharma R P 2016 Study of broadband tunable properties of surface plasmon resonances of noble metal nanoparticles using Mie scattering theory: plasmonic perovskite interaction *Plasmonics* 11 713–9.

74 Xu H and Käll M 2002 Modeling the optical response of nanoparticle-based surface plasmon resonance sensors *Sensors Actuators B Chem.* 87 244–9.

75 Lopatynskyi A M, Lopatynska O G, Guo L J and Chegel V I 2011 Localized surface plasmon resonance biosensor – part I: theoretical study of sensitivity – extended Mie approach *IEEE Sensors J.* 11 361–9.

76 Duque J S, Blandón J S and Riascos H 2017 Localized plasmon resonance in metal nanoparticles using Mie theory *J. Phys. Conf. Ser.* 850 012017.

77 Abid N, Khan A M, Shujait S, Chaudhary K, Ikram M, Imran M, Haider J, Khan M, Khan Q and Maqbool M 2022 Synthesis of nanomaterials using various top-down and bottom-up approaches, influencing factors, advantages, and disadvantages: a review *Adv. Colloid Interf. Sci.* 300 102597.

78 Bloch K, Pardesi K, Satriano C and Ghosh S 2021 Bacteriogenic platinum nanoparticles for application in nanomedicine *Front. Chem.* 9 624344.

79 Cruz J N 2023 *Nanobiomaterials: Perspectives for Medical Applications in the Diagnosis and Treatment of Diseases* (Materials Research Forum LLC).

80 Amabilino D B 2016 *Supramolecular Chemistry at Surfaces* (Royal Society of Chemistry).

81 Ahmed M 2020 Nanomaterial Synthesis *Polymer Science and Nanotechnology* ed R Narain (Elsevier) pp 361–99.

82 Wang Y, Pan J-A, Wu H and Talapin D V 2019 Direct wavelength-selective optical and electron-beam lithography of functional inorganic nanomaterials *ACS Nano* 13 13917–31.

83 Wang Y, Fedin I, Zhang H and Talapin D V 2017 Direct optical lithography of functional inorganic nanomaterials *Science* 357 385–8.

84 Ahmad T, Wani I A, Ahmed J and Al-Hartomy O A 2014 Effect of gold ion concentration on size and properties of gold nanoparticles in TritonX-100 based inverse microemulsions *Appl. Nanosci.* 4 491–8.

85 Wani I A, Ganguly A, Ahmed J and Ahmad T 2011 Silver nanoparticles: ultrasonic wave assisted synthesis, optical characterization and surface area studies *Mater. Lett.* 65 520–2.

86 Wani I A, Khatoon S, Ganguly A, Ahmed J, Ganguli A K and Ahmad T 2010 Silver nanoparticles: large scale solvothermal synthesis and optical properties *Mater. Res. Bull.* 45 1033–8.

87 Polte J, Ahner T T, Delissen F, Sokolov S, Emmerling F, Thünemann A F and Kraehnert R 2010 Mechanism of gold nanoparticle formation in the classical citrate synthesis method derived from coupled in situ XANES and SAXS evaluation *J. Am. Chem. Soc.* 132 1296–301.

88 Grzelczak M, Pérez-Juste J, Mulvaney P and Liz-Marzán L M 2008 Shape control in gold nanoparticle synthesis *Chem. Soc. Rev.* 37 1783–91.

89 Frens G 1973 Controlled nucleation for the regulation of the particle size in monodisperse gold suspensions *Nat. Phys. Sci.* 241 20–2.

90 Kimling J, Maier M, Okenve B, Kotaidis V, Ballot H and Plech A 2006 Turkevich method for gold nanoparticle synthesis revisited *J. Phys. Chem. B* 110 15700–7.

91 Kumar S, Gandhi K S and Kumar R 2007 Modeling of formation of gold nanoparticles by citrate method *Ind. Eng. Chem. Res.* 46 3128–36.

92 Li C, Li D, Wan G, Xu J and Hou W 2011 Facile synthesis of concentrated gold nanoparticles with low size-distribution in water: temperature and pH controls *Nanoscale Res. Lett.* 6 440.

93 Rahman M R, Saleh F S, Okajima T and Ohsaka T 2011 pH dependence of the size and crystallographic orientation of the gold nanoparticles prepared by seed-mediated growth *Langmuir* 27 5126–35.

94 Xia H, Bai S, Hartmann J and Wang D 2010 Synthesis of monodisperse quasi-spherical gold nanoparticles in water via silver(I)-assisted citrate reduction *Langmuir* 26 3585–9.

95 Patungwasa W and Hodak J H 2008 pH tunable morphology of the gold nanoparticles produced by citrate reduction *Mater. Chem. Phys.* 108 45–54.

96 Volkert A A, Subramaniam V and Haes A J 2011 Implications of citrate concentration during the seeded growth synthesis of gold nanoparticles *Chem. Commun.* 47 478–80.

97 Zabetakis K, Ghann W E, Kumar S and Daniel M-C 2012 Effect of high gold salt concentrations on the size and polydispersity of gold nanoparticles prepared by an extended Turkevich–Frens method *Gold Bull.* 45 203–11.

98 Ojea-Jiménez I, Bastús N G and Puntes V 2011 Influence of the sequence of the reagents addition in the citrate-mediated synthesis of gold nanoparticles *J. Phys. Chem. C* 115 15752–7.

99 Ojea-Jiménez I, Romero F M, Bastús N G and Puntes V 2010 Small gold nanoparticles synthesized with sodium citrate and heavy water: insights into the reaction mechanism *J. Phys. Chem. C* 114 1800–4.

100 Brust M, Walker M, Bethell D, Schiffrin D J and Whyman R 1994 Synthesis of thiol-derivatised gold nanoparticles in a two-phase liquid–liquid system *J. Chem. Soc. Chem. Commun.* 0 801–2.

101 Hostetler M J, Green S J, Stokes J J and Murray R W 1996 Monolayers in three dimensions: synthesis and electrochemistry of ω-functionalized alkanethiolate-stabilized gold cluster compounds *J. Am. Chem. Soc.* 118 4212–3.

102 Li Y, Zaluzhna O, Xu B, Gao Y, Modest J M and Tong Y J 2011 Mechanistic insights into the Brust-Schiffrin two-phase synthesis of organo-chalcogenate-protected metal nanoparticles *J. Am. Chem. Soc.* 133 2092–5.

103 Brust M, Fink J, Bethell D, Schiffrin D J and Kiely C 1995 Synthesis and reactions of functionalised gold nanoparticles *J. Chem. Soc. Chem. Commun.* 16 1655–6.

104 Templeton A C, Wuelfing W P and Murray R W 2000 Monolayer-protected cluster molecules *Acc. Chem. Res.* 33 27–36.

105 Kanaras A G, Kamounah F S, Schaumburg K, Kiely C J and Brust M 2002 Thioalkylated tetraethylene glycol: a new ligand for water soluble monolayer protected gold clusters *Chem. Commun.* 20 2294–5.

106 Di Pasqua A J, Mishler R E, Ship Y-L, Dabrowiak J C and Asefa T 2009 Preparation of antibody-conjugated gold nanoparticles *Mater. Lett.* 63 1876–9.

107 Leontowich A F G, Calver C F, Dasog M and Scott R W J 2010 Surface properties of water-soluble glycine-cysteamine-protected gold clusters *Langmuir* 26 1285–90.

108 Busby M, Chiorboli C and Scandola F 2006 Relaxation dynamics and transient behavior of small arenethiol passivated gold nanoparticles *J. Phys. Chem. B* 110 6020–6.

109 Gupta R K, Srinivasan M P and Dharmarajan R 2011 Synthesis of short chain thiol capped gold nanoparticles, their stabilization and immobilization on silicon surface *Colloids Surf. A Physicochem. Eng. Asp.* 390 149–56.

110 Pattabi R M and Pattabi M 2009 Synthesis and characterization of thiosalicylic acid stabilized gold nanoparticles *Spectrochim. Acta A Mol. Biomol. Spectrosc.* 74 195–9.

111 Moores A, Goettmann F, Sanchez C and Floch P L 2004 Phosphinine stabilised gold nanoparticles; synthesis and immobilisation on mesoporous materials *Chem. Commun.* 24 2842–3.

112 Shem P M, Sardar R and Shumaker-Parry J S 2009 One-step synthesis of phosphine-stabilized gold nanoparticles using the mild reducing agent 9-BBN *Langmuir* 25 13279–83.

113 Pettibone J M and Hudgens J W 2011 Gold cluster formation with phosphine ligands: etching as a size-selective synthetic pathway for small clusters *ACS Nano* 5 2989–3002.

114 Gautier C, Taras R, Gladiali S and Bürgi T 2008 Chiral 1, 1′-binaphthyl-2, 2′-dithiol-stabilized gold clusters: size separation and optical activity in the UV–vis *Chirality Pharmacol. Biol. Chem. Consequences Mol. Asymmetry* 20 486–93.

115 De Silva N, Ha J-M, Solovyov A, Nigra M M, Ogino I, Yeh S W, Durkin K A and Katz A 2010 A bioinspired approach for controlling accessibility in calix [4] arene-bound metal cluster catalysts *Nat. Chem.* 2 1062–8.

116 Ha J-M, Solovyov A and Katz A 2009 Synthesis and characterization of accessible metal surfaces in calixarene-bound gold nanoparticles *Langmuir* 25 10548–53.

117 Villa A, Wang D, Su D, Veith G M and Prati L 2010 Using supported Au nanoparticles as starting material for preparing uniform Au/Pd bimetallic catalysts *Phys. Chem. Chem. Phys.* 12 2183–9.

118 Sardar R and Shumaker-Parry J S 2009 9-BBN induced synthesis of nearly monodisperse ω-functionalized alkylthiol stabilized gold nanoparticles *Chem. Mater.* 21 1167–9.

119 Xu C, Sun L, Kepley L J, Crooks R M and Ricco A J 1993 Molecular interactions between organized, surface-confined monolayers and vapor-phase probe molecules. 6. In-situ FT-IR external reflectance spectroscopy of monolayer adsorption and reaction chemistry *Anal. Chem.* 65 2102–7.

120 Kumar A, Mandal S, Selvakannan P R, Pasricha R, Mandale A B and Sastry M 2003 Investigation into the interaction between surface-bound alkylamines and gold nanoparticles *Langmuir* 19 6277–82.

121 Porta F, Krpetić Ž, Prati L, Gaiassi A and Scarì G 2008 Gold-ligand interaction studies of water-soluble aminoalcohol capped gold nanoparticles by NMR *Langmuir* 24 7061–4.

122 Wangoo N, Suri C R and Shekhawat G 2008 Interaction of gold nanoparticles with protein: a spectroscopic study to monitor protein conformational changes *Appl. Phys. Lett.* 92 133104.

123 Wangoo N, Bhasin K K, Mehta S K and Suri C R 2008 Synthesis and capping of water-dispersed gold nanoparticles by an amino acid: bioconjugation and binding studies *J. Colloid Interface Sci.* 323 247–54.

124 Griffin F and Fitzmaurice D 2007 Preparation and thermally promoted ripening of water-soluble gold nanoparticles stabilized by weakly physisorbed ligands *Langmuir* 23 10262–71.

125 Gittins D I and Caruso F 2001 Spontaneous phase transfer of nanoparticulate metals from organic to aqueous media *Angew. Chem. Int. Ed.* 40 3001–4.

126 Yu A, Liang Z, Cho J and Caruso F 2003 Nanostructured electrochemical sensor based on dense gold nanoparticle films *Nano Lett.* 3 1203–7.

127 Gandubert V J and Lennox R B 2005 Assessment of 4-(Dimethylamino)pyridine as a capping agent for gold nanoparticles *Langmuir* 21 6532–9.

128 Barlow B C and Burgess I J 2007 Electrochemical evaluation of 4-(Dimethylamino)pyridine adsorption on polycrystalline gold *Langmuir* 23 1555–63.

129 Lange H, Maultzsch J, Meng W, Mollenhauer D, Paulus B, Peica N, Schlecht S and Thomsen C 2011 Adsorption behavior of 4-Methoxypyridine on gold nanoparticles *Langmuir* 27 7258–64.

130 Li G, Li D, Zhang L, Zhai J and Wang E 2009 One-step synthesis of folic acid protected gold nanoparticles and their receptor-mediated intracellular uptake *Chem. Eur. J.* 15 9868–73.

131 Jena B K and Raj C R 2007 Synthesis of flower-like gold nanoparticles and their electrocatalytic activity towards the oxidation of methanol and the reduction of oxygen *Langmuir* 23 4064–70.

132 Yavuz M S, Li W and Xia Y 2009 Facile synthesis of gold icosahedra in an aqueous solution by reacting HAuCl4 with *N*-vinyl pyrrolidone *Chem. Eur. J.* 15 13181–7.

133 Shon Y-S and Choo H 2003 Organic reactions of monolayer-protected metal nanoparticles *Comptes Rendus Chim.* 6 1009–18.

134 Geng M, Zhang Y, Huang Q, Zhang B, Li Q, Li W and Li J 2010 Functionalization of C60 with gold nanoparticles *Carbon* 48 3570–4.

135 Demartini A, Alloisio M, Cuniberti C, Dellepiane G, Jadhav S A, Thea S, Giorgetti E, Gellini C and Muniz-Miranda M 2009 Polydiacetylene-functionalized noble metal nanocages *J. Phys. Chem. C* 113 19475–81.

136 Jadhav S A 2012 Functional self-assembled monolayers (SAMs) of organic compounds on gold nanoparticles *J. Mater. Chem.* 22 5894–9.

137 Giersig M and Mulvaney P 2002 Preparation of ordered colloid monolayers by electrophoretic deposition *Langmuir* 9(12) 3408–13.

138 Giljohann D A, Seferos D S, Daniel W L, Massich M D, Patel P C and Mirkin C A 2010 Gold nanoparticles for biology and medicine *Angew. Chem. Int. Ed. Engl.* 49 3280–94.

139 Kim D, Park S, Lee J H, Jeong Y Y and Jon S 2007 Antibiofouling polymer-coated gold nanoparticles as a contrast agent for in vivo X-ray computed tomography imaging *J. Am. Chem. Soc.* 129 7661–5.

140 Daniel M-C, Grow M E, Pan H, Bednarek M, Ghann W E, Zabetakis K and Cornish J 2011 Gold nanoparticle-cored poly(propyleneimine) dendrimers as a new platform for multifunctional drug delivery systems *New J. Chem.* 35 2366–74.

141 Templeton A C, Hostetler M J, Kraft C T and Murray R W 1998 Reactivity of monolayer-protected gold cluster molecules: steric effects *J. Am. Chem. Soc.* 120 1906–11.

142 Milette J, Toader V, Reven L and Lennox R B 2011 Tuning the miscibility of gold nanoparticles dispersed in liquid crystals via the thiol-for-DMAP reaction *J. Mater. Chem.* 21 9043–50.

143 Rucareanu S, Maccarini M, Shepherd J L and Lennox R B 2008 Polymer-capped gold nanoparticles by ligand-exchange reactions *J. Mater. Chem.* 18 5830–4.

144 Rucareanu S, Gandubert V J and Lennox R B 2006 4-(*N*,*N*-dimethylamino) pyridine-protected Au nanoparticles: versatile precursors for water- and organic-soluble gold nanoparticles *Chem. Mater.* 18 4674–80.

145 Zhao P, Li N and Astruc D 2013 State of the art in gold nanoparticle synthesis *Coord. Chem. Rev.* 257 638–65.

146 Reetz M T, Helbig W, Quaiser S A, Stimming U, Breuer N and Vogel R 1995 Visualization of surfactants on nanostructured palladium clusters by a combination of STM and high-resolution TEM *Science* 267 367–9.

147 Reetz M T and Helbig W 1994 Size-selective synthesis of nanostructured transition metal clusters *J. Am. Chem. Soc.* 116 7401–2.

148 Huang C-J, Chiu P-H, Wang Y-H, Chen K-L, Linn J-J and Yang C-F 2006 Electrochemically controlling the size of gold nanoparticles *J. Electrochem. Soc.* 153 D193.

149 Dai X and Compton R G 2006 Direct electrodeposition of gold nanoparticles onto indium tin oxide film coated glass: application to the detection of arsenic(III) *Anal. Sci. Int. J. Jpn. Soc. Anal. Chem.* 22 567–70.

150 Angerstein-Kozlowska H, Conway B E, Hamelin A and Stoicoviciu L 1987 Elementary steps of electrochemical oxidation of single-crystal planes of Au Part II. A chemical and structural basis of oxidation of the (111) plane *J. Electroanal. Chem. Interfacial Electrochem.* 228 429–53.

151 Sheridan E, Hjelm J and Forster R J 2007 Electrodeposition of gold nanoparticles on fluorine-doped tin oxide: control of particle density and size distribution *J. Electroanal. Chem.* 608 1–7.

152 Aziz M A, Patra S and Yang H 2008 A facile method of achieving low surface coverage of Au nanoparticles on an indium tin oxide electrode and its application to protein detection *Chem. Commun.* 38 4607–9.

153 Gamero M and Alonso C 2010 Deposition of nanostructurated gold on *n*-doped silicon substrate by different electrochemical methods *J. Appl. Electrochem.* 40 175.

154 Zhang H, Xu J-J and Chen H-Y 2008 Shape-controlled gold nanoarchitectures: synthesis, superhydrophobicity, and electrocatalytic properties *J. Phys. Chem. C* 112 13886–92.

155 Sobri S and Roy S 2008 Nucleation studies of gold on carbon electrodes *J. Eng. Sci. Technol.* 3 62–70.

156 Othman S H, El-Deab M S, Okajima T and Ohsaka T 2009 Novel procedure for the fabrication of gold nanostructures enriched in Au (1 1 0) facet orientation *Electrochem. Commun.* 11 1273–6.

157 Arihara K, Ariga T, Takashima N, Arihara K, Okajima T, Kitamura F, Tokuda K and Ohsaka T 2003 Multiple voltammetric waves for reductive desorption of cysteine and 4-mercaptobenzoic acid monolayers self-assembled on gold substrates *Phys. Chem. Chem. Phys.* 5 3758–61.

158 Zhou Y-G, Rees N V, Pillay J, Tshikhudo R, Vilakazi S and Compton R G 2011 Gold nanoparticles show electroactivity: counting and sorting nanoparticles upon impact with electrodes *Chem. Commun.* 48 224–6.

159 Zhou Y, Zheng X, Wu L, Zhi J and Xu M 2011 An easy method for manufacture of gold nanoparticles on a glassy carbon surface *J. Mater. Sci.* 46 1139–42.

160 Komsiyska L and Staikov G 2008 Electrocrystallization of Au nanoparticles on glassy carbon from $HClO_4$ solution containing $[AuCl4]^-$ *Electrochim. Acta* 54 168–72.

161 Qu L and Dai L 2005 Substrate-enhanced electroless deposition of metal nanoparticles on carbon nanotubes *J. Am. Chem. Soc.* 127 10806–7.

162 Hung V W-S and Kerman K 2011 Gold electrodeposition on carbon nanotubes for the enhanced electrochemical detection of homocysteine *Electrochem. Commun.* 13 328–30.

163 Afraz A, Rafati A A and Najafi M 2014 Optimization of modified carbon paste electrode with multiwalled carbon nanotube/ionic liquid/cauliflower-like gold nanostructures for simultaneous determination of ascorbic acid, dopamine and uric acid *Mater. Sci. Eng. C* 44 58–68.

164 Qiu C, Zhou H, Cao B, Sun L and Yu T 2013 Raman spectroscopy of morphology-controlled deposition of Au on graphene *Carbon* 59 487–94.

165 Lee J, Shim S, Kim B and Shin H S 2011 Surface-enhanced Raman scattering of single-and few-layer graphene by the deposition of gold nanoparticles *Chem. Eur. J.* 17 2381–7.

166 Wayman C M and Darby T P 1975 Nucleation and growth of gold films on graphite: II. The effect of substrate temperature *J. Cryst. Growth* 28 53–67.

167 Cherevko S and Chung C-H 2011 Direct electrodeposition of nanoporous gold with controlled multimodal pore size distribution *Electrochem. Commun.* 13 16–9.

168 Yu C S-S, Lee C-L and Wang C R C 1997 Gold nanorods: electrochemical synthesis and optical properties *J. Phys. Chem. B* 101 6661–4.

169 Walker C H, St. John J V and Wisian-Neilson P 2001 Synthesis and size control of gold nanoparticles stabilized by poly (methylphenylphosphazene) *J. Am. Chem. Soc.* 123 3846–7.

170 Wang X, Li G, Chen T, Yang M, Zhang Z, Wu T and Chen H 2008 Polymer-encapsulated gold-nanoparticle dimers: facile preparation and catalytical application in guided growth of dimeric ZnO-nanowires *Nano Lett.* 8 2643–7.

171 Giljohann D A, Seferos D S, Patel P C, Millstone J E, Rosi N L and Mirkin C A 2007 Oligonucleotide loading determines cellular uptake of DNA-modified gold nanoparticles *Nano Lett.* 7 3818–21.

172 Wang Z, Kanaras A G, Bates A D, Cosstick R and Brust M 2004 Enzymatic DNA processing on gold nanoparticles *J. Mater. Chem.* 14 578–80.

173 Guo S, Wang L and Wang E 2007 Templateless, surfactantless, simple electrochemical route to rapid synthesis of diameter-controlled 3D flowerlike gold microstructure with "clean" surface *Chem. Commun.* 44 3163–5.

174 Shan J and Tenhu H 2007 Recent advances in polymer protected gold nanoparticles: synthesis, properties and applications *Chem. Commun.* 110(10) 4580–98.

175 Newkome G R and Shreiner C 2010 Dendrimers derived from 1→ 3 branching motifs *Chem. Rev.* 110(10) 6338–442.

176 Caminade A-M, Turrin C-O, Laurent R, Ouali A and Delavaux-Nicot B 2011 *Dendrimers: Towards Catalytic, Material and Biomedical Uses* (John Wiley & Sons).

177 Esumi K, Miyamoto K and Yoshimura T 2002 Comparison of PAMAM–Au and PPI–Au nanocomposites and their catalytic activity for reduction of 4-nitrophenol *J. Colloid Interface Sci.* 254 402–5.

178 Esumi K, Houdatsu H and Yoshimura T 2004 Antioxidant action by gold- Pamam dendrimer nanocomposites *Langmuir* 20 2536–8.

179 Endo T, Yoshimura T and Esumi K 2005 Synthesis and catalytic activity of gold–silver binary nanoparticles stabilized by PAMAM dendrimer *J. Colloid Interface Sci.* 286 602–9.

180 Kim T H, Choi H S, Go B R and Kim J 2010 Modification of a glassy carbon surface with amine-terminated dendrimers and its application to electrocatalytic hydrazine oxidation *Electrochem. Commun.* 12 788–91.

181 Yancey D F, Carino E V and Crooks R M 2010 Electrochemical synthesis and electrocatalytic properties of Au@ Pt dendrimer-encapsulated nanoparticles *J. Am. Chem. Soc.* 132 10988–9.

182 Daniel M-C, Ruiz J, Nlate S, Blais J-C and Astruc D 2003 Nanoscopic assemblies between supramolecular redox active metallodendrons and gold nanoparticles: synthesis, characterization, and selective recognition of $H_2PO_4^-$, HSO_4^-, and adenosine-5′-triphosphate (ATP^{2-}) anions *J. Am. Chem. Soc.* 125 2617–28.

183 Kaewtong C, Jiang G, Ponnapati R, Pulpoka B and Advincula R 2010 Redox nanoreactor dendrimer boxes: in situ hybrid gold nanoparticles via terthiophene and carbazole peripheral dendrimer oxidation *Soft Matter* 6 5316–9.

184 Balogh L, Valluzzi R, Laverdure K S, Gido S P, Hagnauer G L and Tomalia D A 1999 Formation of silver and gold dendrimer nanocomposites *J. Nanopart. Res.* 1 353–68.

185 Garcia M E, Baker L A and Crooks R M 1999 Preparation and characterization of dendrimer-gold colloid nanocomposites *Anal. Chem.* 71 256–8.

186 Chechik V and Crooks R M 1999 Monolayers of thiol-terminated dendrimers on the surface of planar and colloidal gold *Langmuir* 15 6364–9.

187 Shan Y, Luo T, Peng C, Sheng R, Cao A, Cao X, Shen M, Guo R, Tomás H and Shi X 2012 Gene delivery using dendrimer-entrapped gold nanoparticles as nonviral vectors *Biomaterials* 33 3025–35.

188 Peng C, Zheng L, Chen Q, Shen M, Guo R, Wang H, Cao X, Zhang G and Shi X 2012 PEGylated dendrimer-entrapped gold nanoparticles for in vivo blood pool and tumor imaging by computed tomography *Biomaterials* 33 1107–19.

189 Kim Y-G, Oh S-K and Crooks R M 2004 Preparation and characterization of 1-2 nm dendrimer-encapsulated gold nanoparticles having very narrow size distributions *Chem. Mater.* 16 167–72.

190 Love C S, Chechik V, Smith D K and Brennan C 2004 Dendron-stabilised gold nanoparticles: generation dependence of core size and thermal stability *J. Mater. Chem.* 14 919–23.

191 Hoffman L W, Andersson G G, Sharma A, Clarke S R and Voelcker N H 2011 New insights into the structure of PAMAM dendrimer/gold nanoparticle nanocomposites *Langmuir* 27 6759–67.

192 Frasconi M, Tortolini C, Botre F and Mazzei F 2010 Multifunctional Au nanoparticle dendrimer-based surface plasmon resonance biosensor and its application for improved insulin detection *Anal. Chem.* 82 7335–42.

193 Namazi H and Fard A M P 2011 Preparation of gold nanoparticles in the presence of citric acid-based dendrimers containing periphery hydroxyl groups *Mater. Chem. Phys.* 129 189–94.

194 Boisselier E, Diallo A K, Salmon L, Ornelas C, Ruiz J and Astruc D 2010 Encapsulation and stabilization of gold nanoparticles with "click" polyethyleneglycol dendrimers *J. Am. Chem. Soc.* 132 2729–42.

195 Jia L, Lv L-P, Xu J-P and Ji J 2011 Phosphorylcholine functionalized dendrimers for the formation of highly stable and reactive gold nanoparticles and their glucose conjugation for biosensing *J. Nanopart. Res.* 13 4075–83.

196 Cho T J, Zangmeister R A, MacCuspie R I, Patri A K and Hackley V A 2011 Newkome-type dendron-stabilized gold nanoparticles: synthesis, reactivity, and stability *Chem. Mater.* 23 2665–76.

197 Jara N, Milán N S, Rahman A, Mouheb L, Boffito D C, Jeffryes C and Dahoumane S A 2021 Photochemical synthesis of gold and silver nanoparticles – a review *Molecules* 26 4585.

198 Lin S K and Cheng W T 2020 Fabrication and characterization of colloidal silver nanoparticle via photochemical synthesis *Mater. Lett.* 261 127077.

199 Eftekhari-Kenzerki Z, Fardid R and Behzad-Behbahani A 2019 Impact of silver nanoparticles on the ultraviolet radiation direct and bystander effects on TK6 cell line *J. Med. Phys.* 44 118.

200 McTiernan C D, Alarcon E I, Hallett-Tapley G L, Murillo-Lopez J, Arratia-Perez R, Netto-Ferreira J C and Scaiano J C 2014 Electron transfer from the benzophenone triplet excited state directs the photochemical synthesis of gold nanoparticles *Photochem. Photobiol. Sci.* 13 149–53.

201 Valandro S R, Poli A L, Neumann M G and Schmitt C C 2019 Photochemical synthesis of Ag and Au nanoparticles using a thioxanthone substituted chitosan as simultaneous photoinitiator and stabilizer *J. Braz. Chem. Soc.* 30 2642–9.

202 Marin M L, McGilvray K L and Scaiano J C 2008 Photochemical strategies for the synthesis of gold nanoparticles from Au(III) and Au(I) using photoinduced free radical generation *J. Am. Chem. Soc.* 130 16572–84.

203 Yuan M, Li Y-B, Guo J, Zhong R-B, Wang L-P and Zhang F 2018 Solvent effects on gold nanoparticle formation from photochemical reduction of Au(III) by UV irradiation *Nucl. Sci. Tech.* 29 158.

204 Zhou Y, Wang C Y, Zhu Y R and Chen Z Y 1999 A novel ultraviolet irradiation technique for shape-controlled synthesis of gold nanoparticles at room temperature *Chem. Mater.* 11 2310–2.

205 Sau T K, Pal A, Jana N R, Wang Z L and Pal T 2001 Size controlled synthesis of gold nanoparticles using photochemically prepared seed particles *J. Nanopart. Res.* 3 257–61.

206 del Pilar R-T M, Díaz-Torres L A and Romero-Servin S 2014 Heparin assisted photochemical synthesis of gold nanoparticles and their performance as SERS substrates *Int. J. Mol. Sci.* 15 19239–52.

207 Grzelczak M and Liz-Marzán L M 2014 The relevance of light in the formation of colloidal metal nanoparticles *Chem. Soc. Rev.* 43 2089–97.

208 Mukha I, Khodko A, Vityuk N, Severynovska O, Pivovarenko V, Kachalova N, Smirnova N and Eremenko A 2020 Light-driven formation of gold/tryptophan nanoparticles *Appl. Nanosci.* 10 2827–33.

209 Rodríguez G C, Gauthier G H, Ladeira L O, Cala J S and Cataño D L 2017 Effect of pH and chloroauric acid concentration on the geometry of gold nanoparticles obtained by photochemical synthesis *J. Phys. Conf. Ser.* 935 (IOP Publishing) 012027.

210 Cheng J, Ge L, Xiong B and He Y 2011 Investigation of pH effect on gold nanorod synthesis *J. Chin. Chem. Soc.* 58 822–7.

211 Unal I S, Demirbas A, Onal I, Ildiz N and Ocsoy I 2020 One step preparation of stable gold nanoparticle using red cabbage extracts under UV light and its catalytic activity *J. Photochem. Photobiol. B* 204 111800.

212 Scaiano J C, Billone P, Gonzalez C M, Marett L, Marin M L, McGilvray K L and Yuan N 2009 Photochemical routes to silver and gold nanoparticles *Pure Appl. Chem.* 81 635–47.

213 Rodríguez-Torres M d P, Díaz-Torres L A, Salas P, Rodríguez-González C and Olmos-López M 2014 UV photochemical synthesis of heparin-coated gold nanoparticles *Gold Bull.* 47 21–31.

214 Bianchi P, Petit G and Monbaliu J-C M 2020 Scalable and robust photochemical flow process towards small spherical gold nanoparticles *React. Chem. Eng.* 5 1224–36.

215 Sanabria-Calaa J A, Conde-Rodrígueza G R, Gauthiera G H, Ladeirab L O, Laverde-Catañoa D A, Peña-Ballesterosa D Y and Merchan-Arenasc D 2018 Gold nanoparticles formation mechanism by photochemical synthesis *Chem. Eng.* 64 https://doi.org/10.3303/CET1864068.

216 Hamaguchi K, Kawasaki H and Arakawa R 2010 Photochemical synthesis of glycine-stabilized gold nanoparticles and its heavy-metal-induced aggregation behavior *Colloids Surf. A Physicochem. Eng. Asp.* 367 167–73.

217 Teixeira P R, Santos M S, Silva A L G, Báo S N, Azevedo R B, Sales M J A and Paterno L G 2016 Photochemically-assisted synthesis of non-toxic and biocompatible gold nanoparticles *Colloids Surf. B Biointerfaces* 148 317–23.

218 Boufi S, Vilar M R, Ferraria A M and do Rego A M B 2013 In situ photochemical generation of silver and gold nanoparticles on chitosan *Colloids Surf. A Physicochem. Eng. Asp.* 439 151–8.

219 Shiraishi Y, Tanaka H, Sakamoto H, Ichikawa S and Hirai T 2017 Photoreductive synthesis of monodispersed Au nanoparticles with citric acid as reductant and surface stabilizing reagent *RSC Adv.* 7 6187–92.

220 Huang Y and Kim D-H 2012 Light-controlled synthesis of gold nanoparticles using a rigid, photoresponsive surfactant *Nanoscale* 4 6312–7.

221 Shang Y, Min C, Hu J, Wang T, Liu H and Hu Y 2013 Synthesis of gold nanoparticles by reduction of HAuCl4 under UV irradiation *Solid State Sci.* 15 17–23.

222 Manna A, Imae T, Yogo T, Aoi K and Okazaki M 2002 Synthesis of gold nanoparticles in a Winsor II type microemulsion and their characterization *J. Colloid Interface Sci.* 256 297–303.

223 Note C, Kosmella S and Koetz J 2006 Poly (ethyleneimine) as reducing and stabilizing agent for the formation of gold nanoparticles in w/o microemulsions *Colloids Surf. A Physicochem. Eng. Asp.* 290 150–6.

224 Mihaly M, Fleancu M C, Olteanu N L, Bojin D, Meghea A and Enachescu M 2012 Synthesis of gold nanoparticles by microemulsion assisted photoreduction method *Comptes Rendus Chim.* 15 1012–21.

225 Ahmad T, Wani I A, Manzoor N, Ahmed J, Kalam A and Al-Shihri A S 2014 Structural characterization, antifungal activity and optical properties of gold nanoparticles prepared by reverse micelles *Adv. Sci. Lett.* 20 1631–6.

226 Pileni M P 1989 *Structure and Reactivity in Reverse Micelles* (Elsevier).

227 Pileni M P 1993 Reverse micelles as microreactors *J. Phys. Chem.* 97 6961–73.

228 Uskoković V and Drofenik M 2005 Synthesis of materials within reverse micelles *Surf. Rev. Lett.* 12 239–77.

229 Luisi P L, Giomini M, Pileni M and Robinson B H 1988 Reverse micelles as hosts for proteins and small molecules *Biochim. Biophys. Acta BBA-Rev. Biomembr.* 947 209–46.

230 Wu M-L, Chen D-H and Huang T-C 2001 Synthesis of Au/Pd bimetallic nanoparticles in reverse micelles *Langmuir* 17 3877–83.

231 Murphy C J, Sau T K, Gole A and Orendorff C J 2005 Surfactant-directed synthesis and optical properties of one-dimensional plasmonic metallic nanostructures *MRS Bull.* 30 349–55.

232 Malik M A, Wani M Y and Hashim M A 2012 Microemulsion method: a novel route to synthesize organic and inorganic nanomaterials: 1st nano update *Arab. J. Chem.* 5 397–417.

233 López-Quintela M A, Tojo C, Blanco M C, Rio L G and Leis J R 2004 Microemulsion dynamics and reactions in microemulsions *Curr. Opin. Colloid Interface Sci.* 9 264–78.

234 Kurihara K, Kizling J, Stenius P and Fendler J H 1983 Laser and pulse radiolytically induced colloidal gold formation in water and in water-in-oil microemulsions *J. Am. Chem. Soc.* 105 2574–9.

235 Hollamby M J, Eastoe J, Chemelli A, Glatter O, Rogers S, Heenan R K and Grillo I 2010 Separation and purification of nanoparticles in a single step *Langmuir* 26 6989–94.

236 Hernandez J, Solla-Gullón J and Herrero E 2004 Gold nanoparticles synthesized in a water-in-oil microemulsion: electrochemical characterization and effect of the surface structure on the oxygen reduction reaction *J. Electroanal. Chem.* 574 185–96.

237 Muralidharan G, Subramanian L, Nallamuthu S K, Santhanam V and Kumar S 2011 Effect of reagent addition rate and temperature on synthesis of gold nanoparticles in microemulsion route *Ind. Eng. Chem. Res.* 50 8786–91.

238 Porta F, Prati L, Rossi M and Scarı` G 2002 Synthesis of Au(0) nanoparticles from W/O microemulsions *Colloids Surf. A Physicochem. Eng. Asp.* 211 43–8.

239 Tianimoghadam S and Salabat A 2018 A microemulsion method for preparation of thiol-functionalized gold nanoparticles *Particuology* 37 33–6.

240 Fustin C-A, Colard C, Filali M, Guillet P, Duwez A-S, Meier M A, Schubert U S and Gohy J-F 2006 Tuning the hydrophilicity of gold nanoparticles templated in star block copolymers *Langmuir* 22 6690–5.

241 Riess G 2003 Micellization of block copolymers *Prog. Polym. Sci.* 28 1107–70.

242 Sakai T and Alexandridis P 2004 Single-step synthesis and stabilization of metal nanoparticles in aqueous pluronic block copolymer solutions at ambient temperature *Langmuir* 20 8426–30.

243 Bakshi M S, Kaura A, Bhandari P, Kaur G, Torigoe K and Esumi K 2006 Synthesis of colloidal gold nanoparticles of different morphologies in the presence of triblock polymer micelles *J. Nanosci. Nanotechnol.* 6 1405–10.

244 Chen S, Guo C, Hu G-H, Wang J, Ma J-H, Liang X-F, Zheng L and Liu H-Z 2006 Effect of hydrophobicity inside PEO- PPO- PEO block copolymer micelles on the stabilization of gold nanoparticles: experiments *Langmuir* 22 9704–11.

245 Khullar P, Mahal A, Singh V, Banipal T S, Kaur G and Bakshi M S 2010 How PEO-PPO-PEO triblock polymer micelles control the synthesis of gold nanoparticles: temperature and hydrophobic effects *Langmuir* 26 11363–71.

246 Liu J, Niu J, Yin L and Jiang F 2011 In situ encapsulation of laccase in nanofibers by electrospinning for development of enzyme biosensors for chlorophenol monitoring *Analyst* 136 4802–8.

247 Ali N and Park S-Y 2008 Micellar structures of poly (styrene-b-4-vinylpyridine) s in THF/toluene mixtures and their functionalization with gold *Langmuir* 24 9279–85.

248 Cuenya B R, Baeck S-H, Jaramillo T F and McFarland E W 2003 Size- and support-dependent electronic and catalytic properties of Au^0/Au^{3+} nanoparticles synthesized from block copolymer micelles *J. Am. Chem. Soc.* 125 12928–34.

249 Li J, Shi L, An Y, Li Y, Chen X and Dong H 2006 Reverse micelles of star-block copolymer as nanoreactors for preparation of gold nanoparticles *Polymer* 47 8480–7.

250 Spatz J P, Mößmer S and Möller M 1996 Mineralization of gold nanoparticles in a block copolymer microemulsion *Chem. Eur. J.* 2 1552–5.

251 Kim D H, Wei A and Won Y-Y 2012 Preparation of super-stable gold nanorods via encapsulation into block copolymer micelles *ACS Appl. Mater. Interfaces* 4 1872–7.

252 Jaramillo T F, Baeck S-H, Cuenya B R and McFarland E W 2003 Catalytic activity of supported Au nanoparticles deposited from block copolymer micelles *J. Am. Chem. Soc.* 125 7148–9.

253 Papp S, Kőrösi L, Gool B, Dederichs T, Mela P, Möller M and Dékány I 2009 Formation of gold nanoparticles in diblock copolymer micelles with various reducing agents: kinetic and thermodynamic studies *J. Therm. Anal. Calorim.* 101 865–72.

254 Goy-López S, Juárez J, Cambón A, Botana J, Pereiro M, Baldomir D, Taboada P and Mosquera V 2010 Block copolymer-regulated synthesis of gold nanocrystals with sharp tips and edges *J. Mater. Chem.* 20 6808–14.

255 Sakai T and Alexandridis P 2005 Mechanism of gold metal ion reduction, nanoparticle growth and size control in aqueous amphiphilic block copolymer solutions at ambient conditions *J. Phys. Chem. B* 109 7766–77.

256 Khullar P, Singh V, Mahal A, Kaur H, Singh V, Banipal T S, Kaur G and Bakshi M S 2011 Tuning the shape and size of gold nanoparticles with triblock polymer micelle structure transitions and environments *J. Phys. Chem. C* 115 10442–54.

257 He P and Urban M W 2005 Phospholipid-stabilized Au–nanoparticles *Biomacromolecules* 6 1224–5.

258 Genç R, Clergeaud G, Ortiz M and O'Sullivan C K 2011 Green synthesis of gold nanoparticles using glycerol incorporated nanosized liposomes *Langmuir* 27 10894–900.

259 Sau T K, Urban A S, Dondapati S K, Fedoruk M, Horton M R, Rogach A L, Stefani F D, Rädler J O and Feldmann J 2009 Controlling loading and optical properties of gold nanoparticles on liposome membranes *Colloids Surf. A Physicochem. Eng. Asp.* 342 92–6.

260 Li S, Al-Misned F A, El-Serehy H A and Yang L 2021 Green synthesis of gold nanoparticles using aqueous extract of *Mentha Longifolia* leaf and investigation of its anti-human breast carcinoma properties in the in vitro condition *Arab. J. Chem.* 14 102931.

261 Al-Radadi N S 2021 Facile one-step green synthesis of gold nanoparticles (AuNp) using licorice root extract: antimicrobial and anticancer study against HepG2 cell line *Arab. J. Chem.* 14 102956.

262 Vijaya Kumar P, Kala M J S and Prakash K S 2019 Green synthesis of gold nanoparticles using *Croton Caudatus Geisel* leaf extract and their biological studies *Mater. Lett.* 236 19–22.

263 Ahmed R H and Mustafa D E 2020 Green synthesis of silver nanoparticles mediated by traditionally used medicinal plants in Sudan *Int. Nano Lett.* 10 1–14.

264 Arunachalam K D, Annamalai S K, Arunachalam A M and Kennedy S 2013 Green synthesis of crystalline silver nanoparticles using *Indigofera aspalathoides*-medicinal plant extract for wound healing applications *Asian J. Chem.* 25 S311.

265 Lakshmanan G, Sathiyaseelan A, Kalaichelvan P T and Murugesan K 2018 Plant-mediated synthesis of silver nanoparticles using fruit extract of *Cleome viscosa* L.: assessment of their antibacterial and anticancer activity *Karbala Int. J. Mod. Sci.* 4 61–8.

266 Vasquez R D, Apostol J G, De Leon J D, Mariano J D, Mirhan C M C, Pangan S S and Zamora E T 2016 Polysaccharide-mediated green synthesis of silver nanoparticles from *Sargassum siliquosum* JG Agardh: assessment of toxicity and hepatoprotective activity *OpenNano* 1 16–24.

267 Kulkarni A P, Srivastava A A and Zunjarrao R S 2012 Plant mediated synthesis of silver nanoparticles and their applications *Int J Pharm. Bio. Sci* 3 121–7.

268 Chandran S P, Chaudhary M, Pasricha R, Ahmad A and Sastry M 2006 Synthesis of gold nanotriangles and silver nanoparticles using Aloevera plant extract *Biotechnol. Prog.* 22 577–83.

269 Thirumurugan A, Jiflin G J, Rajagomathi G, Tomy N A, Ramachandran S and Jaiganesh R 2010 Biotechnological synthesis of gold nanoparticles of *Azadirachta indica* leaf extract *Int. J. Biol. Technol.* 1 75–7.

270 Keshavarzi M, Davoodi D, Pourseyedi S and Taghizadeh S 2018 The effects of three types of alfalfa plants (Medicago sativa) on the biosynthesis of gold nanoparticles: an insight into phytomining *Gold Bull.* 51 99–110.

271 Rao S V 2011 Picosecond nonlinear optical studies of gold nanoparticles synthesised using coriander leaves (*Coriandrum sativum*) *J. Mod. Opt.* 58 1024–9.

272 Shankar S S, Rai A, Ahmad A and Sastry M 2005 Controlling the optical properties of lemongrass extract synthesized gold nanotriangles and potential application in infrared-absorbing optical coatings *Chem. Mater.* 17 566–72.

273 Ankamwar B 2010 Biosynthesis of gold nanoparticles (green-gold) using leaf extract of *Terminalia catappa E-J. Chem.* 7 1334–9.

274 Elia P, Zach R, Hazan S, Kolusheva S, Porat Z and Zeiri Y 2014 Green synthesis of gold nanoparticles using plant extracts as reducing agents *Int. J. Nanomedicine* 9 4007–21.

275 Thangamani N and Bhuvaneshwari N 2019 Green synthesis of gold nanoparticles using *Simarouba glauca* leaf extract and their biological activity of micro-organism *Chem. Phys. Lett.* 732 136587.

276 Goia D and Matijević E 1999 Tailoring the particle size of monodispersed colloidal gold *Colloids Surf. A Physicochem. Eng. Asp.* 146 139–52.

277 Eskandari-Nojehdehi M, Jafarizadeh-Malmiri H and Jafarizad A 2018 Microwave accelerated green synthesis of gold nanoparticles using gum Arabic and their physico-chemical properties assessments *Z. Für Phys. Chem.* 232 325–43.

278 Jafarizad A, Safaee K and Ekinci D 2017 Green synthesis of gold nanoparticles using aqueous extracts of *Ziziphus Jujuba* and gum arabic *J. Clust. Sci.* 28 2765–77.

279 Shih C-M, Shieh Y-T and Twu Y-K 2009 Preparation of gold nanopowders and nanoparticles using chitosan suspensions *Carbohydr. Polym.* 78 309–15.

280 Valentina Y, Rather M Y and Yogamoorthi A 2019 Biosynthesis of gold nanoparticle using cell-free extract of clinical isolates *Staphylococcus Aureus* and *Escherichia coli Malays. J. Med. Biol. Res.* 6 107–16.

281 Sodaiezade H 2020 Gold bioaccumulation by filamentous soil cyanobacteria isolated from mouteh goldmine (Isfahan Province, Iran) *J. Phycol. Res.* 4 469–80.

282 Patil M P, Kang M, Niyonizigiye I, Singh A, Kim J-O, Seo Y B and Kim G-D 2019 Extracellular synthesis of gold nanoparticles using the marine bacterium *Paracoccus haeundaensis* BC74171T and evaluation of their antioxidant activity and antiproliferative effect on normal and cancer cell lines *Colloids Surf. B Biointerfaces* 183 110455.

283 Abdel-Kareem M M and Zohri A A 2018 Extracellular mycosynthesis of gold nanoparticles using *Trichoderma hamatum*: optimization, characterization and antimicrobial activity *Lett. Appl. Microbiol.* 67 465–75.

284 Camas M, Sazak Camas A and Kyeremeh K 2018 Extracellular synthesis and characterization of gold nanoparticles using *Mycobacterium* sp. BRS2A-AR2 isolated from the aerial roots of the Ghanaian mangrove plant, *Rhizophora racemosa Indian J. Microbiol.* 58 214–21.

285 Khalil A T, Ovais M, Iqbal J, Ali A, Ayaz M, Abbas M, Ahmad I and Devkota H P 2021 Microbes-mediated synthesis strategies of metal nanoparticles and their potential role in cancer therapeutics *Semin. Cancer Biol.* 86 693–705.

286 Syed B, Prasad N M and Satish S 2016 Endogenic mediated synthesis of gold nanoparticles bearing bactericidal activity *J. Microsc. Ultrastruct.* 4 162–6.

287 Molnár Z, Bódai V, Szakacs G, Erdélyi B, Fogarassy Z, Sáfrán G, Varga T, Kónya Z, Tóth-Szeles E, Szűcs R and Lagzi I 2018 Green synthesis of gold nanoparticles by thermophilic filamentous fungi *Sci. Rep.* 8 1–12.

288 Gholami-Shabani M, Shams-Ghahfarokhi M, Gholami-Shabani Z, Akbarzadeh A, Riazi G, Ajdari S, Amani A and Razzaghi-Abyaneh M 2015 Enzymatic synthesis of gold nanoparticles using sulfite reductase purified from *Escherichia coli*: a green eco-friendly approach *Process Biochem.* 50 1076–85.

289 Vágó A, Szakacs G, Sáfrán G, Horvath R, Pécz B and Lagzi I 2016 One-step green synthesis of gold nanoparticles by mesophilic filamentous fungi *Chem. Phys. Lett.* 645 1–4.

290 Du L, Xian L and Feng J-X 2011 Rapid extra-/intracellular biosynthesis of gold nanoparticles by the fungus *Penicillium* sp *J. Nanopart. Res.* 13 921–30.

291 Zhang X, He X, Wang K and Yang X 2011 Different active biomolecules involved in biosynthesis of gold nanoparticles by three fungus species *J. Biomed. Nanotechnol.* 7 245–54.

292 Mukherjee P, Ahmad A, Mandal D, Senapati S, Sainkar S R, Khan M I, Ramani R, Parischa R, Ajayakumar P V, Alam M, Sastry M and Kumar R 2001 Bioreduction of AuCl(4)(−) ions by the fungus, *Verticillium* sp. and surface trapping of the gold nanoparticles Formed D.M. and S.S. thank the Council of Scientific and Industrial Research (CSIR), Government of India, for financial assistance *Angew. Chem. Int. Ed. Engl.* 40 3585–8.

293 Gericke M and Pinches A 2006 Biological synthesis of metal nanoparticles *Hydrometallurgy* 83 132–40.

294 Gericke M and Pinches A 2006 Microbial production of gold nanoparticles *Gold Bull.* 39 22–8.

295 Shankar S S, Ahmad A, Pasricha R and Sastry M 2003 Bioreduction of chloroaurate ions by geranium leaves and its endophytic fungus yields gold nanoparticles of different shapes *J. Mater. Chem.* 13 1822–6.

296 Ahmad A, Senapati S, Khan M I, Kumar R and Sastry M 2005 Extra-/intracellular biosynthesis of gold nanoparticles by an alkalotolerant fungus, *Trichothecium* sp *J. Biomed. Nanotechnol.* 1 47–53.

297 Qu Y, You S, Zhang X, Pei X, Shen W, Li Z, Li S and Zhang Z 2018 Biosynthesis of gold nanoparticles using cell-free extracts of *Magnusiomyces ingens* LH-F1 for nitrophenols reduction *Bioprocess Biosyst. Eng.* 41 359–67.

298 Yaqoob A A, Umar K and Ibrahim M N M 2020 Silver nanoparticles: various methods of synthesis, size affecting factors and their potential applications–a review *Appl. Nanosci.* 10 1369–78.

299 Jung J H, Cheol O H, Soo Noh H, Ji J H and Soo Kim S 2006 Metal nanoparticle generation using a small ceramic heater with a local heating area *J. Aerosol Sci.* 37 1662–70.

300 Raffi M, Rumaiz A K, Hasan M M and Shah S I 2007 Studies of the growth parameters for silver nanoparticle synthesis by inert gas condensation *J. Mater. Res.* 22 3378–84.

301 Harra J, Mäkitalo J, Siikanen R, Virkki M, Genty G, Kobayashi T, Kauranen M and Mäkelä J M 2012 Size-controlled aerosol synthesis of silver nanoparticles for plasmonic materials *J. Nanopart. Res.* 14 870.

302 Fumitaka M, Kohno J, Takeda Y, Kondow T and Sawabe H 2000 Formation and size control of silver nanoparticles by laser ablation in aqueous solution *J. Phys. Chem. B* 104 9111–7.

303 Alhamid M Z, Hadi B S and Khumaeni A 2019 Synthesis of silver nanoparticles using laser ablation method utilizing Nd:YAG laser *AIP Conf. Proc.* 2202 020013.

304 Rafique M, Rafique M S, Kalsoom U, Afzal A, Butt S H and Usman A 2019 Laser ablation synthesis of silver nanoparticles in water and dependence on laser nature *Opt. Quant. Electron.* 51 179.

305 Chen S-F and Zhang H 2012 Aggregation kinetics of nanosilver in different water conditions *Adv. Nat. Sci. Nanosci. Nanotechnol.* 3 035006.

306 Dang T M D, Le T T T, Fribourg-Blanc E and Dang M C 2012 Influence of surfactant on the preparation of silver nanoparticles by polyol method *Adv. Nat. Sci. Nanosci. Nanotechnol.* 3 035004.

307 Patil R S, Kokate M R, Jambhale C L, Pawar S M, Han S H and Kolekar S S 2012 One-pot synthesis of PVA-capped silver nanoparticles their characterization and biomedical application *Adv. Nat. Sci. Nanosci. Nanotechnol.* 3 015013.

308 Dong P, Feng J, Zhang D, Li C, Shi Q and Xie X 2020 In situ synthesis of amply antimicrobial silver nanoparticle (AgNP) by polyzwitterionic copolymers bearing hydroxyl groups *React. Funct. Polym.* 153 104609.

309 Aadil K R, Pandey N, Mussatto S I and Jha H 2019 Green synthesis of silver nanoparticles using acacia lignin, their cytotoxicity, catalytic, metal ion sensing capability and antibacterial activity *J. Environ. Chem. Eng.* 7 103296.

310 Wani I A, Khatoon S, Ganguly A, Ahmed J and Ahmad T 2013 Structural characterization and antimicrobial properties of silver nanoparticles prepared by inverse microemulsion method *Colloids Surf. B Biointerfaces* 101 243–50.

311 Manna A, Imae T, Iida M and Hisamatsu N 2001 Formation of silver nanoparticles from a *N*-hexadecylethylenediamine silver nitrate complex *Langmuir* 17 6000–4.

312 Demchenko V, Riabov S, Sinelnikov S, Radchenko O, Kobylinskyi S and Rybalchenko N 2020 Novel approach to synthesis of silver nanoparticles in interpolyelectrolyte complexes based on pectin, chitosan, starch and their derivatives *Carbohydr. Polym.* 242 116431.

313 Huo H, Li Z, Tan T, Chen L, Ungar G, Hoogenboom R, Zhang Q and Liu F 2021 Asymmetric incorporation of silver nanoparticles in polymeric assemblies by coassembly of tadpole-like nanoparticles and amphiphilic block copolymers *Macromol. Rapid Commun.* 42 2100354.

314 Dong X, Ji X, Wu H, Zhao L, Li J and Yang W 2009 Shape control of silver nanoparticles by stepwise citrate reduction *J. Phys. Chem. C* 113 6573–6.

315 Hebeish A A, El-Rafie M H, Abdel-Mohdy F A, Abdel-Halim E S and Emam H E 2010 Carboxymethyl cellulose for green synthesis and stabilization of silver nanoparticles *Carbohydr. Polym.* 82 933–41.

316 Radziuk D, Skirtach A, Sukhorukov G, Shchukin D and Möhwald H 2007 Stabilization of silver nanoparticles by polyelectrolytes and poly (ethylene glycol) *Macromol. Rapid Commun.* 28 848–55.

317 Pillai Z S and Kamat P V 2004 What factors control the size and shape of silver nanoparticles in the citrate ion reduction method *J. Phys. Chem. B* 108 945–51.

318 Kassaee M Z, Mohammadkhani M, Akhavan A and Mohammadi R 2011 In situ formation of silver nanoparticles in PMMA via reduction of silver ions by butylated hydroxytoluene *Struct. Chem.* 22 11–5.

319 Kainourgios P, Tziveleka L-A, Kartsonakis I A, Ioannou E, Roussis V and Charitidis C A 2021 Silver nanoparticles grown on cross-linked poly (methacrylic acid) microspheres: synthesis, characterization, and antifungal activity evaluation *Chemosensors* 9 152.

320 Alavi M and Rai M 2019 Recent progress in nanoformulations of silver nanoparticles with cellulose, chitosan, and alginic acid biopolymers for antibacterial applications *Appl. Microbiol. Biotechnol.* 103 8669–76.

321 Oliveira M M, Ugarte D, Zanchet D and Zarbin A J G 2005 Influence of synthetic parameters on the size, structure, and stability of dodecanethiol-stabilized silver nanoparticles *J. Colloid Interface Sci.* 292 429–35.

322 Luo C, Zhang Y, Zeng X, Zeng Y and Wang Y 2005 The role of poly(ethylene glycol) in the formation of silver nanoparticles *J. Colloid Interface Sci.* 288 444–8.
323 Hsu S L-C and Wu R-T 2007 Synthesis of contamination-free silver nanoparticle suspensions for micro-interconnects *Mater. Lett.* 61 3719–22.
324 Chen M, Feng Y-G, Wang X, Li T-C, Zhang J-Y and Qian D-J 2007 Silver nanoparticles capped by oleylamine: formation, growth, and self-organization *Langmuir* 23 5296–304.
325 Natsuki J and Abe T 2011 Synthesis of pure colloidal silver nanoparticles with high electroconductivity for printed electronic circuits: the effect of amines on their formation in aqueous media *J. Colloid Interface Sci.* 359 19–23.
326 Natsuki J, Natsuki T and Abe T 2013 Low molecular weight compounds as effective dispersing agents in the formation of colloidal silver nanoparticles *J. Nanopart. Res.* 15 1483.
327 Kim D, Jeong S and Moon J 2006 Synthesis of silver nanoparticles using the polyol process and the influence of precursor injection *Nanotechnology* 17 4019–24.
328 Sun Y, Mayers B and Xia Y 2003 Transformation of silver nanospheres into nanobelts and triangular nanoplates through a thermal process *Nano Lett.* 3 675–9.
329 Im S H, Lee Y T, Wiley B and Xia Y 2005 Large-scale synthesis of silver nanocubes: the role of HCl in promoting cube perfection and monodispersity *Angew. Chem. Int. Ed.* 44 2154–7.
330 Tao A, Sinsermsuksakul P and Yang P 2006 Polyhedral silver nanocrystals with distinct scattering signatures *Angew. Chem. Int. Ed.* 45 4597–601.
331 Wiley B J, Chen Y, McLellan J M, Xiong Y, Li Z-Y, Ginger D and Xia Y 2007 Synthesis and optical properties of silver nanobars and nanorice *Nano Lett.* 7 1032–6.
332 Wiley B J, Im S H, Li Z-Y, McLellan J, Siekkinen A and Xia Y 2006 Maneuvering the surface plasmon resonance of silver nanostructures through shape-controlled synthesis *J. Phys. Chem. B* 110 15666–75.
333 Métraux G S and Mirkin C A 2005 Rapid thermal synthesis of silver nanoprisms with chemically tailorable thickness *Adv. Mater.* 17 412–5.
334 Zaheer Z and Rafiuddin 2011 Multi-branched flower-like silver nanoparticles: preparation and characterization *Colloids Surf. A Physicochem. Eng. Asp.* 384 427–31.
335 Cong F-Z, Wei H, Tian X-R and Xu H-X 2012 A facile synthesis of branched silver nanowire structures and its applications in surface-enhanced Raman scattering *Front. Phys.* 7 521–6.
336 Shchukin D G, Radtchenko I L and Sukhorukov G B 2003 Photoinduced reduction of silver inside microscale polyelectrolyte capsules *ChemPhysChem* 4 1101–3.
337 Jin R, Charles Cao Y, Hao E, Métraux G S, Schatz G C and Mirkin C A 2003 Controlling anisotropic nanoparticle growth through plasmon excitation *Nature* 425 487–90.
338 Malval J-P, Jin M, Balan L, Schneider R, Versace D-L, Chaumeil H, Defoin A and Soppera O 2010 Photoinduced size-controlled generation of silver nanoparticles coated with carboxylate-derivatized thioxanthones *J. Phys. Chem. C* 114 10396–402.
339 Sato-Berrú R, Redón R, Vázquez-Olmos A and Saniger J M 2009 Silver nanoparticles synthesized by direct photoreduction of metal salts. Application in surface-enhanced Raman spectroscopy *J. Raman Spectrosc.* 40 376–80.
340 Kumar Ghosh S, Kundu S, Mandal M, Nath S and Pal T 2003 Studies on the evolution of silver nanoparticles in micelle by UV-photoactivation *J. Nanopart. Res.* 5 577–87.
341 Huang L, Zhai M L, Long D W, Peng J, Xu L, Wu G Z, Li J Q and Wei G S 2008 UV-induced synthesis, characterization and formation mechanism of silver nanoparticles in alkalic carboxymethylated chitosan solution *J. Nanopart. Res.* 10 1193–202.
342 Król-Gracz A, Michalak E, Nowak P M and Dyonizy A 2011 Photo-induced chemical reduction of silver bromide to silver nanoparticles *Cent. Eur. J. Chem.* 9 982–9.

343 Omrani A A and Taghavinia N 2012 Photo-induced growth of silver nanoparticles using UV sensitivity of cellulose fibers *Appl. Surf. Sci.* 258 2373–7.

344 Rosa R M, Silva J C, Sanches I S and Henriques C 2017 Simultaneous photo-induced cross-linking and silver nanoparticle formation in a PVP electrospun wound dressing *Mater. Lett.* 207 145–8.

345 Caillosse E, Zaier M, Mezghani M, Hajjar-Garreau S, Vidal L, Lougnot D and Balan L 2020 Photo-induced self-assembly of silver nanoparticles for rapid generation of first and second surface mirrors *ACS Appl. Nano Mater.* 3 6531–40.

346 Kumar V, Gundampati R K, Singh D K, Bano D, Jagannadham M V and Hasan S H 2016 Photoinduced green synthesis of silver nanoparticles with highly effective antibacterial and hydrogen peroxide sensing properties *J. Photochem. Photobiol. B* 162 374–85.

347 Kumar V, Singh D K, Mohan S, Gundampati R K and Hasan S H 2017 Photoinduced green synthesis of silver nanoparticles using aqueous extract of *Physalis angulata* and its antibacterial and antioxidant activity *J. Environ. Chem. Eng.* 5 744–56.

348 Kumar V, Singh D K, Mohan S and Hasan S H 2016 Photo-induced biosynthesis of silver nanoparticles using aqueous extract of *Erigeron bonariensis* and its catalytic activity against acridine orange *J. Photochem. Photobiol. B* 155 39–50.

349 Liu J, Zhang X, Yu M, Li S and Zhang J 2012 Photoinduced silver nanoparticles/nanorings on plasmid DNA scaffolds *Small* 8 310–6.

350 Kumar V, Gundampati R K, Singh D K, Jagannadham M V, Sundar S and Hasan S H 2016 Photo-induced rapid biosynthesis of silver nanoparticle using aqueous extract of *Xanthium strumarium* and its antibacterial and antileishmanial activity *J. Ind. Eng. Chem.* 37 224–36.

351 Johans C, Clohessy J, Fantini S, Kontturi K and Cunnane V J 2002 Electrosynthesis of polyphenylpyrrole coated silver particles at a liquid–liquid interface *Electrochem. Commun.* 4 227–30.

352 Zhang Y, Chen F, Zhuang J, Tang Y, Wang D, Wang Y, Dong A and Ren N 2002 Synthesis of silver nanoparticles via electrochemical reduction on compact zeolite film modified electrodes *Chem. Commun.* 23 2814–5.

353 Ma H, Yin B, Wang S, Jiao Y, Pan W, Huang S, Chen S and Meng F 2004 Synthesis of silver and gold nanoparticles by a novel electrochemical method *ChemPhysChem* 5 68–75.

354 Starowicz M, Stypuła B and Banaś J 2006 Electrochemical synthesis of silver nanoparticles *Electrochem. Commun.* 8 227–30.

355 Khaydarov R A, Khaydarov R R, Gapurova O, Estrin Y and Scheper T 2009 Electrochemical method for the synthesis of silver nanoparticles *J. Nanopart. Res.* 11 1193–200.

356 Nasretdinova G R, Fazleeva R R, Mukhitova R K, Nizameev I R, Kadirov M K, Ziganshina A Y and Yanilkin V V 2015 Electrochemical synthesis of silver nanoparticles in solution *Electrochem. Commun.* 50 69–72.

357 Kuntyi O I, Kytsya A R, Mertsalo I P, Mazur A S, Zozula G I, Bazylyak L I and Topchak R V 2019 Electrochemical synthesis of silver nanoparticles by reversible current in solutions of sodium polyacrylate *Colloid Polym. Sci.* 297 689–95.

358 Kuntyi O, Mazur A, Kytsya A, Karpenko O, Bazylyak L, Mertsalo I, Pokynbroda T and Prokopalo A 2020 Electrochemical synthesis of silver nanoparticles in solutions of rhamnolipid *Micro Nano Lett.* 15 802–7.

359 Nadagouda M N, Speth T F and Varma R S 2011 Microwave-assisted green synthesis of silver nanostructures *Acc. Chem. Res.* 44 469–78.

360 Anjana V N, Joseph M, Francis S, Joseph A, Koshy E P and Mathew B 2021 Microwave assisted green synthesis of silver nanoparticles for optical, catalytic, biological and electrochemical applications *Artif. Cells Nanomedicine Biotechnol.* 49 438–49.

361 Polshettiwar V, Nadagouda M N and Varma R S 2009 Microwave-assisted chemistry: a rapid and sustainable route to synthesis of organics and nanomaterials *Aust. J. Chem.* 62 16–26.

362 Chen J, Wang J, Zhang X and Jin Y 2008 Microwave-assisted green synthesis of silver nanoparticles by carboxymethyl cellulose sodium and silver nitrate *Mater. Chem. Phys.* 108 421–4.

363 Navaladian S, Viswanathan B, Varadarajan T K and Viswanath R P 2008 Microwave-assisted rapid synthesis of anisotropic Ag nanoparticles by solid state transformation *Nanotechnology* 19 045603.

364 Sreeram K J, Nidhin M and Nair B U 2008 Microwave assisted template synthesis of silver nanoparticles *Bull. Mater. Sci.* 31 937–42.

365 Komarneni S, Li D, Newalkar B, Katsuki H and Bhalla A S 2002 Microwave- polyol process for Pt and Ag nanoparticles *Langmuir* 18 5959–62.

366 Yin H, Yamamoto T, Wada Y and Yanagida S 2004 Large-scale and size-controlled synthesis of silver nanoparticles under microwave irradiation *Mater. Chem. Phys.* 83 66–70.

367 Tsuji M, Matsumoto K, Jiang P, Matsuo R, Hikino S, Tang X-L and Kamarudin K S N 2008 The role of adsorption species in the formation of Ag nanostructures by a microwave-polyol route *Bull. Chem. Soc. Jpn.* 81 393–400.

368 Hu B, Wang S-B, Wang K, Zhang M and Yu S-H 2008 Microwave-assisted rapid facile "green" synthesis of uniform silver nanoparticles: self-assembly into multilayered films and their optical properties *J. Phys. Chem. C* 112 11169–74.

369 Pal A, Shah S and Devi S 2009 Microwave-assisted synthesis of silver nanoparticles using ethanol as a reducing agent *Mater. Chem. Phys.* 114 530–2.

370 Zhao X, Xia Y, Li Q, Ma X, Quan F, Geng C and Han Z 2014 Microwave-assisted synthesis of silver nanoparticles using sodium alginate and their antibacterial activity *Colloids Surf. A Physicochem. Eng. Asp.* 444 180–8.

371 Seku K, Gangapuram B R, Pejjai B, Kadimpati K K and Golla N 2018 Microwave-assisted synthesis of silver nanoparticles and their application in catalytic, antibacterial and antioxidant activities *J. Nanostructure Chem.* 8 179–88.

372 Khan A, El-Toni A M, Alrokayan S, Alsalhi M, Alhoshan M and Aldwayyan A S 2011 Microwave-assisted synthesis of silver nanoparticles using poly-*N*-isopropylacrylamide/acrylic acid microgel particles *Colloids Surf. A Physicochem. Eng. Asp.* 377 356–60.

373 Peng H, Yang A and Xiong J 2013 Green, microwave-assisted synthesis of silver nanoparticles using bamboo hemicelluloses and glucose in an aqueous medium *Carbohydr. Polym.* 91 348–55.

374 Adhikari L, Larm N E, Bhawawet N and Baker G A 2018 Rapid microwave-assisted synthesis of silver nanoparticles in a Halide-free deep eutectic solvent *ACS Sustain. Chem. Eng.* 6 5725–31.

375 Dzido G, Markowski P, Małachowska Jutsz A, Prusik K and Jarzębski A B 2015 Rapid continuous microwave-assisted synthesis of silver nanoparticles to achieve very high productivity and full yield: from mechanistic study to optimal fabrication strategy *J. Nanopart. Res.* 17 27.

376 Barani H and Mahltig B 2022 Microwave-assisted synthesis of silver nanoparticles: effect of reaction temperature and precursor concentration on fluorescent property *J. Clust. Sci.* 33 101–11.

377 Kumar A, Vemula P K, Ajayan P M and John G 2008 Silver-nanoparticle-embedded antimicrobial paints based on vegetable oil *Nat. Mater.* 7 236–41.

378 Fayaz A M, Balaji K, Girilal M, Yadav R, Kalaichelvan P T and Venketesan R 2010 Biogenic synthesis of silver nanoparticles and their synergistic effect with antibiotics: a study against gram-positive and gram-negative bacteria *Nanomed. Nanotechnol. Biol. Med.* 6 103–9.

379 Ahmed S and Ikram S 2015 Silver nanoparticles: one pot green synthesis using *Terminalia arjuna* extract for biological application *J Nanomed Nanotechnol* 6 1–6.

380 Abou El-Nour K M, Eftaiha A, Al-Warthan A and Ammar R A 2010 Synthesis and applications of silver nanoparticles *Arab. J. Chem.* 3 135–40.

381 Lee S H and Jun B-H 2019 Silver nanoparticles: synthesis and application for nanomedicine *Int. J. Mol. Sci.* 20 865.

382 Klaus-Joerger T, Joerger R, Olsson E and Granqvist C-G 2001 Bacteria as workers in the living factory: metal-accumulating bacteria and their potential for materials science *Trends Biotechnol.* 19 15–20.

383 Arokiyaraj S, Vincent S, Saravanan M, Lee Y, Oh Y K and Kim K H 2017 Green synthesis of silver nanoparticles using *Rheum palmatum* root extract and their antibacterial activity against *Staphylococcus aureus* and *Pseudomonas aeruginosa Artif. Cells Nanomedicine Biotechnol.* 45 372–9.

384 Patra S, Mukherjee S, Barui A K, Ganguly A, Sreedhar B and Patra C R 2015 Green synthesis, characterization of gold and silver nanoparticles and their potential application for cancer therapeutics *Mater. Sci. Eng. C* 53 298–309.

385 Slawson R M, Trevors J T and Lee H 1992 Silver accumulation and resistance in *Pseudomonas stutzeri Arch. Microbiol.* 158 398–404.

386 Klaus T, Joerger R, Olsson E and Granqvist C-G 1999 Silver-based crystalline nanoparticles, microbially fabricated *Proc. Natl. Acad. Sci.* 96 13611–4.

387 Pugazhenthiran N, Anandan S, Kathiravan G, Udaya Prakash N K, Crawford S and Ashokkumar M 2009 Microbial synthesis of silver nanoparticles by *Bacillus* sp *J. Nanopart. Res.* 11 1811.

388 Tsibakhashvili N Y, Kirkesali E I, Pataraya D T, Gurielidze M A, Kalabegishvili T L, Gvarjaladze D N, Tsertsvadze G I, Frontasyeva M V, Zinicovscaia I I, Wakstein M S, Khakhanov S N, Shvindina N V and Shklover V Y 2011 Microbial synthesis of silver nanoparticles by *Streptomyces glaucus* and *Spirulina platensis Adv. Sci. Lett.* 4 3408–17.

389 El-Shanshoury A, E-R R, ElSilk S E and Ebeid M E 2011 Extracellular biosynthesis of silver nanoparticles using *Escherichia coli* ATCC 8739, *Bacillus subtilis* ATCC 6633, and *Streptococcus thermophilus* ESh1 and their antimicrobial activities *ISRN Nanotechnol.* 2011 1–7.

390 Shahverdi A R, Minaeian S, Shahverdi H R, Jamalifar H and Nohi A-A 2007 Rapid synthesis of silver nanoparticles using culture supernatants of Enterobacteria: a novel biological approach *Process Biochem.* 42 919–23.

391 Saravanan M, Vemu A K and Barik S K 2011 Rapid biosynthesis of silver nanoparticles from *Bacillus megaterium* (NCIM 2326) and their antibacterial activity on multi drug resistant clinical pathogens *Colloids Surf. B Biointerfaces* 88 325–31.

392 Priyadarshini S, Gopinath V, Priyadharsshini N M, MubarakAli D and Velusamy P 2013 Synthesis of anisotropic silver nanoparticles using novel strain, *Bacillus flexus* and its biomedical application *Colloids Surf. B Biointerfaces* 102 232–7.

393 Samadi N, Golkaran D, Eslamifar A, Jamalifar H, Fazeli M R and Mohseni F A 2009 Intra/extracellular biosynthesis of silver nanoparticles by an autochthonous strain of *Proteus mirabilis* isolated from photographic waste *J. Biomed. Nanotechnol.* 5 247–53.

394 Das V L, Thomas R, Varghese R T, Soniya E V, Mathew J and Radhakrishnan E K 2014 Extracellular synthesis of silver nanoparticles by the Bacillus strain CS 11 isolated from industrialized area *3 Biotech* 4 121–6.

395 Ibrahim S, Ahmad Z, Manzoor M Z, Mujahid M, Faheem Z and Adnan A 2021 Optimization for biogenic microbial synthesis of silver nanoparticles through response surface methodology, characterization, their antimicrobial, antioxidant, and catalytic potential *Sci. Rep.* 11 770.

396 Sastry M, Ahmad A, Islam Khan M and Kumar R 2003 Biosynthesis of metal nanoparticles using fungi and actinomycete *Curr. Sci.* 85 162–70.

397 Mandal D, Bolander M E, Mukhopadhyay D, Sarkar G and Mukherjee P 2006 The use of microorganisms for the formation of metal nanoparticles and their application *Appl. Microbiol. Biotechnol.* 69 485–92.

398 Mukherjee P, Ahmad A, Mandal D, Senapati S, Sainkar S R, Khan M I, Parishcha R, Ajaykumar P V, Alam M and Kumar R 2001 Fungus-mediated synthesis of silver nanoparticles and their immobilization in the mycelial matrix: a novel biological approach to nanoparticle synthesis *Nano Lett.* 1 515–9.

399 Bhainsa K C and D'souza S F 2006 Extracellular biosynthesis of silver nanoparticles using the fungus *Aspergillus fumigatus Colloids Surf. B Biointerfaces* 47 160–4.

400 Naqvi S Z H, Kiran U, Ali M I, Jamal A, Hameed A, Ahmed S and Ali N 2013 Combined efficacy of biologically synthesized silver nanoparticles and different antibiotics against multidrug-resistant bacteria *Int. J. Nanomedicine* 8 3187.

401 Ingle A, Rai M, Gade A and Bawaskar M 2009 *Fusarium solani*: a novel biological agent for the extracellular synthesis of silver nanoparticles *J. Nanopart. Res.* 11 2079.

402 Syed A, Saraswati S, Kundu G C and Ahmad A 2013 Biological synthesis of silver nanoparticles using the fungus *Humicola* sp. and evaluation of their cytoxicity using normal and cancer cell lines *Spectrochim. Acta - Part Mol. Biomol. Spectrosc.* 114 144–7.

403 Owaid M N, Raman J, Lakshmanan H, Al-Saeedi S S S, Sabaratnam V and Abed I A 2015 Mycosynthesis of silver nanoparticles by *Pleurotus cornucopiae* var. citrinopileatus and its inhibitory effects against Candida sp *Mater. Lett.* 153 186–90.

404 Al-Bahrani R, Raman J, Lakshmanan H, Hassan A A and Sabaratnam V 2017 Green synthesis of silver nanoparticles using tree oyster mushroom *Pleurotus ostreatus* and its inhibitory activity against pathogenic bacteria *Mater. Lett.* 186 21–5.

405 Vigneshwaran N, Ashtaputre N M, Varadarajan P V, Nachane R P, Paralikar K M and Balasubramanya R H 2007 Biological synthesis of silver nanoparticles using the fungus *Aspergillus flavus Mater. Lett.* 61 1413–8.

406 Mohammed Fayaz A, Balaji K, Kalaichelvan P T and Venkatesan R 2009 Fungal based synthesis of silver nanoparticles – an effect of temperature on the size of particles *Colloids Surf. B Biointerfaces* 74 123–6.

407 Feroze N, Arshad B, Younas M, Afridi M I, Saqib S and Ayaz A 2020 Fungal mediated synthesis of silver nanoparticles and evaluation of antibacterial activity *Microsc. Res. Tech.* 83 72–80.

408 Gudikandula K, Vadapally P and Singara Charya M A 2017 Biogenic synthesis of silver nanoparticles from white rot fungi: their characterization and antibacterial studies *OpenNano* 2 64–78.

409 Tyagi S, Tyagi P K, Gola D, Chauhan N and Bharti R K 2019 Extracellular synthesis of silver nanoparticles using entomopathogenic fungus: characterization and antibacterial potential *SN Appl. Sci.* 1 1545.

410 Elegbede J A, Lateef A, Azeez M A, Asafa T B, Yekeen T A, Oladipo I C, Adebayo E A, Beukes L S and Gueguim-Kana E B 2018 Fungal xylanases-mediated synthesis of silver nanoparticles for catalytic and biomedical applications *IET Nanobiotechnol.* 12 857–63

411 Alavi M 2022 Bacteria and fungi as major bio-sources to fabricate silver nanoparticles with antibacterial activities *Expert Rev. Anti-Infect. Ther.* 20 897–906.

412 Apte M, Sambre D, Gaikawad S, Joshi S, Bankar A, Kumar A R and Zinjarde S 2013 Psychrotrophic yeast *Yarrowia lipolytica* NCYC 789 mediates the synthesis of antimicrobial silver nanoparticles via cell-associated melanin *AMB Express* 3 32.

413 Mourato A, Gadanho M, Lino A R and Tenreiro R 2011 Biosynthesis of crystalline silver and gold nanoparticles by extremophilic yeasts *Bioinorg. Chem. Appl.* 2011 546074.

414 Abdeen S, Geo S, Sukanya S, Praseetha P K and Dhanya R P 2014 Biosynthesis of silver nanoparticles from actinomycetes for therapeutic applications *Int. J. Nano Dimens.* 5 155–62.

415 Karthik L, Kumar G, Kirthi A V, Rahuman A A and Rao K B 2014 *Streptomyces* sp. LK3 mediated synthesis of silver nanoparticles and its biomedical application *Bioprocess Biosyst. Eng.* 37 261–7.

416 Golinska P, Wypij M, Ingle A P, Gupta I, Dahm H and Rai M 2014 Biogenic synthesis of metal nanoparticles from actinomycetes: biomedical applications and cytotoxicity *Appl. Microbiol. Biotechnol.* 98 8083–97.

417 Balagurunathan R, Radhakrishnan M, Rajendran R B and Velmurugan D 2011 Biosynthesis of gold nanoparticles by actinomycete Streptomyces viridogens strain HM10 *Indian J. Biochem. Biophys.* 48 331–5.

418 Chauhan R, Kumar A and Abraham J 2013 A biological approach to the synthesis of silver nanoparticles with *Streptomyces* sp JAR1 and its antimicrobial activity *Sci. Pharm.* 81 607.

419 Sukanya M K, Saju K A, Praseetha P K and Sakthivel G 2013 Therapeutic potential of biologically reduced silver nanoparticles from actinomycete cultures *J. Nanosci.* 2013 940719.

420 Prakasham R S, Buddana S K, Yannam S K and Guntuku G S 2012 Characterization of silver nanoparticles synthesized by using marine isolate *Streptomyces albidoflavus J. Microbiol. Biotechnol.* 22 614–21.

421 Salem S S, EL-Belely E F, Niedbała G, Alnoman M M, Hassan S E-D, Eid A M, Shaheen T I, Elkelish A and Fouda A 2020 Bactericidal and in-vitro cytotoxic efficacy of silver nanoparticles (Ag-NPs) fabricated by endophytic actinomycetes and their use as coating for the textile fabrics *Nanomaterials* 10 2082.

422 Hamed A A, Kabary H, Khedr M and Emam A N 2020 Antibiofilm, antimicrobial and cytotoxic activity of extracellular green-synthesized silver nanoparticles by two marine-derived actinomycete *RSC Adv.* 10 10361–7.

423 Sowani H, Mohite P, Munot H, Shouche Y, Bapat T, Kumar A R, Kulkarni M and Zinjarde S 2016 Green synthesis of gold and silver nanoparticles by an actinomycete *Gordonia amicalis* HS-11: mechanistic aspects and biological application *Process Biochem.* 51 374–83.

424 Subbaiya R, Saravanan M, Priya A R, Shankar K R, Selvam M, Ovais M, Balajee R and Barabadi H 2017 Biomimetic synthesis of silver nanoparticles from *Streptomyces atrovirens* and their potential anticancer activity against human breast cancer cells *IET Nanobiotechnol.* 11 965–72.

425 Abd-Elnaby H M, Abo-Elala G M, Abdel-Raouf U M and Hamed M M 2016 Antibacterial and anticancer activity of extracellular synthesized silver nanoparticles from marine *Streptomyces rochei* MHM13 *Egypt. J. Aquat. Res.* 42 301–12.

426 Mabrouk M, Elkhooly T A and Amer S K 2021 Actinomycete strain type determines the monodispersity and antibacterial properties of biogenically synthesized silver nanoparticles *J. Genet. Eng. Biotechnol.* 19 57.

427 Khan I, Anburajan P, Kumar G, Yoon J-J, Bahuguna A, de Moura A G L, Pugazhendhi A, Kim S-H and Kang S-C 2021 Comparative effect of silver nanoparticles (AgNPs) derived from actinomycetes and henna on biohydrogen production by *Clostridium beijerinckii* (KTCC1737) *Int. J. Energy Res.* 45 17269–78.

428 Xie J, Lee J Y, Wang D I and Ting Y P 2007 Silver nanoplates: from biological to biomimetic synthesis *ACS Nano* 1 429–39.

429 Kathiraven T, Sundaramanickam A, Shanmugam N and Balasubramanian T 2015 Green synthesis of silver nanoparticles using marine algae *Caulerpa racemosa* and their antibacterial activity against some human pathogens *Appl. Nanosci.* 5 499–504.

430 Govindaraju K, Kiruthiga V, Kumar V G and Singaravelu G 2009 Extracellular synthesis of silver nanoparticles by a marine alga, Sargassum wightii Grevilli and their antibacterial effects *J. Nanosci. Nanotechnol.* 9 5497–501.

431 Castro L, Blázquez M L, Muñoz J A, González F and Ballester A 2013 Biological synthesis of metallic nanoparticles using algae *IET Nanobiotechnol.* 7 109–16.

432 Prasad T N V K V, Kambala V S R and Naidu R 2013 Phyconanotechnology: synthesis of silver nanoparticles using brown marine algae *Cystophora moniliformis* and their characterisation *J. Appl. Phycol.* 25 177–82.

433 Sinha S N, Paul D, Halder N, Sengupta D and Patra S K 2015 Green synthesis of silver nanoparticles using fresh water green alga *Pithophora oedogonia* (Mont.) Wittrock and evaluation of their antibacterial activity *Appl. Nanosci.* 5 703–9.

434 Arya A, Mishra V and Chundawat T S 2019 Green synthesis of silver nanoparticles from green algae (*Botryococcus braunii*) and its catalytic behavior for the synthesis of benzimidazoles *Chem. Data Collect.* 20 100190.

435 Chugh D, Viswamalya V S and Das B 2021 Green synthesis of silver nanoparticles with algae and the importance of capping agents in the process *J. Genet. Eng. Biotechnol.* 19 126.

436 Aboelfetoh E F, El-Shenody R A and Ghobara M M 2017 Eco-friendly synthesis of silver nanoparticles using green algae (*Caulerpa serrulata*): reaction optimization, catalytic and antibacterial activities *Environ. Monit. Assess.* 189 349.

437 Acharya D, Satapathy S, Thathapudi J J, Somu P and Mishra G 2022 Biogenic synthesis of silver nanoparticles using marine algae *Cladophora glomerata* and evaluation of apoptotic effects in human colon cancer cells *Mater. Technol.* 37 569–80.

438 Danaei M, Motaghi M M, Naghmachi M, Amirmahani F and Moravej R 2021 Green synthesis of silver nanoparticles (AgNPs) by filamentous algae extract: comprehensive evaluation of antimicrobial and anti-biofilm effects against nosocomial pathogens *Biologia (Bratisl.)* 76 3057–69.

439 Durán N, Marcato P D, Alves O L, De Souza G I H and Esposito E 2005 Mechanistic aspects of biosynthesis of silver nanoparticles by several *Fusarium oxysporum* strains *J Nanobiotechnol* 3 1–7.

440 Mukherjee P, Roy M, Mandal B P, Dey G K, Mukherjee P K, Ghatak J, Tyagi A K and Kale S P 2008 Green synthesis of highly stabilized nanocrystalline silver particles by a non-pathogenic and agriculturally important fungus *T. asperellum Nanotechnology* 19 75103.

441 Ahmad A, Mukherjee P, Senapati S, Mandal D, Khan M I, Kumar R and Sastry M 2003 Extracellular biosynthesis of silver nanoparticles using the fungus *Fusarium oxysporum Colloids Surf. B Biointerfaces* 28 313–8.

442 Korbekandi H, Ashari Z, Iravani S and Abbasi S 2013 Optimization of biological synthesis of silver nanoparticles using *Fusarium oxysporum Iran. J. Pharm. Res. IJPR* 12 289.

443 Roopan S M, Madhumitha G, Rahuman A A, Kamaraj C, Bharathi A and Surendra T V 2013 Low-cost and eco-friendly phyto-synthesis of silver nanoparticles using *Cocos nucifera* coir extract and its larvicidal activity *Ind. Crop. Prod.* 43 631–5.

444 Mukunthan K S, Elumalai E K, Patel T N and Murty V R 2011 *Catharanthus roseus*: a natural source for the synthesis of silver nanoparticles *Asian Pac. J. Trop. Biomed.* 1 270.

445 Shukla V K, Singh R P and Pandey A C 2010 Black pepper assisted biomimetic synthesis of silver nanoparticles *J. Alloys Compd.* 507 L13–6.

446 Mondal N K, Chowdhury A, Dey U, Mukhopadhya P, Chatterjee S, Das K and Datta J K 2014 Green synthesis of silver nanoparticles and its application for mosquito control *Asian Pac. J. Trop. Dis.* 4 S204–10.

447 Anuj S A and Ishnava K B 2013 Plant mediated synthesis of silver nanoparticles by using dried stem powder of *Tinospora cordifolia*, its antibacterial activity and comparison with antibiotics *Int J Pharm. Bio. Sci* 4 849–63.

448 Gardea-Torresdey J L, Gomez E, Peralta-Videa J R, Parsons J G, Troiani H and Jose-Yacaman M 2003 *Alfalfa sprouts*: a natural source for the synthesis of silver nanoparticles *Langmuir* 19 1357–61.

449 Shankar S S, Ahmad A and Sastry M 2003 Geranium leaf assisted biosynthesis of silver nanoparticles *Biotechnol. Prog.* 19 1627–31.

450 Li S, Shen Y, Xie A, Yu X, Qiu L, Zhang L and Zhang Q 2007 Green synthesis of silver nanoparticles using *Capsicum annuum* L. extract *Green Chem.* 9 852–8.

451 Kasthuri J, Veerapandian S and Rajendiran N 2009 Biological synthesis of silver and gold nanoparticles using apiin as reducing agent *Colloids Surf. B Biointerfaces* 68 55–60.

452 Dwivedi A D and Gopal K 2010 Biosynthesis of silver and gold nanoparticles using *Chenopodium album* leaf extract *Colloids Surf. A Physicochem. Eng. Asp.* 369 27–33.

453 Küp F Ö, Çoşkunçay S and Duman F 2020 Biosynthesis of silver nanoparticles using leaf extract of *Aesculus hippocastanum* (horse chestnut): evaluation of their antibacterial, antioxidant and drug release system activities *Mater. Sci. Eng. C* 107 110207.

454 Vinodhini S, Vithiya B S M and Prasad T A A 2022 Green synthesis of silver nanoparticles by employing the *Allium fistulosum*, *Tabernaemontana divaricate* and *Basella alba* leaf extracts for antimicrobial applications *J. King Saud Univ. - Sci.* 34 101939.

455 Loo Y Y, Chieng B W, Nishibuchi M and Radu S 2012 Synthesis of silver nanoparticles by using tea leaf extract from *Camellia sinensis Int. J. Nanomedicine* 7 4263–7.

456 Krishnaraj C, Jagan E G, Rajasekar S, Selvakumar P, Kalaichelvan P T and Mohan N 2010 Synthesis of silver nanoparticles using *Acalypha indica* leaf extracts and its antibacterial activity against water borne pathogens *Colloids Surf. B Biointerfaces* 76 50–6.

457 Ahmed M J, Murtaza G, Mehmood A and Bhatti T M 2015 Green synthesis of silver nanoparticles using leaves extract of *Skimmia laureola*: characterization and antibacterial activity *Mater. Lett.* 153 10–3.

458 Sathyavathi R, Krishna M B, Rao S V, Saritha R and Rao D N 2010 Biosynthesis of silver nanoparticles using *Coriandrum sativum* leaf extract and their application in nonlinear optics *Adv. Sci. Lett.* 3 138–43.

459 Veerasamy R, Xin T Z, Gunasagaran S, Xiang T F W, Yang E F C, Jeyakumar N and Dhanaraj S A 2011 Biosynthesis of silver nanoparticles using mangosteen leaf extract and evaluation of their antimicrobial activities *J. Saudi Chem. Soc.* 15 113–20.

460 Yasir M, Singh J, Tripathi M K, Singh P and Shrivastava R 2017 Green synthesis of silver nanoparticles using leaf extract of common arrowhead houseplant and its anticandidal activity *Pharmacogn. Mag.* 13 S840–4.

461 Kumar V, Singh S, Srivastava B, Bhadouria R and Singh R 2019 Green synthesis of silver nanoparticles using leaf extract of *Holoptelea integrifolia* and preliminary investigation of its antioxidant, anti-inflammatory, antidiabetic and antibacterial activities *J. Environ. Chem. Eng.* 7 103094.

462 Singh R, Hano C, Nath G and Sharma B 2021 Green biosynthesis of silver nanoparticles using leaf extract of *Carissa carandas* L. and their antioxidant and antimicrobial activity against human pathogenic bacteria *Biomolecules* 11 299.

463 Azmi S N H, Al-Jassasi B M H, Al-Sawafi H M S, Al-Shukaili S H G, Rahman N and Nasir M 2021 Optimization for synthesis of silver nanoparticles through response surface methodology using leaf extract of *Boswellia sacra* and its application in antimicrobial activity *Environ. Monit. Assess.* 193 497.

464 Yassin M T, Mostafa A A-F, Al-Askar A A and Al-Otibi F O 2022 Facile green synthesis of silver nanoparticles using aqueous leaf extract of *Origanum majorana* with potential bioactivity against multidrug resistant bacterial strains *Crystals* 12 603.

465 Das S, Das J, Samadder A, Bhattacharyya S S, Das D and Khuda-Bukhsh A R 2013 Biosynthesized silver nanoparticles by ethanolic extracts of *Phytolacca decandra*, *Gelsemium sempervirens*, *Hydrastis canadensis* and *Thuja occidentalis* induce differential cytotoxicity through G2/M arrest in A375 cells *Colloids Surf. B Biointerfaces* 101 325–36.

466 Roy K, Sarkar C K and Ghosh C K 2014 Green synthesis of silver nanoparticles using fruit extract of *Malus domestica* and study of its antimicrobial activity *Dig. J. Nanomater. Biostruct.* 9 1137–47.

467 Gogoi N, Babu P J, Mahanta C and Bora U 2015 Green synthesis and characterization of silver nanoparticles using alcoholic flower extract of *Nyctanthes arbortristis* and in vitro investigation of their antibacterial and cytotoxic activities *Mater. Sci. Eng. C* 46 463–9.

468 Pattanayak S, Mollick M M R, Maity D, Chakraborty S, Dash S K, Chattopadhyay S, Roy S, Chattopadhyay D and Chakraborty M 2017 Butea monosperma bark extract mediated green synthesis of silver nanoparticles: characterization and biomedical applications *J. Saudi Chem. Soc.* 21 673–84.

469 Omran B A, Nassar H N, Fatthallah N A, Hamdy A, El-Shatoury E H and Sh E-G N 2018 Waste upcycling of *Citrus sinensis* peels as a green route for the synthesis of silver nanoparticles *Energy Sources Part Recovery Util. Environ. Eff.* 40 227–36.

470 Rolim W R, Pelegrino M T, de Araújo L B, Ferraz L S, Costa F N, Bernardes J S, Rodigues T, Brocchi M and Seabra A B 2019 Green tea extract mediated biogenic synthesis of silver nanoparticles: characterization, cytotoxicity evaluation and antibacterial activity *Appl. Surf. Sci.* 463 66–74.

471 Hasnain M S, Javed M N, Alam M S, Rishishwar P, Rishishwar S, Ali S, Nayak A K and Beg S 2019 Purple heart plant leaves extract-mediated silver nanoparticle synthesis: optimization by Box-Behnken design *Mater. Sci. Eng. C* 99 1105–14.

472 Dakshayani S S, Marulasiddeshwara M B, Kumar M N S, Golla R, Kumar P R, Devaraja S and Hosamani R 2019 Antimicrobial, anticoagulant and antiplatelet activities of green synthesized silver nanoparticles using Selaginella (Sanjeevini) plant extract *Int. J. Biol. Macromol.* 131 787–97.

473 Khabeeri O M, Al Thabaiti S A and Khan Z 2020 *Citrus sinensis* peel waste assisted synthesis of AgNPs: effect of surfactant on the nucleation and morphology *SN Appl. Sci.* 2 2038.

474 Alsakhawy S A, Baghdadi H H, El Shenawy M A and El-Hosseiny L S 2022 Antibacterial activity of silver nanoparticles phytosynthesized by citrus fruit peel extracts *BioNanoScience* 12 1106–15.

475 Manik U P, Nande A, Raut S and Dhoble S J 2020 Green synthesis of silver nanoparticles using plant leaf extraction of *Artocarpus heterophylus* and *Azadirachta indica* *Results Mater.* 6 100086.

476 Alharbi N S, Alsubhi N S and Felimban A I 2022 Green synthesis of silver nanoparticles using medicinal plants: characterization and application *J. Radiat. Res. Appl. Sci.* 15 109–24.

477 Aisida S O, Ugwu K, Akpa P A, Nwanya A C, Ejikeme P M, Botha S, Ahmad I, Maaza M and Ezema F I 2019 Biogenic synthesis and antibacterial activity of controlled silver nanoparticles using an extract of *Gongronema Latifolium* *Mater. Chem. Phys.* 237 121859.

478 Raghavendra G M, Jung J and Seo J 2016 Step-reduced synthesis of starch-silver nanoparticles *Int. J. Biol. Macromol.* 86 126–8.

479 Jung J, Raghavendra G M, Kim D and Seo J 2018 One-step synthesis of starch-silver nanoparticle solution and its application to antibacterial paper coating *Int. J. Biol. Macromol.* 107 2285–90.

480 Vanamudan A and Sudhakar P P 2016 Biopolymer capped silver nanoparticles with potential for multifaceted applications *Int. J. Biol. Macromol.* 86 262–8.

481 Maryan A S and Gorji M 2016 Synthesize of nano silver using cellulose or glucose as a reduction agent: the study of their antibacterial activity on polyurethan fibers *Bulg. Chem. Commun.* 47 151–5.

482 Chen G, Yan L, Wan X, Zhang Q and Wang Q 2019 In situ synthesis of silver nanoparticles on cellulose fibers using D-glucuronic acid and its antibacterial application *Materials* 12 3101.

483 Hong P, Liu Y, Peng W, Zhang J and Ruan R 2016 Green synthesis and stability evaluation of Ag nanoparticles using bamboo hemicellulose *Bioresources* 11 385–99.

484 Amin M, Iram F, Iqbal M S, Saeed M Z, Raza M and Alam S 2013 Arabinoxylan-mediated synthesis of gold and silver nanoparticles having exceptional high stability *Carbohydr. Polym.* 92 1896–900.

485 Hu S and Hsieh Y-L 2016 Silver nanoparticle synthesis using lignin as reducing and capping agents: a kinetic and mechanistic study *Int. J. Biol. Macromol.* 82 856–62.

486 Zhi J, Wang Y, Lu Y, Ma J and Luo G 2006 In situ preparation of magnetic chitosan/Fe_3O_4 composite nanoparticles in tiny pools of water-in-oil microemulsion *React. Funct. Polym.* 66 1552–8.

487 Loo A L, Pineda M G, Saade H, Treviño M E and López R G 2008 Synthesis of magnetic nanoparticles in bicontinuous microemulsions. Effect of surfactant concentration *J. Mater. Sci.* 43 3649–54.

488 Foroughi F, Hassanzadeh-Tabrizi S A and Amighian J 2015 Microemulsion synthesis and magnetic properties of hydroxyapatite-encapsulated nano $CoFe_2O_4$ *J. Magn. Magn. Mater.* 382 182–7.

489 Wang X, Zhang P, Gao J, Chen X and Yang H 2015 Facile synthesis and magnetic properties of Fe_3C/C nanoparticles via a sol–gel process *Dyes Pigments* 112 305–10.

490 Alagiri M, Ponnusamy S and Muthamizhchelvan C 2012 Synthesis and characterization of NiO nanoparticles by sol–gel method *J. Mater. Sci. Mater. Electron.* 23 728–32.

491 Atrak K, Ramazani A and Taghavi Fardood S 2018 A novel sol–gel synthesis and characterization of $MgFe_2O_4$@γ–Al_2O_3 magnetic nanoparticles using tragacanth gel and its application as a magnetically separable photocatalyst for degradation of organic dyes under visible light *J. Mater. Sci. Mater. Electron.* 29 6702–10.

492 Chen R, Wang W, Zhao X, Zhang Y, Wu S and Li F 2014 Rapid hydrothermal synthesis of magnetic $Co_xNi_{1-x}Fe_2O_4$ nanoparticles and their application on removal of Congo red *Chem. Eng. J.* 242 226–33.

493 Peng J, Hojamberdiev M, Xu Y, Cao B, Wang J and Wu H 2011 Hydrothermal synthesis and magnetic properties of gadolinium-doped $CoFe_2O_4$ nanoparticles *J. Magn. Magn. Mater.* 323 133–7.

494 Sun J, Zhou S, Hou P, Yang Y, Weng J, Li X and Li M 2007 Synthesis and characterization of biocompatible Fe_3O_4 nanoparticles *J. Biomed. Mater. Res. A* 80A 333–41.

495 Massart R and Cabuil V 1987 Effect of some parameters on the formation of colloidal magnetite in alkaline-medium-yield and particle-size control *J. Chim. Phys. Phys.-Chim. Biol.* 84 967–73.

496 Wu W, He Q and Jiang C 2008 Magnetic iron oxide nanoparticles: synthesis and surface functionalization strategies *Nanoscale Res. Lett.* 3 397.

497 Majidi S, Zeinali Sehrig F, Farkhani S M, Soleymani Goloujeh M and Akbarzadeh A 2016 Current methods for synthesis of magnetic nanoparticles *Artif. Cells Nanomedicine Biotechnol.* 44 722–34.

498 Iida H, Takayanagi K, Nakanishi T and Osaka T 2007 Synthesis of Fe_3O_4 nanoparticles with various sizes and magnetic properties by controlled hydrolysis *J. Colloid Interface Sci.* 314 274–80.

499 Refait P and Génin J-M R 1993 The oxidation of ferrous hydroxide in chloride-containing aqueous media and pourbaix diagrams of green rust one *Corros. Sci.* 34 797–819.

500 Olowe A A and Génin J M R 1991 The mechanism of oxidation of ferrous hydroxide in sulphated aqueous media: importance of the initial ratio of the reactants *Corros. Sci.* 32 965–84.

501 Kang Y S, Risbud S, Rabolt J F and Stroeve P 1996 Synthesis and characterization of nanometer-size Fe_3O_4 and γ-Fe_2O_3 particles *Chem. Mater.* 8 2209–11.

502 Feltin N and Pileni M P 1997 New technique for synthesizing iron ferrite magnetic nanosized particles *Langmuir* 13 3927–33.

503 López Pérez J A, López Quintela M A, Mira J, Rivas J and Charles S W 1997 Advances in the preparation of magnetic nanoparticles by the microemulsion method *J. Phys. Chem. B* 101 8045–7.

504 Okoli C, Boutonnet M, Mariey L, Järås S and Rajarao G 2011 Application of magnetic iron oxide nanoparticles prepared from microemulsions for protein purification *J. Chem. Technol. Biotechnol.* 86 1386–93.

505 Lakshmanan R, Okoli C, Boutonnet M, Järås S and Rajarao G K 2014 Microemulsion prepared magnetic nanoparticles for phosphate removal: time efficient studies *J. Environ. Chem. Eng.* 2 185–9.

506 Vidal-Vidal J, Rivas J and López-Quintela M A 2006 Synthesis of monodisperse maghemite nanoparticles by the microemulsion method *Colloids Surf. A Physicochem. Eng. Asp.* 288 44–51.

507 Vestal C R and Zhang Z J 2002 Synthesis of $CoCrFeO_4$ nanoparticles using microemulsion methods and size-dependent studies of their magnetic properties *Chem. Mater.* 14 3817–22.

508 Santra S, Tapec R, Theodoropoulou N, Dobson J, Hebard A and Tan W 2001 Synthesis and characterization of silica-coated iron oxide nanoparticles in microemulsion: the effect of nonionic surfactants *Langmuir* 17 2900–6.

509 Najafi A and Nematipour K 2017 Synthesis and magnetic properties evaluation of monosized FeCo alloy nanoparticles through microemulsion method *J. Supercond. Nov. Magn.* 30 2647–53.

510 Pillai V and Shah D O 1996 Synthesis of high-coercivity cobalt ferrite particles using water-in-oil microemulsions *J. Magn. Magn. Mater.* 163 243–8.

511 Asab G, Zereffa E A and Abdo Seghne T 2020 Synthesis of silica-coated Fe_3O_4 nanoparticles by microemulsion method: characterization and evaluation of antimicrobial activity *Int. J. Biomater.* 2020 e4783612.

512 Mollarasouli F, Zor E, Ozcelikay G and Ozkan S A 2021 Magnetic nanoparticles in developing electrochemical sensors for pharmaceutical and biomedical applications *Talanta* 226 122108.

513 Abbas M, Parvatheeswara Rao B, Naga S M, Takahashi M and Kim C 2013 Synthesis of high magnetization hydrophilic magnetite (Fe_3O_4) nanoparticles in single reaction – surfactantless polyol process *Ceram. Int.* 39 7605–11.

514 Oh A H, Park H-Y, Jung Y-G, Choi S-C and An G S 2020 Synthesis of Fe_3O_4 nanoparticles of various size via the polyol method *Ceram. Int.* 46 10723–8.

515 Iacovita C, Fizeşan I, Pop A, Scorus L, Dudric R, Stiufiuc G, Vedeanu N, Tetean R, Loghin F and Stiufiuc R 2020 In vitro intracellular hyperthermia of iron oxide magnetic nanoparticles, synthesized at high temperature by a polyol process *Pharmaceutics* 12 424.

516 Couto G G, Klein J J, Schreiner W H, Mosca D H, de Oliveira A J A and Zarbin A J G 2007 Nickel nanoparticles obtained by a modified polyol process: synthesis, characterization, and magnetic properties *J. Colloid Interface Sci.* 311 461–8.

517 Jeyadevan B, Shinoda K, Justin R J, Matsumoto T, Sato K, Takahashi H, Sato Y and Tohji K 2006 Polyol process for Fe-based hard(fct-FePt) and soft(FeCo) magnetic nanoparticles *IEEE Trans. Magn.* 42 3030–5.

518 Cheng C, Xu F and Gu H 2011 Facile synthesis and morphology evolution of magnetic iron oxide nanoparticles in different polyol processes *New J. Chem.* 35 1072–9.

519 Liu C, Wu X, Klemmer T, Shukla N, Yang X, Weller D, Roy A G, Tanase M and Laughlin D 2004 Polyol process synthesis of monodispersed FePt nanoparticles *J. Phys. Chem. B* 108 6121–3.

520 Julio Jimenez L F, Herrera Barros A and Caamaño De Ávila Z I 2019 Synthesis and characterization of $Fe_{55}Co_{45}$ magnetic nanoparticles by polyol and green chemistry method *Results Phys.* 15 102785.

521 Li D and Komarneni S 2006 Microwave-assisted polyol process for synthesis of Ni nanoparticles *J. Am. Ceram. Soc.* 89 1510–7.

522 Joseyphus R J, Shinoda K, Kodama D and Jeyadevan B 2010 Size controlled Fe nanoparticles through polyol process and their magnetic properties *Mater. Chem. Phys.* 123 487–93.

523 Wang Y, Nkurikiyimfura I and Pan Z 2015 Sonochemical synthesis of magnetic nanoparticles *Chem. Eng. Commun.* 202 616–21.

524 Hassanjani-Roshan A, Vaezi M R, Shokuhfar A and Rajabali Z 2011 Synthesis of iron oxide nanoparticles via sonochemical method and their characterization *Particuology* 9 95–9.

525 Hee Kim E, Sook Lee H, Kook Kwak B and Kim B-K 2005 Synthesis of ferrofluid with magnetic nanoparticles by sonochemical method for MRI contrast agent *J. Magn. Magn. Mater.* 289 328–30.

526 Dang F, Enomoto N, Hojo J and Enpuku K 2009 Sonochemical synthesis of monodispersed magnetite nanoparticles by using an ethanol–water mixed solvent *Ultrason. Sonochem.* 16 649–54.

527 Aliramaji S, Zamanian A and Sohrabijam Z 2015 Characterization and synthesis of magnetite nanoparticles by innovative sonochemical method *Procedia Mater. Sci.* 11 265–9.

528 Yadav V K, Ali D, Khan S H, Gnanamoorthy G, Choudhary N, Yadav K K, Thai V N, Hussain S A and Manhrdas S 2020 Synthesis and characterization of amorphous iron oxide nanoparticles by the sonochemical method and their application for the remediation of heavy metals from wastewater *Nanomaterials* 10 1551.

529 Bazzaz Dilmaghani A, Monajjemzadeh F, Afshar Mogaddam M R and Farajzadeh M A 2022 Sonochemical synthesis of deep eutectic solvent-coated magnetic nanoparticles and their application in magnetic dispersive micro solid phase extraction–dispersive liquid–liquid microextraction of polycyclic aromatic hydrocarbons from mascara *Microchem. J.* 181 107665.

530 Moradnia F, Taghavi Fardood S, Ramazani A, Min B, Joo S W and Varma R S 2021 Magnetic $Mg_{0.5}Zn_{0.5}FeMnO_4$ nanoparticles: green sol-gel synthesis, characterization, and photocatalytic applications *J. Clean. Prod.* 288 125632.

531 Jasrotia R, Singh V P, Sharma B, Verma A, Puri P, Sharma R and Singh M 2020 Sol-gel synthesized Ba-Nd-Cd-In nanohexaferrites for high frequency and microwave devices applications *J. Alloys Compd.* 830 154687.

532 Elayakumar K, Manikandan A, Dinesh A, Thanrasu K, Kanmani Raja K, Thilak Kumar R, Slimani Y, Jaganathan S K and Baykal A 2019 Enhanced magnetic property and antibacterial biomedical activity of Ce^{3+} doped $CuFe_2O_4$ spinel nanoparticles synthesized by sol-gel method *J. Magn. Magn. Mater.* 478 140–7.

533 Fotukian S M, Barati A, Soleymani M and Alizadeh A M 2020 Solvothermal synthesis of $CuFe_2O_4$ and Fe_3O_4 nanoparticles with high heating efficiency for magnetic hyperthermia application *J. Alloys Compd.* 816 152548.

534 Zhang H, Wu J-R, Wang X, Li X-S, Wu M-X, Liang F and Yang Y-W 2019 One-pot solvothermal synthesis of carboxylatopillar[5]arene-modified Fe_3O_4 magnetic nanoparticles for ultrafast separation of cationic dyes *Dyes Pigments* 162 512–6.

535 Kalam A, Al-Sehemi A G, Assiri M, Du G, Ahmad T, Ahmad I and Pannipara M 2018 Modified solvothermal synthesis of cobalt ferrite ($CoFe_2O_4$) magnetic nanoparticles photocatalysts for degradation of methylene blue with H_2O_2/visible light *Results Phys.* 8 1046–53.

536 Chen Y, Zhang J, Wang Z and Zhou Z 2019 Solvothermal synthesis of size-controlled monodispersed superparamagnetic iron oxide nanoparticles *Appl. Sci.* 9 5157.

537 Muthukumar K, Lakshmi D S, Acharya S D, Natarajan S, Mukherjee A and Bajaj H C 2018 Solvothermal synthesis of magnetic copper ferrite nano sheet and its antimicrobial studies *Mater. Chem. Phys.* 209 172–9.

538 Dai X, Deng Y, Peng X and Jin Y 2017 Quantum-dot light-emitting diodes for large-area displays: towards the dawn of commercialization *Adv. Mater.* 29 1607022.

539 Pu Y, Cai F, Wang D, Wang J-X and Chen J-F 2018 Colloidal synthesis of semiconductor quantum dots toward large-scale production: a review *Ind. Eng. Chem. Res.* 57 1790–802.

540 Gao X, Yang L, Petros J A, Marshall F F, Simons J W and Nie S 2005 In vivo molecular and cellular imaging with quantum dots *Curr. Opin. Biotechnol.* 16 63–72.

541 Mattoussi H, Mauro J M, Goldman E R, Anderson G P, Sundar V C, Mikulec F V and Bawendi M G 2000 Self-assembly of CdSe–ZnS quantum dot bioconjugates using an engineered recombinant protein *J. Am. Chem. Soc.* 122 12142–50.

542 Peng X, Manna L, Yang W, Wickham J, Scher E, Kadavanich A and Alivisatos A P 2000 Shape control of CdSe nanocrystals *Nature* 404 59–61.

543 Wu S, Dou J, Zhang J and Zhang S 2012 A simple and economical one-pot method to synthesize high-quality water soluble CdTe QDs *J. Mater. Chem.* 22 14573–8.

544 Jamieson T, Bakhshi R, Petrova D, Pocock R, Imani M and Seifalian A M 2007 Biological applications of quantum dots *Biomaterials* 28 4717–32.

545 Park J, Dvoracek C, Lee D K H, Galloway J F, Bhang D H C, Pomper P M G and Searson P P C 2011 CuInSe/ZnS core/shell NIR quantum dots for biomedical imaging *Small Weinh. Bergstr. Ger.* 7 3148.

546 Jiang P, Zhu C-N, Zhang Z-L, Tian Z-Q and Pang D-W 2012 Water-soluble Ag_2S quantum dots for near-infrared fluorescence imaging in vivo *Biomaterials* 33 5130–5.

547 Huang H, Susha A S, Kershaw S V, Hung T F and Rogach A L 2015 Control of emission color of high quantum yield $CH_3NH_3PbBr_3$ perovskite quantum dots by precipitation temperature *Adv. Sci.* 2 1500194.

548 Wang D, Liu J, Chen J-F and Dai L 2016 Surface functionalization of carbon dots with polyhedral oligomeric silsesquioxane (POSS) for multifunctional applications *Adv. Mater. Interfaces* 3 1500439.

549 Deng Z, Cao L, Tang F and Zou B 2005 A new route to zinc-blende CdSe nanocrystals: mechanism and synthesis *J. Phys. Chem. B* 109 16671–5.

550 Acharya K P, Khon E, O'Conner T, Nemitz I, Klinkova A, Khnayzer R S, Anzenbacher P and Zamkov M 2011 Heteroepitaxial growth of colloidal nanocrystals onto substrate films via hot-injection routes *ACS Nano* 5 4953–64.

551 Song W-S, Lee H-S, Lee J C, Jang D S, Choi Y, Choi M and Yang H 2013 Amine-derived synthetic approach to color-tunable InP/ZnS quantum dots with high fluorescent qualities *J. Nanopart. Res.* 15 1–10.

552 Jiang P, Tian Z-Q, Zhu C-N, Zhang Z-L and Pang D-W 2012 Emission-tunable near-infrared Ag_2S quantum dots *Chem. Mater.* 24 3–5.

553 Zhu C-N, Jiang P, Zhang Z-L, Zhu D-L, Tian Z-Q and Pang D-W 2013 Ag_2Se quantum dots with tunable emission in the second near-infrared window *ACS Appl. Mater. Interfaces* 5 1186–9.

554 Protesescu L, Yakunin S, Bodnarchuk M I, Krieg F, Caputo R, Hendon C H, Yang R X, Walsh A and Kovalenko M V 2015 Nanocrystals of cesium lead halide perovskites ($CsPbX_3$, X= Cl, Br, and I): novel optoelectronic materials showing bright emission with wide color gamut *Nano Lett.* 15 3692–6.

555 Wang A, Yan X, Zhang M, Sun S, Yang M, Shen W, Pan X, Wang P and Deng Z 2016 Controlled synthesis of lead-free and stable perovskite derivative Cs_2SnI_6 nanocrystals via a facile hot-injection process *Chem. Mater.* 28 8132–40.

556 Pradhan N and Efrima S 2003 Single-precursor, one-pot versatile synthesis under near ambient conditions of tunable, single and dual band fluorescing metal sulfide nanoparticles *J. Am. Chem. Soc.* 125 2050–1.

557 LaMer V K and Dinegar R H 1950 Theory, production and mechanism of formation of monodispersed hydrosols *J. Am. Chem. Soc.* 72 4847–54.

558 Liu T-Y, Li M, Ouyang J, Zaman M B, Wang R, Wu X, Yeh C-S, Lin Q, Yang B and Yu K 2009 Non-injection and low-temperature approach to colloidal photoluminescent PbS nanocrystals with narrow bandwidth *J. Phys. Chem. C* 113 2301–8.

559 Du Y, Xu B, Fu T, Cai M, Li F, Zhang Y and Wang Q 2010 Near-infrared photoluminescent Ag_2S quantum dots from a single source precursor *J. Am. Chem. Soc.* 132 1470–1.

560 Henglein A 2002 Photochemistry of colloidal cadmium sulfide. 2. Effects of adsorbed methyl viologen and of colloidal platinum *ACS Publ.* 86(13) 2291–3.

561 Rossetti R, Nakahara S and Brus L E 1983 Quantum size effects in the redox potentials, resonance Raman spectra, and electronic spectra of CdS crystallites in aqueous solution *J. Chem. Phys.* 79 1086–8.

562 Rogach A L, Katsikas L, Kornowski A, Su D, Eychmüller A and Weller H 1996 Synthesis and characterization of thiol-stabilized CdTe nanocrystals *Berichte Bunsenges. Für Phys. Chem.* 100 1772–8.

563 Yakoubi A, Chaabane T B, Aboulaich A, Mahiou R, Balan L, Medjahdi G and Schneider R 2016 Aqueous synthesis of Cu-doped CdZnS quantum dots with controlled and efficient photoluminescence *J. Lumin.* 175 193–202.

564 Thomas D, Lee H O, Santiago K C, Pelzer M, Kuti A, Treadwell L J and Bahoura M 2021 Europium doping of cadmium selenide (CdSe) quantum dots via rapid microwave synthesis for optoelectronic applications *Dalton Trans.* 51 264–73.

565 Darr J A, Zhang J, Makwana N M and Weng X 2017 Continuous hydrothermal synthesis of inorganic nanoparticles: applications and future directions *Chem. Rev.* 117 11125–238.

566 Ji W, Liu S, Zhang H, Wang R, Xie W and Zhang H 2017 Ultrasonic spray processed, highly efficient all-inorganic quantum-dot light-emitting diodes *ACS Photonics* 4 1271–8.

567 Katelnikovas A, Steponaviciute M, Ezerskyte E, Drabavicius A and Klimkevicius V 2022 Ultrasound-induced synthesis of all-inorganic lead perovskite quantum dots: fast, simple, and highly reproducible *Mater. Today Chem.* 26 101163.

568 Xue J, Li H, Liu J, Wang Y, Liu Y, Sun D, Wang W, Huang L and Tang J 2019 Facile synthesis of silver sulfide quantum dots by one pot reverse microemulsion under ambient temperature *Mater. Lett.* 242 143–6.

569 Manjunatha C, Hari Krishna R and Ashoka S 2021 Chapter 2 - Green Synthesis of Inorganic Nanoparticles Using Microemulsion Methods *Green Sustainable Process for Chemical and Environmental Engineering and Science* ed Inamuddin, R Boddula, M I Ahamed and A M Asiri (Elsevier) pp 41–67.

570 Tarkas H S, Marathe D M, Mahajan M S, Muntaser F, Patil M B, Tak S R and Sali J V 2017 Synthesis of tin monosulfide (SnS) nanoparticles using surfactant free microemulsion (SFME) with the single microemulsion scheme *Mater. Res. Express* 4 025018.

571 He Y, Lu H-T, Sai L-M, Su Y-Y, Hu M, Fan C-H, Huang W and Wang L-H 2008 Microwave synthesis of water-dispersed CdTe/CdS/ZnS core-shell-shell quantum dots with excellent photostability and biocompatibility *Adv. Mater.* 20 3416–21.

572 Duan J, Song L and Zhan J 2009 One-pot synthesis of highly luminescent CdTe quantum dots by microwave irradiation reduction and their Hg^{2+}-sensitive properties *Nano Res.* 2 61–8.

573 Wang L, Yin Y, Jain A and Zhou H S 2014 Aqueous phase synthesis of highly luminescent, nitrogen-doped carbon dots and their application as bioimaging agents *Langmuir* 30 14270–5.

574 Dai M-Q, Zheng W, Huang Z and Yung L-Y L 2012 Aqueous phase synthesis of widely tunable photoluminescence emission CdTe/CdS core/shell quantum dots under a totally ambient atmosphere *J. Mater. Chem.* 22 16336–45.

575 Borse V, Sadawana M and Srivastava R 2016 CdTe quantum dots: aqueous phase synthesis, stability studies and protein conjugation for development of biosensors *Nanophotonics VI* 9884 (SPIE) 245–56.

576 Qian H, Li L and Ren J 2005 One-step and rapid synthesis of high quality alloyed quantum dots (CdSe–CdS) in aqueous phase by microwave irradiation with controllable temperature *Mater. Res. Bull.* 40 1726–36.

577 Mrad M, Ben Chaabane T, Rinnert H, Lavinia B, Jasniewski J, Medjahdi G and Schneider R 2020 Aqueous synthesis for highly emissive 3-mercaptopropionic acid-capped AIZS quantum dots *Inorg. Chem.* 59 6220–31.

578 Xu Y, Hao J, Niu X, Qi S, Chen H, Wang K, Chen X and Yi T 2016 Seed-mediated growth approach for rapid synthesis of high-performance red-emitting CdTe quantum dots in aqueous phase and their application in detection of highly reactive oxygen species *Chem. Eng. J.* 299 201–8.

579 Su Q, Gan L and Yang X 2021 Achieving room temperature phosphorescence in aqueous phase through rigidifying the triplet state and information encryption *Appl. Surf. Sci.* 566 150726.

580 Tan J, Zhang J, Li W, Zhang L and Yue D 2016 Synthesis of amphiphilic carbon quantum dots with phosphorescence properties and their multifunctional applications *J. Mater. Chem. C* 4 10146–53.

581 Fu Y-S, Du X-W, Kulinich S A, Qiu J-S, Qin W-J, Li R, Sun J and Liu J 2007 Stable aqueous dispersion of ZnO quantum dots with strong blue emission via simple solution route *J. Am. Chem. Soc.* 129 16029–33.

582 Li H, Shih W Y and Shih W-H 2007 Non-heavy-metal ZnS quantum dots with bright blue photoluminescence by a one-step aqueous synthesis *Nanotechnology* 18 205604.

583 Zheng Y, Yang Z and Ying J Y 2007 Aqueous synthesis of glutathione-capped ZnSe and $Zn_{1-x}Cd_xSe$ alloyed quantum dots *Adv. Mater.* 19 1475–9.

584 Chen G, Tian F, Zhang Y, Zhang Y, Li C and Wang Q 2014 Tracking of transplanted human mesenchymal stem cells in living mice using near-infrared Ag_2S quantum dots *Adv. Funct. Mater.* 24 2481–8.

585 Yang H-Y, Zhao Y-W, Zhang Z-Y, Xiong H-M and Yu S-N 2013 One-pot synthesis of water-dispersible Ag_2S quantum dots with bright fluorescent emission in the second near-infrared window *Nanotechnology* 24 055706.

586 Wang C, Wang Y, Xu L, Zhang D, Liu M, Li X, Sun H, Lin Q and Yang B 2012 Facile aqueous-phase synthesis of biocompatible and fluorescent Ag_2S nanoclusters for bioimaging: tunable photoluminescence from red to near infrared *Small* 8 3137–42.

587 Ren Q, Ma Y, Zhang S, Ga L and Ai J 2021 One-step synthesis of water-soluble silver sulfide quantum dots and their application to bioimaging *ACS Omega* 6 6361–7.

588 Lu F, Ju W, Zhao N, Zhao T, Zhan C, Wang Q, Fan Q and Huang W 2020 Aqueous synthesis of PEGylated Ag_2S quantum dots and their in vivo tumor targeting behavior *Biochem. Biophys. Res. Commun.* 529 930–5.

589 Mustafa S M, Barzinjy A A, Hamad A H and Hamad S M 2022 Biosynthesis of quantum dots and their usage in solar cells: insight from the novel researches *Int. Nano Lett.* 12 139–51.

590 Labrenz M, Druschel G K, Thomsen-Ebert T, Gilbert B, Welch S A, Kemner K M, Logan G A, Summons R E, Stasio G D, Bond P L, Lai B, Kelly S D and Banfield J F 2000 Formation of sphalerite (ZnS) deposits in natural biofilms of sulfate-reducing bacteria *Science* 290 1744–7.

591 Bai H-J, Zhang Z-M and Gong J 2006 Biological synthesis of semiconductor zinc sulfide nanoparticles by immobilized *Rhodobacter sphaeroides Biotechnol. Lett.* 28 1135–9.

592 Prasad K and Jha A K 2010 Biosynthesis of CdS nanoparticles: an improved green and rapid procedure *J. Colloid Interface Sci.* 342 68–72.

593 Sweeney R Y, Mao C, Gao X, Burt J L, Belcher A M, Georgiou G and Iverson B L 2004 Bacterial biosynthesis of cadmium sulfide nanocrystals *Chem. Biol.* 11 1553–9.

594 Mi C, Wang Y, Zhang J, Huang H, Xu L, Wang S, Fang X, Fang J, Mao C and Xu S 2011 Biosynthesis and characterization of CdS quantum dots in genetically engineered *Escherichia coli J. Biotechnol.* 153 125–32.

595 Gallardo-Benavente C, Carrión O, Todd J D, Pieretti J C, Seabra A B, Durán N, Rubilar O, Pérez-Donoso J M and Quiroz A 2019 Biosynthesis of CdS quantum dots mediated by volatile sulfur compounds released by antarctic *Pseudomonas fragi Front. Microbiol.* 10 https://doi.org/10.3389/fmicb.2019.01866.

596 Qin Z, Yue Q, Liang Y, Zhang J, Zhou L, Hidalgo O B and Liu X 2018 Extracellular biosynthesis of biocompatible cadmium sulfide quantum dots using *Trametes versicolor J. Biotechnol.* 284 52–6.

597 Seshadri S, Saranya K and Kowshik M 2011 Green synthesis of lead sulfide nanoparticles by the lead resistant marine yeast, *Rhodosporidium diobovatum Biotechnol. Prog.* 27 1464–9.

598 Öcal N, Ceylan A and Duman F 2020 Intracellular biosynthesis of PbS quantum dots using *Pseudomonas aeruginosa* ATCC 27853: evaluation of antibacterial effects and DNA cleavage activities *World J. Microbiol. Biotechnol.* 36 147.

599 Kumar S A, Ansary A A, Ahmad A and Khan M I 2007 Extracellular biosynthesis of CdSe quantum dots by the fungus, *Fusarium oxysporum J. Biomed. Nanotechnol.* 3 190–4.

600 Cao K, Chen M-M, Chang F-Y, Cheng Y-Y, Tian L-J, Li F, Deng G-Z and Wu C 2020 The biosynthesis of cadmium selenide quantum dots by *Rhodotorula mucilaginosa* PA-1 for photocatalysis *Biochem. Eng. J.* 156 107497.

601 Suresh A K, Doktycz M J, Wang W, Moon J-W, Gu B, Meyer H M, Hensley D K, Allison D P, Phelps T J and Pelletier D A 2011 Monodispersed biocompatible silver sulfide nanoparticles: facile extracellular biosynthesis using the γ-proteobacterium, *Shewanella oneidensis Acta Biomater.* 7 4253–8.

602 Raouf Hosseini M and Nasiri Sarvi M 2015 Recent achievements in the microbial synthesis of semiconductor metal sulfide nanoparticles *Mater. Sci. Semicond. Process.* 40 293–301.

603 Liu J, Zheng D, Zhong L, Gong A, Wu S and Xie Z 2021 Biosynthesis of biocompatibility Ag_2Se quantum dots in *Saccharomyces cerevisiae* and its application *Biochem. Biophys. Res. Commun.* 544 60–4.

604 Fatimah I, Sahroni I, Muraza O and Doong R 2020 One-pot biosynthesis of SnO_2 quantum dots mediated by *Clitoria ternatea* flower extract for photocatalytic degradation of rhodamine B *J. Environ. Chem. Eng.* 8 103879.

605 Dai H 2002 Carbon nanotubes: opportunities and challenges *Surf. Sci.* 500 218–41.

606 Dineshkumar B, Krishnakumar K, Bhatt A R, Paul D, Cherian J, John A and Suresh S 2015 Single-walled and multi-walled carbon nanotubes based drug delivery system: cancer therapy: a review *Indian J. Cancer* 52 262.

607 Cassell A M, Raymakers J A, Kong J and Dai H 1999 Large scale CVD synthesis of single-walled carbon nanotubes *J. Phys. Chem. B* 103 6484–92.

608 Kukovecz Á, Kozma G and Kónya Z 2013 Multi-Walled Carbon Nanotubes *Springer Handbook of Nanomaterials* Springer Handbooks ed R Vajtai (Berlin, Heidelberg: Springer) pp 147–88.

609 Liew K M, Wong C H, He X Q, Tan M J and Meguid S A 2004 Nanomechanics of single and multiwalled carbon nanotubes *Phys. Rev. B* 69 115429.

610 Lim S, Luo Z, Shen Z and Lin J 2010 Plasma-assisted synthesis of carbon nanotubes *Nanoscale Res. Lett.* 5 1377.

611 Iijima S 1991 Helical microtubules of graphitic carbon *Nature* 354 56.

612 Arora N and Sharma N N 2014 Arc discharge synthesis of carbon nanotubes: comprehensive review *Diam. Relat. Mater.* 50 135–50.

613 Purohit R, Purohit K, Rana S, Rana R S and Patel V 2014 Carbon nanotubes and their growth methods *Procedia Mater. Sci.* 6 716–28.

614 Farhat S and Scott C D 2006 Review of the arc process modeling for fullerene and nanotube production *J. Nanosci. Nanotechnol.* 6 1189–210.

615 Ando Y 2010 Carbon nanotube: the inside story *J. Nanosci. Nanotechnol.* 10 3726–38.

616 Mohammad M I, Moosa A A, Potgieter J H and Ismael M K 2013 Carbon nanotubes synthesis via arc discharge with a Yttria catalyst *Int. Sch. Res. Not.* 2013 e785160.

617 Wang X K, Lin X W, Dravid V P, Ketterson J B and Chang R P H 1995 Carbon nanotubes synthesized in a hydrogen arc discharge *Appl. Phys. Lett.* 66 2430–2.

618 Rouf S A, Usman Z, Masood H T, Majeed A M, Sarwar M, Abbas W, Rouf S A, Usman Z, Masood H T, Majeed A M, Sarwar M and Abbas W 2021 *Synthesis and Purification of Carbon Nanotubes* (IntechOpen).

619 Roch A, Jost O, Schultrich B and Beyer E 2007 High-yield synthesis of single-walled carbon nanotubes with a pulsed arc discharge technique *Phys. Status Solidi B* 244 3907–10.

620 Joseph Berkmans A, Jagannatham M, Rohit Reddy D and Haridoss P 2015 Synthesis of thin bundled single walled carbon nanotubes and nanohorn hybrids by arc discharge technique in open air atmosphere *Diam. Relat. Mater.* 55 12–5.

621 Zhao S, Hong R, Luo Z, Lu H and Yan B 2011 Carbon nanostructures production by AC arc discharge plasma process at atmospheric pressure *J. Nanomater.* 2011 14.

622 Parkansky N, Boxman R L, Alterkop B, Zontag I, Lereah Y and Barkay Z 2004 Single-pulse arc production of carbon nanotubes in ambient air *J. Phys. Appl. Phys.* 37 2715.

623 Tsai Y Y, Su J S, Su C Y and He W H 2009 Production of carbon nanotubes by single-pulse discharge in air *J. Mater. Process. Technol.* 209 4413–6.

624 Gang X, Shenli J, Jian X and Zongqian S 2007 Analysis of the carbon nano-structures formation in liquid arcing *Plasma Sci. Technol.* 9 770.

625 Corbella C, Portal S, Zolotukhin D B, Martinez L, Lin L, Kundrapu M N and Keidar M 2019 Pulsed anodic arc discharge for the synthesis of carbon nanomaterials *Plasma Sources Sci. Technol.* 28 045016.

626 Zhang D, Ye K, Yao Y, Liang F, Qu T, Ma W, Yang B, Dai Y and Watanabe T 2019 Controllable synthesis of carbon nanomaterials by direct current arc discharge from the inner wall of the chamber *Carbon* 142 278–84.

627 Thess A, Lee R, Nikolaev P, Dai H, Petit P, Robert J, Xu C, Lee Y H, Kim S G, Rinzler A G, Colbert D T, Scuseria G E, Tománek D, Fischer J E and Smalley R E 1996 Crystalline ropes of metallic carbon nanotubes *Science* 273 483–7.

628 Niu K-Y, Sun J, Yang J and Du X-W 2012 The synthesis of carbon nanotubes by pulsed-laser ablation of a nickel/carbon composite target in ethanol or ambient air *Sci. Adv. Mater.* 4 463–6.

629 Chrzanowska J, Hoffman J, Małolepszy A, Mazurkiewicz M, Kowalewski T A, Szymanski Z and Stobinski L 2015 Synthesis of carbon nanotubes by the laser ablation method: effect of laser wavelength *Phys. Status Solidi B* 252 1860–7.

630 Li W Z, Xie S S, Qian L X, Chang B H, Zou B S, Zhou W Y, Zhao R A and Wang G 1996 Large-scale synthesis of aligned carbon nanotubes *Science* 274 1701–3.

631 Rathinavel S, Priyadharshini K and Panda D 2021 A review on carbon nanotube: an overview of synthesis, properties, functionalization, characterization, and the application *Mater. Sci. Eng. B* 268 115095.

632 Kang J, Li J, Du X, Shi C, Zhao N and Nash P 2008 Synthesis of carbon nanotubes and carbon onions by CVD using a Ni/Y catalyst supported on copper *Mater. Sci. Eng. A* 475 136–40.

633 Ding E-X, Jiang H, Zhang Q, Tian Y, Laiho P, Hussain A, Liao Y, Wei N and Kauppinen E I 2017 Highly conductive and transparent single-walled carbon nanotube thin films from ethanol by floating catalyst chemical vapor deposition *Nanoscale* 9 17601–9.

634 Ahmad S, Liao Y, Hussain A, Zhang Q, Ding E-X, Jiang H and Kauppinen E I 2019 Systematic investigation of the catalyst composition effects on single-walled carbon nanotubes synthesis in floating-catalyst CVD *Carbon* 149 318–27.

635 Zhou W, Ding L and Liu J 2009 Role of catalysts in the surface synthesis of single-walled carbon nanotubes *Nano Res.* 2 593–8.

636 He M, Chernov A I, Obraztsova E D, Jiang H, Kauppinen E I and Lehtonen J 2013 Synergistic effects in FeCu bimetallic catalyst for low temperature growth of single-walled carbon nanotubes *Carbon* 52 590–4.

637 Escobar M, Moreno M S, Candal R J, Marchi M C, Caso A, Polosecki P I, Rubiolo G H and Goyanes S 2007 Synthesis of carbon nanotubes by CVD: effect of acetylene pressure on nanotubes characteristics *Appl. Surf. Sci.* 254 251–6.

638 Yang Y, Zhang H and Yan Y 2019 Synthesis of CNTs on stainless steel microfibrous composite by CVD: effect of synthesis condition on carbon nanotube growth and structure *Compos. Part B Eng.* 160 369–83.

639 Kuwana K and Saito K 2005 Modeling CVD synthesis of carbon nanotubes: nanoparticle formation from ferrocene *Carbon* 43 2088–95.

640 Allaedini G, Tasirin S M and Aminayi P 2015 Synthesis of CNTs via chemical vapor deposition of carbon dioxide as a carbon source in the presence of NiMgO *J. Alloys Compd.* 647 809–14.

641 Verma B, Sewani H and Balomajumder C 2020 Synthesis of carbon nanotubes via chemical vapor deposition: an advanced application in the management of electroplating effluent *Environ. Sci. Pollut. Res.* 27 14007–18.

642 Kumar U and Yadav B C 2019 Synthesis of carbon nanotubes by direct liquid injection chemical vapor deposition method and its relevance for developing an ultra-sensitive room temperature based CO_2 sensor *J. Taiwan Inst. Chem. Eng.* 96 652–63.

643 Wu S-L, Chen C-M, Kuo J-H and Wey M-Y 2020 Synthesis of carbon nanotubes with controllable diameter by chemical vapor deposition of methane using Fe@Al_2O_3 core–shell nanocomposites *Chem. Eng. Sci.* 217 115541.

644 Prasek J, Drbohlavova J, Chomoucka J, Hubalek J, Jasek O, Adam V and Kizek R 2011 Methods for carbon nanotubes synthesis – review *J. Mater. Chem.* 21 15872–84.

645 Meyyappan M, Delzeit L, Cassell A and Hash D 2003 Carbon nanotube growth by PECVD: a review *Plasma Sources Sci. Technol.* 12 205.

646 Chen M, Chen C-M and Chen C-F 2002 Preparation of high yield multi-walled carbon nanotubes by microwave plasma chemical vapor deposition at low temperature *J. Mater. Sci.* 37 3561–7.

647 Teo K B K, Lee S-B, Chhowalla M, Semet V, Binh V T, Groening O, Castignolles M, Loiseau A, Pirio G, Legagneux P, Pribat D, Hasko D G, Ahmed H, Amaratunga G A J and Milne W I 2003 Plasma enhanced chemical vapour deposition carbon nanotubes/nanofibres – how uniform do they grow *Nanotechnology* 14 204.

648 Boskovic B O, Stolojan V, Khan R U A, Haq S and Silva S R P 2002 Large-area synthesis of carbon nanofibres at room temperature *Nat. Mater.* 1 165–8.

649 Hussain S, Kovacevic E, Amade R, Berndt J, Pattyn C, Dias A, Boulmer-Leborgne C, Ammar M-R and Bertran-Serra E 2018 Plasma synthesis of polyaniline enrobed carbon nanotubes for electrochemical applications *Electrochim. Acta* 268 218–25.

650 Park D, Kim Y H and Lee J K 2003 Pretreatment of stainless steel substrate surface for the growth of carbon nanotubes by PECVD *J. Mater. Sci.* 38 4933–9.

651 Gangele A, Sharma C S and Pandey A K 2017 Synthesis of patterned vertically aligned carbon nanotubes by PECVD using different growth techniques: a review *J. Nanosci. Nanotechnol.* 17 2256–73.

652 Cui L, Chen J, Yang B, Sun D and Jiao T 2015 RF-PECVD synthesis of carbon nanowalls and their field emission properties *Appl. Surf. Sci.* 357 1–7.

653 Shoukat R and Khan M I 2017 Growth of nanotubes using IC-PECVD as benzene carbon carrier *Microsyst. Technol.* 23 5447–53.

654 Tailleur A, Achour A, Djouadi M A, Le Brizoual L, Gautron E and Tristant P 2012 PECVD low temperature synthesis of carbon nanotubes coated with aluminum nitride *Surf. Coat. Technol.* 211 18–23.

655 Park S, Gupta A P, Yeo S J, Jung J, Paik S H, Mativenga M, Kim S H, Shin J H, Ahn J S and Ryu J 2018 Carbon nanotube field emitters synthesized on metal alloy substrate by PECVD for customized compact field emission devices to be used in X-ray source applications *Nanomaterials* 8 378.

656 Yerlanuly Y, Christy D, Van Nong N, Kondo H, Alpysbayeva B, Nemkayeva R, Kadyr M, Ramazanov T, Gabdullin M, Batryshev D and Hori M 2020 Synthesis of carbon nanowalls on the surface of nanoporous alumina membranes by RI-PECVD method *Appl. Surf. Sci.* 523 146533.

657 Krishna A, Gecil Evangeline T, Aravinda L S, Sharath Kumar N, Niranjan Reddy K, Balashanmugam N and Mamilla R S 2020 Synthesis and thermal simulations of novel

encapsulated CNT multifunctional thin-film based nanomaterial of SiO_2-CNT and TiN-CNT by PVD and PECVD techniques for thermal applications *Diam. Relat. Mater.* 109 108029.

658 Sacco L N, van Ginkel H J and Vollebregt S 2022 Synthesis of Carbon nanofibers (CNFs) by PECVD using Ni catalyst printed by spark ablation *2022 IEEE 22nd International Conference on Nanotechnology (NANO)* pp 128–31.

659 Ankit K, Evangeline T G, Aravinda L S, Kumar N S, Sankar M R, Nagahanumaiah K, Reddy N and Balashanmugam N 2022 Encapsulation of CNT Films on Silicon Wafer by DLC Synthesized by PECVD for Application as a Thermal Interface Material *Advances in Micro and Nano Manufacturing and Surface Engineering: Proceedings of AIMTDR 2021* (Singapore: Springer Nature Singapore) pp 95–106.

660 Viculis L M, Mack J J and Kaner R B 2003 A chemical route to carbon nanoscrolls *Science* 299 1361–1.

661 Novoselov K S, Geim A K, Morozov S V, Jiang D, Zhang Y, Dubonos S V, Grigorieva I V and Firsov A A 2004 Electric field effect in atomically thin carbon films *Science* 306 666–9.

662 Datta S S, Strachan D R, Khamis S M and Johnson A T C 2008 Crystallographic etching of few-layer graphene *Nano Lett.* 8 1912–5.

663 Gass M H, Bangert U, Bleloch A L, Wang P, Nair R R and Geim A K 2008 Free-standing graphene at atomic resolution *Nat. Nanotechnol.* 3 676–81.

664 Shukla A, Kumar R, Mazher J and Balan A 2009 Graphene made easy: high quality, large-area samples *Solid State Commun.* 149 718–21.

665 Hiura H, Ebbesen T W, Fujita J, Tanigaki K and Takada T 1994 Role of sp^3 defect structures in graphite and carbon nanotubes *Nature* 367 148–51.

666 Lu X, Yu M, Huang H and Ruoff R S 1999 Tailoring graphite with the goal of achieving single sheets *Nanotechnology* 10 269.

667 Bernhardt T M, Kaiser B and Rademann K 1998 Formation of superperiodic patterns on highly oriented pyrolytic graphite by manipulation of nanosized graphite sheets with the STM tip *Surf. Sci.* 408 86–94.

668 Liu F 2021 Mechanical exfoliation of large area 2D materials from vdW crystals *Prog. Surf. Sci.* 96 100626.

669 Kumar R, Sahoo S, Joanni E, Singh R K, Tan W K, Kar K K and Matsuda A 2019 Recent progress in the synthesis of graphene and derived materials for next generation electrodes of high performance lithium ion batteries *Prog. Energy Combust. Sci.* 75 100786.

670 Jiang F, Yu Y, Feng A and Song L 2018 Effects of ammonia on graphene preparation via microwave assisted intercalation exfoliation method *Ceram. Int.* 44 12763–6.

671 Ma H, Shen Z, Yi M, Ben S, Liang S, Liu L, Zhang Y, Zhang X and Ma S 2017 Direct exfoliation of graphite in water with addition of ammonia solution *J. Colloid Interface Sci.* 503 68–75.

672 Parvez K, Yang S, Feng X and Müllen K 2015 Exfoliation of graphene via wet chemical routes *Synth. Met.* 210 123–32.

673 Guan Z, Shen X, Yu R, Wang Z and Chen L 2016 Chemical intercalation of solvated sodium ions in graphite *Electrochim. Acta* 222 1365–70.

674 Baig Z, Mamat O, Mustapha M, Mumtaz A, Munir K S and Sarfraz M 2018 Investigation of tip sonication effects on structural quality of graphene nanoplatelets (GNPs) for superior solvent dispersion *Ultrason. Sonochem.* 45 133–49.

675 Asgar H, Deen K M, Riaz U, Rahman Z U, Shah U H and Haider W 2018 Synthesis of graphene via ultra-sonic exfoliation of graphite oxide and its electrochemical characterization *Mater. Chem. Phys.* 206 7–11.

676 Bakhshandeh R and Shafiekhani A 2018 Ultrasonic waves and temperature effects on graphene structure fabricated by electrochemical exfoliation method *Mater. Chem. Phys.* 212 95–102.

677 Hernandez Y, Nicolosi V, Lotya M, Blighe F M, Sun Z, De S, McGovern I T, Holland B, Byrne M, Gun'Ko Y K, Boland J J, Niraj P, Duesberg G, Krishnamurthy S, Goodhue R, Hutchison J, Scardaci V, Ferrari A C and Coleman J N 2008 High-yield production of graphene by liquid-phase exfoliation of graphite *Nat. Nanotechnol.* 3 563–8.

678 Narayan R and Kim S O 2015 Surfactant mediated liquid phase exfoliation of graphene *Nano Converg.* 2 20.

679 Liu W W and Wang J N 2011 Direct exfoliation of graphene in organic solvents with addition of NaOH *Chem. Commun.* 47 6888–90.

680 Guardia L, Fernández-Merino M J, Paredes J I, Solís-Fernández P, Villar-Rodil S, Martínez-Alonso A and Tascón J M D 2011 High-throughput production of pristine graphene in an aqueous dispersion assisted by non-ionic surfactants *Carbon* 49 1653–62.

681 Haar S, El Gemayel M, Shin Y, Melinte G, Squillaci M A, Ersen O, Casiraghi C, Ciesielski A and Samorì P 2015 Enhancing the liquid-phase exfoliation of graphene in organic solvents upon addition of *n*-Octylbenzene *Sci. Rep.* 5 16684.

682 Xu Y, Cao H, Xue Y, Li B and Cai W 2018 Liquid-phase exfoliation of graphene: an overview on exfoliation media, techniques, and challenges *Nanomaterials* 8 942.

683 Arao Y, Mori F and Kubouchi M 2017 Efficient solvent systems for improving production of few-layer graphene in liquid phase exfoliation *Carbon* 118 18–24.

684 Lotya M, Hernandez Y, King P J, Smith R J, Nicolosi V, Karlsson L S, Blighe F M, De S, Wang Z, McGovern I T, Duesberg G S and Coleman J N 2009 Liquid phase production of graphene by exfoliation of graphite in surfactant/water solutions *J. Am. Chem. Soc.* 131 3611–20.

685 Wu Z-S, Ren W, Gao L, Zhao J, Chen Z, Liu B, Tang D, Yu B, Jiang C and Cheng H-M 2009 Synthesis of graphene sheets with high electrical conductivity and good thermal stability by hydrogen arc discharge exfoliation *ACS Nano* 3 411–7.

686 Wang G, Wang B, Park J, Wang Y, Sun B and Yao J 2009 Highly efficient and large-scale synthesis of graphene by electrolytic exfoliation *Carbon* 47 3242–6.

687 Ang P K, Wang S, Bao Q, Thong J T L and Loh K P 2009 High-throughput synthesis of graphene by intercalation–exfoliation of graphite oxide and study of ionic screening in graphene transistor *ACS Nano* 3 3587–94.

688 Pingale A D, Owhal A, Katarkar A S, Belgamwar S U and Rathore J S 2021 Facile synthesis of graphene by ultrasonic-assisted electrochemical exfoliation of graphite *Mater. Today Proc.* 44 467–72.

689 Liu F, Wang C, Sui X, Riaz M A, Xu M, Wei L and Chen Y 2019 Synthesis of graphene materials by electrochemical exfoliation: recent progress and future potential *Carbon Energy* 1 173–99.

690 Zhang Y H, Chen Z Y, Ge X M, Liang Y J, Hu S K, Sui Y P and Yu G H 2018 A waterless cleaning method of the Cu foil for CVD graphene growth *Mater. Lett.* 211 258–60.

691 Min B H, Kim D W, Kim K H, Choi H O, Jang S W and Jung H-T 2014 Bulk scale growth of CVD graphene on Ni nanowire foams for a highly dense and elastic 3D conducting electrode *Carbon* 80 446–52.

692 Wang J J, Zhu M Y, Outlaw R A, Zhao X, Manos D M, Holloway B C and Mammana V P 2004 Free-standing subnanometer graphite sheets *Appl. Phys. Lett.* 85 1265–7.

693 Wang J, Zhu M, Outlaw R A, Zhao X, Manos D M and Holloway B C 2004 Synthesis of carbon nanosheets by inductively coupled radio-frequency plasma enhanced chemical vapor deposition *Carbon* 42 2867–72.

694 Hiramatsu M, Shiji K, Amano H and Hori M 2004 Fabrication of vertically aligned carbon nanowalls using capacitively coupled plasma-enhanced chemical vapor deposition assisted by hydrogen radical injection *Appl. Phys. Lett.* 84 4708–10.

695 Zhu M, Wang J, Holloway B C, Outlaw R A, Zhao X, Hou K, Shutthanandan V and Manos D M 2007 A mechanism for carbon nanosheet formation *Carbon* 45 2229–34.

696 Wang C, Yang S, Wang Q, Wang Z and Zhang J 2008 Super-low friction and super-elastic hydrogenated carbon films originated from a unique fullerene-like nanostructure *Nanotechnology* 19 225709.

697 Muñoz R and Gómez-Aleixandre C 2013 Review of CVD synthesis of graphene *Chem. Vap. Depos.* 19 297–322.

698 Hussain A, Mehdi S M, Abbas N, Hussain M and Naqvi R A 2020 Synthesis of graphene from solid carbon sources: a focused review *Mater. Chem. Phys.* 248 122924.

699 Wang J, Ren Z, Hou Y, Yan X, Liu P, Zhang H, Zhang H and Guo J 2020 A review of graphene synthesis at low temperatures by CVD methods *New Carbon Mater.* 35 193–208.

700 Huet B and Raskin J-P 2018 Role of Cu foil in-situ annealing in controlling the size and thickness of CVD graphene domains *Carbon* 129 270–80.

701 Shin J H, Kim S H, Kwon S S and Park W I 2018 Direct CVD growth of graphene on three-dimensionally-shaped dielectric substrates *Carbon* 129 785–9.

702 Palneedi H, Park J H, Maurya D, Peddigari M, Hwang G-T, Annapureddy V, Kim J-W, Choi J-J, Hahn B-D, Priya S, Lee K J and Ryu J 2018 Laser irradiation of metal oxide films and nanostructures: applications and advances *Adv. Mater.* 30 1705148.

703 Zhao Y, Han Q, Cheng Z, Jiang L and Qu L 2017 Integrated graphene systems by laser irradiation for advanced devices *Nano Today* 12 14–30.

704 Kumar R, Singh R K, Singh D P, Joanni E, Yadav R M and Moshkalev S A 2017 Laser-assisted synthesis, reduction and micro-patterning of graphene: recent progress and applications *Coord. Chem. Rev.* 342 34–79.

705 Kumar R, Joanni E, Singh R K, da Silva E T S G, Savu R, Kubota L T and Moshkalev S A 2017 Direct laser writing of micro-supercapacitors on thick graphite oxide films and their electrochemical properties in different liquid inorganic electrolytes *J. Colloid Interface Sci.* 507 271–8.

706 Tseng S-F 2018 Picosecond laser micropatterning of graphene films for rapid heating chips *Appl. Surf. Sci.* 450 380–6.

707 Kasischke M, Subaşı E, Bock C, Pham D-V, Gurevich E L, Kunze U and Ostendorf A 2019 Femtosecond laser patterning of graphene electrodes for thin-film transistors *Appl. Surf. Sci.* 478 299–303.

708 Al-Nimr M A 1997 Heat transfer mechanisms during short-duration laser heating of thin metal films *Int. J. Thermophys.* 18 1257–68.

709 Jiang J, Lin Z, Ye X, Zhong M, Huang T and Zhu H 2014 Graphene synthesis by laser-assisted chemical vapor deposition on Ni plate and the effect of process parameters on uniform graphene growth *Thin Solid Films* 556 206–10.

710 Tu R, Liang Y, Zhang C, Li J, Zhang S, Yang M, Li Q, Goto T, Zhang L, Shi J, Li H, Ohmori H, Kosinova M and Basu B 2018 Fast synthesis of high-quality large-area graphene by laser CVD *Appl. Surf. Sci.* 445 204–10.

711 Wang K, Tai G, Wong K H, Lau S P and Guo W 2011 Ni induced few-layer graphene growth at low temperature by pulsed laser deposition *AIP Adv.* 1 022141.

712 Xiong W, Zhou Y S, Hou W J, Jiang L J, Gao Y, Fan L S, Jiang L, Silvain J F and Lu Y F 2014 Direct writing of graphene patterns on insulating substrates under ambient conditions *Sci. Rep.* 4 4892.

713 Mikheev K G, Zonov R G, Mogileva T N, Fateev A E and Mikheev G M 2021 Optical anisotropy of laser-induced graphene films *Opt. Laser Technol.* 141 107143.

714 Lee S, Toney M F, Ko W, Randel J C, Jung H J, Munakata K, Lu J, Geballe T H, Beasley M R, Sinclair R, Manoharan H C and Salleo A 2010 Laser-synthesized epitaxial graphene *ACS Nano* 4 7524–30.

715 Wei D, Mitchell J I, Tansarawiput C, Nam W, Qi M, Ye P D and Xu X 2013 Laser direct synthesis of graphene on quartz *Carbon* 53 374–9.

716 Rao C N R, Matte H S S R and Subrahmanyam K S 2013 Synthesis and selected properties of graphene and graphene mimics *Acc. Chem. Res.* 46 149–59.

717 Kumar N, Salehiyan R, Chauke V, Joseph Botlhoko O, Setshedi K, Scriba M, Masukume M and Sinha Ray S 2021 Top-down synthesis of graphene: a comprehensive review *FlatChem* 27 100224.

2

Gold and Silver Nanoparticles for Biomedical Applications

Ajit Khosla[1], Irshad A. Wani[2], and Mohammad N. Lone[3]

[1] *School of Advanced Materials and Nanotechnology, Xidian University, Xian, Shaanxi, China*
[2] *Department of Chemistry, Govt. Degree College Anantnag, Anantnag, University of Kashmir, Jammu and Kashmir, India*
[3] *Department of Chemistry, Central University of Kashmir, Ganderbal, Jammu and Kashmir, India*

2.1 Biomedical Applications of Gold Nanoparticles

The clinical applications of gold and silver nanoparticles (NPs) have been significantly influenced by their electronic structure, which plays a central role in shaping their utility in the medical field. These NPs possess a higher number of electrons per atom on their surface, due to which they absorb X-rays almost 1000 times more than absorbed by the soft tissues of the animal body [1]. This property of Au and AgNPs enables them to be used in cancer cell labeling and in enhanced radiotherapy [2]. Both gold and silver nanoparticles exhibit a unique optical characteristic referred to as surface plasmon resonance (SPR). This feature allows them to absorb and scatter light within the near-infrared (NIR) range (650–900 nm) [3]. This SPR is contingent upon the size and shape of the NPs, surface ligands, solvent, etc., and creates a localized enhanced electric field at the Au and AgNP surfaces, which are then converted into heat [4, 5]. The exceptional light-sensitizing properties of gold and silver nanoparticles, specifically through localized surface plasmon resonance (LSPR), make them highly effective for applications in medical fields, such as photothermal therapy [5, 6]. The remarkable chemical and physical attributes of gold and silver nanoparticles position them as outstanding candidates for a multitude of applications, including but not limited to diagnosis, detection, drug delivery, biolabeling, biological sensing, and imaging [7–9]. The dominant properties like size- and shape-dependent optical properties, substantial surface area-to-volume ratio, good biocompatibility, and minimal toxicity have made them an important tool in biotechnology [10].

The exciting physicochemical properties of AuNPs have facilitated their profound use in biomedical engineering [11]. The biocompatibility and colloidal stability of AuNPs have been increased by functionalization with various organic moieties, making them more beneficial for biomedical applications. Further, the optical properties of AuNPs can be toned by creating various morphologies such as Au nanorods [12], nanostars [13], nanocages [14], and nanoshells [15]. The optical characteristics of gold nanoparticles can be deliberately manipulated by creating assembled structures and adjusting the particle spacing within these assemblies [16, 17].

Keeping in view the potential of AuNPs in biomedical field, this section will discuss the latest advances of AuNPs for use in the biomedical field. Some of the important biomedical applications of AuNPs are discussed below.

Metallic, Magnetic and Carbon-Based Nanomaterials: Synthesis and Biomedical Applications, First Edition.
Ajit Khosla, Irshad A. Wani, and Mohammad N. Lone.

2.1.1 Biomedical Imaging

Owing to the tunable optical properties of AuNPs and their desirable sizes, they have received immense interest for use in biomedical imaging. They have employed biomedical imaging techniques such as dark-field microscopy [18], magnetic resonance imaging (MRI) [19, 20], photoacoustic imaging [21, 22], X-ray computer tomography (CT) [20], and fluorescence imaging [23]. Gold NPs, owing to their least toxicity and tunable SPR to NIR region, have overcome the limitations of organic fluorophores like narrow excitation and broad emission spectra and quantum dots such as toxicity due to leaching of toxic metal ions for use in color-based biomedical imaging. Gold NPs are also immune to blinking and photobleaching. The good scattering properties of AuNPs as compared to conventional fluorophores and SPR in the IR region have made them a favorite candidate for dark-field cell imaging applications.

The iodinated contrast agents in CT undergo rapid secretion through kidneys, which results in their shorter circulation times, and creates problems in recording sensitive CT imaging. However, the nanosized Au provides longer circulation times and better tissue specificity after proper surface modification. Gold NPs offer better absorption than iodine, and their low interference with bone and tissue contributes to improved contrast at lower X-ray doses [24]. Although the excretion route is the same for both AuNPs and iodinated contrast agents, their slow clearance rate results in prolonged imaging times. Further, AuNPs induce better contrast as compared to iodinated contrast agents attributed to their high atomic number (79) and K-edge value (80.7 KeV). The optimal size and surface chemistry are the important factors that affect the target efficiency and circulation times. For example, AuNPs of the size of 2 nm and hydrodynamic diameters of 9–10 nm functionalized with PEGylated (tetra ethylene glycol) were found to have longer circulation times and accumulated efficiently in tumors as compared to other charged ligands [25]. However, they also accumulated significantly in reticuloendothelial systems such as liver and kidneys. Such retention in reticuloendothelial systems results in health hazards that hamper the clinical translation of such NPs. The retention of these PEGylated AuNPs is due to their size greater than the kidney filtration threshold of around 5.5 nm [26]. Sun et al. have reported the glycol chitosan (GC) polymers (GC-AuNPs) modified AuNPs as probes for the dual optical/CT imaging of cancer cells [27]. GC AuNPs were then conjugated with matrix metalloproteinase (MMP) peptides, resulting in MMP-GC-AuNPs for fluorescence optical imaging (Figure 2.1A). X-ray absorption showed enhanced contrast results while using MMP-GC-AuNPs compared to commercially available iodine contrast agents (eXIA^{TM}160) while visualizing as spots in phantom CT imaging. Further, the gold particle surface MMP-GC-AuNPs quenched the NIR fluorescence by 95% compared to fluorescence resonance energy transfer (FRET) and 90% by Cy5.5 labeled MMP peptide probe. MMP-GC-AuNPs were injected into mice at high concentrations (200 μL and 300 mg/kg) for dual CT/optical imaging of the HT-29 tumor cells. After 2 hours of injection, the tumor cells were delineated from the normal cells, and the HU (HU Value-25) value was found to increase five times after 24 hours of injection compared to normal tissues. These MMP-GC-AuNPs exhibited longer circulation in the body, thereby facilitating tumor targeting due to enhanced permeability and retention (EPR) effect as compared to iodine based contrast agents. The increase in near-infrared fluorescence (NIRF) signal was detected in HT-29 tumor of the test mice after one hour of intravenous injection of MMP-GC-AuNPs. The maximum NIRF signal was detected after four hours of injection, with the help of which the tumor was easily detected. The NIRF was retained for 48 hours. Histological assay of the dissected mice proved the NIRF signal initiated from MMP-GC-AuNPs. Confirmation of the presence of MMP-GC-AuNPs within tumor cells was established through dark-field microscopy and fluorescence microscopy using antibody-stained MMP enzymes. Thus, an AuNP-based imaging probe proved to be a good tool for imaging the tumor tissues

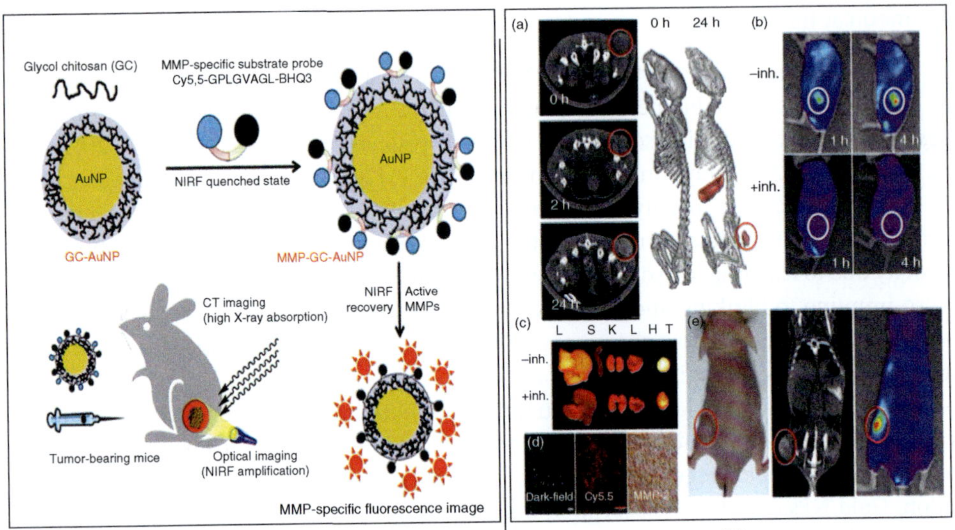

(A) (B)

Figure 2.1 (A) Illustration depicting GC-coated (GC-AuNPs) and MMP peptide probe-conjugated gold nanoparticles (MMP-GC-AuNPs) serving as multimodal CT/optical imaging agents for cancer imaging. (B) (a) Cross-sectional CT scans of tumors before (0 h) and after (2 h, 24 h) tail-vein administration of GC-AuNPs, along with their 3D reconstructed images. (b) NIRF tomographic images of HT-29 tumor-bearing mice postinjection of MMP-GC-AuNPs with and without an inhibitor (blue: low intensity, red: high intensity). (c) NIRF images of excised organs 4 h postinjection, including liver (L), spleen (S), kidney (K), lung (L), heart (H), and tumor (T). (d) Dark-field (left), fluorescence (middle), and MMP-2 enzyme-stained (right) microscopic images of excised HT-29 tumor tissue, with a scale bar indicating 50 mm. (e) Dual CT/optical imaging of the same HT-29 tumor-bearing mouse model 24 h postinjection. Source: (b) Reproduced with permission from Sun et al. [27]/John Wiley & Sons.

using CT and optical imaging (Figure 2.1B). Peng et al. [28] have reported the synthesis of amine-terminated poly(amidoamine) dendrimer-coated AuNPs and improved with PEG monomethyl ether ($G_5.NH_2$-$mPEG_{20}$) for applications in *in vivo* CT imaging. The stability of PEGylated gold dendrimer-encapsulated nanoparticles (Au DENPs) was maintained within a pH range of 5–8 and a temperature range of 0–50 °C. The PEGylated gold dendrimer-encapsulated nanoparticles (Au DENPs) were employed for CT imaging of the blood pools in mice and rats. Following intravenous injection via the tail, the presence of these nanoparticles could be readily visualized within the parenchyma of rats, attributed to the high attenuation coefficient of parenchymal tissue. Furthermore, the PEGylated Au DENPs facilitated blood pool imaging in mice, allowing for the easy identification of various veins, including pulmonary veins, hepatic veins, and inferior vena cava in CT images. Such observations cannot be recorded in mice or rats when omnipaque-iodine was used with concentration like that of PEGylated Au DENPs. The excellent biocompatibility of PEGylated Au DENPs prompted the authors to use these NPs for their applications in *in vivo* CT imaging of the xenograft of mouse tumor model (SPC A1). These PEGylated Au DENPs were injected via two models viz, intraperitoneal and intravenous, and the tumor sites exhibited higher CT values than the tumor before injection. PEGylated Au DENPs diffused into the entire tumor area, and the tumor margins even turned brighter and clearer. The CT values were found to be high even after six hours of injection. Thus, the results indicated that PEGylated Au DENPs could be easily transported to the tumor site irrespective of the administration route. Further, the biodistribution results suggested that the spleen has a higher uptake of AuNPs than the liver and kidney. The findings indicated that PEGylated gold dendrimer-encapsulated nanoparticles (Au DENPs) have the capability to evade the reticuloendothelial system (RES) of the liver and traverse the renal filtration system. Additionally, tumor tissues exhibited absorption of AuNPs at concentrations of 2.3 and 4.7 mg/g for intraperitoneal and intravenous injections, respectively. PEGylated gold nanoparticles (AuNPs) have been employed as promising contrast agents in spectral photon counting computer tomography (SPCCT), also recognized as K-edge imaging [29]. *In vivo*, SPCCT was conducted to quantitatively measure the concentration of Au nanoparticles (NPs) as contrast agents in various organs of white rabbits over time. The experimental animal underwent three scans on the day of Au nanoparticle (NPs) injection – once before injection (T0) and twice after injection (T1 and T2). The quantification of contrast agent biodistribution was conducted by scanning the animal at one week (W_1), one month (M_1), and six months (M_6) postinjection CT images revealed a high concentration of Au nanoparticles (NPs) in the blood at eight minutes postinjection (4.76 mg/mL). The concentration values gradually decreased over time, reaching negligible levels in the blood after one week and one month postinjection (0.16 ± 0.42 mg/mL. Whereas Au concentration was low initially (T1 and T2) in other organs like liver, spleen, and kidney. The concentration increased with time, as evident in CT scans after one week, one month, and six months postinjection. The helical scan by SPCCT of the whole body revealed the AuNPs present in the lymph nodes in the same concentration as that of spleen. The Au content in the muscles and brain was found to be negligible due to the large size of PEGylated AuNPs. The effect of the m-PEGylated AuNPs of different sizes (4–152 nm) has been investigated as contrast agents in CT [30]. The contrast generation with CT was investigated for AuNPs of diverse sizes using clinical CT, SPCCT, and micro-CT systems. No difference in CT contrast was observed for AuNPs of various sizes using any of the test CT systems. However, *in vivo* imaging results demonstrated that AuNPs of sizes equal to or smaller than 15 nm have longer circulation times, whereas AuNPs of larger sizes accumulated in the spleen and liver more swiftly.

AuNPs after being modified with elements such as ^{124}I [31], ^{125}I [32], Gd [33], ^{18}F [34], ^{111}In [35], and ^{64}Cu [36] targeting ligands, have been explored as nanocarriers of signaling agents for MRI, positron emission tomography (PET), and single-photon emission computed tomography (SPECT). The main disadvantages of these imaging techniques such as PET and SPECT are the poor spatial solution, uptake of nonspecific radiotracers by the surrounding tissues, long scan times, etc. Such difficulties have been solved by employing multimodal imaging agents. For example, AuNPs have been labeled with ^{125}I and ^{111}In isotopes and functionalized with MMP9 cleavable to act as contrast agent multispectral SPECT imaging with MMP cleavable peptide [35]. This AuNP-based contrast agent showed a distinct contrast between tumors with different MMP9 expression levels. A significant difference in AuNP accumulation was recorded between tumors with high and low MMP9 expression. Au nanocages (Figure 2.2A) have been investigated for pharmacokinetic and *in vivo* cancer targeting using PET imaging techniques by functionalizing them with radioactive $^{64}Cu^{2+}$ ions [37]. The comparison in radiolabeling efficiency was made between PEGylated Au nanocages of two different sizes labeled with $^{64}Cu^{2+}$ ions. Additionally, the passive targeting ability of Au nanocages was assessed through the EPR effect in the EMT-6 mouse mammary tumor model, utilizing direct imaging with CT/PET. Biodistribution studies have unveiled that 55 nm ^{64}Cu DOTA PEG-AuNCs exhibit swift clearance from the blood of C57BL/6, primarily attributed to their rapid uptake by the reticuloendothelial system of the liver and spleen. However, the hepatic accumulation demonstrated a swift reduction, dropping from 60% ID/g one-hour postinjection to 41.6% ID/g after 24 hours of injection. However, the blood retention of 30 nm ^{64}Cu-DOTA-PEG-AuNCs was found to be six times higher than that of 55 nm ^{64}Cu-DOTA-PEG-AuNCs. The spleen exhibited rapid clearance for both 30 and 55 nm ^{64}Cu-DOTA-PEG-AuNCs, retaining less than 50% of the initial concentration

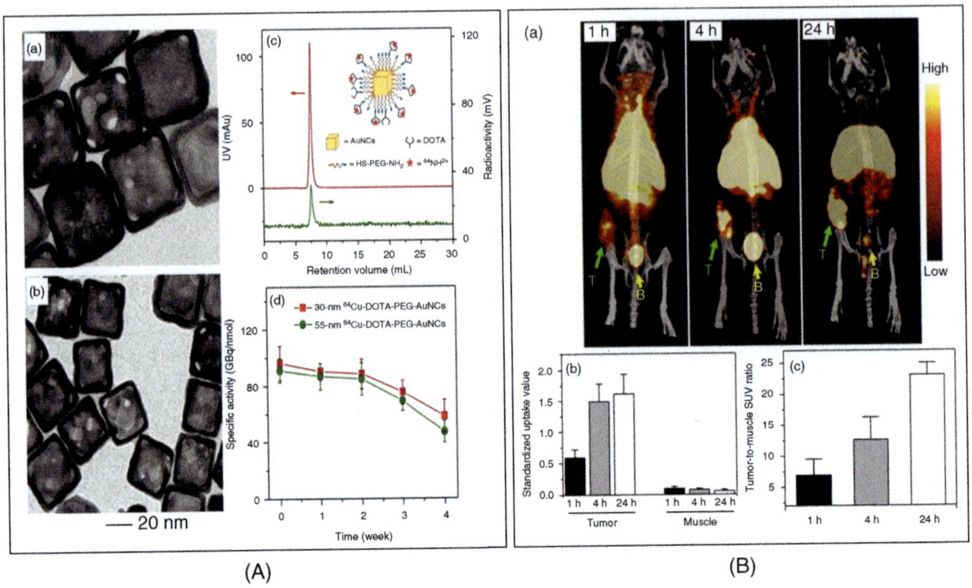

Figure 2.2 (A) Displays representative transmission electron microscopy pictures of DOTA-PEG-AuNCs, showcasing (a) an edge length of 54.5 (±4.4) nm and (b) 30.3 (±4.2) nm. Additionally, (c) fast protein liquid chromatography analysis is presented for the 30 nm 64Cu-DOTA-PEG-AuNCs, and (d) longitudinal specific activities of the 64Cu-DOTA-PEG-AuNCs after incubation at 4 °C for varying durations. (B) (a) PET/CT images illustrate the 30 nm 64Cu-DOTA-PEG-AuNCs in a mouse with an EMT-6 tumor at 1, 4, and 24 h postinjection (3.7 MBq injection/mouse), where T denotes tumor and B denotes the bladder. (b) Standardized uptake values (SUV) in tumor and muscle regions obtained from PET/CT images captured at different time points. (c) Tumor-to-muscle SUV ratios derived from PET/CT images taken at different time intervals. Source: (a, b) Reproduced with permission from Wang et al. [37]/American Chemical Society.

24 hours postinjection. Kidney accumulation was observed to be three times higher for 55 nm ^{64}Cu-DOTA-PEG-AuNCs compared to 30 nm ^{64}Cu-DOTA-PEG-AuNCs during the period from 1 to 24 hours postinjection. The uptake of 30 nm ^{64}Cu-DOTA-PEG-AuNCs by the tumor demonstrated a fourfold increase in comparison to the findings for 55 nm ^{64}Cu-DOTA-PEG-AuNCs. Initial tumor uptake measured 2.68%ID/g after 1 hour of injection, progressively escalating to 7.2% ID/g after 4 hours and eventually reaching 7.9% ID/g after 24 hours. Although 55 nm AuNPs initially displayed higher tumor contrast, the contrast ratio of 30 nm AuNPs was twice that of 55 nm AuNPs, possibly due to higher tumor uptake of smaller-sized AuNPs. The swift localization of 30 nm 64Cu-DOTA-PEG-AuNCs is evident from the PET/CT images of the tumors just one hour postinjection (Figure 2.2B). The standardized uptake values (SUVs) of the AuNPs around the tumor show a significant increase from 1 to 24 hours, which is consistent with the biodistribution profile (Figure 2.2B). Kircher et al. [38] introduced a novel class of nanoparticles known as magnetic particle resonances (MPRs), which are based on gold nanoparticles. These multifunctional nanoparticles integrate MRI, photoacoustic imaging, and Raman imaging. The primary purpose of these nanoparticles is to accurately define the margins of brain tumors in living mice. MPRs, comprising a gold nanoparticle core, a Raman-active layer, a silica shell, and a gadolinium coating, were identified through all three modalities with picomolar sensitivity. Upon injection into mice with glioblastoma, these MPRs accumulated and were retained in tumors, with no noticeable accumulation in the surrounding tissues. The utilization of this triple modality has facilitated enhanced imaging of brain tumors and their resection. Numerous research papers have been published, highlighting the significance of functionalized gold nanoparticles for various *in vivo*

imaging applications [39–43]. The outcomes suggest that gold nanoparticles (AuNPs) can serve as highly effective contrast agents for various *in vivo* imaging applications.

In summary, this section explores the application of gold nanoparticles (AuNPs) in biomedical imaging, highlighting their tunable optical properties and favorable sizes. Several biomedical imaging techniques, such as dark-field microscopy, MRI, photoacoustic imaging, X-ray CT, and fluorescence imaging, have harnessed the unique features of AuNPs. These nanoparticles present benefits like minimal toxicity, SPR in the NIR range, resistance to blinking and photobleaching, and excellent scattering properties, making them well-suited for dark-field cell imaging.

In the context of CT imaging, gold nanoparticles (AuNPs) outperform iodinated contrast agents by showcasing prolonged circulation times and enhanced tissue specificity post-surface modification. The effective application of AuNPs in this domain heavily relies on their optimal size and surface chemistry, influencing target efficiency and circulation durations. Illustrative instances of this encompass the use of PEGylated AuNPs for simultaneous optical and CT imaging of cancer cells and the employment of amine-terminated poly(amidoamine) dendrimer-coated AuNPs for *in vivo* CT imaging purposes. PEGylated AuNPs display notable biocompatibility, extended circulation times, and precise tumor targeting, underscoring the adaptability and promise of AuNPs in advancing biomedical imaging techniques.

Furthermore, AuNPs have been investigated as carriers for signaling agents, including isotopes ^{124}I, ^{125}I, Gd, ^{18}F, ^{111}In, and ^{64}Cu, in MRI, PET, and SPECT. The development of multimodal imaging agents that integrate various imaging modalities has been a focus. For example, AuNPs labeled with ^{125}I and ^{111}In isotopes, functionalized with matrix metalloproteinase (MMP)-cleavable peptides, serve as multispectral SPECT imaging contrast agents, offering distinctive contrast between tumors with varying MMP9 expression levels. Additionally, Au nanocages functionalized with radioactive $^{64}Cu^{2+}$ ions have been explored for PET imaging, showcasing their potential for passive targeting through the EPR effect in tumor models.

Moreover, gold nanoparticles (AuNPs) have played a pivotal role in the advancement of multimodal imaging nanoparticles (MPRs), which integrate gold NPs with Raman-active layers, silica shells, and gadolinium coatings to precisely delineate brain tumor margins in living mice. These MPRs exhibit detectability through MRI, photoacoustic imaging, and Raman imaging with heightened sensitivity and specificity. This innovative approach enhances the imaging and resection of brain tumors. To sum up, this section underscores the adaptability and efficacy of functionalized AuNPs as contrast agents in various *in vivo* imaging applications, demonstrating their potential across diverse imaging modalities within the biomedical field.

2.1.2 AuNPs in Therapeutics

Metallic drugs are currently employed in the field of medicine to clinically treat a range of diseases [44–46]. Numerous metal-based drugs have undergone clinical investigation and are extensively applied, including platinum-derived drugs like cisplatin, tetraplatin, and carboplatin for combating cancer, gold for addressing arthritis, bismuth for treating gastrointestinal and infectious diseases, and gallium for managing cancer-associated hypercalcemia [47–53]. Good biocompatibility of bulk gold, as well as good bioactivity like antibacterial, anticancer, and antirheumatic effects, have prompted the research community to investigate AuNPs in order to treat a variety of diseases [54, 55]. The unique properties of AuNPs, high stability, and distinct optical properties such as SPR have resulted in their use in medicine such as therapeutic agents and drug delivery [56]. AuNP-based medicines have various advantages over conventional therapies such as long circulation times inside the biosystems and longer shelf life. As such, AuNPs have been shown to have various therapeutic effects on various diseases. The antioxidant, antiangiogenic, and anti-inflammatory of AuNPs have made them a versatile candidate against various diseases such as rheumatoid arthritis, cancer, obesity, and muscular degeneration. These diseases are associated with high blood pressure and leaky vasculature, which makes it easy to pass AuNPs into the diseased tissues. Thus, such AuNPs can either reach the target by passive mechanism or can be specially functionalized to possess anti-angiogenic effects. The fractured vasculature at the disease site has pores with a diameter of 200–400 nm, which allow the AuNPs in this size range to easily pass through them to the diseased tissue site. The uptake, biodistribution, localization, and pharmacokinetics of AuNPs vary with their size and shape. The cytotoxic effect of citrate-capped AuNPs of different sizes has been studied by Vechia et al. [57] using various normal and cancerous cell lines like murine fibroblasts (NIH3T3), human erythrocytes, melanoma cells (B16F10) and human cervix carcinoma cells (HeLa). Spherical AuNPs of three sizes viz., 10, 20, and 30 nm have been synthesized by the Turkevisch method. The studies have revealed that AuNPs of 20 nm size exert a higher cytotoxic effect on cancerous HeLa cells, whereas they are least cytotoxic to noncancerous cells. However, 10 nm-sized AuNPs were found to show higher cytotoxicity against B16F10 than 20 and 30 nm-sized AuNPs. HeLa cells exhibited lower IC50 (IC50 2.1 μg/mL) values and hence were found to be more sensitive with AuNPs as compared to B16F10 cells (IC50 > 70 μg/mL). Thus, the cytotoxic effect of AuNPs was found to be dependent on cell type, concentration of AuNPs, and size. Huang et al. [58] conducted a study examining the impact of citrate-capped gold nanoparticles (AuNPs) with three distinct sizes (3, 13, and 45 nm) on rabbit articular chondrocytes. The results demonstrated 13 nm AuNPs of 2 nM concentration induced cell death of chondrocytes as evidenced by flow cytometric analysis (Figure 2.3a–c) and fluorescence imaging. The apoptotic characteristics like mitochondrial damage, oozing out of phosphatidylserine, and nuclear contents were noticed (Figure 2.3d–h). Gold NPs of the size of 3 and 45 nm of the same concentration did not induce cytotoxicity in the chondrocytes. Further, 13 nm AuNPs were found to induce reactive oxygen species (ROS) independent apoptosis as the ROS scavengers proved to be ineffective in preventing cytotoxicity. The results indicated the size-dependent toxicity of chondrocytes. In some cases, AuNPs have been modified with the therapeutic agents directly, such as drug molecules or any other biomolecule, with known inhibitory effects on cell signaling [59–61]. In other cases, AuNPs are encapsulated with such molecules, which facilitate the active targeting of AuNPs toward specific cancerous cells/tissues [62]. Such an association occurs easily on the AuNP surface due to interaction such as thiolation or ionic interaction between oppositely charged entities such as AuNP surface and biomolecules. Such enhanced functionalization affects the physicochemical properties of AuNPs, which in turn affect their mobility, safety, and biocompatibility. Functionalization also prevents nonspecific interaction of AuNPs with proteins, which otherwise result in the creation of protein coronas, leading to premature clearance of AuNPs due to opsonization by phagocytic cells. Even the surface charge affects the behavior of AuNPs in biological systems. AuNPs with neutral surfaces are unreactive and have been found to have a higher rate of opsonization. Even AuNPs with hydrophilic surfaces behave differently than AuNPs with hydrophobic surfaces [63]. For instance, PEGylation of AuNPs improves the biocompatibility of AuNPs by removing the neutrality of AuNP surfaces. The additional functionalization of PEGylated gold nanoparticles (AuNPs) with molecules like polyarginine results in their visualization both on the cell membrane and within the kidney COS-1 cells of the African green monkey [63]. Yassiri et al. [64] have documented the synthesis of mangiferin-encapsulated radioactive AuNPs by using Au-198 gold precursor. The resulting MGF-^{198}AuNPs were used as the therapeutic agent for prostate tumors (PC-3). *In vivo* therapeutic investigations have demonstrated an 80% retention of the injected dose for up to 24 hours postinjection. A fivefold reduction in tumor growth was recorded after three weeks of postinjection of MGF-^{198}AuNPs. Deng et al. [65] have formulated multifunctional AuNPs to deliver anticancer agents like doxorubicin, AS1411, and anti-221 simultaneously to improve

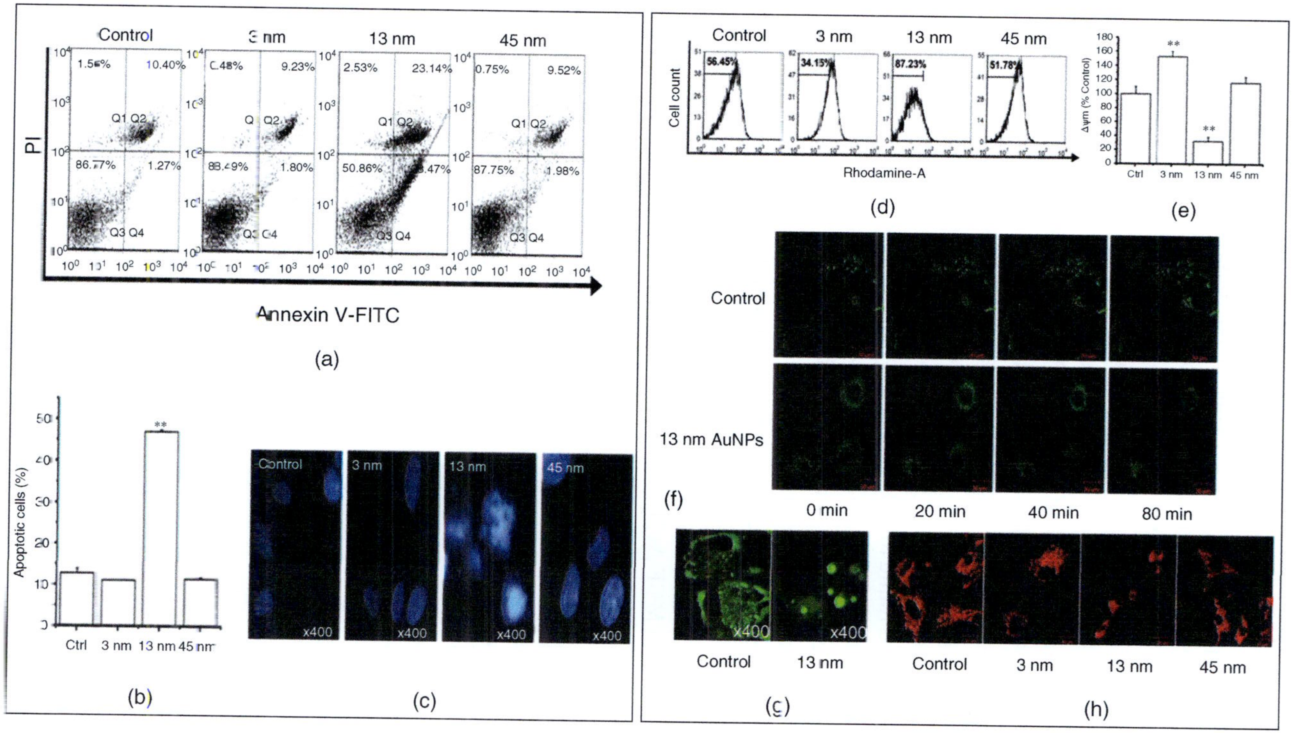

Figure 2.3 Illustrates the induction of apoptosis in rabbit articular chondrocytes by 13 nm AuNPs. (a) Flow cytometry (FCM) analysis depicts annexin V/PI staining for chondrocytes treated with 3, 13, and 45 nm AuNPs. (b) Statistical outcomes from three independent FCM analyses indicating $^{**}P < 0.01$ compared with the control. (c) Fluorescence imaging of chondrocytes stained with Hoechst 33258 after a 24-hour treatment with 3, 13, and 45 nm AuNPs, respectively, with an original magnification of ×400. Further, it is demonstrated that 13 nm AuNPs induce apoptosis through mitochondrial damage. (d) FCM analysis using Rho 123 staining shows the loss of $\Delta\Psi m$ induced by 3, 13, and 45 nm AuNPs. (e) Statistical results from three independent FCM analyses indicate $^{**}P < 0.01$ compared with the control. (f) Time-lapse imaging captures cells stained with Rho 123 after treatment with 13 nm AuNPs for 8 hours, with a scale bar of 20 μm. (g) Fluorescence imaging reveals chondrocytes transfected with GFP-Cyt.c after an 8-hour treatment with 13 nm AuNPs, with an original magnification of ×400. (h) Confocal imaging of chondrocytes stained with Mitotracker Red 633 after 8-hour treatment with 3, 13, and 45 nm AuNPs, respectively, with a scale bar of 20 μm. Source: (a–h) Huang et al. [58]/Springer Nature/CC BY 4.0.

the treatment efficacy of doxorubicin drug-resistant leukemia (K562DR). Here, 40 nm AuNPs were linked with targeting moieties like folate and As141 receptors and therapeutic agents like doxorubicin and anti-mi RNA molecules (anti-221). Folate moieties recognize folate receptors and AS141 aptamers recognize NCL receptors that are overexpressed cancer cell surface. The cellular uptake of AuNPs takes place via receptor-mediated endocytosis. AuNP conjugates, once shuttled inside the cells, shed the cargo (AS1411 aptamer, anti-221, and Dox), which kills the cancerous cell through synergistic action. AS1411 aptamer and anti-221 collectively induced the suppression of endogenous NCL and miT-221 and prevented leukemogenesis by sensitizing the cancerous cells to doxorubicin. Aboyewa et al. [66] have studied the antitumor efficacy of AuNPs synthesized using honey bush tree (HB-AuNPs) and Mangifera (MGF) encapsulated AuNPs in combination with doxorubicin (DOX) and Ayurvedic medicine. The mean size of HB-AuNPs and MGF-AuNPs was recorded to be 20 and 26 nm, respectively. The stability of AuNPs was tested in various biological media and buffers such as BSA, FBS, DPBS, DMEM, and RPMI1640, and the SPR spectra of AuNPs were found to be intact even after 24 hours of incubation in various media. The efficacy of these AuNPs was evaluated against cancer cell lines including Caco-2, U87, and PC-3, as well as noncancerous cell lines such as MCF-12A. The NP dosage was varied from 15.62 to 1000 μg/mL. The MCF-12A cell line was found to be 100% viable when treated with both HB-AuNPs and MGF-AuNPs at all the test concentrations (15.62–250 μg/mL). Even at higher concentrations, the viability of the MCF-12A cell line remained unaffected by the presence of AuNPs. However, both types of AuNPs affected the cancerous cell lines differently. The cell viability of U87 and PC-3 cancerous cell lines was significantly reduced after treatment with both types of AuNPs in the range of 31.25–1000 μg/mL. Whereas Caco-2 cell lines were affected only at higher concentrations of AuNPs (500–1000 μg/mL). The results suggested that Caco-2 cell lines were found to be less susceptible to both HB-AuNPs and MGF-AuNPs than other cell lines. U87 cell line was found to be most susceptible to AuNPs with IC50 values of 85.9 and 121.4 μg/mL for MGF-AuNPs and HB-AuNPs, respectively. Whereas IC^{50} values of AuNPs for other cell lines were found to be significantly higher (1000 μg/mL). Gold nanoparticles modified with two thiol ligands, specifically thioglycolic acid (TGA) and glutathione (GSH), have been utilized as carriers for delivering the anticancer drug 5-fluorouracil (5-FU) [67]. The gold nanoparticles (AuNPs) exhibited sizes within the range of 9–17 nm and were employed with varying ligand/5-fluorouracil (5-FU) molar ratios to assess their anticancer efficacy against a colorectal cancer cell line. A peak loading capacity of 29.34% was attained at the 5-FU/TGA molar ratio of 1 : 1, while a maximum drug loading capacity of 20.95% was achieved for 5-FU/GSH at the molar ratio of 1 : 2. The release of the drug from the functionalized AuNPs was observed to be contingent on the pH level. The drug was released slowly from GSH-NP as compared to TGA-AuNPs at pH of 7 and 8. Upon a decrease in pH from 7 to 5, GSH-NP showed no significant difference in release pattern, whereas significantly slower drug release was observed to occur from TGA-NPs. Further decrease in pH to 3 results in faster release of drug from both the types of AuNPs, and the complete release was noticed after 4 hours at pH of 3. Thus, from this drug release study, GSH-AuNPs were chosen for further cytotoxic studies against cancer cell lines due to the slow drug release profile. It was observed that cells treated with 5-FU/GSH-AuNPs showed significant apoptosis and cell death (39.33%) as compared to cells treated with drug 5-FU alone (19.33%). Thus, the results confirmed the increase in anticancer activity of the 5-FU drug against the colorectal cancer cells when combined with functionalized AuNPs as compared to the free drug alone. Thus, localized drug delivery is one of the applications that AuNPs have been used effectively to allow for selective and localized targeting.

In summary, this section delves into the utilization of metal-based drugs, with a specific focus on gold nanoparticles (AuNPs), in the clinical treatment of diverse diseases within the realm of medicine. Extensive research has been conducted on various metal-based drugs, encompassing platinum derivatives for cancer therapy, gold for arthritis, bismuth for gastrointestinal conditions, and gallium for cancer-related hypercalcemia. The intrinsic biocompatibility and bioactivity of bulk gold have spurred investigations into the therapeutic potential of AuNPs, given their distinct properties, such as high stability and unique optical characteristics, including SPR. These attributes render AuNPs suitable for medicinal applications, functioning as therapeutic agents and drug delivery vehicles. Noteworthy advantages of AuNP-based medicines include prolonged circulation times and extended shelf life, surpassing conventional therapies.

Furthermore, the section underscores the critical influence of size and shape on the uptake, biodistribution, localization, and pharmacokinetics of AuNPs. Studies have elucidated the size-dependent toxicity of differently sized citrate-capped AuNPs on diverse cell lines, highlighting the need for careful consideration in their application. Moreover, efforts have been directed toward functionalizing AuNPs with therapeutic agents, either directly or through encapsulation, to augment their targeting capabilities. The surface properties of AuNPs, including charge and hydrophilicity, play pivotal roles in dictating their behavior within biological systems.

Illustrative studies, such as the synthesis of mangiferin-encapsulated radioactive AuNPs for prostate tumor therapy, multifunctional AuNPs delivering anticancer agents for leukemia treatment, and the exploration of honeybush tree and mangiferin-encapsulated AuNPs combined with doxorubicin and Ayurvedic medicine, exemplify the versatility of AuNPs in medical applications. These investigations showcase the potential of AuNPs for localized drug delivery, demonstrating enhanced targeting capabilities and increased therapeutic efficacy across a spectrum of diseases.

2.1.3 Gold Nanoparticles as Biosensors

The exceptional optical characteristics of gold nanoparticles (AuNPs), combined with their favorable biocompatibility and inert properties, render them an ideal choice for biosensing applications. This segment will offer an in-depth account of employing functionalized gold nanoparticles (AuNPs) for the detection of diverse biomolecular components.

Gold NPs modified with various functional moieties have been used to detect proteins. For instance, Schoefield et al. [68] have employed (2-mercaptoethyl R-D mannopyranoside), a mannose derivative functionalized AuNPs for the detection of lectin concanavalin A (Con A). when the Con A was added to 16 nm-sized AuNPs sol, aggregation occurred which resulted in the shifting of the SPR band and simultaneous color change of the AuNP sol. Mannose stabilized AuNPs showed good specificity with LOD of 0.04 μM against Con A. Narain et al. [69] have used biotinylated AuNPs for the detection of streptavidin. Biotinylated AuNPs were synthesized by RAFT polymerization. The interaction between streptavidin and the biotin ligands present on the surface of gold nanoparticles (AuNPs) was subsequently explored. Spectroscopic monitoring was employed to observe the alterations upon the addition of streptavidin to two distinct types of biotinylated gold nanoparticles (GNP 5 and GNP 6). The introduction of streptavidin to biotinylated gold nanoparticles (AuNPs) led to particle aggregation, whereas no such aggregation was observed when streptavidin was added to non-biotinylated AuNPs. Additionally, there was no discernible change in absorbance over time in the latter case. A bathochromic shift of a few nanometers was observed spectroscopically for the biotinylated AuNPs upon the addition of streptavidin. The findings suggested that the aggregation of biotinylated gold nanoparticles (AuNPs) and the consequent shifts in the SPR position were induced by the binding of biotin ligands with streptavidin. Chandrawati et al. [70] have employed gold nanoparticles (AuNPs) functionalized with peptides for assessing the activity of blood coagulation factor XIII. The changes due to the interaction between the functionalized AuNPs and blood coagulation factor were monitored spectroscopically. Surface functionalization of 15 nm AuNPs was done using two peptide substrates viz., Cys-Ala-Leu-Asn-AsnGly-Gln-Gly

(CALNNGQG) and Cys-Ala-Leu-Asn-Asn-Gly-Lys-Gly (CALNNGKG). AuNPs functionalized with CALNNGQG were observed to exhibit stability at physiological salt concentrations, maintaining a steady SPR peak at 522 nm. In contrast, CALNNGKG demonstrated a notable redshift of 40 nm in the SPR peak, likely attributed to the robust interaction between the amino groups of lysine residues and the surface of gold nanoparticles (AuNPs). Nonfunctionalized AuNPs showed instant aggregation at the same physiological conditions due to the charge screening effect. Later, stable dispersions of lysine-modified AuNPs were synthesized using polylysine (PLL) as an alternative substrate, which showed a characteristic and stable SPR peak at 522 nm. Subsequently, CALNNGQG- and lysine-functionalized gold nanoparticles (AuNPs) were subjected to incubation with factor XIII in the presence of $CaCl_2$ and thrombin, essential for the activation of factor XIII, within a buffer at pH 7.4. Upon the introduction of 0.5 U/mL Factor XIII, a noteworthy red shift in the SPR peak of the AuNP combination occurred, transitioning from 522 to 559 nm. This shift was attributed to the establishment of e-(g-glutamyl)-lysine isopeptide bonds between the lysine residue of PLL and the glutamine residues of CALNNGQG, resulting in crosslinking and altering the interparticle distance between gold nanoparticles (AuNPs). Nevertheless, exposing the AuNP system to only one component of the assay, such as factor XIII or $CaCl_2$ and thrombin, did not result in any shift of the SPR peak. This observation suggests that particle aggregation is contingent upon the activation of factor XIII, rather than the presence of the individual components alone. Chen et al. [71] developed gold nanoparticles (AuNPs) that were functionalized with polypeptides to facilitate the identification of matrix metalloproteinase matrilysin (MMP-7), serving as an indicative marker for salivary gland cancer. The peptide used for functionalization of AuNPs is called JR2EC, which contains two recognition sites for MMP-7. The immobilization of JR2EC on the AuNP surface was facilitated by the introduction of cysteine residue at position 22 of the polypeptide. The stability of gold nanoparticles (AuNPs) immobilized with polypeptides was observed to be maintained at physiological salt concentrations. Nevertheless, the colloidal stability of the AuNP suspension decreased upon the addition of MMP-7, attributed to the hydrolysis of the immobilized peptide on the surface of AuNPs. MMP-7 rapidly triggered the aggregation of gold nanoparticles (AuNPs), causing a noticeable shift in the SPR peak. The addition of 4 μg/mL of MMP-7 to the polypeptide-functionalized gold nanoparticle (AuNP) solution led to a shift in the SPR from 521 to 560 nm, and there was no apparent change in peak intensity. Further, SPR peak intensity increases linearly with MMP-7 concentration up to 8 minutes. However, exponential decay occurred for longer incubation periods of 20 minutes. A nearly linear relationship was identified up to a concentration of 2 μg/mL of MMP-7, noticeable at 4 minutes of incubation time. However, beyond this point, a saturation in the shift occurred. Indeed, this sensor utilizing gold nanoparticles (AuNPs) demonstrates considerable potential for the diagnostic identification of salivary gland cancer. A highly sensitive immunoassay based on antibody-coated AuNPs has been developed by Thanh et al. [72] for the recognition of antibodies. The assay operates through the aggregation of antibodies with antigen-coated silver nanoparticles (AgNPs), leading to a change in the SPR absorption at 520 nm. A color change from red to purple has been reported within 30 minutes. After 2 hours of assay, precipitation of the large-sized AgNPs was observed with clear supernatant. Initially, AgNPs showed an SPR peak at 520 nm, which red-shifted to longer wavelengths (600–700 nm) after 60 minutes. transmission electron microscopy (TEM) images revealed the initial particle size of protein-coated NPs to be 10 nm. With time, aggregation was recorded with large islands of AuNPs seen after 30 minutes of assay, and large micrometric aggregates were seen after 2 hours. Further, this gold NP-based assay recorded a sensitivity as low as 1 μg/mL for antibodies which is comparable to that reported by ELISA technique.

Colorimetric assays based on gold nanoparticles (AuNPs) have been employed to detect diverse oligonucleotide sequences, which play a crucial role in early diagnosis and applications related to drug delivery. For instance, He et al. [73] have produced a DNA probe based on chemically modified oligonucleotide-capped AuNPs. The effective integration of the oligonucleotide onto the gold nanoparticle (AuNP) surface was confirmed through a DNA melting assay. The 12-mer oligonucleotide was used as a probe material on the AuNP surface, exhibiting a complementary sequence to half of the target DNA. The combination of S_3: Au and S_4: Au conjugate produced SPR at 534 nm, whereas the addition of linking nucleotide with a sequence complementary to S_3 and S_4 resulted in flocculation of the Au colloid with a red shift of the SPR peak along with a decline in intensity. These changes in intensity and peak shifting occurred due to the formation of NP aggregates. Recently, Karakus et al. [74] have developed AuNP-based colorimetric sensor for detecting SARS-CoV-2 spike antigen. Antibody-capped AuNPs (AuNP-mAb), when exposed to SARS-CoV-2 spike antigen, resulted in irreversible aggregation of AuNPs due to antigen-antibody interaction. Various quantities of SARS-CoV-2 spike antigen were subjected to incubation with 200 μL of gold nanoparticle (AuNP)-monoclonal antibody (mAb) probe for a duration of 10 minutes. Incubation resulted in the visible color change of the AuNP solution with a corresponding spectral shift of 25 nm. The changes were confirmed by TEM and dynamic light scattering (DLS) studies. TEM images revealed the antigen-antibody interaction at the AuNP surface, resulting in aggregation. DLS analysis indicated that the hydrodynamic size of gold nanoparticles (AuNPs) expanded from 34 to 264 nm after introducing the SARS-CoV-2 spike antigen. UV-Visible spectroscopic studies showed the red shifting of the SPR peak and a decrease in peak intensity due to antigen–antibody interactions (Figure 2.4A). The red shift occurred initially upon the introduction of 259 ng/mL of the antigen, and a linear relationship was observed between the wavelength and the concentration of the SARS-CoV-2 spike antigen (Figure 2.4B). The selectivity of the AuNP-based probe was confirmed by the addition of different types of spike antigens (1 μg/mL) such as *Streptococcus pneumonia* antigen, MERS-CoV antigen, and H1N1 antigen which resulted in no significant change in the color or spectral properties of the AuNP probe system. The schematic representation of the recognition of SARS-CoV-2 spike antigen by AuNP-mAb is shown in Figure 2.4C. Yrad et al. [75] have employed dextran-capped AuNPs as labels for the visually detecting dengue-1 RNA employing the lateral flow technique. The detection principle relies on the nucleic acid sandwich mechanism, utilizing gold nanoparticles (NPs) as probes for identifying a variety of biocomponents such as DNA [76], whole bacterial cells [77–79], antibiotics [80], biomarkers [81], virus particles [82].

In brief, this section focuses on the application of functionalized gold nanoparticles (AuNPs) in biosensing. The distinctive optical characteristics, biocompatibility, and inert properties of gold nanoparticles (AuNPs) render them appropriate choices for applications in biosensing. Various functional moieties have been employed to modify AuNPs for the identification of biomolecular components. Several examples are provided, illustrating the use of functionalized AuNPs for the detection of specific proteins. Functionalization with mannose derivatives enabled the detection of lectin Concanavalin A, while biotinylated AuNPs were utilized for streptavidin detection. Peptide-functionalized AuNPs were employed to determine blood coagulation factor XIII activity, demonstrating changes in spectral properties upon interaction. Polypeptide-functionalized AuNPs were used to detect matrix metalloproteinase matrilysin, indicating potential for salivary gland cancer diagnosis. Additionally, antibody-coated AuNPs were developed for a highly sensitive immunoassay to recognize antibodies. Colorimetric assays based on gold nanoparticles (AuNPs) have been utilized to detect essential oligonucleotide sequences, which play a vital role in early diagnosis and drug delivery applications. Chemically modified oligonucleotide-capped AuNPs were used in a DNA probe-based assay, showcasing changes in spectral properties upon target DNA interaction. Another study developed an AuNP-based colorimetric sensor detecting SARS-CoV-2 spike antigen, demonstrating visible color changes and spectral shifts upon antigen–antibody interaction.

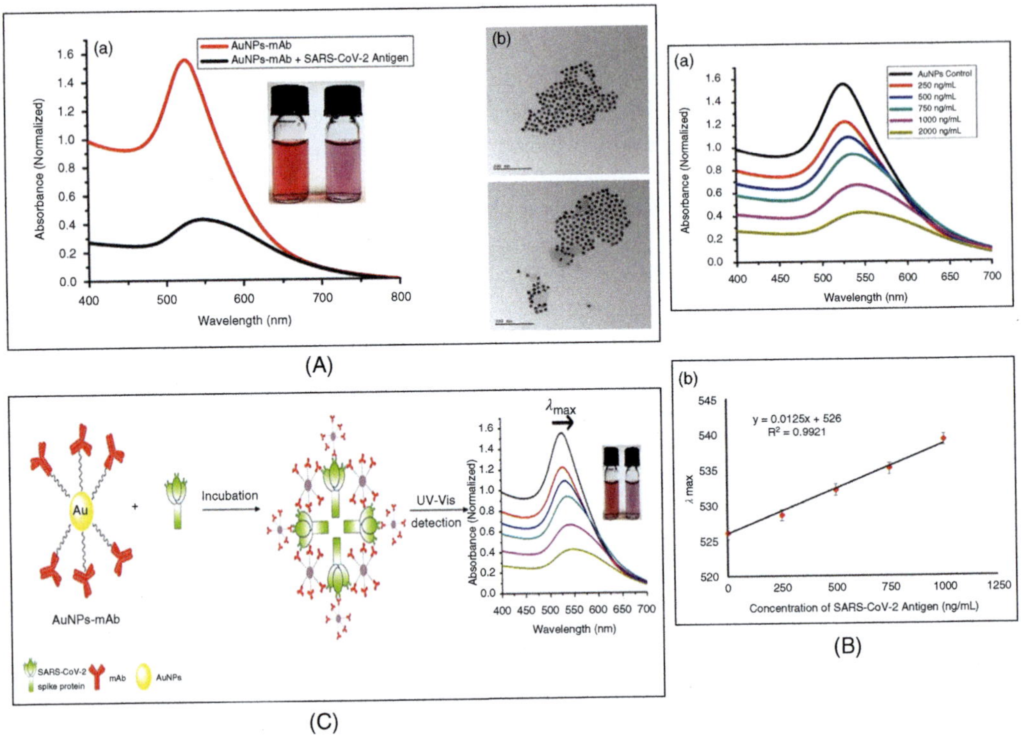

Figure 2.4 (A) Presents (a) the absorption spectra of AuNPs-mAb, comparing the absence (red line) and presence (black line) of 2 mg/mL of the SARS-CoV-2 spike antigen. Additionally, (b) TEM images of AuNPs-mAb postincubation with the SARS-CoV-2 spike antigen are provided, with a scale bar of 200 nm. (B) (a) The UV-Vis spectrum of AuNP-mAb is displayed in response to varying concentrations of the SARS-CoV-2 spike antigen (250, 500, 750, 1000, and 2000 ng/mL), while (b) shows the redshifted wavelength of the LSPR peak with different concentrations of the SARS-CoV-2 spike antigen (250, 500, 750, and 1000 ng/mL). (C) Illustrates the UV-Vis detection mechanism of the SARS-CoV-2 spike antigen based on the aggregation of AuNPs-mAb on the surface of the SARS-CoV-2 spike antigen. Source: (a–c) Reproduced with permission from Karakus [74]/Elsevier.

Furthermore, the section highlights the use of gold nanoparticles as probes for the detection of various biocomponents, including DNA, whole bacterial cells, antibiotics, biomarkers, and virus particles. The versatility of functionalized AuNPs in biosensing applications is emphasized, showcasing their potential for specific and sensitive detection in various diagnostic scenarios.

2.1.4 Gold Nanoparticles in Photothermal Therapy

In various applications, gold nanoparticles (AuNPs) have been employed for photothermal therapy (PTT), aiming to eliminate cancerous cells [83]. Photothermal therapy stands out as a minimally invasive technique for cancer treatment, characterized by high precision and selectivity [84]. The unique optical properties of AuNPs such as SPR enhance their absorption and scattering properties. Hence, they have been used as a promising candidate to produce the localized heating effect once irradiated with laser energy [85]. Gold NPs also offer other favorable properties for PTT such as good biocompatibility, ease of surface modifications, non-photobleaching, and ease of production of AuNPs of variable sizes and shapes such as nanorods, nanoshells, nanospheres, nanostars, nanocages, etc. [86]. Visible light offers poor penetration due to low absorption coefficient by the water molecules present in the body, so it is not suitable for PTT and imaging of the cancer cells present inside the body [87]. Therefore, the penetration depth and the peak absorption wavelength of the NPs must fall within the NIR range (600–1400 nm). Within this range, water and other components including hemoglobin and melanin have low absorption coefficients. Hence, NIR radiation could shine deep inside the body, unlike visible light. Among the various types of nanomaterials, AuNPs offer the possibility of tuning the maximum wavelength to a longer wavelength in the NIR range due to SPR. The absorption wavelength of AuNPs can be adjusted by various modifications such as changing the size and shape which directly impacts the SPR band. Hence, depending on the location of the tumors inside the body, the peak absorption of AuNPs can be tuned. Further, to minimize the long-term toxicity and to enhance the specificity to minimize the damage to healthy tissues, AuNPs have been conjugated with various inorganic/organic/polymeric molecules and biomolecular moieties such as SiO_2 [88], Prussian blue [89], CTAB [90], MOFs [91], and zeolitic imidazole framework (ZIF) [92].

PEGylated Au nanorods have been used by Pop et al. [93] in PTT to decrease the tumor volume using murine melanoma model. Gold nanorods have been shown to attain ablative temperatures quickly upon exposure to NIR light. Temperatures within the ablation range of 55–65 °C were maintained by varying the AuNP concentration and LED flux density. Further, with the help of Au nanorods, LED was found to be sufficient to induce ablation in tumor models up to the depth of 1.5 cm. PEGylated Au nanorods with an aspect ratio of 4.6 have been used to study the pharmacokinetics of GNRs after intravenous and intratumor administration [94]. The concentration of GNR was found to be similar around tumor regions in both I.V. and I.T. administration. The antitumor effect of GNRs was assessed by injecting about 1.4 mg/kg of GNRs every three weeks to mice bearing EAC tumors. After exposing the tumor to IR lasers, the temperature surged to 79 °C in the central region of the tumor and decreased to 41 °C at the tumor boundary. The AuNRs coupled with IR radiation arrested the growth of the tumor, without any change in tumor size, whereas tumor of the untreated mice kept on growing until the size reached 6.3-fold the initial size. Gold nanorods were linked with anti-epidermal growth factor receptor (anti-EGFR) and exposed to both malignant and nonmalignant epithelial cells, as outlined by El-Sayed et al. [95]. Subsequently, the cells were exposed to red laser light at 800 nm. The observation revealed that malignant cells were eradicated with half the laser energy compared to nonmalignant cells. Later, Mustafa et al. [96] again used PEGylated AuNRs for *in vivo* PTT of deep tissue malignancies using NIR laser. The administration of PEGylated AuNRs both directly and intravenously induced the inhibition of tumor growth in mice after 13 days of injection (Figure 2.5A). The average tumor growth showed a reduction of over 96% in the case of directly treated HSC-3 xenografts and over 74% for intravenously treated HSC-3 xenografts compared to the control tumor. Thermal transient measurement showed that rapid heating occurred within the first three minutes, and a Newtonian cooling behavior was observed after the removal of the NIR source Figure 2.5B. Gold NRs have been recently used for the photoablation of various types of malignant and nonmalignant tumors [97–99].

In this section, the emphasis is on utilizing gold nanoparticles (AuNPs) in photothermal therapy (PTT) to eradicate cancerous cells. Photothermal therapy is underscored as a minimally invasive approach to cancer treatment, characterized by its high precision and selectivity. The distinctive optical properties of AuNPs, particularly SPR, render them well-suited for generating a localized heating effect upon exposure to laser energy. The favorable properties of gold nanoparticles for PTT, including good biocompatibility, ease of surface modifications, and various shapes and sizes, are discussed. The section emphasizes the importance of using NIR radiation for PTT and imaging within the body due to its deeper penetration compared to visible light. AuNPs are noted for their ability to tune the maximum absorption wavelength to the NIR range, which is crucial for effective treatment.

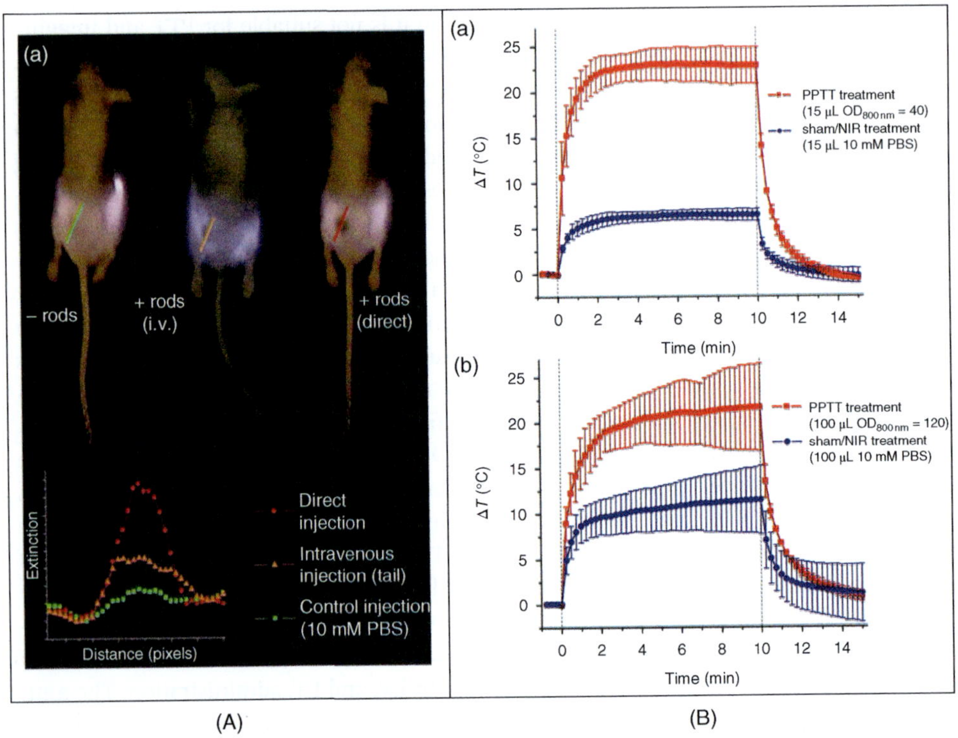

Figure 2.5 (A) Displays near-infrared (NIR) transmission images of mice before photothermal therapy (PPTT) interventions. The inset provides intensity line scans of NIR extinction at tumor sites for control (j), intravenous (N), and direct (d) administration of PEGylated gold nanorods. In the control group, mice were injected interstitially with 15 μL of 10 mM phosphate-buffered saline (PBS) alone. Directly administered mice received interstitial injections of 15 μL PEGylated gold nanorods (ODk = 800 = 40, 2 min accumulation), while intravenously administered mice received 100 μL PEGylated gold nanorod (ODk = 800 = 120, 24 h accumulation) injections. (B) Thermal transient measurements of HSC-3 tumor interstitia are presented during (a) direct and (b) intravenous near-infrared PPTT (j) and sham/NIR (d) treatments using PEGylated gold nanorods. Direct PPTT treatments involved the administration of 15 μL PEGylated gold nanorods (ODk = 800 = 40, 2 min accumulation) followed by 10 min of 0.9–1.1 W/cm^2 NIR laser exposure. Intravenous PPTT treatments included the administration of 100 μL PEGylated gold nanorods (ODk = 800 = 120, 24 h accumulation) followed by 10 min of 1.7–1.9 W/cm^2 NIR laser exposure. Sham/NIR treatments consisted of the administration of 15 μL 10 mM PBS and NIR laser exposure with comparable time and power density. Reported errors are presented as standard deviation. Source: Reproduced with permission from Dickerson [96]/Elsevier.

To address long-term toxicity and enhance specificity while minimizing damage to healthy tissues, AuNPs are conjugated with various inorganic/organic/polymeric molecules and biomolecular moieties. The text then provides an example of PEGylated Au nanorods used in PTT to decrease tumor volume in a murine melanoma model. The nanorods are shown to induce ablation in tumor models up to a depth of 1.5 cm. Additionally, the pharmacokinetics of PEGylated Au nanorods are discussed, highlighting their concentration around tumor regions after both intravenous and intratumor administration. The antitumor effect of these nanorods is explored, showing a significant reduction in tumor growth when exposed to infrared (IR) lasers.

The section also mentions the conjugation of gold nanorods with anti-EGFR for the elimination of malignant cells with lower laser energy compared to nonmalignant cells. Further studies using PEGylated Au nanorods for *in vivo* PTT of deep tissue malignancies are described, indicating a substantial decrease in tumor growth after injection. The thermal behavior of the nanorods during laser exposure is detailed, showcasing rapid heating and Newtonian cooling after the removal of the NIR source. Lastly, gold nanorods are mentioned for the photoablation of various types of malignant and nonmalignant tumors, highlighting their potential in cancer therapy.

2.2 Biomedical Applications of Silver Nanoparticles

Silver nanoparticles find diverse applications in biomedical fields, including but not limited to antimicrobial coatings, drug delivery, the detection and diagnosis of pathogenic microbes, tissue regeneration, and therapeutics. The important biomedical application of AgNPs is discussed in this section.

2.2.1 As Antimicrobial Agents

Silver nanoparticles have been instrumental in advancing novel antimicrobial materials, particularly for integration into medical devices. Additionally, they have been included as additives in tooth implants, bone cement, catheters, and bone fractures to deter biofilm formation and impede pathogen growth. Silver nanoparticles have been utilized to address the growing challenge of multidrug resistance in pathogens. They demonstrate potent antimicrobial activity against both gram-positive and gram-negative bacteria. Sondi et al. [100] conducted an investigation into the antimicrobial activity of silver nanoparticles (AgNPs) using *Escherichia coli* as a gram-negative bacterial model. Silver nanoparticles of variable concentrations were used for the antimicrobial study. The bactericidal action of AgNPs was confirmed electron microscopically by the formation of pits on the surface of bacteria. Bacterial cells exposed to silver nanoparticles (AgNPs) exhibited heightened permeability of the cell membrane, ultimately leading to cell death. The concentration of AgNPs was identified as a determining factor in their antimicrobial activity. Morones et al. [101] explored the size dependence of the antimicrobial activity of silver nanoparticles (AgNPs) within the size range of 1–100 nm. Their observations indicated that AgNPs exhibited size-dependent antimicrobial activity, with smaller-sized AgNPs (1–10 nm) displaying greater interaction with the bacterial cell membrane compared to larger-sized AgNPs. Pal et al. [102] conducted a study on the antimicrobial activity of silver nanoparticles (AgNPs) with varying sizes against *E. coli*. TEM images of AgNP-treated AgNPs showed changes in microbial membrane which resulted in bacterial death (Figure 2.6). It was observed that truncated Ag nanoplates displayed the highest biocidal action as compared to rod-shaped and spherical AgNPs. The authors concluded that [103] plane of Ag nanoplanes interacts most effectively with bacterial cell membrane. Silver nanoparticles with sizes ranging from 17.9 to 59.6 nm were produced through the utilization of a surfactant obtained from *Trianthema decandra*, and these nanoparticles were applied for antimicrobial purposes [104]. The antimicrobial effectiveness of silver nanoparticles (AgNPs) was evaluated against eight different microorganisms through the Kirby–Bauer method, which indicated the varied susceptibility of these microbes against AgNPs. The antimicrobial susceptibility was recorded to be higher for gram-negative bacteria than for gram-positive bacteria. Ahmad et al. [105] utilized a biosynthetic approach involving the cell extract of *Candida albicans* to synthesize AgNPs. The resulting AgNPs were both small (30 nm) and highly crystalline, boasting a surface area of 18.9 m^2/g. These fabricated AgNPs exhibited notable antimicrobial efficacy against two distinct microorganisms, namely *Staphylococcus aureus* and *E. coli*. The assessment of antimicrobial activity was conducted through growth curve studies and disc diffusion assays. The biocidal effectiveness of the AgNPs was more pronounced against *E. coli*, a gram-negative bacterium, compared to *S. aureus*, a gram-positive bacterium. Notably, the Minimum Inhibitory Concentration at 80% (MIC_{80}) values

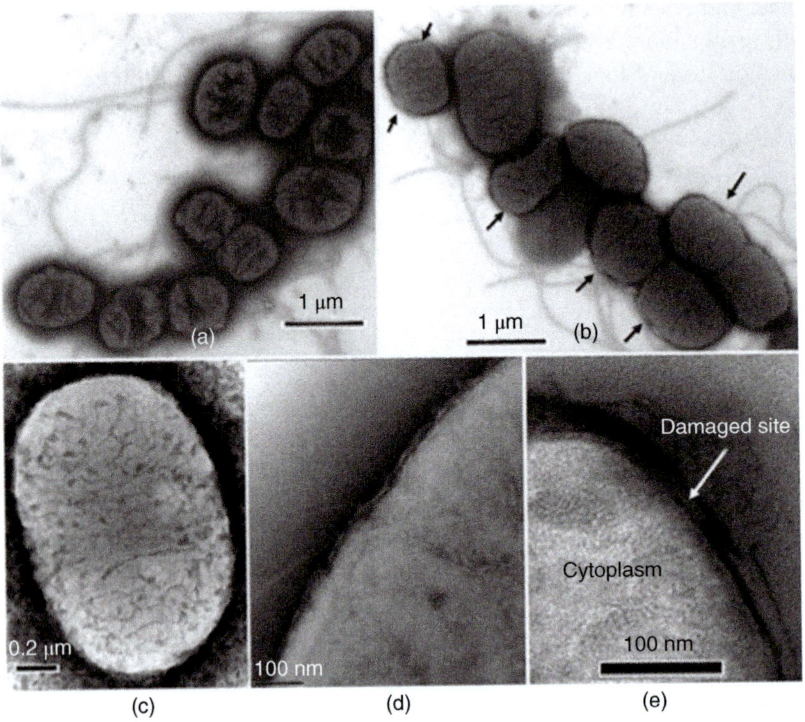

Figure 2.6 Electron energy loss spectroscopy (EFTEM) images depicting *E. coli* cells are presented as follows: (a) *E. coli* that has not undergone any treatment, where flagella is visible. (b) *E. coli* cultivated on agar plates supplemented with silver ions ($AgNO_3$). Arrows highlight partially damaged membranes. Despite this damage, these cells remain viable. (c) *E. coli* subjected to treatment with triangular silver nanoplates. Dark irregular pits, representing silver nanoparticles, are evident on the cell surface. (d) *E. coli* is treated with spherical silver nanoparticles. (e) A magnified view of a segment of the bacterial cell membrane treated with triangular silver nanoparticles reveals damage in multiple locations. Source: Reproduced with permission from Pal [102]/American Society For Microbiology.

indicated a lower concentration (8 μg/mL) for *E. coli* compared to *S. aureus* (32 μg/mL). Additionally, significant and well-defined zones of inhibition (ZOI) were observed surrounding discs infused with AgNPs, measuring 24 mm for *E. coli* and 19 mm for *S. aureus*. Wani et al. [106] conducted the synthesis of AgNPs within microemulsions utilizing three distinct surfactants: CTAB, Tergitol, and Triton X-100. The variation in surfactant type led to differences in particle size and morphology. The antimicrobial efficacy of the AgNPs was evaluated against the bacterium *E. coli* and the fungal yeast *C. albicans*. In both cases, the antimicrobial activity was observed to be size-dependent. Specifically, the smallest Ag nanodiscs (8 nm), synthesized using Triton X-100 surfactant, exhibited the highest antimicrobial activity. Following this, 30 nm-sized Ag nanospheres (Tergitol) and then 30 nm AgNPs (CTAB) showed decreasing levels of antimicrobial effectiveness. The Minimum Inhibitory Concentration (MIC) values for AgNPs were recorded as higher for *E. coli* compared to *C. albicans*. In the disc diffusion assay, the results demonstrated larger ZOI for AgNPs against *C. albicans* compared to *E. coli*. This suggests that C. albicans, the yeast, displayed greater susceptibility to AgNPs of different sizes in comparison to the bacterium *E. coli*. Bindu and Umadevi [107] successfully synthesized silver nanoparticles (AgNPs) utilizing leaf extract from *Hibiscus cannabinus*. The synthesis involved varying concentrations of Ag^+ ions, specifically 8 mM, 5 mM, and 10^{-2} M. The synthesized AgNPs were then subjected to antimicrobial testing against several microbial species, including *E. coli*, *Proteus mirabilis*, and *Shigella flexneri*, revealing significant antimicrobial activity. Interestingly, the antimicrobial activity was observed to be higher for AgNPs synthesized at lower Ag^+ ion concentrations compared to higher concentrations. This heightened efficacy at lower concentrations was attributed to the formation of smaller-sized AgNPs. The results imply that the size of the AgNPs influenced their antimicrobial activity, with smaller nanoparticles exhibiting increased effectiveness against the tested microbial species. Silver NPs have been synthesized using polysaccharide extract from four marine microalgae [108]. The polysaccharide acted as both a reducing agent and a stabilizing agent. Silver NPs were evaluated for antimicrobial activity against microbes on the fabrics. The antimicrobial activity depended upon the size of AgNPs, the type of fabric and the type of algal species used. The gram-positive bacterium showed lower antimicrobial activity in comparison to the gram-negative bacterium. The gram-positive bacterium was found to absorb a lower quantity of Ag^+ ions than the gram-negative bacterium, suggesting a greater interaction between Ag^+ ions and gram-negative bacterium. Silver NPs have been synthesized using *Lycopersicon esculentum* (red tomato) aqueous extract [109]. Spherical silver nanoparticles (AgNPs) of various concentrations underwent testing for their antimicrobial efficacy against *E. coli*. Notably, these silver NPs exhibited a distinct ZOI against *E. coli*, showcasing a MIC of 50 μg/mL. Additionally, AgNPs capped with polyvinyl alcohol (PVA) were synthesized through a chemical reduction method employing different pH values. The alteration of pH from 6 to 12 during synthesis resulted in a reduction in particle size, decreasing from 26 to 10 nm. The impact of these PVA-capped AgNPs on both *E. coli* and *Pseudomonas* sp. was assessed using the Kirby–Bauer disc diffusion method. The antimicrobial effectiveness of the AgNPs displayed an incremental trend with the elevation of pH in the silver colloidal solution against both bacterial species. Specifically, the maximum ZOI was observed at pH 12, while the minimum was recorded at pH 6 (refer to Figure 2.7A). This highlights the influential role of pH in modulating the antimicrobial activity of AgNPs against *E. coli* and *Pseudomonas* sp. Furthermore, the fungicidal potential of these AgNPs was examined against *Aspergillus niger* and *Penicillium* sp. Interestingly, *Penicillium* sp. exhibited greater susceptibility to the AgNPs compared to *A. niger*. The antifungal activity demonstrated an escalation with the increase in pH of the AgNP solution, as evidenced by the ZOI (refer to Figure 2.7B). This observation underscores the influence of pH on the antifungal efficacy of AgNPs, with higher pH values contributing to increased inhibitory effects against both *A. niger* and *Penicillium* sp. Thus, both the fungal species exhibited higher ZOI at pH 12 due to smaller-sized AgNP obtained at this pH. Chahar et al. [111] conducted a comparative analysis of the antimicrobial activity of silver nanoparticles (AgNPs) synthesized through a biological route using *Berginea ligulata*, *Rheum australe*, and *Selinum vaginatum*, and commercially available AgNPs prepared via a chemical route using gelatin, PVA, and PVP as stabilizers. The biosynthetic route produced smaller-sized AgNPs with higher antimicrobial activity than chemically synthesized AgNPs. The antimicrobial activity was evaluated against three microbial pathogens viz., *Salmonella typhi*, *Bacillus cerus*, and *Pseudomonas aeruginosa*. The order of biocidal action of the AgNPs was found to be: (AgNPs) *Spartina vaginatum* > (AgNPs) *Rhododendron australe* > (AgNPs) *Bergenia ligulata* > (AgNPs) gelatin > (AgNPs) PVP > (AgNPs) PVA. The biocidal action was observed to be the direct function of particle size. The order of sensitivity of microbial species to AgNPs was found to be; *P. aeruginosa* > *B. cerus* > *S. typhi*. Silver nanoparticles of three different morphologies viz., nanospheres, nanodiscs, and nanotriangles were synthesized by Cheon et al. [112] and evaluated the morphological dependence of their antimicrobial action against *E. coli*, *S. aureus*, and *P. aeruginosa*. The order of antimicrobial action of AgNPs against the test microbes from the ZOI values was recorded as spherical AgNPs > disk AgNPs > triangular AgNPs. Moreover, the antimicrobial activity was found to increase with the rise in the concentration of silver nanospheres (AgNSs) and silver nanodiscs (AgNDs) from 0.1 to 0.7 mg/mL against *E. coli*. However, AgNTs displayed the least antimicrobial activity at all concentrations. Silver nanoparticles synthesized using a biosynthetic

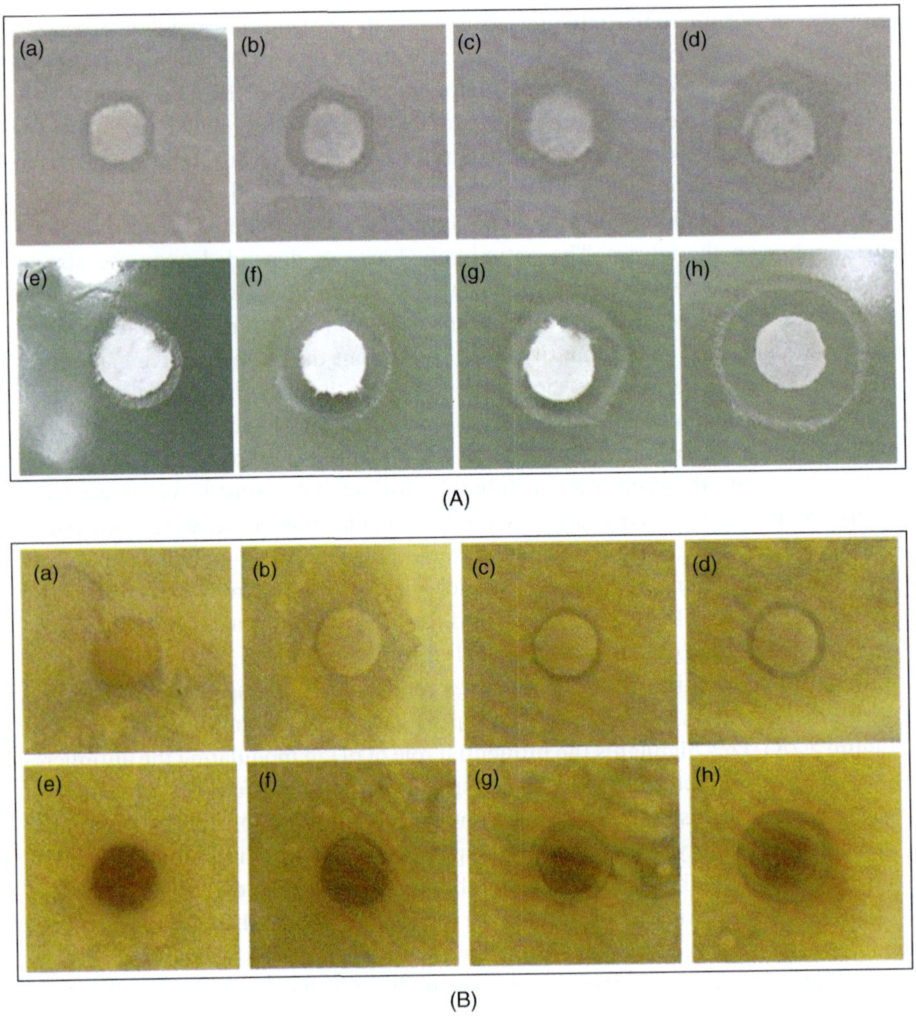

Figure 2.7 (A) Displays representative outcomes illustrating the antibacterial effectiveness of silver nanoparticles (AgNPs) against *Escherichia coli* (a–d) and *Pseudomonas* sp. (e–h) under varying pH conditions: (a, e) pH 6, (b, f) pH 8, (c, g) pH 10, and (d, h) pH 12. (B) Provides representative findings demonstrating the fungicidal activity of AgNPs against *Aspergillus niger* (a–d) and *Penicillium* sp. (e–h) under different pH conditions: (a, e) pH 6, (b, f) pH 8, (c, g) pH 10, and (d, h) pH 12. Source: Reproduced with permission from Ajitha [110]/Elsevier.

route using microbial cultural supernatants of *Bacillus amyloliquefaciens* and *Bacillus subtillis* have demonstrated notable antimicrobial efficacy against a variety of microbial species including gram-negative bacteria: *E. coli*, *P. aeruginosa*, *Salmonella*, gram-positive bacterial: *S. aureus*, *Streptococcus pyogenes*, and the fungus *C. albicans* [103]. Silver nanoparticles synthesized using the reducing agents from natural sources such as extracts of fruits such as apples, oranges, and vegetables such as garlic, reddish, potatoes, red pepper, and white onion have been used for the assessment of their antimicrobial activity against various microbes such as *S. aureus, Bacillus cereus, E. coli,* and *Candida krusei*. The particle size of AgNPs ranged from 9 to 30 nm. Silver NPs obtained using potato extract showed the highest antimicrobial activity [113]. Numerous research reports have been published on the antimicrobial activity of silver nanoparticles (AgNPs) synthesized through diverse chemical and green synthetic routes [114–121].

In summary, this section emphasizes the utilization of silver nanoparticles (AgNPs) in the creation of innovative antimicrobial materials for medical devices. The integration of AgNPs in diverse medical applications, such as tooth implants, bone cement, catheters, and bone fractures, aims to deter biofilm formation and hinder pathogen growth. The adoption of silver nanoparticles is specifically underscored as a strategy to address the growing concern of multidrug resistance in pathogens.

Researchers, including Sondi et al., Morones et al., and Pal et al., have investigated the antimicrobial properties of silver nanoparticles (AgNPs) against both gram-positive and gram-negative bacteria. Their studies reveal that the antimicrobial effects of AgNPs are dependent on both concentration and size. Electron microscopy analyses confirm the bactericidal action of AgNPs, as they cause damage to bacterial cell membranes.

The synthesis of silver nanoparticles (AgNPs) through diverse methods and sources, encompassing biosynthetic routes, surfactants, and plant extracts, is extensively discussed. The antimicrobial efficacy of these synthesized AgNPs is assessed against various microbial species, revealing variations in susceptibility between gram-positive and gram-negative bacteria. The section underscores the significance of particle size and morphology in influencing the antimicrobial properties of AgNPs.

Studies conducted by Ahmad et al., Wani et al., Bindu and Umadevi, and others highlight the effectiveness of AgNPs synthesized through different methods, including biosynthetic routes, in inhibiting the growth of both bacteria and fungi. Additionally, a comparison between AgNPs synthesized through biological and chemical routes is addressed, emphasizing the smaller size and heightened antimicrobial activity of biosynthesized AgNPs.

Furthermore, the section delves into the morphological dependence of antimicrobial activity, exploring the impact of AgNPs with distinct shapes (nanospheres, nanodiscs, and nanotriangles). Cheon et al. contribute to this aspect, reporting an enhanced antimicrobial activity with increased concentration. Investigations involving AgNPs synthesized from microbial cultural supernatant and natural reducing agents found in fruits and vegetables also demonstrate their effectiveness against various microbial species.

The section concludes by noting the abundance of research reports that delve into the antimicrobial activity of silver nanoparticles (AgNPs) synthesized through diverse chemical and green synthetic routes. It underscores the wide array of applications and the significant potential of AgNPs in effectively combating microbial infections.

2.2.2 Silver NPs in Wound Care Applications

Nanotechnology has enabled the production of novel wound-healing systems that are effective as well as economical. For example, AgNPs are highly effective in killing pathogenic microbes and hence have displayed the potential wound-healing abilities when incorporated in wound dressings. Because of their substantial surface area to volume ratio, ease of surface functionalization with various active biomolecules like DNA and drugs, and intrinsic antimicrobial activity against a broad spectrum of microorganisms, silver nanoparticles (AgNPs) have found application as topical agents in wound healing. Silver nanoparticles have been used in the form of Ag nanocomposites. Ag nanocomposites are formed by the incorporation of AgNPs into gels, colloids, and polymeric materials. The collaborative action of AgNPs, along with phytochemicals and various other beneficial biomolecules, works synergistically to accelerate the wound-healing process. For instance, Al-shmgani et al. [122] synthesized AgNPs using *Catharanthus roseus* leaf extract and evaluated them for wound-healing applications. The wound-healing potential of AgNPs was assessed using the excision wound model on male albino mice. Mice treated with AgNPs exhibited better healing activity than those treated with both positive and negative controls. Further, AgNP-treated wounds exhibited no pus formation or bleeding during treatment, whereas notable inflammation was observed in wound treated with positive control. Significant wound closure and a notable reduction in wound size were observed in wounds treated with silver nanoparticles (AgNPs) compared to those treated with the positive control. Silver nanoparticle-treated wounds displayed 98% of closure at the conclusion of the experiment as compared to 85% closure in case of wounds treated with positive control. ε-polylysine/AgNP nanocomposite (EPL-g-butyl@AgNPs) was synthetized by Dai et al. [123], which exhibited synergistic antimicrobial effects. The Ag nanocomposite was found to exhibit good aqueous stability and was effective against both *S. aureus* (gram-positive) and *P. aeruginosa* (gram-negative) bacteria. The antibacterial activity of AgNPs and EPL-g-butyl@AgNPs composites was assessed by *in vivo* study using diabetic rats with wounds. Rat wounds were subjected to treatment with EPL-g-butyl@AgNPs and AgNPs at the same concentrations of 28 μg/mL. It was observed that, after nine days of treatment, the sizes of the wounds became smaller in rats treated with EPL-g-butyl@AgNPs and AgNPs compared to those in the model and blank control groups. Furthermore, after 18 days of treatment, wounds treated with EPL-g-butyl@AgNPs exhibited a faster healing process than those treated with AgNPs alone. Histological examination of the wounds revealed that both EPL-g-butyl@AgNPs and AgNPs treatments resulted in fewer inflammatory cells compared to the model and blank control groups. Chitosan, composed of *N*-acetylglycosamine and glucosamine, is a biocompatible natural biopolymer is used as a wound-healing material in the form of sponges, scaffolds, hydrogels, and membranes. Although it finds its usage in wound-healing applications, due to the lack of its significant antimicrobial activity, treatment of infectious wounds with chitosan is limited. However, the incorporation of AgNPs in chitosan makes it one of the wonderful antibacterial dressing materials due to the high antimicrobial activity of AgNPs. For instance, chitosan-capped AgNPs have been synthesized by Oryan et al. [124] and investigated the effects of its topical application on burn wound healing. The results have demonstrated the chitosan-capped AgNPs-treated wounds exhibited lower inflammation as evident by the lower levels of interleukin 1B and neutrophil counts. Chitosan-AgNP treated also exhibited rapid reepithelization with complete reepithelization of the wound lesions achieved on the seventh day of post-wound treatment. The wound treated with Chitosan-AgNP also exhibited improved granulation, tissue formation, and enhanced basic fibroblast growth factor on the seventh day of post wounding. Based on the histological findings and biochemical assays, it was concluded that AgNp-composite materials stimulated the healing of the burn wound by reducing the length of the repair phases. AgNPs have also been incorporated in gelatin to form Ag nanocomposite and utilized for wound-healing applications. For example. Nanogelatin composite material composed of gelatin, AgNPs, and curcumin has been synthesized by Khanh et al. [125] for wound-healing applications. The nanocomposite was made by incorporating AgNPs in a gelatin and encapsulated with curcumin for use in wound care applications. The authors concluded that the GelCurAg nanocomposite demonstrated exceptional antimicrobial activity, suggesting its potential use as a biocompatible material for treating infectious wounds. Wu et al. [126] have synthesized AgNPs and assembled them on the surface of bacterial cellulose (BC) nanofibers forming AgNP-BC nanocomposite and studied them for wound-healing applications. The results suggested a significant reduction in wound flora of wounds treated with Ag-BC nanocomposite material. Furthermore, the wound showed faster reepithelization and thicker dermis and epidermis as compared to wounds treated with BC alone and untreated control wounds. The findings indicated that the Ag-BC nanocomposite has the potential to alleviate inflammation and enhance the healing process of scaffold wounds. Silver nanoparticles (AgNPs) were synthesized by Sood and Chopra [127] using *Ocimum sanctum* leaf extract. The resultant AgNPs were then integrated into a carbopol gel base, creating an Ag nanogel.

The resulting Ag nanogel has been evaluated for *in vivo* wound-healing applications using rat model. The application of the AgNP-based gel topically demonstrated a quicker rate of wound healing compared to both the control and standard treatments. Furthermore, the wound treated with AgNP-gel formulation took less time for closure as compared to standard and control. Tannic acid (TA) stabilized AgNPs were synthesized by Orlowski et al. [128], which displayed potent antibacterial action against *P. aeruginosa*, *S. aureus*, and *E. coli*. TA-stabilized AgNPs were found to stimulate the movement of keratinocytes *in vitro*. TA stabilized of the size of 33 and 46 nm when used for wound healing in splint wounds displayed better wound closure ability, faster epithelization, angiogenesis, and granulation tissue formation. The silver nanoparticles (AgNPs) induced the expression of cytokines, such as VEGF-α, PDG.

F-β, and TGF-β1, crucial for effective wound healing, when compared to the control. Nevertheless, the 13 nm-sized silver nanoparticles (AgNPs) triggered a robust inflammatory response through the wound-healing process in damaged skin. Mehwish et al. [129] have synthesized *Moringa oleifera* seed polysaccharide-AgNP composite (MOS-PS-AgNPs) for wound dressing applications. MOS-PS-AgNPs were spherical in appearance (Figure 2.8A) with no cytotoxicity toward mouse fibroblasts. *In vivo* experiments were conducted to assess the wound-healing efficacy of MOS-PS-AgNPs in male rat models. The wound-healing process was found to be dose-dependent. A substantial decrease in wound area has been observed after nine days of treatment in high dose (MOS-PS-AgNPs)-treated group (91%) then, followed by medium dose (63%), low dose-treated groups (51%) positive control (42%), and negative control group (30%). Complete wound healing was observed after 10 days of wound treatment without any scars (Figure 2.8B). Microscopic pathological alterations during distinct stages of wound healing in various mouse groups are illustrated in Figure 2.8C. The images show that considerable neutrophil and inflammatory cells are present in the wound area in all the groups. The negative control showed severe tissue damage and fibrosis. Granular formation and extracellular matrix were noted in MOS-PS-AgNPs-treated groups in comparison to other experimental groups. The formation of the new epidermis and granulation tissue was displayed by MOS-PS-AgNPs-treated groups in the wounded area in comparison to other experimental groups. Nejad et al. [130] have synthesized antimicrobial AgNPs impregnated in zwitterionic dopamine methacrylamide hydrogels (ZWDO) for use in wound-healing applications. Electron microscopy images have revealed that AgNPs of the size of 20 nm were homogenously incorporated into the hydrogel network. The antimicrobial activity of ZWDO hydrogels alone and ZWDO-AgNP composites was assessed by disc diffusion assay. The results suggested that no zones of inhibition were recorded for ZWDO alone, whereas ZWDO-AgNP composites exhibited significant ZOI against the microbes with an increase in diameters of ZOI of the order of 157%, 148%, and 129% for *E. coli*, *S. aureus*, and *P. aeruginosa* respectively. Further, ZWDO-AgNP-composite hydrogel exhibited bactericidal action against the test microbes, as evident from the OD measurements at 600 nm. The OD values of all the bacterial strains were found to decrease after treatment with ZWDO-AgNP-composite hydrogel. Whereas no change in OD was recorded for the bacterial species treated with ZWDO hydrogels alone. The ZWDO-AgNP hydrogel was then tested for its wound-healing ability by applying the ZWDO hydrogel and ZWDO-AgNP-composite hydrogel on the separate groups of rats inflicted with back wounds. Whereas no dressing was applied to the third group of rats taken as control. After 15 days of dressing, the wounds were assessed, and it was observed that the percentage of size reduction in the area of the wounds for the three groups of rats was 59% for control, 80% for ZWDO, and 98% for ZWDO-AgNPs hydrogel groups. The findings suggested that the ZWDO-AgNPs hydrogel demonstrated notable wound-healing capabilities compared to other groups. Numerous research reports have been published on the application of silver nanoparticles (AgNPs) in wound-healing contexts [131–137].

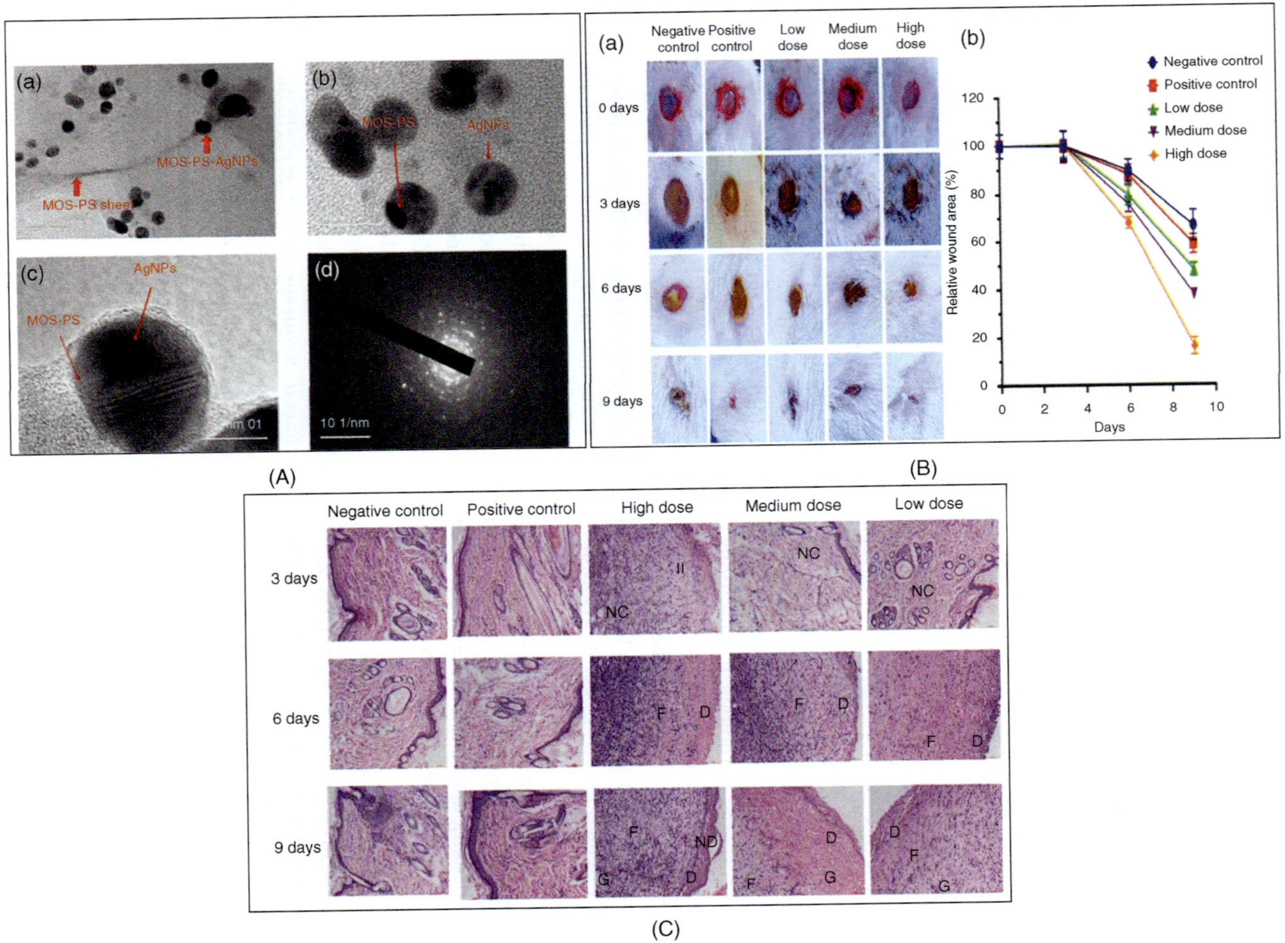

Figure 2.8 (A) Showcases the characterization of MOS-PS-AgNPs through transmission electron micrograph. In (B) the wound-healing appearance (a) and healing rate (b) are documented at various time intervals following treatment with MOS-PS-AgNPs at different concentrations (low dose: 25 μg/mL, medium dose: 50 μg/mL, and high dose: 100 μg/mL), in comparison to the negative control (treated with water) and positive control (treated with betadine) groups. The values represent the mean ± SD of four replicates, and the results reveal a significant ($p < 0.05$) difference compared to the control group. (C) Presents the histological examination of treated wounds on days 3, 6, and 9 through H and E staining. The micrograph was captured at a scale bar of 100 nm, illustrating features such as the infiltration of inflammatory cells (II), neutrophil cells (NC), fibroblasts (F), dermis (D), neodermis (ND), and granulation tissue (G). Source: Reproduced with permission from Mehwish [129]/Elsevier.

This section delves into the application of nanotechnology in wound healing, specifically focusing on the utilization of silver nanoparticles (AgNPs) due to their antimicrobial properties. The integration of AgNPs into various wound dressing materials is explored, highlighting their effectiveness and economic feasibility in wound-healing systems. AgNPs, renowned for their substantial antimicrobial activity against a diverse range of microorganisms, are underscored for their potential in wound-healing applications.

The discussion revolves around the incorporation of AgNPs into wound dressings, resulting in the formation of Ag nanocomposites. These composites are created by integrating AgNPs into gels, colloids, and polymeric materials. Several studies are cited to substantiate the effectiveness of AgNPs in wound healing. For instance, Al-shmgani et al. synthesized AgNPs using *Catharanthus roseus* leaf extract, showcasing enhanced wound-healing activity in mice. Another study by Dai et al. focused on ε-polylysine/AgNP nanocomposites, revealing synergistic antimicrobial effects effective against both gram-positive and gram-negative bacteria. *In vivo* studies with diabetic rats further supported the superior wound-healing properties of these nanocomposites.

The text also explores the incorporation of AgNPs into diverse materials like chitosan, gelatin, and BC. Instances include chitosan-capped AgNPs synthesized by Oryan et al., demonstrating improved burn wound healing characterized by reduced inflammation and rapid reepithelization. Additionally, AgNPs assembled on BC nanofibers (AgNP-BC nanocomposite) exhibited reduced wound flora, faster reepithelization, and thicker dermis and epidermis.

Various methods of synthesizing AgNPs for wound-healing applications are covered, encompassing the use of plant extracts, biosynthetic routes, and natural reducing agents. Examples include AgNPs synthesized using *O. sanctum* leaf extract, gelatin-encapsulated AgNPs with curcumin, and tannic acid-stabilized AgNPs. These synthesized AgNPs demonstrated effective antibacterial activity and stimulated the expression of cytokines involved in wound healing.

The section concludes by highlighting the synthesis of AgNPs using *Moringa oleifera* seed polysaccharide for wound dressing applications, which exhibited dose-dependent wound healing. Additionally, AgNPs impregnated in zwitterionic dopamine methacrylamide hydrogels showcased significant antimicrobial activity and wound-healing capabilities. The comprehensive review acknowledges numerous research reports supporting the use of AgNPs in wound-healing applications, emphasizing their potential to develop novel and effective wound dressings.

2.2.3 Silver Nanoparticles in Dentistry

Silver has been employed in dentistry since the early nineteenth century owing to the antimicrobial properties exhibited by silver ions [138]. Nonetheless, the emergence of nanotechnology in the twenty-first century has spotlighted the application of silver in the form of silver nanoparticles (AgNPs) in dentistry, attributed to the gradual release of silver ions from the surface of AgNPs [139–142]. Several studies have demonstrated the potential of silver nanoparticles (AgNPs) in managing the growth of various pathogenic microbes, including those associated with oral pathogens [143–145]. Silver NPs have been vainly studied for their use in areas such as oral microbiology and dental prosthesis. The oral cavity is home to around 70 microbial which include protozoa, fungi, bacteria, and viruses [146, 147]. The microbial composition in the oral cavity plays a pivotal role in comprehending the etiology of oral diseases, directly influencing the health and disease status of an individual. The balance of microbiota is vital for maintaining the oral and systemic health of an individual [146]. The most abundant of the microorganisms for a healthy oral microbiota belong to *Proteobacteria*, *Firmicutes*, *Actinobacteria phyla*, and *Fusobacteria* [148]. Silver NPs have shown a strong antimicrobial action in numerous species of various gram-positive as well as gram-negative bacteria [117, 122, 126, 143, 144]. The research findings indicate an inverse relationship between the size of silver nanoparticles (AgNPs) and their antimicrobial activity. Smaller-sized AgNPs are more effective in preventing biofilm formation and are effective against various oral pathogens such as *Streptococcus mutans* and *Streptococcus orallis*. Cabal et al. [149] have used soda-lime glass-coated AgNPs for *in vitro* inhibition of the biofilm of *Streptococcus oralist*. The three strains of *S. oralist* were grown on titanium alloy discs, discs coated with glass, glass plates coated with AgNPs, and uncoated titanium alloy discs. The alteration in bacterial colonization and biofilm formation was visualized by SEM. SEM images revealed bacterial cells were densely colonized on the uncoated titanium alloy discs and glass-coated discs contained after 24 hours of incubation. However, AgNP-coated glass discouraged the adhesion of the bacterial colonies on the surface significantly. As a result, only a few bacterial colonies appeared to be located inside the porous coating. The results unequivocally demonstrated that silver nanoparticles (AgNPs) prevented bacterial adhesion on the surface of the glass and hindered the formation of biofilms. The toxicity of the AgNPs was further assessed by examining the release profile of Ag^+ ions from the AgNPs. The recorded observation indicated the concentration of Ag^+ ions in the supernatant was around c. 7 ppm, which is lower (30 ppm) than the concentration at which Ag^+ ions become cytotoxic to human fibroblasts [150]. Glass ionomer cement (GIC) finds application in pediatric dentistry for therapeutic oral health care, primarily due to its ability to release and store fluoride. The fluoride release capacity of GIC positions it as an effective anti-caries agent attributed to the inhibition of bacterial growth by fluoride ions. Nevertheless, to sustain its anti-caries efficacy, the GIC material requires periodic refilling with fluoride. Therefore, there is a need to impregnate GIC material with an antimicrobial agent, which provides GIC cement with long-lasting antimicrobial properties for combating oral diseases effectively. In this direction, several studies have been published that report the impregnation of GIC cement with AgNPs, and the resultant material exhibited notable antimicrobial action against both gram-positive and gram-negative microbes. [151, 152]. The antimicrobial activity observed in the GIC-AgNP material composite was attributed to the release of Ag^+ ions from the silver nanoparticles (AgNPs) into the cement matrix. This release effectively inhibited the formation of bacterial biofilms and served as a preventive measure against dental caries. Barot et al. [153] have synthesized AgNP-immobilized halloysite nanotubes (HNT) based resins as fillers and incorporated them in dental resin composites composed of Bis-GMA/TEGDMA with a ratio of 70/30 in varied percentages. The mechanical and antimicrobial properties of the resulting Ag/HNTNP-based dental resin composite were compared with the normal composite containing conventional glass fillers. The results suggested that incorporating HNT/AgNPs into dental resin enhanced the antimicrobial properties of the resin. Notably, the composite with 5% HNT/Ag, both without glass fillers and with glass fillers, demonstrated outstanding antimicrobial activity and superior mechanical properties. Importantly, these composites exhibited no cytotoxicity toward the tested cell line (NIH_3T_3). Cataldi et al. [154] have synthesized a Chitlac-AgNP by stabilizing AgNPs with a polyelectrolyte solution containing Chitlac. Chitilac-AgNPs were subsequently administered onto BisGMA/TEGDMA methacrylic thermosets to evaluate their antibacterial efficacy and cytotoxicity, with the aim of comparing the outcomes to those of uncoated BisGMA/TEGDMA methacrylic thermosets. The evaluation of the proliferative and adhesive potential of human gingival fibroblasts (HGFs) was conducted on both AgNP-coated and uncoated BisGMA/TEGDMA methacrylic thermosets, employing *Streptococcus mitis* as a cultural model system. The viability and quantification of the test microbe were carried out using fluorescence intensity and SEM. The images suggested that chitilac alone coated thermosets contained *S. mitis* microbes in enormous quantities, with FI_{live} and FI_{dead} as 16.18 and 1.8, respectively. However, in the same microbe, when cultured on a chitilac-AgNP-coated thermoset, the amount of living bacterial cells decreased, and the dead cells increased with FI_{live} and FI_{dead} of the order of 4.68 and 6.38, respectively. When the cultures are inoculated with saliva, the amount of detectable living cells further decreases. Low density culture (LDC) cytotoxic assay of the thermoset coated with chitilac and chitilac-AgNPs showed the minimal cytotoxic effect exerted by the thermoset on HGFs after 24 hours as was evident by the lower lactate dehydrogenase (LDH) values. LDH values increased from 8.9 and 7.8-fold to 27.9 and 22-fold for chitilac thermosets and chitilac-Ag thermosets after 48 hours. Nevertheless, the presence of *S. mitis* significantly decreased the release of LDH by human fibroblasts in both instances. Additionally, the introduction of saliva into the cultural medium further diminished the LDH values. Furthermore, a notable reduction in the secretion of collagen type I was observed on both surfaces during the co-culture. The findings indicate that chitilac-AgNPs exhibit promising potential as antimicrobial materials in the realm of restorative dentistry, displaying minimal cytotoxicity against eukaryotic cells. Silver nanoparticles (AgNPs) of two different sizes were synthesized by Cristobal et al. [155] and assessed for their antimicrobial activity and their impact on the adhesion of *S. mutans* to the surfaces of brackets and wires used in orthodontic therapies. The silver nanoparticles exhibited robust antimicrobial and antiadherence properties against *S. mutans*. The outcomes underscore the significant potential of employing AgNPs for managing white spot lesions in orthodontic treatments. There are several research reports on the utilization of silver nanoparticles (AgNPs) for application across various fields within dentistry like orthodontics [139, 141, 156], endodontics [157, 158], periodontics [159, 160], and prosthodontics [161–163].

This section delves into the historical use of silver in dentistry, tracing its roots back to the nineteenth century, primarily driven by the antimicrobial activity of silver ions. Transitioning to the twenty-first century, the discussion highlights the contemporary application of silver in the form of silver nanoparticles (AgNPs) in dentistry. Emphasis is placed on the gradual release of silver ions from AgNPs, showcasing their potential in effectively managing the growth of various pathogenic microbes, including those associated with oral pathogens. The oral cavity, harboring a diverse microflora comprising approximately 70 microbial species, is deemed essential for comprehending the etiology of oral diseases. The balance of microbiota is crucial for maintaining both oral and systemic health. Silver nanoparticles (AgNPs) exhibit strong antimicrobial effects against numerous species of both gram-positive and gram-negative bacteria. Research suggests an inverse correlation between the size of AgNPs and their antimicrobial activity, with smaller-sized AgNPs demonstrating greater effectiveness in inhibiting biofilm formation and targeting oral pathogens like *S. mutans* and *S. orallis*. Multiple studies are cited to support the evidence of AgNPs' efficacy in preventing biofilm formation. For instance, soda-lime glass-coated AgNPs demonstrated *in vitro* inhibition of *S. orallis* biofilm. SEM images revealed that AgNP-coated glass significantly discouraged bacterial adhesion and inhibited biofilm formation. The toxicity of silver nanoparticles (AgNPs) was evaluated by examining the release profile of Ag^+ ions, confirming concentrations below cytotoxic levels.

The integration of silver nanoparticles (AgNPs) into dental materials, including GIC commonly utilized in pediatric dentistry, is examined. GIC impregnated with AgNPs showcased antimicrobial activity against both gram-positive and gram-negative microbes. This antimicrobial effect is attributed to the release of Ag^+ ions from AgNPs, which effectively inhibits bacterial biofilm formation and serves as a preventive measure against dental caries. Innovative approaches, such as synthesizing AgNP-immobilized halloysite nanotubes (HNT) for dental resin composites, are discussed. The mechanical and antimicrobial properties of the resulting Ag/HNTNP-based dental resin composite were compared with conventional composites, highlighting enhanced antimicrobial properties without cytotoxicity. Further studies involve synthesizing Chitlac-AgNP and applying it to methacrylic thermosets to assess antibacterial activity. The results suggest reduced bacterial adhesion and cytotoxicity, presenting Chitlac-AgNPs as promising antimicrobial materials in restorative dentistry. Other research reports, including the synthesis of AgNPs with different sizes for orthodontic therapies and their potential in various fields of dentistry, such as endodontics, periodontics, and prosthodontics, are briefly mentioned. The overall theme emphasizes the versatility and potential of AgNPs in addressing microbial challenges in various dental applications.

2.3 Toxicity of Gold and Silver Nanoparticles

Gold and silver NPs have emerged as prominent nanomaterials with unique physicochemical properties and multifaceted applications across various disciplines, including biomedicine and engineering [93, 95, 159, 164]. Their versatile nature has sparked extensive research to harness their potential in drug delivery, medical imaging, biosensing, catalysis, and antimicrobial therapies, among other innovative applications [65, 165–167]. As these nanoparticles continue to find wider usage, it becomes increasingly crucial to undertake comprehensive investigations into their effects on biological systems and evaluate their potential toxicity. In this section, we embark on a thorough exploration of the existing literature pertaining to the toxicity of gold and silver nanoparticles, aiming to understand the factors that govern their adverse impacts on diverse biological entities. We delve into how particle size, shape, surface modifications, and other physicochemical characteristics intricately influence their interactions with cells, tissues, and organs. By gaining a deeper understanding of the potential hazards associated with these nanoparticles, we can make informed and responsible decisions when employing them in various fields.

2.3.1 Toxicity of Gold Nanoparticles

Several factors influence the toxicity of gold nanoparticles (AuNPs) in biological systems, including but not limited to particle size, morphology, surface modifications, or functionalization. Gold nanoparticles (AuNPs) have been modified with various specific biomolecules, including chemical moieties, peptides, amino acids, and carbohydrate molecules, which increased their biocompatibility for diagnostic or therapeutic applications. However, this surface functionalization directly affects the interaction of AuNPs in biological systems [168]. Hence, it is crucial to investigate the cellular toxicity of gold nanoparticles (AuNPs) in relation to their surface functionalization. As an illustration, Su et al. [169] investigated the influence of surface charge density (SCD) of gold nanoparticles (AuNPs) on the interaction with HeLa cells. They synthesized a range of peptide-ligated AuNPs designated in a series from GNP01 to GNP17. The peptide ligands were used to synthesize the AuNPs with different SCD, which varied from positive in decreasing order from GNP01-GNP08 through zero (GNP09) to negative in increasing order from GNP10-GNP17. The interplay between HeLa cells and arrays of gold nanoparticles (AuNPs) resulted in the cellular uptake of AuNPs, quantitatively evaluated using inductively coupled plasma mass spectrometry (ICP-MS) to analyze Au content in HeLa cells. The results revealed a substantial increase in the cellular uptake of AuNPs, notably at GNP05. The cellular uptake decreased drastically from GNP06 to GNP17, and negligible uptake was observed in this range of AuNP arrays. Therefore, the conclusion drawn was that the cellular uptake of gold nanoparticles (AuNPs) primarily occurred for those AuNPs with high positive values of SCD and was negligible for those AuNP arrays with lower positive values or zero or negative values of SCD. The interaction between gold nanoparticles (AuNPs) and HeLa cells was additionally explored through zeta potential measurements, taking into account the effective SCD and considering the shielding effect of the peptide ligands, as well as surface solvation in solution. The trend between the cellular uptake of AuNPs and effective SCD was found to be almost similar but with different transition points like GNP05 for cellular uptake and GNP for effective SCD. Wang et al. [164] have studied the cytotoxicity of Au nanorods (AuNRs) synthesized using CTAB and compared its cytotoxicity with AuNRs coated with the positively charged poly(diallyl dimethyl ammonium chloride) (PDDAC) and negatively charged PSS (poly(sodium 4-styrene sulfonate)) and HS-PEG-COOH (thiol-terminated polyethylene glycol with carboxyl end groups)-coated rods. The findings suggested that CTAB (cetyltrimethylammonium bromide)-coated gold nanoparticles (AuNPs) caused cell death by creating defects in the cell membrane. In contrast, gold nanorods (AuNRs) coated with PDDAC (poly(diallyldimethylammonium chloride)), PSS (poly(sodium 4-styrene sulfonate)), and HS-PEG-COOH (thiol-terminated polyethylene glycol with carboxyl end groups) exhibited good biocompatibility toward the cells. TEM images depicted a noticeable alteration in the membrane structure of cells subjected to CTAB-coated gold nanorods (AuNRs). In this situation, the nanorods (NRs) coated with CTAB aggregated on the cell membrane, and a portion of these NRs penetrated and entered the cell membrane. The assessment of cell viability for cells exposed to different nanorods (NRs) was conducted using a dead-live assay after 24 hours of exposure. The findings indicated a higher percentage of dead cells in the group treated with CTAB-coated NRs than in the other NR groups (Figure 2.9A). FESEM images revealed the formation of cell defects resulting in cell injury. However, PDDAC and PSS-coated AuNRs formed fewer aggregates on the cell surface and rather got internalized in the cells due to the process of endocytosis (Figure 2.9B). The internalization of gold nanorods (AuNRs) coated with PDDAC (poly(diallyldimethylammonium chloride)) and PSS (poly(sodium 4-styrene sulfonate)) induced the development of additional endosomes and lysosomes in comparison to the control cells, as observed in TEM images (Figure 2.9C). A few PEG-AuNRs were found to be either internalized or adhered to the cells, possibly due to the orientation of the negatively charged functional group toward the aqueous environment, which prevented the PEG-AuNRs from coming close to the cells. The variation in cytotoxicity between gold nanorods (AuNRs) coated with CTAB and AuNRs coated with other charged ligands was assigned to the nature and packing structure of the surface molecules rather than their surface charge. The toxicity of the CTAB-coated AuNRs was assigned to the CTAB bilayer on the AuNRs, which induced cell damage and resulted in cell death. Moreover, the toxicity associated with CTAB-coated gold nanorods (AuNRs) was attributed to the CTAB-AuNRs rather than free CTAB molecules. However, contrary opposite conclusions were published by Alkilany et al. [170] while investigating the toxicity of CTAB-coated Au nanorods on human colon carcinoma cells (HT-29). Their conclusion was that the toxicity against the tested cell line (HT-29) was attributed to free CTAB molecules in the solution rather than CTAB-coated gold nanorods. CTAB molecules with a concentration of 200 nM lowered the cell viability by up to 50%. The cytotoxicity of the CTAB-coated AuNRs was found to decrease substantially by overcoating with polymers like PAA (negatively charged) and PAH (positively charged). The reduction in cytotoxicity due to overcoating was ascribed to the reduction in cell exposure to the CTAB molecules on the AuNR surfaces. The decrease in cytotoxicity was attributed to the findings from inductively coupled plasma mass spectrometry (ICP-MS) studies, indicating that the cellular uptake of gold nanorods (AuNRs) by HT-29 cells depended on the surface chemistry of the AuNRs and ranged from 50 to 2300. Further, AuNRs, when coming in contact with the biological media, were found to bear the same surface charge possibly due to adsorption of the bovine serum albumin molecules on the surface of AuNRs. In a separate investigation conducted by the same research group, CTAB-stabilized AuNRs were prepared and their surfaces were modified with various other surfactant molecules to synthesized positively charged and negatively charged AuNRs [171]. The stability and biocompatibility of these AuNRs were enhanced by polyacrylic acid (PAA), a negatively charged polyelectrolyte marked as PAA-Au Poly(acrylic acid) (PAA)-coated gold nanorods (AuNRs) were subsequently coated with poly(allylamine hydrochloride) (PAH) to generate positively charged AuNRs, designated as PAH-AuNRs. The toxicity of PAA-coated gold nanorods (PAA-AuNRs) and PAH-coated gold nanorods (PAH-AuNRs) was assessed against cultured bovine

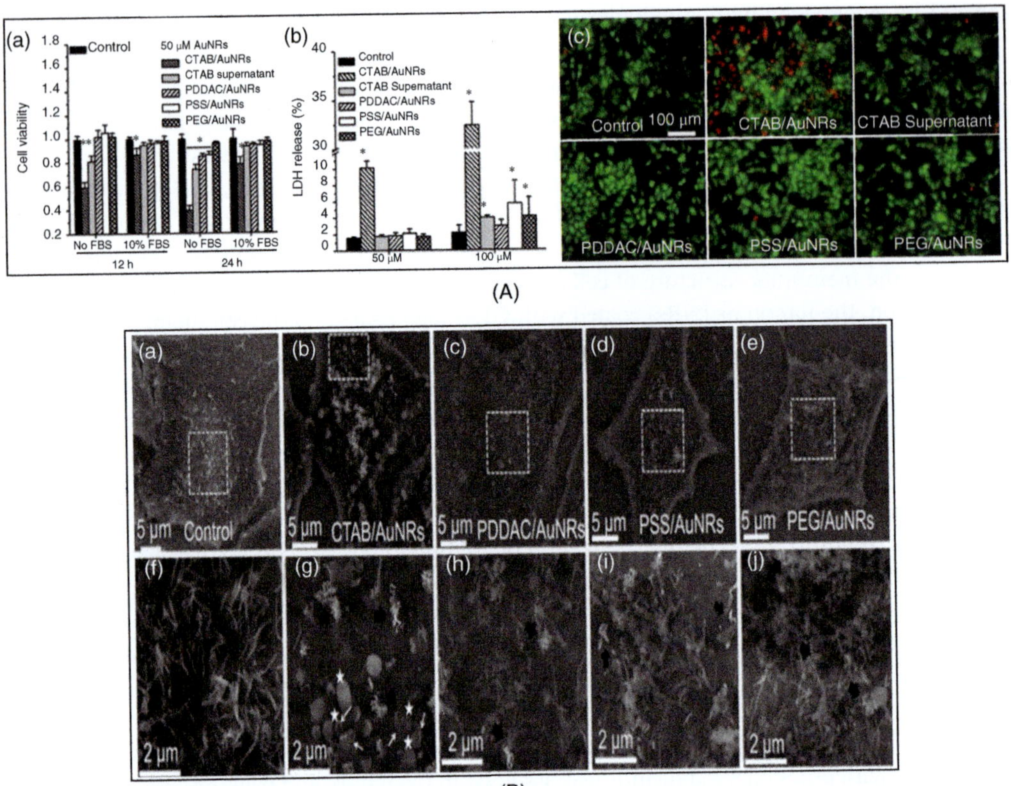

Figure 2.9 (A) Illustrates the impact of the surface characteristics of AuNRs on A549 cells. Subfigure (a) depicts the influence of 50 mM AuNRs on A549 cell viability in both serum-free and complete medium with 10% FBS, assessed through a CCK-8 assay. Subfigure (b) demonstrates the cytotoxicity evaluation of 50 mM AuNRs and CTAB supernatant from the CTAB–AuNRs solution in a serum-free medium, using LDH release. Subfigure (c) presents the results of a live–dead assay conducted after 24 hours of exposure. The asterisk (*) denotes statistically significant effects of AuNRs on cells ($p < 0.05$). The results are presented as mean values and standard deviations based on three replicates. (B) Environmental scanning electron microscope (ESEM) images reveal cytoplasm membrane structures in different conditions. Subfigure (a) displays untreated A549 cells in a serum-free medium, while subfigures (b–e) show A549 cells exposed to 50 mM AuNRs with various surface coatings (CTAB, PDDAC, PSS, and HS-PEG-COOH) for 12 hours. The upper figures (a–e) present whole cell images and their magnified versions are displayed below (f–j). In subfigures (g–j), black arrowheads, white stars, and white arrowheads indicate AuNR aggregates, blebs, and induced defects in the membranes, respectively. (C) Features transmission electron microscope (TEM) images of cytoplasm membranes and intracellular ultrastructures. Similar to Figure 2.3, subfigure (a) shows untreated A549 cells in a serum-free medium, while subfigures (b–e) present A549 cells exposed to 50 mM AuNRs with different surface coatings for 12 hours. The above figures (a–e) represent whole cell images, and their magnified versions are shown below (f–j). In subfigures (g–j), black arrowheads indicate lysosomes or AuNRs within lysosomes, while white arrowheads signify AuNRs inserted in the cell membrane. Source: Reproduced with permission from Wang [164]/Royal Society of Chemistry.

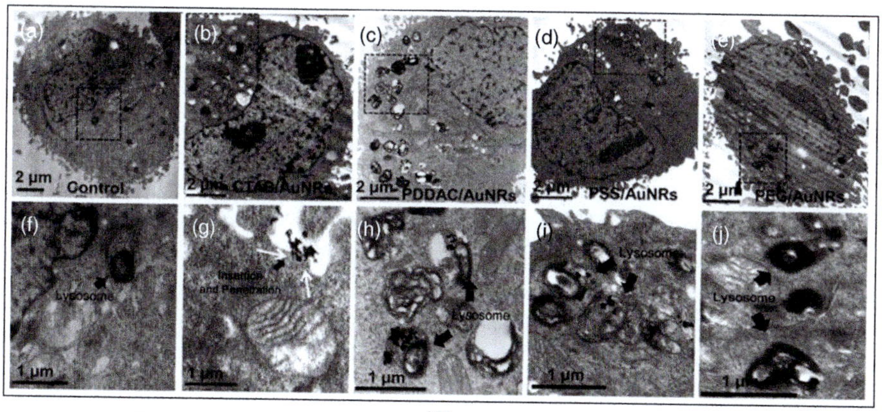

(C)

Figure 2.9 *(Continued)*

aortic endothelial cells (BAEC) through an *in vitro* viability test. Dose-dependent cytotoxicity was recorded by both types of NRs against BAEC. Later, both types of AuNRs were then evaluated for toxicity against endothelial cells in aortic rings. PAH-AuNRs induced significant cell death in endothelial cells at 1 nM concentration. However, toxicity evaluation of AuNRs against smooth muscle cells in aortic rings indicated no significant cell death. Therefore, gold nanorods (AuNRs) were observed to exhibit toxicity toward endothelial cells and not to smooth muscles. In another study, CTAB from AuNRs were replaced from the surface of AuNRs with phosphatidylcholine (PC) by extraction from the AuNR solution into the chloroform phase containing phosphatidylcholine (PC) [172]. Both CTAB-AuNRs and PC-AuNRs were subsequently assessed for cytotoxicity against HeLa cells. It was observed that PC-passivated AuNRs induced the least cytotoxicity to the HeLa cells. Whereas significant cytotoxicity was recorded by twice centrifuged CTAB-capped AuNR solution. The results indicated that the toxicity of CTAB-AuNRs can be significantly diminished by replacing the CTAB with alternative stabilizing agents, which are more biocompatible and less toxic to the cells. Sodium citrate is another stabilizer that influences the toxicity of AuNPs. For instance, citrate-stabilized AuNRs were found to reduce the spread of human alveolar type-II cell lines like A549 and NCIH441. These AuNRs also reduced the viability of cells in the MTT assay. However, the uptake behavior of AuNRs in endothelial or epithelial cells [173]. Among the various capping agents, polyethylene glycol (PEG) was found to be the most biocompatible, which reduces the nonspecific binding of the biomolecules on the AuNR surfaces and also avoids the recognition by macrophages [174]. PEG modified AuNPs also possess prolonged blood circulation and improved permeability and retention compared to other capping agents.

The toxicity of AuNPs varies with the shape of NPs. Gold NPs of various shapes were synthesized by various research groups and investigated for their toxicity on the various cell lines both *in vitro* and *in vivo*. For illustration, Wang et al. [175] have synthesized AuNPs of three different shapes, viz., Au nanorods, Au nanocages, and Au nanohexapods and assessed their toxicity and photothermal efficiency. It was observed that the Au nanohexapods possessed lower toxicity, higher cellular uptake, and comparable photothermal efficiency relative to Au nanorods and nanocages. ICP-MS studies have revealed the photothermal conversion efficiency per atom based on the number of Au atoms per nanostructure in all three types of Au nanostructures. The efficiency of photothermal conversion was observed to be the highest for gold nanocages, followed by gold hexapods and then gold nanorods. The *in vitro* cell toxicity of Au nanostructures was also assessed against the breast MDA-MB-435 breast cancer cells [175]. CTAB-coated gold nanorods exhibited significant toxicity.

However, the toxicity was mitigated by replacing CTAB from the surface of gold nanorods with PEG at the same concentration. PVP-coated gold nanocages also exhibited significant toxicity at higher concentrations. The toxicity of Au nanocages was mitigated by substituting PVP with PEG. However, the PEGylated Au hexapods displayed no toxicity at all against the cell line. Based on the molecular simulations, one of the studies revealed that spherical AuNPs possess the highest internalization due to the least membrane binding energies rate, followed by cubic AuNPs and disc-shaped AuNPs [176]. Nangia et al. [177] have done an analysis of the rate of internalization of the AuNPs of different shapes on the basis of MD simulations and charge density. The authors have concluded that rice-shaped AuNPs possess the highest rates of internalization as compared to Au nanospheres, Au nanopyramids, Au nanocones, Au nanorods, and Au nanocubes. Wang et al. [178] have revealed the effect of engineered AuNPs of different shapes on the growth and development of zebrafish embryos. The results indicated that AuNPs exhibit toxicity as a function of size. The AuNRs displayed the least toxicity; no significant malfunction was observed in the zebrafish embryos incubated with AuNRs. In contrast, Au nanospheres were found to be most toxic to the zebrafish embryos. Karakoçaka et al. [179] have investigated the biocompatibility of AuNP of various shapes such as Au nanospheres of various sizes, Au nanocubes, and Au nanorods on retinal pigment epithelial cell line (APRE-19) using MTT assay. The results indicated that Au nanospheres (size: 50 and 100 nm) and Au nanocubes with size of 50 nm did not show any toxicity toward the cell line. Whereas Au nanospheres with sizes between 5 and 30 nm and Au nanorods adversely decreased the cell viability even at lower concentrations (0.05 mg/mL).

The impact of surface hydrophobicity on the toxicity of gold nanoparticles (AuNPs) has been explored. Chompoosor et al. [180] produced a series of AuNPs altered with quaternary ammonium salts with varying chain lengths ranging from C_1 to C_6. The acute toxicity of this series of AuNPs was investigated by mitochondrial activity assay and the long-term toxicity by generation of ROS species and DNA damage. The toxicity results have been found to be dependent upon the chain length. The results have suggested that acute cellular toxicity and ROS production increased with an increase in hydrophobicity. However, DNA damage decreased with an increase in AuNP hydrophobicity. Uz et al. [181] investigated the effect of PEG density on PEGylated gold nanoparticles (AuNPs) concerning cell cycle arrest, DNA damage, and apoptosis. Gold NPs of two different core diameters were synthesized, and each gold NP core was coated with PEG of three different densities. It was noted that protein absorption was controlled by PEG hydrodynamic volumes and steric hindrance. In contrast, cellular uptake and cell viability varied with AuNP core diameter and PEG hydrodynamic volumes. Gold nanoparticles (AuNPs) with various PEG grafting densities induced significant cell cycle arrest and DNA damage in $CaCo_2$ and PC3 cells, with no observed apoptosis. However, no severe cell damage was noticed on 3T3 cells by AuNPs at one particular grafting density (~0.65 chains/nm^2). The results indicate that carefully choosing the grafting density can result in a discriminatory behavior of PEGylated AuNPs between healthy (3T3) and cancerous cells. The effect of PEGylated AuNPs on the viability of skeletal muscle cells was explored by Leite et al. [182]. It was witnessed that PEGylated AuNPs, when co-cultured with differential skeletal muscle cells (C_2C_{12}), caused no change in cell viability. However, an increase in mitochondrial membrane potential and intracellular ATP levels has been observed. An increase in the values of INF-c and TGF-b1 was also observed, which suggested the potential ability of AuNPs to activate fibrosis and inflammation, which can result in cellular susceptibility to death stimuli. The researchers investigated the cellular uptake and intracellular trafficking of 15 nm AuNPs coated with citrate and PEG when exposed to human alveolar epithelial cells (A549) [183]. The results suggested that more PEGylated AuNPs were found in the cytosol. However, the preferred site of AuNP localization was found to be vesicles. PEGylated AuNPs were found to be internalized in small vesicles, and the cellular uptake was mostly caused by endocytosis. Citrate-stabilized AuNPs were found to be localized mostly in large-sized vesicles, and the uptake occurred by micropinocytosis. Larger exposure times resulted in the accumulation of both types of AuNPs in larger vesicles, which suggested the intracellular trafficking of the NPs. The findings suggest that the surface coatings of AuNPs play a crucial role in the cellular uptake and trafficking of these nanoparticles. Numerous research studies support the notion that PEGylation is a widely adopted method to mitigate toxicity and enhance the biological compatibility of AuNPs [184, 185].

The size of AuNPs is a crucial parameter that influences their toxicity in biological systems. Gold NPs of different sizes can be taken up by biological cells through various processes, and the uptake mechanism depends on the particle size [186, 187]. A research group has observed that AuNPs are taken up by biological cells through receptor-mediated endocytosis facilitated by the nonspecific adsorption of serum proteins on the surface of the AuNPs [188]. Gold nanoparticles (AuNPs) were synthesized in three size groups: 10, 25, and 50 nm. The internalization studies revealed that 50 nm-sized AuNPs were most easily taken up by cells, followed by 25 nm AuNPs and 10 nm AuNPs [188]. The findings suggest that the cellular uptake of AuNPs is influenced by their size [189–191]. Sabine et al. [192] investigated the size-dependent toxicity of AuNPs using four different cell lines. They systematically studied the toxicity of AuNPs stabilized by triphenyl phosphine derivatives, such as triphenylphosphine monosulfonate (TPPMS) and tris-sulfonated triphenylphosphine (TPPTS), with AuNP sizes ranging from 0.8 to 15 nm against the cell lines L929, J774A1, HeLa, and SKMel-28. The findings substantiated the size-dependent toxicity of AuNPs, where the 1.4 nm-sized AuNPs exhibited the highest toxicity, as indicated by IC50 values ranging from 30 mmol/L to 46 mmol/L. Interestingly, Au nanoclusters, which were stabilized by the same ligand and varied in size (0.8, 1.2, and 1.8 nm), demonstrated significantly lower toxicity, approximately six to eight times less than that observed for 1.4 nm-sized AuNPs. The calculated IC50 values for these Au clusters fell within the range of 140 to 250 mmol/L. However, 15 nm-sized AuNPs displayed no toxicity at all against all four cell lines up to the concentrations of 6300 μm. The toxicity assessment of 1.4 nm-sized AuNPs revealed that it was independent of the type of ligand used for their stabilization. Furthermore, co-culturing 1.4 nm AuNPs with test cell lines for 12 hours led to cell necrosis. In contrast, under similar conditions, 1.2 nm AuNPs-induced cell apoptosis. Another study by the same research group focused on investigating the toxicity of 1.4 nm AuNPs based on their ligand chemistry [193]. The authors observed that pretreatment of 1.4 nm-sized AuNPs with agents like glutathione, TPPMS, and *N*-acetylcysteine reduces their toxicity. Gold NPs of a similar size but capped with glutathione also do not induce oxidative stress. Hence, much less toxicity was observed with 1.4 nm AuNPs coated with glutathione. A separate research team explored the impact of AuNPs with three distinct sizes (1.5, 4, and 14 nm) on various aspects, including cell viability, DNA methylation, pluripotency, and neuronal differentiation of human embryonic stem cells (hESCs) [194]. The findings revealed that exposure to 1.5 nm AuNPs led to cell death in hESC-derived neural progenitor cells. hESC cells exposed to 1.5 nm thiolated AuNPs exhibited loss of cohesiveness and detachment even at concentrations as low as 0.1 μg/mL. In contrast, gold NPs of other sizes showed minimal toxic effects on hESC cells, even at concentrations as high as 10 μg/mL. Additionally, treatment with thiolated AuNPs at a concentration of 10 μg/mL for 24 hours resulted in a significant decrease in global DNA methylation and a subsequent increase in global hydroxymethylation in hESC DNA. Yen et al. [195] also explored the impact of AuNP particle size on the toxicity to J774 A1 macrophages. They have observed that AuNPs demonstrated size-dependent toxicity and the toxicity decreased with an increase in particle size. Further, increase in the concentration of AuNPs resulted in an increase in toxicity due to uptake of AuNPs by cells in higher amounts. AuNPs with concentration below 10 ppm showed no significant toxicity. However, significant toxicity to J774 A1 macrophages was recorded at the

AuNP concentration of 10 ppm. Treatment of macrophages with AuNPs resulted in an increase in the size of the macrophage cells with spread morphology. Further upregulation of IL-1, IL-6, and TNF-a genes occurred in macrophages treated with AuNPs at 1 ppm concentration for 3 hours. The results indicate that treatment of macrophages with AuNPs, even at a low concentration of 1 ppm prompted the immunological response event, though no significant cytotoxicity was induced. No particle size-dependent cytotoxicity was recorded by Freese et al. [173] after conducting experiments involving the incubation of gold nanoparticles (AuNPs) with three distinct sizes (10, 11, and 25 nm) alongside human cerebral microvascular endothelial (hcMEC/D3) cells and human dermal microvascular endothelial (HDMEC) cells. The observed toxicity was determined to be dependent on the dosage administered. Significant adverse effects on cell viability for both cell types were noted at elevated concentrations of AuNPs, specifically at 500 μM and 1000 μM. Additionally, serum protein-coated citrate-stabilized AuNPs with a size of 17 nm exhibited toxicity toward A549 cells. The assessment of toxicity was conducted through LDH, MTT assays, and ATP depletion measurements [196]. Numerous studies have been conducted to assess the impact of particle size on the toxicity of mammalian cells and tissues [197–200]. Toxicity induced by gold nanoparticles (AuNPs) has been attributed to oxidative stress, depletion of intracellular oxidative pool, generation of ROS, and reactive nitrogen species, necrosis, release of nitric oxide, and inhibition of vascular endothelial growth factor (VEGF)-induced cell proliferation [193, 201–204].

In conclusion, the toxicity of gold nanoparticles (AuNPs) in biological systems is impacted by various factors, including particle size, shape, and surface modifications. Although the surface functionalization of AuNPs with specific biomolecules can enhance their biocompatibility for diagnostic or therapeutic applications, it also modifies their interactions within biological systems.

Studies indicate that the cellular toxicity of gold nanoparticles (AuNPs) is contingent upon their SCD. Higher positive SCD values are associated with increased cellular uptake of AuNPs, while lower positive or negative SCD values result in negligible cellular uptake. Zeta potential measurements have confirmed the correlation between cellular uptake and effective SCD after considering the shielding effect of peptide ligands.

AuNPs of different shapes have been investigated for their toxicity, and size-dependent effects have been observed. Smaller AuNPs generally exhibit higher toxicity compared to larger ones. Gold nanorods coated with CTAB were observed to trigger cell death by causing damage to the cell membrane, while PDDAC, PSS, and HS-PEG-COOH-coated AuNRs displayed good biocompatibility. Surface hydrophobicity also plays a role in the toxicity of AuNPs. Hydrophobic AuNPs were found to induce higher cellular toxicity, but replacing hydrophobic stabilizing agents with biocompatible ones like PEG can reduce the toxicity. Overall, the size of gold nanoparticles (AuNPs) plays a crucial role in determining their toxicity, with smaller NPs often showing heightened toxicity. The entry into cells, internal processes, and movement within cells of gold nanoparticles (AuNPs) display differences influenced by both their size and surface coatings. PEGylation is a common method used to reduce toxicity and improve biocompatibility. AuNPs have been found to induce toxicity through various mechanisms, including oxidative stress, ROS production, necrosis, and inhibition of cell proliferation.

In summary, understanding the factors governing the toxicity of AuNPs in biological systems is crucial for safe and effective biomedical applications. Further research in this area is essential to ensure the safe use of AuNPs in various biomedical and nanomedicine applications.

2.3.2 Toxicity of Silver Nanoparticles

Silver NPs have received considerable attention as antimicrobial agents for various engineering applications [205, 206]. Considering the diverse potential applications, it is imperative to assess the potential toxicity of silver nanoparticles (AgNPs) on biological systems. Silver NP-coated fabrics were produced by Lee et al. and evaluated for their antimicrobial activity. Further, the AgNP-treated fabrics were evaluated for their safety for skin irritations against the guinea pigs. The results indicated that AgNP-based cotton fabrics were safe and caused no skin irritation to guinea pigs [207]. Kim et al. conducted a study on the toxicity of silver nanoparticles (AgNPs) in human hepatoma cells. The results revealed that silver NPs aggregated in the cytoplasm and nuclei of the treated cells, leading to cytotoxicity through the induction of oxidative stress [208]. Ahmad et al. [209] conducted a study on the impact of both polysaccharide-functionalized and non-functionalized silver nanoparticles (AgNPs) on DNA damage in mouse embryonic stem (mES) cells and mouse embryonic fibroblasts. Both types of AgNPs were found to elicit cell death, but the cellular response to DNA damage varied depending on the surface chemistry of the AgNPs. Polysaccharide-coated AgNPs induced more pronounced DNA damage compared to their uncoated counterparts. Furthermore, the distribution of polysaccharide-coated AgNPs within the cells was noted to be more uniform than that of the uncoated AgNPs. The AgNPs coated with polysaccharides showed a greater induction of apoptosis compared to the uncoated AgNPs [209]. Silver NPs of the sizes of 15 and 100 nm were evaluated for *in vitro* toxicity in BRL 3A rat liver cells [210]. The findings revealed that the toxicity of AgNPs exhibited a dose-dependent pattern, with elevated LDG levels observed at higher concentrations. Notably, significant cytotoxicity was observed within the concentration range of 10–50 μg/mL. When comparing silver nanoparticles (AgNPs) of sizes 100 and 15 nm at concentrations of 25 and 50 μg/mL, the larger 100 nm-sized AgNPs demonstrated higher toxicity. The MTT assay further confirmed considerable cytotoxic effects of AgNPs in the range of 5–50 μg/mL on rat liver cells, with EC50 values of 24 μg/mL for 15 nm AgNPs and 19 μg/mL for 100 nm AgNPs. Exposure of rat liver cells to AgNPs resulted in reduction of GSH levels and ROS accumulation inside the cells, which resulted in a compromised cellular antioxidant defense mechanism. The effect of 45 nm-sized AgNPs on the exposure of coronary endothelial cells (CEC) has been studied by Hernandez et al. [211]. The impact of AgNPs on CEC was examined by exposing CEC to escalating concentrations of AgNPs (0.1–100 μg/mL) for a duration of 24 hours. The results demonstrated that low concentrations of AgNPs (1–10 μg/mL) inhibited the proliferation of CEC as compared to cell cultures grown in absence of AgNPs. However, significant cell proliferation was observed by culturing CEC with AgNPs at high concentrations (50–100 μg/mL). Cell proliferation was explained to the increased production of NO at higher AgNP concentrations. The neurotoxic effect of AgNP on the rat hippocampal CA1 neurons was investigated by Zhaowei et al. [212]. Silver nanoparticles (AgNPs) spanning sizes from 32 to 380 nm decreased peak amplitude ${I_{Na}}^+$ which resulted in a decline in Na^+ concentration. The alteration in I_{Na+} probably led to a change in the shape of the action potential, resulting in observed neural dysfunction. The intraperitoneal injection of 25 nm-sized silver nanoparticles (AgNPs) led to changes in gene expression, oxidative stress from free radicals, and apoptosis in the brain tissues of the examined mice [213]. Oral administration of AgNPs solution to rats resulted in atropic and degenerative changes in the ventral and dorsal surfaces of the tongue and salivary glands [214]. Many reports have been published on the *in vitro* and *in vivo* toxicity of AgNPs [215–218].

Several research papers have highlighted that the toxicity of silver nanoparticles is contingent upon their concentration, environmental factors, size, and exposure time [219]. Larger silver nanoparticles exhibit lower cytotoxicity compared to smaller ones due to their higher surface area per unit mass and the increased release of Ag^+ ions. The cytotoxicity of silver nanoparticles larger than 10 nm is predominantly influenced by the direct interaction between the nanoparticles and macrophages [220]. In experiments involving sizes ranging from 10 to 100 nm and concentrations between 5 and 10 mg/mL, silver nanoparticles (AgNPs) significantly affect mitochondrial functions [221]. Moreover, *in vitro* studies have indicated that silver nanoparticles have an impact on reproduction and can induce morphological abnormalities in non-mammalian animals [222]. The cytotoxicity of silver nanoparticles is affected by various factors including surface chemistry, crystallinity, shape, size, and capping agents. Additionally, environmental factors such as pH, ionic strength, and the presence of complexing agents also play a role [223].

Exposure of humans to silver nanoparticles (AgNPs) takes place through diverse pathways, encompassing dermal contact, inhalation, oral administration, blood circulation, and vaginal administration [224–226]. The inhalation of silver nanoparticles (AgNPs) has been associated with respiratory toxicities, including pulmonary inflammation, airway hyperresponsiveness, and asthma [227, 228]. Research indicates that the liver is notably sensitive to the inhalation of silver nanoparticles (AgNPs), and a proposed occupational exposure limit of $0.19\,\mu g/m^3$ has been suggested for AgNPs [229]. The ingestion of silver nanoparticles (AgNPs) constitutes a significant route of exposure, with adults estimated to have a daily intake ranging from 20 to 80 μg. *In vivo* studies have shown that AgNPs can elicit various adverse effects, such as gastrointestinal inflammation, oxidative stress, alterations in the structure of intestinal villi, disruptions in gut microbiota, and liver injuries [230–232]. Even though intact skin acts as a barrier to silver nanoparticles (AgNPs), specific nanoproducts used in wound healing and infection prevention may enable the direct entry of AgNPs into subcutaneous tissue and the bloodstream. Consequently, this can lead to allergic dermatitis and eye irritation [233]. Within the realm of biomedicine, it is essential to acknowledge iatrogenic exposure to silver nanoparticles (AgNPs). Administering medications containing silver nanoparticles (AgNPs) intranasally can lead to their distribution in diverse organs, including the liver, spleen, kidney, lung, nasal cavity, olfactory bulb, and lateral ventricle. This distribution has been observed in animal experiments [234]. Accumulation of silver nanoparticles (AgNPs) in these organs may lead to nasal mucosa erosion and hemosiderin deposition in the spleen. These effects are linked to oxidative stress in the nose and macrophage-mediated destruction of erythrocytes in the spleen, respectively [234]. The systemic administration of silver nanoparticles (AgNPs) has been observed to primarily concentrate in the liver and spleen, causing inflammatory infiltration and necrosis [235].

In summary, silver nanoparticles (AgNPs) have been extensively investigated for their antimicrobial properties and applications in various bioengineering fields. However, their potential impact on biological systems raises concerns. Numerous studies have indicated that AgNPs can lead to cytotoxicity, oxidative stress, DNA damage, and apoptosis across different cell types. The toxicity of AgNPs is influenced by factors such as concentration, size, surface chemistry, and environmental conditions. Particularly, the size of AgNPs is a crucial determinant in their cytotoxicity, as smaller nanoparticles generally exhibit greater toxicity compared to larger ones. Moreover, the cytotoxic effects are attributed, in part, to the release of Ag^+ ions from the nanoparticles. The interaction between silver nanoparticles (AgNPs) and macrophages has been identified as a critical factor influencing their cytotoxicity, particularly for nanoparticles larger than 10 nm. Human exposure to AgNPs can occur through various routes, including dermal contact, inhalation, oral administration, and blood circulation. Inhalation of AgNPs has been associated with respiratory toxicities, while oral exposure can result in gastrointestinal inflammation, oxidative stress, and structural changes in the intestines. Although intact skin serves as a barrier against AgNPs, certain nano-products utilized in wound healing and infection prevention can facilitate their direct entry into subcutaneous tissue and blood circulation, leading to adverse effects.

In conclusion, the toxicity of silver nanoparticles (AgNPs) is influenced by various factors, emphasizing the need for further research to enhance our understanding of their potential risks. Implementing appropriate safety measures in the diverse applications of AgNPs is crucial to ensure their responsible use.

2.4 Conclusion and Future Perspectives

In conclusion, the chapter underscores the multifaceted applications of AuNPs and AgNPs in the biomedical field. AuNPs exhibit significant potential in biomedical imaging, offering advantages in various imaging techniques and multimodal imaging. Their therapeutic applications span diverse diseases, emphasizing their unique properties and the importance of size and shape in influencing their behavior. Furthermore, the biosensing applications of functionalized AuNPs showcase their versatility in specific and sensitive detection. On the other hand, AgNPs are highlighted for their antimicrobial properties, playing a pivotal role in addressing multidrug resistance and wound healing. The chapter concludes by emphasizing the need for continued research to understand and harness the potential of these nanoparticles for safe and effective biomedical applications. The chapter lays the foundation for several future perspectives in the application of gold nanoparticles (AuNPs) and silver nanoparticles (AgNPs) in biomedical contexts. These prospects include:

Advanced imaging technologies. The continuous evolution of imaging technologies will likely see increased integration with AuNPs, opening avenues for even more precise and sensitive imaging. Future research may explore novel imaging techniques or improvements in existing ones, with a focus on enhancing contrast and resolution.

Therapeutic innovations. The therapeutic potential of AuNPs is likely to be a focal point for future research. This could encompass the exploration and development of innovative drug delivery strategies, innovative therapeutic agents, or advancements in targeting specific diseases. Tailoring the size, shape, and surface modifications of AuNPs for optimized therapeutic outcomes will be an area of exploration.

Biocompatibility enhancement. Further research will likely delve into strategies to enhance the biocompatibility of AuNPs. This could involve refining surface modifications, exploring new biocompatible coatings, or investigating alternative synthesis methods to minimize toxicity and improve overall safety in biological systems.

Biosensing advancements. Future developments in biosensing are anticipated to leverage the unique properties of AuNPs for even more sophisticated and sensitive detection methods. Research may focus on expanding the range of detectable biomolecules, enhancing the speed and accuracy of detection, and adapting these technologies for point-of-care applications.

Antimicrobial materials and wound healing. The application of AgNPs in antimicrobial materials and wound healing is likely to see further refinements. Future research may explore new materials for incorporating AgNPs, optimizing release profiles, and addressing challenges related to potential cytotoxicity. There could also be investigations into incorporating AgNPs into 3D-printed structures or smart materials for advanced wound care.

Safety and toxicity studies. As the use of nanoparticles in biomedical applications expands, there will be a growing emphasis on safety and toxicity studies. Future research will likely explore in-depth mechanisms of cellular interactions, long-term effects, and potential cumulative toxicity. This is crucial for establishing comprehensive guidelines and ensuring the safe use of these nanoparticles.

Multifunctional nanoparticles. The development of multifunctional nanoparticles, capable of serving multiple purposes simultaneously, is an exciting avenue. Future research may aim to create nanoparticles that integrate imaging, therapeutic, and biosensing capabilities within a single platform, providing a holistic approach to biomedical applications.

Clinical translation. Bridging the gap between laboratory research and clinical applications will be a key focus. Future studies may involve more extensive preclinical trials, addressing regulatory challenges and working toward the incorporation of these nanoparticles into routine clinical practice.

In summary, the future perspectives of this chapter revolve around refining and expanding the applications of AuNPs and AgNPs in biomedical settings, with a strong emphasis on safety, efficacy, and sustainability. Continued interdisciplinary collaboration between materials scientists, biologists, and clinicians will play a pivotal role in driving these advancements.

References

1 Popovtzer R, Agrawal A, Kotov N A, Popovtzer A, Balter J, Carey Thomas E and Kopelman R 2008 Targeted gold nanoparticles enable molecular CT imaging of cancer *Nano Lett.* 8 4593–6.

2 Hainfeld J F, Dilmanian F A, Slatkin D N and Smilowitz H M 2008 Radiotherapy enhancement with gold nanoparticles *J. Pharm. Pharmacol.* 60 977–85.

3 Weissleder R 2001 A clearer vision for in vivo imaging *Nat. Biotechnol.* 19 316–7.

4 Dykman L and Khlebtsov N 2012 Gold nanoparticles in biomedical applications: recent advances and perspectives *Chem. Soc. Rev.* 41 2256–82.

5 Burduşel A-C, Gherasim O, Grumezescu A M, Mogoantă L, Ficai A and Andronescu E 2018 Biomedical applications of silver nanoparticles: an up-to-date overview *Nanomaterials* 8 681.

6 Huang X and El-Sayed M A 2011 Plasmonic photo-thermal therapy (PPTT) *Alex. J. Med.* 47 1–9.

7 Haider A and Kang I-K 2015 Preparation of silver nanoparticles and their industrial and biomedical applications: a comprehensive review *Adv. Mater. Sci. Eng.* 2015 e165257.

8 Shanmuganathan R, Karuppusamy I, Saravanan M, Muthukumar H, Ponnuchamy K, Ramkumar V S and Pugazhendhi A 2019 Synthesis of silver nanoparticles and their biomedical applications – a comprehensive review *Curr. Pharm. Des.* 25 2650–60.

9 Anil Bansal S, Kumar V, Karimi J, Pal Singh A and Kumar S 2020 Role of gold nanoparticles in advanced biomedical applications *Nanoscale Adv.* 2 3764–87.

10 Zhang X 2015 Gold nanoparticles: recent advances in the biomedical applications *Cell Biochem. Biophys.* 72 771–5.

11 Kim E Y, Kumar D, Khang G and Lim D-K 2015 Recent advances in gold nanoparticle-based bioengineering applications *J. Mater. Chem. B* 3 8433–44.

12 Kimling J, Maier M, Okenve B, Kotaidis V, Ballot H and Plech A 2006 Turkevich method for gold nanoparticle synthesis revisited *J. Phys. Chem. B* 110 15700–7.

13 Kumar P S, Pastoriza-Santos I, Rodríguez-González B, de Abajo F J G and Liz-Marzán L M 2007 High-yield synthesis and optical response of gold nanostars *Nanotechnology* 19 015606.

14 Skrabalak S E, Chen J, Sun Y, Lu X, Au L, Cobley C M and Xia Y 2008 Gold nanocages: synthesis, properties, and applications *Acc. Chem. Res.* 41 1587–95.

15 Lal S, Westcott S L, Taylor R N, Jackson J B, Nordlander P and Halas N J 2002 Light interaction between gold nanoshells plasmon resonance and planar optical waveguides *J. Phys. Chem. B* 106 5609–12.

16 Funston A M, Novo C, Davis T J and Mulvaney P 2009 Plasmon coupling of gold nanorods at short distances and in different geometries *Nano Lett.* 9 1651–8.

17 Lin S, Li M, Dujardin E, Girard C and Mann S 2005 One-dimensional plasmon coupling by facile self-assembly of gold nanoparticles into branched chain networks *Adv. Mater.* 17 2553–9.

18 Rosman C, Pierrat S, Henkel A, Tarantola M, Schneider D, Sunnick E, Janshoff A and Sönnichsen C 2012 A new approach to assess gold nanoparticle uptake by mammalian cells: combining optical dark-field and transmission electron microscopy *Small* 8 3683–90.

19 Deboutttière P-J, Roux S, Vocanson F, Billotey C, Beuf O, Favre-Réguillon A, Lin Y, Pellet-Rostaing S, Lamartine R, Perriat P and Tillement O 2006 Design of gold nanoparticles for magnetic resonance imaging *Adv. Funct. Mater.* 16 2330–9.

20 Alric C, Taleb J, Le Duc G, Mandon C, Billotey C, Le Meur-Herland A, Brochard T, Vocanson F, Janier M, Perriat P, Roux S and Tillement O 2008 Gadolinium chelate coated gold nanoparticles as contrast agents for both X-ray computed tomography and magnetic resonance imaging *J. Am. Chem. Soc.* 130 5908–15.

21 Li W and Chen X 2015 Gold nanoparticles for photoacoustic imaging *Nanomedicine* 10 299–320.

22 Mallidi S, Larson T, Tam J, Joshi P P, Karpiouk A, Sokolov K and Emelianov S 2009 Multiwavelength photoacoustic imaging and plasmon resonance coupling of gold nanoparticles for selective detection of cancer *Nano Lett.* 9 2825–31.

23 Yu K-K, Li K, Lu C-Y, Xie Y-M, Liu Y-H, Zhou Q, Bao J-K and Yu X-Q 2020 Multifunctional gold nanoparticles as smart nanovehicles with enhanced tumour-targeting abilities for intracellular pH mapping and in vivo MR/fluorescence imaging *Nanoscale* 12 2002–10.

24 Hainfeld J F, Slatkin D N, Focella T M and Smilowitz H M 2006 Gold nanoparticles: a new X-ray contrast agent *Br. J. Radiol.* 79(939) 248–53.

25 Arvizo R R, Miranda O R, Moyano D F, Walden C A, Giri K, Bhattacharya R, Robertson J D, Rotello V M, Reid J M and Mukherjee P 2011 Modulating pharmacokinetics, tumor uptake and biodistribution by engineered nanoparticles *PLoS One* 6 e24374.

26 Longmire M, Choyke P L and Kobayashi H 2008 Clearance properties of nano-sized particles and molecules as imaging agents: considerations and caveats *Nanomedicine* 3 703–17.

27 Sun I-C, Eun D-K, Koo H, Ko C-Y, Kim H-S, Yi D K, Choi K, Kwon I C, Kim K and Ahn C-H 2011 Tumor-targeting gold particles for dual computed tomography/optical cancer imaging *Angew. Chem. Int. Ed.* 50 9348–51.

28 Peng C, Zheng L, Chen Q, Shen M, Guo R, Wang H, Cao X, Zhang G and Shi X 2012 PEGylated dendrimer-entrapped gold nanoparticles for in vivo blood pool and tumor imaging by computed tomography *Biomaterials* 33 1107–19.

29 Si-Mohamed S, Cormode D P, Bar-Ness D, Sigovan M, Naha P C, Langlois J-B, Chalabreysse L, Coulon P, Blevis I, Roessl E, Erhard K, Boussel L and Douek P 2017 Evaluation of spectral photon counting computed tomography K-edge imaging for determination of gold nanoparticle biodistribution in vivo *Nanoscale* 9 18246–57.

30 Dong Y C, Hajfathalian M, Maidment P S N, Hsu J C, Naha P C, Si-Mohamed S, Breuilly M, Kim J, Chhour P, Douek P, Litt H I and Cormode D P 2019 Effect of gold nanoparticle size on their properties as contrast agents for computed tomography *Sci. Rep.* 9 14912.

31 Cormode D P, Roessl E, Thran A, Skajaa T, Gordon R E, Schlomka J P, Fuster V, Fisher E A, Mulder W J M, Proksa R and Fayad Z A 2010 Atherosclerotic plaque composition: analysis with multicolor CT and targeted gold nanoparticles *Radiology* 256 774–82.

32 Kim Y-H, Jeon J, Hong S H, Rhim W-K, Lee Y-S, Youn H, Chung J-K, Lee M C, Lee D S, Kang K W and Nam J-M 2011 Tumor targeting and imaging using cyclic RGD-PEGylated gold nanoparticle probes with directly conjugated iodine 125 *Small* 7 2052–60.

33 Moriggi L, Cannizzo C, Dumas E, Mayer C R, Ulianov A and Helm L 2009 Gold nanoparticles functionalized with gadolinium chelates as high-relaxivity MRI contrast agents *J. Am. Chem. Soc.* 131 10828–9.

34 Key J and Leary J F 2014 Nanoparticles for multimodal in vivo imaging in nanomedicine *Int. J. Nanomedicine* 9 711–26.

35 Black K C L, Akers W J, Sudlow G, Xu B, Laforest R and Achilefu S 2014 Dual-radiolabeled nanoparticle SPECT probes for bioimaging *Nanoscale* 7 440–4.

36 Zhao Y, Sultan D, Detering L, Cho S, Sun G, Pierce R, Wooley K L and Liu Y 2014 Copper-64-alloyed gold nanoparticles for cancer imaging: improved radiolabel stability and diagnostic accuracy *Angew. Chem.* 126 160–3.

37 Wang Y, Liu Y, Luehmann H, Xia X, Brown P, Jarreau C, Welch M and Xia Y 2012 Evaluating the pharmacokinetics and in vivo cancer targeting capability of Au nanocages by positron emission tomography imaging *ACS Nano* 6 5880–8.

38 Kircher M F, de la Zerda A, Jokerst J V, Zavaleta C L, Kempen P J, Mittra E, Pitter K, Huang R, Campos C, Habte F, Sinclair R, Brennan C W, Mellinghoff I K, Holland E C and Gambhir S S 2012 A brain tumor molecular imaging strategy using a new triple-modality MRI-photoacoustic-Raman nanoparticle *Nat. Med.* 18 829–34.

39 Marquardt R M, Nafiujjaman M, Kim T H, Chung S-J, Hadrick K, Kim T and Jeong J-W 2022 A mouse model of endometriosis with nanoparticle labeling for in vivo photoacoustic imaging *Reprod. Sci.* 29 2947–59.

40 Wang R, Deng J, He D, Yang E, Yang W, Shi D, Jiang Y, Qiu Z, Webster T J and Shen Y 2019 PEGylated hollow gold nanoparticles for combined X-ray radiation and photothermal therapy in vitro and enhanced CT imaging in vivo *Nanomedicine Nanotechnol. Biol. Med.* 16 195–205.

41 Loynachan C N, Soleimany A P, Dudani J S, Lin Y, Najer A, Bekdemir A, Chen Q, Bhatia S N and Stevens M M 2019 Renal clearable catalytic gold nanoclusters for in vivo disease monitoring *Nat. Nanotechnol.* 14 883–90.

42 Khademi S, Sarkar S, Shakeri-Zadeh A, Attaran N, Kharrazi S, Ay M R, Azimian H and Ghadiri H 2019 Targeted gold nanoparticles enable molecular CT imaging of head and neck cancer: an in vivo study *Int. J. Biochem. Cell Biol.* 114 105554.

43 Kotcherlakota R, Nimushakavi S, Roy A, Yadavalli H C, Mukherjee S, Haque S and Patra C R 2019 Biosynthesized gold nanoparticles: in vivo study of near-infrared fluorescence (NIR)-based bio-imaging and cell labeling applications *ACS Biomater Sci. Eng.* 5 5439–52.

44 Boros E, Dyson P J and Gasser G 2020 Classification of metal-based drugs according to their mechanisms of action *Chem* 6 41–60.

45 Aderibigbe B A 2017 Metal-based nanoparticles for the treatment of infectious diseases *Molecules* 22 1370.

46 Yu Y, Xu Q, He S, Xiong H, Zhang Q, Xu W, Ricotta V, Bai L, Zhang Q, Yu Z, Ding J, Xiao H and Zhou D 2019 Recent advances in delivery of photosensitive metal-based drugs *Coord. Chem. Rev.* 387 154–79.

47 Ghosh S 2019 Cisplatin: the first metal based anticancer drug *Bioorganic Chem.* 88 102925.

48 Rixe O, Ortuzar W, Alvarez M, Parker R, Reed E, Paull K and Fojo T 1996 Oxaliplatin, tetraplatin, cisplatin, and carboplatin: Spectrum of activity in drug-resistant cell lines and in the cell lines of the national cancer institute's anticancer drug screen panel *Biochem. Pharmacol.* 52 1855–65.

49 Sanderson B J S, Ferguson L R and Denny W A 1996 Mutagenic and carcinogenic properties of platinum-based anticancer drugs *Mutat. Res. Mol. Mech. Mutagen.* 355 59–70.

50 Di Pasqua A J, Goodisman J and Dabrowiak J C 2012 Understanding how the platinum anticancer drug carboplatin works: from the bottle to the cell *Inorganica Chim. Acta* 389 29–35.

51 Finkelstein A E, Walz D T, Batista V, Mizraji M, Roisman F and Misher A 1976 Auranofin. New oral gold compound for treatment of rheumatoid arthritis *Ann. Rheum. Dis.* 35 251–7.

52 Krusius F E, Markkanen A and Peltola P 1970 Plasma levels and urinary excretion of gold during routine treatment of rheumatoid arthritis *Ann. Rheum. Dis.* 29 232–5.

53 Chisholm M A, Mulloy A L and Taylor A T 1996 Acute management of cancer-related hypercalcemia *Ann. Pharmacother.* 30 507–13.

54 Mieszawska A J, Mulder W J M, Fayad Z A and Cormode D P 2013 Multifunctional gold nanoparticles for diagnosis and therapy of disease *Mol. Pharm.* 10 831–47.

55 Shanmugasundaram T, Radhakrishnan M, Gopikrishnan V, Kadirvelu K and Balagurunathan R 2017 Biocompatible silver, gold and silver/gold alloy nanoparticles for enhanced cancer therapy: in vitro and in vivo perspectives *Nanoscale* 9 16773–90.

56 Bowman M-C, Ballard T E, Ackerson C J, Feldheim D L, Margolis D M and Melander C 2008 Inhibition of HIV fusion with multivalent gold nanoparticles *J. Am. Chem. Soc.* 130 6896–7.

57 Vechia I C D, Steiner B T, Freitas M L, dos Santos Pedroso Fidelis G, Galvani N C, Ronchi J M, Possato J C, Fagundes M Í, Rigo F K, Feuser P E, de Araújo P H H and Machado-de-Ávila R A 2020 Comparative cytotoxic effect of citrate-capped gold nanoparticles with different sizes on noncancerous and cancerous cell lines *J. Nanopart. Res.* 22 133.

58 Huang H, Quan Y, Wang X and Chen T 2016 Gold nanoparticles of diameter 13 nm induce apoptosis in rabbit articular chondrocytes *Nanoscale Res. Lett.* 11 249.

59 Tsai L-C, Hsieh H-Y, Lu K-Y, Wang S-Y and Mi F-L 2016 EGCG/gelatin-doxorubicin gold nanoparticles enhance therapeutic efficacy of doxorubicin for prostate cancer treatment *Nanomedicine* 11 9–30.

60 Zhang J, Mou L and Jiang X 2020 Surface chemistry of gold nanoparticles for health-related applications *Chem. Sci.* 11 923–36.

61 Vinhas R, Cordeiro M, Carlos F F, Mendo S, Fernandes A R, Figueiredo S and Baptista P V 2015 Gold nanoparticle-based theranostics: disease diagnostics and treatment using a single nanomaterial *Nanobiosensors Dis. Diagn.* 4 11.

62 Sibuyi N R S, Thovhogi N, Gabuza K B, Meyer M D, Drah M, Onani M O, Skepu A, Madiehe A M and Meyer M 2017 Peptide-functionalized nanoparticles for the selective induction of apoptosis in target cells *Nanomedicine* 12 1631–45.

63 Oh E, Delehanty J B, Sapsford K E, Susumu K, Goswami R, Blanco-Canosa J B, Dawson P E, Granek J, Shoff M, Zhang Q, Goering P L, Huston A and Medintz I L 2011 Cellular uptake and fate of PEGylated gold nanoparticles is dependent on both cell-penetration peptides and particle size *ACS Nano* 5 6434–48.

64 Al-Yasiri A Y, Khoobchandani M, Cutler C S, Watkinson L, Carmack T, Smith C J, Kuchuk M, Loyalka S K, Lugão A B and Katti K V 2017 Mangiferin functionalized radioactive gold nanoparticles (MGF-198AuNPs) in prostate tumor therapy: green nanotechnology for production, in vivo tumor retention and evaluation of therapeutic efficacy *Dalton Trans.* 46 14561–71.

65 Deng R, Ji B, Yu H, Bao W, Yang Z, Yu Y, Cui Y, Du Y, Song M, Liu S, Meguellati K and Yan F 2019 Multifunctional gold nanoparticles overcome microRNA regulatory network mediated-multidrug resistant leukemia *Sci. Rep.* 9 5348.

66 Aboyewa J A, Sibuyi N R S, Meyer M and Oguntibeju O O 2021 Gold nanoparticles synthesized using extracts of cyclopia intermedia, commonly known as honeybush, amplify the cytotoxic effects of doxorubicin *Nanomaterials* 11 132.

67 Safwat M A, Soliman G M, Sayed D and Attia M A 2016 Gold nanoparticles enhance 5-fluorouracil anticancer efficacy against colorectal cancer cells *Int. J. Pharm.* 513 648–58.

68 Schofield C L, Haines A H, Field R A and Russell D A 2006 Silver and gold glyconanoparticles for colorimetric bioassays *Langmuir* 22 6707–11.

69 Narain R, Housni A, Gody G, Boullanger P, Charreyre M-T and Delair T 2007 Preparation of Biotinylated Glyconanoparticles via a Photochemical Process and Study of Their Bioconjugation to Streptavidin. ACS Publications.

70 Chandrawati R and Stevens M M 2014 Controlled assembly of peptide-functionalized gold nanoparticles for label-free detection of blood coagulation factor XIII activity *Chem. Commun.* 50 5431–4.

71 Chen P, Selegård R, Aili D and Liedberg B 2013 Peptide functionalized gold nanoparticles for colorimetric detection of matrilysin (MMP-7) activity *Nanoscale* 5 8973–6.

72 Thanh N T K and Rosenzweig Z 2002 Development of an aggregation-based immunoassay for anti-protein a using gold nanoparticles *Anal. Chem.* 74 1624–8.

73 He L, Musick M D, Nicewarner S R, Salinas F G, Benkovic S J, Natan M J and Keating C D 2000 Colloidal Au-enhanced surface plasmon resonance for ultrasensitive detection of DNA hybridization *J. Am. Chem. Soc.* 122 9071–7.

74 Karakuş E, Erdemir E, Demirbilek N and Liv L 2021 Colorimetric and electrochemical detection of SARS-CoV-2 spike antigen with a gold nanoparticle-based biosensor *Anal. Chim. Acta* 1182 338939.

75 Yrad F M, Castañares J M and Alocilja E C 2019 Visual detection of dengue-1 RNA using gold nanoparticle-based lateral flow biosensor *Diagnostics* 9 74.

76 Khalil I, Yehye W A, Julkapli N M, Rahmati S, Sina A A I, Basirun W J and Johan M R 2019 Graphene oxide and gold nanoparticle based dual platform with short DNA probe for the PCR free DNA biosensing using surface-enhanced Raman scattering *Biosens. Bioelectron.* 131 214–23.

77 Zou Y, Liang J, She Z and Kraatz H-B 2019 Gold nanoparticles-based multifunctional nanoconjugates for highly sensitive and enzyme-free detection of *E. coli* K12 *Talanta* 193 15–22.

78 Vu Q K, Tran Q H, Vu N P, Anh T-L, Dang T T L, Matteo T and Nguyen T H H 2021 A label-free electrochemical biosensor based on screen-printed electrodes modified with gold nanoparticles for quick detection of bacterial pathogens *Mater. Today Commun.* 26 101726.

79 Lin D, Pillai R G, Lee W E and Jemere A B 2019 An impedimetric biosensor for *E. coli* O157:H7 based on the use of self-assembled gold nanoparticles and protein G *Microchim. Acta* 186 169.

80 Zahra Q u A, Luo Z, Ali R, Khan M I, Li F and Qiu B 2021 Advances in gold nanoparticles-based colorimetric aptasensors for the detection of antibiotics: an overview of the past decade *Nanomaterials* 11 840.

81 Alba-Patiño A, Russell S M, Borges M, Pazos-Pérez N, Álvarez-Puebla R A and de la Rica R 2020 Nanoparticle-based mobile biosensors for the rapid detection of sepsis biomarkers in whole blood *Nanoscale Adv.* 2 1253–60.

82 Toubanaki D K, Margaroni M, Prapas A and Karagouni E 2020 Development of a nanoparticle-based lateral flow strip biosensor for visual detection of whole nervous necrosis virus particles *Sci. Rep.* 10 6529.

83 Riley R S and Day E S 2017 Gold nanoparticle-mediated photothermal therapy: applications and opportunities for multimodal cancer treatment *WIREs Nanomed. Nanobiotechnol.* 9 e1449.

84 Norouzi H, Khoshgard K and Akbarzadeh F 2018 In vitro outlook of gold nanoparticles in photo-thermal therapy: a literature review *Lasers Med. Sci.* 33 917–26.

85 Cheng X, Sun R, Yin L, Chai Z, Shi H and Gao M 2017 Light-triggered assembly of gold nanoparticles for photothermal therapy and photoacoustic imaging of tumors in vivo *Adv. Mater.* 29 1604894.

86 Liu Y, Crawford B M and Vo-Dinh T 2018 Gold nanoparticles-mediated photothermal therapy and immunotherapy *Immunotherapy* 10 1175–88.

87 Yang W, Liang H, Ma S, Wang D and Huang J 2019 Gold nanoparticle based photothermal therapy: development and application for effective cancer treatment *Sustain. Mater. Technol.* 22 e00109.

88 Sharma P, Brown S C, Singh A, Iwakuma N, Pyrgiotakis G, Krishna V, Knapik J A, Barr K, Moudgil B M and Grobmyer S R 2010 Near-infrared absorbing and luminescent gold speckled silica nanoparticles for photothermal therapy *J. Mater. Chem.* 20 5182–5.

89 Jing L, Liang X, Deng Z, Feng S, Li X, Huang M, Li C and Dai Z 2014 Prussian blue coated gold nanoparticles for simultaneous photoacoustic/CT bimodal imaging and photothermal ablation of cancer *Biomaterials* 35 5814–21.

90 Grabinski C, Schaeublin N, Wijaya A, D'Couto H, Baxamusa S H, Hamad-Schifferli K and Hussain S M 2011 Effect of gold nanorod surface chemistry on cellular response *ACS Nano* 5 2870–9.

91 Huang J, Xu Z, Jiang Y, Law W, Dong B, Zeng X, Ma M, Xu G, Zou J and Yang C 2021 Metal organic framework-coated gold nanorod as an on-demand drug delivery platform for chemo-photothermal cancer therapy *J. Nanobiotechnology* 19 219.

92 Fang L, Wang W, Liu Y, Xie Z and Chen L 2017 Zeolitic imidazole framework coated Au nanorods for enhanced photothermal therapy and stability *Dalton Trans.* 46 8933–7.

93 Popp M K, Oubou I, Shepherd C and Nager Z 2014 Anderson C and Pagliaro L 2014 Photothermal therapy using gold nanorods and near-infrared light in a murine melanoma model increases survival and decreases tumor volume *J. Nanomater.* 126 126.

94 El-Sayed M A, Shabaka A A, El-Shabrawy O A, Yassin N A, Mahmoud S S, El-Shenawy S M, Al-Ashqar E, Eisa W H, Farag N M, El-Shaer M A, Salah N and Al-Abd A M 2013 Tissue distribution and efficacy of gold nanorods coupled with laser induced photoplasmonic therapy in Ehrlich carcinoma solid tumor model *PLoS One* 8 e76207.

95 Huang X, El-Sayed I H, Qian W and El-Sayed M A 2006 Cancer cell imaging and photothermal therapy in the near-infrared region by using gold nanorods *J. Am. Chem. Soc.* 128 2115–20.

96 Dickerson E B, Dreaden E C, Huang X, El-Sayed I H, Chu H, Pushpanketh S, McDonald J F and El-Sayed M A 2008 Gold nanorod assisted near-infrared plasmonic photothermal therapy (PPTT) of squamous cell carcinoma in mice *Cancer Lett.* 269 57–66.

97 Bagley A F, Hill S, Rogers G S and Bhatia S N 2013 Plasmonic photothermal heating of intraperitoneal tumors through the use of an implanted near-infrared source *ACS Nano* 7 8089–97.

98 Deinavizadeh M, Kiasat A, Hooshmand N, Shafiei M, Sabaeian M, Mirzajani R, Zahraei S M, Labouta H I and El-Sayed M A 2021 Smart NIR-light and pH responsive doxorubicin-loaded GNRs@SBA-15-SH nanocomposite for chemo-photothermal therapy of cancer *Nanophotonics* 10 3303–19.

99 Ali M R K, Wu Y and El-Sayed M A 2019 Gold-nanoparticle-assisted plasmonic photothermal therapy advances toward clinical application *J. Phys. Chem. C* 123 15375–93.

100 Sondi I and Salopek-Sondi B 2004 Silver nanoparticles as antimicrobial agent: a case study on *E. coli* as a model for gram-negative bacteria *J. Colloid Interface Sci.* 275 177–82.

101 Morones J R, Elechiguerra J L, Camacho A, Holt K, Kouri J B, Ramírez J T and Yacaman M J 2005 The bactericidal effect of silver nanoparticles *Nanotechnology* 16 2346.

102 Pal S, Tak Y K and Song J M 2007 Does the antibacterial activity of silver nanoparticles depend on the shape of the nanoparticle? A study of thegram-negative bacterium *Escherichia coli Appl. Environ. Microbiol.* 73 1712–20.

103 Ghiuţă I, Cristea D, Croitoru C, Kost J, Wenkert R, Vyrides I, Anayiotos A and Munteanu D 2018 Characterization and antimicrobial activity of silver nanoparticles, biosynthesized using Bacillus species *Appl. Surf. Sci.* 438 66–73.

104 Geethalakshmi R and Sarada D V L 2013 Characterization and antimicrobial activity of gold and silver nanoparticles synthesized using saponin isolated from *Trianthema decandra* L *Ind. Crop. Prod.* 51 107–15.

105 Ahmad T, Wani I A, Manzoor N, Ahmed J and Asiri A M 2013 Biosynthesis, structural characterization and antimicrobial activity of gold and silver nanoparticles *Colloids Surf. B Biointerfaces* 107 227–34.

106 Wani I A, Khatoon S, Ganguly A, Ahmed J and Ahmad T 2013 Structural characterization and antimicrobial properties of silver nanoparticles prepared by inverse microemulsion method *Colloids Surf. B Biointerfaces* 101 243–50.

107 Bindhu M R and Umadevi M 2013 Synthesis of monodispersed silver nanoparticles using *Hibiscus cannabinus* leaf extract and its antimicrobial activity *Spectrochim. Acta A Mol. Biomol. Spectrosc.* 101 184–90.

108 El-Rafie H M, El-Rafie M H and Zahran M K 2013 Green synthesis of silver nanoparticles using polysaccharides extracted from marine macro algae *Carbohydr. Polym.* 96 403–10.

109 Maiti S, Krishnan D, Barman G, Ghosh S K and Laha J K 2014 Antimicrobial activities of silver nanoparticles synthesized from *Lycopersicon esculentum* extract *J. Anal. Sci. Technol.* 5 40.

110 Ajitha B, Reddy Y A K and Sreedhara Reddy P 2015 Enhanced antimicrobial activity of silver nanoparticles with controlled particle size by pH variation *Powder Technol.* 269 110–7.

111 Chahar V, Sharma B, Shukla G, Srivastava A and Bhatnagar A 2018 Study of antimicrobial activity of silver nanoparticles synthesized using green and chemical approach *Colloids Surf. Physicochem. Eng. Asp.* 554 149–55.

112 Cheon J Y, Kim S J, Rhee Y H, Kwon O H and Park W H 2019 Shape-dependent antimicrobial activities of silver nanoparticles *Int. J. Nanomedicine* 14 2773–80.

113 Wasilewska A, Klekotka U, Zambrzycka M, Zambrowski G, Święcicka I and Kalska-Szostko B 2023 Physico-chemical properties and antimicrobial activity of silver nanoparticles fabricated by green synthesis *Food Chem.* 400 133960.

114 Aygün A, Özdemir S, Gülcan M, Yalçın M S, Uçar M and Şen F 2022 Characterization and antioxidant-antimicrobial activity of silver nanoparticles synthesized using Punica granatum extract *Int. J. Environ. Sci. Technol.* 19 2781–8.

115 Al-Ogaidi M A Z and Rasheed B G 2022 Enhancement of antimicrobial activity of silver nanoparticles using lasers *Lasers Manuf. Mater. Process.* 9 610–21.

116 Alduraihem N S, Bhat R S, Al-Zahrani S A, Elnagar D M, Alobaid H M and Daghestani M H 2023 Anticancer and antimicrobial activity of silver nanoparticles synthesized from pods of *Acacia nilotica Processes* 11 301.

117 Kemala P, Idroes R, Khairan K, Ramli M, Jalil Z, Idroes G M, Tallei T E, Helwani Z, Safitri E, Iqhrammullah M and Nasution R 2022 Green synthesis and antimicrobial activities of silver nanoparticles using *Calotropis gigantea* from Ie Seu-Um Geothermal Area, Aceh Province, Indonesia *Molecules* 27 5310.

118 Marinescu L, Ficai D, Ficai A, Oprea O, Nicoara A I, Vasile B S, Boanta L, Marin A, Andronescu E and Holban A-M 2022 Comparative antimicrobial activity of silver nanoparticles obtained by wet chemical reduction and solvothermal methods *Int. J. Mol. Sci.* 23 5982.

119 Singla S, Jana A, Thakur R, Kumari C, Goyal S and Pradhan J 2022 Green synthesis of silver nanoparticles using *Oxalis griffithii* extract and assessing their antimicrobial activity *OpenNano* 7 100047.

120 Mahendran N, Anand B, Rajarajan M, Muthuvel A and Mohana V 2022 Green synthesis, characterization and antimicrobial activities of silver nanoparticles using *Cissus quadarangularis* leaf extract *Mater. Today Proc.* 49 2620–3.

121 Ozturk D, Ozguven A, Yonten V and Ertas M 2022 Green synthesis, characterization and antimicrobial activity of silver nanoparticles using *Ornithogalum narbonense* L *Inorg. Nano-Met. Chem.* 52 329–41.

122 Al-Shmgani H S A, Mohammed W H, Sulaiman G M and Saadoon A H 2017 Biosynthesis of silver nanoparticles from *Catharanthus roseus* leaf extract and assessing their antioxidant, antimicrobial, and wound-healing activities *Artif. Cells Nanomedicine Biotechnol.* 45 1234–40.

123 Dai X, Guo Q, Zhao Y, Zhang P, Zhang T, Zhang X and Li C 2016 Functional silver nanoparticle as a benign antimicrobial agent that eradicates antibiotic-resistant bacteria and promotes wound healing *ACS Appl. Mater. Interfaces* 8 25798–807.

124 Oryan A, Alemzadeh E, Tashkhourian J and Nami Ana S F 2018 Topical delivery of chitosan-capped silver nanoparticles speeds up healing in burn wounds: a preclinical study *Carbohydr. Polym.* 200 82–92.

125 Loan Khanh L, Thanh Truc N, Tan Dat N, Thi Phuong Nghi N, van Toi V, Thi Thu Hoai N, Ngoc Quyen T, Loan T T T and Hiep N T 2019 Gelatin-stabilized composites of silver nanoparticles and curcumin: characterization, antibacterial and antioxidant study *Sci. Technol. Adv. Mater.* 20 276–90.

126 Wu J, Zheng Y, Wen X, Lin Q, Chen X and Wu Z 2014 Silver nanoparticle/bacterial cellulose gel membranes for antibacterial wound dressing: investigation in vitro and in vivo *Biomed. Mater.* 9 035005.

127 Sood R and Chopra D S 2018 Optimization of reaction conditions to fabricate *Ocimum sanctum* synthesized silver nanoparticles and its application to nano-gel systems for burn wounds *Mater. Sci. Eng. C* 92 575–89.

128 Orlowski P, Zmigrodzka M, Tomaszewska E, Ranoszek-Soliwoda K, Czupryn M, Antos-Bielska M, Szemraj J, Celichowski G, Grobelny J and Krzyzowska M 2018 Tannic acid-modified silver nanoparticles for wound healing: the importance of size *Int. J. Nanomedicine* 13 991–1007.

129 Mehwish H M, Liu G, Rajoka M S R, Cai H, Zhong J, Song X, Xia L, Wang M, Aadil R M, Inam-Ur-Raheem M, Xiong Y, Wu H, Amirzada M I, Zhu Q and He Z 2021 Therapeutic potential of *Moringa oleifera* seed polysaccharide embedded silver nanoparticles in wound *Int. J. Biol. Macromol.* 184 144–58.

130 GhavamiNejad A, Park C H and Kim C S 2016 In situ synthesis of antimicrobial silver nanoparticles within antifouling Zwitterionic hydrogels by catecholic redox chemistry for wound healing application *Biomacromolecules* 17 1213–23.

131 Tra Thanh N, Ho Hieu M, Tran Minh Phuong N, Do Bui Thuan T, Nguyen Thi Thu H, Thai V P, Do Minh T, Nguyen Dai H, Vo V T and Nguyen Thi H 2018 Optimization and characterization of electrospun polycaprolactone coated with gelatin-silver nanoparticles for wound healing application *Mater. Sci. Eng. C* 91 318–29.

132 Gunasekaran T, Nigusse T and Dhanaraju M D 2011 Silver nanoparticles as real topical bullets for wound healing *J. Am. Coll. Clin. Wound Spec.* 3 82–96.

133 Tian J, Wong K K, Ho C-M, Lok C-N, Yu W-Y, Che C-M, Chiu J-F and Tam P K 2007 Topical delivery of silver nanoparticles promotes wound healing *ChemMedChem* 2 129–36.

134 Rigo C, Ferroni L, Tocco I, Roman M, Munivrana I, Gardin C, Cairns W R, Vindigni V, Azzena B and Barbante C 2013 Active silver nanoparticles for wound healing *Int. J. Mol. Sci.* 14 4817–40.

135 Tang Q, Chen C, Jiang Y, Huang J, Liu Y, Nthumba P M, Gu G, Wu X, Zhao Y and Ren J 2020 Engineering an adhesive based on photosensitive polymer hydrogels and silver nanoparticles for wound healing *J. Mater. Chem. B* 8 5756–64.

136 Ji X, Li R, Liu G, Jia W, Sun M, Liu Y, Luo Y and Cheng Z 2021 Phase separation-based electrospun Janus nanofibers loaded with *Rana chensinensis* skin peptides/silver nanoparticles for wound healing *Mater. Des.* 207 109864.

137 Pansara C, Mishra R, Mehta T, Parikh A and Garg S 2020 Formulation of chitosan stabilized silver nanoparticle-containing wound healing film: in vitro and in vivo characterization *J. Pharm. Sci.* 109 2196–205.

138 Peng J J-Y, Botelho M G and Matinlinna J P 2012 Silver compounds used in dentistry for caries management: a review *J. Dent.* 40 531–41.

139 Venugopal A, Muthuchamy N, Tejani H, Gopalan A-I, Lee K-P, Lee H-J and Kyung H M 2016 Incorporation of silver nanoparticles on the surface of orthodontic microimplants to achieve antimicrobial properties *Korean J. Orthod.* 47 3–10.

140 Fernandes G L, Delbem A C B, Do Amaral J G, Gorup L F, Fernandes R A, De Souza N F N, Souza J A S, Monteiro D R, Hunt A M A, Camargo E R and Barbosa D B 2018 Nanosynthesis of silver-calcium glycerophosphate: promising association against oral pathogens *Antibiotics* 7 52.

141 Zhang N, Melo M A S, Antonucci J M, Lin N J, Lin-Gibson S, Bai Y and Xu H H K 2016 Novel dental cement to combat biofilms and reduce acids for orthodontic applications to avoid enamel demineralization *Materials* 9 413.

142 Cheng L, Zhang K, Zhou C-C, Weir M D, Zhou X-D and Xu H H K 2016 One-year water-ageing of calcium phosphate composite containing nano-silver and quaternary ammonium to inhibit biofilms *Int. J. Oral Sci.* 8 172–81.

143 Lu Z, Rong K, Li J, Yang H and Chen R 2013 Size-dependent antibacterial activities of silver nanoparticles against oral anaerobic pathogenic bacteria *J. Mater. Sci. Mater. Med.* 24 1465–71.

144 Panpaliya N P, Dahake P T, Kale Y J, Dadpe M V, Kendre S B, Siddiqi A G and Maggavi U R 2019 In vitro evaluation of antimicrobial property of silver nanoparticles and chlorhexidine against five different oral pathogenic bacteria *Saudi Dent. J.* 31 76–83.

145 Kandhan T S, Roy A, Lakshmi T and Rajeshkumar S 2019 Green synthesis of Rosemary oleoresin mediated silver nanoparticles and its effect on oral pathogens *Res. J. Pharm. Technol.* 12 5379.

146 Zarco M, Vess T and Ginsburg G 2012 The oral microbiome in health and disease and the potential impact on personalized dental medicine *Oral Dis.* 18 109–20.

147 Sampaio-Maia B, Caldas I M, Pereira M L, Pérez-Mongiovi D and Araujo R 2016 Chapter Four – The Oral Microbiome in Health and Its Implication in Oral and Systemic Diseases *Advances in Applied Microbiology*, vol 97, ed S Sariaslani and G M Gadd (Academic Press) pp 171–210.

148 Wang L and Ganly I 2014 The oral microbiome and oral cancer *Clin. Lab. Med.* 34 711–9.

149 Cabal B, Cafini F, Esteban-Tejeda L, Alou L, Bartolomé J F, Sevillano D, López-Piriz R, Torrecillas R and Moya J S 2012 Inhibitory effect on in vitro *Streptococcus oralis* biofilm of a soda-lime glass containing silver nanoparticles coating on titanium alloy *PLoS One* 7 e42393.

150 Panácek A, Kolár M, Vecerová R, Prucek R, Soukupová J, Krystof V, Hamal P, Zboril R and Kvítek L 2009 Antifungal activity of silver nanoparticles against Candida spp *Biomaterials* 30 6333–40.

151 Porenczuk A, Grzeczkowicz A, Maciejewska I, Gołaś M, Piskorska K, Kolenda A, Gozdowski D, Kopeć-Swoboda E, Granicka L and Olczak-Kowalczyk D 2019 An initial evaluation of cytotoxicity, genotoxicity and antibacterial effectiveness of a disinfection liquid containing silver nanoparticles alone and combined with a glass-ionomer cement and dentin bonding systems *Adv. Clin. Exp. Med. Off. Organ Wroclaw Med. Univ.* 28 75–83.

152 Paiva L, Fidalgo T K S, da Costa L P, Maia L C, Balan L, Anselme K, Ploux L and Thiré R M S M 2018 Antibacterial properties and compressive strength of new one-step preparation silver nanoparticles in glass ionomer cements (NanoAg-GIC) *J. Dent.* 69 102–9.

153 Barot T, Rawtani D and Kulkarni P 2020 Physicochemical and biological assessment of silver nanoparticles immobilized Halloysite nanotubes-based resin composite for dental applications *Heliyon* 6 e03601.

154 Cataldi A, Gallorini M, Di Giulio M, Guarnieri S, Mariggiò M A, Traini T, Di Pietro R, Cellini L, Marsich E and Sancilio S 2016 Adhesion of human gingival fibroblasts/*Streptococcus mitis* co-culture on the nanocomposite system Chitlac-nAg *J. Mater. Sci. Mater. Med.* 27 88.

155 Espinosa-Cristóbal L F, López-Ruiz N, Cabada-Tarín D, Reyes-López S Y, Zaragoza-Contreras A, Constandse-Cortéz D, Donohué-Cornejo A, Tovar-Carrillo K, Cuevas-González J C and Kobayashi T 2018 Antiadherence and antimicrobial properties of silver nanoparticles against *Streptococcus mutans* on brackets and wires used for orthodontic treatments *J. Nanomater.* 2018 e9248527.

156 Ghorbanzadeh R, Pourakbari B and Bahador A 2015 Effects of baseplates of orthodontic appliances with in situ generated silver nanoparticles on cariogenic bacteria: a randomized, double-blind cross-over clinical trial *J. Contemp. Dent. Pract.* 16 291–8.

157 Jonaidi-Jafari N, Izadi M and Javidi P 2016 The effects of silver nanoparticles on antimicrobial activity of ProRoot mineral trioxide aggregate (MTA) and calcium enriched mixture (CEM) *J Clin Exp Dent* 8 e22–6.

158 Mendes M S, Resende L D, Pinto C A, Raldi D P, Cardoso F G and Habitante S M 2017 Radiopacity of mineral trioxide aggregate with and without inclusion of silver nanoparticles *J. Contemp. Dent. Pract.* 18 448–51.

159 Niska K, Knap N, Kędzia A, Jaskiewicz M, Kamysz W and Inkielewicz-Stepniak I 2016 Capping agent-dependent toxicity and antimicrobial activity of silver nanoparticles: an in vitro study. Concerns about potential application in dental practice *Int. J. Med. Sci.* 13 772–82.

160 Bapat R A, Joshi C P, Bapat P, Chaubal T V, Pandurangappa R, Jnanendrappa N, Gorain B, Khurana S and Kesharwani P 2019 The use of nanoparticles as biomaterials in dentistry *Drug Discov. Today* 24 85–98.

161 Besinis A, Hadi S D, Le H R, Tredwin C and Handy R D 2017 Antibacterial activity and biofilm inhibition by surface modified titanium alloy medical implants following application of silver, titanium dioxide and hydroxyapatite nanocoatings *Nanotoxicology* 11 327–38

162 Fu C, Zhang X, Savino K, Gabrys P, Gao Y, Chaimayo W, Miller B L and Yates M Z 2016 Antimicrobial silver-hydroxyapatite composite coatings through two-stage electrochemical synthesis *Surf. Coat. Technol.* 301 13–9.

163 Pokrowiecki R, Zaręba T, Szaraniec B, Pałka K, Mielczarek A, Menaszek E and Tyski S 2017 In vitro studies of nanosilver-doped titanium implants for oral and maxillofacial surgery *Int. J. Nanomedicine* 12 4285.

164 Wang L, Jiang X, Ji Y, Bai R, Zhao Y, Wu X and Chen C 2013 Surface chemistry of gold nanorods: origin of cell membrane damage and cytotoxicity *Nanoscale* 5 8384–91.

165 El-Kurdi R and Patra D 2019 Gold and silver nanoparticles in resonance Rayleigh scattering techniques for chemical sensing and biosensing: a review *Microchim. Acta* 186 1–18.

166 Anker J N, Hall W P, Lyandres O, Shah N C, Zhao J and Van Duyne R P 2008 Biosensing with plasmonic nanosensors *Nat. Mater.* 7(6) 442–53.

167 Baghayeri M, Mahdavi B, Hosseinpor-Mohsen Abadi Z and Farhadi S 2018 Green synthesis of silver nanoparticles using water extract of Salvia leriifolia: Antibacterial studies and applications as catalysts in the electrochemical detection of nitrite *Appl. Organomet. Chem.* 32 e4057.

168 Kim S T, Saha K, Kim C and Rotello V M 2013 The role of surface functionality in determining nanoparticle cytotoxicity *Acc. Chem. Res.* 46 681–91.

169 Su G, Zhou H, Mu Q, Zhang Y, Li L, Jiao P, Jiang G and Yan B 2012 Effective surface charge density determines the electrostatic attraction between nanoparticles and cells *J. Phys. Chem. C* 116 4993–8.

170 Alkilany A M, Nagaria P K, Hexel C R, Shaw T J, Murphy C J and Wyatt M D 2009 Cellular Uptake and Cytotoxicity of Gold Nanorods: Molecular Origin of Cytotoxicity and Surface Effects *Small* 5 701–8.

171 Alkilany A M, Shatanawi A, Kurtz T, Caldwell R B and Caldwell R W 2012 Toxicity and cellular uptake of gold nanorods in vascular endothelium and smooth muscles of isolated rat blood vessel: importance of surface modification *Small Weinh. Bergstr. Ger.* 8 1270–8.

172 Takahashi H, Niidome Y, Niidome T, Kaneko K, Kawasaki H and Yamada S 2006 Modification of gold nanorods using phosphatidylcholine to reduce cytotoxicity *Langmuir* 22 2–5.

173 Freese C, Uboldi C, Gibson M I, Unger R E, Weksler B B, Romero I A, Couraud P O and Kirkpatrick C J 2012 Uptake and cytotoxicity of citrate-coated gold nanospheres: comparative studies on human endothelial and epithelial cells *Part. Fibre Toxicol.* 9.

174 Anon 2006 PEG-modified gold nanorods with a stealth character for in vivo applications *J. Control. Release* 114 343–7.

175 Wang Y, Black K C L, Luehmann H, Li W, Zhang Y, Cai X, Wan D, Liu S-Y, Li M, Kim P, Li Z-Y, Wang L V, Liu Y and Xia Y 2013 Comparison study of gold nanohexapods, nanorods, and nanocages for photothermal cancer treatment *ACS Nano* 7 2068–77.

176 Li Y, Kröger M and Liu W K 2015 Shape effect in cellular uptake of PEGylated nanoparticles: comparison between sphere, rod, cube and disk *Nanoscale* 7 16631–46.

177 Nangia S and Sureshkumar R 2012 Effects of nanoparticle charge and shape anisotropy on translocation through cell membranes *Langmuir* 28 17666–71.

178 Wang Z, Xie D, Liu H, Bao Z and Wang Y 2016 Toxicity assessment of precise engineered gold nanoparticles with different shapes in zebrafish embryos *RSC Adv.* 6 33009–13.

179 Karakoçak B B, Raliya R, Davis J T, Chavalmane S, Wang W-N, Ravi N and Biswas P 2016 Biocompatibility of gold nanoparticles in retinal pigment epithelial cell line *Toxicol. In Vitro* 37 61–9.

180 Chompoosor A, Saha K, Ghosh P S, Macarthy D J, Miranda O R, Zhu Z-J, Arcaro K F and Rotello V M 2010 The role of surface functionality on acute cytotoxicity, ROS generation and DNA damage by cationic gold nanoparticles *Small* 6 2246–9.

181 Uz M, Bulmus V and Alsoy Altinkaya S 2016 Effect of PEG grafting density and hydrodynamic volume on gold nanoparticle–cell interactions: an investigation on cell cycle, apoptosis, and DNA damage *Langmuir* 32 5997–6009.

182 Leite P E C, Pereira M R, do Santos C A, N, Campos A P C, Esteves T M and Granjeiro J M 2015 Gold nanoparticles do not induce myotube cytotoxicity but increase the susceptibility to cell death *Toxicol. In Vitro* 29 819–27.

183 Brandenberger C, Mühlfeld C, Ali Z, Lenz A-G, Schmid O, Parak W J, Gehr P and Rothen-Rutishauser B 2010 Quantitative evaluation of cellular uptake and trafficking of plain and polyethylene glycol-coated gold nanoparticles *Small* 6 1669–78.

184 Cho W-S, Cho M, Jeong J, Choi M, Cho H-Y, Han B S, Kim S H, Kim H O, Lim Y T, Chung B H and Jeong J 2009 Acute toxicity and pharmacokinetics of 13 nm-sized PEG-coated gold nanoparticles *Toxicol. Appl. Pharmacol.* 236 16–24.

185 Soenen S J, Manshian B B, Abdelmonem A M, Montenegro J-M, Tan S, Balcaen L, Vanhaecke F, Brisson A R, Parak W J, De Smedt S C and Braeckmans K 2014 The cellular interactions of PEGylated gold nanoparticles: effect of PEGylation on cellular uptake and cytotoxicity *Part. Part. Syst. Charact.* 31 794–800.

186 Jiang W, Kim B Y S, Rutka J T and Chan W C W 2008 Nanoparticle-mediated cellular response is size-dependent *Nat. Nanotechnol.* 3 145–50.

187 Ahn S, Seo E, Kim K H and Lee S J 2015 Physical property control on the cellular uptake pathway and spatial distribution of nanoparticles in cells *J. Biomed. Nanotechnol.* 11 1051–70.

188 Ma X, Wu Y, Jin S, Tian Y, Zhang X, Zhao Y, Yu L and Liang X-J 2011 Gold nanoparticles induce autophagosome accumulation through size-dependent nanoparticle uptake and lysosome impairment *ACS Nano* 5 8629–39.

189 Cheng X, Tian X, Wu A, Li J, Tian J, Chong Y, Chai Z, Zhao Y, Chen C and Ge C 2015 Protein corona influences cellular uptake of gold nanoparticles by phagocytic and nonphagocytic cells in a size-dependent manner *ACS Appl. Mater. Interfaces* 7 20568–75.

190 Coradeghini R, Gioria S, García C P, Nativo P, Franchini F, Gilliland D, Ponti J and Rossi F 2013 Size-dependent toxicity and cell interaction mechanisms of gold nanoparticles on mouse fibroblasts *Toxicol. Lett.* 217 205–16.

191 Wang S-H, Lee C-W, Chiou A and Wei P-K 2010 Size-dependent endocytosis of gold nanoparticles studied by three-dimensional mapping of plasmonic scattering images *J. Nanobiotechnology* 8 33.

192 Pan Y, Neuss S, Leifert A, Fischer M, Wen F, Simon U, Schmid G and Brandau W D W J 2007 Size-dependent cytotoxicity of gold nanoparticles *Small* 3 1941–9.

193 Pan Y, Leifert A, Ruau D, Neuss S, Bornemann J, Schmid G, Brandau W, Simon U and Jahnen-Dechent W 2009 Gold nanoparticles of diameter 1.4 nm trigger necrosis by oxidative stress and mitochondrial damage *Small* 5 2067–76.

194 Senut M-C, Zhang Y, Liu F, Sen A, Ruden D M and Mao G 2016 Size-dependent toxicity of gold nanoparticles on human embryonic stem cells and their neural derivatives *Small* 12 631–46.

195 Yen H-J, Hsu S and Tsai C-L 2009 Cytotoxicity and immunological response of gold and silver nanoparticles of different sizes *Small* 5 1553–61.

196 Choi S Y, Jeong S, Jang S H, Park J, Park J H, Ock K S, Lee S Y and Joo S-W 2012 In vitro toxicity of serum protein-adsorbed citrate-reduced gold nanoparticles in human lung adenocarcinoma cells *Toxicol. In Vitro* 26 229–37.

197 Yao M, He L, McClements D J and Xiao H 2015 Uptake of Gold Nanoparticles by Intestinal Epithelial Cells: Impact of Particle Size on Their Absorption, Accumulation, and Toxicity *J. Agric. Food Chem.* 63 8044–9.

198 Taylor U, Barchanski A, Garrels W, Klein S, Kues W, Barcikowski S and Rath D 2012 Toxicity of Gold Nanoparticles on Somatic and Reproductive Cells *Nano-Biotechnology for Biomedical and Diagnostic Research Advances in Experimental Medicine and Biology* ed E Zahavy, A Ordentlich, S Yitzhaki and A Shafferman (Dordrecht: Springer Netherlands) pp 125–33.

199 Xia Q, Li H, Liu Y, Zhang S, Feng Q and Xiao K 2017 The effect of particle size on the genotoxicity of gold nanoparticles *J. Biomed. Mater. Res. A* 105 710–9.

200 Lasagna-Reeves C, Gonzalez-Romero D, Barria M A, Olmedo I, Clos A, Ramanujam V M S, Urayama A, Vergara L, Kogan M J and Soto C 2010 Bioaccumulation and toxicity of gold nanoparticles after repeated administration in mice *Biochem. Biophys. Res. Commun.* 393 649–55.

201 Li J J, Hartono D, Ong C-N, Bay B-H and Yung L-Y L 2010 Autophagy and oxidative stress associated with gold nanoparticles *Biomaterials* 31 5996–6003.

202 Jia H Y, Liu Y, Zhang X J, Han L, Du L B, Tian Q and Xu Y C 2008 Potential oxidative stress of gold nanoparticles by induced-NO releasing in serum *J. Am. Chem. Soc.* 131 40–1.

203 Bhattacharya R, Mukherjee P, Xiong Z, Atala A, Soker S and Mukhopadhyay D 2004 Gold nanoparticles inhibit VEGF165-induced proliferation of HUVEC cells *Nano Lett.* 4 2479–81.

204 Staroverov S A, Aksinenko N M, Gabalov K P, Vasilenko O A, Vidyasheva I V, Shchyogolev S Y and Dykman L A 2009 Effect of gold nanoparticles on the respiratory activity of peritoneal macrophages *Gold Bull.* 42 153–6.

205 Arora S, Jain J, Rajwade J M and Paknikar K M 2008 Cellular responses induced by silver nanoparticles: in vitro studies *Toxicol. Lett.* 179 93–100.

206 Chadeau E, Oulahal N, Dubost L, Favergeon F and Degraeve P 2010 Anti-Listeria innocua activity of silver functionalised textile prepared with plasma technology *Food Control* 21 505–12.

207 YeonáLee H, KunáPark H, MiáLee Y and BumáPark S 2007 A practical procedure for producing silver nanocoated fabric and its antibacterial evaluation for biomedical applications *Chem. Commun.* 2959–61.

208 Kim S, Choi J E, Choi J, Chung K-H, Park K, Yi J and Ryu D-Y 2009 Oxidative stress-dependent toxicity of silver nanoparticles in human hepatoma cells *Toxicol. In Vitro* 23 1076–84.

209 Ahamed M, Karns M, Goodson M, Rowe J, Hussain S M, Schlager J J and Hong Y 2008 DNA damage response to different surface chemistry of silver nanoparticles in mammalian cells *Toxicol. Appl. Pharmacol.* 233 404–10.

210 Hussain S M, Hess K L, Gearhart J M, Geiss K T and Schlager J J 2005 In vitro toxicity of nanoparticles in BRL 3A rat liver cells *Toxicol. In Vitro* 19 975–83.

211 Rosas-Hernández H, Jiménez-Badillo S, Martínez-Cuevas P P, Gracia-Espino E, Terrones H, Terrones M, Hussain S M, Ali S F and González C 2009 Effects of 45-nm silver nanoparticles on coronary endothelial cells and isolated rat aortic rings *Toxicol. Lett.* 191 305–13.

212 Liu Z, Ren G, Zhang T and Yang Z 2009 Action potential changes associated with the inhibitory effects on voltage-gated sodium current of hippocampal CA1 neurons by silver nanoparticles *Toxicology* 264 179–84.

213 Rahman M F, Wang J, Patterson T A, Saini U T, Robinson B L, Newport G D, Murdock R C, Schlager J J, Hussain S M and Ali S F 2009 Expression of genes related to oxidative stress in the mouse brain after exposure to silver-25 nanoparticles *Toxicol. Lett.* 187 15–21.

214 Soliman F S, Hegazy E, Youssef M and Ghali L 2023 Study the potential impacts of chronic oral administration of silver nanoparticles on the lingual mucous membrane and salivary glands of albino rats *Dent. Sci. Updat.* 4 65–76.

215 Sajid Mohammed Z, Lataef R, Alwash S W, Jassim Sultan S, Mihdy Mohammed N and Abed A S 2022 Investigation on the possible cutaneous toxicity of silver nanoparticles in Swiss albino rats *J. Nanostructures* 12 754–60.

216 Babaei M, Tayemeh M B, Jo M S, Yu I J and Johari S A 2022 Trophic transfer and toxicity of silver nanoparticles along a phytoplankton-zooplankton-fish food chain *Sci. Total Environ.* 842 156807.

217 Sibiya A, Gopi N, Jeyavani J, Mahboob S, Al-Ghanim K A, Sultana S, Mustafa A, Govindarajan M and Vaseeharan B 2022 Comparative toxicity of silver nanoparticles and silver nitrate in freshwater fish *Oreochromis mossambicus*: a multi-biomarker approach *Comp. Biochem. Physiol. Part C Toxicol. Pharmacol.* 259 109391.

218 Skóra B, Piechowiak T and Szychowski K A 2022 Epidermal growth factor-labeled liposomes as a way to target the toxicity of silver nanoparticles into EGFR-overexpressing cancer cells in vitro *Toxicol. Appl. Pharmacol.* 443 116009.

219 Akter M, Sikder M T, Rahman M M, Ullah A K M A, Hossain K F B, Banik S, Hosokawa T, Saito T and Kurasaki M 2018 A systematic review on silver nanoparticles-induced cytotoxicity: physicochemical properties and perspectives *J. Adv. Res.* 9 1–16.

220 Pratsinis A, Hervella P, Leroux J-C, Pratsinis S E and Sotiriou G A 2013 Toxicity of silver nanoparticles in macrophages *Small* 9 2576–84.

221 Arora S, Jain J, Rajwade J M and Paknikar K M 2009 Interactions of silver nanoparticles with primary mouse fibroblasts and liver cells *Toxicol. Appl. Pharmacol.* 236 310–8.

222 Ema M, Okuda H, Gamo M and Honda K 2017 A review of reproductive and developmental toxicity of silver nanoparticles in laboratory animals *Reprod. Toxicol.* 67 149–64.

223 Marambio-Jones C and Hoek E M V 2010 A review of the antibacterial effects of silver nanomaterials and potential implications for human health and the environment *J. Nanopart. Res.* 12 1531–51.

224 Wei L, Lu J, Xu H, Patel A, Chen Z-S and Chen G 2015 Silver nanoparticles: synthesis, properties, and therapeutic applications *Drug Discov. Today* 20 595–601.

225 Bourquin J, Milosevic A, Hauser D, Lehner R, Blank F, Petri-Fink A and Rothen-Rutishauser B 2018 Biodistribution, clearance, and long-term fate of clinically relevant nanomaterials *Adv. Mater.* 30 1704307.

226 Reidy B, Haase A, Luch A, Dawson K A and Lynch I 2013 Mechanisms of silver nanoparticle release, transformation and toxicity: a critical review of current knowledge and recommendations for future studies and applications *Materials* 6 2295–350.

227 Seiffert J, Buckley A, Leo B, Martin N G, Zhu J, Dai R, Hussain F, Guo C, Warren J, Hodgson A, Gong J, Ryan M P, Zhang J J, Porter A, Tetley T D, Gow A, Smith R and Chung K F 2016 Pulmonary effects of inhalation of spark-generated silver nanoparticles in Brown-Norway and Sprague–Dawley rats *Respir. Res.* 17 85.

228 Seiffert J, Hussain F, Wiegman C, Li F, Bey L, Baker W, Porter A, Ryan M P, Chang Y, Gow A, Zhang J, Zhu J, Tetley T D and Chung K F 2015 Pulmonary toxicity of instilled silver nanoparticles: influence of size, coating and rat strain *PLoS One* 10 e0119726.

229 Weldon B A, Faustman E M, Oberdörster G, Workman T, Griffith W C, Kneuer C and Yu I J 2016 Occupational exposure limit for silver nanoparticles: considerations on the derivation of a general health-based value *Nanotoxicology* 10 945–56.

230 Fröhlich E E and Fröhlich E 2016 Cytotoxicity of nanoparticles contained in food on intestinal cells and the gut microbiota *Int. J. Mol. Sci.* 17 509.

231 van der Zande M, Vandebriel R J, Van Doren E, Kramer E, Herrera Rivera Z, Serrano-Rojero C S, Gremmer E R, Mast J, Peters R J B, Hollman P C H, Hendriksen P J M, Marvin H J P, Peijnenburg A A C M and Bouwmeester H 2012 Distribution, elimination, and toxicity of silver nanoparticles and silver ions in rats after 28-day oral exposure *ACS Nano* 6 7427–42.

232 Loeschner K, Hadrup N, Qvortrup K, Larsen A, Gao X, Vogel U, Mortensen A, Lam H R and Larsen E H 2011 Distribution of silver in rats following 28 days of repeated oral exposure to silver nanoparticles or silver acetate *Part. Fibre Toxicol.* 8 18.

233 Hadrup N, Sharma A K and Loeschner K 2018 Toxicity of silver ions, metallic silver, and silver nanoparticle materials after in vivo dermal and mucosal surface exposure: a review *Regul. Toxicol. Pharmacol.* 98 257–67.

234 Genter M B, Newman N C, Shertzer H G, Ali S F and Bolon B 2012 Distribution and systemic effects of intranasally administered 25 nm silver nanoparticles in adult mice *Toxicol. Pathol.* 40 1004–13.

235 Recordati C, De Maglie M, Bianchessi S, Argentiere S, Cella C, Mattiello S, Cubadda F, Aureli F, D'Amato M, Raggi A, Lenardi C, Milani P and Scanziani E 2016 Tissue distribution and acute toxicity of silver after single intravenous administration in mice: nano-specific and size-dependent effects *Part. Fibre Toxicol.* 13 12.

3

Biomedical Applications of Iron-Oxide-based Magnetic Nanoparticles

Ajit Khosla[1], Irshad A. Wani[2], and Mohammad N. Lone[3]

[1] *School of Advanced Materials and Nanotechnology, Xidian University, Xian, Shaanxi, China*
[2] *Department of Chemistry, Govt. Degree College Anantnag, Anantnag, University of Kashmir, Jammu and Kashmir, India*
[3] *Department of Chemistry, Central University of Kashmir, Ganderbal, Jammu and Kashmir, India*

3.1 Biomedical Applications of MNPs

In the realm of medicine, nanoparticles find application across three primary categories: metallic, nonmetallic, and composite nanoparticles. Among these, the most frequently utilized are metal and metal-oxide nanoparticles [1]. In the realm of metal-oxide nanoparticles, particular attention is directed toward ferrites, specifically superparamagnetic iron oxide nanoparticles (SPIONs), owing to their distinct magnetic properties and compatibility with biological systems. Iron oxide nanoparticles, overall, have recently gained significant attention due to their exceptional characteristics, such as superparamagnetism, a high surface-to-volume ratio, a larger surface area, and practical separation methods. The distinctive feature of superparamagnetic nanoparticles lies in their responsiveness to magnetic fields, losing their magnetism once the external field is removed. This trait, combined with good colloidal stability, holds considerable clinical importance as it reduces the risk of nanoparticle aggregation in the bloodstream [2]. Additionally, the biodegradability of SPIONs is considered a crucial aspect for their clinical application [3]. Among the various iron oxides in this group are hematite (α-Fe_2O_3), maghemite (γ-Fe_2O_3) and magnetite (Fe_3O_4) [4]. From a medical perspective, magnetite is particularly noteworthy as a highly promising SPION, mainly attributable to its distinctive magnetic, catalytic, and biochemical characteristics [5]. Fe_3O_4 nanoparticles offer a significant surface area, aiding in the absorption and immobilization of molecules or pharmaceuticals [6]. Enhancing the absorbent capacity of these nanoparticles is easily achievable through straightforward modifications. Additionally, adjusting their size and structure is readily attainable through various synthesis methods [7]. The flexibility of Fe_3O_4 nanoparticles facilitates the creation of versatile platforms, where coatings can be applied to magnetite nanoparticles using additional compounds like SiO_2 or even biodegradable polymers [8, 9].

Therapeutic magnetite nanoparticles (MNPs) typically consist of three essential components: (a) a magnetite core, which acts as a carrier for therapeutic agents, (b) a coated MNP core facilitating beneficial interactions within the body, and (c) a therapeutic payload with a specific in vivo function. Additionally, a particle may include a ligand that identifies specific receptors on the outer surface of cancer cells [10]. Fascinatingly, these configurations are subject to modification, and MNPs, whether coated or not, can be deposited onto alternative nanostructures such as layered double hydroxides (LDHs). Additionally, it is possible to coat Fe_3O_4 NPs with LDHs, producing nanostructures with diverse geomorphology and properties [11]. The flexible synthesis of

Metallic, Magnetic and Carbon-Based Nanomaterials: Synthesis and Biomedical Applications, First Edition.
Ajit Khosla, Irshad A. Wani, and Mohammad N. Lone.

nanostructured multifunctional platforms has generated substantial interest in the application of magnetite in nanomedicine. The development of core–shell nanostructures not only modifies the properties of iron oxides, but recent studies have also indicated that surface modifications with diverse surfactants can markedly influence the magnetic properties and transverse relaxation rate of water molecules in hydrophilic magnetic nanoparticle colloids [12, 13].

In recent decades, there has been widespread research on magnetic nanoparticles, specifically paramagnetic metal oxides like iron oxide nanoparticles (such as magnetite (Fe_3O_4) or maghemite (γ-Fe_2O_3)), exploring a range of biomedical applications. These applications encompass magnetic resonance imaging (MRI), multimodal imaging, drug delivery for transporting clinically important pharmaceutical compounds, and additional medical therapies like magnetic hyperthermia. The following sections delve into the diverse and significant biomedical applications of some essential MNPs.

3.1.1 MNPs as Imaging Contrast Agents for Biological Targets

MRI serves as a clinical diagnostic imaging technique wherein hydrogen atoms within an organism's body are exposed to a magnetic field, and their reactions vary depending on the surrounding environment. This approach is utilized for the diagnosis of various diseases [14]. When exposed to a strong magnetic field, protons undergo processing and alignment with the magnetic field. Following this, a radiofrequency pulse is administered to induce perturbation. The perturbed proton undergoes relaxation processes, namely T_1 (longitudinal or spin-lattice relaxation) and T_2 (transverse or spin-spin relaxation), eventually returning to the equilibrium state [15]. Variations in these relaxation processes are harnessed to produce images with diverse contrasts, emphasizing differences between tissues in pathology or anatomy. Employing different exogenous contrast agents enhances visibility, resulting in clearer contrast images. Traditionally, contrast agents in MRI imaging have involved the use of various paramagnetic materials such as Gd, Eu, Nd, and Mn, which play a role in influencing and modifying relaxation times [16]. For example, contrast agents based on Gd have been employed, exhibiting relatively low relaxivity values [17]. Gadolinium (Gd) chelates commonly decrease T_1 relaxation times at low concentrations, while T_2 shortening is observed primarily at high concentrations, raising potential concerns about toxicity [18]. The release of Gd^{3+} has been found to induce nephrogenic systemic fibrosis (NPS) in test animals [19]. Therefore, the main drawback of some of these materials is that they may induce severe adverse effects on the animal body such as impaired renal function and deposition in various tissues and organs. However, magnetic metal oxides have been recently used as a favorite contrast agent and important MRI probe due to their strong contrast effect on the surrounding tissues. In comparison to conventional paramagnetic contrast agents, superparamagnetic nanoparticles (SPNPs) have been employed as a new class of contrast agents, which has been seen to shorten both T_1 and T_2 relaxation values at low concentrations. Tailoring the size and appropriate functionalization of these SPNPs can be done to meet the specific requirements of target tissues. These nanoparticles induce shorter T_2 values, which result from the significant difference in magnetic susceptibility between MNPs and the surrounding medium. This disparity leads to the diffusion of protons across the magnetic field gradient and the dephasing of magnetic moments [20]. SPIONs have been studied as contrast agents in MRI for the diagnosis and treatment of cancerous tumors. For instance, SPIONs were synthesized by Kim et al. [21] by the sonochemical method and were covered with oleic acid surfactant to obtain ferrofluid for use as MRI contrast agents. The SPIONs, coated with oleic acid, exhibited easy dispersion in a chitosan solution, and their observed size was determined to be 65 nm. The hysteresis curve of synthesized SPIONs showed no coercive force with the saturation magnetization of 83 emu/g. the ferrofluids were prepared with varying Fe concentrations and were found to be stable for 30 days. T_1 and T_2 weighted MRI images of ferrofluid were compared with the standard Resovist and water. T_1 images of both ferrofluid and Resovist were brighter than T_1 images of water, whereas T_2 images were darker than T_2 images of water. T_1 and T_2 weighted images of both ferrofluid and Resovist were found to have the same contrasts. Chee et al. [22] conducted the synthesis of ultrasmall superparamagnetic iron oxide nanoparticles (USPIONs) by developing a library comprising 21 peptide ligands, aiming to create self-assembled monolayers on the SPIONs' surface. The evaluation of these peptides-coated USPIONs involved comparisons based on criteria such as colloidal stability, biocompatibility, magnetic performance, and dispersion in circulatory fluids. Rigorous selection criteria were applied, considering factors like the absence of cellular cytotoxicity, stability against electrolyte aggregation, no nonspecific binding to cells, and enhancement in in vivo MRI contrast. Among the candidates, USPIONs coated with bisphosphorylated peptide 2PG-S*VVVT-PEG4-ol demonstrated superior biocompatibility, minimal cytotoxicity, and nonspecific adhesion to cells. These USPIONs showcased enhanced magnetic resonance (MR) properties, with r_1 and r_2 relaxivity values of 2.4 mM^1 s^{-1} and 217.8 mM^{-1} s^{-1}, along with a higher relaxivity ratio (r_2/r_1) exceeding 90 when compared to commercially available contrast agents. Additionally, 2PG-S*VVVT-PEG4-ol-coated USPIONs were applied as a T_2 contrast agent, resulting in improved contrast enhancement of liver tumors. The stability of peptide-coated USPIONs was assessed under various conditions, considering both single peptide-coated USPIONs and mixed peptide coatings. The evaluation included measuring the normalized aggregation parameter based on UV absorption before and after incubation in different conditions. The stability of single peptide-coated USPIONs varied based on the peptide and ligand design. Notably, bisphosphorylated peptide 2PG-S*VVVT-PEG4-ol demonstrated the highest stability across all tested conditions, resisting aggregation even in high NaCl concentrations and elevated temperatures. The introduction of a short hydrophilic ethylene glycol moiety at the peptide's head played a crucial role in preventing electrolyte-induced aggregation. Mixed peptide coatings were also explored, and combinations of mono or bisphosphorylated peptides with the ligand (PO_3H_2)-S-C_{16}-OH demonstrated stability. The peptidol 2PG-S*VVVT-ol combined with the ligand (PO_3H_2)-S-C_{16}-OH provided optimal performance for mixed peptide-coated USPIONs. Additional evaluations included assessing cell toxicity and nonspecific binding, revealing that two highly stable USPION preparations, namely 2PG-S*VVVT-PEG4-ol and 2PG-S*VVVT-ol: (PO_3H_2)-S-C11-EG3-OH, exhibited no toxicity to PLC/PRF5 cells and minimal nonspecific binding (Figure 3.1A). In vivo experiments underscored the efficacy of 2PG-S*VVVT-PEG4-ol-coated USPIONs in contrast-enhanced T_2-weighted MRI. The nanoparticles demonstrated a robust and sustained reduction in the MRI signal in the liver. Furthermore, the peptide-coated USPIONs were assessed for imaging liver tumors, revealing notable contrast enhancement and enhanced delineation of liver tumor contours (Figure 3.1B). Overall, the study suggests that the bisphosphorylated peptide 2PG-S*VVVT-PEG4-ol and its mixed peptide coatings, especially when combined with (PO_3H_2)-S-C_{16}-OH, are promising candidates for in vitro and in vivo applications, including liver cancer imaging. The nanoparticles demonstrated stability, low toxicity, and specific binding in experimental settings, showcasing their potential for biomedical applications.

Lymphotropic SPIONS were used by Harisinghani et al. [23] to detect lymph node metastasis. The SPIONS were able to detect the otherwise undetectable nodal metastasis in combination with high-resolution MRI upon entering the lymph nodes via interstitial lymphatic fluid. The sensitivity of MRI to lymphotropic SPIONS was found to be around 90.5% as compared to 35.4% for conventional MRI. Further, the diagnostic specificity of lymphotropic SPIONs increased from 90.4% to 97.8%. Two types of SPIONs, namely ferumoxide, which are partially dextran-coated SPIONs and Ferumoxtran-10, and complete dextran-coated SPIONs were used by Varallyay et al. [24] for the delineation of intracranial tumors using MRI and compare the results with gadolinium chelate.

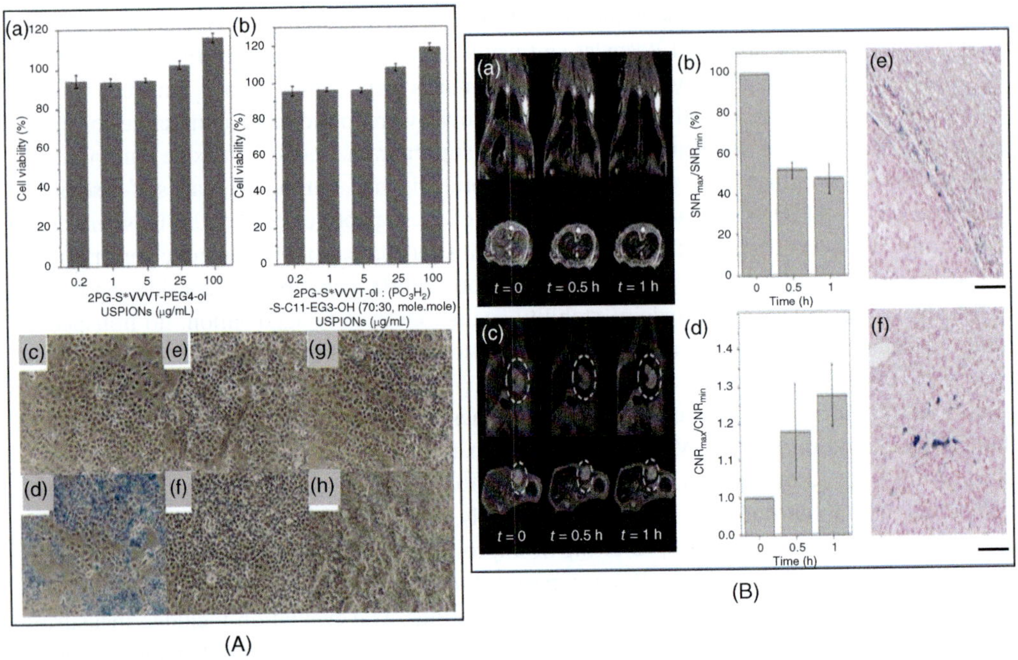

Figure 3.1 (A) Depicts the in vitro assessment of peptide-coated USPIONs. The cell viability assay conducted with PLC/PRF5 cells showed no evidence of cell toxicity at the tested concentration of 100 μg/mL of Fe after 48 hours of incubation at 37 °C for both (a) 2PG-S*VVVT-PEG4-ol and (b) 2PG-S*VVVT-ol:(PO_3H_2)-SC_{11}-EG_3-OH (70:30, mole:mole)-coated USPIONs. The cell nonspecific binding test was performed with PLC/PRF5 cells and the control sample. (c) demonstrated cells incubated without USPIONs. Overnight incubation at 37 °C with (d) TMAOH-coated USPIONs. (e, f) 2PG-S*VVVT-PEG4-ol, and (g, h) 2PG-S*VVVT-ol:(PO_3H_2)-S-C11-EG3-OH (70:30, mole:mole)-coated USPIONs at the concentrations of 10 μg/mL (e, g) and 100 μg/mL (d, f, h) of Fe showed no detectable nonspecific binding. The cells were stained with Prussian blue reagent and counterstained with Nuclear Fast Red solution, and the results are presented as the average and standard deviation of three repeated experiments. (B) Illustrates the in vivo MRI with the contrast agent 2PG-S*VVVT-PEG4-ol-coated USPIONs. (a) In vivo MR images in the coronal (upper) and transverse (lower) planes of NCr nude mice at 0, 0.5, and 1 hour after intravenous injection of 2PG-S*VVVT-PEG4-ol-coated USPIONs. (b) Quantification of liver contrast at 0, 0.5, and 1 hour after the accumulation of peptide-coated USPIONs in NCr nude mice at a dose of 1.0 mg Fe/kg. (c) In vivo MR images in the coronal (upper) and transverse (lower) planes of the orthotopic xenograph liver tumor model at 0, 0.5, and 1 hour after intravenous injection of 2PG-S*VVVT-PEG4-ol-coated USPIONs. The dashed circle indicates the tumor location. (d) Quantification of the contrast-to-noise ratio of tumor-to-liver contrast at 0, 0.5, and 1 hour after the accumulation of 2PG-S*VVVT-PEG4-ol-coated USPIONs at a dose of 2.0 mg Fe/kg. (e, f) Histopathological analysis of the liver of mice after 1 hour post intravenous injection of 2PG-S*VVVT-PEG4-ol-coated USPIONs at a dose of 2.0 mg Fe/kg. The scale bar is 100 μm. Source: Reproduced with permission from Chee et al. [22]/American Chemical Society.

The results showed that ferumoxide administration caused no significant change in T_1 or T_2 intensity, whereas shortening of either T_1 or T_2 occurs in patients administered with ferumoxtran-10, indicating the penetration of iron oxide NPS into tumor. Signal enhancement with ferumoxtran-10 was found to be comparable to gadolinium chelate but with more variations. Further partial dextran coating of ferumoxide caused rapid opsonization resulting in a short half-life of the particle in blood. Histological examination showed the cellular uptake occurred by parenchymal cells at the margins of the tumors. Nahrendorf et al. [25] synthesized VINP-28, VCAM-1 targeting magnetic fluorescent molecule made up of iron oxide core for use as a contrast agent in cardiovascular disease imaging. Vascular cell adhesion molecule 1, VCAM-1, is an immunoglobin protein that acts as a biomarker for atherosclerosis and is expressed on the endothelial cells stimulated by cytokines. Iron oxide NP-based VINP-28 particles were used to target selectively macrophages and endothelial cells, which overexpress VCAM-1. The in vivo MRI findings in correlation with fluorescence reflectance images showed a strong fluorescent signal in aortic atheromata arising from VINP-28. Chen et al. [26] synthesized a nanoconjugate of Iron oxide nanoparticles (IONPs) and dopamine, which was easily conjugated with human serum albumin (HSA). The IONPs coated with HAS (HAS-IONPs) were subsequently dual-labeled with 64Cu-DOTA and Cy5.5. These labeled nanoparticles were evaluated in xenograft mouse models to investigate their in vivo behavior using positron emission tomography (PET)/NIRF/MRI tri-modality imaging (Figure 3.2A). Upon injecting HAS-IONPs into tumor-infected mice at the dose of 10 mg Fe/kg, near-infrared fluorescence (NIRF) images showed the clear delineation of tumor after 1 hour of injection with an increase in tumor/muscle ratio from 1.98 ± 0.20 at 1 hour to 3.08 ± 0.28 at 18 hours. Further PET images showed the elevation in tumor uptake of HAS-IONPs from 5.46 ± 0.64 at 1 hour to $8.45 \pm 0.86\%$ ID/g after 18 hours. Further PET results showed a comparatively much higher tumor/muscle ratio than NIRF. MRI scans displayed a clear signal drop of $29.9 \pm 4.2\%$ pre- and post-18 hours of injection. *Ex vivo* PET and NIRF studies of tumor allografts displayed substantial intensities from tumors. Further PET/NIRF images showed a substantial accumulation of HAS-IONPs in the liver compared to other organs (Figure 3.2B).

Iron oxide-polymer core–shell NPs with an iron oxide core diameter of 55 ± 6 nm have been synthesized by Hayashi et al. [27] and explored for application in MRI-guided magnetic thermochemotherapy. The schematic representation of the synthesis of polymer core–shell nanoparticles containing iron oxide and doxorubicin (iron oxide-DOX) is depicted diagrammatically in Figure 3.3A. DOX loaded Core–shell nanoparticles were cultured with HAC-2 tumor cells which is human ovary cancer cell line. The presence of core–shell nanoparticles induced the expression of foliate receptors from the tumor cells after 24 hours, with the bright field and fluorescent images showing no leakage of DOX molecules from the core–shell particles into the cultural medium (Figure 3.3B(a–c)). Unmodified iron oxide-polymer core–shell NPs were not observed in the vicinity of tumor cells, which indicated that foliate receptors of HAC-2 tumor cells exhibit specific binding to the folate receptors present on the surface of the core–shell NPs. Further MRI scans of the tumor after intraperitoneal injection of DOX-loaded core–shell NPs at the optimum hypothermia dosage of 4 μg/g after 24 hours of injection showed a significant reduction in the MRI signal with respect to the neighboring tissues, which remained almost unchanged. No significant variation in MRI signal was recorded with tumors loaded with unmodified core–shell iron oxide NPs (Figure 3.3B(d–g)). MRI and histological analysis indicated DOX-loaded iron oxide core–shell NPs were located specifically in tumor cells, which was also evident by hematoxylin and eosin (HE)/iron (Berlin blue) staining analysis of the tumor sections (Figure 3.3B(h,i)). In yet another study, chemoferroptosis combination therapy was investigated in tumor-induced mice by MRI using a combined polypeptide vesicle loaded with drug cisplatin and SPIONs (Pt and Fe_3O_4 at PP) [28]. The examination of the tumor accumulation of Pt and Fe_3O_4 at PP through MR imaging provided T_2-weighted MR images with transverse relaxation (r_2) values of 271.5 mM s^{-1}, indicating the effectiveness of Fe_3O_4 nanoparticles as contrast agents for MRI. Additionally, after the injection of Pt and Fe_3O_4 at PP in mice, there was an observed enhancement in T_2 contrast at tumor sites compared to the preinjection state. It has been observed that r_2 values of SPIONs increase with an increase in size. It has been reported that spin-spin relaxation time (T_2) weighted MRI signal intensity increases with a corresponding change in MR image color from white to black with an increase in the size of SPIONs from 4 nm to 12 nm [29]. However, a limiting value of r_2 is reached as the size of SPIONS reaches a threshold due to the occurrence of a static dephasing regime [30]. This size-dependent relaxation of SPIONS resulting in poor contrast in MRI has somewhat limited

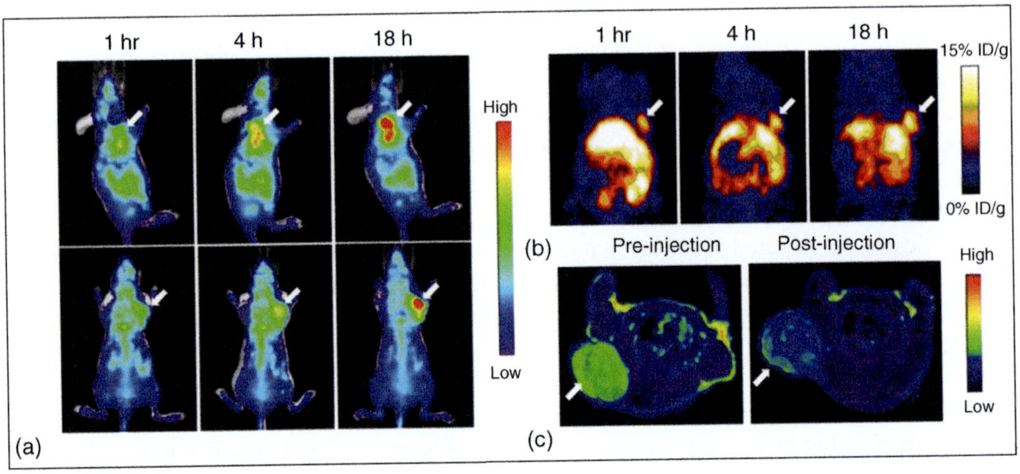

(A)

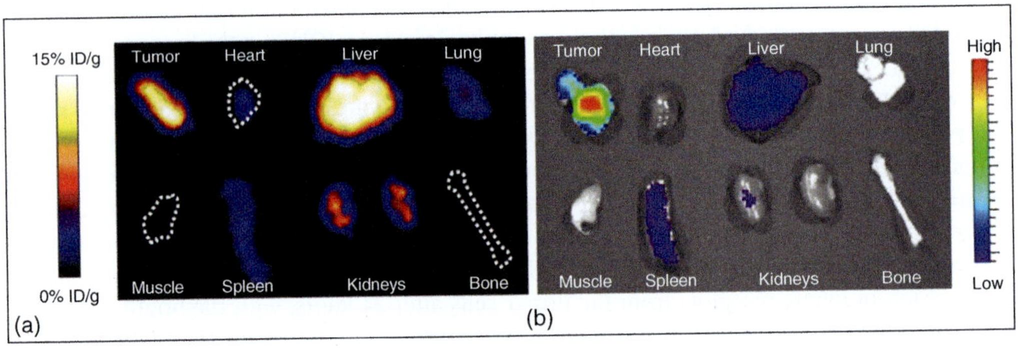

(B)

Figure 3.2 (A) Depicts the following: (a) In vivo near-infrared fluorescence (NIRF) images of a mouse administered with HSA-IONPs. The images were captured at 1 hour, 4 hours, and 18 hours post-injection. (b) In vivo positron emission tomography (PET) imaging results of a mouse injected with HSA-IONPs. The images were obtained at 1 hour, 4 hours, and 18 hours post-injection. (c) Magnetic resonance imaging (MRI) images were acquired both before and 18 hours after injection. (B) Illustrates: (a) *Ex vivo* PET imaging focusing on the tumor and major organs. (b) *Ex vivo* NIRF imaging targeting the tumor and major organs. Source: Reproduced with permission from Xie et al. [26]/Elsevier.

their use for advanced MR imaging [31, 32]. Therefore, new types of magnetic materials known as magnetism-engineered iron oxide (MEIO) nanoparticles/nanoprobes have been developed to increase the magnetic contrast of MRI by significant reduction of spin-spin relaxation time (T_2) of the protons [33]. Such a series of MEIO nanoprobes have been developed by doping iron oxide (Fe_3O_4) by divalent transition metal cations such as Mn, Co, Fe, or Ni to produce MFe_2O_4 spinal structures [33]. This series of metal-doped MEI nanoparticles have been found to induce MRI signal enhancement effects. For instance, manganese-doped MEI (Mn Fe_2O_4), which has the highest magnetization value of 110 emu/g, exhibited the prominent MR contrast effect with R_2 value of 358 l/mmol/s (Figure 3.4A). In case of other metal-doped MEI NPs with diminished magnetization values, shown in brackets, R_2 values were found to decrease viz., 218 for $FeFe_2O_4$ (101 emu/g), 172 for $CoFe_2O_4$ (99 emu/g), 152 $NiFe_2O_4$ (85 emu/g) and 62 l/mmol/s for CLIO. Further r_2 values, which is a direct reflection of MR contrast enhancement, were also found to increase gradually with a size of MEIO nanoparticles from 78 mM^{-1} s^{-1} (4 nm), 106 mM^{-1} s^{-1} (6 nm), 130 mM^{-1} s^{-1} (9 nm), and 218 mM^{-1} s^{-1} (12 nm). Mn Fe_2O_4 was used in conjugation with Herceptin for in vivo imaging of small-sized cancers in mice at the dosage of 20 mg/kg. MRI images captured before injection, along with those obtained at one hour and two hours post-injection, revealed a transition in the tumor color from red to blue in the color-coded MRI. This alteration signifies a distinct and unequivocal detection of the tumor site in the MRI images (Figure 3.4B). The R_2 value experienced a rise from 25% one hour post-injection to 34% two hours after injection at the tumor tissue. In contrast, the CLIO-Herceptin conjugate, when administered in tumor-bearing mice, demonstrated an increase in R_2 at the tumor site of less than 5% [33]. Zinc ferrite (Zn_xFe_3-xO_4) octapods have been synthesized by Yang et al. [34] with variable zinc doping ratios as a contrast agent for MRI. It was noticed that both saturation magnetization and r_2 values showed an increased trend with an increase in the Zn dopant concentration with the r_2 value of 989.1 mM^{-1} s^{-1} reported at ($x = 0.44$) and the T_2 relaxivity, which is 10.2 times greater than reported for the commercial sample. These octapods demonstrated effective contrast in in vivo MRI imaging for detecting hepatic and small metastatic tumors at low doses. The T_2 relaxivity was observed to be influenced by the surface environment or coating of the MNP surfaces [35, 36]. Various reports have been published on the use of magnetic iron oxide NPs in imaging applications [37–42].

In conclusion, MRI stands as a valuable clinical diagnostic tool that leverages the behavior of hydrogen atoms in the body under a magnetic field. The technique relies on the T_1 and T_2 relaxation processes of these atoms to generate images that provide contrasting information about various tissues and pathologies. Despite the historical use of traditional paramagnetic materials like Gd as contrast agents, they have limitations, including low relaxivity values and potential toxicity. Recently, SPNPs, particularly SPIONs, have emerged as promising contrast agents for MRI. The customizable size and functionalization of SPNPs make them adaptable to specific tissue requirements, displaying shorter T_1 and T_2 relaxation times at low concentrations and resulting in robust contrast effects. SPIONs have demonstrated potential applications in cancerous tumor diagnosis and treatment, detection of lymph node metastasis, and delineation of intracranial tumors.

MEIO nanoparticles, involving the doping of Fe_3O_4 with transition metal cations, have been developed to enhance the magnetic contrast of MRI. Among them, manganese-doped MEIO ($MnFe_2O_4$) exhibited the most robust contrast effect. Other magnetic materials, such as zinc ferrite octapods, also show promise as MRI contrast agents. It is noteworthy that the surface environment and coating of magnetic nanoparticles can influence their T_2 relaxivity.

In summary, the continuous development of novel magnetic materials and nanoparticles has expanded the capabilities of MRI by improving contrast enhancement and detection capabilities, paving the way for advanced imaging techniques and applications.

3.1.2 MNPs in Drug Delivery Applications for Cancer Therapy

Over the past few decades, significant advancements have been made in drug delivery systems, involving the integration of the therapeutic effects of drugs with external magnetic systems and imaging technologies like MRI, magnetic particle imaging (MPI), or near-infrared optical imaging (NIR) [43]. Such a sophisticated system allows the precise delivery of the drugs, which increases its efficiency. Such a combination of therapy with imaging techniques is known as theranostics [44]. A proficient theranostic system can be devised by creating a unified vehicle that possesses the ability to adsorb, transport, and release drugs, coupled with simultaneous imaging capabilities, thereby augmenting the targeting properties [45]. To fulfill these objectives, MNPs hold a distinct advantage for use in theranostic-based drug delivery systems, primarily owing to their superparamagnetic properties and biocompatibility [45]. They have also been advantageous due to their function in imaging as contrast agents with different formulations [46, 47]. The therapeutic efficiency of MNPs was also enhanced due to the local heating produced by the NPs inside

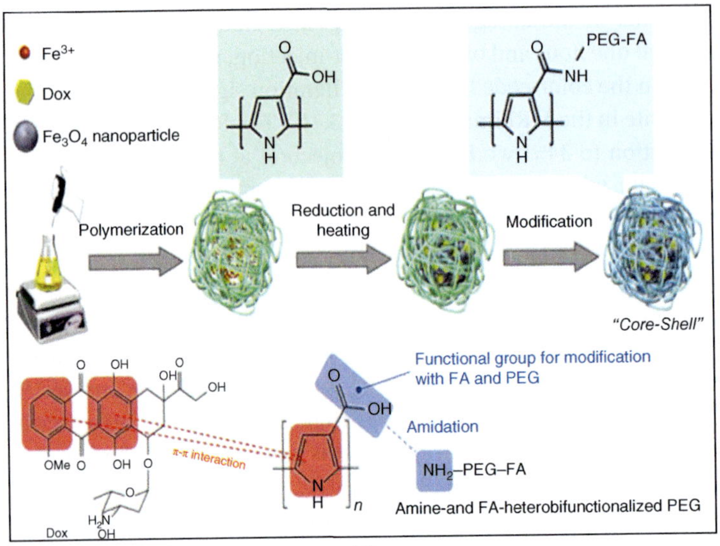

(A)

Figure 3.3 (A) Outline of the Core–Shell Synthesis Scheme, while (B) demonstrates cellular uptake and in vivo MRI-guided magnetic thermochemotherapy. The details are as follows: (a) Core–shell cellular uptake: Images from left to right include bright field, fluorescence, and overlay images. (b) Photograph and T_2-weighted MR images of mice following Core–Shell injection. The red square in the photograph marks the area subjected to MRI scanning, while yellow circles in the MRI images indicate tumor sites. (c) HE- and iron-stained tumor sections for mice with and without Core–Shell injection. The blue-stained area in the iron-stained sections represents the region where iron ions, that is, the core–shells, are present. (d) Thermal images of mice with and without Mag or Core–Shell injection under the influence of AMF. White circles highlight abdominal tumor sites. (e) Changes in mouse abdominal temperature with and without Mag or Core–Shell injection under the influence of AMF. (f) Weights of intraperitoneal tumors excised from mouse abdomens 42 days after treatment, with significance denoted by $^{*}p < 0.05$, $^{**}p < 0.01$, $^{***}p < 0.005$. (g) Changes in body weight post-treatment, with significance denoted by $^{*}p < 0.05$, $^{**}p < 0.01$, $^{***}p < 0.005$. (h and i) depict HE-, TUNEL-, and iron-stained as well as undyed tumor sections of untreated mice and mice 24 hours after magnetic thermochemotherapy, respectively. Surviving tumor tissues are enclosed in a yellow dotted line. In TUNEL-stained sections, brown and blue areas indicate dead and surviving tissues, respectively. In iron-stained sections, the blue-stained area indicates the region where iron ions, that is, the core–shells, are present. In undyed sections, red-stained areas indicate the presence of Dox. The scale bar is 800 μm. Source: Reproduced with permission from Hayashi et al. [27] and Gao et al. [28]/American Chemical Society.

the biological systems upon application of the external magnetic field [48]. Indeed, this localized heating process is known as magnetic hyperthermia (MH), and it effectively induces the death of cancer cells due to their heightened sensitivity to elevated temperatures (42–49 °C) compared to normal healthy cells [49]. Such a thermal stimulus can act as a trigger to prompt the MNPs to release the loaded drugs and hence can increase the efficiency of the treatment by acting in synergy with thermotherapy [50]. Cancers are usually treated with conventional therapeutic methods such as radiotherapy and radiotherapy and surgeries [51]. The localized cancers are usually eradicated with radiotherapy or surgery, whereas in some cases chemotherapeutic agents are used to treat cancers, but the main drawback of chemotherapeutic agents is their lack of specificity [52]. Chemotherapeutic compounds usually destroy simultaneously both healthy as well as cancerous cells, thus giving rise to various health complications. These compounds attach cells in a particular cell cycle phase independent of the cell type. The problem of the nonspecificity of chemotherapeutic agents has been addressed by loading the compounds onto some specific carrier agents for targeted drug delivery [53]. In this context, endeavors have been undertaken to create

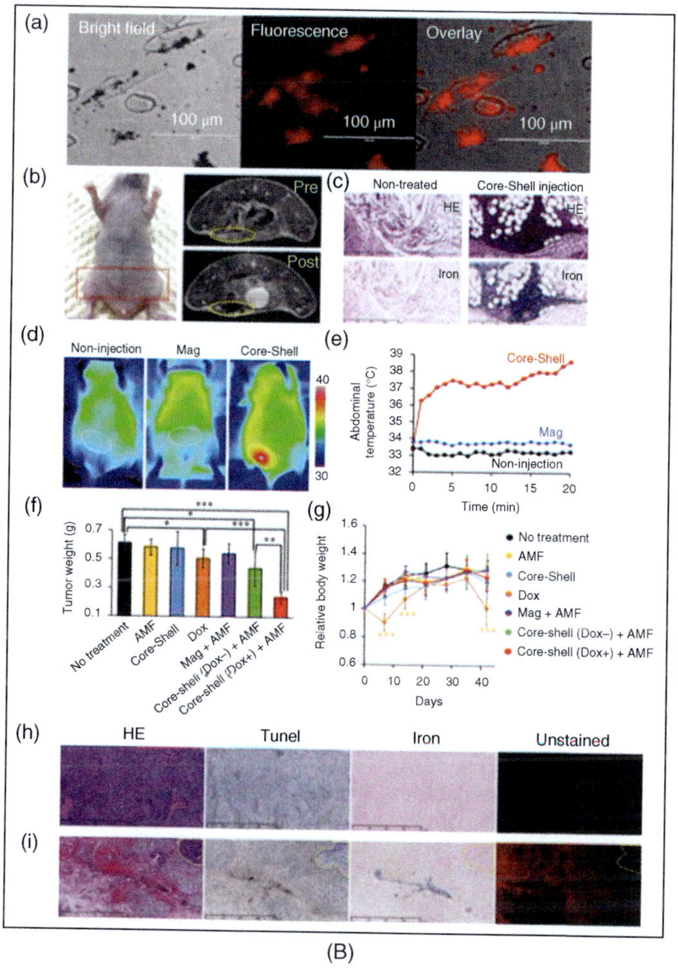

Figure 3.3 (*Continued*)

diverse nanoparticle-based drug delivery systems aimed at delivering therapeutic agents specifically to pathological cells. Among the array of nanoparticles, magnetic nanoparticles, including IONPs, have demonstrated promise as viable candidates for the targeted and precise delivery of anticancer drugs to tumors [54]. Magnetic nanoparticles used for this purpose are either single-core nanosystems consisting of individual IONPs or multicore nanosystems consisting of magnetic nano assemblies embedded in a macromolecular matrix [55]. To provide an example, magnetic nanoparticles incorporating IONPs cross-linked with glyoxal have been successfully synthesized using the microemulsion route. This synthesis aims to enable the loading and delivery of doxorubicin (DOX) [56]. The effective integration and distribution of magnetic IONPs within the GMS have been validated through energy-dispersive spectroscopy and elemental mapping, affirming the presence of robust and uniform Fe signals within the GMS microspheres. The magnetic properties of GMS microspheres were consequently inherited from IONPs, showcasing superparamagnetism. It was noted that the drug loading and release profile is intricately governed by the extent of swelling in the polymer matrix, a parameter controlled by the degree of cross-linking and the quantity of MNPs incorporated into the polymer matrix. A nanocomposite of silica nanospheres (80–150 nm) and IONPs (maghemite) core (10–12 nm) with different maghemite content (Si/Fe ratio of 58 and

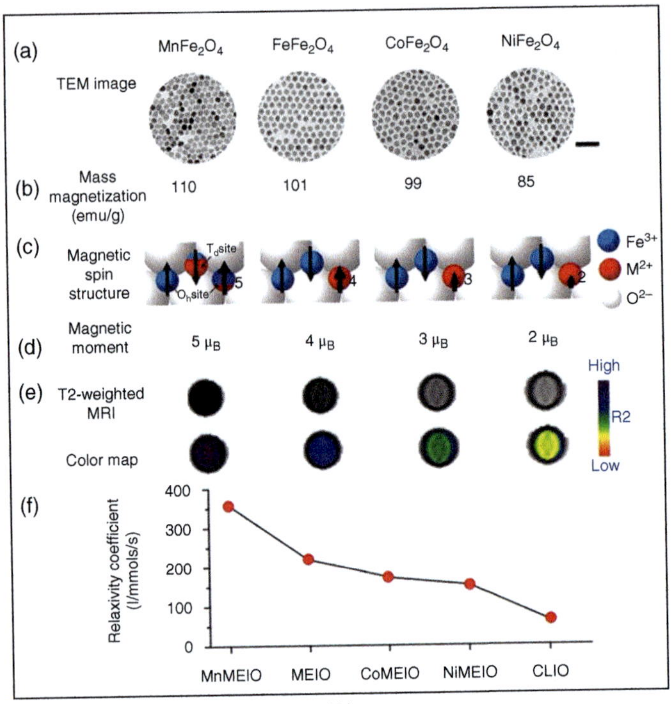

Figure 3.4 (A): Magnetism-engineered iron oxide (MEIO) nanoparticles and effects of their magnetic spin on MRI. (a) Transmission electron microscopy (TEM) images depicting $MnFe_2O_4$ (MnMEIO), Fe_3O_4 (MEIO), $CoFe_2O_4$ (CoMEIO), and $NiFe_2O_4$ (NiMEIO) nanoparticles, all synthesized to be below 12 nm with narrow size distributions (standard deviation ≤ 8%). Scale bar: 50 nm. (b) Mass magnetization values of MFe_2O_4. (c, d) Diagrams illustrating spin alignments of magnetic ions in spinel structures under an external magnetic field and the magnetic spin moment of MFe_2O_4 nanoparticles. In face-centered cubic lattices of oxygen, magnetic spins at Oh sites align in parallel with the external magnetic field, whereas those at Td sites align antiparallel. $MnFe_2O_4$ exhibits the highest mass magnetization value, with a magnetic spin magnitude of 5 mB. (e, f) T_2-weighted spin-echo MR images, their color maps, and relaxivity (R_2) of a series of MEIO nanoparticles at 1.5 T. In (f), The R_2 of CLIO is also presented for comparison. Consistent with mass magnetization results, MnMEIO demonstrates the strongest MR contrast effect (i.e., darkest MR image and violet in color map) and the highest R_2 coefficient. Mass magnetization value, MR contrast, and R_2 coefficient gradually decrease as M^{2+} changes from Mn^{2+} to Fe^{2+}, to Co^{2+}, and to Ni^{2+}. (B) In Vivo MR Detection of Cancer Using Magnetic Nanoparticle–Herceptin Conjugates. (a–f) Color maps of T2-weighted MR images of a mouse implanted with the cancer cell line NIH3T6.7 at different time points after injection of MnMEIO-Herceptin conjugates or CLIO-Herceptin conjugates (preinjection (a, d); and one hour (b, e) or two hours (c, f) after injection). In a–c, gradual color changes at the tumor site from red (low R_2) to blue (high R_2) indicate progressive targeting by MnMEIO-Herceptin conjugates. In contrast, almost no change is seen in the mouse treated with CLIO-Herceptin conjugate (d–f). (g) Plot of R_2 change versus time. In the mouse treated with MnMEIO-Herceptin conjugate (squares), significant R_2 changes (up to 34%) were observed with time after treatment. In contrast, R2 changed by ≤5% after treatment with CLIO-Herceptin conjugate (dots) and by ≤13% after treatment with 12-nm–MEIO–Herceptin conjugate (triangles).
(h) *Ex vivo* MR images (i–iii) of explanted tumors (eight hours) and their color maps (iv–vi). The tumor explanted after treatment with MnMEIO-Herceptin conjugate (i) is dark; that explants following CLIO-Herceptin conjugate treatment (ii) or no treatment (iii) show no contrast. Consistently, in the image color-coded according to R2, the tumor explanted after MnMEIO-Herceptin conjugate treatment is blue (iv), whereas that after CLIO-Herceptin conjugate treatment (v) or no treatment (vi) is red. Source: Reproduced with permission from Lee et al. [33]/Springer Nature.

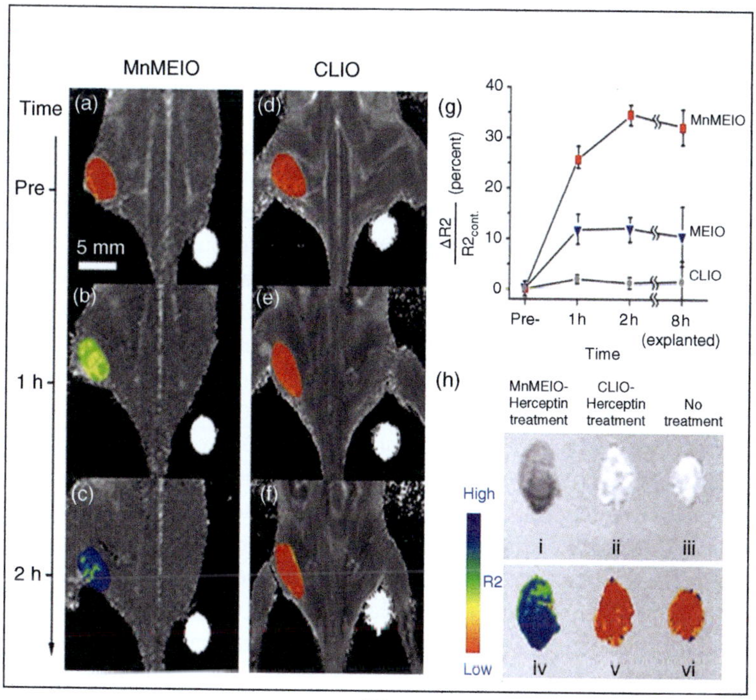

(B)

Figure 3.4 (*Continued*)

20) has been synthesized and used for the delivery of prednisolone drug in magnetic field [57]. The presence of maghemite NPs in the silica structure was confirmed by XRD and Mossbauer spectroscopy and the nanocomposite exhibited superparamagnetic behavior. The magnetic silica nanoparticles displayed a slower drug release than the drug-free ones. Inside the silica nanospheres, the prednisolone drug underwent a polymorphic transition from polymorph I to polymorph II, which showed a slower release profile than polymorph I or polymorph II in the free state. In another study, MNPs were embedded in bioactive glass (BG) and the resulting nanocomposite was loaded with doxorubicin to check its loading efficiency at various concentrations against the human osteosarcoma cell line (MG63) invitro [58]. Magnetization curves depict the superparamagnetic behavior of MNP/BG composite particles, and increase in saturation magnetization occurred due to increase in iron oxide content in the BG structure. MNP-BG were found to be nontoxic at various concentrations and the DOX released from the nanocomposite resulted in decrease in cell viability of the MG63 cell line. Fluorescent microscopy studies revealed decrease in cell proliferation of the MG63 cell line with FITC-labeled Fe_3O_4-BG particles at concentrations of 250 and 500 μg/mL. Further, the cell proliferation decreased with increase in the content of iron oxide in Fe_3O_4-BG particles. The decrease in cell proliferation was attributed to the change in cell homeostasis by the degradation of BG inside cells, resulting in the release of ionic dissolution products inside cells. Cell death was also reported to occur due to free radicals generated by IONPs, which resulted in damage of cellular components and, finally, cell death. Drug loading efficiency also varied with the amount of iron oxide content inside amine-modified Fe_3O_4-BG particles. The drug loading efficiency was observed to be highest for NH_2-0.5-Fe_3O_4/BG (76%) as compared to NH_2-0.25-Fe_3O_4/BG (70%) and NH_2-1-Fe_3O_4/BG (76%). IONPs coated with silica shells were synthesized and further grafted with poly(N isopropylacrylamide-co-3-(methacryloxypropyl)trimethoxysilane), PNIPAM,

for controlled drug delivery of test drug, Ibuprofen [59]. The IONP-based polymer hybrid nano drug carrier showed a very good magnetization saturation value (19.5 emu/g) and high surface area (505 m^2/g). The resulting NP-polymer hybrid formed a thermoresponsive drug delivery vehicle with temperature-controlled drug release. At lower temperatures (below 25 °C) only 20% of the drug was released after 16 hours as compared to the 80% release occurred at higher temperature (above 40 °C). Iron-oxide-based mesoporous magnetic microparticles (IO-MMMs) have been synthesized by El-boubbou et al. [60] and utilized for the drug loading and delivery of hydrophilic (Dox or Daun)) and hydrophobic (Tam) chemotherapeutic drugs. IO-MMMs were found to have a high surface area (240 m^2/g) and high biocompatibility, which made them ideal candidates for drug delivery. Cytotoxicity studies have confirmed that drugs loaded on IO-MMMs in combined form (Daun/Tam at IO-MMMs) specifically reduced the viability of breast cancer cell lines (MCF7 and KAIMRC1) as well as colorectal cancer cells (HCT8) significantly up to 10-fold and were least toxic to the normal noncancerous cell lines. Both the drugs in free state were toxic to both normal and cancer cells at equivalent concentrations. Magnetic response of the IO-MMMs was investigated by incubating the cancer cell line (KAIMRC1 cells) with IO-MMMs and applying the static magnetic field. Surprisingly, it was observed that cancer cells, when incubated, exhibited alignment with the magnetic field within petri dishes. Conversely, in the absence of a magnetic field, no such alignment was detected. The magnetic field effects demonstrated the possibility of concentrating and spatially locating IONP-based microparticles inside cells for magnetically induced hyperthermia during drug delivery. Cellular uptake of drug-loaded IO-MMMs by flow cytometry suggested a significant uptake of sevenfold after 2 hours of incubation and then reduced to twofold after 24 hours in comparison to normal primary cells. The results confirmed an enhanced uptake of IONPs by cancerous cells in comparison to normal cells. This enhanced cellular uptake of IONPs by cancer cells was attributed to the high metabolic activity of the tumor cells in comparison to the normal cells. Cellular uptake and internalization was also ascertained by electron microscopy. SEM images showed the IO-MMMs particles were found attached mainly to the cell membranes, with some particles located on the periphery and the particles were also found to be wrapped by the cell membranes. Representative transmission electron microscope (TEM) images also demonstrate the adherence of IO-MMMs particles to the cellular membrane, followed by internalization either individually or in clusters through the plasma membrane. Many reports are available that show the enhanced drug delivery of IONP-based polymeric particles/gels against various cancer cell lines [61, 62].

Attari et al. [63] synthesized arginine-capped IONPs and subsequently coupled them with the methotrexate (MTX) drug for the purpose of targeting cancerous cells. The average diameter of the resulting IONPs-Arg-MTX was found to be 27 nm. The resulting NPs showed good magnetic properties with Ms of 28 emu/g. The NPs exerted a minimal hemolytic effect on HRBCs, which confirmed their biocompatibility. In vitro cytotoxicity assessments were carried out using different concentrations against breast cancer cell lines, namely 4T1 and MCF-7. The findings indicated that nanocarriers based on MNPs led to a reduction in the cell viability of cancer cells in comparison to free methotrexate (MTX) with IC_{50} values of 230 and 200 nM after 48 hours and 72 hours, respectively, against MCF-7 cancer cell line, whereas IC_{50} values of 380 and 265 nM were recorded against 4T1 cell line after 48 hours and 72 hours. The IC_{50} values of the free MTX were found to be much higher (450 and 400 nM for MCF-7; 390 and 575 nM for 4T1 cell line) than MNP-based nanocarriers. The release profile of the nanocarrier based on MNPs was investigated in the presence and absence of the enzyme proteinase K. Drug release was found to be faster (55% after 12 hours) in the presence of the enzyme as compared to the absence of the enzyme (22%) because the peptide bond underwent enzymatic cleavage between MTX and NPs. Huang et al. [64] synthesized SPIONs with coatings of polyethylene glycol (PEG) and polyethyleneimine (PEI). Following this, the particles were further modified with folic acid (FA) and loaded with the model anticancer drug doxorubicin (DOX), aiming to enable targeted drug delivery specifically to anticancer cells. The hydrodynamic sizes of the SPIONs, FA-SPIONs, and DA at FA-SPIONs were measured to be 23 nm, 40 nm, and 67 nm, respectively. pH-responsive drug release profile indicated the total DOX release of 90% occurred after 48 hours at pH of 5.4, which was found to be remarkably higher compared to pH of 7. In vitro cytotoxicity results by MTT assay suggested the high toxicity of DOX at FA-SPIONs toward the MCF-7 cancer cells compared to free DOX, whereas FA-SPIONs demonstrated the least cytotoxicity. It was documented that DOX at FA-SPIONs entered cancer cells through receptor-mediated endocytosis, while free DOX entered cells through passive diffusion. The introduction of a magnetic field was observed to amplify the cytotoxicity of DOX at FA-SPIONs against MCF-7 cells, indicating a potential enhancement in the interaction between DOX at FA-SPIONs and MCF-7 cells facilitated by the magnetic field. Moreover, fluorescence microscopy images revealed a significantly brighter red fluorescence in cells treated with DOX at FA-SPIONs compared to those treated with free DOX, suggesting a higher concentration of DOX in the former group. Additionally, the application of a magnetic field led to an augmentation in the fluorescence of MCF cells treated with DOX at FA-SPIONs compared to cells without a magnetic field. Biodistribution findings in mice indicated a notable accumulation of DOX at FA-SPIONs in the tumor, approximately 6.9 times higher in the presence of a magnetic field compared to the injection of DOX at FA-SPIONs in the absence of a magnetic field. A reduction in iron content was noted in the spleen, lungs, and liver of mice treated with DOX at FA-SPIONs in the presence of a magnetic field compared to those treated in the absence of a magnetic field. Furthermore, the in vivo antitumor activity was assessed by administering DOX at FA-SPIONs to mice with tumors. A significant decrease in tumor size was observed in mice treated with DOX at FA-SPIONs in the presence of a magnetic field, and the tumors displayed substantial necrosis. Quan et al. [65] synthesized hyaluronic acid-shelled (HAS)-coated IONPs as a drug delivery platform for anticancer therapy. The delivery system was created by incorporating 0.5 mg of doxorubicin (Dox) and 1 mg of IONPs into 10 mg of HAS matrices. The hydrodynamic size of DOX at HSA-IONPs was found to be 50.8 nm. Biodistribution studies showed the drug uptake by the 4T1 tumor increased to 15.6 and 7.8 times using DOX at HSA-IONPs after one hour and four hours, respectively, compared to the free Dox used alone. The higher tumor accumulation of nanoformulation was attributed to the interaction of the SHA sheath with the glycoproteins, gp60, albondin receptor and SPARC, that is, secreted protein, acidic, rich in cysteine. Many such research reports have been published that have successfully evaluated the application of IONPs as carriers of drugs for cancer therapy [66–70].

Utilizing active targeting strategies has become a common approach to enhance the accumulation and uptake of anticancer drugs with the aid of MNPs. In these scenarios, a variety of biomolecules or molecular self-assemblies, including exosomes, aptamers, antibodies, and peptides, are attached to the external surface of the MNPs. For instance, Jia et al. [71] employed exosomes that were modified with a neuropilin-1-targeted peptide (RGERPPR, RGE) and loaded with MNPs/curcumin (RGE-Exo-SPION/Cur) to specifically target glioma cells. MNP-modified exosomes were found to easily pass through the blood barrier and reach the targeted cancer cells. Fluorescence intensity results indicated that RGE-Exo-SPION/Cur managed to reach the tumor area earlier and stayed longer inside tumor cells at higher concentrations as compared to control. Further Imaging results suggested a strong enhancement in imaging signal for RGE-Exo-SPION/Cur groups compared to those without SPIONs. SPIONs and curcumin were seen to act synergistically with each other to exert a stronger inhibitory effect on tumor growth. Park et al. [72] have synthesized magneto nanoclusters composed of oleic-acid-stabilized SPIONPs and amphiphilic phospholipids conjugated with HER2-specific aptide (APT_{HER2}) to obtain APT_{HER2}-MNP. APT_{HER2}–MNPs were then loaded with anticancer drug, Docetaxel (DXT) to evaluate the antitumor efficacy of the resulting nanosystem, DOX at APT_{HER2}-MNP, in the NIH3T6.7 xenograft tumor model. Results from cell binding experiments revealed that APTHER2-MNP retained a notable affinity for binding to HER2-overexpressing NIH3T6.7 cells. DTX at APTHER2-MNCs were loaded with 3.2% DTX, and 80% of the DTX drug was released gradually over a period of 10 days. Further, DTX at APT_{HER2}-MNCs showed higher toxicity to the tumor cells in comparison to the free DTX and the toxicity of APT_{HER2}-MNCs was found to be the lowest. Tumor growth was reduced substantially (50%) by DTX at APT_{HER2}-MNCs as compared to the free DTX at the same dose of 2 mg/kg, which did not show the appreciable inhibition of tumor growth. However, free DTX significantly inhibited tumor growth (67%) at higher concentrations of 10 mg/kg, but such as dos free drug also showed significant cell toxicity. Magnetic lipocomplexes were synthesized by incorporating magnetic IONPs and DOX into lipid materials for enhanced stimulated targeted drug delivery [73]. IONPs were present in the core of lipocomplexes surrounded by lipid bilayer and loaded with DOX. The resulting DOX-loaded magnetic lipocomplexes displayed enhanced antitumor activity against the mouse colorectal carcinoma cell line (CT26). In vivo tumor imaging also showed the suppression of tumor growth by magnetically sensitive DOX at magnetic lipocomplexes. Oleic acid IONPs further modified with chitosan (IONPs at OA-CS NPs) have been synthesized and utilized for the precise delivery of drug, 5-floro uracil to the cancer site [74]. The release profile of the drug-loaded nanoparticles (IONPs at OA-CS-5FLU NPs) was investigated under varying pH conditions. The results have shown that the adsorbed drug was released slowly at the physiological pH of 7.4, with only 62% of the drug released over a period of 500 minutes. However, in order to replicate the physiological conditions of cancer cells, the drug release profile was also examined at a pH of 5.2. The results demonstrated a faster drug release, with over 73% of the drug released within 500 minutes. Further, the cell toxicity studies of the IONPs at OA-CS-5FLU NPs were also studied using an MTT assay against various cancer cell lines like HeLa S3, A549, and MCF-7. The investigations into cell viability revealed that the drug-loaded nanoparticles demonstrated heightened toxicity against A549 and HeLa S3 cells, showcasing IC50 values of 12.9 and 23 μg/mL, respectively. The cell viability of all the cancer cell lines did not change on treatment with the IONPs at OA-CS NPs only. Further fluorescent microscopy images revealed the significant cell death of A549, HeLa S3 cells occurred at concentrations of 15 and 40 μg/mL. The cell death resulted due to the modification in the cell morphologies and nuclear fragmentation. A biocompatible nanocarrier composed of biodegradable biopolymer, IONPs (Fe_3O_4) and doxorubicin (DOXI) ((PVLPEG-PVLDOXI-PCL-PHOS-Fe_3O_4) have been synthesized to produce the magnetic vector for drug internalization under magnetic field [75]. Acylhydrazine linker was used to create the mild acidic condition required to release drug at the cancerous site. The resulting magnetic nanocarrier was found to be superparamagnetic in nature and exhibited excellent tumor penetration. The study of drug delivery induced by a magnetic field involved incubating magnetic nanocapsules with the mouse mammary gland cancer cell line (4T). Flow cytometry analysis indicated a clear shift in intensity over cells when comparing the results with and without magnet in Allophycocyanin (APC) channels. The results suggested an increased internalization of nanocapsules when exposed to a magnetic field. Furthermore, confocal laser scanning microscopy images illustrated that the internalization of DOXI in 4T cells showed a proportional increase with both the duration and intensity of the magnetic field. Further, enhanced antitumor activity was observed with DOXI-conjugated nanocapsules due to prolonged retention of nanocarriers in the nucleus.

In conclusion, the integration of MNPs into drug delivery systems has undergone significant advancements in recent decades. The fusion of therapeutic drug effects with external magnetic systems and imaging technologies such as MRI, MPI, or NIR has given rise to a sophisticated paradigm known as theranostics. This innovative system enables precise drug delivery, ultimately enhancing treatment effectiveness. MNPs, with their superparamagnetic properties and biocompatibility, are particularly favored in theranostic-based drug delivery systems. They serve a dual role as contrast agents for imaging and facilitators of localized heating, termed MH, effectively targeting and eliminating cancer cells due to their heightened sensitivity to elevated temperatures. The incorporation of MNPs into different carrier agents, such as silica nanoparticles, BG, or polymeric hybrids, has enabled targeted drug delivery to pathological cells. Various studies have demonstrated enhanced drug delivery, cytotoxicity against cancer cells, and improved tumor accumulation using MNPs as drug carriers. Active targeting strategies involving the attachment of biomolecules or molecular self-assemblies to MNPs have also been employed to enhance the accumulation and uptake of anticancer drugs. In summary, the incorporation of MNPs in drug delivery systems shows considerable potential for improving the specificity, efficacy, and imaging capabilities of cancer treatments.

3.1.3 Magnetic Nanoparticles in Hyperthermia Therapy

In hyperthermia therapy, cancer cells undergo exposure to temperatures ranging from 41 °C to 45 °C, resulting in cell damage through various mechanisms. These include increased membrane cell permeability, induction of protein denaturation, damage to the tumor vasculature, and disruption of nucleic acid and protein synthesis [76]. Hyperthermia can be categorized into three types:

(a) Whole-body hyperthermia, which involves applying heat to the entire body using methods such as hot water, electric blankets, and hot wax.

(b) Specific organ hyperthermia, where heat is applied to a particular organ affected by cancer.

(c) Local hyperthermia, wherein heat is directly applied to cancerous cells or tumors [77]. In local hyperthermia, heat is generated through the use of electromagnetic waves like ultrasound, microwaves, or radio waves. This process involves placing heat-generating instruments either on the surface of the skin or inside the skin, especially for superficial cancer, or by implanting them deep within the target cancer cells [78]. All three types of hyperthermia treatments come with various disadvantages. Nonspecific heat treatment poses a risk of affecting healthy cells due to high temperatures. Additionally, challenges include difficulty in controlling and measuring temperature at the cellular level, as well as nonuniformity in the applied heat at the targeted region, where the heat delivery may be limited [79, 80]. Even though local hyperthermia is deemed superior in comparison to the other two types due to its enhanced heat uniformity and better control over the exposed area, it is not without its drawbacks. A significant limitation is its high invasiveness, particularly for regions with deep-seated cancers, and it demonstrates a shallow penetration depth, typically within the range of a few centimeters [81, 82]. These side effects of local hyperthermia can be minimized using nanotechnology. Nanotechnology makes use of nanomaterials such as magnetic nanoparticles in local hyperthermia [83]. Magnetic nanoparticles are transported to tumor regions and externally heated through the application of an alternating magnetic field. This approach is less invasive, offers superior localized heating, and is particularly suitable for deep-seated cancer targets.

Two main characteristics that describe the magnetic characteristics of MNPs are the fine size effects and surface effects [84]. Fine size effects reflect whether the MNPs are single-domain or multidomain based on structural considerations. Surface effects is associated with several parameters such as breakdown of the symmetry of the crystals on the particle surface, surface strain, oxidation, and dangling bonds. Indeed, the ratio of surface atoms to core atoms increases with a decrease in particle size. Consequently, the magnetic properties of MNPs diverge significantly from those of bulk materials. This difference is attributed to the influence of size and surface effects, leading to variations in parameters such as magnetic anisotropy, saturation magnetization, magnetic moment per atom, coercivity, and Curie temperature [85]. Indeed, MNPs with sizes falling within the range of 10–50 nm are widely employed in various medical applications [86].

Exactly, within this size range, MNPs undergo a transition to a single domain phase and achieve superparamagnetic behavior when reaching a specific temperature known as the blocking temperature. In the superparamagnetic state, MNPs emulate the characteristics of a giant superparamagnetic atom, displaying a significant magnetic moment, zero coercivity, and magnetic remanence when subjected to a magnetic field [87]. In MH applications, achieving a high saturation magnetization (Ms) in MNPs is crucial. This high Ms is necessary to generate a significant amount of heat within tumor cells when subjected to an alternating magnetic field (AMF). Larger Ms values also enhance control over the movement of MNPs within the bloodstream under the influence of an external magnetic field. Additionally, maintaining the superparamagnetic behavior of MNPs is essential for ensuring good colloidal stability in these applications [88]. At temperatures below the blocking temperature, MNPs in the superparamagnetic state achieve colloidal stability by losing their magnetism. Additionally, in the presence of an external magnetic field, smaller MNPs demonstrate reduced dipole-dipole interactions, thereby minimizing particle aggregation. This phenomenon is attributed to the fact that the dipole-dipole interaction energy scales as r^6, where r denotes the interparticle distance [89]. Indeed, the heat generation capability of MNPs in the presence of an external magnetic field is influenced by both high saturation magnetization (Ms) and superparamagnetic behavior. However, there are additional factors that contribute to this capability, including anisotropy, as well as the amplitude and frequency of the applied magnetic field [90]. IONPs stand out as highly versatile options for MH applications thanks to their exceptional biocompatibility, biodegradability, and straightforward synthesis process [91]. Certainly, the characteristics of IONPs can be easily customized by modifying the synthetic methods and adjusting the surface structures [92]. Even the iron atoms released from IONPs inside the animal body by metabolic activity can be easily absorbed and stored inside the body [93]. Naked IONPs were synthesized using the coprecipitation method. Subsequently, they were passivated with Bovine Serum Albumin (BSA) to generate BSA-coated MNPs, denoted as MNP-A, with an average diameter of 28 nm [94]. The application of MNP-A in thermal therapy was assessed by evaluating their surface absorption rates (SAR) with other already published reports. MNP-A exhibited an enhanced SAR of 36 W/g iron oxide in comparison to citrate-coated IONPs (MNP-B), which displayed SAR values of 5.2 W/g iron oxide at an AMF amplitude of 6.3 kAm-1 and a frequency of 400 kHz, as reported in another source [95]. The thermal ablation ability of MNP-A was evaluated by incubating them with Hela cells with NPs for two hours at the dosage of 4 mg/mL, followed by exposure to the magnetic field (400 KHz, 6.3 kA/m). The cytotoxicity results indicated the complete cell death of the Hela cells due to thermal ablation. No cell death was detected in the control cells, specifically the Hela cells treated with MNP-A but not exposed to the external magnetic field. Cysteine-coated MNPs (Cys at MNPs) of an average diameter of 8 nm have been synthesized by Hayashi et al. [96] and evaluated for their hyperthermic property. The resulting Cys at MNPs demonstrated a saturation magnetization of 24 emu/g, along with zero coercivity and remanent magnetization at 300 K. The hyperthermic properties of MNPs were evaluated by dispersing the NPs in agar phantom at the concentration of 10 mg/mL and then applying the alternating current (AC) magnetic field, AMF (230 KHz, 100 Oe). The temperature of the phantom containing Cys at MNPs augmented from 37 °C to 42 °C after 6 minutes of exposure, which is an effective temperature for hyperthermia and then increased to 44 °C after 15 minutes of exposure. The heat generated remained constant afterward, with the SAR value of 156 W/g. Kossatz et al. [97] synthesized superparamagnetic IONPs known as MF66. Subsequently, these IONPs were functionalized with the Nucant multivalent pseudopeptide, N6L (MF66-N6L), doxorubicin (DOX) (MF66-DOX), and a combination of N6L and DOX (MF66-N6LDOX). The cytotoxic potential of these functionalized IONPs was then evaluated in a breast adenocarcinoma cell line (MDA-MB-231). This assessment included harnessing their MH property in tandem with the delivery of the anticancer drug. The intramural injection of magnetic nanoparticle formulations, followed by MH treatment, led to a substantial decrease in tumor growth. After 28 days of treatment, a notable reduction of up to 40% in the tumor volume was observed compared to the initial volume (Vt_0). After the MH treatment, there was a macroscopic occurrence of eschars around the tumor area, leading to a subsequent reduction in tumor volume. In contrast, no similar effects were observed in the test animals that did not undergo hyperthermia treatment. Beola et al. [98] emphasized the importance of SPION concentration in influencing the cell death pathway initiated by AMF stimulation. At lower concentrations, the AMF stimulus resulted in a decrease in B-cell lymphoma 2 (Bcl-2) mRNA expression and an elevation of the Bax/Bcl-2 ratio, leading to the activation of the intrinsic cell death pathway through mitochondria permeabilization. Conversely, at higher concentrations, AMF stimulation induced caspase-8 activation without a simultaneous increase in the Bax/Bcl-2 ratio, suggesting a predominant activation of the extrinsic death pathway. In their investigation, Adamiano et al. [99] explored the consequences of MH induced by two different superparamagnetic nanomaterials: iron-doped hydroxyapatite (FeHA) and IONPs coated with amorphous calcium phosphate (Mag at CaP). Their results demonstrated that, despite FeHA exhibiting more efficient cellular internalization, Mag at CaP was more proficient in inducing apoptosis. The disparity in effectiveness was suggested to arise from variances in physicochemical properties, particle–particle interactions, or the rate of AMF absorption between the two materials. This study underscored the crucial role of the physicochemical characteristics of nanoparticles not only in determining treatment effectiveness but also in shaping the underlying mechanisms of action. Soleymani et al. [100] conducted the synthesis of hyaluronic acid-capped iron oxide nanoparticles (HA at IONPs) and explored their heat generation capability under various strengths of alternating magnetic fields. The study systematically compared the heat generation performance of HA at IONPs with commercially available IONPs. Notably, the intrinsic power loss of HA at IONPs was determined to be 44 times higher than that of the commercially available IONPs. Furthermore, the HA at IONPs exhibited not only high heat efficiency but also demonstrated good biocompatibility. Additionally, their specific affinity for CD44 receptor-overexpressing cancer cells was highlighted, suggesting their potential as promising candidates for therapeutic applications. Mondal et al. [101] conducted the synthesis of IONPs and hydroxyapatite-coated IONPs (Hap at IONPs) with the aim of utilizing them in magnetic hyperthermia-based cancer therapy. The IONPs displayed a robust magnetization, with a saturation magnetization of 83.2 emu/g. Interestingly, the application of a hydroxyapatite coating to IONPs resulted in a reduction of the saturation magnetization to 40.6 emu/g at 300 K and 1.5 T. Despite this decrease, the saturation magnetization remained sufficient for MH treatment, surpassing 20 emu/g, a rarity in such applications. Both types of nanoparticles, IO and IO-HAp, exhibited superparamagnetic behavior, as evidenced by their low coercivity and remanent magnetization. The synthesized IONPs and hydroxyapatite-coated IONPs (HAp at IONPs) nanomaterials were subjected to testing to assess their heat generation capabilities in an AMF. The hyperthermia experiment initiated at a room temperature of 25 °C demonstrated that the IONPs experienced a rapid temperature increase, reaching up to 66 °C, while the HAp at IONPs achieved a lower temperature of 54 °C. The presence of a hydroxyapatite nanoparticle coating over the IONPs acted as an insulator, resulting in a less pronounced temperature rise.

Maintaining temperatures between 42 °C and 47 °C for an extended duration is crucial for effective cancer cell eradication. The heating efficiency of the synthesized superparamagnetic HAp at IONPs was evaluated using MG-63 cells during a 30-minute incubation in the hyperthermia field. Outcomes from the trypan blue assay conducted after hyperthermia treatment indicated that all cells retained the stain, underscoring the nanoparticles' efficacy in hyperthermia. Additionally, a TEM analysis of the treated MG-63 cells revealed cell arrest in the necrosis phase, potentially attributed to heightened levels of reactive oxygen species (ROS) induced by the heat generated through MH.

Sapareto et al. [102] proposed the utilization of Cumulative Equivalent Minutes as a method for computing a thermal dose in MH applications. This involves the introduction of biocompatible nanoparticles with moderate size and shape at a reasonable dosage. The conversion of all thermal exposure into "equivalent minutes" is achieved through the application of the following equation:

$$t_{43} = \sum_{t=0}^{t=\text{final}} \mathrm{R}^{(43-\bar{\mathrm{T}})} \Delta \mathrm{t} \tag{3.1}$$

In the context of MH, t_{43} represents the equivalent time at a temperature of 43 °C, and T denotes the average temperature over a specific time interval ΔT. The parameter R holds a value of 0.5 when the temperature exceeds 43 °C and 0.25 when it falls below 42 °C.

Creating MNPs with high SAR is crucial for minimizing the required thermal dose in various applications [103]. SAR serves as a critical parameter for assessing the power deposition and heating efficiency of MNPs. The determination of SAR involves suspending a known concentration of the material in a liquid with a known heat capacity and exposing it to an AMF of known strength and frequency [104]. The ensuing temperature change is then observed over a specific time duration. In essence, SAR quantifies the efficiency with which MNPs can convert magnetic energy into heat, as expressed by the following equation [105]:

$$\mathrm{SAR} = \frac{\text{absorbed power}}{\text{mass of MNPs}} \tag{3.2}$$

Two approaches, namely calorimetric and magnetometric methods, are employed to measure the SAR. In the calorimetric method, the SAR value of a sample is determined by monitoring the temperature changes over time and subsequently calculating the time derivative of the temperature at the initial point. The SAR value of the sample is given by the following equation:

$$\mathrm{SAR} = \frac{C_{p,s}}{m_{\mathrm{MNPs}}} \left| \frac{d\mathrm{T}}{dt} \right|_{t=0} \tag{3.3}$$

The accuracy of SAR values in calorimetry is influenced by several factors, including magnetic field gradient, sample volume, thermal loss, and the measurement methodology [106, 107]. The positioning of the temperature sensor is critical when dealing with a large sample volume, as rapid heating can lead to high heat dissipation and affect the precision of the SAR values [107]. Moreover, a sample possessing a high magnetic value has the potential to diminish the intensity of the applied field, resulting in underestimated SAR values. The precision of SAR value measurement is also contingent on the initial value slope, with the linear slope region being constrained, except in adiabatic processes [108]. Therefore, it is important to consider these factors when conducting calorimetry experiments to ensure accurate and precise measurement of SAR values. By carefully controlling these variables, researchers can obtain reliable data that can be used to study the thermodynamic properties of materials and their potential applications.

Magnetometric methods for SAR measurements entail measuring the dynamic magnetization (Mt) of a sample and subsequently calculating the integration of this Mt with respect to the applied magnetic field [109]:

$$\mathrm{SAR} = \frac{f}{c} \oint \vec{\mathrm{M}}_{\mathrm{t}} . d\vec{\mathrm{H}}_{\mathrm{t}} \tag{3.4}$$

Here, where f represents the frequency and c denotes the weight concentration of MNPs.

In clinical applications of MNPs, it is essential to consider two parameters: the frequency of the applied magnetic field and the optimal AC magnetic field. Increasing both frequency and field amplitude can result in increased heating of MNPs [110]. However, there are certain limits to this. For example, Hergt et al. [110] found that 500 kHz and 10 kA/m are suitable for MH applications. Similarly, Hergt et al. [111] recommended 400 kHz and 11.7 kA/m for the same purpose. Hilger et al. [112] proposed 10 kA/m and 400 kHz to treat breast cancer. In order to calculate the SAR values, Rosensweig [113] presented a formula that considers the unique characteristics of MNPs, encompassing factors such as size, crystal structure, shape, magnetic susceptibility, and saturation magnetization. This formula also integrates the frequency (f) and applied strength (H):

$$\mathrm{SAR} = \frac{8\pi^3 \mu_0{}^2 M_s{}^2 r^3 H_0{}^2 f^2 \tau}{3\rho k_\mathrm{B} T(1 + 2\pi f\tau)^2} \tag{3.5}$$

Equation 3.5 expresses the correlation between the SAR and the applied magnetic field (H0), frequency (f), and the density of MNPs: SAR = $\mu_0 \rho H_0 f$. Here, μ_0 signifies the permeability of free space, with a value of $4\pi \times 10^{-7}$ N/A^{-2}, while ρ represents the density of the MNP. As noted by Hergt et al. [110], this equation emphasizes that the SAR is contingent on the product of H0 and f.

The dispersion or aggregation of MNPs is a crucial factor during the application of in vivo MH [87]. Uniform dispersion of MNPs enhances heating performance, whereas aggregated MNPs can impede Brown relaxation mechanisms and elevate dipolar interaction among MNPs [85]. However, highly viscous or solid environments can immobilize MNPs and decrease their heating potential. Additionally, when MNPs are immersed in biological media, they can form a protein corona due to protein-MNP interactions, which can modify protein functionality and biocompatibility, potentially resulting in immunogenicity [114, 115]. Hence, it is vital to comprehend protein-MNP interactions and the retention time of the protein corona for the effective utilization of MNPs in MH applications [116]. Research on magnetic materials indicates that improvements in their SAR performance can be achieved by manipulating factors such as size, shape, composition, and coating [117–119]. The summarized results in Table 3.1 from studies on both coated and uncoated MNPs for MH applications against various cancer cell lines hold importance. These findings contribute to the effective implementation of in vivo MH, emphasizing the need to understand the behavior of MNPs in different environments and their interactions with biological molecules.

In conclusion, hyperthermia therapy emerges as a promising strategy for cancer treatment, involving the exposure of cancer cells to temperatures ranging from 41 °C to 45 °C. This therapeutic approach inflicts damage on cancer cells through diverse mechanisms, including heightened cell membrane permeability, protein denaturation, impairment of tumor vasculature, and interference with nucleic acid and protein synthesis. The administration of hyperthermia comes in various forms, such as whole-body hyperthermia, specific organ hyperthermia, or local hyperthermia, each presenting its distinct advantages and disadvantages. In the realm of hyperthermia methods, local hyperthermia is distinguished by its exceptional heat uniformity and precision in targeting specific areas. However, its application is constrained in treating deeply situated cancers due to its invasive nature and limited penetration depth. Nanotechnology, particularly the utilization of MNPs, emerges as a promising solution to address the limitations of local hyperthermia and enhance its effectiveness. The magnetic properties of MNPs, influenced by factors like fine size effects and surface effects, play a crucial role in determining their heat generation capabilities. MNPs in the superparamagnetic state, characterized by high saturation magnetization (Ms), exhibit remarkable heat generation when subjected to an AMF. Nanoparticles within the size range of 10–50 nm are commonly employed for medical applications. Various types of MNPs, including BSA-coated MNPs, cysteine-coated MNPs, Nucant multivalent pseudopeptide-functionalized MNPs, hyaluronic acid-capped MNPs, and hydroxyapatite-coated MNPs, have demonstrated promising results in MH applications. These MNPs exhibit favorable characteristics such as good biocompatibility, biodegradability, and efficient heat generation when exposed to an AMF. The

Table 3.1 Summary of the studies on both coated and uncoated magnetic nanoparticles (MNPs) for magnetic hyperthermia applications against various cancer cell lines.

S. No	Magnetic nanoparticle used	Preparation method employed	Magnetic properties of MNPs		Alternating magnetic field (KHz)	Tested cell lines	Reference
			Ms (emu/g)	SAR (W/g)			
1.	Protein-coated Fe_3O_4 (12 nm)	Coprecipitation	—	36	400	Hela cells	[94]
2.	Uncoated IONPs	Spray coprecipitation	68.8	69.8	150	KB and L929 cells	[120]
	Chitosan coated IONPs		25.6	64.6			
	Starch-coated IONPs		16.4	62.3			
3.	allyl-Fe_3O_4, size = 8 nm	One-pot condensation and thiol-ene click reaction	24	156	230	Glioma 261	[96]
4.	Oleic acid-capped IONPs (Fe_3O_4-OA-MN) (5 nm)	Coprecipitation method	34.39	33.5	265	MCF7	[121]
	Polyethylene glycol capped IONPs (Fe_3O_4-PEG-MN) (10 nm)	Coprecipitation method	37.37	28.3	265	MCF7	
	Uncoated IONPs (Fe_3O_4-MN) (20 nm)	Coprecipitation method	59.7	38.4	265	MCF7	
5.	Iron oxide nanocubes (13–40 nm)	Thermal decomposition	78–95	2452	520	KB cancer cells	[122]
6.	Chitosan nanofibers-IONP composites (E-CHS-Fe_3O_4)	Coprecipitation	16.3	—	750–1150	HFL1 cells	[123]
7.	Dextran-coated Fe_3O_4 NPs	—	68.0	46	4 MA/m	Dendritic cells	[124]

8.	Folate-coated iron oxide nanoparticles	—	—	13 ± 3	108	HeLa and KB cells	[125]
9.	Maghemite NPs embedded in mesoporous silica	Massart method	—	—	100	A549, Saos-2, HepG2 cells	
10.	SPIONs	Sol gel	2.5	20	141	—	[126]
	Hyaluronic acid-coated Fe_3O_4	Hydrothermal method	11.3–29.6	27.0–62.1	120 kHz	MDA-MB-231 cell	[100]
	Oleic acid-capped spherical IONPs	Thermal decomposition	101.5	189.6	380	—	[127]
	Oleic acid-capped cubic IONPs		107.3	356.2			
	IONPs (20 nm)	—	50	57.06 ± 0.78	351	M2 macrophage cells	[128]

effectiveness of MH treatment hinges on factors such as the concentration of MNPs, frequency, and amplitude of the applied magnetic field. SAR values serve as metrics for measuring the power deposition and heating efficiency of MNPs.

Accurate measurement of SAR values is achieved through calorimetric and magnetometric methods, taking into account variables like magnetic field gradient, thermal loss, sample volume, and measurement methodology. In summary, the optimization of magnetic field parameters and the development of MNPs with high SAR values are pivotal for enhancing the efficacy of MH therapy. Ongoing research and advancements in nanotechnology hold significant potential for advancing the outcomes of cancer treatment through hyperthermia therapy utilizing MNPs.

3.1.4 Magnetic Nanoparticles in Gene Therapy

Magnetic nanoparticles have drawn considerable attention in the realm of gene therapy due to their distinctive properties and versatile applications. These nanoparticles display superparamagnetism, signifying their magnetization exclusively in the presence of an external magnetic field. Moreover, their compact size enables efficient cellular uptake and biodistribution, often complemented by coatings with biocompatible materials to enhance stability and address potential toxicity concerns. A primary application of magnetic nanoparticles in gene therapy is their notable role in targeted delivery [129, 130]. By functionalizing their surface with specific ligands or antibodies, researchers can selectively bind and deliver therapeutic genes or proteins to specific cells or tissues [131]. This targeted approach offers advantages such as increased accumulation at the desired site, reduced off-target effects, and enhanced therapeutic efficacy. In the domain of gene therapy, magnetic nanoparticles have been utilized to enhance the efficiency of delivering genes [132]. Researchers have enhanced the cellular uptake and transfection efficiency of therapeutic genes by conjugating gene vectors or plasmids to magnetic nanoparticles [132, 133].

Gene therapy stands out as a highly promising therapeutic approach that entails introducing an external gene into living cells to produce therapeutic proteins, offering potential solutions for various severe diseases, including cancer and genetic disorders [129, 130]. Yet, the effectiveness of gene therapy hinges on the proficient delivery of sizable and delicate DNA molecules into the nucleus of targeted cells, safeguarding them from degradation by nucleases. To surmount this challenge, considerable research has concentrated on creating nonviral vectors that are not only safe but also efficient, ensuring the protection and controlled release of genetic material at the intended site [134, 135]. Taking inspiration from magnetic drug delivery systems, which enable targeted accumulation of therapeutics using magnetic fields, various researchers have explored the utilization of MNPs in gene delivery [136]. The concept of "magnetofection" was introduced to characterize the delivery of nucleic acids facilitated by a magnetic field, where magnetic nanoparticles serve as associated vectors. Integrating magnetic nanoparticles into the transfection process under the influence of a magnetic field presents notable advantages throughout the gene delivery process [137].

It is commonly acknowledged that nucleic acids carry a negative surface charge [138]. Consequently, when utilized in gene transfection, MNPs need to undergo surface modification to enable compacting with genes or gene/carrier complexes, while also enhancing stability and biocompatibility [133]. Two principal methods are employed for loading genes onto MNPs. The first approach encompasses coating cationic substances onto the surface of MNPs, allowing them to bind with negatively charged genes through electrostatic absorption [139]. Polyethylenimine (PEI), characterized by its numerous amino groups, is frequently utilized for gene delivery [140]. As an illustration, polyethylenimine (PEI) can be employed as a coating on the surface of 11-mercaptoundecanoic acid-modified Au/Ni/Si nanospears using a layer-by-layer technique. This procedure results in the development of positively charged MNPs with the capability to load genes efficiently. Furthermore, PEI can directly modify the surface of iron oxide nanoparticles, establishing a core–shell structure that effectively binds with siRNA, leading to the formation of well-dispersed nanoparticles. Other polycations like poly-arginine, poly-L-lysine, and chitosans can also be surface-modified onto MNPs to facilitate the successful binding of genes [141, 142]. The alternative approach entails compacting gene/carrier complexes, with the surface charge not being the decisive factor in this strategy [131, 133]. Studies on various MNPs with different surface charges have confirmed this observation [132]. In this particular approach, genes usually bind to polycations initially, following which MNPs with an opposite charge are introduced to the gene/carrier complexes through electrostatic adsorption. MNPs featuring diverse charges, including modifications with citric acid, carboxylmethyl-dextran (CMD), g-aminopropyltriethoxy silane (APTES), betaine, and PEI, have shown the capability to compact genes similarly and improve gene transfection efficiency when a magnetic field is applied [131].

The sugar calyx and lipid bilayer of cell membranes possess negative charges, allowing cationic magnetic gene carriers to facilitate both cell binding and absorption [143]. However, while circulating, these carriers may interact with negatively charged serum proteins, potentially impeding cellular uptake [144]. The electrostatic interaction results in the adsorption of serum proteins on the surface of nanoparticles, leading to an increase in particle size and a shift in surface charge from positive to negative [145]. As a consequence, MNPs might encounter challenges in binding to negatively charged cell membranes and achieving internalization [146, 147]. In certain instances, the absorption of serum proteins can potentially reduce the binding of MNPs to DNA [148]. However, applying a magnetic field has been demonstrated to aid cationic gene carriers in overcoming interference from serum [149]. When a magnetic field is applied, the cellular uptake of MNPs/DNA complexes appears to be largely unaffected by serum interference. Notably, in specific cases, the cellular uptake of PEI-PAAIO/pEGFP complexes has been observed to increase by more than sevenfold during serum incubation [149, 150]. Subsequent investigations have verified that the stability of pDNA/BPEI-SPION complexes, formed by combining BPEI with superparamagnetic iron oxide nanoparticles to facilitate pDNA complexation, extends beyond 24 hours. In contrast, unshielded pDNA undergoes degradation within a 24-hour timeframe when exposed to DNase [151]. Taken together, these studies collectively illustrate that MNPs not only augment cellular uptake but also safeguard gene stability in the presence of serum proteins. Magnetofection has proven to markedly accelerate the accumulation of MNP/gene complexes on the cell surface, thereby reducing the cellular uptake time [152]. Y. Nie et al. [148] showcased the swift enrichment of MNPs/gene complexes on the cell surface through magnetofection, facilitating their rapid uptake into target cells without causing harm. The velocity at which MNPs reach the target cell is intricately linked to the strength and duration of the magnetic field, along with the extent and proximity of the magnetic field [153]. Broadly speaking, cellular uptake of nanoparticles involves multiple mechanisms, including phagocytosis, macropinocytosis, clathrin mediated endocytosis, and caveolin mediated endocytosis [154, 155]. The cell uptake mechanism of magnetofection revealed that the internalization pathway of MNPs is similar to that of PEI polyplexes, involving clathrin-mediated endocytosis in human lung epithelial cells and liver cancer cells, or caveolin-mediated endocytosis in human cervical cancer cells [156]. Later studies have substantiated that different configurations of MNPs are internalized via clathrin-mediated endocytosis in both baby hamster kidney cells and HeLa cells [142, 157]. Together, these investigations indicate that magnetofection facilitates cellular uptake without influencing the entry pathway of nanoparticles.

Morishita et al. [158] have developed the distinctive vector "HVJ-E (hemagglutinating virus of Japan-envelope)," which facilitates the swift transfer of plasmid DNA, oligonucleotides, and proteins into cells through cell fusion. In this study, the researchers associated HVJ-E with magnetic nanoparticles to potentially enhance its transfection efficiency under the influence of a magnetic force. Specifically, they utilized maghemite particles with an average size of 29 nm, controlled by a magnetic force. In an in vitro cell culture system, a notable improvement in transfection efficiency was observed with the HVJ-E technology, achieved by combining a mix of magnetite particles with protamine sulfate, resulting in a cationic surface charge on the maghemite particles. This approach led to a reduction in the required HVJ titer.

The introduction of magnetic nanoparticles facilitated the association of HVJ-E with the cell membrane through a magnetic force. Additionally, the authors discovered that maghemite particles, surface-coated with heparin (but not protamine sulfate), enhanced transfection efficiency when directly injected into the mouse liver in an in vivo model. The study revealed that the size and surface chemistry of magnetic particles could be customized to meet specific physical and biological characteristics.

Overall, the authors demonstrated that HVJ-E-based gene transfer can be improved by magnetic nanoparticles with various surface modifications, including adjustments in size or charge. These findings hold promise for addressing fundamental limitations in in vivo gene therapy. The researchers further aimed to explore whether targeted delivery using adenovirus (Ad) complexed with biodegradable magnetically responsive nanoparticles (MRNP) could enhance the safety and efficiency of these vectors. Chorny et al. [159] employed fluorescently labeled MNPs created from polylactide incorporating iron oxide. These MNPs were surface-modified with the D1 domain of the coxsackievirus and adenovirus receptor (CAR), serving as an affinity linker. The goal was to augment transgene expression in vascular cells that lacked the CAR in culture. The authors assessed the cellular uptake of MNPs and the expression of a green fluorescent protein reporter transgene in cultured endothelial and smooth muscle cells. Fluorimetric assays utilized specific excitation/emission wavelengths of 540 nm/575 nm for endothelial cells and 485 nm/535 nm for smooth muscle cells. The study revealed that the stable association of adenovirus (Ad) with MNPs resulted in the formation of MNP-Ad complexes. Exposure to a high gradient magnetic field facilitated rapid binding kinetics of these complexes to cells, leading to significantly elevated gene transfer levels compared to free vector or nonmagnetic control treatments. Mechanistic insights obtained from multiple regression analysis indicated that MNP-Ad-mediated transduction differed from that of free Ad, with the complexes identified as the primary contributors to gene transfer under magnetic conditions. Crucially, the magnetically enhanced transduction did not induce any adverse effects on cell viability or growth kinetics. The authors successfully demonstrated that the affinity complexation of adenoviral vectors with biodegradable MNPs has the potential to enhance gene delivery, presenting a promising approach to broaden the applicability of viral gene therapy strategies. Namgung et al. [151] conducted a study where they developed thermally cross-linked iron oxide nanoparticles (TCL-SPION). These SPIONs were modified by attaching a polyethylene glycol moiety and covalently linking them with branched polyethylenimine (BPEI). The resulting nanomaterial demonstrated high efficiency in magnetofection, even in the presence of serum-conditioned media.

Surface modification of TCL-SPION nanoparticles involved conjugating low molecular weight BPEI to the carboxylic groups of the magnetic nanoparticles through NHS-mediated amidation. The conjugation ratios of BPEI/SPION were determined for various combinations, and the magnetic properties of the resulting BPEI-SPION complexes exhibited sufficient magnetization for effective magnetofection.

Unmodified SPION nanoparticles initially had a negative zeta potential due to carboxylic groups on their surface, hindering interaction with negatively charged nucleic acids. The successful synthesis of hybrid BPEI-SPION nanoparticles was confirmed by converting the negative surface charge of unmodified SPION into a positive charge through covalent linkage with BPEI. Various polyplexes, including BPEI-SPION/pDNA and BPEI/pDNA, were assessed at different N/P ratios to analyze surface charge changes. BPEI 1.8K/pDNA polyplexes exhibited a negative surface charge at low N/P ratios, indicating insufficient complexation. In contrast, BPEI-SPION/pDNA complexes displayed a positive surface charge even at low N/P ratios, gradually increasing until reaching a plateau at an N/P ratio of 5. The multiple conjugations of BPEI to SPION led to a higher charge density and the formation of stable complexes with pDNA, even at low N/P ratios. The successful conjugation of BPEI to SPION was verified through a gel retardation assay, confirming the complexation of BPEI-SPION with pDNA. Unmodified SPION nanoparticles exhibited spherical particles with an approximate diameter of 5 nm, and the conjugation of BPEI slightly increased the nanoparticle size to around 10 nm. The addition of pDNA to SPION did not significantly alter the nanoparticle morphology. However, the interaction between BPEI-coated SPION and pDNA resulted in the formation of cloud-like particles composed of several discrete nanoparticles with a diameter of 50 nm.

To protect pDNA from enzymatic degradation in serum, BPEI-SPION magnetoplexes were tested. These magnetoplexes effectively shielded pDNA from degradation by serum nucleases, contrasting with the degradation observed in naked pDNA within 24 hours. This demonstrated the efficient protection of pDNA provided by BPEI-SPION complexes, even at a low N/P ratio of 2. The study also delved into the impact of magnetofection on gene delivery. It was observed that magnetofection facilitated the swift accumulation and sedimentation of magnetoplexes at the cell surface, resulting in an improved transfection efficiency when compared to samples without the application of a magnetic field. Additionally, the effectiveness of magnetofection in transfecting primary vascular endothelial cells (HUVEC) was assessed. Delivery of the IL-10 gene, known for its anti-inflammatory properties, to HUVEC, was accomplished using BPEI-SPION magnetoplexes.

Subsequently, the conditioned media containing IL-10 were added to TNF-a-treated HUVEC, leading to a noteworthy inhibition of PAI-1 expression, a prothrombotic protein induced by TNF-a. These findings highlight the potential of magnetofection as an efficient strategy for therapeutic gene delivery. Mu et al. [160] utilized polydopamine-coated iron oxide nanoparticles (Fe_3O_4 at PDA) with augmented photothermal capabilities for the effective loading and delivery of siRNA. Subsequently, these nanoparticles underwent an additional coating with mesenchymal stem cell (MSC) membranes. The study demonstrated the retention of photothermal functionality and MR imaging capability inherited from Fe_3O_4 at PDA in the MSC membrane-coated Fe_3O_4 at PDA-siRNA nanoparticles (Fe_3O_4 at PDA-siRNA at MSCs). These synthesized nanocomplexes exhibited excellent siRNA delivery abilities in DU145 cells. Furthermore, significant inhibition of endogenous Plk1 gene expression and the induction of apoptosis in DU145 cells was observed upon delivering siRNA targeting the Plk1 gene using Fe_3O_4 at PDA-siRNA at MSCs nanoparticles. Remarkably, the study demonstrated remarkable antitumor efficacy in a DU145 xenograft mice model through the combination of photothermal treatment and gene silencing. The potential of Fe_3O_4 at PDA-siRNA at MSCs nanoparticles as carriers for gene and photothermal therapy has been highlighted through preliminary in vitro and in vivo studies. In examining the expression of Plk1, a proto-oncogene with elevated expression in tumor cells, DU145 cells were investigated after the delivery of siPlk1 via Fe_3O_4 at PDA-siPlk1 at MSCs nanoparticles. The Plk1 mRNA expression level in DU145 cells was assessed 24 hours post-transfection using qRT-polymerase chain reaction (PCR). Significantly reduced Plk1 mRNA expression was observed in cells treated with Fe_3O_4 at PDA-siPlk1 at MSCs NPs or Lipo2K/siPlk1, in comparison to naked siPlk1 or the control group treated with Fe_3O_4 at PDA-siRNA at MSCs nanoparticles, as illustrated in Figure 3.5a. Fe_3O_4 at PDA-siPlk1 at MSCs nanoparticles demonstrated a notable reduction in Plk1 mRNA expression, approximately 50% lower compared to the control group. Subsequently, both qRT-PCR and western blot analyses confirmed a substantial decrease in Plk1 protein expression in DU145 cells treated with Fe_3O_4 at PDA-siPlk1 at MSCs nanoparticles, in contrast to Fe_3O_4 at PDA-siRNA at MSCs nanoparticles. This aligns with the mRNA expression data presented in Figure 3.5b, establishing a consistent reduction in Plk1 expression associated with apoptosis in cancer cells. Following transfection with various formulations, DU145 cells were stained with Annexin V-FITC and PI and then analyzed using flow cytometry. As depicted in Figure 3.5c, treatment with Fe_3O_4 at PDA-siPlk1 at MSCs nanoparticles resulted in 22.9% cell apoptosis (comprising both early and late apoptotic cells), a notably higher percentage compared to the 1.1% observed with Fe_3O_4 at PDA-siRNA at MSCs nanoparticles after 72 hours of treatment. These findings underscore the ability of Fe_3O_4 at PDA-siPlk1 at MSCs nanoparticles to induce apoptosis in DU145 cells through Plk1 gene silencing. It is noteworthy to mention that gene editing has been transformed by the Clustered Regularly Interspaced Short Palindromic Repeats-associated protein (CRISPR/Cas9) system. Nevertheless, the use of viral delivery systems comes with substantial side effects, and the direct delivery of naked DNA presents practical challenges. To overcome these limitations, there is a compelling need for nonviral delivery methods for CRISPR/Cas9 components in prospective therapeutic applications. Addressing this demand, Rohiwal et al. [161] synthesized Polyethylenimine (PEI) MNPs and formed complexes with CRISPR/Cas9 to achieve efficient delivery of plasmid DNA encoding the CRISPR-Cas system in vitro using eukaryotic cells. The MNPs synthesized displayed a narrow size distribution and adequate colloidal stability, featuring an average particle size of approximately 20 nm. Notably, these MNPs exhibited comparable effectiveness in both homology-directed repair and nonhomologous end-joining events in comparison to the conventional lipofectamine transfection method. This nonviral delivery strategy employing PEI-MNPs holds significant promise for improving the safety and applicability of gene editing by facilitating the delivery of plasmids encoding CRISPR/Cas9 and template DNA. Furthermore, the study explored the application of magnetofection, wherein PEI-MNPs were combined with an applied magnetic field to enhance transfection efficiency. The introduction of a magnetic field led to a marginal increase in transfection efficiency, suggesting the accelerated movement and accumulation of magnetoplexes at the cell surface. In conclusion, this investigation underscored the potential of PEI-MNPs as a nonviral delivery system for CRISPR/Cas9 and template DNA, providing enhanced safety and efficiency for gene editing applications. Numerous scientific reports exist regarding the use of MNPs for the effective delivery of therapeutic genes [162, 163].

To sum up, magnetic nanoparticles have emerged as a highly promising tool in the realm of gene therapy. Their distinctive properties, including superparamagnetism and small size, render them well-suited for the targeted delivery of therapeutic genes or proteins to specific cells or tissues. Through surface functionalization of magnetic nanoparticles with specific ligands or antibodies, researchers can selectively bind and deliver genetic material, providing advantages such as increased accumulation at the desired site, reduced off-target effects, and enhanced therapeutic efficacy. The modification of magnetic nanoparticle surfaces with cationic substances or gene/carrier complexes enables efficient compacting and binding of genes, thereby improving cellular uptake and transfection efficiency. Furthermore, the application of a magnetic field during magnetofection has proven effective in overcoming serum interference, improving cellular uptake, and maintaining gene stability in the presence of serum proteins. Combining magnetic nanoparticles with gene vectors or carriers has been demonstrated to enhance transgene expression and increase gene delivery efficiency, presenting a promising avenue to broaden the spectrum of gene therapy strategies. Additionally, the amalgamation of magnetic nanoparticles with other gene delivery systems, such as HVJ-E or MSC membranes, has exhibited heightened transfection efficiency and potential for combined photothermal therapy and gene silencing. In summary, magnetic nanoparticles provide a versatile platform for targeted gene delivery, showing substantial potential in addressing challenges associated with effective gene therapy. Ongoing research and

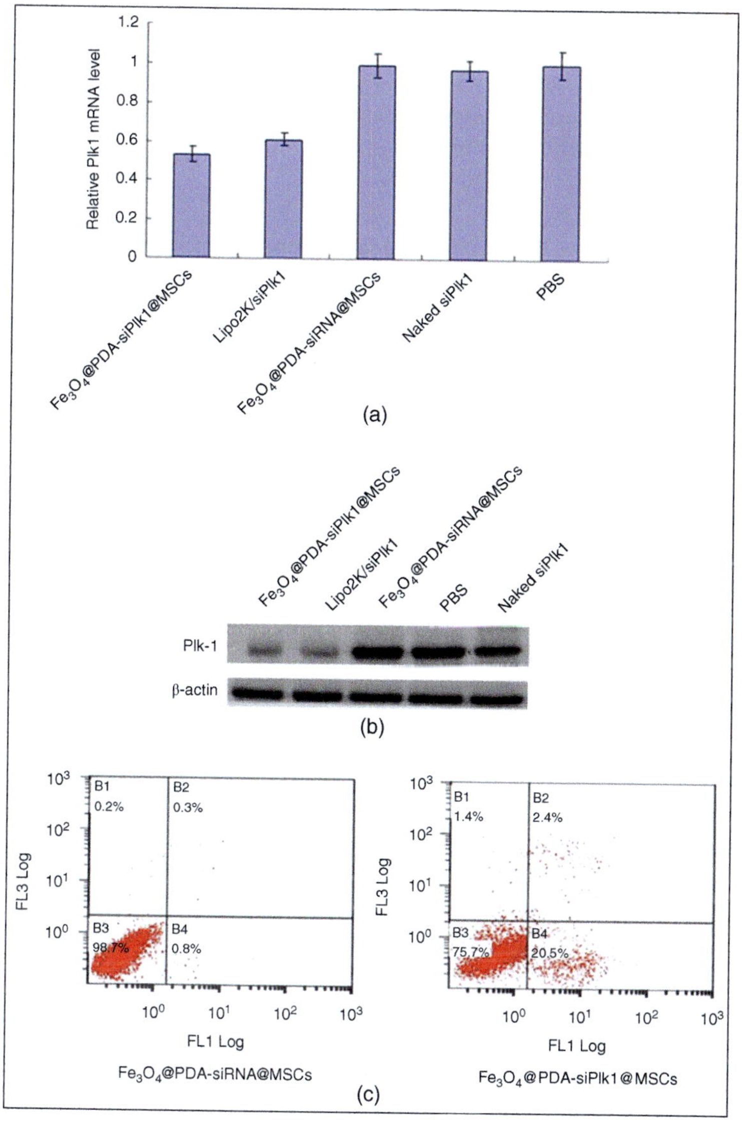

Figure 3.5 Illustrates the effectiveness of in vitro gene silencing on DU145 cells. The relative Plk1 mRNA (a) and Plk1 protein (b) expression in DU145 cells were assessed after treatment with PBS, naked siPlk1, Fe_3O_4 at PDA-siRNA at MSCs, Lipo2K/siPlk1, or Fe_3O_4 at PDA-siPlk1 at MSCs, as determined through qRT-PCR and western blot analysis. Additionally, (c) depicts the apoptosis of DU145 cells treated with Fe_3O_4 at PDA-siRNA at MSCs or Fe_3O_4 at PDA-siPlk1 at MSCs, as measured by flow cytometry. Source: Reproduced from Mu et al. [160]/with permission of American Chemical Society.

development in this domain will contribute to advancing gene therapy as a highly effective and safe therapeutic approach for various diseases.

3.1.5 Magnetic Nanoparticles in Biosensing and Detection

Magnetic nanoparticles have become a highly promising asset in biosensing and detection, owing to their distinct characteristics and wide-ranging applications. Comprising magnetic materials like iron oxide or cobalt ferrite, these nanoparticles exhibit magnetic traits, facilitating effortless manipulation and control within diverse biological systems. With their compact size, extensive surface area, and amenability to functionalization, they prove to be excellent options for the precise and selective identification of biomarkers, pathogens, and environmental pollutants. In this section, we explore the utilization of magnetic nanoparticles in biosensing, focusing on their application in detection of disease biomarkers and pathogens.

3.1.5.1 Detection of Disease Biomarkers

In the realm of cancer biosensing with MNPs, a crucial phase in biosensor advancement entails anchoring particular biological components onto MNPs that have been functionalized. Cancer biomarkers, which include molecules or structures associated with disease advancement, provide valuable prognostic information regarding the occurrence and outcomes of cancer [164]. Detecting these biomarkers, ranging from small entities like peptides, aptamers, and DNA to larger proteins, holds the utmost significance [165]. The modified surface of MNPs features linker groups that aid in the attachment of complementary biomolecules. Diverse bioconjugation methods encompass physical interactions like electrostatic and hydrophilic–hydrophobic interactions, as well as affinity interactions, in addition to chemical interactions such as covalent bonds. These techniques allow for easy immobilization of biomolecules on MNPs or transducers [166]. Particular interactions, like the biotin-avidin and antigen–antibody systems, have proven to be valuable in the design of biosensors [167]. After the conjugation of MNPs with specific biomarkers, they are utilized in the concluding stage of the sensing process. Over the past few decades, there has been notable attention and progress in the detection of cancer biomarkers.

Numerous effective sensing techniques have been created for the identification of cancer biomarkers in plasma, blood, or afflicted tissues. These methods include electrophoresis, optical approaches such as electrochemiluminescence, fluorescence, colorimetric assays, surface plasmon resonance, and surface-enhanced Raman spectroscopy (SERS). Immunological methods like enzyme-linked immunosorbent assay (ELISA) and PCR are also employed. Additionally, techniques like microcantilevers and electrochemical assays have found application in this context [168, 169]. The choice of labeling approach for biosensing events depends on the specific method and may involve electroactive molecules, fluorescent labels, enzymes, or nano-/micro-particles.

The integration of MNPs into biosensing systems can be accomplished through two primary strategies: direct labeling and indirect labeling. In the direct labeling approach, MNPs are attached to the transducing element through affinity recognition reactions, such as complementary DNA sequences or streptavidin-biotin interactions. For example, the sensor surface can be modified with magnetic particles that have been functionalized with single-stranded DNA, acting as capture probes. Upon contact with complementary oligonucleotides, hybridization occurs, leading to a physical and/or chemical response [170]. In the indirect labeling approach, a principle similar to ELISA, known as sandwich immunoassays, is employed. In this technique, primary antibodies that complement the target protein are anchored to the sensing surface. This is followed by an affinity reaction with a solution containing the biomarker. Subsequently, secondary antibodies labeled with biotin are introduced, facilitating an affinity reaction when the sensor surface comes into contact with the streptavidin-labeled MNP solution [170]. For a comprehensive overview, the sensing methodologies for cancer detection can be categorized according to the detection principles into three main groups: electrochemical, optical, and magnetic.

Malignant neoplasms, commonly known as cancer, rank second as the primary cause of mortality globally, following cardiovascular diseases. These neoplasms are accountable for millions of deaths worldwide [75]. The timely identification of cancer is pivotal in substantially improving the chances of successful treatment and full recovery. Since cancer cells originate from the body's own cells, early-stage malignant transformations present considerable challenges for analysis.

However, the heightened expression of numerous genes in cancerous tissues offers an alternative pathway for analytical purposes and therapeutic exploration. Tumor markers are essential in the identification of cancer and clinical screening [77]. Consequently, endeavors to detect cancer at the earliest possible stage intensify the demand for detection methods, particularly in the realm of biosensors [78]. A unique scenario unfolds in later stages when cancer becomes evident, and metastases occur due to the dissemination of cancer cells from the primary tumor. Circulating tumor cells play a crucial role, and the early detection of these cells provides an opportunity for effective treatment. The use of MNPs in the detection of cancer biomarkers is increasingly favored, attributed to the specific binding capacity of antibody-modified magnetic particles (MPs). These MP-based systems facilitate the capture of target markers or malignant cells, efficiently isolating them from complex sample matrices. This is particularly beneficial when addressing cancer cells, which often coexist with a large surplus of normal cells. Furthermore, MP-based systems provide a high preconcentration effect, making them valuable in detecting extremely low concentrations within complex samples such as whole blood, biopsies, blood serum, cell lysates, cultivation media, and paraffin-embedded tissue sections. As cancer markers typically lack complete selectivity, the identification of a single target is often insufficient for dependable screening, diagnosis, or treatment monitoring. Hence, the use of a panel of biomarkers enhances the sensitivity of diagnostic methods and diminishes the likelihood of false positive or false negative outcomes.

Indeed, utilizing a panel of biomarkers enhances the sensitivity of diagnostic methods and decreases the probability of false positive or false negative results. For example, the research team led by Rusling has devised a microfluidic platform designed for the simultaneous detection of multiple protein targets. They have introduced a modular microfluidics configuration able to concurrently identify proinflammatory cytokines IL-6 and IL-8 [171]. Streptavidin-coupled MPs (1 μm), modified with both biotinylated horseradish peroxidase (HRP) and antibody 2 (Ab_2), were employed to capture analytes and enable their magnetic separation from undiluted calf serum samples. The magnetoimmunocomplex was captured in the microfluidic device using corresponding immobilized antibody 1 (Ab_1) on an 8-electrode array. Amperometry was employed for detection using HQ/H_2O_2 substrates at −0.2 V versus an Ag/AgCl reference electrode, achieving a limit of detection (LOD) of 5 fg/mL for IL-6 and 7 fg/mL for IL-8 in undiluted calf serum. Each analysis of a single sample took 45 minutes, and the device allowed the simultaneous measurement of up to eight analytes. A comparable device was utilized for detecting prostate cancer biomarkers, specifically prostate-specific antigen (PSA) and prostate membrane-specific antigen (PMSA) [172]. In this case, a paramagnetic composite of Fe_3O_4/reduced graphene oxide (rGO)/Ab_2 served as an artificial alternative to the HRP enzymatic label. The immunocomplex was introduced into a microfluidic chamber, captured by Ab_1-rGO on a solid-phase extraction (SPE) platform (rGO modified with MNPs), and the intrinsic HRP-like activity of Fe_3O_4 nanoparticles (NPs) was measured using H_2O_2 as the electron acceptor from the electrode. The reduced reagent cost of US$0.85 for a single 2-protein assay, along with reduced instrumentation requirements, renders this sensor format suitable for broader utilization. Magnetic nanoparticles, in conjunction with sandwich immunosensors, were also utilized for the detection of p53-specific autoantibodies [173]. Autoantibodies targeting tumor-associated antigens have shown significant promise as markers for reliable cancer diagnosis, prognosis, and therapy. In the analysis of human sera from individuals at high risk of colorectal cancer and those already diagnosed, a disposable SPE biosensor was employed. MNPs modified with the HaloTag fusion protein were utilized for target capture, and a sandwich immunoassay was conducted using HRP-labeled anti-human antibody 2 (Ab_2), followed by magnetic immobilization on the SPE platform. The preference for disposable formats in clinical settings is motivated by the need to prevent contamination from potentially infectious samples. Additionally, for the detection of estrogen receptor alpha (ERα), a breast cancer biomarker, an inexpensive and disposable microfluidic electrochemical array was employed [174]. The anti-ERα antibody 2 (Ab_2) on horseradish peroxidase/magnetic particles (HRP/MPs) captured the target, and the complete complex was attached to DNA-responsive elements on gold nanoparticles (AuNPs)-modified glassy carbon electrode (GCE). Amperometric measurements employing common substrates, HQ/H_2O_2, yielded a LOD of 5 fg/mL for ERα. The simultaneous detection of alpha-fetoprotein (AFP) and carcinoembryonic antigen (CEA) was accomplished using metal-coded probes and applying stripping square-wave voltammetry (SWV) for quantification on a single electrode [175]. Apoferritins labeled with metals, Cd-Apo and Pb-Apo, were anchored on poly(lysine) along with their corresponding antibodies (anti-AFP or anti-CEA) utilizing gold nanoparticles (AuNPs). Employing Dynabeads antibody 1 (Ab_1)-modified MPs facilitated magnetic separation, followed by the introduction of anti-AFP-AuNP-Cd-Apo and anti-CEA-AuNP-Pb-Apo to create sandwich immunocomplexes. Stripping SWV was conducted on a bismuth-film-modified GCE, detecting the signal for AFP ($\sim Cd^{2+}$) at −0.78 V and CEA ($\sim Pb^{2+}$) at −0.53 V, with LODs of 4 pg/mL for both markers. Utilizing apoferritin-based containers, capable of holding around 4500 iron atoms, enhances the method's sensitivity due to their high capacity for metal labels. A refined approach incorporated dopamine self-polymerized Fe_3O_4 nanoparticles (NPs) coated with a dual-template molecularly imprinted polymer to simultaneously enrich both AFP and CEA. Recombinant apoferritin loaded with metal ions was linked with a graphene-gold nanoparticle (AuNP) composite and corresponding antibodies, leading to enhanced LODs of 0.3 pg/mL for AFP and 0.35 pg/mL for CEA [176]. Electrochemical scanning techniques facilitate the straightforward separation of various metal labels, paving the way for uncomplicated multianalyte sensing platforms.

A novel homogeneous protocol, employing a controlled release system, was utilized for the identification of squamous cell carcinoma antigen (SCCA). This method utilized magnetic mesoporous Fe_3O_4 as a nanocontainer and aminated polystyrene microspheres as a molecular gate [177]. The pores of Fe_3O_4, negatively charged and conjugated with antibodies, were filled with toluidine blue and sealed with positively charged aminated polystyrene microspheres. Upon capturing SCCA by the antibodies, it displaced the polystyrene microsphere on the surface, leading to the release of toluidine blue from the pores. The devoid nanocontainers were subsequently magnetically extracted, and the toluidine blue in the supernatant was quantified using SWV, achieving a LOD of 0.33 pg/mL. Performing homogeneous assays in solution or pseudohomogeneous formats on sensing surfaces eliminates the requirement for separation steps, streamlining assay complexity and accelerating the response. Conversely, heterogeneous formats usually yield enhanced sensitivity. The research team, under the leadership of Pingarron, utilized MPs modified with a specific capture antibody (Ab_1) for the selection and separation of the analyte from the sample matrix. Subsequently, horseradish peroxidase-labeled antibody (Ab_2) was employed for amperometric measurement. The entire immunoassay could be conducted as a one-step procedure by simultaneously incubating the sample with both Ab_1-MP and Ab_2-HRP [178]. The resulting magnetic immunocomplex was isolated, subjected to washing, and magnetically captured for the amperometric measurement of HQ/H_2O_2. The paramagnetic characteristics of substances were beneficial in modification stages, especially when working with intricate composite materials formed through multi-step procedures. In the precise identification of CEA, corallite-shaped magnetic hybrid $Fe_3O_4/MnO_2/Pt$ nanocomposites were employed as labels for amperometric detection [179]. Antibody 1 (Ab_1) was attached to a GCE enhanced with amine-providing graphene sheets, and simultaneously, antibody 2 (Ab_2) was linked to the nanocomposite. CEA was captured by Ab_1 on the electrode surface, and the Ab_2-$Fe_3O_4/MnO_2/Pt$ nanocomposite was incubated with the sensor without the need for any magnetic separation steps. The catalytic features of the nanocomposite constituents, combined with multiple amplifications and facilitated electron transfer through graphene sheets, allowed the sensor to identify CEA with an LOD of

0.16 pg/mL and an extensive linear range up to 20 ng/mL. This indicates successful performance, even in human serum. A comparable LOD for CEA at 0.3 pg/mL was attained through the use of magnetite-silica-polystyrene ($Fe_3O_4/SiO_2/PS$) nanocomposites for capture and subsequent magnetic separation. [180]. Mesoporous silica nanospheres incorporating CdTe quantum dots (QDs), in conjunction with antibody 2 (Ab_2), functioned as the electrochemiluminescence (ECL) label, delivering elevated quantities of CdTe QDs through their encapsulation within the mesoporous silica. The immunoassay consisted of two stages: the capture of the target by $Fe_3O_4/SiO_2/PS$-Ab_1 and subsequent incubation with the ECL label (Ab_2). Following the ultimate magnetic separation and washing, the immunocomplex was adsorbed onto a GCE, and ECL was gauged in a tripropylamine solution. The assay duration was slightly extended to 2.5 hours, attributed to the additional steps in the process. For the detection of PSA, ECL quenching via resonance energy transfer was utilized [181]. Magnetic separation was once again employed exclusively for probe preparation. MNPs of Fe_3O_4, coated with MnO_2 nanosheets and functionalized with Ab_2, were utilized as the quenching probe. The presence of Fe_3O_4/MnO_2 impeded the excitation of nitrogen-doped graphene quantum dots (NGQDs), resulting in a reduction in electrochemiluminescence (ECL) intensity. NGQDs were utilized for signal generation and loaded onto nanostructured $Ni(OH)_2$ to form a composite with enhanced ECL behavior. This composite was modified with Ab_1 and applied onto a gold nanoparticle (AuNP)-modified GCE. Following PSA binding to Ab_1, the Fe_3O_4/MnO_2-Ab_2 probe finalized the sandwich immunocomplex on the electrode, causing a reduction in electrochemiluminescence (ECL) intensity due to resonance energy transfer to the quencher probe. The achieved LOD was 5 fg/mL. ECL measurements are integrated into various clinical immunoanalyzers, and their combination with magnetic separation further enhances the LODs. Qiu et al. devised an electrode immobilization-free and homogeneous electrochemical method for the sensitive detection of thrombin in human serums [182]. The method relies on the target-promoted proximity binding-induced hybridization chain reaction (HCR) to generate MNP-decorated DNA polymers. The binding of thrombin by two aptamers increases their local concentrations, prompting a strand displacement reaction and initiating HCR between labeled hairpins. Streptavidin-modified MNPs bind to the biotin moieties, forming MNPs/DNA polymers. The accumulated MNPs/DNA polymers on AuNP-deposited electrodes result in a highly amplified current for detecting thrombin within the concentration range of 5 pM to 50 nM, with a detection limit of 1.1 pM. The technique showcases the specific identification of low levels of thrombin in diluted human serums, indicating its capability for monitoring protein biomarkers in disease diagnosis. The assay is enzyme-free, uncomplicated, and effective, bypassing the need for enzyme engagement and electrode incubation. It provides heightened sensitivity by employing two specific aptamers and signal amplification through HCR. The adaptability of aptamers suggests the prospect of extending this method to identify various other protein biomarkers. Upadhyay et al. [183] introduced an immunoassay technique known as s-MARSA (magnetic nanoparticle enrichment and enzyme-mediated signal amplification) for detecting Apolipoprotein E (ApoE) in cerebrospinal fluid (CSF). ApoE serves as a biomarker with a correlation inversely linked to midbrain dopamine concentration, rendering it pertinent for monitoring the side effects of pharmacotherapy in patients with Parkinson's disease and Schizophrenia. The s-MARSA method presents several advantages compared to the conventional ELISA technique. It boasts a lower detection limit of 5 fg/mL and a more extensive linear detection range, covering from 5 fg/mL to 4 μg/mL. The analysis can be concluded within an hour, utilizing only a small sample volume (2 μL) of CSF. The outcomes obtained through s-MARSA exhibit a robust correlation with those derived from ELISA. Moreover, the method underwent validation with 53 clinical samples, revealing clear linearity across various dopamine states in Parkinson's patients, as well as physiological levels in healthy individuals. The established s-MARSA method demonstrates significant potential for clinical application in monitoring the pharmacotherapy of patients with Parkinson's disease and Schizophrenia.

In conclusion, the field of cancer biosensing utilizing MNPs has witnessed significant advancements in the detection of cancer biomarkers. Various strategies for immobilizing specific biological elements onto functionalized MNPs have been employed, enabling the detection of small molecules, peptides, aptamers, DNA, and proteins. The integration of MNPs into biosensing systems can be achieved through direct labeling, where MNPs are immobilized at the transducing element through affinity recognition reactions, or indirect labeling, which employs sandwich immunoassays. Homogeneous assays in solution or pseudohomogeneous formats eliminate the need for separation steps, while heterogeneous formats generally provide better sensitivity. The use of microfluidics in combination with MNPs allows for enhanced separation efficiency. MPs have also been utilized as labels in various sensing platforms, such as amperometric detection, electrochemiluminescence, and resonance energy transfer, resulting in low limits of detection and broad linear ranges.

In addition to cancer biomarker detection, MNPs have been applied in the detection of other biomarkers such as thrombin in human serums and Apolipoprotein E (ApoE) making MNPs suitable for monitoring the pharmacotherapy of other diseases as well, such as Parkinson's disease and Schizophrenia patients. Overall, the integration of MNPs into biosensing systems has shown great promise in improving the detection of disease biomarkers, offering enhanced sensitivity, specificity, and efficiency. These advancements contribute to the development of diagnostic methods and therapeutic monitoring, facilitating early detection and effective treatment of various diseases.

3.1.5.2 MNPs in the Detection of Pathogens

The use of MNPs has gained significant attention in the field of pathogen detection due to their unique properties and versatile applications. MNPs offer several advantages, including high surface area, easy functionalization with biomolecules, and magnetic responsiveness, making them ideal candidates for sensitive and rapid detection of pathogens. The integration of MNPs with various detection techniques, such as magnetic immunoassays, biosensors, and magnetic particle spectroscopy, has shown great potential in the detection of a wide range of pathogens, including viruses and bacteria. This section aims to provide an overview of the recent developments and applications of MNPs in pathogen detection, highlighting their role in improving sensitivity, selectivity, and speed of detection assays. The combination of MNPs with advanced detection platforms holds great promise for addressing the challenges in pathogen detection and enhancing our capabilities in diagnosing and controlling infectious diseases.

Detection of Viruses Assays conducted outside of laboratory settings play a vital role in the swift detection, identification, and subsequent treatment of potentially dangerous substances. In this context, the utilization of MNPs is of great importance as they have the potential to significantly reduce the LOD for pathogenic viruses and bacteria. Even a single microbe has the potential to trigger the onset of disease, underscoring the significance of achieving highly sensitive detection. An example of this is the creation of a chronoamperometric magnetoimmunosensor utilizing gold nanoparticles (AuNPs) for the detection of the influenza virus [184]. The viruses were captured and concentrated through specific antibody-conjugated magnetic nanoparticles (Ab_1-MNPs), leading to the creation of a sandwich complex involving MNPs, viruses, and AuNPs. This complex was magnetically captured and deposited onto a carbon surface, where the reduction of H^+ into H_2 occurred catalytically by the deposited AuNPs. The resultant current signal exhibited a direct proportionality to the virus titer. The complete analytical procedure was completed within 2.5 hours, achieving a LOD of 18 hemagglutination units (HAU). In an alternative instance, a digital electrochemical ELISA for the avian influenza virus H7N9 was created, employing a microelectrode array that was enhanced with a monolayer of gold nanoparticles (AuNPs) [185]. Bifunctional fluorescence magnetic nanoparticles (bi-FMNs) underwent modification through the coimmobilization of specific antibodies (Ab_2) and alkaline phosphatase (ALP). The viruses captured on bi-FMNs were then magnetically separated and loaded onto the microelectrode array. The separation of individual immunocomplexes of bi-FMNs, each containing a single virus, onto microelectrodes was monitored using fluorescence microscopy. Finally, ALP catalyzed the hydrolysis of p-aminophenyl phosphate (PAPP) to p-aminophenol, leading to the reduction of Ag^+ to Ag. The resulting signal was measured using linear sweep voltammetry to determine the virus concentration. By manipulating the ratio of bifunctional fluorescence magnetic nanoparticles (bi-FMNs) to target concentration, it was ensured that, statistically, only one virus particle was captured per bi-FMN. This approach facilitated the feasibility of single virus analysis. The digital ELISA proved effective in detecting avian influenza virus H7N9 in chicken serum and liver samples, achieving a LOD of 7.8 fg/mL. Notably, this was accomplished without the need for intricate sample pretreatment or advanced instrumentation. In a different scenario involving hepatitis B virus surface antigen (HBsAg), an electrochemiluminescence (ECL) immunosensor was devised [186]. The method employed specific antibodies (Ab_1) immobilized on MNPs, while Ab_2 was attached to a polyamidoamine (PAMAM) dendrimer carrying QDs. The electrochemiluminescence (ECL) signal emitted by the QDs bound to the electrode surface served to quantify the analyte. This strategy incorporated multiple signal enhancement techniques, utilizing MNPs for preconcentration, improving electron transfer between the nanoprobe and the electrode, and facilitating efficient loading of QDs through the dendrimer. Consequently, the sensor achieved a LOD of 0.8 fg/mL within a one-hour analysis time, successfully detecting virus particles in human serum. Additionally, an ECL approach using electroluminescent nanospheres based on a poly(styrene/acrylamide) copolymer doped with CdSe/ZnS QDs and MNPs was developed for the sensitive detection of the Ebola virus [187]. The electroluminescent nanospheres demonstrated an approximately 85-fold amplification of electrochemiluminescence (ECL) intensities compared to the QDs, thereby enhancing sensitivity. The LOD achieved was 5 pg/mL, and the linear range extended up to 30 ng/mL for the Ebola virus. The analysis was completed within two hours, eliminating the need for complex sample pretreatment. The urgency highlighted by the Ebola outbreak in Africa underscored the necessity for rapid point-of-care assay tools to facilitate swift screening and treatment initiation for confirmed positive patients. These examples collectively highlight the potential of MNPs in developing highly sensitive and efficient detection assays for various viral pathogens, addressing the critical demand for rapid and precise diagnostic tools. It's important to note that norovirus, highly contagious and causing gastroenteritis, poses a significant risk to both elderly individuals and young children. To address this public health concern, there is an urgent need for sensitive and rapid detection methods. In this study, a novel detection approach called localized surface plasmon resonance (LSPR)-amplified magnetofluoroimmunoassay was developed for detecting norovirus in biocomplex media without compromising the fluorescence response [188]. The detection approach involved creating a hybrid nanocomposite composed of gold nanoparticles (AuNPs) and MNPs. These AuNP/MNP hybrid nanocomposites were linked with antibodies (Ab) specific to norovirus genogroup II, as well as CdSeS QDs. This composite facilitated the separation of the target virus from impurities using a magnetic field and induced a robust LSPR effect. Norovirus particles were introduced to the Ab-conjugated AuNP/MNP composite and Ab-CdSeS QDs, and subsequently, the Au/MNPs-Ab-virus-Ab-CdSeS QDs complex was separated using an external magnetic field. The developed method demonstrated the capability to detect norovirus-like particles in the concentration range of 1 pg/mL to 5 ng/mL, achieving a LOD of 0.48 pg/mL in feces. The method successfully detected clinical samples of feces containing norovirus from patients, achieving an LOD of 84 RNA copies/mL. The detection remained unaffected by impurities present in the feces solution. This sensing system demonstrated applicability for on-site norovirus detection in feces samples. The LSPR-amplified magnetofluoroimmunoassay exhibited high sensitivity, selectivity, and simplicity in detecting norovirus, with a sensitivity approximately 1000 times higher than that of a commercially available ELISA kit. When applied to clinical samples, the LOD for genogroup II norovirus was 98 RNA copies/ml for GII.2 and 84 RNA copies/ml for GII.3. Despite varying sensitivities for different genotypes, this detection system reliably detected norovirus even in feces specimens containing impurities. Due to its outstanding sensitivity, selectivity, and minimal interference from impurities in clinical samples, this detection method is well-suited for on-site norovirus detection. The study by Zhong et al. [189] centers on the rapid and sensitive detection of SARS-CoV-2 through the use of functionalized MNPs. In their research, MNPs were conjugated with antibodies targeting the SARS-CoV-2 spike protein, serving as sensors. Experiments were conducted on ACS spectra and MPS signals to analyze detection sensitivity and the LOD. The method involved using MNPs as sensors, detecting the ratio of the third to first harmonics of the MPS signal. A proof of concept was demonstrated using 100 nm-sized polystyrene beads conjugated with SARS-CoV-2 spike protein to mimic the virus. The achieved LOD with this approach was approximately 0.084 nM (5.9 femtomoles) for detecting the mimic SARS-CoV-2. The specificity of the method relies on the specificity of the spike protein antibody binding to SARS-CoV-2 spike proteins. The LOD and specificity stand as crucial parameters in diagnostic tests, and the current method exhibits a LOD comparable to other documented approaches for SARS-CoV-2 detection. Ongoing experiments are in place to assess the method's specificity and enhance the LOD through optimization of the measurement setup and sample preparation. The paper notes that, while the proposed approach meets the specified criteria for diagnostic tests, except for being equipment-free, the measurement setup can be constructed as a point-of-care device at low cost. Moreover, the method holds potential for visualizing the spatial distribution of SARS-CoV-2 in vivo when combined with multicolor MPI. Multiple research groups have underscored the role of MNPs in detecting various virus types [189–192].

In summary, the use of MNPs in viral pathogen detection holds substantial promise for developing assays that are both highly sensitive and rapid. Various studies have demonstrated the successful development of biosensors and immunoassays using MNPs for the detection of viruses such as influenza, avian influenza, hepatitis B, Ebola, norovirus, and SARS-CoV-2. These assays employ different strategies, including magnetoimmunosensors, digital electrochemical ELISAs, electrochemiluminescence immunosensors, and LSPR-amplified magnetofluoroimmunoassays. MNPs functionalized with specific antibodies enable the capture and separation of target viruses from complex samples, while various labeling techniques enhance the detection signals. These assays have demonstrated remarkable sensitivity, with low LODs ranging from femtograms to picograms per milliliter, enabling the detection of viruses in clinical samples and biocomplex media. The development of MNPs-based assays addresses the critical need for rapid and accurate diagnostic tools for viral pathogens, facilitating timely screening, treatment initiation, and disease management. Further research and optimization are expected to enhance the performance and applicability of these assays, ultimately enabling their deployment as cost-effective and equipment-free point-of-care devices. The role of MNPs in virus detection continues to be a focus of research, offering potential advancements in the field of viral diagnostics.

Detection of Bacteria An impedance biosensor with high sensitivity was developed to detect *Listeria monocytogenes*, a harmful bacterium [193]. The biosensor employed biotinylated monoclonal antibody 1 (Ab_1) immobilized on streptavidin-modified MNPs. Through this setup, Listeria cells were effectively separated from the sample matrix. The captured microbes were then labeled with

Ab_2-urease-AuNPs conjugate, resulting in the formation of sandwich complexes. The enzyme urease, which is present in the conjugate, initiated the hydrolysis of urea into ammonia and hydrogen carbonate. This process resulted in an elevation of the ionic strength in the surrounding medium. This change in ionic strength was quantified using an interdigitated microelectrodes array. The biosensor exhibited remarkable reusability, with the capability to be reused over 50 times. In terms of sensitivity, the biosensor demonstrated an LOD of 300 colony-forming units per milliliter (CFU/mL) for detecting Listeria monocytogenes in both pure culture and spiked lettuce samples. Detailed TEM images provided clear evidence of the binding between both particle types and the surface of the Listeria cells. Praiwi et al. [194] have created an electrochemical immunoassay with high sensitivity for the detection of Salmonella enterica serovar Typhimurium. This particular foodborne pathogen is notorious for causing infections even at remarkably low concentrations. The assay integrated DNA biobarcode labels and utilized the technique of differential pulse anodic stripping voltammetry (DPASV), aided by silver (Ag) enhancement. In this approach, the biobarcodes comprised latex spheres modified with ferromagnetic Fe_3O_4 particles, each bearing specific antibodies and DNA sequences. The immunoassay was conducted by introducing biobarcode particles, Salmonella cells, and biotin-conjugated antibodies into well plates. After the binding of antigens and antibodies, excess antibodies were eliminated through magnetic collection, and the remaining biobarcode particles were exposed to avidin-coated screen-printed electrodes. This enabled the formation of an avidin-biotin bond, facilitating the removal of excess biobarcode particles. The quantification of the remaining biobarcode particles on the electrode was carried out by measuring silver ions (Ag^+) using DPASV. The assay demonstrated minimal response to *Escherichia coli* and achieved an LOD of 12 colony-forming units per milliliter (CFU/mL) in a buffer solution. The LOD for both heat-killed and whole-cell *Salmonella Typhimurium* in milk, green bean sprouts, and raw eggs fell within the range of 13 to 26 colony-forming units per milliliter (CFU/mL). Remarkably, this electrochemical immunoassay achieved the lowest reported LOD for Salmonella without the need for pre-enrichment of the samples. Additionally, a highly sensitive electrochemical biosensor platform was developed for the detection of Mycobacterium tuberculosis antigen by modifying a GCE with graphene quantum dot (GQD)-coated Fe_3O_4 at Ag core–shell nanostructures (Fe_3O_4 at Ag/GQD) [195]. The synthesis of the Fe_3O_4 at Ag/GQD nanocomposite involved coating Fe_3O_4 nanoparticles with a thin layer of silver (Ag), followed by the deposition of GQDs on the surface. This composite provided several advantages for the biosensor platform. Firstly, Fe_3O_4 nanoparticles increased the surface-to-volume ratio of the electrode, providing more binding sites for the immobilization of antibodies. This increased the sensitivity of the biosensor by allowing more antigens to be captured and detected. Secondly, the Ag layer on the Fe_3O_4 nanoparticles enhanced the electrical conductivity of the electrode, improving the electron transfer between the electrode and the analyte. This resulted in more efficient electrochemical signal transduction, leading to higher sensitivity and lower detection limits. Finally, the GQDs served as a substrate for the immobilization of antibodies, enabling a higher loading of antibodies onto the electrode. This increased the binding capacity of the biosensor, enabling the detection of lower concentrations of the target antigen. The AuNPs conjugated to the anti CFP-10 antibody served as a signal amplification label. The AuNPs provided a large surface area for the immobilization of antibodies, allowing for more efficient binding of the target antigen. Additionally, the AuNPs exhibited excellent catalytic properties, enhancing the electrochemical signal generated during the detection process. In summary, the amalgamation of Fe_3O_4 at Ag/GQD nanocomposites and gold nanoparticles (AuNPs) linked to antibodies produced a highly sensitive biosensor platform for the detection of *Mycobacterium tuberculosis* antigen. This platform has the potential to be used for the early diagnosis of tuberculosis, providing a rapid and accurate method for screening and monitoring the disease. The immunosensor had a wide range of concentrations it could detect, from 0.005 to 500 μg/mL, and it was able to detect very low concentrations with an LOD of 0.33 ng/mL. This low LOD suggests that the biosensor is highly sensitive and capable of detecting very low levels of the target antigen. Xu et al. [196] explored an impedimetric immunosensor for the swift detection of *E. coli* O157:H7 and *S. typhimurium* in food, employing nano magnetic beads (MBs) and a screen-printed interdigitated microelectrode (SP-IDME). Specific biotinylated antibodies (Ab) were used to capture the target bacteria, and these antibodies were applied to modify streptavidin-coated MBs. The MBs-Ab-cell complexes were then labeled using glucose oxidase (GOx)-Ab conjugates. The resulting complex, comprised of MBs-Ab-cell-Ab-GOx, was combined with a glucose solution, initiating an enzymatic reaction that led to the production of gluconic acid. This acid caused an increase in the ion strength of the solution, resulting in a reduction in the solution's impedance as measured on the screen-printed interdigitated microelectrode (SP-IDME). The immunosensor successfully detected *E. coli* O157:H7 and *S. typhimurium* in pure culture samples, with a detection range spanning from 102 to 106 colony-forming units per milliliter (CFU/mL). Additionally, the immunosensor demonstrated the detection of *E. coli* O157:H7 in ground beef and *S. typhimurium* in chicken rinse water. The determined LODs for the two bacteria in food were 2.05×10^3 cfu/g and 1.04×10^3 cfu/mL, respectively. It is noteworthy that this immunosensor employed a bare electrode for impedance measurement without the need for any surface modifications. This characteristic contributes to its cost-effectiveness, reproducibility, ease of operation, and simplicity of preservation. Enhanced sensitivity in detecting *E. coli* was achieved by utilizing MNPs conjugated with Ab_1, effectively separating bacteria from broth samples [197]. In the next phase of the assay, Ab_2-AuNPs were employed to create the sandwich complex. Notably, this process only took 45 minutes to reach an LOD of 10 colony-forming units per milliliter (CFU/mL) and a linear detection range extending up to 10^6 CFU/mL. In the subsequent stage of the assay, Ab_2-AuNPs were employed to form the sandwich complex. This method achieved a LOD of 10 colony-forming units per milliliter (CFU/mL) and demonstrated a linear detection range up to 106 CFU/mL, all within a brief duration of 45 minutes. Benefiting from its simplicity and rapid processing, this biosensor proved effective for on-site screening of bacterial contamination in beef, vegetables, juices, and water systems. A comparative analysis was conducted between microparticles (MPs, 2.8 μm) and MNPs, 300 nm, both utilizing the same anti-Salmonella capture antibody (Ab_1) [198]. The captured microorganisms were noted to create aggregates in both the magnetic micro and nanosized particles, with a more noticeable impact seen in MNPs, primarily due to the multivalency of the magnetic carriers and bacteria. The detection of Ab_2-HRP was accomplished through amperometric measurement using H_2O_2/HQ substrates. The electrochemical immunosensor successfully differentiated pathogenic *S. typhimurium*, with carriers of MPs and MNPs demonstrating LODs of 540 and 290 CFU/mL, respectively, within a total time of one hour. Through an eight-hour pre-enrichment step, the detection of 1 CFU in 25 mL of milk (equivalent to 0.04 CFU/mL) was successfully achieved. It is important to highlight that a decrease in amperometric responses was observed when working with whole milk, especially when employing MNPs. This observation indicates a possible matrix effect, possibly due to the greater surface area-to-volume ratio associated with the smaller size of MNPs. This factor may lead to heightened nonspecific adsorption in complex media, such as whole milk. Despite the preconcentration phase, this method provided quicker results than traditional cultivation methods, which typically take more than two days for the detection of *Salmonella*. Another research effort described the creation of an electrochemical immunoassay employing iron oxide/gold core/shell nanoparticles (Fe at Au) conjugated with anti-Salmonella antibodies [199]. Fe at Au nanoparticles underwent modification with different self-assembled monolayers and were characterized using methods such as ultraviolet-visible spectrometry, TEM, and voltammetric techniques.

The detection of *S. typhimurium* on screen-printed carbon electrodes was accomplished through square-wave anodic stripping voltammetry, utilizing CdS nanocrystals. A calibration curve was generated within the concentration range of 1×10^1 to 1×10^6 cells/mL, and the determined LOD was 13 cells/mL. A cost-effective and straightforward immunosensor was created for the swift detection of *Staphylococcus aureus* B-1266, employing Fe_3O_4–SiO_2–NH_2 nanoparticles as the direct signal label [200]. The process of the electrochemical immunoassay included exposing bacteria to an excess of magnetite nanoparticles, followed by magnetic separation to eliminate the unbound nanoparticles. Afterward, the labeled immunocomplex was developed on the surface of a planar electrode, and the electrochemical response was gauged from the magnetite nanoparticles present within the immunocomplex. The electrochemical immunosensor allowed for the selective and accurate detection of *S. aureus* within the range of 10–10^5 colony-forming units per milliliter (CFU/mL), exhibiting a relative standard deviation below 10%. The determined LOD for *S. aureus* was 8.7 CFU/mL.

The development of a sensitive, rapid, and cost-effective assay aimed to detect *E. coli*, *S. aureus*, and *S. typhimurium* in potable water and apple juice. The assay utilized electrical impedance spectroscopy alongside screen-printed interdigitated electrodes, incorporating magnetite nanoparticles functionalized with melittin (MLT) [201]. The assay successfully detected *E. coli* at concentrations as low as 1 CFU/mL in water and 3.5 CFU/mL in apple juice within 25 minutes, without the necessity for sample preparation. The MNPs employed in the assay were modified with carboxyl groups, displaying a negative charge of −28 mV at pH 7.0. TEM analysis (Figure 3.6A(a)) confirmed an average particle diameter of 8.5 nm, while electron diffraction patterns (Figure 3.6A(b)) verified the typical structure and composition of magnetite. The MNPs underwent functionalization through the interaction between the carboxylated particles and the cationic amine groups in MLT. MNPs-MLT demonstrated robust interactions with bacterial cells, as depicted in the TEM image (Figure 3.6A(c)), facilitating their separation through magnetic precipitation. The analytical approach utilized precipitation with MNPs-MLT and impedance measurements using silver screen-printed interdigitated electrodes (Ag SPIDEs) for detecting *S. aureus*, *S. typhimurium*, and *E. coli*. The capacitance versus frequency spectra acquired with the MNPs-MLT biosensor illustrated a connection with bacteria concentration, showing increased capacitance at higher concentrations (Figure 3.6B). The calibration curves exhibited a linear correlation between capacitance change and bacteria concentration for *E. coli* and *S. aureus*. In contrast, the capacitance spectra for *S. typhimurium* remained consistent across concentrations, allowing it to be distinguished from the blank spectrum. Further experiments with *S. typhimurium* concentrations below 1 CFU/mL were not reproducible, suggesting the need for an alternative dilution approach using fluorescence microscopy for reliable measurements. The evaluation of *E. coli* in artificially contaminated samples demonstrated no significant matrix effect in potable water, although a reduced detection range was observed compared to pure cultures. In the analysis of apple juice, a dilution factor of 5 facilitated successful detection within the range of 3.5–30 CFU/mL. Lower dilution factors resulted in bacterial overgrowth and saturation of the biosensor at low concentrations. Recovery percentages ranged from 89.5% to 97.5%, indicating the potential to determine low bacterial concentrations in food samples. This approach, utilizing high-affinity peptides, highlights its appropriateness for quantifying pathogenic bacteria in food. Numerous additional publications have explored the application of magnetic nanoparticles for the detection of bacterial pathogens [202–204].

In summary, the advancement of nanomaterial-based biosensors and immunoassays exhibits considerable potential in achieving sensitive detection of bacterial pathogens. These methodologies, leveraging MNPs, specific antibodies, and diverse labeling techniques, present advantages such as high sensitivity, swift detection, and the prospect of on-site screening. Impedance biosensors have demonstrated remarkable sensitivity, showcasing low LODs within the CFU/mL range,

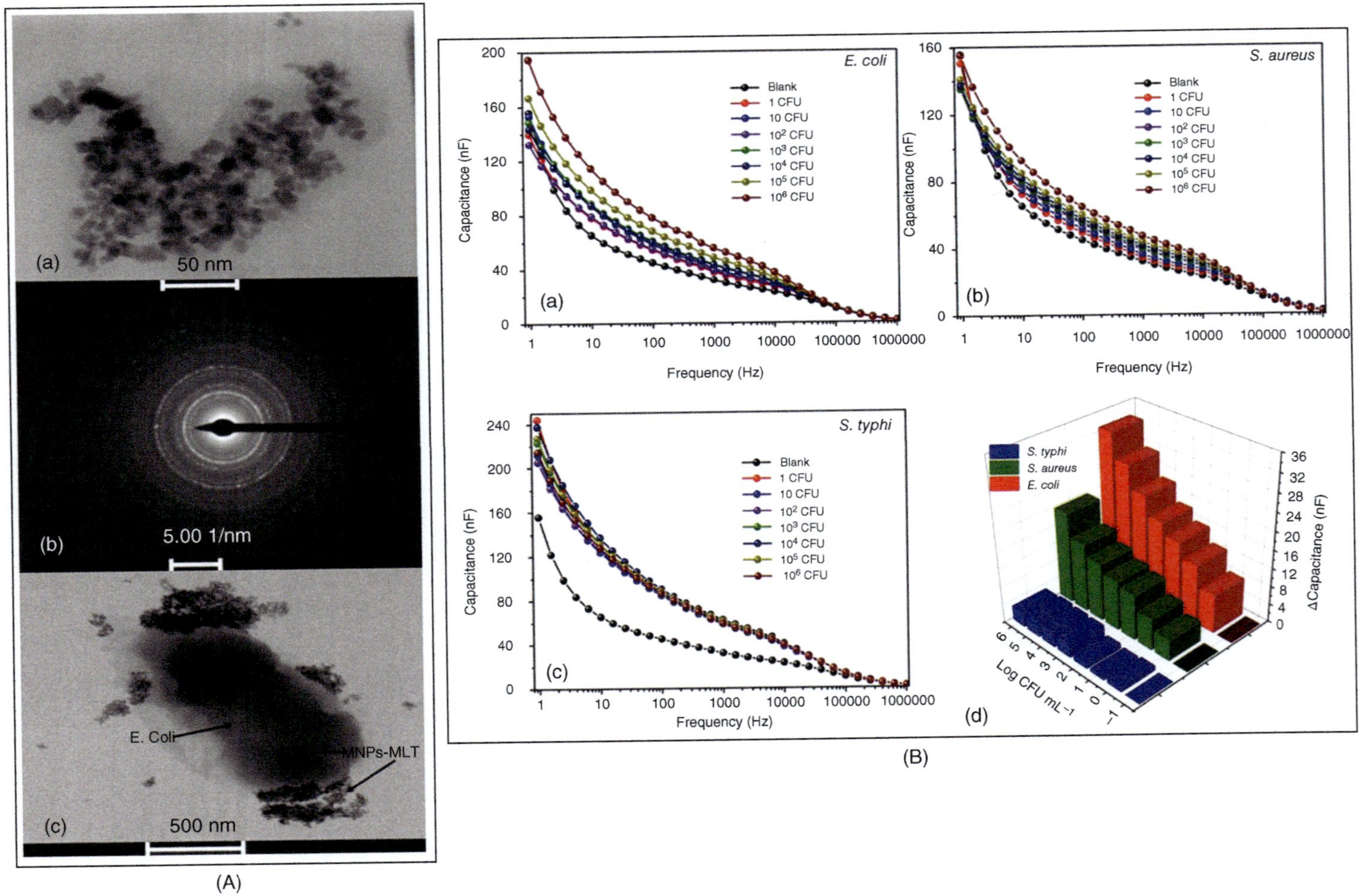

Figure 3.6 (A) Presents images obtained through transmission electron microscopy (TEM), including (a) the synthesized MNPs, (b) selected area electron diffraction (SAED) patterns of the MNPs, and (c) a TEM image depicting the complex of MNPs with MLT-*E. coli*. In Figure (B), the capacitance spectrum for the MNP-MLT system is illustrated for *E. coli* (a), *S. aureus* (b), and *S. typhimurium* (c). The accompanying column graph exhibits the calibration curves for *E. coli*, *S. aureus*, and *S. typhimurium* at 10 kHz (d). Source: Reproduced with permission from Wilson et al. [201] and Tietze et al. [210]/Elsevier.

rendering them suitable for the detection of bacteria in intricate samples like food matrices. Electrochemical immunoassays, employing nanomaterials like GQDs and iron oxide/gold core/shell nanoparticles (Fe at Au), have exhibited enhanced sensitivity by improving binding capacity and electrochemical signal transduction. The application of MNPs functionalized with specific antibodies has enabled the selective detection of bacteria, utilizing techniques such as electrical impedance spectroscopy or capacitance measurements. These nanomaterial-based biosensors hold substantial potential for dependable and precise screening methods in domains such as food safety, environmental monitoring, and disease diagnosis. Ongoing research and development are anticipated to propel the progress and commercialization of these biosensors for practical applications in the near future.

3.2 Toxicity of MNPs to Biological Systems

Acquiring knowledge about the potential toxicity of MNPs is crucial and relies on various factors such as their dimensions, form, composition, surface alterations, concentration, amount, distribution within the body, accessibility to biological systems, dissolvability, immunogenicity, and the way the body processes them [205]. Due to their elevated surface-to-volume ratio, MNPs are more prone to interact with bodily elements when inhaled, absorbed, or ingested [206]. Additionally, their capacity to traverse cellular barriers and resist biodegradation enhances their potential toxicity. When undertaking toxicity studies, it is essential to explore diverse aspects, including acute and chronic toxicity mechanisms, degradation of particle products, stimulation of defensive and inflammatory responses, metabolic effects, long-term toxicity, and elimination in cellular and animal models [207, 208]. This section concentrates on assessing the toxicity parameters of iron oxide MNPs by reviewing studies conducted in cell lines, aquatic organisms, and rodent animal models. The findings from these investigations offer valuable insights that can inform future research in toxicity analysis, particularly when employing animal models. The ultimate goal of this research is to guarantee the secure and advantageous application of MNPs in human contexts.

3.2.1 Biodistribution and Metabolism of MNPs

The distribution throughout the body and the duration of circulation of MNPs are affected by their surface chemistry and the method of delivery [209]. MNPs exceeding 200 nm in size are generally captured by the spleen via mechanical filtration, whereas those smaller than 10 nm can be eliminated through renal clearance [17]. Hence, the ideal size range for administration in specific applications is considered to be 10–100 nm. Biodistribution studies indicate that these particles predominantly accumulate in the liver (80–90%), with secondary accumulation in the spleen (5–8%) and bone marrow (1–2%) [210]. The interaction, biodistribution, internalization mechanism, metabolism, and potential toxicity of MNPs depend on their surface modifications [211]. Prior investigations have comprehensively evaluated the toxicity of iron oxide nanoparticles with varying sizes and surface functionalization at the cellular level [212, 213]. ICR mice were subjected to low ($4.89 \times 10^5 \pm 2.37 \times 10^4/cm^3$) and high concentrations ($9.34 \times 10^5 \pm 5.11 \times 10^4/cm^3$) of fluorescent magnetic nanoparticles (FMNPs) for a duration of four weeks using a nose-only exposure system [214]. The analysis of biodistribution showed that the FMNPs were mainly present in the liver, testis, spleen, lung, and brain, suggesting their capability to traverse the blood-brain barrier and blood-testis barrier. In a separate study, the biodistribution and elimination of MNPs were examined after the injection of MNP suspensions (10 mg Fe/mL) through the tail vein) [211]. The findings indicated that the distribution of iron and its elimination were intricate processes affected by fluctuating iron concentrations in various tissues over time. Despite this, the introduction of iron did not lead to persistent alterations in liver enzyme levels or oxidative stress. In another study, male Fisher 344 rats with intracerebral brain tumors underwent injections of polyethylene glycol-coated magnetic nanoparticles (PEG-MNPs) through burr holes drilled in their skulls [215]. This research explores the biodistribution patterns of polyethylene glycol-coated magnetic nanoparticles (PEG-MNPs) in elimination organs, namely the liver, lung, spleen, and kidney, employing a rat model. The findings demonstrate that PEG-MNPs exhibit an extended circulation lifetime, enabling enhanced magnetic targeting of brain tumors. At the one-hour mark, the concentrations of PEG-MNPs in the liver and spleen were significantly reduced (approximately 12-fold and 2.5-fold, respectively) compared to starch-coated MNPs (D). Although the liver concentrations of PEG-MNPs consistently stayed lower than D from 1 to 60 hours, spleen values continued to rise and were notably higher at later time points, as confirmed by histological observations. Over the 60-hour assessment, there was limited to no dispersion of PEG-MNPs observed in the lung or kidney. The research additionally verified the improved and specific magnetic targeting of brain tumors with PEG-MNPs (12 mg Fe/kg) after one hour, achieving nanoparticle delivery of up to 1.0% of the injected dose per gram of tumor tissue—an enhancement of 15 times compared to targeted D (0.07% injected dose/g tissue). MRI and histological examinations visually affirmed the improved targeting and indicated that passive mechanisms have a restricted role in nanoparticle retention in tissues. The research proposed that liver biodistribution was comparatively less noteworthy compared to other MNPs investigated, whereas the distribution in the spleen and the prolonged accumulation of PEG-MNPs raised apprehensions about toxicity. Hence, the research team proposed a prolonged study to evaluate potential risks. Additionally, alterations in the biodistribution of iron in various body tissues over time were observed in ICR mice following an intraperitoneal injection of 3-aminopropyl-triethoxysilane-coated magnetic nanoparticles (APTS-MNPs), with analysis points at 1, 2, 3, and 4 weeks [216]. Liver and kidney functions were assessed, and subsequent pathological analysis involved the dissection of the heart, liver, spleen, lung, kidney, testis, and brain. The distribution of iron in various body tissues exhibited temporal changes, with a larger proportion of the injected iron localizing in the liver and spleen compared to other tissues. Following APTS-MNPs injection, the serum indicated an increase in AST and LDH levels. Histological analyses of selected tissues revealed no noticeable abnormal alterations. Ongoing assessments indicated no changes in liver and kidney function, suggesting the safe applicability of APTS-MNPs for in vivo purposes. Nevertheless, the research concluded that additional inquiries at a molecular level were necessary. The attachment of MNPs to external structures can trigger varied reactions, impacting their absorption and processing contingent on the organ or cell type. A separate investigation illustrated that CD301b molecules distinctly labeled different groups of MNPs in diverse organs, indicating their participation in systemic metabolism, encompassing adipose tissue, skeletal muscle, colon, and pancreas [217]. This research delved into the expression and role of CD301b+ MNPs in diverse tissues. Findings revealed the presence of CD301b+ cells within subsets of MNPs in peripheral organs, including epididymal fat, colon, and peritoneum. These CD301b+ MNPs displayed a phenotype characterized by elevated expression of MHC class II and CD11b, intermediate to high expression of CD11c, and moderate expression of F4/80. Notably, the expression of CD301b did not align with other M2 macrophage markers like CD206 and Arginase1, suggesting that CD301b+ MNPs represent a unique cell subset. Elimination of CD301b+ MNPs in mice resulted in weight loss and changes in overall energy balance. The absence of CD301b+ MNPs led to decreased food intake, reduced energy expenditure, and a transition toward lipid oxidation. Furthermore, depletion of CD301b+ MNPs caused hypoglycemia and heightened insulin sensitivity in animals fed both regular chow and a high-fat diet. These results indicate that CD301b+ MNPs contribute to the regulation of metabolism and energy homeostasis, irrespective of their interaction with the central nervous system.

3.2.2 In vitro Toxicity Studies of MNPs

Spherical-shaped Fe_3O_4 MNPs measuring an average size of 72.6 ± 0.6 nm underwent thorough characterization and were examined in vitro to evaluate their influence on the cellular membrane microstructure and function of red blood cells, utilizing eryptosis indices as an analytical measure [218]. The findings indicated that at a concentration of 25 μg/mL, Fe_3O_4-MNPs caused considerable impairment to erythrocytes. Subsequent in vivo experiments on 8-week-old female CDfi IGS rats revealed that Fe_3O_4 MNPs at a concentration of 12 mg/kg induced apoptosis in circulating erythrocytes. The presence of Fe_3O_4 MNPs significantly affected the mechanical properties of erythrocytes, resulting in pathological alterations in cell membranes, abnormal cytosolic calcium levels, oxidative stress, and ultimately programmed cell death, observed both in in vitro and in vivo conditions. In an independent investigation utilizing a BeWob30 placental cell barrier model, the toxicity of different nanoparticles, including uncoated Fe_3O_4, Na-oleate-coated Fe_3O_4 (OC-Fe_3O_4), and fluorescent rhodamine-labeled silica nanoparticles (Fl-25 SiO_2, Fl-50 SiO_2), was assessed [219]. Iron oxide nanoparticles displayed cytotoxicity at lower concentrations and shorter exposure durations when contrasted with silica particles. Similarly, the cytotoxic profiles of MNPs without coating and those coated with n-octyltriethoxysilane were compared in PC12 (rat pheochromocytoma) and ReNcell at VM (human neural stem cells). The coated MNPs exhibited diminished cytotoxic effects compared to their uncoated counterparts, with variations observed across different cell types and the presence of serum influencing nanoparticle toxicity [220]. In an alternative investigation, untreated MNPs with a size range of approximately 20–60 nm were applied to treat NRK-52E cell lines (rat, kidney). The outcomes revealed heightened expression of rat-related proteins along with increased levels of glutathione-related and chaperone proteins. These changes acted as protective mechanisms for NRK-52E cell lines, mitigating apoptosis and reducing ROS levels [221]. A comparative assessment of toxicity was conducted on four distinct types of MNPs, which included uncoated-magnetite MNPs (cationic), starch-magnetite MNPs, dextran-magnetite MNPs, and uncoated-maghemite MNPs (anionic) using rat pheochromocytoma PC12 cells. The findings revealed diverse interactions with cells and varying cytotoxic effects. Uncoated-magnetite particles remained on the external surface of cells without inducing cytotoxicity up to 0.1 mg/mL, whereas starch and dextran-coated particles exhibited some reduction in cell viability at higher concentrations [222]. Additionally, anionic maghemite nanoparticles were investigated in human prostatic tumor cells (PC3). These particles were taken up by the cells and accumulated within vesicles. Upon exposure to an alternating current (AC) field, these nanoparticles generated intracellular heat, leading to effective cell destruction through MH without causing detectable cytotoxicity [223]. Both investigations suggested that maghemite MNPs are safe and do not cause notable toxicity, even during prolonged exposure. In a fascinating study, the effects of magnetically stimulated human primary adipose-derived stem cells (ADSCs) were examined using MNPs with a size of 6.6 nm, carrying an average of 21 picograms of iron per cell [224]. The ADSCs were subjected to various magnetic field (MF) intensities, ranging from low intensity (0.5 mT) to higher intensities (14.7 and 21.6 mT). Interestingly, the type, duration, intensity, and frequency of the MF had distinct effects on the differentiation of the ADSCs. Short intermittent exposure (two days) to low-intensity MF facilitated adipogenesis, while prolonged exposure (seven days) induced osteogenesis. The introduction of MNPs into ADSCs resulted in a shift toward an osteoblastic lineage, suggesting the potential use of MF-exposed ADSCs-MNPs for osteoblastic conversion in cell therapies targeting systemic osteoporosis. Further, in vivo studies are required to assess the feasibility of this potential therapeutic approach for osteoporosis treatment. The possible cytotoxicity of MNPs was explored using a human foreskin fibroblast (HFF2) cell line, and the MNPs were synthesized through an environmentally friendly and one-pot method. Various concentrations of both uncoated and amino acid-coated MNPs were evaluated over a 72-hour duration. Importantly, no cytotoxic effects were observed, indicating the biocompatibility of the MNPs with the cells [225]. Furthermore, the investigation was expanded to include MNPs coated with a triblock copolymer (PEO-polyurethane-PEO) of various chain lengths, assessing their impact on diverse cell types such as HUVECs, human retinal pigment epithelial cells, C4-2 metastatic human prostate adenocarcinoma cells, and PC3 prostate cancer cells. The results revealed that short-chain copolymers exhibited greater cytotoxicity compared to their long-chain counterparts, and the ascribed cause of toxicity was linked to the polymer coating rather than the magnetite core. This observation is advantageous as it maintains a higher magnetic component for larger magnetic moments [226]. Additional studies involving dextran- and PVA-coated iron oxide MNPs on brain-derived endothelial cells and microglial cells did not show any signs of cytotoxicity [227]. Likewise, Ferumoxtran-10, an ultrasmall dextran-coated superparamagnetic iron oxide particle, demonstrated no toxic effects on human monocyte-macrophage interactions [228]. The manipulation of the shape and size of nanoparticles, coated with different concentrations of PVA and subjected to stirring, revealed that augmenting the r-ratio could improve the biocompatibility of nanoparticles, resulting in reduced cellular toxicity effects [229]. For enhanced surface functionalization and modification of surface charge properties, MNPs underwent coating with dextran, aminodextran, heparin, and dimercaptosuccinic acid (DMSA). The interaction of these coated nanoparticles with HeLa cells demonstrated that cationic dextran-coated MNPs were effectively internalized by the cells without adversely affecting cell viability, making them promising candidates for in vivo biomedical applications such as cell tracking via MRI. Conversely, anionic DMSA-coated nanoparticles exhibited less efficient cellular uptake, while neutral-charged nanoparticles were not taken up at all [230]. Studies involving DMSA-coated MNPs on NCTC 1469 nonparenchymal hepatocytes revealed no noteworthy impacts on cell viability, cell cycle, oxidative stress, or apoptosis. This underscores the importance of the particular cell line and its response to the coating [231]. Both investigations featuring DMSA-coated MNPs highlight that apparently similar coatings can result in varied outcomes, contingent on the specific cell line being examined.

3.2.3 In vivo Toxicity Studies of MNPs

The engagements of MNPs within biological systems are intricate and continually developing [232, 233]. The size of MNPs becomes pivotal upon their introduction into the body. MNPs with a dimension smaller than 10 nm are eliminated through renal extravasation, while larger particles surpassing 200 nm are ensnared by the spleen [234]. Therefore, MNPs ranging in size from approximately 10–100 nm are deemed most suitable for internal use. Collectively, factors such as size, shape, surface charge, and stability play pivotal roles in determining the overall behavior of MNPs within the organism. To engineer nanoparticles with targeted capabilities for specific cells and tissues, it is imperative to grasp their pharmacokinetic aspects, including absorption, distribution, metabolism, and excretion in the organism's system. A thorough understanding of the metabolism and clearance criteria of MNPs is crucial for assessing potential toxicity. Upon entry into the body, MNPs can be absorbed through interactions with cells and various biological components [235, 236]. Following this, the MNPs are distributed among various organs and may undergo additional metabolic processes [237]. A critical focus is placed on scrutinizing how MNPs interact with and disrupt physiological iron metabolism as they degrade within the body. The excessive accumulation of intracellular iron can lead to the impairment of essential cellular components, including proteins and nucleic acids. This constitutes a vital biosafety phenomenon that necessitates thorough investigation [238].

To comprehensively assess the various aspects of MNPs, it is crucial to start from the fundamental level of the ecological pyramid, acknowledging the substantial role played by small organisms in ecological balance and food chains. These model organisms can serve as valuable indicators for testing specific criteria in research. In a study that investigated the impact of Fe_2O_3 magnetic nanoparticles and arsenate (As (V)), synergistic effects were noted in Ceriodaphnia dubia (water flea). The bioaccumulation of MNPs rose with concentration, reaching its peak at 20 mg/L within six hours, but subsequently declined due to the settling and aggregation of nanoparticles. Intriguingly, nano-Fe_2O_3 alone exhibited no toxicity or mortality across all tested concentrations. However, when combined with As (V), the toxicity was intensified [239]. Similarly, during a toxicity assessment comparing two types of MNPs (MNP and MNP integrated with zeolite) in mussels (*Mytilus galloprovincialis*) over 1, 3, and 7 days, both variants of MNPs induced changes in animal physiology at specific concentrations of iron oxide NPs. Mussels exposed to these particles showed indications of oxidative stress, and the addition of zeolite to iron oxides led to additional cellular damage and increased oxidative effects by the seventh day [240]. Another investigation concentrated on the size impacts of hematite nanoparticles (NPs) in *E. coli* cells. The adsorption process of hematite NPs, with sizes of 76 and 96 nm, achieved equilibrium at a quicker rate (30–40 minutes) in comparison to smaller NPs (60–90 minutes). The descending order of size impact rate in mg Fe/L was 98 nm > 76 nm > 53 nm > 26 nm. This research lays the foundation for modeling the thermodynamics and kinetics of the adsorption of various NPs on microorganisms [241]. A study investigated the influence of iron oxide MNPs on the symbiotic relationship between the fungus Arbuscular mycorrhizal and plants. Fe_3O_4 MNPs were determined to be toxic to fungi at higher concentrations, leading to adverse effects on the mutual interaction between fungi and plants. This toxicity had implications for soil carbon accumulation, phosphorous cycling, crop yield, and soil fertility [242]. Concerning MNP coatings for stability and compatibility, a study employed iron oxide nanoparticles (NPs) of identical size (5–6 nm) with various coatings (ascorbate, citrate, dextran, and polyvinyl-pyrrolidone) on water flea (Daphnia magna). Each coating had distinct effects, with polyvinyl-pyrrolidone (PVP) exhibiting the lowest acute toxicity, highest colloidal stability, and a low rate of accumulation and adsorption. Toxicity may arise from reduced colloidal stability, ion release from the core material, and the generation of ROS. The study suggested that negative charge carrier coatings of ascorbate and citrate NPs are of lesser significance for in vivo testing compared to in vitro assessments [243]. Similarly, in the assessment of the toxicity of DMSA-coated and unfunctionalized MNPs (fresh and aged uncoated iron oxide NPs) on three aquatic organisms (green algae, duckweed, and water fleas), it was noted that green algae exhibited the highest sensitivity. Daphnia demonstrated minor effects with uncoated NPs over a wide concentration range [244]. In a research project employing Drosophila (fruit fly) as a model organism, the mechanistic role of orally administered nanoparticles (NPs) was examined. As these NPs entered the gut, they successfully traversed the peritrophic membrane, triggering apoptosis. This gut-related toxicity led to developmental delays, reduced pupa counts, impaired fly hatching, and weight loss. After adult flies emerged from larvae exposed to nanoparticles (NPs), subsequent scrutiny revealed anomalies in both phenotype and genotype. These anomalies encompassed irregularities in sensory organs, other body parts, and indications of impaired larval crawling or adult climbing. The observed alterations in phenotypic and behavioral evaluations following NP exposure were linked to disruptions in signaling pathways, including Notch, Wnt, and the estimated glomerular filtration rate (EGFR) [245]. The findings of this study emphasize the necessity for additional research encompassing cell signaling, proteomics, and genomics to elucidate the underlying mechanisms of NP-induced toxicity.

To evaluate the impact of prenatal exposure to iron oxide MNPs concerning dose dependency and surface charge, male and female Crl:CD1(ICR) (CD1) mice were administered intraperitoneal injections of positively charged polyethyleneimine-Fe_2O_3-NPs (PEI-NPs) and negatively charged poly(acrylic acid)-Fe_2O_3-NPs (PAA-NPs) at low and high doses of 10 and 100 mg/kg, respectively. The examinations were conducted on days 8, 9, and 10 of gestation. The outcomes showed that a low MNP dose, regardless of charge, did not produce any harmful effects. However, exposure to high doses led to charge-dependent fetal loss and morphological alterations in the uteri (for both charges) and testes (for positive charge only) of the surviving offspring [246]. In a separate in vivo study employing a chick embryo chorioallantoic membrane model (CAM), embryos were administered injections of Fe_3O_4/salicylic acid nanoparticles featuring sizes spanning from 60.3 nm to 79.9 nm. Following a 24-hour period, the histological scrutiny of the heart, liver, kidney, and lungs indicated that the tissue deposits of MNPs were well-tolerated and exhibited biocompatibility with the embryos and chickens. Furthermore, intravenous administration of Fe_3O_4/salicylic acid NPs of sizes between 50 and 100 nm demonstrated minimal risk of embolism and exhibited prolonged intravascular persistence, suggesting their potential as efficient magnetically targeted agents [247]. When compared to rodent and chicken models, aquatic organisms such as fish emerge as a promising in vivo model for exploring the potential toxicity of MNPs. Fish present several advantages, including convenient MNP administration through waterborne exposure, generation of a large biological sample size from a single parent, thereby reducing inter-species variation, and a significant reduction in the necessity to sacrifice higher vertebrates, aligning with the 3R principle (replacement, reduction, and refinement). An investigation that compared the toxicity of iron oxide nanoparticles with iron salts in blackfish (Capoeta fusca) revealed that $Fe(NO_3)_3$ exhibited the highest acute toxicity, followed by $FeCl_3$, Fe_3O_4, and $FeSO_4$ NPs. Exposure to Fe_3O_4 NPs and iron salts induced histopathological abnormalities in the gills and intestine, including the presence of aneurysms, hyperplasia, edema, fusion of lamellae, lamellar synechiae, and necrosis in the gills. Comparatively, Fe_3O_4 NPs exhibited higher tissue uptake than iron salts, with the liver demonstrating the highest bioconcentration factors for $FeCl_3$, Fe_3O_4 NPs, and $FeSO_4$, while the gills showed the highest bioconcentration factors for $Fe(NO_3)_3$ [248]. Recently, a study by Malhotra et al. [249] showcased alterations in behavior and biochemical parameters in zebrafish exposed to uncoated Fe_3O_4 MNPs (15 nm). Zebrafish exposed to low concentrations (1 ppm) did not display notable abnormalities, whereas higher concentrations (10 ppm) led to changes in behavior, neurotransmitter levels, and stress hormones in the brain. Intriguingly, the modification of the surface of Fe_3O_4 MNPs using inert materials such as carbon was identified to mitigate their in vivo toxicity [250]. In an alternate investigation, MNPs featuring distinct cores composed of ferrite and manganese ferrite underwent examination for their toxicity in both in vitro and in vivo settings. Both categories of MNPs, specifically pegylated and varying in sizes from 3 nm to 14–20 nm, displayed diminished survival rates in zebrafish embryos when associated with the manganese-based MNPs. Conversely, the iron oxide MNPs did not lead to mortality or abnormal hatching. Subsequently, the PEGylated iron oxide MNP, characterized by a cubic shape and a size of 20 nm, was selected for further assessment in male Balb/c mice through tail vein injection. The findings of this study indicated the absence of toxicity indications and showcased excellent MRI contrast [251]. The in vitro biocompatibility and toxicity of different coated single-core and multicore MNPs were evaluated using human hepatocellular carcinoma (Hep G2) and human colorectal adenocarcinoma (Caco-2) cells. Additionally, the in vivo assessment was conducted using a Xenopus laevis embryo model [252]. In vitro, the MNPs showed no cytotoxicity, but variations in internalization and accumulation were noted depending on the particle size. Xenopus laevis embryos were identified as a rapid, cost-effective, and high-throughput alternative for toxicity assessment before proceeding to investigations involving nanotherapeutics and rodent models. Another investigation delved into the in vitro toxicity and biodegradation of $ZnxFe_3O_4$ on mammalian cell lines and extended its examination to in vivo studies using Xenopus laevis embryos [253]. The Zn-Fe_3O_4 MNPs did not induce toxicity in mammalian cell lines. However, in Xenopus laevis embryos, short-term exposure resulted in abnormal phenotypes and elevated expression of metal response proteins and metal transporters. Notably, gene expression returned to baseline levels following prolonged exposure, suggesting a reduction in nanoparticle absorption due to metabolic, distribution, and excretion processes. In an interesting investigation, the toxicity of magnetic iron oxide nanoparticles (MIONs) synthesized with carob leaf extract was assessed in specific brain areas of Wistar rats. The MIONs were found to cause neuronal degeneration in the hippocampus and striatum areas of the brain, and they disrupted iron homeostasis in the striatum and midbrain [254]. To enhance prostate cancer MRI, biocompatibility assessments were conducted on MNPs coated with prostate-specific membrane antigen (PSMA) and linked with the J591 antibody. The in vitro investigation demonstrated increased cellular iron absorption without compromising the viability of prostate cancer cells. In in vivo examinations using tumor-bearing NOD/SCID mice, specific binding of PSMA-J591-coated MNPs was observed, leading to enhanced MRI contrast for the localization and detection of tumors [255]. In an independent investigation, the intravenous application of sub-lethal amounts of sodium oleate-coated Fe_3O_4 (OC-Fe_3O_4) nanoparticles in female Wistar rats did not result in noteworthy alterations in body weight. Nonetheless, elevated doses revealed slight liver tissue damage, and exposure to the MIONs caused neuronal degeneration in the hippocampus and striatum regions of the brain [256]. Moreover, an in vivo examination assessed the toxicity of $CuFe_2O_4$ nanoparticles in zebrafish larvae, demonstrating dose and time-dependent impacts such as delayed hatching, mechanical membrane damage, and significant apoptosis in different body regions. The research underscored that acute toxicity induced by ROS escalated with elevated nanoparticle concentrations, resulting in the suppression of catalase (CAT) activity and heightened oxidative stress [257].

To sum up, acquiring a thorough comprehension of the potential toxicity associated with MNPs is crucial, especially given their potential applications across diverse fields, including medicine. The toxicity of MNPs is contingent on various factors like size, shape, surface modification, concentration, dosage, biodistribution, bioavailability, solubility, immunogenicity, and pharmacokinetics. The significant surface-to-volume ratio of MNPs heightens the probability of interaction with biological elements when inhaled, absorbed, or ingested, thereby magnifying their potential for toxicity.

The biodistribution and metabolism of MNPs are crucial aspects to assess their potential toxicity. The size of MNPs plays a crucial role in determining their behavior within the body, with the preferred size range for administration falling approximately between 10 and 100 nm. Research on biodistribution has indicated that MNPs tend to accumulate predominantly in the spleen, liver, and bone marrow. Modifications to the surface of MNPs impact how they interact, distribute, undergo internalization, metabolize, and potentially induce toxicity. Investigations into toxicity using cell lines and model organisms, including zebrafish, drosophila, and mussels, have yielded valuable insights into the safety and potential hazards of MNPs in vivo.

Examinations into toxicity should cover a range of factors, encompassing acute and chronic toxicity mechanisms, degradation of particle products, triggering defensive and inflammatory responses, metabolic effects, long-term toxicity, and elimination, particularly when using cellular and animal models. This section focused on analyzing the toxicity parameters of iron oxide MNPs by reviewing research conducted in cell lines, aquatic organisms, and rodent animal models. The knowledge obtained from these investigations offers valuable insights and can guide future research in toxicity analysis, especially in the use of animal models, to ensure the safe and beneficial application of MNPs in human contexts.

In summary, the assessment of MNP toxicity using various in vivo and in vitro models has shed light on the factors that influence their biocompatibility. These studies are crucial for developing safe and effective MNP-based applications in the fields of medicine and beyond. Nevertheless, additional research and inquiry are necessary to comprehensively grasp the fundamental mechanisms of MNP-induced toxicity and guarantee their secure utilization in diverse biomedical applications. By considering these aspects and conducting comprehensive toxicity evaluations, MNPs can be harnessed for beneficial applications while minimizing potential adverse effects in living systems.

3.3 Conclusion and Future Perspectives

The comprehensive exploration of iron oxide-based MNPs in biomedical applications presented in this chapter underscores the multifaceted potential and evolving landscape of this burgeoning field. The diverse applications of MNPs, spanning imaging, drug delivery, hyperthermia therapy, gene therapy, and biosensing, reveal a promising trajectory for advancing diagnostics and therapeutics in medicine. As we conclude this chapter, we consolidate key insights and outline future perspectives, emphasizing the significance of ongoing research and collaborative efforts.

3.3.1 Imaging Applications

MNPs, especially SPIONs, show significant potential as contrast agents in MRI. Their adjustable properties and heightened contrast effects offer advantages over conventional paramagnetic materials. Subsequent research should emphasize the improvement of surface modifications for MNPs and the exploration of MEIO nanoparticles to enhance MRI contrast. Progress in MNPs could pave the way for innovative imaging techniques with heightened sensitivity and specificity.

3.3.2 Drug Delivery Systems

MNPs in drug delivery systems, forming theranostic platforms, demonstrate remarkable progress. Their ability to combine therapy and imaging enhances treatment precision and efficacy, particularly in cancer treatment. Future research avenues include exploring advanced carrier agents, optimizing drug release kinetics, and further investigating active targeting strategies. Harnessing MNPs' potential in drug delivery systems may revolutionize personalized medicine.

3.3.3 Hyperthermia Therapy

MNPs play a pivotal role in hyperthermia therapy by efficiently inducing controlled heating in cancer cells. Surface-functionalized MNPs exhibit notable biocompatibility and heat generation capabilities. Research should prioritize optimizing MNPs for hyperthermia applications, considering concentration, size, and surface modifications. Advancements in MF parameters may enhance the precision and effectiveness of hyperthermia therapy.

3.3.4 Gene Therapy

MNPs offer a versatile platform for targeted gene delivery, overcoming challenges in gene therapy. Surface modifications and magnetofection techniques enhance their role in efficient transfection and gene expression. Continued efforts should focus on refining magnetic vectors, exploring new surface modifications, and investigating combined therapies. MNPs hold the potential to reshape gene therapy strategies for improved therapeutic outcomes.

3.3.5 Biosensing and Detection

MNPs significantly advance cancer biosensing and detection of pathogens by providing high sensitivity and specificity. Their integration into various sensing platforms offers diverse applications in diagnostic methods. Ongoing research should explore innovative strategies for MNPs in biosensing, expanding applications beyond cancer to diseases like Parkinson's and schizophrenia. Advancements may lead to rapid, cost-effective point-of-care devices.

3.3.6 Toxicity and Biodistribution

Comprehensive assessment of MNPs' toxicity is vital for safe applications. Factors such as size, surface modifications, and biodistribution influence their interaction with biological systems. Future studies should delve deeper into understanding the long-term effects, immunogenicity, and metabolism of MNPs. Research efforts must continue to refine toxicity assessments, ensuring the safe utilization of MNPs in diverse biomedical applications.

In conclusion, the journey through the biomedical applications of MNPs reveals a dynamic and promising field with transformative potential. Future endeavors should be marked by interdisciplinary collaboration, addressing current challenges, and exploring novel avenues for innovation. The continued exploration of MNPs will undoubtedly contribute to the evolution of diagnostics and therapeutics, with the ultimate goal of enhancing patient outcomes and advancing the frontiers of medical science.

References

1 Yu Z, Gao L, Chen K, Zhang W, Zhang Q, Li Q and Hu K 2021 Nanoparticles: a new approach to upgrade cancer diagnosis and treatment *Nanoscale Res. Lett.* 16 88.

2 Akbarzadeh A, Samiei M and Davaran S 2012 Magnetic nanoparticles: preparation, physical properties, and applications in biomedicine *Nanoscale Res. Lett.* 7 1–13.

3 Pistone A, Iannazzo D, Panseri S, Montesi M, Tampieri A and Galvagno S 2014 Hydroxyapatite-magnetite-MWCNT nanocomposite as a biocompatible multifunctional drug delivery system for bone tissue engineering *Nanotechnology* 25 425701.

4 Mahmoudi M, Sant S, Wang B, Laurent S and Sen T 2011 Superparamagnetic iron oxide nanoparticles (SPIONs): development, surface modification and applications in chemotherapy *Adv. Drug Deliv. Rev.* 63 24–46.

5 Natarajan S, Harini K, Gajula G P, Sarmento B, Neves-Petersen M T and Thiagarajan V 2019 Multifunctional magnetic iron oxide nanoparticles: diverse synthetic approaches, surface modifications, cytotoxicity towards biomedical and industrial applications *BMC Mater.* 1 1–22.

6 Radoń A, Hawelek L, Lukowiec D, Kubacki J and Wlodarczyk P 2019 Dielectric and electromagnetic interference shielding properties of high entropy (Zn, Fe, Ni, Mg, Cd) Fe_2O_4 ferrite *Sci. Rep.* 9 1–13.

7 Roca A G, Gutiérrez L, Gavilán H, Brollo M E F, Veintemillas-Verdaguer S and del Puerto Morales M 2019 Design strategies for shape-controlled magnetic iron oxide nanoparticles *Adv. Drug Deliv. Rev.* 138 68–104.

8 Farazi R, Vaezi M R, Molaei M J, Saeidifar M and Behnam-Ghader A A 2018 Effect of pH and temperature on doxorubicin hydrochloride release from magnetite/graphene oxide nanocomposites *Mater. Today Proc.* 5 15726–32.

9 Keshavarz H, Khavandi A, Alamolhoda S and Naimi-Jamal M R 2020 pH-sensitive magnetite mesoporous silica nanocomposites for controlled drug delivery and hyperthermia *RSC Adv.* 10 39008–16.

10 Kievit F M and Zhang M 2011 Surface engineering of iron oxide nanoparticles for targeted cancer therapy *Acc. Chem. Res.* 44 853–62.

11 Prasad C, Tang H and Liu W 2018 Magnetic Fe_3O_4 based layered double hydroxides (LDHs) nanocomposites (Fe_3O_4/LDHs): recent review of progress in synthesis, properties and applications *J. Nanostructure Chem.* 8 393–412.

12 Stepanov A, Burilov V, Pinus M, Mustafina A, Rümmeli M H, Mendez R G, Amirov R, Lukashenko S, Zvereva E and Katsuba S 2014 Water transverse relaxation rates in aqueous dispersions of superparamagnetic iron oxide nanoclusters with diverse hydrophilic coating *Colloids Surf. Physicochem. Eng. Asp.* 443 450–8.

13 Daou T J, Grenèche J M, Pourroy G, Buathong S, Derory A, Ulhaq-Bouillet C, Donnio B, Guillon D and Begin-Colin S 2008 Coupling agent effect on magnetic properties of functionalized magnetite-based nanoparticles *Chem. Mater.* 20 5869–75.

14 Vlaardingerbroek M T and Boer J A 2013 *Magnetic Resonance Imaging: Theory and Practice* (Berlin, Heidelberg: Springer Science & Business Media).

15 Grover V P B, Tognarelli J M, Crossey M M E, Cox I J, Taylor-Robinson S D and McPhail M J W 2015 Magnetic resonance imaging: principles and techniques: lessons for clinicians *J. Clin. Exp. Hepatol.* 5 246–55 https://doi.org/10.1016/j.jceh.2015.08.001.

16 Rümenapp C, Gleich B and Haase A 2012 Magnetic nanoparticles in magnetic resonance imaging and diagnostics *Pharm. Res.* 29 1165–79 https://doi.org/10.1007/s11095-012-0711-y.

17 Mittal A, Roy I and Gandhi S 2022 Magnetic nanoparticles: an overview for biomedical applications *Magnetochemistry* 8 107.

18 Terreno E, Castelli D D, Viale A and Aime S 2010 Challenges for molecular magnetic resonance imaging *Chem. Rev.* 110 3019–42 https://doi.org/10.1021/cr100025t.

19 Forghani R 2016 Adverse effects of gadolinium-based contrast agents: changes in practice patterns *Top. Magn. Reson. Imaging* 25 163 https://doi.org/10.1097/RMR.0000000000000095.

20 Sun C, Lee J S H and Zhang M 2008 Magnetic nanoparticles in MR imaging and drug delivery *Adv. Drug Deliv. Rev.* 60 1252–65 https://doi.org/10.1016/j.addr.2008.03.018.

21 Hee Kim E, Sook Lee H, Kook Kwak B and Kim B-K 2005 Synthesis of ferrofluid with magnetic nanoparticles by sonochemical method for MRI contrast agent *J. Magn. Magn. Mater.* 289 328–30 https://doi.org/10.1016/j.jmmm.2004.11.093.

22 Chee H L, Gan C R R, Ng M, Low L, Fernig D G, Bhakoo K K and Paramelle D 2018 Biocompatible peptide-coated ultrasmall superparamagnetic iron oxide nanoparticles for in vivo contrast-enhanced magnetic resonance imaging *ACS Nano* 12 6480–91 https://doi.org/10.1021/acsnano.7b07572.

23 Harisinghani M G, Barentsz J, Hahn P F, Deserno W M, Tabatabaei S, van de Kaa C H, de la Rosette J and Weissleder R 2003 Noninvasive detection of clinically occult lymph-node metastases in prostate cancer *N. Engl. J. Med.* 348 2491–9 https://doi.org/10.1056/NEJMoa022749.

24 Varallyay P, Nesbit G, Muldoon L L, Nixon R R, Delashaw J, Cohen J I, Petrillo A, Rink D and Neuwelt E A 2002 Comparison of two superparamagnetic viral-sized iron oxide particles ferumoxides and ferumoxtran-10 with a gadolinium chelate in imaging intracranial tumors *Am. J. Neuroradiol.* 23 510–9.

25 Nahrendorf M, Jaffer F A, Kelly K A, Sosnovik D E, Aikawa E, Libby P and Weissleder R 2006 Noninvasive vascular cell adhesion molecule-1 imaging identifies inflammatory activation of cells in atherosclerosis *Circulation* 114 1504–11 https://doi.org/10.1161/CIRCULATIONAHA.106.646380.

26 Xie J, Chen K, Huang J, Lee S, Wang J, Gao J, Li X and Chen X 2010 PET/NIRF/MRI triple functional iron oxide nanoparticles *Biomaterials* 31 3016–22 https://doi.org/10.1016/j.biomaterials.2010.01.010.

27 Hayashi K, Sato Y, Sakamoto W and Yogo T 2017 Theranostic nanoparticles for MRI-guided thermochemotherapy: "tight" clustering of magnetic nanoparticles boosts relaxivity and heat-generation power *ACS Biomater. Sci. Eng.* 3 95–105 https://doi.org/10.1021/acsbiomaterials.6b00536.

28 Gao Z, He T, Zhang P, Li X, Zhang Y, Lin J, Hao J, Huang P and Cui J 2020 Polypeptide-based theranostics with tumor-microenvironment-activatable cascade reaction for chemo-ferroptosis combination therapy *ACS Appl. Mater. Interfaces* 12 20271–80 https://doi.org/10.1021/acsami.0c03748.

29 Jun Y, Huh Y-M, Choi J, Lee J-H, Song H-T, Kim S, Kim S, Yoon S, Kim K-S and Shin J-S 2005 Nanoscale size effect of magnetic nanocrystals and their utilization for cancer diagnosis via magnetic resonance imaging *J. Am. Chem. Soc.* 127 5732–3.

30 Joos A, Löwa N, Wiekhorst F, Gleich B and Haase A 2017 Size-dependent MR relaxivities of magnetic nanoparticles *J. Magn. Magn. Mater.* 427 122–6 https://doi.org/10.1016/j.jmmm.2016.11.021.

31 Weissleder R, Lee A S, Fischman A J, Reimer P, Shen T, Wilkinson R, Callahan R J and Brady T J 1991 Polyclonal human immunoglobulin G labeled with polymeric iron oxide: antibody MR imaging *Radiology* 181 245–9.

32 Tanimoto A and Kuribayashi S 2006 Application of superparamagnetic iron oxide to imaging of hepatocellular carcinoma *Eur. J. Radiol.* 58 200–16.

33 Lee J-H, Huh Y-M, Jun Y, Seo J, Jang J, Song H-T, Kim S, Cho E-J, Yoon H-G and Suh J-S 2007 Artificially engineered magnetic nanoparticles for ultra-sensitive molecular imaging *Nat. Med.* 13 95–9.

34 Yang L, Sun C, Lin H, Gong X, Zhou T, Deng W-T, Chen Z and Gao J 2019 Sensitive contrast-enhanced magnetic resonance imaging of orthotopic and metastatic hepatic tumors by ultralow doses of zinc ferrite octapods *Chem. Mater.* 31 1381–90 https://doi.org/10.1021/acs.chemmater.8b04760.

35 Liu X L, Wang Y T, Ng C T, Wang R, Jing G Y, Yi J B, Yang J, Bay B H, Yung L-Y L, Fan D D, Ding J and Fan H M 2014 Coating engineering of $MnFe_2O_4$ nanoparticles with superhigh T2 relaxivity and efficient cellular uptake for highly sensitive magnetic resonance imaging *Adv. Mater. Interfaces* 1 1300069 https://doi.org/10.1002/admi.201300069.

36 Gao Z, Hou Y, Zeng J, Chen L, Liu C, Yang W and Gao M 2017 Tumor microenvironment-triggered aggregation of antiphagocytosis 99mTc-labeled Fe_3O_4 nanoprobes for enhanced tumor imaging in vivo *Adv. Mater.* 29 1701095 https://doi.org/10.1002/adma.201701095.

37 Ma D, Chen J, Luo Y, Wang H and Shi X 2017 Zwitterion coated ultrasmall iron oxide nanoparticles for enhanced T1-weighted magnetic resonance imaging applications *J. Mater. Chem. B* 5 7267–73 https://doi.org/10.1039/C7TB01588G.

38 Sundaresan V, Menon J U, Rahimi M, Nguyen K T and Wadajkar A S 2014 Dual-responsive polymer-coated iron oxide nanoparticles for drug delivery and imaging applications *Int. J. Pharm.* 466 1–7 https://doi.org/10.1016/j.ijpharm.2014.03.016.

39 Li J, Shi X and Shen M 2014 Hydrothermal synthesis and functionalization of iron oxide nanoparticles for MR imaging applications *Part. Part. Syst. Charact.* 31 1223–37 https://doi.org/10.1002/ppsc.201400087.

40 Peng X-H, Qian X, Mao H and Wang A Y 2008 Targeted magnetic iron oxide nanoparticles for tumor imaging and therapy *Int. J. Nanomedicine* 3 311.

41 Ma D, Shi M, Li X, Zhang J, Fan Y, Sun K, Jiang T, Peng C and Shi X 2020 Redox-sensitive clustered ultrasmall iron oxide nanoparticles for switchable T2/T1-weighted magnetic resonance imaging applications *Bioconjug. Chem.* 31 352–9 https://doi.org/10.1021/acs.bioconjchem.9b00659.

42 Jeon M, Halbert M V, Stephen Z R and Zhang M 2021 Iron oxide nanoparticles as T1 contrast agents for magnetic resonance imaging: fundamentals, challenges, applications, and prospectives *Adv. Mater.* 33 1906539 https://doi.org/10.1002/adma.201906539.

43 Qiao Y, Wan J, Zhou L, Ma W, Yang Y, Luo W, Yu Z and Wang H 2019 Stimuli-responsive nanotherapeutics for precision drug delivery and cancer therapy *Wiley Interdiscip. Rev. Nanomed. Nanobiotechnol.* 11 e1527.

44 Dadfar S M, Roemhild K, Drude N I, von Stillfried S, Knüchel R, Kiessling F and Lammers T 2019 Iron oxide nanoparticles: diagnostic, therapeutic and theranostic applications *Adv. Drug Deliv. Rev.* 138 302–25.

45 Xie L, Jin W, Chen H and Zhang Q 2019 Superparamagnetic iron oxide nanoparticles for cancer diagnosis and therapy *J. Biomed. Nanotechnol.* 15 215–416.

46 Reimer P, Schuierer G, Balzer T and Peters P E 1995 Application of a superparamagnetic iron oxide (Resovis®) for MR imaging of human cerebral blood volume *Magn. Reson. Med.* 34 694–7.

47 Taboada E, Rodríguez E, Roig A, Oró J, Roch A and Muller R N 2007 Relaxometric and magnetic characterization of ultrasmall iron oxide nanoparticles with high magnetization. Evaluation as potential T1 magnetic resonance imaging contrast agents for molecular imaging *Langmuir* 23 4583–8 https://doi.org/10.1021/la063415s.

48 Blanco-Andujar C, Walter A, Cotin G, Bordeianu C, Mertz D, Felder-Flesch D and Begin-Colin S 2016 Design of iron oxide-based nanoparticles for MRI and magnetic hyperthermia *Nanomedicine* 11 1889–910 https://doi.org/10.2217/nnm-2016-5001.

49 Lucena G N, dos Santos C C, Pinto G C, Amantéa B E, Piazza R D, Jafelicci Júnior M and Marques R F C 2020 Surface engineering of magnetic nanoparticles for hyperthermia and drug delivery *Med. Devices Sens.* 3 e10100 https://doi.org/10.1002/mds3.10100.

50 Salmani M M, Hashemian M, Yekta H J, Nejad M G, Saber-Samandari S and Khandan A 2020 Synergic effects of magnetic nanoparticles on hyperthermia-based therapy and controlled drug delivery for bone substitute application *J. Supercond. Nov. Magn.* 33 2809–20 https://doi.org/10.1007/s10948-020-05530-1.

51 Pérez-Herrero E and Fernández-Medarde A 2015 Advanced targeted therapies in cancer: drug nanocarriers, the future of chemotherapy *Eur. J. Pharm. Biopharm.* 93 52–79 https://doi.org/10.1016/j.ejpb.2015.03.018.

52 Estanqueiro M, Amaral M H, Conceição J and Sousa Lobo J M 2015 Nanotechnological carriers for cancer chemotherapy: the state of the art *Colloids Surf. B Biointerfaces* 126 631–48 https://doi.org/10.1016/j.colsurfb.2014.12.041.

53 Cole A J, Yang V C and David A E 2011 Cancer theranostics: the rise of targeted magnetic nanoparticles *Trends Biotechnol.* 29 323–32 https://doi.org/10.1016/j.tibtech.2011.03.001.

54 Singh A and Sahoo S K 2014 Magnetic nanoparticles: a novel platform for cancer theranostics *Drug Discov. Today* 19 474–81 https://doi.org/10.1016/j.drudis.2013.10.005.

55 Kumar C S S R and Mohammad F 2011 Magnetic nanomaterials for hyperthermia-based therapy and controlled drug delivery *Adv. Drug Deliv. Rev.* 63 789–808 https://doi.org/10.1016/j.addr.2011.03.008.

56 Sumitha N S, Prakash P, Nair B N and Sailaja G S 2021 Degradation-dependent controlled delivery of doxorubicin by glyoxal cross-linked magnetic and porous chitosan microspheres *ACS Omega* 6 21472–84 https://doi.org/10.1021/acsomega.1c02303.

57 Szegedi Á, Trendafilova I, Mihály J, Lázár K, Németh P, Momekov G, Momekova D, Marinov L, Nikolova I and Popova M 2020 New insight on prednisolone polymorphs in mesoporous silica/maghemite nanocomposites *J. Drug Deliv. Sci. Technol.* 60 102092 https://doi.org/10.1016/j.jddst.2020.102092.

58 Sabouri Z, Labbaf S, Karimzadeh F, Baharlou-Houreh A, McFarlane T V and Esfahani M N 2021 Fe_3O_4/bioactive glass nanostructure: a promising therapeutic platform for osteosarcoma treatment *Biomed. Mater.* 16 035016.

59 Peralta M E, Jadhav S A, Magnacca G, Scalarone D, Mártire D O, Parolo M E and Carlos L 2019 Synthesis and in vitro testing of thermoresponsive polymer-grafted core-shell magnetic mesoporous silica nanoparticles for efficient controlled and targeted drug delivery *J. Colloid Interface Sci.* 544 198–205 https://doi.org/10.1016/j.jcis.2019.02.086.

60 El-Boubbou K, Ali R, Al-Zahrani H, Trivilegio T, Alanazi A H, Khan A L, Boudjelal M and AlKushi A 2019 Preparation of iron oxide mesoporous magnetic microparticles as novel multidrug carriers for synergistic anticancer therapy and deep tumor penetration *Sci. Rep.* 9 1–20 https://doi.org/10.1038/s41598-019-46007-z.

61 El-Boubbou K, Ali R, Al-Humaid S, Alhallaj A, Lemine O M, Boudjelal M and AlKushi A 2021 Iron oxide mesoporous magnetic nanostructures with high surface area for enhanced and selective drug delivery to metastatic cancer cells *Pharmaceutics* 13 553 https://doi.org/10.3390/pharmaceutics13040553.

62 Zou Y, Li D, Wang Y, Ouyang Z, Peng Y, Tomás H, Xia J, Rodrigues J, Shen M and Shi X 2020 Polyethylenimine nanogels incorporated with ultrasmall iron oxide nanoparticles and doxorubicin for MR imaging-guided chemotherapy of tumors *Bioconjug. Chem.* 31 907–15 https://doi.org/10.1021/acs.bioconjchem.0c00036.

63 Attari E, Nosrati H, Danafar H and Kheiri Manjili H 2019 Methotrexate anticancer drug delivery to breast cancer cell lines by iron oxide magnetic based nanocarrier *J. Biomed. Mater. Res. A* 107 2492–500 https://doi.org/10.1002/jbm.a.36755.

64 Huang Y, Mao K, Zhang B and Zhao Y 2017 Superparamagnetic iron oxide nanoparticles conjugated with folic acid for dual target-specific drug delivery and MRI in cancer theranostics *Mater. Sci. Eng. C* 70 763–71 https://doi.org/10.1016/j.msec.2016.09.052.

65 Quan Q, Xie J, Gao H, Yang M, Zhang F, Liu G, Lin X, Wang A, Eden H S, Lee S, Zhang G and Chen X 2011 HSA coated iron oxide nanoparticles as drug delivery vehicles for cancer therapy *Mol. Pharm.* 8 1669–76 https://doi.org/10.1021/mp200006f.

66 Marcu A, Pop S, Dumitrache F, Mocanu M, Niculite C M, Gherghiceanu M, Lungu C P, Fleaca C, Ianchis R, Barbut A, Grigoriu C and Morjan I 2013 Magnetic iron oxide nanoparticles as drug delivery system in breast cancer *Appl. Surf. Sci.* 281 60–5 https://doi.org/10.1016/j.apsusc.2013.02.072.

67 Estelrich J, Escribano E, Queralt J and Busquets M A 2015 Iron oxide nanoparticles for magnetically-guided and magnetically-responsive drug delivery *Int. J. Mol. Sci.* 16 8070–101 https://doi.org/10.3390/ijms16048070.

68 Kayal S and Ramanujan R V 2010 Doxorubicin loaded PVA coated iron oxide nanoparticles for targeted drug delivery *Mater. Sci. Eng. C* 30 484–90 https://doi.org/10.1016/j.msec.2010.01.006.

69 Laurent S, Saei A A, Behzadi S, Panahifar A and Mahmoudi M 2014 Superparamagnetic iron oxide nanoparticles for delivery of therapeutic agents: opportunities and challenges *Expert Opin. Drug Deliv.* 11 1449–70 https://doi.org/10.1517/17425247.2014.924501.

70 Gaihre B, Khil M S, Lee D R and Kim H Y 2009 Gelatin-coated magnetic iron oxide nanoparticles as carrier system: drug loading and in vitro drug release study *Int. J. Pharm.* 365 180–9 https://doi.org/10.1016/j.ijpharm.2008.08.020.

71 Jia G, Han Y, An Y, Ding Y, He C, Wang X and Tang Q 2018 NRP-1 targeted and cargo-loaded exosomes facilitate simultaneous imaging and therapy of glioma in vitro and in vivo *Biomaterials* 178 302–16 https://doi.org/10.1016/j.biomaterials.2018.06.029.

72 Park J, Park S, Kim S, Lee I-H, Saw P E, Lee K, Kim Y-C, Kim Y-J, Farokhzad O C, Jeong Y Y and Jon S 2013 HER2-specific aptide conjugated magneto-nanoclusters for potential breast cancer imaging and therapy *J. Mater. Chem. B* 1 4576–83 https://doi.org/10.1039/C3TB20613K.

73 Choi G-E, Kang M-S, Kim Y-J, Yoon J J and Jeong Y-I 2019 Magnetically responsive drug delivery using doxorubicin and iron oxide nanoparticle-incorporated lipocomplexes *J. Nanosci. Nanotechnol.* 19 675–9 https://doi.org/10.1166/jnn.2019.15910.

74 Ayyanaar S, Balachandran C, Bhaskar R C, Kesavan M P, Aoki S, Raja R P, Rajesh J, Webster T J and Rajagopal G 2020 ROS-responsive chitosan coated magnetic iron oxide nanoparticles as potential vehicles for targeted drug delivery in cancer therapy *Int. J. Nanomedicine* 15 3333–46 https://doi.org/10.2147/IJN.S249240.

75 Ganivada M N, Rao N V, Dinda H, Kumar P, Das Sarma J and Shunmugam R 2014 Biodegradable magnetic nanocarrier for stimuli responsive drug release *Macromolecules* 47 2703–11 https://doi.org/10.1021/ma500384m.

76 Fratila R M and Fuente J M D L 2018 *Nanomaterials for Magnetic and Optical Hyperthermia Applications* (Amsterdam: Elsevier).

77 Pérez-Hernández M 2019 Chapter 8 - Mechanisms of Cell Death Induced by Optical Hyperthermia *Nanomaterials for Magnetic and Optical Hyperthermia Applications* ed R M Fratila and J M De La Fuente (Amsterdam: Elsevier) pp 201–28 https://doi.org/10.1016/B978-0-12-813928-8.00008-9.

78 Dhavalikar R, Bohórquez A C and Rinaldi C 2019 Chapter 10 - Image-Guided Thermal Therapy Using Magnetic Particle Imaging and Magnetic Fluid Hyperthermia *Nanomaterials for Magnetic and Optical Hyperthermia Applications* ed R M Fratila and J M De La Fuente (Amsterdam: Elsevier) pp 265–86 https://doi.org/10.1016/B978-0-12-813928-8.00010-7.

79 Gel'vich E A and Mazokhin V N 2001 Technical aspects of electromagnetic hyperthermia in medicine *Crit. Rev. Biomed. Eng.* 29 77–97 https://doi.org/10.1615/CritRevBiomedEng.v29.i1.30.

80 Kok H P and Crezee J 2017 Progress and Future Directions in Hyperthermia Treatment Planning *2017 First IEEE MTT- Int. Microw. Bio Conf. IMBIOC* pp 1–4 https://doi.org/10.1109/IMBIOC.2017.7965793.

81 Szasz O and Szasz A 2016 Heating, efficacy and dose of local hyperthermia *Open J. Biophys.* 6 10–8.

82 Obaidat I M, Narayanaswamy V, Alaabed S, Sambasivam S and Muralee Gopi C V 2019 Principles of magnetic hyperthermia: a focus on using multifunctional hybrid magnetic nanoparticles *Magnetochemistry* 5 67.

83 Beik J, Abed Z, Ghoreishi F S, Hosseini-Nami S, Mehrzadi S, Shakeri-Zadeh A and Kamrava S K 2016 Nanotechnology in hyperthermia cancer therapy: from fundamental principles to advanced applications *J. Controlled Release* 235 205–21.

84 Obaidat I M, Issa B and Haik Y 2015 Magnetic properties of magnetic nanoparticles for efficient hyperthermia *Nanomaterials* 5 63–89.

85 Carrey J, Mehdaoui B and Respaud M 2011 Simple models for dynamic hysteresis loop calculations of magnetic single-domain nanoparticles: application to magnetic hyperthermia optimization *J. Appl. Phys.* 109 083921 https://doi.org/10.1063/1.3551582.

86 Mehdaoui B, Meffre A, Carrey J, Lachaize S, Lacroix L-M, Gougeon M, Chaudret B and Respaud M 2011 Optimal size of nanoparticles for magnetic hyperthermia: a combined theoretical and experimental study *Adv. Funct. Mater.* 21 4573–81 https://doi.org/10.1002/adfm.201101243.

87 Shaterabadi Z, Nabiyouni G and Soleymani M 2018 Physics responsible for heating efficiency and self-controlled temperature rise of magnetic nanoparticles in magnetic hyperthermia therapy *Prog. Biophys. Mol. Biol.* 133 9–19 https://doi.org/10.1016/j.pbiomolbio.2017.10.001.

88 Hedayatnasab Z, Abnisa F and Daud W M A W 2017 Review on magnetic nanoparticles for magnetic nanofluid hyperthermia application *Mater. Des.* 123 174–96.

89 Salunkhe A B, Khot V M and Pawar S H 2014 Magnetic hyperthermia with magnetic nanoparticles: a status review *Curr. Top. Med. Chem.* 14 572–94.

90 Bañobre-López M, Teijeiro A and Rivas J 2013 Magnetic nanoparticle-based hyperthermia for cancer treatment *Rep. Pract. Oncol. Radiother.* 18 397–400 https://doi.org/10.1016/j.rpor.2013.09.011.

91 Colombo M, Carregal-Romero S, Casula M F, Gutiérrez L, Morales M P, Bohm I B, Heverhagen J T, Prosperi D and Parak W J 2012 Biological applications of magnetic nanoparticles *Chem. Soc. Rev.* 41 4306–34 https://doi.org/10.1039/C2CS15337H.

92 Rachakatla R S, Balivada S, Seo G-M, Myers C B, Wang H, Samarakoon T N, Dani R, Pyle M, Kroh F O, Walker B, Leaym X, Koper O B, Chikan V, Bossmann S H, Tamura M and Troyer D L 2010 Attenuation of mouse melanoma by A/C magnetic field after delivery of bi-magnetic nanoparticles by neural progenitor cells *ACS Nano* 4 7093–104 https://doi.org/10.1021/nn100870z.

93 Ruiz A, Gutiérrez L, Cáceres-Vélez P R, Santos D, Chaves S B, Fascineli M L, Garcia M P, Azevedo R B and Morales M P 2015 Biotransformation of magnetic nanoparticles as a function of coating in a rat model *Nanoscale* 7 16321–9 https://doi.org/10.1039/C5NR03780H.

94 Samanta B, Yan H, Fischer N O, Shi J, Jerry D J and Rotello V M 2008 Protein-passivated Fe_3O_4 nanoparticles: low toxicity and rapid heating for thermal therapy *J. Mater. Chem.* 18 1204–8 https://doi.org/10.1039/B718745A.

95 Sheparovych R, Sahoo Y, Motornov M, Wang S, Luo H, Prasad P N, Sokolov I and Minko S 2006 Polyelectrolyte stabilized nanowires from Fe_3O_4 nanoparticles via magnetic field induced self-assembly *Chem. Mater.* 18 591–3 https://doi.org/10.1021/cm051274u.

96 Hayashi K, Ono K, Suzuki H, Sawada M, Moriya M, Sakamoto W and Yogo T 2010 One-pot biofunctionalization of magnetic nanoparticles via thiol–ene click reaction for magnetic hyperthermia and magnetic resonance imaging *Chem. Mater.* 22 3768–72 https://doi.org/10.1021/cm100810g.

97 Kossatz S, Grandke J, Couleaud P, Latorre A, Aires A, Crosbie-Staunton K, Ludwig R, Dähring H, Ettelt V, Lazaro-Carrillo A, Calero M, Sader M, Courty J, Volkov Y, Prina-Mello A, Villanueva A, Somoza Á, Cortajarena A L, Miranda R and Hilger I 2015 Efficient treatment of breast cancer xenografts with multifunctionalized iron oxide nanoparticles combining magnetic hyperthermia and anti-cancer drug delivery *Breast Cancer Res.* 17 66 https://doi.org/10.1186/s13058-015-0576-1.

98 Beola L, Asín L, Roma-Rodrigues C, Fernández-Afonso Y, Fratila R M, Serantes D, Ruta S, Chantrell R W, Fernandes A R, Baptista P V, de la Fuente J M, Grazú V and Gutiérrez L 2020 The intracellular number of magnetic nanoparticles modulates the apoptotic death pathway after magnetic hyperthermia treatment *ACS Appl. Mater. Interfaces* 12 43474–87 https://doi.org/10.1021/acsami.0c12900.

99 Adamiano A, Wu V M, Carella F, Lamura G, Canepa F, Tampieri A, Iafisco M and Uskoković V 2019 Magnetic calcium phosphates nanocomposites for the intracellular hyperthermia of cancers of bone and brain *Nanomedicine* 14 1267–89 https://doi.org/10.2217/nnm-2018-0372.

100 Soleymani M, Velashjerdi M, Shaterabadi Z and Barati A 2020 One-pot preparation of hyaluronic acid-coated iron oxide nanoparticles for magnetic hyperthermia therapy and targeting CD44-overexpressing cancer cells *Carbohydr. Polym.* 237 116130 https://doi.org/10.1016/j.carbpol.2020.116130.

101 Mondal S, Manivasagan P, Bharathiraja S, Santha Moorthy M, Nguyen V T, Kim H H, Nam S Y, Lee K D and Oh J 2017 Hydroxyapatite coated iron oxide nanoparticles: a promising nanomaterial for magnetic hyperthermia cancer treatment *Nanomaterials* 7 426 https://doi.org/10.3390/nano7120426.

102 Sapareto S A and Dewey W C 1984 Thermal dose determination in cancer therapy *Int. J. Radiat. Oncol.* 10 787–800 https://doi.org/10.1016/0360-3016(84)90379-1.

103 Giustini A J, Petryk A A, Cassim S M, Tate J A, Baker I and Hoopes P J 2010 Magnetic nanoparticle hyperthermia in cancer treatment *Nano LIFE* 01 17–32 https://doi.org/10.1142/S1793984410000067.

104 Chang D, Lim M, Goos J A C M, Qiao R, Ng Y Y, Mansfeld F M, Jackson M, Davis T P and Kavallaris M 2018 Biologically targeted magnetic hyperthermia: potential and limitations *Front. Pharmacol.* 9 831 https://www.frontiersin.org/articles/10.3389/fphar.2018.00831 (accessed June 8, 2023).

105 Périgo E A, Hemery G, Sandre O, Ortega D, Garaio E, Plazaola F and Teran F J 2015 Fundamentals and advances in magnetic hyperthermia *Appl. Phys. Rev.* 2 041302 https://doi.org/10.1063/1.4935688.

106 Huang S, Wang S-Y, Gupta A, Borca-Tasciuc D-A and Salon S J 2012 On the measurement technique for specific absorption rate of nanoparticles in an alternating electromagnetic field *Meas. Sci. Technol.* 23 035701 https://doi.org/10.1088/0957-0233/23/3/035701.

107 Wang S-Y, Huang S and Borca-Tasciuc D-A 2013 Potential sources of errors in measuring and evaluating the specific loss power of magnetic nanoparticles in an alternating magnetic field *IEEE Trans. Magn.* 49 255–62 https://doi.org/10.1109/TMAG.2012.2224648.

108 Natividad E, Castro M, Goglio G, Andreu I, Epherre R, Duguet E and Mediano A 2012 New insights into the heating mechanisms and self-regulating abilities of manganite perovskite nanoparticles suitable for magnetic fluid hyperthermia *Nanoscale* 4 3954–62 https://doi.org/10.1039/C2NR30667K.

109 Garaio E, Sandre O, Collantes J-M, Garcia J A, Mornet S and Plazaola F 2014 Specific absorption rate dependence on temperature in magnetic field hyperthermia measured by dynamic hysteresis losses (ac magnetometry) *Nanotechnology* 26 015704 https://doi.org/10.1088/0957-4484/26/1/015704.

110 Hergt R and Dutz S 2007 Magnetic particle hyperthermia—biophysical limitations of a visionary tumour therapy *J. Magn. Magn. Mater.* 311 187–92 https://doi.org/10.1016/j.jmmm.2006.10.1156.

111 Hergt R, Hiergeist R, Hilger I, Kaiser W A, Lapatnikov Y, Margel S and Richter U 2004 Maghemite nanoparticles with very high AC-losses for application in RF-magnetic hyperthermia *J. Magn. Magn. Mater.* 270 345–57 https://doi.org/10.1016/j.jmmm.2003.09.001.

112 Dutz S, Hergt R, Mürbe J, Müller R, Zeisberger M, Andrä W, Töpfer J and Bellemann M E 2007 Hysteresis losses of magnetic nanoparticle powders in the single domain size range *J. Magn. Magn. Mater.* 308 305–12 https://doi.org/10.1016/j.jmmm.2006.06.005.

113 Rosensweig R E 2002 Heating magnetic fluid with alternating magnetic field *J. Magn. Magn. Mater.* 252 370–4 https://doi.org/10.1016/S0304-8853(02)00706-0.

114 Iacovita C, Florea A, Dudric R, Pall E, Moldovan A I, Tetean R, Stiufiuc R and Lucaciu C M 2016 Small versus large iron oxide magnetic nanoparticles: hyperthermia and cell uptake properties *Molecules* 21 1357 https://doi.org/10.3390/molecules21101357.

115 Yallapu M M, Chauhan N, Othman S F, Khalilzad-Sharghi V, Ebeling M C, Khan S, Jaggi M and Chauhan S C 2015 Implications of protein corona on physico-chemical and biological properties of magnetic nanoparticles *Biomaterials* 46 1–12 https://doi.org/10.1016/j.biomaterials.2014.12.045.

116 Iacovita C, Florea A, Scorus L, Pall E, Dudric R, Moldovan A I, Stiufiuc R, Tetean R and Lucaciu C M 2019 Hyperthermia, cytotoxicity, and cellular uptake properties of manganese and zinc ferrite magnetic nanoparticles synthesized by a polyol-mediated process *Nanomaterials* 9 1489 https://doi.org/10.3390/nano9101489.

117 Yoo D, Lee J-H, Shin T-H and Cheon J 2011 Theranostic magnetic nanoparticles *Acc. Chem. Res.* 44 863–74 https://doi.org/10.1021/ar200085c.

118 Liu X L and Fan H M 2014 Innovative magnetic nanoparticle platform for magnetic resonance imaging and magnetic fluid hyperthermia applications *Curr. Opin. Chem. Eng.* 4 38–46 https://doi.org/10.1016/j.coche.2013.12.010.

119 Fatima H, Charinpanitkul T and Kim K-S 2021 Fundamentals to apply magnetic nanoparticles for hyperthermia therapy *Nanomaterials* 11 1203 https://doi.org/10.3390/nano11051203.

120 Kim D-H, Kim K-N, Kim K-M and Lee Y-K 2009 Targeting to carcinoma cells with chitosan- and starch-coated magnetic nanoparticles for magnetic hyperthermia *J. Biomed. Mater. Res. A* 88A 1–11 https://doi.org/10.1002/jbm.a.31775.

121 Ghosh R, Pradhan L, Devi Y P, Meena S S, Tewari R, Kumar A, Sharma S, Gajbhiye N S, Vatsa R K, Pandey B N and Ningthoujam R S 2011 Induction heating studies of Fe_3O_4 magnetic nanoparticles capped with oleic acid and polyethylene glycol for hyperthermia *J. Mater. Chem.* 21 13388–98 https://doi.org/10.1039/C1JM10092K.

122 Guardia P, Di Corato R, Lartigue L, Wilhelm C, Espinosa A, Garcia-Hernandez M, Gazeau F, Manna L and Pellegrino T 2012 Water-soluble iron oxide nanocubes with high values of specific absorption rate for cancer cell hyperthermia treatment *ACS Nano* 6 3080–91 https://doi.org/10.1021/nn2048137.

123 Lin T-C, Lin F-H and Lin J-C 2012 In vitro feasibility study of the use of a magnetic electrospun chitosan nanofiber composite for hyperthermia treatment of tumor cells *Acta Biomater.* 8 2704–11 https://doi.org/10.1016/j.actbio.2012.03.045.

124 Asín L, Ibarra M R, Tres A and Goya G F 2012 Controlled cell death by magnetic hyperthermia: effects of exposure time, field amplitude, and nanoparticle concentration *Pharm. Res.* 29 1319–27 https://doi.org/10.1007/s11095-012-0710-z.

125 Sonvico F, Mornet S, Vasseur S, Dubernet C, Jaillard D, Degrouard J, Hoebeke J, Duguet E, Colombo P and Couvreur P 2005 Folate-conjugated iron oxide nanoparticles for solid tumor targeting as potential specific magnetic hyperthermia mediators: synthesis, physicochemical characterization, and in vitro experiments *Bioconjug. Chem.* 16 1181–8 https://doi.org/10.1021/bc050050z.

126 Le Renard P-E, Lortz R, Senatore C, Rapin J-P, Buchegger F, Petri-Fink A, Hofmann H, Doelker E and Jordan O 2011 Magnetic and in vitro heating properties of implants formed in situ from injectable formulations and containing superparamagnetic iron oxide nanoparticles (SPIONs) embedded in silica microparticles for magnetically induced local hyperthermia *J. Magn. Magn. Mater.* 323 1054–63 https://doi.org/10.1016/j.jmmm.2010.12.003.

127 Bauer L M, Situ S F, Griswold M A and Samia A C S 2016 High-performance iron oxide nanoparticles for magnetic particle imaging – guided hyperthermia (hMPI) *Nanoscale* 8 12162–9 https://doi.org/10.1039/C6NR01877G.

128 Wang W, Li F, Li S, Hu Y, Xu M, Zhang Y, Khan M I, Wang S, Wu M, Ding W and Qiu B 2021 M2 macrophage-targeted iron oxide nanoparticles for magnetic resonance image-guided magnetic hyperthermia therapy *J. Mater. Sci. Technol.* 81 77–87 https://doi.org/10.1016/j.jmst.2020.11.058.

129 Mykhaylyk O, Antequera Y S, Vlaskou D and Plank C 2007 Generation of magnetic nonviral gene transfer agents and magnetofection in vitro *Nat. Protoc.* 2 2391–411 https://doi.org/10.1038/nprot.2007.352.

130 Luo T, Liang H, Jin R and Nie Y 2019 Virus-inspired and mimetic designs in non-viral gene delivery *J. Gene Med.* 21 e3090 https://doi.org/10.1002/jgm.3090.

131 Bi Q, Song X, Hu A, Luo T, Jin R, Ai H and Nie Y 2020 Magnetofection: magic magnetic nanoparticles for efficient gene delivery *Chin. Chem. Lett.* 31 3041–6 https://doi.org/10.1016/j.cclet.2020.07.030.

132 Wang X, Zhou L, Ma Y and Gu H 2009 Charged magnetic nanoparticles for enhancing gene transfection *IEEE Trans. Nanotechnol.* 8 142–7 https://doi.org/10.1109/TNANO.2009.2013946.

133 Ma Y, Wang X and Gu H 2012 Enhancement of the efficiency of PEI/liposome transfection by magnetofectins formed via electrostatic self-assembly *Chin. Sci. Bull.* 57 4005–11 https://doi.org/10.1007/s11434-012-5184-1.

134 Chen X, Chen Y, Xin H, Wan T and Ping Y 2020 Near-infrared optogenetic engineering of photothermal nano CRISPR for programmable genome editing *Proc. Natl. Acad. Sci.* 117 2395–405 https://doi.org/10.1073/pnas.1912220117.

135 Liang H, Chen X, Jin R, Ke B, Barz M, Ai H and Nie Y 2020 Integration of indocyanine green analogs as near-infrared fluorescent carrier for precise imaging-guided gene delivery *Small* 16 1906538 https://doi.org/10.1002/smll.201906538.

136 Plank C, Anton M, Rudolph C, Rosenecker J and Krötz F 2003 Enhancing and targeting nucleic acid delivery by magnetic force *Expert Opin. Biol. Ther.* 3 745–58 https://doi.org/10.1517/14712598.3.5.745.

137 Xie L, Jiang W, Nie Y, He Y, Jiang Q, Lan F, Wu Y and Gu Z 2013 Low aggregation magnetic polyethyleneimine complexes with different saturation magnetization for efficient gene transfection in vitro and in vivo *RSC Adv.* 3 23571–81 https://doi.org/10.1039/C3RA43588A.

138 Kim J, Mirando A C, Popel A S and Green J J 2017 Gene delivery nanoparticles to modulate angiogenesis *Adv. Drug Deliv. Rev.* 119 20–43 https://doi.org/10.1016/j.addr.2016.11.003.

139 Liu G, Xie J, Zhang F, Wang Z, Luo K, Zhu L, Quan Q, Niu G, Lee S, Ai H and Chen X 2011 N-Alkyl-PEI-functionalized iron oxide nanoclusters for efficient siRNA delivery *Small* 7 2742–9 https://doi.org/10.1002/smll.201100825.

140 Pandey A P and Sawant K K 2016 Polyethylenimine: a versatile, multifunctional non-viral vector for nucleic acid delivery *Mater. Sci. Eng. C* 68 904–18 https://doi.org/10.1016/j.msec.2016.07.066.

141 Veiseh O, Kievit F M, Mok H, Ayesh J, Clark C, Fang C, Leung M, Arami H, Park J O and Zhang M 2011 Cell transcytosing poly-arginine coated magnetic nanovector for safe and effective siRNA delivery *Biomaterials* 32 5717–25.

142 Arsianti M, Lim M, Marquis C P and Amal R 2010 Polyethylenimine based magnetic iron-oxide vector: the effect of vector component assembly on cellular entry mechanism, intracellular localization, and cellular viability *Biomacromolecules* 11 2521–31.

143 Vaughan H J, Green J J and Tzeng S Y 2020 Cancer-targeting nanoparticles for combinatorial nucleic acid delivery *Adv. Mater.* 32 1901081 https://doi.org/10.1002/adma.201901081.

144 Nagase K, Hasegawa M, Ayano E, Maitani Y and Kanazawa H 2019 Effect of polymer phase transition behavior on temperature-responsive polymer-modified liposomes for siRNA transfection *Int. J. Mol. Sci.* 20 430 https://doi.org/10.3390/ijms20020430.

145 Geppert M, Petters C, Thiel K and Dringen R 2012 The presence of serum alters the properties of iron oxide nanoparticles and lowers their accumulation by cultured brain astrocytes *J. Nanoparticle Res.* 15 1349 https://doi.org/10.1007/s11051-012-1349-8.

146 Arsianti M, Lim M, Marquis C P and Amal R 2010 Assembly of polyethylenimine-based magnetic iron oxide vectors: insights into gene delivery *Langmuir* 26 7314–26 https://doi.org/10.1021/la9041919.

147 Liu Y, Wu W, Wang Y, Han S, Yuan Y, Huang J, Shuai X and Peng Z 2021 Recent development of gene therapy for pancreatic cancer using non-viral nanovectors *Biomater. Sci.* 9 6673–90 https://doi.org/10.1039/D1BM00748C.

148 Xie L, Jiang Q, He Y, Nie Y, Yue D and Gu Z 2015 Insight into the efficient transfection activity of a designed low aggregated magnetic polyethyleneimine/DNA complex in serum-containing medium and the application in vivo *Biomater. Sci.* 3 446–56 https://doi.org/10.1039/C4BM00317A.

149 Sun S-L, Lo Y-L, Chen H-Y and Wang L-F 2012 Hybrid polyethylenimine and polyacrylic acid-bound iron oxide as a magnetoplex for gene delivery *Langmuir* 28 3542–52 https://doi.org/10.1021/la204529u.

150 Zhang T-Y, Wu J-H, Xu Q-H, Wang X-R, Lu J, Hu Y, Jo J, Yamamoto M, Ling D, Tabata Y and Gao J-Q 2017 Design of magnetic gene complexes as effective and serum resistant gene delivery systems for mesenchymal stem cells *Int. J. Pharm.* 520 1–13 https://doi.org/10.1016/j.ijpharm.2017.01.041.

151 Namgung R, Singha K, Yu M K, Jon S, Kim Y S, Ahn Y, Park I-K and Kim W J 2010 Hybrid superparamagnetic iron oxide nanoparticle-branched polyethylenimine magnetoplexes for gene transfection of vascular endothelial cells *Biomaterials* 31 4204–13 https://doi.org/10.1016/j.biomaterials.2010.01.123.

152 Plank C, Schillinger U, Scherer F, Bergemann C, Rémy J-S, Krötz F, Anton M, Lausier J and Rosenecker J 2003 The magnetofection method: using magnetic force to enhance gene delivery *Biol. Chem.* 384 737–47 https://doi.org/10.1515/BC.2003.082.

153 Furlani E P and Xue X 2012 A model for predicting field-directed particle transport in the magnetofection process *Pharm. Res.* 29 1366–79 https://doi.org/10.1007/s11095-012-0681-0.

154 Foroozandeh P and Aziz A A 2018 Insight into cellular uptake and intracellular trafficking of nanoparticles *Nanoscale Res. Lett.* 13 339 https://doi.org/10.1186/s11671-018-2728-6.

155 Gao H, Xiong Y, Zhang S, Yang Z, Cao S and Jiang X 2014 RGD and interleukin-13 peptide functionalized nanoparticles for enhanced glioblastoma cells and neovasculature dual targeting delivery and elevated tumor penetration *Mol. Pharm.* 11 1042–52 https://doi.org/10.1021/mp400751g.

156 Huth S, Lausier J, Gersting S W, Rudolph C, Plank C, Welsch U and Rosenecker J 2004 Insights into the mechanism of magnetofection using PEI-based magnetofectins for gene transfer *J. Gene Med.* 6 923–36 https://doi.org/10.1002/jgm.577.

157 Ota S, Takahashi Y, Tomitaka A, Yamada T, Kami D, Watanabe M and Takemura Y 2013 Transfection efficiency influenced by aggregation of DNA/polyethylenimine max/magnetic nanoparticle complexes *J. Nanoparticle Res.* 15 1653 https://doi.org/10.1007/s11051-013-1653-y.

158 Morishita N, Nakagami H, Morishita R, Takeda S, Mishima F, BungoTerazono S, Nishijima Y and Kaneda N T 2005 Magnetic nanoparticles with surface modification enhanced gene delivery of HVJ-E vector *Biochem. Biophys. Res. Commun.* 334 1121–6 https://doi.org/10.1016/j.bbrc.2005.06.204.

159 Chorny M, Fishbein I, Alferiev I and Levy R J 2009 Magnetically responsive biodegradable nanoparticles enhance adenoviral gene transfer in cultured smooth muscle and endothelial cells *Mol. Pharm.* 6 1380–7 https://doi.org/10.1021/mp900017m.

160 Mu X, Li J, Yan S, Zhang H, Zhang W, Zhang F and Jiang J 2018 siRNA delivery with stem cell membrane-coated magnetic nanoparticles for imaging-guided photothermal therapy and gene therapy *ACS Biomater. Sci. Eng.* 4 3895–905 https://doi.org/10.1021/acsbiomaterials.8b00858.

161 Rohiwal S S, Dvorakova N, Klima J, Vaskovicova M, Senigl F, Slouf M, Pavlova E, Stepanek P, Babuka D, Benes H, Ellederova Z and Stieger K 2020 Polyethylenimine based magnetic nanoparticles mediated non-viral CRISPR/Cas9 system for genome editing *Sci. Rep.* 10 4619 https://doi.org/10.1038/s41598-020-61465-6.

162 Jin L, Wang Q, Chen J, Wang Z, Xin H and Zhang D 2019 Efficient delivery of therapeutic siRNA by Fe_3O_4 magnetic nanoparticles into oral cancer cells *Pharmaceutics* 11 615 https://doi.org/10.3390/pharmaceutics11110615.

163 Gong M, Liu H, Sun N, Xie Y, Yan F and Cai L 2020 Polyethylenimine-dextran-coated magnetic nanoparticles loaded with miR-302b suppress osteosarcoma in vitro and in vivo *Nanomedicine* 15 711–23 https://doi.org/10.2217/nnm-2019-0218.

164 Hosu O, Selvolini G, Cristea C and Marrazza G 2018 Electrochemical immunosensors for disease detection and diagnosis *Curr. Med. Chem.* 25 4119–37 https://doi.org/10.2174/0929867324666170727104429.

165 Sapsford K E, Algar W R, Berti L, Gemmill K B, Casey B J, Oh E, Stewart M H and Medintz I L 2013 Functionalizing nanoparticles with biological molecules: developing chemistries that facilitate nanotechnology *Chem. Rev.* 113 1904–2074 https://doi.org/10.1021/cr300143v.

166 Kouassi G K and Irudayaraj J 2006 Magnetic and gold-coated magnetic nanoparticles as a DNA sensor *Anal. Chem.* 78 3234–41 https://doi.org/10.1021/ac051621j.

167 Chen Y-T, Kolhatkar A G, Zenasni O, Xu S and Lee T R 2017 Biosensing using magnetic particle detection techniques *Sensors* 17 2300 https://doi.org/10.3390/s17102300.

168 Wu L and Qu X 2015 Cancer biomarker detection: recent achievements and challenges *Chem. Soc. Rev.* 44 2963–97 https://doi.org/10.1039/C4CS00370E.

169 Hosu O, Florea A, Cristea C and Sandulescu R 2019 Chapter 6 – Functionalized Advanced Hybrid Materials for Biosensing Applications *Advanced Biosensors for Health Care Applications* ed Inamuddin, R Khan, A Mohammad and A M Asiri (Amsterdam: Elsevier) pp 171–207 https://doi.org/10.1016/B978-0-12-815743-5.00006-8.

170 Liu G, Li R-W, Chen Y, Wu Y, Yang X, Yi X and Liu Y 2015 Magnetic nanoparticle for biomedicine applications *J. Nanotechnol. Nanomed. Nanobiotechnol.* 2 1–7. doi: 10.24966/NTMB-2044/100003.

171 Otieno B A, Krause C E, Latus A, Chikkaveeraiah B V, Faria R C and Rusling J F 2014 On-line protein capture on magnetic beads for ultrasensitive microfluidic immunoassays of cancer biomarkers *Biosens. Bioelectron.* 53 268–74 https://doi.org/10.1016/j.bios.2013.09.054.

172 Sharafeldin M, Bishop G W, Bhakta S, El-Sawy A, Suib S L and Rusling J F 2017 Fe_3O_4 nanoparticles on graphene oxide sheets for isolation and ultrasensitive amperometric detection of cancer biomarker proteins *Biosens. Bioelectron.* 91 359–66 https://doi.org/10.1016/j.bios.2016.12.052.

173 Garranzo-Asensio M, Guzman-Aranguez A, Povés C, Fernández-Aceñero M J, Torrente-Rodríguez R M, Ruiz-Valdepeñas Montiel V, Domínguez G, Frutos L S, Rodríguez N, Villalba M, Pingarrón J M, Campuzano S and Barderas R 2016 Toward liquid biopsy: determination of the humoral immune response in cancer patients using HaloTag fusion protein-modified electrochemical bioplatforms *Anal. Chem.* 88 12339–45 https://doi.org/10.1021/acs.analchem.6b03526.

174 Uliana C V, Peverari C R, Afonso A S, Cominetti M R and Faria R C 2018 Fully disposable microfluidic electrochemical device for detection of estrogen receptor alpha breast cancer biomarker *Biosens. Bioelectron.* 99 156–62 https://doi.org/10.1016/j.bios.2017.07.043.

175 Wang D, Li T, Gan N, Zhang H, Long N, Hu F, Cao Y, Jiang Q and Jiang S 2015 Electrochemical coding for multiplexed immunoassays of biomarkers based on bio-based polymer-nanotags *Electrochimica Acta* 163 238–45 https://doi.org/10.1016/j.electacta.2015.02.145.

176 Wang D, Gan N, Zhang H, Li T, Qiao L, Cao Y, Su X and Jiang S 2015 Simultaneous electrochemical immunoassay using graphene–Au grafted recombinant apoferritin-encoded metallic labels as signal tags and dual-template magnetic molecular imprinted polymer as capture probes *Biosens. Bioelectron.* 65 78–82 https://doi.org/10.1016/j.bios.2014.09.085.

177 Gao J, Guo Z, Yu S, Su F, Ma H, Du B, Wei Q and Pang X 2015 A novel controlled release system-based homogeneous immunoassay protocol for SCCA using magnetic mesoporous Fe_3O_4 as a nanocontainer and aminated polystyrene microspheres as a molecular gate *Biosens. Bioelectron.* 66 141–5 https://doi.org/10.1016/j.bios.2014.10.078.

178 Eletxigerra U, Martinez-Perdiguero J, Merino S, Barderas R, Torrente-Rodríguez R M, Villalonga R, Pingarrón J M and Campuzano S 2015 Amperometric magnetoimmunosensor for ErbB2 breast cancer biomarker determination in human serum, cell lysates and intact breast cancer cells *Biosens. Bioelectron.* 70 34–41 https://doi.org/10.1016/j.bios.2015.03.017.

179 Wu D, Ma H, Zhang Y, Jia H, Yan T and Wei Q 2015 Corallite-like magnetic Fe_3O_4@MnO_2@Pt Nanocomposites as multiple signal amplifiers for the detection of carcinoembryonic antigen *ACS Appl. Mater. Interfaces* 7 18786–93 https://doi.org/10.1021/acsami.5b05443.

180 Dong H, Han T-T, Ren L-L and Ding S-N 2017 Novel sandwich-structured electrochemiluminescence immunosensing platform via CdTe quantum dots-embedded mesoporous silica nanospheres as enhanced signal labels and Fe_3O_4@SiO_2@PS nanocomposites as magnetic separable carriers *J. Electroanal. Chem.* 806 32–40 https://doi.org/10.1016/j.jelechem.2017.10.038.

181 Zhu W, Saddam Khan M, Cao W, Sun X, Ma H, Zhang Y and Wei Q 2018 $Ni(OH)_2$/NGQDs-based electrochemiluminescence immunosensor for prostate specific antigen detection by coupling resonance energy transfer with Fe_3O_4@MnO_2 composites *Biosens. Bioelectron.* 99 346–52 https://doi.org/10.1016/j.bios.2017.08.005.

182 Qiu F, Gan X, Jiang B, Yuan R and Xiang Y 2021 Electrode immobilization-free and sensitive electrochemical sensing of thrombin via magnetic nanoparticle-decorated DNA polymers *Sens. Actuators B Chem.* 331 129395 https://doi.org/10.1016/j.snb.2020.129395.

183 Upadhyay N, Tripathi M, Chaddha R K, Ramachandran R, Elavarasi A, Hariprasad G and Elangovan R 2023 Development of sensitive magnetic nanoparticle assisted rapid sandwich assay(s-MARSA) to monitor Parkinson's disease and Schizophrenia pharmacotherapy *Anal. Biochem.* 667 115082 https://doi.org/10.1016/j.ab.2023.115082.

184 Sayhi M, Ouerghi O, Belgacem K, Arbi M, Tepeli Y, Ghram A, Anik Ü, Österlund L, Laouini D and Diouani M F 2018 Electrochemical detection of influenza virus H_9N_2 based on both immunomagnetic extraction and gold catalysis using an immobilization-free screen printed carbon microelectrode *Biosens. Bioelectron.* 107 170–7 https://doi.org/10.1016/j.bios.2018.02.018.

185 Wu Z, Guo W-J, Bai Y-Y, Zhang L, Hu J, Pang D-W and Zhang Z-L 2018 Digital single virus electrochemical enzyme-linked immunoassay for ultrasensitive H_7N_9 avian influenza virus counting *Anal. Chem.* 90 1683–90 https://doi.org/10.1021/acs.analchem.7b03281.

186 Babamiri B, Hallaj R and Salimi A 2018 Ultrasensitive electrochemiluminescence immunosensor for determination of hepatitis B virus surface antigen using CdTe@CdS-PAMAM dendrimer as luminescent labels and Fe_3O_4 nanoparticles as magnetic beads *Sens. Actuators B Chem.* 254 551–60 https://doi.org/10.1016/j.snb.2017.07.016.

187 Wu Z, Hu J, Zeng T, Zhang Z-L, Chen J, Wong G, Qiu X, Liu W, Gao G F, Bi Y and Pang D-W 2017 Ultrasensitive ebola virus detection based on electroluminescent nanospheres and immunomagnetic separation *Anal. Chem.* 89 2039–48 https://doi.org/10.1021/acs.analchem.6b04632.

188 Takemura K, Lee J, Suzuki T, Hara T, Abe F and Park E Y 2019 Ultrasensitive detection of norovirus using a magnetofluoroimmunoassay based on synergic properties of gold/magnetic nanoparticle hybrid nanocomposites and quantum dots *Sens. Actuators B Chem.* 296 126672 https://doi.org/10.1016/j.snb.2019.126672.

189 Zhong J, Rösch E L, Viereck T, Schilling M and Ludwig F 2021 Toward rapid and sensitive detection of SARS-CoV-2 with functionalized magnetic nanoparticles *ACS Sens.* 6 976–84 https://doi.org/10.1021/acssensors.0c02160.

190 Fang Y, Wang Y, Zhu L, Liu H, Su X, Liu Y, Chen Z, Chen H and He N 2023 A novel cartridge for nucleic acid extraction, amplification and detection of infectious disease pathogens with the help of magnetic nanoparticles *Chin. Chem. Lett.* 34 108092 https://doi.org/10.1016/j.cclet.2022.108092.

191 Wang C, Wang C, Wang X, Wang K, Zhu Y, Rong Z, Wang W, Xiao R and Wang S 2019 Magnetic SERS strip for sensitive and simultaneous detection of respiratory viruses *ACS Appl. Mater. Interfaces* 11 19495–505 https://doi.org/10.1021/acsami.9b03920.

192 Barrios-Gumiel A, Sepúlveda-Crespo D, Jiménez J L, Gómez R, Muñoz-Fernández M Á and de la Mata F J 2019 Dendronized magnetic nanoparticles for HIV-1 capture and rapid diagnostic *Colloids Surf. B Biointerfaces* 181 360–8 https://doi.org/10.1016/j.colsurfb.2019.05.050.

193 Chen Q, Lin J, Gan C, Wang Y, Wang D, Xiong Y, Lai W, Li Y and Wang M 2015 A sensitive impedance biosensor based on immunomagnetic separation and urease catalysis for rapid detection of Listeria monocytogenes using an immobilization-free interdigitated array microelectrode *Biosens. Bioelectron.* 74 504–11 https://doi.org/10.1016/j.bios.2015.06.007.

194 Pratiwi F W, Rijiravanich P, Somasundrum M and Surareungchai W 2013 Electrochemical immunoassay for Salmonella Typhimurium based on magnetically collected Ag-enhanced DNA biobarcode labels *Analyst* 138 5011–8 https://doi.org/10.1039/C3AN00606A.

195 Tufa L T, Oh S, Tran V T, Kim J, Jeong K-J, Park T J, Kim H-J and Lee J 2018 Electrochemical immunosensor using nanotriplex of graphene quantum dots, Fe_3O_4, and Ag nanoparticles for tuberculosis *Electrochimica Acta* 290 369–77 https://doi.org/10.1016/j.electacta.2018.09.108.

196 Xu M, Wang R and Li Y 2016 Rapid detection of Escherichia coli O157:H7 and Salmonella Typhimurium in foods using an electrochemical immunosensor based on screen-printed interdigitated microelectrode and immunomagnetic separation *Talanta* 148 200–8 https://doi.org/10.1016/j.talanta.2015.10.082.

197 Wang Y and Alocilja E C 2015 Gold nanoparticle-labeled biosensor for rapid and sensitive detection of bacterial pathogens *J. Biol. Eng.* 9 16 https://doi.org/10.1186/s13036-015-0014-z.

198 Brandão D, Liébana S, Campoy S, Alegret S and Isabel Pividori M 2015 Immunomagnetic separation of Salmonella with tailored magnetic micro and nanocarriers. A comparative study *Talanta* 143 198–204 https://doi.org/10.1016/j.talanta.2015.05.035.

199 Freitas M, Viswanathan S, Nouws H P A, Oliveira M B P P and Delerue-Matos C 2014 Iron oxide/gold core/shell nanomagnetic probes and CdS biolabels for amplified electrochemical immunosensing of Salmonella typhimurium *Biosens. Bioelectron.* 51 195–200 https://doi.org/10.1016/j.bios.2013.07.048.

200 Kozitsina A, Svalova T, Malysheva N, Glazyrina Y, Matern A and Rusinov V 2017 Determination of Staphylococcus aureus B-1266 by an enzyme-free electrochemical immunosensor incorporating magnetite nanoparticles *Anal. Lett.* 50 924–35 https://doi.org/10.1080/00032719.2016.1204312.

201 Wilson D, Materón E, Ibáñez-Redín G, Faria R C, Correa D S and Oliveira O N 2019 Erratum to "Electrical detection of pathogenic bacteria in food samples using information visualization methods with a sensor based on magnetic nanoparticles functionalized with antimicrobial peptides" [Talanta 194 (2019) 611–618] *Talanta* 200 562 https://doi.org/10.1016/j.talanta.2019.03.085.

202 Chen C-T, Yu J-W and Ho Y-P 2019 Identification of bacteria in juice/lettuce using magnetic nanoparticles and selected reaction monitoring mass spectrometry *J. Food Drug Anal.* 27 575–84 https://doi.org/10.1016/j.jfda.2018.09.006.

203 Zou D, Jin L, Wu B, Hu L, Chen X, Huang G and Zhang J 2019 Rapid detection of Salmonella in milk by biofunctionalised magnetic nanoparticle cluster sensor based on nuclear magnetic resonance *Int. Dairy J.* 91 82–8 https://doi.org/10.1016/j.idairyj.2018.11.011.

204 Chattopadhyay S, Sabharwal P K, Jain S, Kaur A and Singh H 2019 Functionalized polymeric magnetic nanoparticle assisted SERS immunosensor for the sensitive detection of S. typhimurium *Anal. Chim. Acta* 1067 98–106 https://doi.org/10.1016/j.aca.2019.03.050.

205 Anik M I, Hossain M K, Hossain I, Mahfuz A M U B, Rahman M T and Ahmed I 2021 Recent progress of magnetic nanoparticles in biomedical applications: a review *Nano Sel.* 2 1146–86 https://doi.org/10.1002/nano.202000162.

206 Iliasov A R, Nizamov T R, Naumenko V A, Garanina A S, Vodopyanov S S, Nikitin A A, Pershina A G, Chernysheva A A, Kan Y, Mogilnikov P S, Metelkina O N, Schetinin I V, Savchenko A G, Majouga A G and Abakumov M A 2021 Non-magnetic shell coating of magnetic nanoparticles as key factor of toxicity for cancer cells in a low frequency alternating magnetic field *Colloids Surf. B Biointerfaces* 206 111931 https://doi.org/10.1016/j.colsurfb.2021.111931.

207 Malhotra N, Lee J-S, Liman R A D, Ruallo J M S, Villaflores O B, Ger T-R and Hsiao C-D 2020 Potential toxicity of iron oxide magnetic nanoparticles: a review *Molecules* 25 3159 https://doi.org/10.3390/molecules25143159.

208 Mou X, Ali Z, Li S and He N 2015 Applications of magnetic nanoparticles in targeted drug delivery system *J. Nanosci. Nanotechnol.* 15 54–62 https://doi.org/10.1166/jnn.2015.9585.

209 Shubayev V I, Pisanic T R and Jin S 2009 Magnetic nanoparticles for theragnostics *Adv. Drug Deliv. Rev.* 61 467–77 https://doi.org/10.1016/j.addr.2009.03.007.

210 Tietze R, Lyer S, Dürr S, Struffert T, Engelhorn T, Schwarz M, Eckert E, Göen T, Vasylyev S, Peukert W, Wiekhorst F, Trahms L, Dörfler A and Alexiou C 2013 Efficient drug-delivery using magnetic nanoparticles — biodistribution and therapeutic effects in tumour bearing rabbits *Nanomedicine Nanotechnol. Biol. Med.* 9 961–71 https://doi.org/10.1016/j.nano.2013.05.001.

211 Jain T K, Reddy M K, Morales M A, Leslie-Pelecky D L and Labhasetwar V 2008 Biodistribution, clearance, and biocompatibility of iron oxide magnetic nanoparticles in rats *Mol. Pharm.* 5 316–27 https://doi.org/10.1021/mp7001285.

212 Vallabani N V S and Singh S 2018 Recent advances and future prospects of iron oxide nanoparticles in biomedicine and diagnostics *3 Biotech* 8 279 https://doi.org/10.1007/s13205-018-1286-z.

213 Mohammed L, Gomaa H G, Ragab D and Zhu J 2017 Magnetic nanoparticles for environmental and biomedical applications: a review *Particuology* 30 1–14 https://doi.org/10.1016/j.partic.2016.06.001.

214 Kwon J-T, Hwang S-K, Jin H, Kim D-S, Minai-Tehrani A, Yoon H-J, Choi M, Yoon T-J, Han D-Y, Kang Y-W, Yoon B-I, Lee J-K and Cho M-H 2008 Body distribution of inhaled fluorescent magnetic nanoparticles in the mice *J. Occup. Health* 50 1–6 https://doi.org/10.1539/joh.50.1.

215 Cole A J, David A E, Wang J, Galbán C J and Yang V C 2011 Magnetic brain tumor targeting and biodistribution of long-circulating PEG-modified, cross-linked starch-coated iron oxide nanoparticles *Biomaterials* 32 6291–301 https://doi.org/10.1016/j.biomaterials.2011.05.024.

216 Wang X, Zhang J, Yang X, Tang Z, Hu Y, Chen B and Tang J 2014 In vivo assessment of hepatotoxicity, nephrotoxicity and biodistribution using 3-aminopropyltriethoxysilane-coated magnetic nanoparticles (APTS-MNPs) in ICR mice *Chin. Sci. Bull.* 59 1800–8 https://doi.org/10.1007/s11434-014-0296-4.

217 Kumamoto Y, Camporez J P G, Jurczak M J, Shanabrough M, Horvath T, Shulman G I and Iwasaki A 2016 CD301b+ mononuclear phagocytes maintain positive energy balance through secretion of resistin-like molecule alpha *Immunity* 45 583–96 https://doi.org/10.1016/j.immuni.2016.08.002.

218 Ran Q, Xiang Y, Liu Y, Xiang L, Li F, Deng X, Xiao Y, Chen L, Chen L and Li Z 2015 Eryptosis indices as a novel predictive parameter for biocompatibility of Fe_3O_4 magnetic nanoparticles on erythrocytes *Sci. Rep.* 5 16209 https://doi.org/10.1038/srep16209.

219 Correia Carreira S, Walker L, Paul K and Saunders M 2015 The toxicity, transport and uptake of nanoparticles in the in vitro BeWo b30 placental cell barrier model used within NanoTEST *Nanotoxicology* 9 66–78 https://doi.org/10.3109/17435390.2013.833317.

220 Ma W, Gehret P M, Hoff R E, Kelly L P and Suh W H 2019 The investigation into the toxic potential of iron oxide nanoparticles utilizing rat pheochromocytoma and human neural stem cells *Nanomaterials* 9 453 https://doi.org/10.3390/nano9030453.

221 Lin Y-R, Kuo C-J, Lin H Y-H, Wu C-J and Liang S-S 2014 A proteomics analysis to evaluate cytotoxicity in NRK-52E cells caused by unmodified nano-Fe_3O_4 *Sci. World J.* 2014 e754721 https://doi.org/10.1155/2014/754721.

222 Alam S, Ahmad R, Pranaw K, Mishra P and Khare S K 2018 Asparaginase conjugated magnetic nanoparticles used for reducing acrylamide formation in food model system *Bioresour. Technol.* 269 121–6 https://doi.org/10.1016/j.biortech.2018.08.095.

223 Wilhelm C, Fortin J-P and Gazeau F 2007 Tumour cell toxicity of intracellular hyperthermia mediated by magnetic nanoparticles *J. Nanosci. Nanotechnol.* 7 2933–7 https://doi.org/10.1166/jnn.2007.668.

224 Labusca L, Herea D D, Danceanu C-M, Minuti A E, Stavilla C, Grigoras M, Gherca D, Stoian G, Ababei G, Chiriac H and Lupu N 2020 The effect of magnetic field exposure on differentiation of magnetite nanoparticle-loaded adipose-derived stem cells *Mater. Sci. Eng. C* 109 110652 https://doi.org/10.1016/j.msec.2020.110652.

225 Nosrati H, Salehiabar M, Attari E, Davaran S, Danafar H and Manjili H K 2018 Green and one-pot surface coating of iron oxide magnetic nanoparticles with natural amino acids and biocompatibility investigation *Appl. Organomet. Chem.* 32 e4069 https://doi.org/10.1002/aoc.4069.

226 Häfeli U O, Riffle J S, Harris-Shekhawat L, Carmichael-Baranauskas A, Mark F, Dailey J P and Bardenstein D 2009 Cell uptake and in vitro toxicity of magnetic nanoparticles suitable for drug delivery *Mol. Pharm.* 6 1417–28 https://doi.org/10.1021/mp900083m.

227 Cengelli F, Maysinger D, Tschudi-Monnet F, Montet X, Corot C, Petri-Fink A, Hofmann H and Juillerat-Jeanneret L 2006 Interaction of functionalized superparamagnetic iron oxide nanoparticles with brain structures *J. Pharmacol. Exp. Ther.* 318 108–16 https://doi.org/10.1124/jpet.106.101915.

228 Müller K, Skepper J N, Posfai M, Trivedi R, Howarth S, Corot C, Lancelot E, Thompson P W, Brown A P and Gillard J H 2007 Effect of ultrasmall superparamagnetic iron oxide nanoparticles (Ferumoxtran-10) on human monocyte-macrophages in vitro *Biomaterials* 28 1629–42 https://doi.org/10.1016/j.biomaterials.2006.12.003.

229 Mahmoudi M, Simchi A, Milani A S and Stroeve P 2009 Cell toxicity of superparamagnetic iron oxide nanoparticles *J. Colloid Interface Sci.* 336 510–8 https://doi.org/10.1016/j.jcis.2009.04.046.

230 Villanueva A, Cañete M, Roca A G, Calero M, Veintemillas-Verdaguer S, Serna C J, Morales M D P and Miranda R 2009 The influence of surface functionalization on the enhanced internalization of magnetic nanoparticles in cancer cells *Nanotechnology* 20 115103 https://doi.org/10.1088/0957-4484/20/11/115103.

231 Mejías R, Gutiérrez L, Salas G, Pérez-Yagüe S, Zotes T M, Lázaro F J, Morales M P and Barber D F 2013 Long term biotransformation and toxicity of dimercaptosuccinic acid-coated magnetic nanoparticles support their use in biomedical applications *J. Controlled Release* 171 225–33 https://doi.org/10.1016/j.jconrel.2013.07.019.

232 Shen C-C, Wang C-C, Liao M-H and Jan T-R 2011 A single exposure to iron oxide nanoparticles attenuates antigen-specific antibody production and T-cell reactivity in ovalbumin-sensitized BALB/c mice *Int. J. Nanomedicine* 6 1229–35 https://doi.org/10.2147/IJN.S21019.

233 Schlachter E K, Widmer H R, Bregy A, Lönnfors-Weitzel T, Vajtai I, Corazza N, Bernau V J, Weitzel T, Mordasini P, Slotboom J, Herrmann G, Bogni S, Hofmann H, Frenz M and Reinert M 2011 Metabolic pathway and distribution of superparamagnetic iron oxide nanoparticles: in vivo study *Int. J. Nanomedicine* 6 1793–800 https://doi.org/10.2147/IJN.S23638.

234 Gupta A K and Gupta M 2005 Synthesis and surface engineering of iron oxide nanoparticles for biomedical applications *Biomaterials* 26 3995–4021 https://doi.org/10.1016/j.biomaterials.2004.10.012.

235 Zavisova V, Koneracka M, Gabelova A, Svitkova B, Ursinyova M, Kubovcikova M, Antal I, Khmara I, Jurikova A, Molcan M, Ognjanović M, Antić B and Kopcansky P 2019 Effect of magnetic nanoparticles coating on cell proliferation and uptake *J. Magn. Magn. Mater.* 472 66–73 https://doi.org/10.1016/j.jmmm.2018.09.116.

236 Quini C C, Próspero A G, Calabresi M F F, Moretto G M, Zufelato N, Krishnan S, Pina D R, Oliveira R B, Baffa O, Bakuzis A F and Miranda J R A 2017 Real-time liver uptake and biodistribution of magnetic nanoparticles determined by AC biosusceptometry *Nanomedicine Nanotechnol. Biol. Med.* 13 1519–29 https://doi.org/10.1016/j.nano.2017.02.005.

237 Poller W C, Pieber M, Boehm-Sturm P, Ramberger E, Karampelas V, Möller K, Schleicher M, Wiekhorst F, Löwa N, Wagner S, Schnorr J, Taupitz M, Stangl K, Stangl V and Ludwig A 2018 Very small superparamagnetic iron oxide nanoparticles: long-term fate and metabolic processing in atherosclerotic mice *Nanomedicine Nanotechnol. Biol. Med.* 14 2575–86 https://doi.org/10.1016/j.nano.2018.07.013.

238 Wei H, Hu Y, Wang J, Gao X, Qian X and Tang M 2021 Superparamagnetic iron oxide nanoparticles: cytotoxicity, metabolism, and cellular behavior in biomedicine applications *Int. J. Nanomedicine* 16 6097–113 https://doi.org/10.2147/IJN.S321984.

239 Hu J, Wang D, Wang J and Wang J 2012 Bioaccumulation of Fe_2O_3(magnetic) nanoparticles in Ceriodaphnia dubia *Environ. Pollut.* 162 216–22 https://doi.org/10.1016/j.envpol.2011.11.016.

240 Taze C, Panetas I, Kalogiannis S, Feidantsis K, Gallios G P, Kastrinaki G, Konstandopoulos A G, Václavíková M, Ivanicova L and Kaloyianni M 2016 Toxicity assessment and comparison between two types of iron oxide nanoparticles in Mytilus galloprovincialis *Aquat. Toxicol.* 172 9–20 https://doi.org/10.1016/j.aquatox.2015.12.013.

241 Zhang W, Rittmann B and Chen Y 2011 Size effects on adsorption of hematite nanoparticles on E. coli cells *Environ. Sci. Technol.* 45 2172–8 https://doi.org/10.1021/es103376y.

242 Cao J, Feng Y, Lin X, Wang J and Xie X 2017 Iron oxide magnetic nanoparticles deteriorate the mutual interaction between arbuscular mycorrhizal fungi and plant *J. Soils Sediments* 17 841–51 https://doi.org/10.1007/s11368-016-1561-8.

243 Baumann J, Köser J, Arndt D and Filser J 2014 The coating makes the difference: acute effects of iron oxide nanoparticles on Daphnia magna *Sci. Total Environ.* 484 176–84 https://doi.org/10.1016/j.scitotenv.2014.03.023.

244 Zhang Y-Q, Dringen R, Petters C, Rastedt W, Köser J, Filser J and Stolte S 2016 Toxicity of dimercaptosuccinate-coated and un-functionalized magnetic iron oxide nanoparticles towards aquatic organisms *Environ. Sci. Nano* 3 754–67 https://doi.org/10.1039/C5EN00222B.

245 Pappus S A and Mishra M 2018 A Drosophila Model to Decipher the Toxicity of Nanoparticles Taken Through Oral Routes *Cellular and Molecular Toxicology of Nanoparticles* ed Q Saquib, M Faisal, A A Al-Khedhairy and A A Alatar (Cham: Springer International Publishing) pp 311–22 https://doi.org/10.1007/978-3-319-72041-8_18.

246 Di Bona K R, Xu Y, Gray M, Fair D, Hayles H, Milad L, Montes A, Sherwood J, Bao Y and Rasco J F 2015 Short- and long-term effects of prenatal exposure to iron oxide nanoparticles: influence of surface charge and dose on developmental and reproductive toxicity *Int. J. Mol. Sci.* 16 30251–68 https://doi.org/10.3390/ijms161226231.

247 Buteică S A, Mihăiescu D E, Rogoveanu I, Mărgăritescu D N and Mîndrilă I 2016 Chick chorioallantoic membrane model as a preclinical tool for nanoparticles biology study *Rom. Biotechnol. Lett.* 21 11688–94.

248 Sayadi M H, Mansouri B, Shahri E, Tyler C R, Shekari H and Kharkan J 2020 Exposure effects of iron oxide nanoparticles and iron salts in blackfish (Capoeta fusca): acute toxicity, bioaccumulation, depuration, and tissue histopathology *Chemosphere* 247 125900 https://doi.org/10.1016/j.chemosphere.2020.125900.

249 Malhotra N, Audira G, Chen J-R, Siregar P, Hsu H-S, Lee J-S, Ger T-R and Hsiao C-D 2020 Surface modification of magnetic nanoparticles by carbon-coating can increase its biosafety: evidences from biochemical and neurobehavioral tests in zebrafish *Molecules* 25 2256 https://doi.org/10.3390/molecules25092256.

250 Malhotra N, Chen J-R, Sarasamma S, Audira G, Siregar P, Liang S-T, Lai Y-H, Lin G-M, Ger T-R and Hsiao C-D 2019 Ecotoxicity assessment of Fe_3O_4 magnetic nanoparticle exposure in adult zebrafish at an environmental pertinent concentration by behavioral and biochemical testing *Nanomaterials* 9 873 https://doi.org/10.3390/nano9060873.

251 Caro C, Egea-Benavente D, Polvillo R, Royo J L, Pernia Leal M and García-Martín M L 2019 Comprehensive toxicity assessment of PEGylated magnetic nanoparticles for in vivo applications *Colloids Surf. B Biointerfaces* 177 253–9 https://doi.org/10.1016/j.colsurfb.2019.01.051.

252 Marín-Barba M, Gavilán H, Gutiérrez L, Lozano-Velasco E, Rodríguez-Ramiro I, Wheeler G N, Morris C J, Morales M P and Ruiz A 2018 Unravelling the mechanisms that determine the uptake and metabolism of magnetic single and multicore nanoparticles in a Xenopus laevis model *Nanoscale* 10 690–704 https://doi.org/10.1039/C7NR06020C.

253 Rivero M, Marín-Barba M, Gutiérrez L, Lozano-Velasco E, Wheeler G N, Sánchez-Marcos J, Muñoz-Bonilla A, Morris C J and Ruiz A 2019 Toxicity and biodegradation of zinc ferrite nanoparticles in Xenopus laevis *J. Nanoparticle Res.* 21 181 https://doi.org/10.1007/s11051-019-4631-1.

254 Fahmy H M, Aly E M, Mohamed F F, Noor N A and Elsayed A A 2020 Neurotoxicity of green- synthesized magnetic iron oxide nanoparticles in different brain areas of wistar rats *Neurotoxicology* 77 80–93 https://doi.org/10.1016/j.neuro.2019.12.014.

255 Tse B W-C, Cowin G J, Soekmadji C, Jovanovic L, Vasireddy R S, Ling M-T, Khatri A, Liu T, Thierry B and Russell P J 2015 PSMA-targeting iron oxide magnetic nanoparticles enhance MRI of preclinical prostate cancer *Nanomedicine* 10 375–86 https://doi.org/10.2217/nnm.14.122.

256 Volkovova K, Handy R D, Staruchova M, Tulinska J, Kebis A, Pribojova J, Ulicna O, Kucharská J and Dusinska M 2015 Health effects of selected nanoparticles in vivo: liver function and hepatotoxicity following intravenous injection of titanium dioxide and Na-oleate-coated iron oxide nanoparticles in rodents *Nanotoxicology* 9 95–105.

257 Ahmad F, Liu X, Zhou Y and Yao H 2015 An in vivo evaluation of acute toxicity of cobalt ferrite ($CoFe_2O_4$) nanoparticles in larval-embryo Zebrafish (Danio rerio) *Aquat. Toxicol.* 166 21–8 https://doi.org/10.1016/j.aquatox.2015.07.003.

4

Biomedical Application of Quantum Dots

Ajit Khosla[1], Irshad A. Wani[2], and Mohammad N. Lone[3]

[1] School of Advanced Materials and Nanotechnology, Xidian University, Xian, Shaanxi, China
[2] Department of Chemistry, Govt. Degree College Anantnag, Anantnag, University of Kashmir, Jammu and Kashmir, India
[3] Department of Chemistry, Central University of Kashmir, Ganderbal, Jammu and Kashmir, India

4.1 Biomedical Applications of QDs

Quantum dots (QDs) are semiconductor nanoparticles composed of group II–VI or III–V elements. Typically, their physical dimensions are smaller than the exciton Bohr radius [1]. When these nanoparticles are exposed to visible light, some of their electrons get excited to a higher energy state. When these excited electrons return to their original (ground) state, they emit photons of a characteristic frequency specific to the material. These unique optoelectronic properties of QDs arise from the quantum confinement effect. QDs optoelectronic behavior can be explained using semiconductor physics and quantum mechanics. Semiconductor systems have two energy bands: the valence band and the conduction band, with the energy gap between them, referred to as the band gap energy. When subjected to optical or electronic excitation, electrons in the valence band become mobile and move to the conduction band, leaving behind a hole. This process, known as electron-hole recombination, occurs after a certain period, approximately 20 ns. It can cause QDs to emit light at specific wavelengths, typically at a rate of 8×10^6. The energy required to excite an electron from the valence band to the conduction band, leading to fluorescence, depends on the size, shape, and composition of the QDs. This size dependence becomes significant when the QD dimensions approach or equal the exciton Bohr radius [2, 3]. Figure 4.1 shows the size-dependent fluorescence emission of QDs upon illumination with UV light.

QDs, particularly those in the 2–6 nm size range, have garnered significant attention because of their size similarity to many biological macromolecules like proteins and nucleic acids [1, 6]. QDs have garnered significant attention in the biomedical research community due to their ability to tune fluorescence emissions, making them an attractive tool for conducting a wide range of biological experiments. QDs possess distinctive electronic properties stemming from quantum size effects, in addition to their substantial extinction coefficients, tunable absorption wavelengths, and broad absorption spectra. These attributes make them highly desirable lumiphores for various biological applications [3]. Recent research has emphasized the development of compact, biocompatible, and less toxic QDs with multifunctional capabilities for advanced imaging and sensing applications in the biomedical field. These QDs possess various advantageous properties, primarily attributed to their continuous absorption profile across different sizes. This property enables them to be excited simultaneously by a single wavelength, allowing their utilization in time-resolved fluorescence bioimaging [7]. This stands in contrast to conventional dyes that necessitate specific excitation

Metallic, Magnetic and Carbon-Based Nanomaterials: Synthesis and Biomedical Applications, First Edition.
Ajit Khosla, Irshad A. Wani, and Mohammad N. Lone.

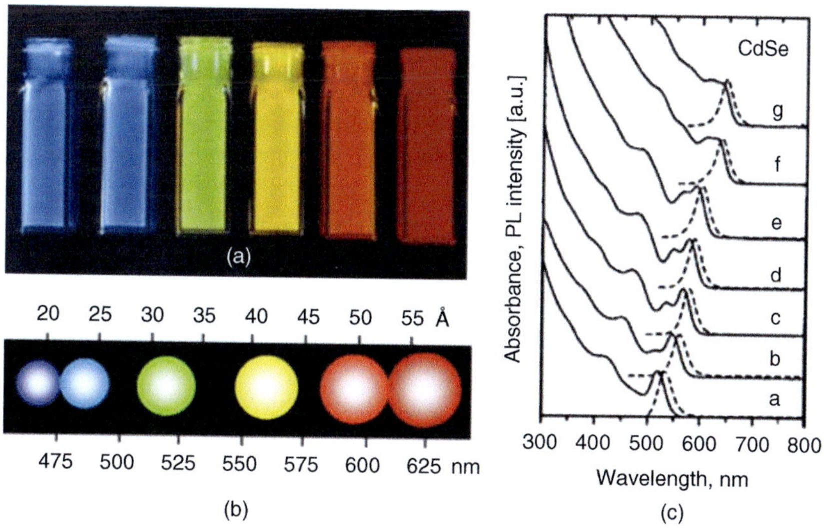

Figure 4.1 (a) Variation in photoluminescence (PL) color based on size. Source: Adapted with permission from Reference [4] Copyright (1997) American Chemical Society. (b) Illustrative representation depicting the relationship between size, color, and PL wavelengths of CdSe-ZnS QDs [5]. (c) Spectra displaying absorption (solid lines) and PL (dashed lines) of CdSe QDs with different sizes. Source: (a) Reproduced with permission from Dabbousi et al. [4]/American Chemical Society, (b) & (c) Talapin et al. [5]/with permission of American Chemical Society and Adapted from Talapin et al. [5].

wavelengths due to their narrow excitation spectra. Traditional dyes also exhibit wide emission spectra, increasing the likelihood of spectral overlap and limiting their utility as spectral fluorescent probes for simultaneously tagging different biomolecules and resolving their spectra. In contrast, QDs have broad absorption spectra, facilitating the simultaneous excitation of QDs of different sizes with a single wavelength [1, 8]. QDs also exhibit narrow emission spectra that can be precisely controlled by altering the core size, composition, or surface coating. This versatility allows for the engineering of QDs to emit light at accurate wavelengths ranging from ultraviolet (UV) to infrared (IR). Their combination of broad absorption and narrow emission spectra makes QDs ideal candidates for multiplexed imaging, enabling the encoding of small molecule libraries, including genes and proteins [9, 10], using various colors and intensities [8]. This approach can generate approximately 10,000–15,000 distinguishable codes. QDs possess a significant Stokes shift, enhancing their detection signal. They exhibit remarkable stability against bleaching [11], making them suitable for monitoring biological processes and tracking biomolecules like proteins. QDs demonstrate exceptional photostability, significantly outperforming organic dyes. For instance, dihydrolipoic acid (DHLA)-capped cadmium selenide-zinc sulfide (CdSe-ZnS) QDs demonstrated approximately 100 times greater stability and nearly 20 times higher brightness compared to rhodamine 6G, maintaining their intensity even after 14 hours of excitation. QDs offer intrinsic brightness, enabling the imaging of individual QDs [8, 12–17]. The optical properties of QDs with organic fluorophores have been summarized in Table 4.1 [3] The prolonged fluorescence lifetime of QDs after excitation proves advantageous in time-gated imaging. Organic dyes exhibit rapid fluorescence emissions upon excitation (< 5 ns), closely matching the short-lived autofluorescence background decay from various naturally occurring species. This close temporal alignment reduces the signal-to-noise ratio. In contrast, QDs exhibit much slower emission decay times, typically ranging from 30 to 100 ns, which significantly surpass the autofluorescence background decay rate (Figure 4.2) [20, 21]. This property, combined with a high photon turnover rate, contributes to an increased signal-to-noise ratio. This approach has been successfully employed to produce images of 3T3 mouse fibroblasts

Table 4.1 A comparison of the typical optical characteristics of green-emitting CdSe/ZnS quantum dots (QDs), the organic fluorophore rhodamine 6G, and the green fluorescent protein (S65T mutant).

Properties	Green-emitting QDs	Rhodamine 6G organic fluorophore	Green fluorescent protein
Emission wavelength [nm]	550	555	509
Full-width at half maximum [nm]	20–40	40–55	35–45
Absorbance profile [nm]	<520	530	487
Quantum yield [%]	20–65	50–90	68
Molar absorptivity [$M^{-1}cm^{-1}$]	10,000–100,000	80,000	39,200
Fluorescence lifetime [ns]	15–20	1–5	3
Other properties	Tuneable emission Photostable Large surface area Continuous absorbance profile	Small Spectral tail	Tuneable emission Genetic incorporation Spectral tail

The information has been compiled from the following sources. Data was compiled from the following References [4, 5, 7, 12, 18, 19].

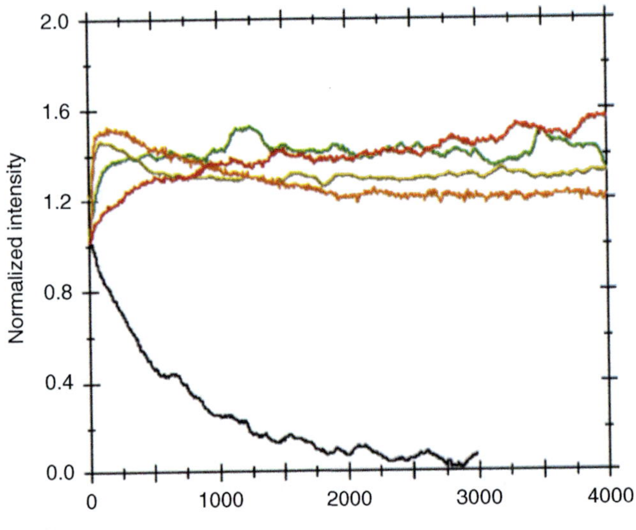

Figure 4.2 The fluorescence intensity over time of silanized nanocrystals and rhodamine 6G at 488 nm is shown. The nanocrystals maintain a consistent emission for up to 4 hours, while the dye fades within 10 minutes. The colors represent the emission of the nanocrystals, while rhodamine 6G is depicted in black. Source: Reproduced with permission from Gerion et al. [20]/American Chemical Society.

with a notably high signal-to-noise ratio [7]. This book chapter aims to offer a comprehensive overview of the recent applications of QDs in the field of biomedicine.

4.1.1 In Bioimaging and Biolabeling

Numerous methods have been devised to link QDs with biomolecules for binding or recognizing specific molecular targets. This process, known as bioconjugation of QDs, can involve either covalent or non-covalent interactions. Functionalization methods for QDs can be categorized into covalent and non-covalent bioconjugation approaches. For covalent cross-linking, various functional groups bearing moieties, including carboxyl groups, thiol groups, amine groups, etc., [12] have been employed. Additionally, azide click chemistry has been harnessed as a strategy for the bioconjugation of nanoparticles [22]. Various non-covalent interactions, such as electrostatic interactions, biotin-avidin binding, and metal chelation, have been effectively utilized for the conjugation of QDs with a wide range of biomolecules. Negatively charged molecules like nucleic acids and proteins can adhere to the positively charged surface of QDs via electrostatic interactions [23, 24]. Metal ion chelating moieties, such as polyhistidine tags fused to proteins or peptides, serve as linkers due to their strong affinity for metal ions like zinc and nickel ions [25]. In some cases, these biomolecules can be directly ligated to the QD surface during the synthesis process [26, 27]. Bioconjugation of QDs has also been successfully achieved through biotin-avidin conjugation [12, 13, 28]. QDs functionalized with biomolecules have proven to be excellent agents for specific cell imaging and labeling. Their high photostability and robust photoluminescence make functionalized QDs ideal for single-particle tracking within cells [11, 29]. Compared to most organic fluorophores, QDs offer more intense and persistent signals. Additionally, QDs have been employed for multicolor imaging, enabling the simultaneous detection and profiling of multiple targets [11, 30, 31]. Beyond imaging, QDs have found applications in intracellular tracking and sensing of biomolecules [32–36], organelle targeting [37], and labeling of cell surface receptors [38–41].

4.1.2 In Cell Imaging

The most frequently employed biomolecules for functionalizing QDs for labeling include proteins (such as antibodies), peptides, aptamers, and other small biomolecules. These ligands serve to target particular cell surface proteins and receptors, which often exhibit overexpression in certain cell types. For instance, biofunctionalized QDs find applications in labeling Her2, EGFR, and prostate-specific membrane antigens for cancer diagnosis, as these antigens commonly exhibit overexpression in specific cancer cell types [13, 42, 43]. Similarly, the arginine-glycine-aspartate (RGD) peptide is a widely used ligand for targeting cancer cells because of its capability to bind to integrins on the cell surface [44, 45]. Nucleic acid aptamers, known for their strong affinity for specific targets, are discovered through an in vitro selection process. This process involves screening a large DNA/RNA library using either whole cells or cell surface proteins as target molecules [46]. Aptamers are typically straightforward to synthesize and manipulate and exhibit greater stability when compared to antibodies. Both DNA and RNA-based aptamers have been utilized to target different cell types, including leukemia cells, prostate cancer cells, and lymphoma cells [47, 48]. Likewise, small molecules like QDs capped with folate acid have been examined as effective options for imaging folate receptors, which exhibit overexpression on various cancer cells [49, 50]. Utilizing these small molecules offers the benefit of controlling the hydrodynamic size of bioconjugated QDs to a specific range, a critical factor for in vivo imaging [51]. The functionalization of QDs for specific labeling primarily involves biomolecules such as proteins (including antibodies), peptides, aptamers, and other small biomolecules [11, 12, 52]. These ligands are utilized to label specific cell surface proteins and receptors that tend to be over-expressed on certain cell types. For instance, biofunctionalized QDs can be employed to label cancer-related antigens like Her2, EGFR, and prostate-specific membrane antigens, as these antigens are often over expressed on certain cancer cells. Similarly, the Arginine-glycine aspartate (RGD) peptide is a popular ligand for cancer cell targeting due to its ability to bind to cell surface integrins. Nucleic acid aptamers, which exhibit high affinity for their targets, are identified through in vitro selection from DNA/RNA libraries using whole cells or cell surface proteins as targets. Aptamers are relatively easy to synthesize and manipulate and are often more stable than antibodies. DNA and RNA-based aptamers have been used to target various cell types, including leukemia cells, prostate cancer cells, and lymphoma cells. Additionally, small molecules like folate acid-capped QDs have been explored for imaging folate receptors that are over-expressed on various cancer cells. The use of these small molecules offers the advantage of limiting the hydrodynamic size of bioconjugated QDs, which is crucial for in vivo imaging. While surface labeling of cells with QDs has been achieved with relative ease, the true challenge lies in the specific intracellular labeling of living cells with QDs. Due to their larger size compared to organic fluorophores, QDs cannot directly enter living cells as easily as most organic fluorophores. Nevertheless, QDs can be introduced into cells through endocytosis, which can be either receptor-mediated or nonspecific. Once internalized, they are typically sequestered within lysosomes and endosomes [11, 12, 52]. However, the presence of these compartments, which keep them isolated from cytoplasmic biomolecules, poses challenges for intracellular labeling of biomolecules with QDs. Initial efforts to deliver QDs into the cytosol involved their conjugation with cell-penetrating peptides or the use of liposome-based transfection agents. Unfortunately, both methods proved ineffective, as they still led to endosomal sequestration and QD aggregation [52]. To address the challenge of delivering QDs into the cytosol, novel approaches have been adopted. Polyethyleneimine (PEI)-based polymers with endosomal-disrupting properties have been employed for internalizing QDs. When these polymers attach to the QD surface, they generate increased osmotic pressure by absorbing protons from the acidic environment. This elevated osmotic pressure disrupts the membrane, allowing QDs to enter the cytosol. However, a notable limitation of this method is the significant toxicity of the polymers to biological systems. To mitigate this toxicity, biodegradable polymeric nanospheres have also been explored [53]. These biodegradable polymers encapsulated QD deliver these QDs into cell cytosol after polymer degradation, allowing QDs to target specific cell types. However, the microinjection method, though tedious, is still considered to be the most efficient type for QD delivery into cell cytosol [31, 53]. This method is considered to be more suitable for quantitative identification of some specific intracellular biomolecules. Multiplex protein tracking, stem cell labeling, gene expression detection, and monitoring intracellular protein interaction dynamics are some of the recent progress of QD-based cell imaging [33, 35, 36, 54]; robust and bright fluorophores QD are quite useful for single-molecule imaging [55, 56] and opens up the possibility of monitoring the real-time complex biological process [57].

4.1.3 In Cell Tracking

A significant study demonstrated the successful labeling of individual blastomeres in Xenopus embryos using QDs enclosed in stable phospholipid micelles, preventing agglomeration. This approach allowed for the labeling of all cell types within the embryo. At the levels necessary for fluorescence visualization (2×10^9 per cell), these QDs were determined to be nontoxic to the cells. However, a concentration of 5×10^9 per cell was found to induce cell abnormalities. These QDs were primarily localized within the injected cells, although unintentional translocation into the nucleus was observed during some stages of embryo development. In a separate study, another research group labeled dictyosteliumdiscoideum with QDs and observed that cell physiology and morphology remained unaffected [11]. The impact of starvation on dictyosteliumdiscoideum was studied by labeling different cell populations with QDs of varying colors. It was noted that QDs exhibited no noticeable loss of fluorescence, enabling longer-term tracking of the cells. When Zebrafish embryo blastomeres were labeled with QDs and co-injected with CFP, it resulted in the transfer of QDs to daughter cells [6]. Some of these daughter cells did not exhibit QD fluorescence, although CFP fluorescence was still observed. This was attributed to QD aggregation and unequal inheritance by the daughter cells. This fluorescence loss and instability of QDs in biological solutions raised concerns for their long-term applications. A recently developed method allows the

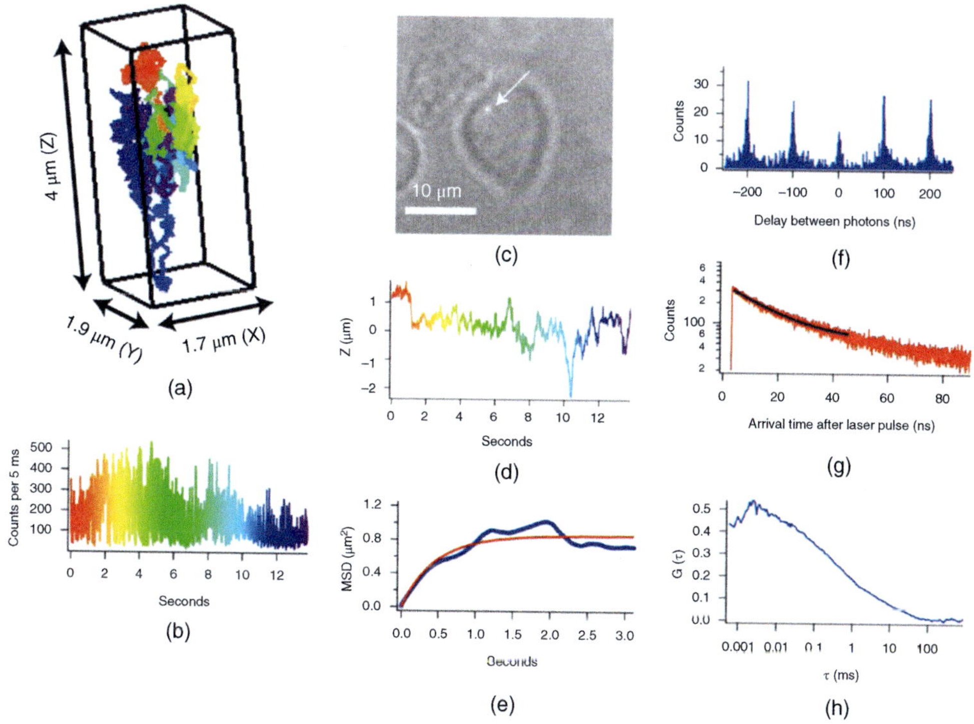

Figure 4.3 (a) The three-dimensional path of a single IgE-FcεRI labeled with a quantum dot (QD) on an unstimulated mast cell at 37 °C, represented using a rainbow color scheme to indicate the temporal progression. (b) The counts utilized for three-dimensional tracking and feedback. (c) A CCD image illustrates the receptor's position relative to the mast cell (arrow pointing to QD). (d) The Z-position of this receptor exhibits over 4 µm of Z-motion, which may go undetected in CCD-based tracking methods. (e) The mean-squared displacement (blue) and its corresponding fit (red). (f) A photon-pair correlation measurement derived from this approximately 14-second trajectory indicates fluorescence photon antibunching. (g) A histogram of photon arrival times in relation to the excitation laser pulse (red) and its exponential fit (black), yielding a 16 ns fluorescence lifetime. (h) An autocorrelation analysis of the photon detection rate obtained during this trajectory, with the decay in the correlation curve primarily influenced by quantum-dot blinking dynamics. Source: Reproduced with permission from Wells et al. [58]/American Chemical Society.

observation of individual proteins labeled with a single-QD in living cells. This technique tracks the three-dimensional molecular motions through active feedback every five milliseconds, employing four overlapping confocal volume elements [58] (Figure 4.3). In contrast to conventional 3D molecular tracking approaches that use charge-coupled device cameras, this method presents several advantages, including an increased Z-tracking range (approximately 10 µm), significantly lower excitation energy (around 15 µW), and the ability to perform time-resolved spectroscopy (e.g., fluorescence correlation spectroscopy or fluorescence lifetime measurements) on the monitored molecules. Particularly noteworthy is the successful demonstration of fluorescence photon antibunching observed in individual QD-labeled proteins within living cells. This demonstration encompassed the monitoring of a singular dye-labeled nucleotide (Cy5-dUTP) at transport rates pertinent to biological processes. The trajectories of receptors on the cell membrane offered insights into the three-dimensional and nanoscale characteristics of cell surface topology. In the later stages of the signal transduction cascade, formations of QD-labeled IgE-FcεRI were identified during ligand-mediated endocytosis and traced as they underwent rapid vesicular transit through the cell at an approximate speed of 950 nm/s. Alongside the substantial kinetic data extracted from these three-dimensional trajectories, individual photon arrival times were precisely recorded with an accuracy of about 400 ps, facilitating the application of various time-resolved analysis methods that are not viable with camera-based tracking techniques.

The study by de Mel et al. [59] introduced an inventive application of combining QDs with specific antibodies for the real-time monitoring of Endothelial Progenitor Cells (EPCs) derived from peripheral blood. This pioneering method sheds light on the potential of QDs to monitor the adhesion of EPCs onto vascular grafts within a bioreactor under pulsatile flow conditions. The research underscores the significance of comprehending physicochemical properties, encompassing surface topography, surface energy, and surface chemistry, in shaping the interaction between EPCs and tissue-engineering scaffolds. Despite established protocols for EPC extraction and cell culture, the behavior of EPCs on tissue-engineered grafts is influenced by scaffold characteristics, fluid dynamics, and the patient's pathophysiological condition. Monitoring the integration and transformation of EPCs into functional endothelium without sacrificing the tissue-engineered construct is considered indispensable. The study addresses the imperative need for live monitoring of EPCs, pivotal for endothelialization and angiogenesis associated with tissue-engineered constructs. The coupling of antibodies with QDs, facilitated by EDC/NHS, emerges as a gentle yet effective method for preserving the integrity of QDs. The study affirms the efficacy of this method through diverse analyses, including FTIR and gel electrophoresis. The research underscores the significance of specificity in cellular uptake, emphasizing the advantages of QDs bound to antibodies over free QDs for targeting cell surface markers. The use of MSA-coated CdTe/CdS nanocrystals guarantees stability in weak acidic solutions, effectively addressing concerns related to thrombogenicity. Hemolytic assays validate the biocompatibility of QDs, indicating their potential introduction into the vasculature for monitoring EPC progress during in vivo studies or direct EPC cellular therapy. That peptide-conjugated CdSe/ZnS QDs encapsulated with PEG were introduced into viable mesenchymal stem cells. Conjugation of Tat peptides to QDs led to an increased photoluminescence (PL) intensity, indicating successful peptide coupling. Fluorescent microscopy confirmed intracellular QD fluorescence in mesenchymal stem cells (MSCs) labeled with QD–Tat peptides, while control experiments with DSPE-PEG-coated QDs showed no intracellular fluorescence, supporting the effectiveness of Tat peptides in facilitating QD internalization into stem cells. Further examination with a laser confocal microscope demonstrated a uniform distribution of QD–Tat peptides throughout the cytoplasm. Multiplexing experiments using two-color QDs successfully bound to Tat peptides and introduced them simultaneously into stem cells. Transmission electron microscopy provided high-resolution images, showing internalized QD–Tat peptides within endosomes, consistent with endocytosis-mediated uptake. Flow cytometry analysis confirmed the high efficiency (>95%) of Tat peptide-conjugated QDs entering MSCs. Engraftment studies in nude mice injected with QD-645-Tat peptide-labeled MSCs revealed the presence of labeled stem cells in various organs, with a higher concentration in the lung. While stem cells were found in other organs, the absence of luminescent spots in the brain, heart, or kidney suggested selective localization of QD–Tat peptide-labeled stem cells. The research team led by Mathias P. Clausen [60] utilized a traditional fluorescence microscope to conduct multicolor single particle tracking (SPT) with QDs. Investigating the individual behaviors of EGFR, CD59, and GM1 clusters within live cells at the single QD level, the study employs spectrally distinct colors for each molecule, achieving extended single-molecule trajectories and facilitating robust analysis at a 25 Hz frame rate. The results reveal unexpected heterogeneity in the plasma membrane, surpassing predictions based on random Brownian diffusion. Variations in diffusion coefficients among the molecular species are observed, with cholesterol depletion using mbCD leading to reduced average diffusion coefficients and decreased heterogeneity for all three types of molecules. The results indicate consistent diffusion coefficients for EGFR and CD59 but significantly slower values for GM1 clusters, which the study clarified as representing GM1 clusters rather than individual molecules. The observed reduction in diffusion coefficients post-cholesterol depletion aligns with established findings. The study validates the single trajectory analysis method through simulated data, demonstrating precision even at short time intervals. While emphasizing the advantages of QDs for simultaneous tracking of multiple molecular species, the research highlights the challenge in controlling QD valence, advocating for improved monovalent QD conjugates. The multi-species approach enables the simultaneous observation of up to four molecular species, allowing detailed comparisons within a single cell or membrane environment. This methodology holds promise for exploring variations in diffusion coefficients within a cell and investigating molecular interactions, such as receptor dimer formation, underscoring the potential of QD-based SPT techniques for comprehensive cell biology investigations. In an investigation by Costa et al. [61], stem cells derived from adipose tissue in goats (g-ADSCs) underwent labeling using luminescent intracytoplasmic nanocrystals QSsto facilitate tracking post-infusion into animal models. The labeled cells, whether fresh or fixed, were administered via tail vein injection into goat mammary glands and mouse testicles and kidneys. Examination of biopsies 30 days after infusion indicated successful labeling and subsequent paraffin processing and fluorescence microscopy validated the sustained effectiveness of the labeling over the long term. Analysis of frozen mammary gland biopsies after 90 days maintained the quality of the labeling. The nanocrystals, passively transported into the cells, demonstrated stability and fluorescence both in vitro and ex vivo, even after fixation. Histological evaluations confirmed the presence of nanocrystal particles in the g-ADSCs within the animal organs, confirming cell migration through the bloodstream. The study underscored the benefits of frozen sections over paraffin-blocked ones for extended tracking and highlighted Qdots' suitability for effective and enduring tracking of g-ADSCs in live tissues. The results imply potential applications in cell therapy, showcasing the robustness and viability of the labeling technique over an extended timeframe. This study by Keller et al. [62] employed non-blinking "giant" quantum dots (gQDs) to investigate the dynamics of IgE-FcεRI receptors in live cells through the use of a confocal-based 3D SPT microscope. The application of gQDs significantly increases the likelihood of observing IgE-FcεRI for durations surpassing one minute by seven times compared to commercially available QDs. To ensure the trajectories in SPT originate from individual gQDs and not aggregates, a time-gated photon-pair correlation analysis is implemented. The prolonged tracking enabled by gQDs facilitates the identification of multiple changes in the diffusion rates of individual IgE-FcεRI receptors occurring over extended (>1 minute) time periods. These alterations are quantified using a time-dependent diffusion coefficient and hidden Markov modeling. The non-blinking characteristic of gQDs is anticipated to play a pivotal role in future live cell 2D and 3D SPT studies, particularly in scenarios involving changes in cellular dynamics over several minutes. The research on single-virus tracking (SVT) and quantum-dot-based SVT (QSVT) encapsulates a paradigm shift in our approach to understanding viral infections at the cellular level. The challenges posed by conventional fluorescent tags, such as their low brightness and poor photostability, have hindered the progress of SVT techniques. However, the advent of QSVT, leveraging the exceptional photostability, high brightness, and tunable color range of QDs, has revolutionized the field. The research team led by Liu et al. [63] highlighted the comprehensive QSVT protocol provided, offering a practical guide for researchers, spanning the intricate processes from virus preparation to image acquisition and data analysis. The intricate details of the protocol underscore its robustness and reliability, positioning QSVT as a transformative tool for real-time monitoring of individual viruses in live cells. The anticipated results from QSVT experiments with influenza A virus (IAV) and Japanese encephalitis virus (JEV) infections provide illuminating insights. In the case of IAV, the dynamic trajectory of the virus, clathrin recruitment, and subsequent cytoskeleton-dependent transport are meticulously elucidated, shedding light on the mechanisms underlying virus entry. Similarly, the study of JEV infection unveils distinct stages, outlining a slow-fast-slow movement pattern from the cell periphery to the perinuclear region along actin filaments and microtubules. These findings not only deepen our understanding of viral infection dynamics but also underscore the versatility and applicability of QSVT in unraveling the intricate interplay between viruses and host cells. In conclusion, the integration of QSVT into the realm of SVT represents a watershed moment, transcending the limitations of conventional tags and offering unparalleled insights into the complex world of viral infections. This chapter advocates for the widespread adoption of QSVT as a cornerstone technique for visualizing and analyzing the dynamic behaviors of individual viruses, ultimately paving the way for innovative discoveries and advancements in our understanding of host-pathogen interactions. Several research reports have been published on the use of QDs in SVT [64, 65].

4.1.4 In Vivo Animal Imaging

The utilization of QDs for whole-body imaging has received limited attention, primarily due to the various challenges encountered when attempting to image an entire animal. The most significant challenge is the potential toxicity associated with QDs, which impacts their applicability in both human and animal contexts. Imaging animal tissues with QDs introduces complexities related to QD autofluorescence upon excitation, as well as issues with QD absorbance and scattering by the animal tissues. Given that tissues exhibit minimal absorption and scattering in the near-infrared (NIR) regions, QDs need to be engineered to emit fluorescence in the NIR range in order to enhance signal strength [66]. The extent of tissue autofluorescence also varies according to the excitation light wavelength. Given the broad absorption spectra of QDs, it is possible to select wavelengths that minimize autofluorescence [67]. Some studies have explored self-illuminating QDs [68] for this purpose. QDs function through bioluminescence resonance energy transfer, which operates in a manner similar to fluorescence resonance energy transfer (FRET), with the key difference being the use of bioluminescent molecules like luciferase as an energy donor. Luciferase is activated by an enzyme, such as coelenterazine, leading to the excitation of QDs. This method offers the advantage of minimizing tissue autofluorescence and background by eradicating the necessity for external excitation light. However, it suffers from a significant drawback due to the introduction of potentially immunogenic substances. The efficacy of this approach relies on the uniform distribution of the enzyme. Enlargement in size of the complex creates challenges for extravasation and penetration into tissues, tumors, or organs [68]. A significant challenge in the in vivo application of QDs is their clearance from the bloodstream. QDs, like many other nanoparticles, often experience widespread uptake by the reticuloendothelial system, leading to a reduction in their concentration in the blood [69]. QDs coated with PEG have proven effective in preventing interactions with opsonins and extending the circulating lifespan of QDs. Nevertheless, the challenge of nonspecific uptake remains unresolved [70–72]. Accumulation after In vivo injection of QDs to various biological targets has been reported such as tumors [73–75], vasculature [66], and several other targets in tissue sections and necropsy [76]. PEG-encapsulated, antibody-conjugated QDs have been used to target prostate-specific membrane antigen [77], a cell surface marker, which can also be expressed for several non-prostate primary carcinomas [78]. The QD particles, once binding to human prostate tumor xenografts, produced signals that were significantly amplified compared to those produced by previously used green fluorescent protein for in vivo imaging of cancer cells [79, 80]. QDs attached to the streptavidin binding site of tumor-specific chimera phase have been synthesized so as to precisely target tumors imaging of vasculature and lymphatics in MDA-MB-435 xenograft tumor systems has been reported. Labeling of such tumor vasculature has been shown to be possible, evidenced by colocalization with lymphatic markers [71]. Imaging of vasculature by type II QDs has been attempted. Upon excitation with light similar to that used to illuminate a surgical field, coronary vasculature could be visualized with the signal-to-noise ratio of 5 : 1 [66]. Using QDs, it was possible to image a coronary vasculature at a depth of 1.5–2.0 mm with resolution enough to identify the named blood vessels [74]. Similarly, it was possible to visualize vasculature in mice through intact skin and adipose tissue 90 mm deep. Imaging of capillaries was also found to be possible through 250 mm od adipose tissue, and the imaging yielded better details at a greater depth by employing power less than needed for FITC-dextran, a traditional organic fluorophore [81]. Studies have shown that after injection into mice, QDs can be colocalized with isosulfan blue, a frequently used lymphatic dye in auxiliary nodes [82]. Identification of SLN in pigs has been reported in pigs using QDs in the GI tract, in the esophagus, in melanoma drainage systems and in pleural space [83, 84]. Once injected, these QDs have been found to be localized rapidly to the SLN and can be imaged in the lung tissues at a depth of up to 5 cm [84]. Although by using QDs, SLN could be identified after resection, even if it is matted and bloody, several studies have reported a slight blue shift of biocompatible IR-emitting QDs [85]. Reportedly, there have been instances of QDs exhibiting a photoluminescent quantum yield slightly below 4%, along with emission spectra broader than visible light [68, 86].

Photodynamic therapy (PDT) has been extensively researched in the therapeutic application of QDs [87]. PDT utilizes light to activate a photosensitizing agent, which subsequently transfers its energy to generate reactive singlet oxygen species (ROS), triggering apoptosis of nearby cancer cells. This method has garnered significant attention for its potential in cancer treatment. The use of QDs, especially in PDT, has been thoroughly investigated due to its potential for more targeted and precise treatment than traditional chemotherapy. Yet, traditional PDT agents frequently encounter constraints such as low singlet oxygen quantum yields, susceptibility to photobleaching, and potential in vivo toxicity. QDs have displayed potential as photosensitizing agents to address these limitations. Different QD variants, such as those composed of CdSe, CdTe, and carbon-based materials, have been investigated for PDT applications, either independently or in conjunction with other photosensitizing agents [88, 89]. Although QDs demonstrate excellent photostability and solubility in water, their clinical application has been impeded by notable toxicity to healthy tissues and inefficient generation of ROS [90]. Ge et al. reported the development of a novel PDT agent based on highly water-dispersible graphene QDs with high singlet oxygen generation. The graphene QDs were synthesized via a hydrothermal technique employing polythiophene derivatives as a carbon source. These QDs displayed a wide absorption band with intense emission peaking at 680 nm and demonstrated enhanced photostability compared to traditional CdTe QDs. In a murine xenograft model of breast cancer, the graphene QDs exhibited a high signal-to-noise ratio for live tumor imaging and proved effective in tumor degradation upon light exposure and QD activation, with no observed regrowth for up to 50 days. Furthermore, the authors showcased the versatility of graphene QDs as multifaceted theranostics, serving as fluorescent, photoacoustic, and thermal agents for simultaneous cancer diagnosis and treatment [91]. Kuo et al. reported on the use of nitrogen-doped graphene-based QDs for dual photodynamic antimicrobial therapy and bioimaging [92]. The nitrogen-doped QDs developed exhibited a higher reactive oxygen species production than nitrogen-free QDs after just 3 minutes of photoexcitation (670 nm laser, 0.1 W/cm^2). Moreover, the study demonstrated that PDT antimicrobial efficacy increased with higher nitrogen-binding compositions. Additionally, the near-infrared (NIR) and excellent photostability properties of graphene QDs facilitated enhanced contrast and imaging of bacteria. QDs can serve as carriers to deliver and monitor the effectiveness of therapeutic payloads, as the attached cargo can quench the particles' fluorescence via FRET. These platforms offer the potential for investigating the pharmacokinetic and pharmacodynamic behavior of chemotherapeutic agents and monitoring disease progression and therapeutic responses. For instance, Minko et al. described a QD carrier with doxorubicin (DOX) conjugated via a pH-sensitive hydrazone bond, along with a

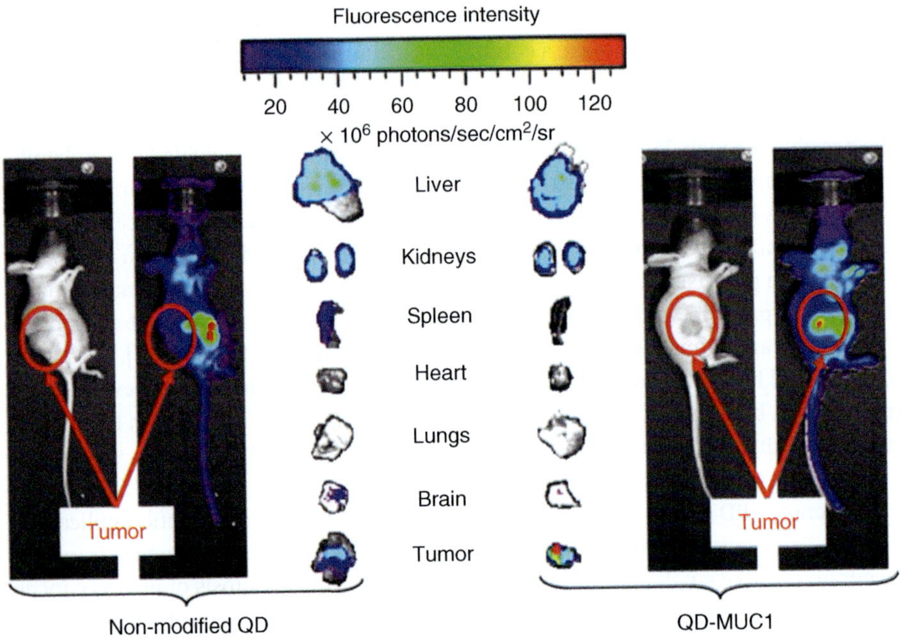

Figure 4.4 In vivo imaging in a nude mouse model 24 hours post intravenous QD injection revealed the real-time biodistribution of both targeted and non-targeted QDs. Fluorescence intensity, indicative of QD accumulation, ranged from blue (lowest intensity) to red (highest intensity). Targeted QDs exhibited notably higher accumulation at the tumor site compared to non-targeted QDs. Source: Reproduced with permission from Savla et al. [93]/Elsevier.

DNA aptamer targeting MUC1. This system aimed to enhance delivery into ovarian cancer cells, mitigate systemic chemotherapy effects through targeted delivery, and offer an effective imaging modality for monitoring drug release. Increased uptake of the QD-MUC1-DOX particles compared to unmodified QDs by DOX-resistant A2780-AD human ovarian cancer cells was confirmed, along with the intracellular release of DOX. In an in vivo study using a nude mouse model, the targeted QD conjugate exhibited higher tumor accumulation and lower accumulation in other organs compared to the non-targeted QDs (Figure 4.4) [93]. QDs represent an exciting prospect for driving advancements in tumor theranostic platforms, with recent studies concentrating on the development of nanocarriers that co-deliver chemotherapeutic agents and QDs. For instance, a combination of paclitaxel and CdTe/CdS/ZnS QDs encapsulated in nanostructured lipid carriers exhibited high encapsulation efficiency, robust drug loading, and effective tumor targeting. Moreover, novel strategies in QD-based therapeutics involve incorporating quercetin onto CdSe/ZnS QDs to engineer potent antibacterial and anticancer nanoplatforms, as well as integrating 5-Fluorouracil onto Mn-ZnS QDs enclosed within chitosan biopolymer and conjugated with folic acid to enhance targeting specificity toward malignant tissues expressing folate receptors [94]. Moreover, QDs have been employed to advance gene therapy by enabling intracellular delivery and imaging of treatment with small interfering RNA (siRNA). SiRNA, consisting of short double-stranded RNA chains, can selectively silence genes through an endogenous pathway called RNA interference (RNAi) [95]. In a 2005 study, CdSe-core QDs, ZnS-shell QDs, and/or siRNA targeting the Lamin A/C gene were encapsulated in a commercially available cationic lipid reagent, Lipofectamine (Invitrogen). Fibroblasts were transfected, and siRNA uptake and silencing were assessed using flow cytometry, Western blotting, and immunofluorescence staining. The findings illustrated in Figure 4.5 revealed a strong association between gene knockdown and fluorescent intensity in cells transfected with

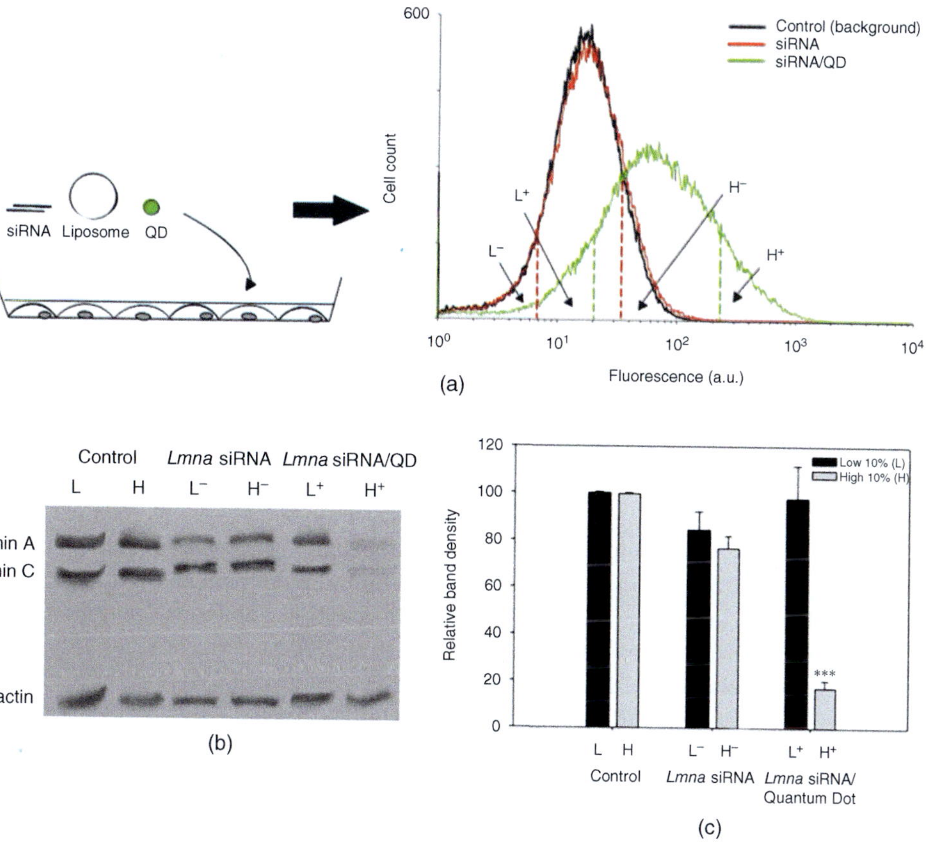

Figure 4.5 (a) Illustration and experimental outcomes depicting the transfection process of fibroblast cells with liposome/siRNA/QD complexes. Flow cytometry was employed to measure fluorescence, and histograms exhibit fluorescence levels in control cells, siRNA-treated cells, and QD/siRNA-treated cells. (b) Western blot analysis of Lamin A/C expression post-transfection, with β-actin acting as the loading control. Results revealed knockdown of the Lamin genes in cells exhibiting high levels of siRNA/QD fluorescence. (c) Densitometry analysis of Western blot bands was also performed. This figure highlights the utilization of Cd/Se core–Zn/S shell QD/siRNA complexes for sorting transfected cell populations and assessing gene knockdown efficiency. Source: (a)–(c) Reproduced with permission from Chen et al. [96]/Oxford University Press and (b) Savla et al. [93]/Oxford University Press/CC BY.

both the QDs and siRNA, indicating QDs' suitability as probes for tracking siRNA delivery. Moreover, the study demonstrated that QD mediated siRNA delivery enhanced in vitro gene silencing efficacy. The formulation delivering both the QDs and siRNA together achieved a 90% silencing efficiency, markedly higher than the control formulation delivering only siRNA, which yielded a 20–30% silencing efficiency [96]. Positively charged QDs have been employed as carriers of siRNA, thereby preventing the requirement for an extra transfection agent like Lipofectamine. In a documented investigation, CdSe/ZnSe QDs modified with L-arginine, both with and without β-cyclodextrin, served as siRNA delivery vectors aimed at silencing the HPV18 E6 gene [97]. The objective of surface modification of QDs with L-arginine was to create a large, positively charged surface area capable of electrostatically binding with negatively charged siRNA. Subsequently, the complexes formed by QD and siRNA were transfected into HeLa cells. Cell uptake and trafficking were monitored using flow cytometry and confocal microscopy, respectively, while efficacy was assessed through RT-PCR and Western blotting. The research illustrated that the complexes were cytocompatible at QD concentrations below 70 μg/mL after 24 hours of exposure. Moreover, the QD and siRNA complexes achieved nearly 80% gene knockdown and approximately 80% suppression of the target protein [97]. In more recent studies, nanocarriers based on cadmium sulfoselenide/zinc sulfide QDs loaded with siRNA have been assessed for their ability to target human telomerase reverse transcriptase (TERT) in central nervous system tumor cells [98]. The findings indicated a high gene transfection efficiency (>80%) in two glioblastoma cell lines. Moreover, the suppression of TERT gene expression markedly hindered the proliferation of glioblastoma cells. Another notable application entails the utilization of QDs and siRNA to investigate the involvement of genes like SOX9 in the chondrogenic differentiation of adult MSCs. Notably, the downregulation of SOX9 impeded the expression of cartilage-specific markers in MSCs, leading to a delay in cartilage repair.

4.1.5 Detection of Pathogens and Toxins

QDs have been employed to detect pathogenic microorganisms and toxins, playing a role in defining the characteristics of these pathogens, including their virulence. A range of pathogenic microorganisms have been targeted, including *Escherichia coli* 0157:H7, *Salmonella Typhi*, *Cryptosporidium parvum*, *Listeria monocytogenes*, and *Giardia lamblia* [98, 99]. In comparison to conventional staining kits, the simultaneous multiplexed labeling of *C. pavrum* and *G. lamblia* achieved a strong signal-to-noise ratio of 17 when using immunofluorescent QD fluorophores. This approach exhibited superior brightness and photostability [99]; however, few reports have questioned the sensitivity of QD-based assay compared to ELISA-based techniques [100]. Bacterial and fungal populations have been labeled using QDs-conjugated wheat germ agglutinin and transferrin [101]. Since human transferrin is strappingly correlated with virulence, hence transferrin-conjugated QDs can provide the test for the presence of pathogenic virulence [102]. In this study, only pathogenic strains of staphylococci were labeled with transferrin-bound QDs, which suggested that they can be used as a quick test for the recognition of hostile and invasive staphylococci. QDs have also been tested for the detection of viruses [103, 104]. Detection of respiratory syncytial virus (RSV) F-protein was possible using a QD-based immunofluorescent technique. Qualitative analysis of the difference in F-protein expression between strains was possible by using the immunofluorescent technique. Detection of Hepatitis B and C viruses has also been demonstrated by employing QD-based in situ hybridization techniques. Multiplexed detection of HCV and HBV with a signal-to-noise ratio of 150 was conceivable by means of printed microarrays of sequences that are complementary top 53 conjugated QDs and hepatitis B and C virus genomes [105]. Likewise, QDs have been used to detect toxins [106, 107]. The labeling of cholera toxin, staphylococcal enterotoxin B (SEB), ricin, and Shiga-like toxin 1 (SLT-1) was carried out using QD immunofluorescence. The study demonstrated the detection of specific toxins at concentrations as low as 3 μg/ml for staphylococcal enterotoxin B (SEB). Furthermore, multiplexed detection was achieved using a mixture of all four toxins. However, challenges related to potential nonspecific binding and cross-reactivity may have arisen due to issues with the antibodies or QD incorporation, necessitating further investigation in this area [6]. The core principle of this detection method hinges on the interplay between antigens and antibodies. The progression of detection is markedly impacted by enhancing the concentration of target bacteria. Modified magnetic beads (MBs), equipped with antibodies, play a pivotal role in both enriching and isolating target bacteria from the sample system through magnetic separation. When combined with luminescent QDs, this technique exhibits the capacity to detect and measure foodborne pathogens accurately.

In the investigation conducted by Dogan et al. [108], a sandwich immunoassay platform was devised, employing chitosan-modified CdTe QDs and Fe_3O_4@Au Magnetic Nanoparticles (MNPs). Through the interaction facilitated by avidin-biotinylated antibodies, CdTe QDs underwent chitosan modification, while MNPs were conjugated with an *E. coli* antibody. The process commenced with MNP-antibody complexes capturing the target bacteria, followed by the integration with QD-antibody complexes to generate MNP-*E coli*-QD conjugates. In this arrangement, the fluorescence spectra of QDs were measured using a hand-held spectrometer. The analysis duration was constrained to 120 minutes, and the proposed method exhibited a working linear range and a limit of detection of 10^2–10^8 and 30 CFU/mL, respectively.

In a study conducted by Martynenko et al. [109], the synthesis of QDs and Magnetic Nanoparticles (MNPs) was exemplified through the utilization of Fe_3O_4 superparamagnetic iron oxide nanoparticles (SIONPs) to form ternary quantum dots (t-QDs). These t-QDs, comprising $CaCO_3$ and AIS/ZnS QDs, established a platform characterized by a photoluminescence band spanning a few hundred nanoseconds and a long lifetime (LT). This platform served as a basis for the conjugation of *Legionella pneumophila*-specific antibodies through Michael's addition, showcasing an innovative synthesis that integrates magnetic and luminescent properties. The resultant complex provides a versatile platform conducive to specific antibody labeling, particularly in the context of detecting *L. pneumophila*. Through the utilization of a layer-by-layer methodology, $CaCO_3$ microbeads with embedded Fe_3O_4 superparamagnetic iron oxide nanoparticles (SIONP) were systematically coated with AIS/ZnS cs-QDs. Subsequently, the surfaces of the resulting $CaCO_3$@ Fe_3O_4@AIS/ZnS microparticles were subjected to conjugation with specifically modified anti-*L pneumophila* polyclonal antibodies. These magneto-fluorescence microbeads were intricately designed to identify and capture targeted bacterial cells. Following a sequence of procedures involving magnetic separation and meticulous washing, the concentration of the identified bacteria could be precisely determined. Significantly, the proposed platform demonstrated a low limit of detection (LOD) for *L. pneumophila* (100 CFU/mL), nearing a threshold of critical importance in cooling water. This innovative methodology underscores the capacity for accurate and sensitive detection of *L. pneumophila* within complex environmental contexts. Utilizing multicolor QDs to label multiple target cells during immune-magnetic separation allows for the simultaneous detection of various bacteria within the same detection system, effectively doubling the detection efficiency. MBs with a substantial volume demonstrate efficiency in enriching target bacteria, preventing their dispersion in the solution. However, maintaining a balance is crucial, as an excessively large volume can result in the blocking effect of small-volume QDs. Moreover, the majority of magnetic separation technologies involve multiple washing steps, which can potentially impact accuracy and detection time. In addressing these challenges, Yin et al. [110] implemented a reverse assaying strategy (RAS) to read the fluorescence signal during the simultaneous detection of *Salmonella enterica* and *E. coli*. The research involved combining numerous QDs with distinct emission wavelengths with specific anti-*S. enterica* and anti-*E. coli* antibodies. Magnetic microbeads (IMBs) were also attached to both types of antibodies. After the immune binding between the QDs, IMBs, and the targeted bacteria, the resulting sandwich complexes underwent magnetic separation for isolation. Subsequently, the remaining QD-antibody conjugates were utilized to quantify fluorescence intensity. The results highlighted that the immunomagnetic separation-QD method achieved a remarkable LOD, reaching as low as 30 colony-forming units per milliliter (CFU/mL) for *E. coli* and 60 CFU mL−1 for *Staphylococcus aureus*. Additionally, the newly devised rapid assessment system (RAS) immunosensors proved effective in swiftly detecting foodborne pathogens in authentic milk samples within just one hour, eliminating the need for multiple washing steps.

Fluorescent QDs have also been used in sandwich assays. The term "sandwich" denotes the capture of target bacteria through the interaction with two different capturing molecules. This arrangement was created by employing maghemite particles and CdTe QDs for the concurrent quantification of three bacterial types: *S. aureus*, *MRSA* (methicillin-resistant *S. aureus*), and *Klebsiella pneumoniae* [111] MNPs were modified with specific gene fragments (fnbA, mecA, and wcaG) and their corresponding oligonucleotides. This led to the formation of a sandwich structure involving MNPs, the three types of target bacteria, and CdTe QDs. Subsequent washing through magnetic separation occurred. The concentration of the target bacteria was determined by evaluating the fluorescence intensity of free CdTe-fnbA, CdTe-mecA, and CdTe-wcaG. The method's LOD was set at 100 colony-forming units per milliliter (CFU/mL). A double-antibody sandwich immunoassay involves capturing target bacteria using two distinct antibodies, making it well-suited for bacteria with both macromolecular protein and polysaccharide components [112]. Song et al. [113] successfully applied this method for the rapid assay of 16 M *Brucella melitensis* (*B. melitensis*) within 105 minutes, for the detection of *C. jejuni*, a one-step detection platform was developed [114]. In this study, CdSe/ZnS QDs modified with antibodies served as fluorescent identification probes, while nanoγ-Fe_2O_3 magnetic nanospheres conjugated with antibodies acted as capture probes. The LOD was determined to be 103 colony-forming units per milliliter (CFU/mL), and the fluorescence intensity of the sandwich immunocomplex showed a linear increase within the range of 103–107 CFU/mL. Importantly, the entire detection process was completed within one hour. CdS QDs, enclosed within in-situ-grown zeolitic imidazolate scaffold-8 (ZIF-8) particles, served as amplification markers for the highly sensitive electrochemical detection of *E. coli* O157:H7 [115]. The carbon glass electrode (GCE) underwent modification with anti-*E. coli* O157:H7 antibody (Ab) and polyaminobenzoic acid (PABA), resulting in the formation of BSA/Ab-PABA/GCE. CdS@ZIF-8, post-modified with polyethyleneimine (PEI) and a specific antibody, demonstrated selective recognition of *E. coli* O157:H7. Following HCl etching, Cd^{2+} ions were released and subsequently analyzed using differential pulse voltammetry to quantify the presence of *E. coli* O157:H7. In 2020, Hunsur Ravikumar, R, and Balakrishna introduced an innovative approach employing a double-antibody immune sandwich technique [116]. In this method, CdSe/CdS core–shell quantum dots (cs-QDs) were paired with antibodies specific to *E. coli*, and the transparent glass surface was modified using polyclonal antibodies through plasma etching. The transparency of the glass played a crucial role in enhancing detection by minimizing the impact of background radiation. The fluorescence intensity of the QDs served as an indicator of the target concentration, resulting in an impressively low detection limit of 50 colony-forming units per milliliter (CFU/mL). Various types of assays/techniques have been employed using QDS for the detection of microbes such as QD-based microfluidic chips, [117] QD-based biosensors, [118] QD-based lateral flow assays [119], and others [120].

4.1.6 Immunolabeling and Molecular Localization

Fluorescence immunolabeling is a commonly employed method for visualizing signal transduction-related molecules and investigating cellular structures. QDs have demonstrated their suitability as probes in this domain due to their durable optical characteristics. In this context, staining of F-actin in fixed cells was carried out using CdSe/ZnS QDs [121]. After this first experiment, a myriad of experiments was performed to localize molecules in cells and tissues using QDs both in fixed and live samples [122]. Immunofluorescence labeling of heat shock 70 protein-mortalin which was latter suggested as a reliable marker for the normal transformed cells, was carried out using QDs so as to obtain different patterns in normal and transformed cells [123] due to the high photostability of QDs, 3-D reconstructed high-resolution image of membrane domain band 3 in erythrocytes was generated by Tokumasu and Dvorak using confocal microscopy by collecting 40 consecutive optical sections [124]. Immunohistochemical (IHC) protocol to detect intracellular antigens in mouse and rat brain tissue sections was developed by Ness et al. by combining QD labeling with conventional enzymatic signal amplification. It was proved that QD IHC labeling had better sensitivity as compared to similar IHC approach employing conventional dyes [125]. Wu et al. developed dependable and specific QD probes for localizing the breast cancer cell surface marker Her2, nuclear antigens in fixed cells, cytoskeleton fibers, live cells, and tissue sections. These QD-based probes offer enhanced sensitivity, brightness, and exceptional photostability when compared to conventional organic dyes. These QDs were employed to label the membrane glycoproteins of breast tumor cells to study the effects of heat stress. These and other studies have proved that QDs are excellent probes with enhanced specificity for studying biological problems [122].

Zdobnova et al. [126] conducted a study that delved into the application of semiconductor QDs coupled with cancer-specific targeting ligands for the fluorescent visualization of cancer cells, with a specific focus on the human epidermal growth factor receptor 2/neu (HER2/neu). The significant overexpression of HER2/neu on the surface of numerous cancer cells underscores its pivotal role as a target for cancer diagnostics. To address this, the study harnessed antibody single-chain variable fragment (scFv) fragments as targeting agents, leveraging their advantages over full-size antibodies, including their smaller size, reduced cross-reactivity, and diminished immunogenicity. In an effort to render the QDs biocompatible and water-soluble, modifications were introduced using mercaptoacetic acid (MAA). The resultant water-soluble QDs maintained their nonaggregate state and retained their fluorescence capabilities. To facilitate the binding of QDs to scFv antibodies, the study employed the barnase-barstar module, creating a versatile "LEGO bricks" system for the design of QD-based constructions. MAA-coated QDs were conjugated with the barstar protein, a crucial component of the barnase-barstar module. This conjugation yielded QD-barstar complexes that exhibited heightened fluorescence quantum yield and increased stability in water solutions compared to their MAA-coated QD counterparts. Analysis of fluorescence spectra revealed a threefold improvement in quantum yield and a red shift in the fluorescence peak for QD-barstar conjugates in contrast to unconjugated QDs. These enhancements were attributed to changes in the QDs polarity in the surrounding environment, neutralization of surface charge, and additional passivation facilitated by protein SH-groups linking to the QD surface. The practical utility of the barnase-barstar module for the targeted delivery of QDs was demonstrated through the visualization of HER2/neu-overexpressing cells. QD-barstar conjugates exhibited efficient staining of HER2/neu on the surface of SKBR-3 human breast cancer cells when subjected to incubation with a 4D5 scFv-dibarnase. The observed minimal nonspecific binding of QD-barstar in isolation further underscored the specificity and efficacy of the QD-antibody conjugates. Overall, the study's findings highlight the successful development of QD-antibody conjugates for the visualization of cancer cells, offering advantages such as high brightness, photostability, and specificity in biological applications. The study by Urban et al. [127] explored the comparative use of colloidal semiconductor QDs and organic fluorophores for super-resolution imaging, specifically in the context of visualizing interactions among neurons. Key findings from the investigation include an assessment of the performance of QDs, particularly bradykinin functionalized CdSe/CdS QDs (BKQDs), in comparison to traditional organic fluorophores, with a specific focus on their application in super-resolution imaging of neurons. In cultures of primary rat hippocampal neuronal glial cells, both neurons and astrocytes expressing BK receptors underwent treatment with BKQDs. Fluorescence imaging depicted in Figure 4.6 verified the successful binding of BKQDs, revealing stationary BKQDs and minimal diffusion (Figure 4.7). The study compared super-resolution imaging using BKQD probes with TAMRA, an organic fluorophore utilized in STORM imaging. Single QD fluorescence displayed bright, blinking BKQDs with a typical signal-to-noise ratio (SNR) intensity ratio of up to 10. Photostability was demonstrated over a 25 s imaging period (Figure 4.6a,b), and super-resolution processing depicted QDs along neuronal cells with a localization error of 50 nm (Figure 4.7e). The reconstructed image disclosed that seemingly single fluorescent spots in diffraction-limited images actually comprised multiple QDs in close proximity, with specific QDs

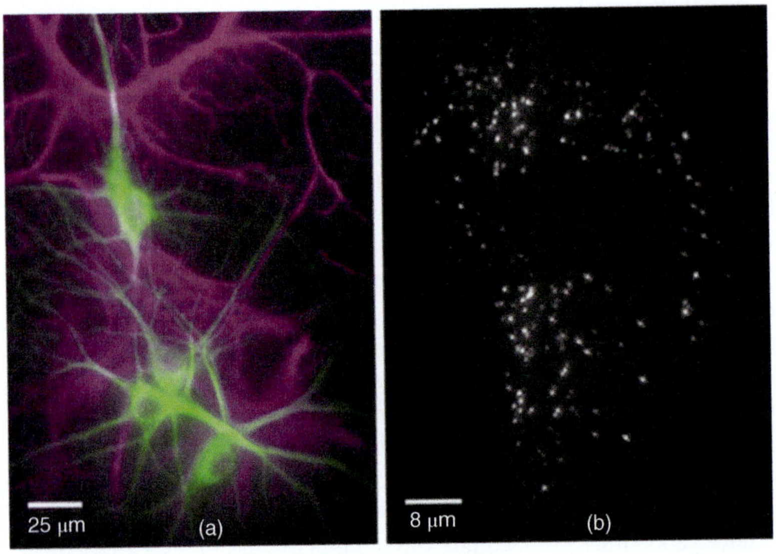

Figure 4.6 Representative images of mixed neuronal glial cultures. In Subfigure (a), a fluorescence image showcases various cell types, emphasizing astrocytes in purple and neurons in green within a hippocampal neuronal-glial culture. Subfigure (b) displays a fluorescence image of neuronal cultures labeled with BKQDs, featuring a single frame from a culture treated with 30 nM BKQDs for 15 minutes. Source: Reproduced with permission from Urban et al. [127]/American Chemical Society.

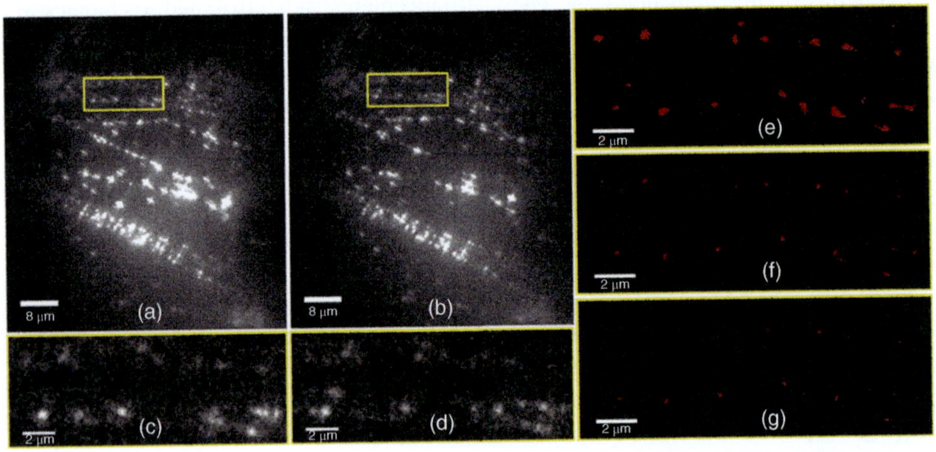

Figure 4.7 The outcomes of STORM imaging on BKQD-treated neuronal cultures. Subfigure (a) exhibits the first frame from a 1000-frame movie capturing a neuronal culture treated with BKQDs, while Subfigure (b) displays the 1000th frame from the same sequence. Subfigures (c) and (d) offer close-ups of the regions marked by yellow boxes in a and b, respectively. Subfigures (e–g) showcase super-resolution reconstruction images of these regions, demonstrating successful localization at precision levels of (e) 50 nm, (f) 30 nm, and (g) 25 nm, highlighting effective localization across various resolution levels. Source: Reproduced with permission from Urban et al. [127]/American Chemical Society.

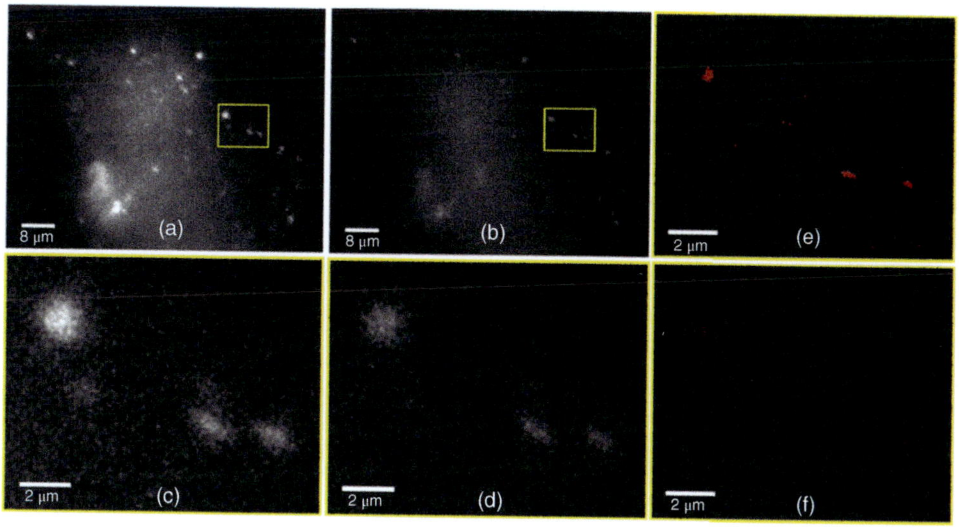

Figure 4.8 The outcomes of STORM imaging for neuronal cultures treated with TAMRA-BK. Subfigure (a) exhibits the first frame of a 1000-frame movie capturing a neuronal culture subjected to TAMRA-BK treatment, while Subfigure (b) showcases the 1000th frame, revealing noticeable photobleaching. Subfigures (c) and (d) offer detailed views of the regions delineated by yellow boxes in A and B, respectively. Subfigures (e–f) present super-resolution reconstructed images of these regions, demonstrating effective localization at a precision of (e) 50 nm and sparse localized points at (f) 30 nm. Source: Reproduced with permission from Urban et al. [127]/American Chemical Society.

localized at 30 nm (F) and even 25 nm (G). In comparison, TAMRA-BK, imaged at four times the fluence, exhibited dimmer, less frequently blinking fluorescent spots (Figure 4.8a,b) and significant photobleaching. Super-resolution processing disclosed large fluorescent blurs resulting from multiple closely spaced fluorescent molecules (Figure 4.8e). TAMRA-BK molecules could only be localized with a 50 nm accuracy, significantly inferior to QDs. At a 30 nm localization error cutoff, substantially fewer TAMRA-BK probes were visible, resulting in the loss of structural information (Figure 4.8f). In practical terms, BKQDs offer advantages over TAMRA, such as simplified imaging conditions in PBS or neurobasal media and enhanced photostability. This is in contrast to TAMRA, which requires oxygen-scavenging buffers and demonstrates a lower SNR and photostability. Simulations shed light on the factors contributing to the superior localization precision of QDs. It is primarily attributed to their faster blinking and increased photostability, with the noteworthy observation that heightened brightness (high SNR) does not significantly contribute to localization precision in densely labeled scenarios. In comparison to other fluorophores like AlexaFluor 647, QDs, especially CdSe/CdS QDs, are anticipated to outperform in terms of blinking, photostability, and brightness.

The research investigation by Yang et al. [128] delved into the application of high-quality CdSe QDs as fluorescent markers for detecting carcino-embryonic antigen (CEA) on HeLa cells. CdSe QDs were synthesized within a paraffin–oleic acid system and utilized for both direct and indirect detection of CEA. The primary antibody (Ab), specifically rabbit anti-CEA8, and the secondary Ab, goat anti-rabbit IgG, were covalently attached to carboxyl functionalized CdSe QDs. The successful application of both QDs–antibody and QDs–IgG probes in labeling HeLa cells underscores the practicality of CdSe QDs for diverse biological applications, including protein probes and cell imaging. Two labeling techniques were employed: direct labeling using QDs–antibody and indirect labeling using QDs–IgG. In the direct method, water-soluble CdSe QDs were conjugated with the primary Ab, followed by incubation with HeLa cells. Illustrated in Figure 4.9. QDs successfully identified the

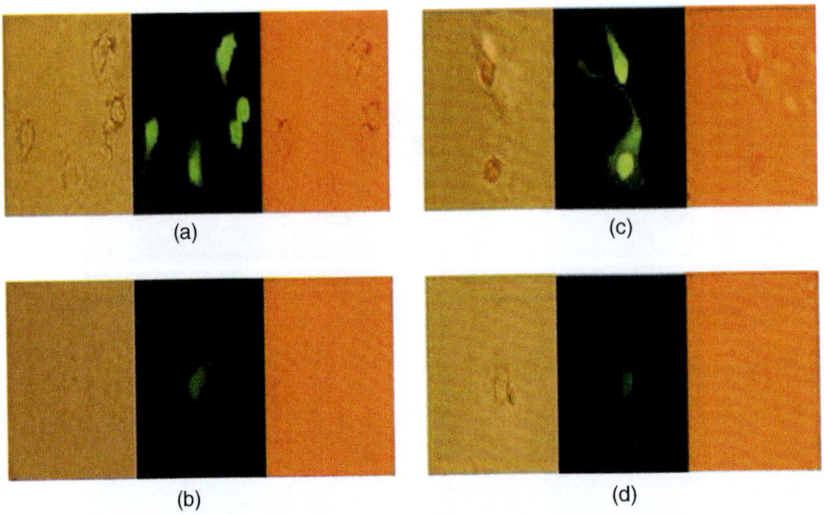

Figure 4.9 Showcase images of cell membranes, both fixed and live, that were directly labeled using QDs–antibody. Panel (a) depicts fixed cells after a 1-hour incubation with QDs–antibody, while panel (b) shows fixed cells incubated with bare QDs for the same duration. Panel (c) presents live cells after 1-hour incubation with QDs–antibody, and panel (d) illustrates live cells incubated with bare QDs for 1 hour. Each set of panels includes phase-contrast images in the left row, fluorescent images in the central row, and overlays of the left and central rows in the right row. Source: Reproduced with permission from Yang et al. [128]/John Wiley & Sons.

targets, generating a vibrant fluorescence presentation over the cell membrane. Unfortunately, this method revealed some non-specificity, as indicated in Figure 4.9b,d. This constraint poses a challenge to its wider application in immuno-assay and immunolabeling. The indirect method utilized secondary Ab IgG linked to QDs, forming probes for labeling HeLa cells after the primary antibody was bound to the cell surface. The findings reveal the effective labeling of HeLa cells (A, D) through QD-IgG probes, highlighting their notable specificity to the targets compared to the direct labeling method. Additionally, Figure 4.10b,e clearly demonstrates the successful elimination of nonspecific adsorption. Analysis of absorption and fluorescence spectra revealed alterations upon conjugation, confirming the formation of QDs–antibody and QDs–IgG conjugates. The study underscores the importance of the antibody-to-QDs ratio in influencing imaging outcomes, with an excess of antibodies utilized to minimize interference from bare QDs. While direct labeling offered simplicity, it exhibited some non-specificity, constraining its broader applications. Conversely, indirect labeling with QDs–IgG probes successfully eliminated nonspecific adsorption, enhancing specificity in HeLa cell labeling.

The study by Zhang et al. [129] focused on the synthesis and application of CdTe–CdS core–shell QDs in biological labeling and cell imaging. These QDs were synthesized in an aqueous solution using a hybrid water-bathing and hydrothermal method, incorporating L-cysteine (L-Cys) as a stabilizer. This approach enhances efficiency in synthesis time and stability compared to using only the hydrothermal method. The successful conjugation of CdTe–CdS QDs with the rabbit anti-CEACAM8 (CD67) antibody demonstrates their effectiveness as fluorescent probes for direct immunolabeling and imaging of HeLa cells. The study underscores the specific identification and binding between the antibody and target antigen, resulting in vibrant fluorescence along the cell membrane. QDs–antibody conjugates exhibit heightened fluorescence intensity compared to bare QDs, attributed to the connection between carboxylic groups on the QD surface and amino groups on the antibody. The slight blue shift in the fluorescence peak with QDs–antibody conjugates

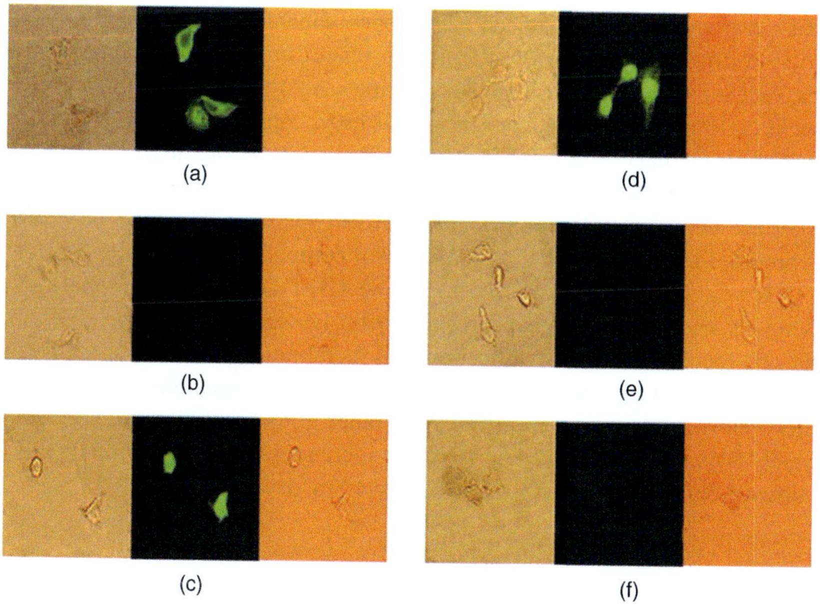

Figure 4.10 Showcase fixed and live cell membranes that underwent indirect labeling using QDs–IgG. Panel (a) illustrates fixed cells following a 1-hour incubation with QDs–IgG, while panel (b) exhibits fixed cells incubated with bare QDs for the same duration. In panel (c), fixed cells are presented after an extended 24-hour incubation with QDs–IgG. Shifting to live cells, panel (d) displays live cells incubated with QDs–IgG for 1 hour, panel (e) showcases live cells incubated with bare QDs for 1 hour, and panel (f) reveals live cells after a 24-hour incubation with QDs–IgG. Each set of panels includes phase-contrast images in the left row, fluorescent images in the central row, and overlays of the left and central rows in the right row. Source: Reproduced with permission from Yang et al. [128]/John Wiley & Sons.

suggests potential adjustments in dipole–dipole interactions and Stokes shifts. Additionally, the research emphasizes the simplicity of the direct labeling method, highlighting the potential of CdTe–CdS QDs as probes for applications in immuno-assays, immunolabeling, and various biological contexts. The water-soluble L-Cys capped CdTe–CdS QDs offer advantages, including heightened sensitivity, straightforward preparation, customizable and narrow emission spectra, and robust photochemical stability compared to organic fluorophores. Mahmoud et al. [130] have devised an advanced methodology for the antibody labeling process using QDs. The refined procedure encompasses crucial steps, including antibody (Ab) reduction, affinity purification, and QD–Ab conjugation. The optimization strategy involves reducing antibodies to generate highly functional 75-kDa heavy–light antibody fragments. Subsequent affinity purification of these fragments, followed by their association with QDs, results in QD–Ab conjugates exhibiting significantly enhanced functionality when compared to conventional methods. This advanced approach is versatile and can be applied to diverse antibody types and nanocrystals. The study compares various reducing agents, demonstrating that reduction with 2-mercaptoethylamine (2-MEA) yields the most functional antibody fragments. Further purification through affinity purification augments the purity of the 75-kDa fragments. Examination of the QD Ab conjugates produced using the optimized procedures in agarose gel confirms the success of the conjugation. Subsequent purification enhances the functional quality of the nanoprobes and eliminates unbound antibodies. Functional assessments on peripheral blood mononuclear cells indicate that the developed nanoprobes adeptly label antigens, resulting in a fourfold increase in fluorescence intensity post-purification. The labeling sensitivity is markedly higher in the optimized nanoprobes compared to those prepared through standard procedures. Dot–blot assays validate the heightened sensitivities of the advanced nanoprobes, capable of detecting less than 10 and 1 ng of antigen, reflecting a more than 100-fold improvement in sensitivity compared to traditional nanoprobes. Hunsur Ravikumar, R, and Balakrishna's 2020 methodology effectively demonstrated the subcellular localization of Sigmar1 in various heart compartments [131]. Using targeted QD labeling with anti-Sigmar1 on ultra-thin sections from adult mouse hearts, QD-TEM images revealed densely labeled QDs on mitochondrial membranes (both inner and outer), lysosomes, and the interface between endoplasmic reticulum/sarcoplasmic reticulum (ER/SR) membranes and mitochondrial membranes (Figure 4.11a,b). For comparison, an isotype control with rabbit IgG and QDs was included (Figure 4.11c,d). The QD-TEM imaging emphasized the predominant presence of endogenous Sigmar1, especially on lysosomes and mitochondrial membranes, with a specific focus on the inner mitochondrial membrane. This study employed QD-mediated immunolabeling, offering a unique insight into Sigmar1's subcellular localization, particularly on the mitochondrial membrane, specifically the inner mitochondrial membrane, in cardiac tissue. Additionally, Sigmar1 was detected on sarcoplasmic/endoplasmic reticulum (S/ER) and lysosomes at the ultrastructural level, as depicted in Figure 4.11a–d.

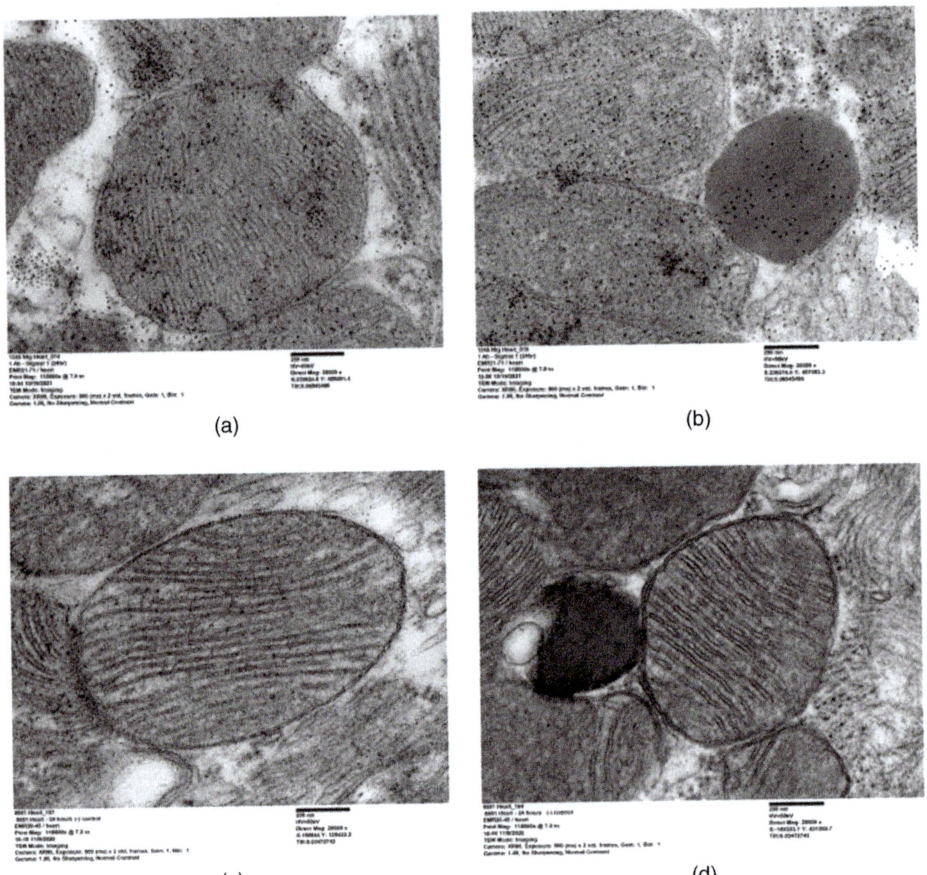

Figure 4.11 QD labeling of Sigmar1 in heart tissues of adult mice. (a,b) Representative TEM images illustrating Sigmar1-Labeled QDs on mitochondria (both outer and inner membranes), the interface between mitochondria and endoplasmic reticulum (ER)/sarcoplasmic reticulum (SR) membranes, and lysosomes. (c,d) TEM image of heart sections with an isotype control labeled with QDs and rabbit IgG. Scale Bar: 200 nm. Source: Reproduced with permission from Aishwarya et al. [131]/MyJoVE Corporation.

4.1.7 Studying Signaling Pathways

Fluorophores play a significant role in monitoring signaling pathways within or between cells. QDs have also found utility in this area. QDs linked to serotonin were applied to target neurotransmitter receptors on cell surfaces. Notably, these QD probes, in addition to recognizing and labeling serotonin-specific neurotransmitters, exhibited a dose-dependent inhibition of serotonin transport [132], thereby making the QD valuable probes for exploration of the mechanism beneath the serotonin transportation. Similarly, QD fluorescent probes have been used to study the dynamics of individual glycerin receptors (GlyRs) in neural membranes [28]. QDs labeled with GlyRs using primary antibody and secondary fragment bridges facilitated the capture of consecutive images for up to 20 minutes. Tracking QDs and their trajectories allowed for the characterization of three GlyR locations and their corresponding diffusion coefficients. Due to the significantly smaller size of QDs (10–15 nm) compared to beads (1 micron), the diffusion coefficients obtained using QDs are notably larger than those obtained with fluorescent beads. QDs have proven to be more suitable as probes for single-molecule experiments in living cells due to their enhanced photostability compared to organic dyes and smaller size relative to beads, bridging the gap between small organic fluorophores and larger beads. Additionally, QDs have been employed to investigate other signaling pathways, such as erbB-HER receptor-mediated transduction [133]. Here, the binding and activation of erbB receptors are done using EGF-conjugated QDs, triggering the internalization of both EGF-QD and its receptor via endocytosis. The high photostability of QDs allowed for studying and understanding the kinetics of the EGF-QD binding and internalization process, which is essential for receptor-targeted therapeutics [133]. QDs have been employed to investigate cell endocytosis. In this study, the efficiency of cell endocytosis was compared using sugar-conjugated QDs of sizes 5, 15, and 50 nm. The results indicated that cell endocytosis was dependent on particle size, with 50 nm particles being the most effective in triggering endocytosis, while 15 nm particles were somewhat less effective, and 5 nm particles were the least effective in inducing cell endocytosis [134] such a QD-based experiment proved to be highly informative for designing artificial molecular delivery systems such as drug and gene delivery.

4.1.8 Tumor Biology Investigation

The assessment of tumor pathology and drug delivery routes to tumors often involves the study of tumor vasculature. The imaging of blood cell distribution in the interstitial spaces has been achieved through the combination of QD imaging with second-harmonic generation (SHG) [135], which has been employed for imaging collagen in normal and cancerous tissues [136, 137]. An evaluation of tissue penetrability can be conducted by infusing QD microbeads with different sizes into tumor vasculature. The outcomes revealed varying distributions of microbeads within intra and extra-vascular compartments, offering valuable insights into the in vivo assessment of drug delivery to tumors. Additionally, QDs of diverse colors have been employed to label and track five distinct cell populations, facilitating the study of tumor cell extravasation and seeding [138, 139]. Similarly, QDs (QDs) have been employed in the study of tumor vasculogenesis by utilizing precursor cells derived from bone marrow. The images generated illustrated the dynamics of blood flow, cell adhesion, and cell rolling [139]. QDs have the potential to be used in assessing cell motility, which is closely associated with metastatic potential [140]. This assessment entails calculating phagokinetic tracks left behind as cells pass over marker layers and ingest them. QDs have been explored for this application, where the incorporation of QDs into the substrate enabled the observation of phagokinetic tracks in both human mammary epithelial cells and non-tumor cells [141].

In the realm of cancer diagnosis, treatment, and monitoring, tumor markers play a crucial role. The clinical effectiveness of traditional detection methods for these markers is hindered by their limited sensitivity. However, the utilization of QDs in the in vitro diagnosis of various tumor markers is on the rise, driven by the distinctive optical properties of QDs, their adaptable functional modifications, and the availability of diverse immunoassay techniques [142]. This shift toward QD-based detection signifies a technological advancement with the potential to enhance the precision and reliability of tumor marker evaluations in clinical settings [143]. Gazouli et al. [144] employed CdSe QDs to identify surface antigens on circulating tumor cells (CTCs) in colorectal cancer, utilizing the fluorescent signal of QDs for CTC detection. This approach highlights the potential of QD-based assays as promising tools for tumor antigen identification. Additionally, Gan and co-workers [145] developed an extremely sensitive electrochemiluminescence immunoassay for detecting carcinoembryonic antigen 19-9 (CA19-9) in serum. Their method incorporated QDs, graphene nanocomposites, and dextran-Fe_3O_4 magnetic nanoparticles. The process involved immobilizing anti-CA19-9 monoclonal antibodies on dextran-Fe_3O_4 magnetic nanoparticles to create capture probes for CA19-9 enrichment. Following this, a multitude of QDs were fixed onto the graphene surface, and an alternative form of anti-CA19-9 monoclonal antibody was applied to the QDs' surface. The dual antibody sandwich technique was employed to detect CA19-9 in serum. Through this method, numerous QDs can be firmly attached to the graphene surface, amplifying the emission signal strength and potentially allowing for extremely sensitive antigen detection. As reported, the minimum detectable limit for CA19-9 using this approach was 0.002 pg/mL. The study by Cao et al. [146] introduced an immunoassay approach for the detection of tumor biomarkers, utilizing QDs and polystyrene microspheres (PSMs) as a platform for immobilizing biomolecules. In this method, carcinoembryonic antigen (CEA) and neuron-specific enolase (NSE) were chosen as models, and two antibodies were immobilized on the surface of PSMs to capture CEA and NSE antigens, respectively. Subsequently, streptavidin-linked QDs were conjugated with another monoclonal antibody targeting CEA and NSE. The identification of tumor markers in serum was accomplished through the double-antibody sandwich method. The minimal detection limits for CEA and NSE were both found to be 0.625 ng/mL, demonstrating sensitivity beyond clinical cutoff levels. In addition to having greater fluorescence intensity and more durable photochemical properties than organic dyes, QDs showcase adjustable emission spectra influenced by their size. Furthermore, a singular excitation wavelength has the capability to activate multiple emission wavelengths, making QDs especially beneficial for the detection of multiple markers simultaneously. Tian et al. [147] introduced a method for simultaneously identifying alpha-fetoprotein (AFP) and carcinoembryonic antigen (CEA) in serum using multicolor QDs. AFP and CEA were labeled with QDs-520 and QDs-620 (the suffix indicating the emission wavelength), respectively. After undergoing ultrafiltration-based purification, a fluorescence polarization immunoassay was implemented. This approach operates on the principle of competition between the free (unlabeled) analyte and the fluorescently labeled antigen for binding sites on antibodies, enabling the detection of AFP and CEA in human serum. The method demonstrated significant reproducibility compared to the traditional enzyme-linked immunosorbent assay, boasting minimal detection limits of 0.28 ng/mL for AFP and 0.36 ng/mL for CEA. Its combination of high sensitivity, specificity, and ease of implementation renders this approach an attractive choice for clinical diagnosis involving AFP and CEA. In an independent investigation, Wu and colleagues [148] employed a combination of MBs and multicolor QDs to simultaneously detect cytokeratin 19 fragment (CYRFA 21-1), NSE, and carcinoembryonic antigen (CEA) in lung cancer cells using the double-antibody sandwich method. MBs were conjugated with monoclonal antibodies specific to CYRFA 21-1, CEA, and NSE, while activated multicolor QDs were coupled with a complementary pair of dithiothreitol-treated monoclonal antibodies. This facilitated the formation of an immune complex capable of capturing the three freely circulating tumor markers in the serum. The concentration of the target proteins was determined by monitoring the fluorescent signal emitted by the QD-conjugated probe, with detection limits of 364 pg/mL for CYRFA21-1, 38 pg/mL for CEA, and 370 pg/mL for NSE in a single assay. Wang et al. [149] developed a multicolor QD-based immuno chromatographic test strip (ICTS) for the simultaneous detection of various tumor markers, including alpha-fetoprotein (AFP) and carcinoembryonic antigen (CEA) in serum. Utilizing the double-antibody sandwich method, they attached QDs-546 to anti-AFP monoclonal antibodies and QDs-620 to anti-CEA monoclonal antibodies. A combination of mouse anti-AFP monoclonal antibodies and mouse anti-CEA monoclonal antibodies was employed to capture the resulting antigen-antibody complexes. The nitrocellulose membrane served as the test line, and the control line featured goat anti-mouse immunoglobulin G (IgG). Following this, the fluorescence intensity of the sample was assessed using an easily accessible test strip reader, allowing for the simultaneous quantitative detection of different tumor markers. Under optimal conditions, the minimum detection thresholds for alpha-fetoprotein (AFP) and carcinoembryonic antigen (CEA) were 3 ng/mL and 2 ng/mL, respectively. This ICTS utilizing multicolor QDs offers a simple and rapid approach for the simultaneous quantitative detection of multiple tumor markers.

4.2 Toxicity of Quantum Dots

The primary concern regarding the use of QDs in biological applications is their potential toxicity. Each individual QD possesses unique physicochemical properties that influence their toxicity. Several factors contribute to the toxicity of QDs, including (1) the size of the QDs, (2) the material used in their preparation, (3) the dosage of QDs administered, (4) the route of QD administration, and (5) the type of capping material used and environmental conditions. Toxicity can also be influenced by residual organic molecules. A review by Hardman in 2006 emphasized the importance of studying the toxicity of QDs, indicating that this concern has been ongoing for the past 15 years and is a critical consideration before introducing QDs for human use. Valizadeh, in 2012, conducted research on QD toxicity in amoebas, plants, and animals exposed to various drugs. Some examples of toxicity studies are discussed below. One of the most commonly used QDs is cadmium selenide (CdSe). Cadmium, a component of CdSe, is known for its toxicity. It tends to accumulate in the spleen and does not readily degrade. Reports indicate that cadmium-based QDs can lead to localized neutrophil inflammation in the lungs. Additionally, the oxidation of cadmium QDs results in the formation of a reduced form of cadmium, which can cause cell death. Another cause of cell death is the generation of singlet oxygen species through the oxidation of lipids and proteins within the cell. Sometimes, lysosomes and endosomes can contribute to the release of heavy metals from QDs. Inhalation exposure to QDs has been associated with toxicity, as QDs are taken up through endocytosis and can accumulate in various cells over weeks to months. Some significant toxic effects include pericardial edema, ocular edema, and spinal curvature. For example, CdSe QDs have been reported to cause acute toxicity, particularly when tested with primary hepatocytes as a liver model. This toxicity has been linked to the core of cadmium selenide (cadmium) binding to the sulfhydryl group of mitochondrial proteins, leading to hepatic injury. The surface oxidation of QDs can produce a reduced form of cadmium, which can be released from the QDs and result in cell death. The core of CdSe QDs has been associated with apoptosis. Coating the QDs with a shell has proven effective in reducing the leaching of core materials through cells and mitigating metal-induced toxicity. Group III and IV QDs exhibit fewer toxic effects and are commonly used in optical probes and biosensors. Potential approaches to address QD toxicity, such as using metal-free or carbon materials in QD preparation, have been explored. Another strategy to reduce QD toxicity is the use of biocompatible functionalization. This involves employing mercapto groups to mitigate the toxicity of QDs. For biological applications, QDs can be encapsulated in phospholipid micelles to maintain their optical properties without surface alterations. Additionally, techniques like conjugating QDs with antibodies have been employed [150, 151].

4.3 Conclusion and Future Outlook

The distinctive optoelectronic properties of QDs have led to a wide range of applications in the fields of nanobiotechnology and nanomedicine. Significant progress has been made in biology by enhancing the biocompatibility of QDs through methods such as polymer encapsulation, water solubility, and biodistribution. QDs have demonstrated several advantages over fluorescent dye molecules, including enhanced stability and brightness. These advantages position them to complement well-established techniques like fluorophores or fluorescent protein fusion, offering benefits like better photostability, single-molecule sensitivity over extended time frames, and improved near-infrared emissions. Biocompatible QDs are also strong candidates for applications in multimodal in vivo imaging. However, challenges related to the limited tissue penetration depth of excitation light and tissue autofluorescence have been addressed through the use of QDs with near-infrared fluorescence. Yet, significant concerns persist regarding QD usage, particularly in in vivo applications. These include toxicity issues, especially for heavy metal ion-based QDs like CdSe, suboptimal clearance of QD particles larger than 5 nm, QD aggregation in capillaries and glomerular spaces, degradation of QD coatings within animal bodies, and potential interactions between QDs and genetic materials. Addressing these concerns requires special attention. As a result, breakthroughs are anticipated with the development of non-blinking and non-toxic QDs, or QDs with biodegradable coatings. These advancements have the potential to revolutionize biomedical imaging, nanodiagnostics, and nanotherapy, with the goal of detecting and treating various health issues. Indeed, QDs have emerged as a versatile tool for improving diagnostic accuracy, fluoroimmunoassays, multiplexed imaging, dual imaging, therapeutic platforms, real-time in vivo and cellular process imaging, as well as tracking of single cells and biological molecules. The design and development of QDs have significantly advanced due to increased interest in inorganic particles over the last decade. Recent progress in synthesis and bioconjugation strategies has led to improved control over QD size, polydispersity, quantum yield, and surface properties. These flexible strategies enable QDs in complex hybrid nanomaterials while providing enhanced colloidal stability and a robust platform for further bioconjugation. However, there is a need to address the investigation, quantification, and control over the spatial distribution of functional groups in literature reports. This becomes more important with increasing biological complexity. While QDs have shown potential across various fields as biological probes, their most promising reports are valid for in vitro and ex vivo applications; QDs have surpassed traditional fluorophores in almost every aspect. Innovative conjugation approaches with biologically active molecules have paved the way for advanced bioimaging techniques. Moreover, the utilization of QDs with ex vivo patient tissues and fluids holds promise for enhancing disease understanding and diagnosis speed and aiding in drug identification. Despite these advances, significant limitations and regulatory issues remain before the application of QDs in medicine and in vivo imaging in human subjects [150]. The challenge arises from inefficient delivery resulting from low biological specificity and poorly controlled biodistribution to target tissues [152]. Furthermore, there is a limited understanding of the biological interaction of QDs reported in the literature. Addressing these limitations will be crucial for the clinical success of QDs. Efforts should focus on developing heavy metal-free QDs to reduce safety concerns associated with heavy metal-based QDs. Enhanced synthesis methods and a more profound comprehension of the structure-function correlation of QDs at biological interfaces are imperative for surmounting these hurdles and advancing QDs toward their clinical promise.

References

1 Weng J and Ren J 2006 Luminescent quantum dots: a very attractive and promising tool in biomedicine *Curr. Med. Chem.* 13 897–909

2 Brus L E 1984 Electron–electron and electron-hole interactions in small semiconductor crystallites: the size dependence of the lowest excited electronic state *J. Chem. Phys.* 80 4403–9

3 Anon 2010 Quantum dots in biomedical applications: advances and challenges *J. Nanophotonics* 4 042503

4 Dabbousi B O, Rodriguez-Viejo J, Mikulec F V, Heine J R, Mattoussi H, Ober R, Jensen K F and Bawendi M G 1997 (CdSe)ZnS core–shell quantum dots: synthesis and characterization of a size series of highly luminescent nanocrystallites *J. Phys. Chem. B* 101 9463–75

5 Talapin D V, Rogach A L, Kornowski A, Haase M and Weller H 2001 Highly luminescent monodisperse CdSe and CdSe/ZnS nanocrystals synthesized in a hexadecylamine–trioctylphosphine oxide–trioctylphospine mixture *Nano Lett.* 1 207–11

6 Jamieson T, Bakhshi R, Petrova D, Pocock R, Imani M and Seifalian A M 2007 Biological applications of quantum dots *Biomaterials* 28 4717–32

7 Dahan M, Laurence T, Pinaud F, Chemla D S, Alivisatos A P, Sauer M and Weiss S 2001 Time-gated biological imaging by use of colloidal quantum dots *Opt. Lett.* 26 825–7

8 Han M, Gao X, Su J Z and Nie S 2001 Quantum-dot-tagged microbeads for multiplexed optical coding of biomolecules *Nat. Biotechnol.* 19 631–5

9 Gao X and Nie S 2003 Doping mesoporous materials with multicolor quantum dots *J. Phys. Chem. B* 107 11575–8

10 Gao X and Nie S 2004 Quantum dot-encoded mesoporous beads with high brightness and uniformity: rapid readout using flow cytometry *Anal. Chem.* 76 2406–10

11 Jaiswal J K, Mattoussi H, Mauro J M and Simon S M 2003 Long-term multiple color imaging of live cells using quantum dot bioconjugates *Nat. Biotechnol.* 21 47–51

12 Chan W C W and Nie S 1998 Quantum dot bioconjugates for ultrasensitive nonisotopic detection *Science* 281 2016–8

13 Wu X, Liu H, Liu J, Haley K N, Treadway J A, Larson J P, Ge N, Peale F and Bruchez M P 2003 Immunofluorescent labeling of cancer marker Her2 and other cellular targets with semiconductor quantum dots *Nat. Biotechnol.* 21 41–6

14 Panchuk-Voloshina N, Haugland R P, Bishop-Stewart J, Bhalgat M K, Millard P J, Mao F, Leung W-Y and Haugland R P 1999 Alexa dyes, a series of new fluorescent dyes that yield exceptionally bright, photostable conjugates *J. Histochem. Cytochem.* 47 1179–88

15 Nirmal M and Brus L 1999 Luminescence photophysics in semiconductor nanocrystals *Acc. Chem. Res.* 32 407–14

16 Alivisatos A P 1996 Perspectives on the physical chemistry of semiconductor nanocrystals *J. Phys. Chem.* 100 13226–39

17 Murray C B, Kagan C R and Bawendi M G 2000 Synthesis and characterization of monodisperse nanocrystals and close-packed nanocrystal assemblies *Annu. Rev. Mater. Sci.* 30 545–610

18 Grecco H E, Lidke K A, Heintzmann R, Lidke D S, Spagnuolo C, Martinez O E, Jares Erijman E A and Jovin T M 2004 Ensemble and single particle photophysical properties (two-photon excitation, anisotropy, FRET, lifetime, spectral conversion) of commercial quantum dots in solution and in live cells *Microsc. Res. Tech.* 65 169–79

19 Noginov M A, Vondrova M, Williams S N, Bahoura M, Gavrilenko V I, Black S M, Drachev V P, Shalaev V M and Sykes A 2005 Spectroscopic studies of liquid solutions of R6G laser dye and Ag nanoparticle aggregates *J. Opt. A Pure Appl. Opt.* 7 S219

20 Gerion D, Pinaud F, Williams S C, Parak W J, Zanchet D, Weiss S and Alivisatos A P 2001 Synthesis and properties of biocompatible water-soluble silica-coated CdSe/ZnS semiconductor quantum dots *J. Phys. Chem. B* 105 8861–71

21 Pinaud F, Michalet X, Bentolila L A, Tsay J M, Doose S, Li J J, Iyer G and Weiss S 2006 Advances in fluorescence imaging with quantum dot bio-probes *Biomaterials* 27 1679–87

22 Zhan N, Palui G, Merkl J-P and Mattoussi H 2016 Bio-orthogonal coupling as a means of quantifying the ligand density on hydrophilic quantum dots *J. Am. Chem. Soc.* 138 3190–201

23 Jiang G, Susha A S, Lutich A A, Stefani F D, Feldmann J and Rogach A L 2009 Cascaded FRET in conjugated polymer/quantum dot/dye-labeled DNA complexes for DNA hybridization detection *ACS Nano* 3 4127–31

24 Clapp A R, Medintz I L, Mauro J M, Fisher B R, Bawendi M G and Mattoussi H 2004 Fluorescence resonance energy transfer between quantum dot donors and dye-labeled protein acceptors *J. Am. Chem. Soc.* 126 301–10

25 Medintz I L, Clapp A R, Brunel F M, Tiefenbrunn T, Tetsuo Uyeda H, Chang E L, Deschamps J R, Dawson P E and Mattoussi H 2006 Proteolytic activity monitored by fluorescence resonance energy transfer through quantum-dot–peptide conjugates *Nat. Mater.* 5 581–9

26 Ma N, Sargent E H and Kelley S O 2009 One-step DNA-programmed growth of luminescent and biofunctionalized nanocrystals *Nat. Nanotechnol.* 4 121–5

27 He X, Gao L and Ma N 2013 One-step instant synthesis of protein-conjugated quantum dots at room temperature *Sci. Rep.* 3 2825

28 Dahan M, Lévi S, Luccardini C, Rostaing P, Riveau B and Triller A 2003 Diffusion dynamics of glycine receptors revealed by single-quantum dot tracking *Science* 302 442–5

29 Zamir E, Lommerse P H, Kinkhabwala A, Grecco H E and Bastiaens P I 2010 Fluorescence fluctuations of quantum-dot sensors capture intracellular protein interaction dynamics *Nat. Methods* 7 295–8

30 Kobayashi H, Hama Y, Koyama Y, Barrett T, Regino C A S, Urano Y and Choyke P L 2007 Simultaneous multicolor imaging of five different lymphatic basins using quantum dots *Nano Lett.* 7 1711–6

31 Delehanty J B, Bradburne C E, Susumu K, Boeneman K, Mei B C, Farrell D, Blanco-Canosa J B, Dawson P E, Mattoussi H and Medintz I L 2011 Spatiotemporal multicolor labeling of individual cells using peptide-functionalized quantum dots and mixed delivery techniques *J. Am. Chem. Soc.* 133 10482–9

32 Barua S and Rege K 2009 Cancer-cell-phenotype-dependent differential intracellular trafficking of unconjugated quantum dots *Small* 5 370–6

33 Lee S, Jung K J, Jung H S and Chang S 2012 Dynamics of multiple trafficking behaviors of individual synaptic vesicles revealed by quantum-dot based presynaptic probe *PLoS One* 7 e38045

34 Biermann B, Sokoll S, Klueva J, Missler M, Wiegert J S, Sibarita J-B and Heine M 2014 Imaging of molecular surface dynamics in brain slices using single-particle tracking *Nat. Commun.* 5 3024

35 Chen M-L, He Y-J, Chen X-W and Wang J-H 2013 Quantum-dot-conjugated graphene as a probe for simultaneous cancer-targeted fluorescent imaging, tracking, and monitoring drug delivery *Bioconjug. Chem.* 24 387–97

36 Clausen M P and Lagerholm B C 2013 Visualization of plasma membrane compartmentalization by high-speed quantum dot tracking *Nano Lett.* 13 2332–7

37 Xu J, Teslaa T, Wu T-H, Chiou P-Y, Teitell M A and Weiss S 2012 Nanoblade delivery and incorporation of quantum dot conjugates into tubulin networks in live cells *Nano Lett.* 12 5669–72

38 Zrazhevskiy P and Gao X 2013 Quantum dot imaging platform for single-cell molecular profiling *Nat. Commun.* 4 1619

39 Ag D, Bongartz R, Dogan L E, Seleci M, Walter J-G, Demirkol D O, Stahl F, Ozcelik S, Timur S and Scheper T 2014 Biofunctional quantum dots as fluorescence probe for cell-specific targeting *Colloids Surf. B Biointerfaces* 114 96–103

40 Jeyadevi R, Sivasudha T, Rameshkumar A, Ananth D A, Aseervatham G S B, Kumaresan K, Kumar L D, Jagadeeswari S and Renganathan R 2013 Enhancement of anti arthritic effect of quercetin using thioglycolic acid-capped cadmium telluride quantum dots as nanocarrier in adjuvant induced arthritic Wistar rats *Colloids Surf. B Biointerfaces* 112 255–63

41 Mansur A A P, Saliba J B and Mansur H S 2013 Surface modified fluorescent quantum dots with neurotransmitter ligands for potential targeting of cell signaling applications *Colloids Surf. B Biointerfaces* 111 60–70

42 Schieber C, Bestetti A, Lim J P, Ryan A D, Nguyen T, Eldridge R, White A R, Gleeson P A, Donnelly P S, Williams S J and Mulvaney P 2012 Conjugation of transferrin to azide-modified CdSe/ZnS core–shell quantum dots using cyclooctyne click chemistry *Angew. Chem.* 124 10675–9

43 Feng L, Long H-Y, Liu R-K, Sun D-N, Liu C, Long L-L, Li Y, Chen S and Xiao B 2013 A quantum dot probe conjugated with Aβ antibody for molecular imaging of Alzheimer's disease in a mouse model *Cell. Mol. Neurobiol.* 33 759–65

44 Narayanan K, Yen S K, Dou Q, Padmanabhan P, Sudhaharan T, Ahmed S, Ying J Y and Selvan S T 2013 Mimicking cellular transport mechanism in stem cells through endosomal escape of new peptide-coated quantum dots *Sci. Rep.* 3 2184

45 Chen Y, Molnar M, Li L, Friberg P, Gan L-M, Brismar H and Fu Y 2013 Characterization of VCAM-1-binding peptide-functionalized quantum dots for molecular imaging of inflamed endothelium *PLoS One* 8 e83805

46 Chen L Q, Xiao S J, Hu P P, Peng L, Ma J, Luo L F, Li Y F and Huang C Z 2012 Aptamer-mediated nanoparticle-based protein labeling platform for intracellular imaging and tracking endocytosis dynamics *Anal. Chem.* 84 3099–110

47 Chu T C, Shieh F, Lavery L A, Levy M, Richards-Kortum R, Korgel B A and Ellington A D 2006 Labeling tumor cells with fluorescent nanocrystal–aptamer bioconjugates *Biosens. Bioelectron.* 21 1859–66

48 Farokhzad O C, Jon S, Khademhosseini A, Tran T-N T, LaVan D A and Langer R 2004 Nanoparticle-aptamer bioconjugates: a new approach for targeting prostate cancer cells *Cancer Res.* 64 7668–72

49 Zhang Y, Liu J-M and Yan X-P 2013 Self-assembly of folate onto polyethyleneimine-coated CdS/ZnS quantum dots for targeted turn-on fluorescence imaging of folate receptor overexpressed cancer cells *Anal. Chem.* 85 228–34

50 Song E-Q, Zhang Z-L, Luo Q-Y, Lu W, Shi Y-B and Pang D-W 2009 Tumor cell targeting using folate-conjugated fluorescent quantum dots and receptor-mediated endocytosis *Clin. Chem.* 55 955–63

51 He X and Ma N 2014 An overview of recent advances in quantum dots for biomedical applications *Colloids Surf. B Biointerfaces* 124 118–31

52 Delehanty J B, Medintz I L, Pons T, Brunel F M, Dawson P E and Mattoussi H 2006 Self-assembled quantum dot–peptide bioconjugates for selective intracellular delivery *Bioconjug. Chem.* 17 920–7

53 Kim B Y S, Jiang W, Oreopoulos J, Yip C M, Rutka J T and Chan W C W 2008 Biodegradable quantum dot nanocomposites enable live cell labeling and imaging of cytoplasmic targets *Nano Lett.* 8 3887–92

54 Yum K, Wang N and Yu M-F 2010 Electrochemically-controlled deconjugation and delivery of single quantum dots into the nucleus of living cells *Small Weinh. Bergstr. Ger.* 6 2109

55 Smith A M and Nie S 2012 Compact quantum dots for single-molecule imaging *JoVE J. Vis. Exp.* e4236

56 Baba K and Nishida K 2012 Single-molecule tracking in living cells using single quantum dot applications *Theranostics* 2 655

57 Yeh H-Y, Yates M V, Mulchandani A and Chen W 2010 Molecular beacon–quantum dot–Au nanoparticle hybrid nanoprobes for visualizing virus replication in living cells *Chem. Commun.* 46 3914–6

58 Wells N P, Lessard G A, Goodwin P M, Phipps M E, Cutler P J, Lidke D S, Wilson B S and Werner J H 2010 Time-resolved three-dimensional molecular tracking in live cells *Nano Lett.* 10 4732–7

59 de Mel A, Oh J-T, Ramesh B and Seifalian A M 2012 Biofunctionalized quantum dots for live monitoring of stem cells: applications in regenerative medicine *Regen. Med.* 7 335–47

60 Clausen M P, Arnspang E C, Ballou B, Bear J E and Lagerholm B C 2014 Simultaneous multi-species tracking in live cells with quantum dot conjugates *PLoS One* 9 e97671

61 Costa C R M, Feitosa M L T, Bezerra D O, Carvalho Y K P, Olivindo R F G, Fernando P B, Silva G C, Silva M L G, Ambrósio C E, Conde Júnior A M, Argolo Neto N M, Costa Silva L M and Carvalho M A M 2017 Labeling of adipose-derived stem cells with quantum dots provides stable and long-term fluorescent signal for ex vivo cell tracking *In Vitro Cell. Dev. Biol. Anim.* 53 363–70

62 Keller A M, Ghosh Y, DeVore M S, Phipps M E, Stewart M H, Wilson B S, Lidke D S, Hollingsworth J A and Werner J H 2014 3-Dimensional tracking of non-blinking 'Giant' quantum dots in live cells *Adv. Funct. Mater.* 24 4796–803

63 Liu H-Y, Wang Z-G, Liu S-L and Pang D-W 2023 Single-virus tracking with quantum dots in live cells *Nat. Protoc.* 18 458–89

64 Yang Y-B, Tang Y-D, Hu Y, Yu F, Xiong J-Y, Sun M-X, Lyu C, Peng J-M, Tian Z-J, Cai X-H and An T-Q 2020 Single virus tracking with quantum dots packaged into enveloped viruses using CRISPR *Nano Lett.* 20 1417–27

65 Wang Z-G, Liu S-L and Pang D-W 2021 Quantum dots: a promising fluorescent label for probing virus trafficking *Acc. Chem. Res.* 54 2991–3002

66 Lim Y T, Kim S, Nakayama A, Stott N E, Bawendi M G and Frangioni J V 2003 Selection of quantum dot wavelengths for biomedical assays and imaging *Mol. Imaging* 2 https://doi.org/10.1162/15353500200302163

67 Frangioni J V 2003 In vivo near-infrared fluorescence imaging *Curr. Opin. Chem. Biol.* 7 626–34

68 So M-K, Xu C, Loening A M, Gambhir S S and Rao J 2006 Self-illuminating quantum dot conjugates for in vivo imaging *Nat. Biotechnol.* 24 339–43

69 Brannon-Peppas L and Blanchette J O 2004 Nanoparticle and targeted systems for cancer therapy *Adv. Drug Deliv. Rev.* 56 1649–59

70 Peracchia M T, Fattal E, Desmaele D, Besnard M, Noel J P, Gomis J M, Appel M, d'Angelo J and Couvreur P 1999 Stealth® PEGylated polycyanoacrylate nanoparticles for intravenous administration and splenic targeting *J. Control. Release* 60 121–8

71 Åkerman M E, Chan W C, Laakkonen P, Bhatia S N and Ruoslahti E 2002 Nanocrystal targeting in vivo *Proc. Natl. Acad. Sci.* 99 12617–21

72 Cai W, Shin D-W, Chen K, Gheysens O, Cao Q, Wang S X, Gambhir S S and Chen X 2006 Peptide-labeled near-infrared quantum dots for imaging tumor vasculature in living subjects *Nano Lett.* 6 669–76

73 Gao X, Cui Y, Levenson R M, Chung L W and Nie S 2004 In vivo cancer targeting and imaging with semiconductor quantum dots *Nat. Biotechnol.* 22 969–76

74 Morgan N Y, English S, Chen W, Chernomordik V, Russo A, Smith P D and Gandjbakhche A 2005 Real time in vivo non-invasive optical imaging using near-infrared fluorescent quantum dots1 *Acad. Radiol.* 12 313–23

75 Chen L, Zurita A J, Ardelt P U, Giordano R J, Arap W and Pasqualini R 2004 Design and validation of a bifunctional ligand display system for receptor targeting *Chem. Biol.* 11 1081–91

76 Ballou B, Lagerholm B C, Ernst L A, Bruchez M P and Waggoner A S 2004 Noninvasive imaging of quantum dots in mice *Bioconjug. Chem.* 15 79–86

77 Dyadyusha L, Yin H, Jaiswal S, Brown T, Baumberg J J, Booy F P and Melvin T 2005 Quenching of CdSe quantum dot emission, a new approach for biosensing *Chem. Commun.* (25) 3201–3

78 Chang S S, Reuter V E, Heston W D W and Gaudin P B 2001 Metastatic renal cell carcinoma neovasculature expresses prostate-specific membrane antigen *Urology* 57 801–5

79 Hoffman R M 2002 Green fluorescent protein imaging of tumour growth, metastasis, and angiogenesis in mouse models *Lancet Oncol.* 3 546–56

80 Yang M, Baranov E, Wang J-W, Jiang P, Wang X, Sun F-X, Bouvet M, Moossa A, Penman S and Hoffman R M 2002 Direct external imaging of nascent cancer, tumor progression, angiogenesis, and metastasis on internal organs in the fluorescent orthotopic model *Proc. Natl. Acad. Sci.* 99 3824–9

81 Larson D R, Zipfel W R, Williams R M, Clark S W, Bruchez M P, Wise F W and Webb W W 2003 Water-soluble quantum dots for multiphoton fluorescence imaging in vivo *Science* 300 1434–6

82 Kim S, Lim Y T, Soltesz E G, De Grand A M, Lee J, Nakayama A, Parker J A, Mihaljevic T, Laurence R G and Dor D M 2004 Near-infrared fluorescent type II quantum dots for sentinel lymph node mapping *Nat. Biotechnol.* 22 93–7

83 Ricciardi S, Carleo F, Jaus M O, Di Martino M, Carbone L, Ricci A and Cardillo G 2021 Malignant pleural mesothelioma nodal status: where are we at? *J. Clin. Med.* 10 5177

84 Soltesz E G, Kim S, Laurence R G, DeGrand A M, Parungo C P, Dor D M, Cohn L H, Bawendi M G, Frangioni J V and Mihaljevic T 2005 Intraoperative sentinel lymph node mapping of the lung using near-infrared fluorescent quantum dots *Ann. Thorac. Surg.* 79 269–77

85 Geho D, Lahar N, Gurnani P, Huebschman M, Herrmann P, Espina V, Shi A, Wulfkuhle J, Garner H, Petricoin E, Liotta L A and Rosenblatt K P 2005 Pegylated, steptavidin-conjugated quantum dots are effective detection elements for reverse-phase protein microarrays *Bioconjug. Chem.* 16 559–66

86 Giepmans B N, Deerinck T J, Smarr B L, Jones Y Z and Ellisman M H 2005 Correlated light and electron microscopic imaging of multiple endogenous proteins using quantum dots *Nat. Methods* 2 743–9

87 Chen W, Ouyang J, Liu H, Chen M, Zeng K, Sheng J, Liu Z, Han Y, Wang L, Li J, Deng L, Liu Y-N and Guo S 2017 Black phosphorus nanosheet-based drug delivery system for synergistic photodynamic/photothermal/chemotherapy of cancer *Adv. Mater.* 29 1603864

88 Park Y I, Kim H M, Kim J H, Moon K C, Yoo B, Lee K T, Lee N, Choi Y, Park W, Ling D, Na K, Moon W K, Choi S H, Park H S, Yoon S-Y, Suh Y D, Lee S H and Hyeon T 2012

Theranostic probe based on lanthanide-doped nanoparticles for simultaneous in vivo dual-modal imaging and photodynamic therapy *Adv. Mater.* 24 5755–61

89 Juzenas P, Chen W, Sun Y-P, Coelho M A N, Generalov R, Generalova N and Christensen I L 2008 Quantum dots and nanoparticles for photodynamic and radiation therapies of cancer *Adv. Drug Deliv. Rev.* 60 1600–14

90 Markovic Z M, Ristic B Z, Arsikin K M, Klisic D G, Harhaji-Trajkovic L M, Todorovic-Markovic B M, Kepic D P, Kravic-Stevovic T K, Jovanovic S P, Milenkovic M M, Milivojevic D D, Bumbasirevic V Z, Dramicanin M D and Trajkovic V S 2012 Graphene quantum dots as autophagy-inducing photodynamic agents *Biomaterials* 33 7084–92

91 Ge J, Jia Q, Liu W, Guo L, Liu Q, Lan M, Zhang H, Meng X and Wang P 2015 Red-emissive carbon dots for fluorescent, photoacoustic, and thermal theranostics in living mice *Adv. Mater.* 27 4169–77

92 Kuo W-S, Chen H-H, Chen S-Y, Chang C-Y, Chen P-C, Hou Y-I, Shao Y-T, Kao H-F, Lilian Hsu C-L, Chen Y-C, Chen S-J, Wu S-R and Wang J-Y 2017 Graphene quantum dots with nitrogen-doped content dependence for highly efficient dual-modality photodynamic antimicrobial therapy and bioimaging *Biomaterials* 120 185–94

93 Savla R, Taratula O, Garbuzenko O and Minko T 2011 Tumor targeted quantum dot-mucin 1 aptamer-doxorubicin conjugate for imaging and treatment of cancer *J. Control. Release* 153 16–22

94 Yang X, Zhang W, Zhao Z, Li N, Mou Z, Sun D, Cai Y, Wang W and Lin Y 2017 Quercetin loading CdSe/ZnS nanoparticles as efficient antibacterial and anticancer materials *J. Inorg. Biochem.* 167 36–48

95 Whitehead K A, Langer R and Anderson D G 2009 Knocking down barriers: advances in siRNA delivery *Nat. Rev. Drug Discov.* 8 129–38

96 Chen A A, Derfus A M, Khetani S R and Bhatia S N 2005 Quantum dots to monitor RNAi delivery and improve gene silencing *Nucleic Acids Res.* 33 e190

97 Li J-M, Zhao M-X, Su H, Wang Y-Y, Tan C-P, Ji L-N and Mao Z-W 2011 Multifunctional quantum-dot-based siRNA delivery for HPV18 E6 gene silence and intracellular imaging *Biomaterials* 32 7978–87

98 Lee L Y, Ong S L, Hu J Y, Ng W J, Feng Y, Tan X and Wong S W 2004 Use of semiconductor quantum dots for photostable immunofluorescence labeling of *Cryptosporidium parvum Appl. Environ. Microbiol.* 70 5732–6

99 Zhu L, Ang S and Liu W-T 2004 Quantum dots as a novel immunofluorescent detection system for *Cryptosporidium parvum* and *Giardia lamblia Appl. Environ. Microbiol.* 70 597–8

100 Tully E, Hearty S, Leonard P and O'Kennedy R 2006 The development of rapid fluorescence-based immunoassays, using quantum dot-labelled antibodies for the detection of *Listeria monocytogenes* cell surface proteins *Int. J. Biol. Macromol.* 39 127–34

101 Kloepfer J A, Mielke R E, Wong M S, Nealson K H, Stucky G and Nadeau J L 2003 Quantum dots as strain- and metabolism-specific microbiological labels *Appl. Environ. Microbiol.* 69 4205–13

102 Modun B, Morrissey J and Williams P 2000 The staphylococcal transferrin receptor: a glycolytic enzyme with novel functions *Trends Microbiol.* 8 231–7

103 Gerion D, Chen F, Kannan B, Fu A, Parak W J, Chen D J, Majumdar A and Alivisatos A P 2003 Room-temperature single-nucleotide polymorphism and multiallele DNA detection using fluorescent nanocrystals and microarrays *Anal. Chem.* 75 4766–72

104 Agrawal A, Tripp R A, Anderson L J and Nie S 2005 Real-time detection of virus particles and viral protein expression with two-color nanoparticle probes *J. Virol.* 79 8625–8

105 Choi A O, Cho S J, Desbarats J, Lovrić J and Maysinger D 2007 Quantum dot-induced cell death involves Fas upregulation and lipid peroxidation in human neuroblastoma cells *J. Nanobiotechnology* 5 1

106 Goldman E R, Anderson G P, Tran P T, Mattoussi H, Charles P T and Mauro J M 2002 Conjugation of luminescent quantum dots with antibodies using an engineered adaptor protein to provide new reagents for fluoroimmunoassays *Anal. Chem.* 74 841–7

107 Goldman E R, Clapp A R, Anderson G P, Uyeda H T, Mauro J M, Medintz I L and Mattoussi H 2004 Multiplexed toxin analysis using four colors of quantum dot fluororeagents *Anal. Chem.* 76 684–8

108 Dogan Ü, Kasap E, Cetin D, Suludere Z, Boyaci I H, Türkyılmaz C, Ertas N and Tamer U 2016 Rapid detection of bacteria based on homogenous immunoassay using chitosan modified quantum dots *Sensors Actuators B Chem.* 233 369–78

109 Martynenko I V, Kusić D, Weigert F, Stafford S, Donnelly F C, Evstigneev R, Gromova Y, Baranov A V, Rühle B, Kunte H-J, Gun'ko Y K and Resch-Genger U 2019 Magneto-fluorescent microbeads for bacteria detection constructed from superparamagnetic Fe_3O_4 nanoparticles and AIS/ZnS quantum dots *Anal. Chem.* 91 12661–9

110 Yin B, Wang Y, Dong M, Wu J, Ran B, Xie M, Joo S W and Chen Y 2016 One-step multiplexed detection of foodborne pathogens: combining a quantum dot-mediated reverse assaying strategy and magnetic separation *Biosens. Bioelectron.* 86 996–1002

111 Cihalova K, Hegerova D, Jimenez A M, Milosavljevic V, Kudr J, Skalickova S, Hynek D, Kopel P, Vaculovicova M and Adam V 2017 Antibody-free detection of infectious bacteria using quantum dots-based barcode assay *J. Pharm. Biomed. Anal.* 134 325–32

112 Lv R, Chen Y, Xia N, Liang Y, He Q, Li M, Qi Z, Lu Y and Zhao S 2019 Development of a double-antibody sandwich ELISA for rapid detection to C-peptide in human urine *J. Pharm. Biomed. Anal.* 162 179–84

113 Song D, Qu X, Liu Y, Li L, Yin D, Li J, Xu K, Xie R, Zhai Y, Zhang H, Bao H, Zhao C, Wang J, Song X and Song W 2017 A rapid detection method of brucella with quantum dots and magnetic beads conjugated with different polyclonal antibodies *Nanoscale Res. Lett.* 12 179

114 He K, Zhang X, Chen L, Zhao R and Wang L 2018 Sensitive detection of *Campylobacter jejuni* using one-step strategy based on functional nanospheres of immunomagnetic capture and quantum dots *Ecotoxicol. Environ. Saf.* 163 274–8

115 Zhong M, Yang L, Yang H, Cheng C, Deng W, Tan Y, Xie Q and Yao S 2019 An electrochemical immunobiosensor for ultrasensitive detection of *Escherichia coli* O157:H7 using CdS quantum dots-encapsulated metal-organic frameworks as signal-amplifying tags *Biosens. Bioelectron.* 126 493–500

116 Ravikumar C H, Shwetharani R and Balakrishna R G 2020 Surface modified glass substrate for sensing *E. coli* using highly stable and luminescent CdSe/CdS core shell quantum dots *J. Photochem. Photobiol. B* 204 111799

117 Hao L, Xue L, Huang F, Cai G, Qi W, Zhang M, Han Q A, Wang Z and Lin J 2020 A microfluidic biosensor based on magnetic nanoparticle separation, quantum dots labeling and mno2 nanoflower amplification for rapid and sensitive detection of *Salmonella typhimurium* *Micromachines* 11(3) 281

118 Zhang Q, Liu Y, Nie Y, Ma Q and Zhao B 2019 Surface plasmon coupling electrochemiluminescence assay based on the use of AuNP@C3N4QD@mSiO2 for the determination of the Shiga toxin-producing *Escherichia coli* (STEC) gene *Microchim. Acta* 186 656

119 Hu J, Tang F, Jiang Y-Z and Liu C 2020 Rapid screening and quantitative detection of Salmonella using a quantum dot nanobead-based biosensor *Analyst* 145 2184–90

120 Yang Q, Farooq U, Chen W, Ullah M W and Wang S 2020 Fluorimetric detection of single pathogenic bacterium in milk and sewage water using pH-sensitive fluorescent carbon dots and MALDI-TOF MS *Microorganisms* 8 53

121 Bruchez M, Moronne M, Gin P, Weiss S and Alivisatos A P 1998 Semiconductor nanocrystals as fluorescent biological labels *Science* 281 2013–6

122 Tokumasu F and Dvorak J 2003 Development and application of quantum dots for immunocytochemistry of human erythrocytes *J. Microsc.* 211(3) 256–61

123 Kaul Z, Yaguchi T, Kaul S C, Hirano T, Wadhwa R and Taira K 2003 Mortalin imaging in normal and cancer cells with quantum dot immuno-conjugates *Cell Res.* 13 503–7

124 Regulacio M D, Win K Y, Lo S L, Zhang S-Y, Zhang X, Wang S, Han M-Y and Zheng Y 2013 Aqueous synthesis of highly luminescent AgInS 2–ZnS quantum dots and their biological applications *Nanoscale* 5 2322–7

125 Ness J M, Akhtar R S, Latham C B and Roth K A 2003 Combined tyramide signal amplification and quantum dots for sensitive and photostable immunofluorescence detection *J. Histochem. Cytochem.* 51 981–7

126 Zdobnova T A, Dorofeev S G, Tananaev P N, Vasiliev R B, Balandin T G, Edelweiss E F, Stremovskiy O A, Balalaeva I V, Turchin I V, Lebedenko E N, Zlomanov V P and Deyev S M 2009 Fluorescent immunolabeling of cancer cells by quantum dots and antibody scFv fragment *J. Biomed. Opt.* 14 021004

127 Urban J M, Chiang W, Hammond J W, Cogan N M B, Litzburg A, Burke R, Stern H A, Gelbard H A, Nilsson B L and Krauss T D 2021 Quantum dots for improved single-molecule localization microscopy *J. Phys. Chem. B* 125 2566–76

128 Yang D, Chen Q, Wang W and Xu S 2008 Direct and indirect immunolabelling of HeLa cells with quantum dots *Lumin. J. Biol. Chem. Lumin.* 23 169–74

129 Zhang H, Sun P, Liu C, Gao H, Xu L, Fang J, Wang M, Liu J and Xu S 2011 L-Cysteine capped CdTe–CdS core–shell quantum dots: preparation, characterization and immuno-labeling of HeLa cells *Luminescence* 26 86–92

130 Mahmoud W, Rousserie G, Reveil B, Tabary T, Millot J-M, Artemyev M, Oleinikov V A, Cohen J H M, Nabiev I and Sukhanova A 2011 Advanced procedures for labeling of antibodies with quantum dots *Anal. Biochem.* 416 180–5

131 Aishwarya R, Abdullah C S, Remex N S, Nitu S, Hartman B, King J and Bhuiyan M S 2022 Visualizing subcellular localization of a protein in the heart using quantum dots-mediated immuno-labeling followed by transmission electron microscopy *J. Vis. Exp.* (187) e64085

132 Rosenthal S J, Tomlinson I, Adkins E M, Schroeter S, Adams S, Swafford L, McBride J, Wang Y, DeFelice L J and Blakely R D 2002 Targeting cell surface receptors with ligand-conjugated nanocrystals *J. Am. Chem. Soc.* 124 4586–94

133 Lidke D S, Nagy P, Heintzmann R, Arndt-Jovin D J, Post J N, Grecco H E, Jares-Erijman E A and Jovin T M 2004 Quantum dot ligands provide new insights into erbB/HER receptor–mediated signal transduction *Nat. Biotechnol.* 22 198

134 Osaki F, Kanamori T, Sando S, Sera T and Aoyama Y 2004 A quantum dot conjugated sugar ball and its cellular uptake. On the size effects of endocytosis in the subviral region *J. Am. Chem. Soc.* 126 6520–1

135 Williams R M, Zipfel W R and Webb W W 2001 Multiphoton microscopy in biological research *Curr. Opin. Chem. Biol.* 5 603–8

136 Zoumi A, Yeh A and Tromberg B J 2002 Imaging cells and extracellular matrix *in vivo* by using second-harmonic generation and two-photon excited fluorescence *Proc. Natl. Acad. Sci.* 99 11014–9

137 Brown E, McKee T, DiTomaso E, Pluen A, Seed B, Boucher Y and Jain R K 2003 Dynamic imaging of collagen and its modulation in tumors in vivo using second-harmonic generation *Nat. Med.* 9 796–800

138 Voura E B, Jaiswal J K, Mattoussi H and Simon S M 2004 Tracking metastatic tumor cell extravasation with quantum dot nanocrystals and fluorescence emission-scanning microscopy *Nat. Med.* 10 993–8

139 Stroh M, Zimmer J P, Duda D G, Levchenko T S, Cohen K S, Brown E B, Scadden D T, Torchilin V P, Bawendi M G and Fukumura D 2005 Quantum dots spectrally distinguish multiple species within the tumor milieu in vivo *Nat. Med.* 11 678–82

140 Partin A W, Schoeniger J S, Mohler J L and Coffey D S 1989 Fourier analysis of cell motility: correlation of motility with metastatic potential *Proc. Natl. Acad. Sci.* 86 1254–8

141 Parak W J, Boudreau R, Le Gros M, Gerion D, Zanchet D, Micheel C M, Williams S C, Alivisatos A P and Larabell C 2002 Cell motility and metastatic potential studies based on quantum dot imaging of phagokinetic tracks *Adv. Mater.* 14 882

142 Fang C-C, Chou C-C, Yang Y-Q, Wei-Kai T, Wang Y-T and Chan Y-H 2018 Multiplexed detection of tumor markers with multicolor polymer dot-based immunochromatography test strip *Anal. Chem.* 90 2134–40

143 Liu B, Jiang B, Zheng Z and Liu T 2019 Semiconductor quantum dots in tumor research *J. Lumin.* 209 61–8

144 Gazouli M, Lyberopoulou A, Pericleous P, Rizos S, Aravantinos G, Nikiteas N, Anagnou N P and Efstathopoulos E P 2012 Development of a quantum-dot-labelled magnetic immunoassay method for circulating colorectal cancer cell detection *World J Gastroenterol: WJG* 18 4419–26

145 Gan N, Zhou J, Xiong P, Li T, Jiang S, Cao Y and Jiang Q 2013 An ultrasensitive electrochemiluminescence immunoassay for carbohydrate antigen 19-9 in serum based on antibody labeled Fe_3O_4 nanoparticles as capture probes and graphene/CdTe quantum dot bionanoconjugates as signal amplifiers *Int. J. Mol. Sci.* 14 10397–411

146 Cao Z, Li H, Lau C and Zhang Y 2011 Cross-talk-free simultaneous fluoroimmunoassay of two biomarkers based on dual-color quantum dots *Anal. Chim. Acta* 698 44–50

147 Tian J, Zhou L, Zhao Y, Wang Y, Peng Y and Zhao S 2012 Multiplexed detection of tumor markers with multicolor quantum dots based on fluorescence polarization immunoassay *Talanta* 92 72–7

148 Wu S, Liu L, Li G, Jing F, Mao H, Jin Q, Zhai W, Zhang H, Zhao J and Jia C 2016 Multiplexed detection of lung cancer biomarkers based on quantum dots and microbeads *Talanta* 156–157 48–54

149 Wang C, Hou F and Ma Y 2015 Simultaneous quantitative detection of multiple tumor markers with a rapid and sensitive multicolor quantum dots based immunochromatographic test strip *Biosens. Bioelectron.* 68 156–62

150 Zhao M-X and Zhu B-J 2016 The research and applications of quantum dots as nano-carriers for targeted drug delivery and cancer therapy *Nanoscale Res. Lett.* 11 207

151 Bachtold A, Fuhrer M S, Plyasunov S, Forero M, Anderson E H, Zettl A and McEuen P L 2000 Scanned probe microscopy of electronic transport in carbon nanotubes *Phys. Rev. Lett.* 84 6082–5

152 Yong K-T 2019 Biophotonic Imaging and Sensing *2019 24th Microoptics Conference (MOC)*, IEEE, pp 20–1

5

Biomedical Applications of Carbon-Based Nanomaterials

Ajit Khosla[1], Irshad A. Wani[2], and Mohammad N. Lone[3]

[1] *School of Advanced Materials and Nanotechnology, Xidian University, Xian, Shaanxi, China*
[2] *Department of Chemistry, Govt. Degree College Anantnag, Anantnag, University of Kashmir, Jammu and Kashmir, India*
[3] *Department of Chemistry, Central University of Kashmir, Ganderbal, Jammu and Kashmir, India*

5.1 Introduction

The concept of nanotechnology originates from the Greek word "nano," which translates to "dwarf." Nanotechnology focuses on materials within the size range of 0.1–100 nm [1]. Nanotechnology draws upon the practical principles of materials science, physics, electronics, and engineering at the submicron or molecular scale. It encompasses a diverse range of materials, including nanoparticles, quantum dots, dendrimers, nanoemulsions, and more. One prominent category among nanomaterials is carbon-based nanomaterials, which consist solely of carbon and are part of the fullerene (C60) family. These carbon-based nanomaterials exist in various forms, including spheres, ellipsoids, and tubes [2, 3]. Carbon-based nanomaterials, such as CNTs, are indeed cylindrical structures with hollow interiors. They are produced through polymerization reactions using carbon-based monomers. Carbon nanotubes (CNTs) fall under the category of carbon allotropes and were initially observed by Iijima in 1991 [4, 5]. CNTs exhibit remarkable properties, including high stiffness in their carbon–carbon (C–C) bonds, with a tensile strength of nearly 150 GPa (Giga Pascals) and a Young's modulus close to 1. These exceptional mechanical characteristics make them ideal for various biomedical applications. Their combination of high tensile strength, substantial surface area, and exceptional absorption capabilities has led to their use as follows.

Controlled Drug Delivery: CNTs can be functionalized to carry and release drugs in a controlled manner, improving the effectiveness of drug delivery systems.

Solubility Enhancement: CNTs can enhance the solubility of poorly water-soluble drugs, making them more bioavailable.

Gene Delivery: They are employed in gene therapy for delivering genetic material to target cells.
Vaccine Delivery: CNTs can serve as carriers for vaccines, aiding in the controlled release of vaccine components.

Enzyme and Hormone Delivery: CNTs have applications in delivering enzymes and hormones, offering precise delivery mechanisms.

Due to these unique properties, CNTs are gaining attention in the field of biomedical research and are contributing to the development of innovative healthcare technologies [6]. Indeed, CNTs and

Metallic, Magnetic and Carbon-Based Nanomaterials: Synthesis and Biomedical Applications, First Edition.
Ajit Khosla, Irshad A. Wani, and Mohammad N. Lone.

other carbon-based nanomaterials play a crucial role in the development of advanced diagnostic tools for the identification of various diseases. These include:

Nanoprobes: CNTs can be functionalized with specific molecules or antibodies that bind to disease markers or biomolecules. They are used as nanoprobes to target and identify disease-related substances in biological samples, aiding in early disease detection.

Nanorobots: Although still in the experimental stage, the concept of nanorobots involves tiny robotic devices made from nanoscale materials like CNTs. These nanorobots can be designed to perform specific tasks, such as targeting and treating cancer cells or delivering drugs to specific locations in the body while minimizing side effects.

Nanosensors: CNTs can be integrated into nanosensors to detect specific molecules or biomarkers associated with diseases. They provide high sensitivity and selectivity, making them valuable tools for rapid and accurate disease diagnosis.

Actuators: In some diagnostic applications, CNTs can serve as actuators, responding to external stimuli or signals. For instance, they can change their properties in the presence of certain biomarkers, providing a signal for disease detection. By leveraging the unique properties of CNTs, these nanoscale technologies hold great promise for improving the sensitivity, specificity, and speed of disease diagnosis, ultimately contributing to more effective healthcare and early intervention [7]. CNTs can exist in two primary structural forms.

Single-walled carbon nanotubes (SWCNTs): These consist of a single layer of carbon atoms arranged in a cylindrical tube. SWCNTs have unique electronic and mechanical properties and are often used in various nanotechnology applications.

Multi-walled carbon nanotubes (MWCNTs): These structures comprise multiple concentric layers of CNTs, one inside the other, resembling a Russian nesting doll. MWCNTs exhibit enhanced mechanical strength and are used in applications requiring higher structural integrity. The synthesis of CNTs typically involves chemical methods and can be done using the following methods:

Chemical Vapor Deposition (CVD): In this process, a carbon-containing gas, such as methane or ethylene, is introduced into a reactor chamber with a substrate, and a catalyst (often iron, nickel, or cobalt) is used to promote the growth of nanotubes. The gas is then subjected to high temperatures, and CNTs grow on the catalyst particles.

Arc Discharge: This method involves using an electric arc discharge between two carbon electrodes in an inert gas atmosphere. The high temperature and pressure conditions in the arc promote the formation of CNTs, which are collected on the electrode.

Laser Ablation: In laser ablation, a high-power laser is directed at a carbon target in the presence of a reactive gas. The intense energy from the laser vaporizes carbon atoms, which then condense to form CNTs.

Each synthesis method has its advantages and limitations, and the choice of method depends on the specific application and desired properties of the CNTs being produced [8]. Graphene is indeed a remarkable two-dimensional nanomaterial composed of carbon atoms that are sp^2 hybridized. It forms a single layer of carbon atoms arranged in a hexagonal lattice, creating a one atom-thick sheet. Graphene is renowned for its extraordinary optoelectronic, thermal, and mechanical properties, which include the following:

Exceptional Electrical Conductivity: Graphene is an excellent conductor of electricity due to the delocalized π-electrons in its hexagonal lattice structure. This property makes it a promising candidate for various electronic and sensor applications.

High Thermal Conductivity: Graphene exhibits exceptionally high thermal conductivity, making it an excellent material for heat management and thermal applications.

Mechanical Strength: Graphene is incredibly strong for its thinness. It has an ultimate tensile strength of over 100 times that of steel, making it an ideal material for reinforcing composites and enhancing mechanical properties.

Transparency: Despite its remarkable electrical conductivity, graphene is transparent, allowing it to be used in transparent conductive coatings for applications like touchscreens and solar cells.

Flexibility: Graphene is flexible and can be bent without breaking. This property is valuable for applications in flexible electronics and wearable devices.

Biocompatibility: Graphene exhibits biocompatibility to some extent, making it suitable for use in biomedical applications such as drug delivery, biosensors, and tissue engineering.

Optical Properties: Graphene displays unique optical properties, including strong light absorption across a wide spectrum, which has applications in photodetectors, photovoltaics, and optical modulators.

Due to these remarkable properties, graphene has garnered significant attention in the fields of materials science, electronics, energy, and beyond, and it holds great promise for a wide range of innovative applications [9–11]. Indeed, rapid advancements in the synthesis and functionalization of graphene have opened up a multitude of outstanding applications across various fields, including composite materials. Some of the notable applications of graphene in composite materials include

Enhanced Mechanical Strength: Graphene, when incorporated into composite materials, significantly improves their mechanical properties. The strong and lightweight nature of graphene enhances the tensile strength and stiffness of composites, making them ideal for applications where structural integrity is crucial, such as aerospace and automotive components.

Improved Thermal Conductivity: The high thermal conductivity of graphene can be harnessed to enhance the heat-dissipating properties of composites. This is particularly valuable in electronic and thermal management applications.

Electrical Conductivity: Graphene's excellent electrical conductivity is beneficial in creating conductive composite materials for use in electrical circuits, sensors, and electromagnetic interference shielding.

Barrier Properties: Graphene's impermeability to gases and liquids can be utilized to enhance the barrier properties of composite materials, making them suitable for packaging applications that require protection against moisture, oxygen, and other environmental factors.

Lightweighting: Graphene's lightweight nature contributes to reducing the overall weight of composite materials, which is crucial in industries like automotive and aerospace to improve fuel efficiency and reduce emissions.

Corrosion Resistance: Incorporating graphene into composites can enhance their resistance to corrosion, making them suitable for applications in harsh environments or where exposure to corrosive agents is a concern.

Antibacterial Properties: Graphene-based composites have shown antibacterial properties, making them suitable for use in medical devices, coatings, and textiles for infection control.

Energy Storage: Graphene composites are explored for use in energy storage devices such as supercapacitors and batteries due to their high surface area and electrical conductivity.

These applications represent just a fraction of the potential uses of graphene in composite materials. As research and development in this field continue to progress, we can expect even more innovative applications to emerge, contributing to the advancement of various industries [12–14], nanoelectronics [15], energy storage such as hydrogen storage, fuel cells and supercapacitors [16–18], catalysis, and sensors [19–22]. Apart from the applications, graphene has shown significant potential for many biomedical applications. Since the first report on the use of graphene oxide (GO) as an efficient nanodrug carrier [23], a lot of research has been done to explore the various biomedical applications of graphene such as drug or gene delivery, imaging, and sensing of biological specimens, antibacterial materials and biocompatible scaffolds for cell culture. These

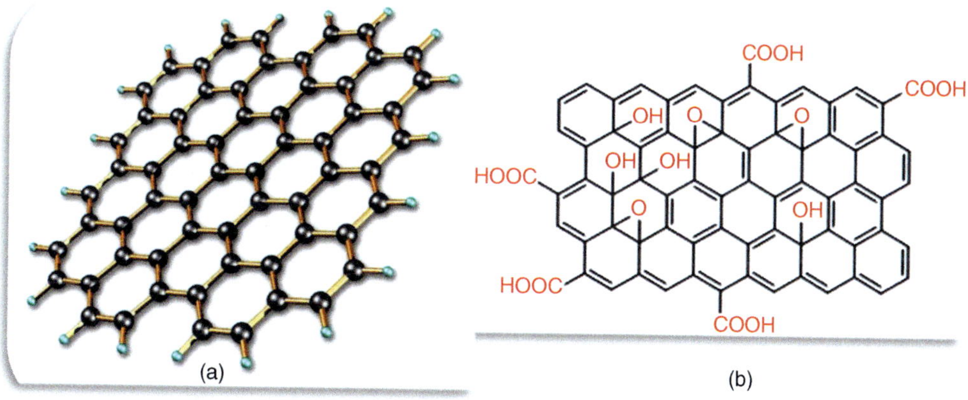

Figure 5.1 Schematic models of (a) graphene and (b) graphene oxide. Source: Adapted from Yaqoob et al. [24].

and various other biomedical properties of graphene and its derivatives are due to its appealing properties such as low cost of production, facile chemical and biological functionalization, high surface area (2630 m^2/g), excellent electronic conductivity (200,000 $cm^2/V\,s$) and thermal conductivity (~5000 W/m/K), and high mechanical strength (Young's modulus, ~1100 Gpa) [21, 22]. Schematic models of graphene and GO are shown in Figure 5.1. This book chapter presents an overview of the recent advances in the biomedical applications of carbon-based nanomaterials with a specific focus on CNTs and graphene.

5.2 Classification of Carbon Nanotubes (CNTs)

Depending upon the structure, CNTs may be classified into two types viz., SWCNTs and MWCNTs [8]. SWCNTs contain the one-atom-thick sheet of graphene rolled into a cylindrical structure with a radius of up to 1 nm. During the synthesis, both the ends of the SWCNT are closed by the rings of carbon–carbon bonds [25]. Whereas MWCNTs consist of few layers (2–10 layers) of graphene sheets with an external diameter equal to or greater than 10 nm [26]. Depending on the arrangement of carbon atoms, MWCNTs give three different structural configurations: zigzag arrangement, armchair arrangement, and chiral or helical arrangement. The degree of chirality in a carbon nanotube is a measure of electronic properties and helps in designing a wide range of electronic instruments. The diameter of the carbon nanotube as well as metallic or semi-metallic properties is also governed by the degree of chirality in a tube [5, 27]. Depending upon the pattern of arrangement of graphite sheets, multi-walled carbon nanotubes (MWNTs) are further classified into two types, viz., Russian-doll model structure and parchment model structure. In the former model, the graphite sheets are arranged in concentric layers above others, such as a sheet of (0, 8) SWCNT enclosed inside another SWCNT of larger diameter (0, 10). In the latter model, a single sheet of graphite is rolled upon itself, which resembles a rolled newspaper or a scroll of parchment paper [28]. Apart from these two structures, another type of CNT has also been described, which is a double-walled carbon nanotube (DWCNT). DWCNTs have a structural resemblance with SWCNTs and have been found to have many applications in the pharmaceutical field [29, 30]. There are also three other prototypic structures of CNTs: carbon nanobuds, carbon nanohorns, and carbon nanotorus (Figure 5.2) [26].

Carbon nanohorns (CNHs) are single-walled carbon nanomaterials (CNMs) that are obtained from single sheet of graphene of 2–3 nm diameter in which a tip of the graphene roll is capped with a five-member ring. CNHs have a cone shape and present a horn-like structure with one narrow

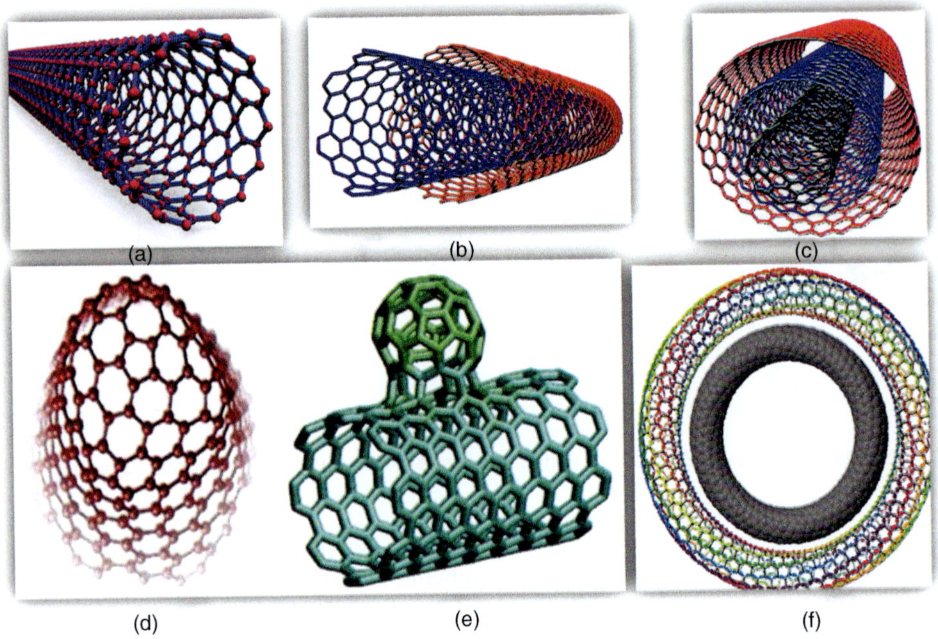

Figure 5.2 Types of carbon nanotubes: (a) SWCNT, (b) DWCNT, (c) MWCNT, (d) carbon nanohorn, (e) carbon nanobud, and (f) carbon nanotorus.

and one broad open end. They show a resemblance with dahlia petals and are often referred to as "dahlia-like" aggregates [31]. The diameter of the CNHs is a function of length and is produced by laser ablation methods [31]. CNHs have been found to have several applications is drug delivery to cancer cells [32–34]. Carbon nanotorus are CNTs with structure bent to form torus or doughnut shape [33]. In contrast, carbon nanobuds are structurally similar to CNTs, with an outgrowth on the external surface. Carbon nanobuds are synthesized from both the CNTs and fullerene with the properties intermediate between the two.

5.3 Biofunctionalization/Biomodification of Carbon Nanotubes (CNTs)

The various chemical modifications of CNTs have been summarized previously [35–37]. These methods can be employed for the direct or indirect modification of CNTs with biomolecules, except for the method involving fluorination of CNTs [38]. Depending on how biomolecules are attached to CNTs, there are three distinct approaches: covalent linkage, which involves forming covalent bonds; noncovalent linkage, which entails physical absorption; and hybrid linkage (Figure 5.3) [39]. The final approach combines noncovalent linkage by attaching an anchor molecule to the CNT, which is then covalently linked to the biomolecule of interest. In the covalent functionalization of CNTs, it involves reacting CNTs with carboxylic acid (—COOH) groups. These —COOH groups are introduced either at the more reactive open ends or the defect sites of both single-walled and multi-walled CNTs through oxidation using strong acids (Figure 5.3a) [35–37]. However, the functionalization of side walls is carried out by either 1,3-dipolar cycloadditions (Figure 5.3d) [39] or nitrene cycloaddition (Figure 5.3b) [39] or arylation using diazonium salts (Figure 5.3c) [40]. In some cases, these functionalized CNTs can be directly linked with biomolecules of interest

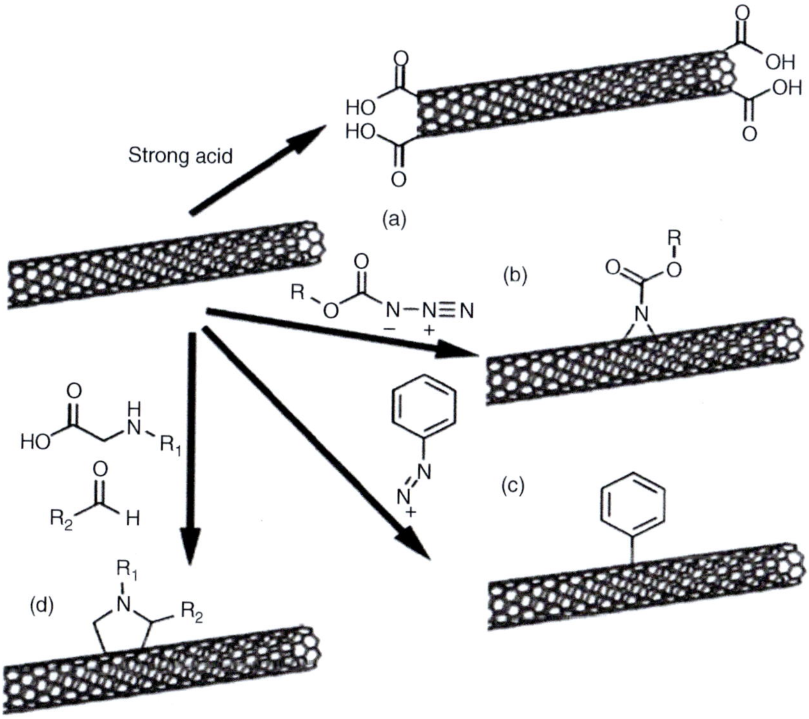

Figure 5.3 Methods for chemical functionalization of carbon nanotubes: (a) oxidation by acid, (b) nitrene cycloaddition, (c) arylation using diazonium salts, and (d) 1,3-dipolar cycloaddition. Source: Reproduced with permission from Yang et al. [39]/IOP Publishing.

[41, 42]. Some instances use standard carbodiimide (EDC) peptide coupling chemistry protocols for the activation of carboxylic acid-functionalized CNTs with N-hydroxy succinimide (NHS), which is then linked with secondary antibodies after activation (Figure 5.4a) [39]. Some other instances involve the reaction of functionalized CNTs with a bifunctional spacer, and the resulting species is then lined with a biomolecule of interest [43]. Recently, DNA has been successfully attached to oxidized SWCNTs in both organic and aqueous solutions. The attachment site of DNA to SWCNTs has been confirmed by connecting DNA-modified CNTs with gold nanoparticles that are also modified with DNA sequences complementary to the attached DNA. Transmission electron microscopy (TEM) images of the resulting hybrid show gold nanoparticles attached to the tips and defects on the sidewalls of the SWCNTs. This approach offers a reliable method for directly assembling CNTs with DNA or creating composites with gold nanoparticles, which can be used in nanoscale devices. A perfluorinated polymer called Nafion has been utilized as a novel stabilizing agent for biomodified CNTs. This has led to the development of CNT-based biosensing devices. When CNT-nafion-modified glassy carbon electrodes are used, they exhibit a stable and robust electrolytic response to hydrogen peroxide. Importantly, the Nafion coating does not interfere with the electrocatalytic properties of the CNTs, making it a valuable advancement for biosensing applications [44]. Azide chemistry has also been used for the biomodification of single-walled and multi-walled CNTs. The functionalization of tips and side walls of the CNTs was done using azide photochemistry so as to produce water-soluble DNA-coated nanotubes using DNA oligonucleotides, which were synthesized in situ form the reactive groups on each photo adduct [45]. Azide thermolysis has also been used successfully in solution to functionalize either fullerenes [46] or the sides of the single-walled CNTs [47].

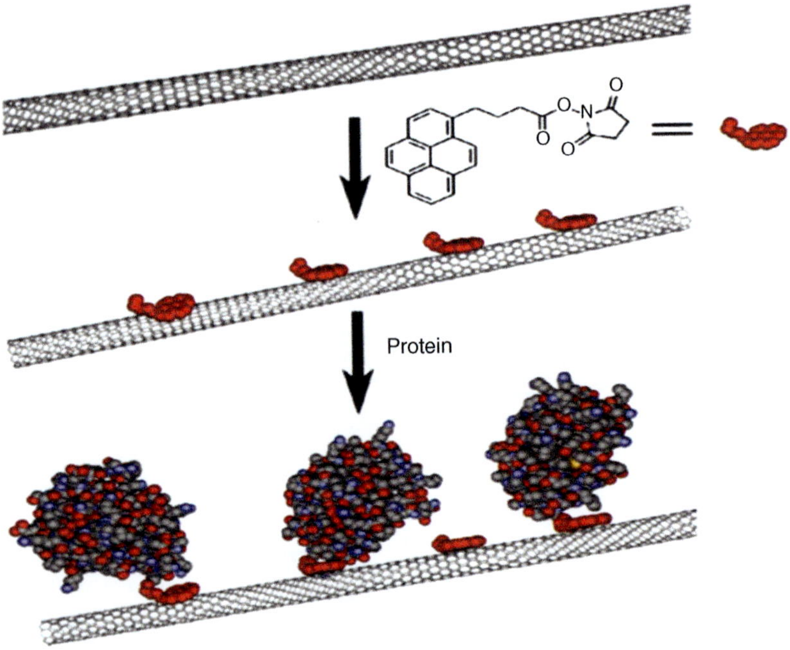

Figure 5.4 Representative of hybrid approach where an NHS-functionalized pyrene molecule is first absorbed onto a carbon nanotube, followed by reaction with the lysine residues of the target protein. Source: Reproduced with permission from Yang et al. [39]/IOP Publishing.

Coupling with carboxylic acid-functionalized CNTs is viewed as one of the simpler methods for chemically modifying CNTs. However, its primary drawback is its lack of specificity. Excessive oxidation of CNTs during this process can considerably change their structure and properties, including their conductivity. Methods that involve side wall modification are considered to be less detrimental to the structure of CNTs. These methods allow for the introduction of functional groups with higher specificity for biomolecules, making them a more favorable choice when precision in modification is required [39]. Noncovalent interactions with CNTs are facilitated by their extensive aromatic hydrophobic surfaces, making them capable of non-covalently interacting with complementary molecules. These interactions can occur on both the external and internal surfaces of the CNTs. Transporting DNA through MWCNTs has been achieved through noncovalent interactions, although the small diameter of CNTs limits the applicability of this method. Additionally, biomolecules like proteins and DNA have a greater affinity for the external surface of CNTs compared to the internal surface. Noncovalent interactions between hydrophobic or aromatic molecules and CNTs can be employed for further functionalization with biomolecules. To prevent nonspecific protein adsorption onto CNTs, a coating of polyethylene glycol (PEG) can be applied through noncovalent interactions. This coating helps avoid unwanted protein binding on CNT surfaces [48–50]. These PEG-coated CNTs can be further modified to provide sites for binding proteins chemically. Another approach employs functionalization of certain polyaromatic compounds adsorbed on the surface of CNTs with certain specific functional groups such as NHS-activated acids [51] for further linking to proteins (Figure 5.4). Although this noncovalent interaction is nonspecific and hydrophobic in nature, it has still been applied for the transport of number of biomolecules [36], which include polysaccharides such as helical amylase [36, 48] and heavy metal doped DNA.

Noncovalent interactions offer several advantages when modifying CNTs with biomolecules, including:

Preservation of CNT Structure: Noncovalent interactions do not significantly alter the structure of CNTs, unlike covalent interactions involving oxidation. This allows for easy comparison of CNT properties, such as conductivity, before and after modification with biomolecules.

Comparative Analysis: Noncovalent modifications enable straightforward evaluation of CNT properties before and after biomolecule attachment, simplifying analysis.

However, noncovalent interactions also come with certain disadvantages as follows:

Lack of Specificity: Noncovalent interactions may lack specificity, potentially leading to unintended binding with molecules other than the target biomolecules.

Potential for Biomolecule Denaturation: Nonspecific interactions could lead to the denaturation or alteration of certain biomolecules upon adsorption onto CNTs, affecting their function or structure.

Encapsulation: In some cases, nonspecific interactions may result in the total encapsulation of CNTs by biomolecules, leading to the loss of certain CNT properties or hindering their intended applications. While noncovalent interactions offer advantages in terms of structural preservation and ease of analysis, researchers must carefully consider the potential issues of specificity and biomolecule behavior to ensure the desired outcomes in CNT-biomolecule interactions [52]. Given the complexity of the chemical composition and structure of biomolecules, the development of methods of chemical modification, which are mild, efficient, and selective, is still an ongoing challenge. Although a number of methods or techniques are used for this purpose [53], the challenges encountered for the functionalization of CNTs are considered like those met in polymer chemistry. In order to understand the interactions, it is better to go through the methods used for the formation of protein-polymer bioconjugates [54]. In general, modification of proteins is done by employing the reactions involving thiol (Cys) functionalities of proteins and lysine/α-N terminus of amines (Figure 5.5) [39]. DNA and RNA are also modified similarly after the incorporation of a thiol-spacer or short amine spacer at the 5-end of the backbone.

5.4 Biomedical Applications of CNTs

The use of CNTs in the biomedical field presents several challenges. Firstly, ensuring safety requires the use of high-purity CNTs to limit the potential release of toxic ions in biological environments. However, preparing high-purity CNT samples on a large scale is challenging, often requiring a compromise between quality and quantity. Additionally, achieving good dispersions of CNTs in solvents, particularly in water, is difficult due to their strong hydrophobicity. Functionalization through chemical oxidation or the addition of dispersing agents or surfactants can help stabilize the suspension. Another strategy to improve dispersion is to limit or avoid drying steps during processing. Furthermore, the proper dispersion of CNTs in a fluid can lead to a significant increase in viscosity, making it challenging to prepare nanocomposite materials with a high volume fraction of well-dispersed CNTs. In terms of electronic structure, SWCNTs can behave as either semiconductors or metals, with a statistical distribution of two-thirds semiconducting and one-third metallic within a random SWCNT population. As the number of concentric walls increases, there is a progressive increase in the population of metallic MWCNTs due to inter-wall interactions. In terms of mechanical performance, CNTs can exhibit a large surface area of up to 1000 m^2/g in the case of DWCNT, a very high aspect ratio and low density, and excellent mechanical resistance and flexibility when the number of walls is low. As described by Mohajeri et al., CNTs are increasingly used

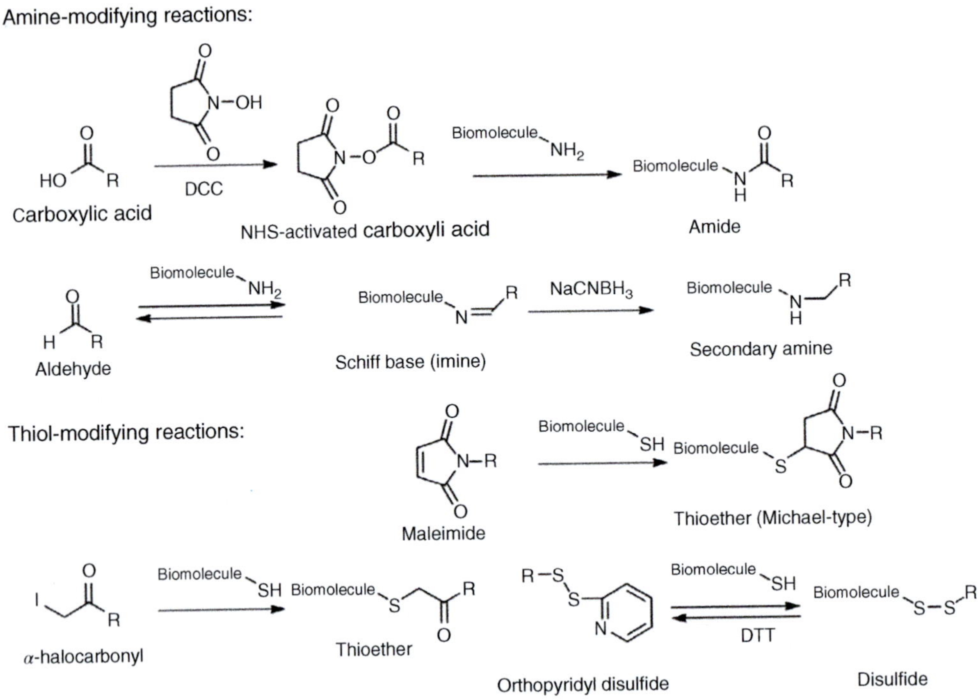

Figure 5.5 Key methods for chemical modification of biomolecules bearing amine. Source: Reproduced with permission from Yang et al. [39]/IOP Publishing.

for biomedical purposes due to their good biocompatibility for diagnostics or treatment of various diseases [55–57].

5.4.1 In Controlled and Targeted Drug Delivery

Paul Eric proposed the concept of the "magic bullet." He envisioned the development of specialized drug delivery systems that could target specific sites in the body for controlled and precise drug delivery, while sparing healthy tissues. This concept has been influential in the field of pharmacology and drug delivery. Indeed, extensive research has been conducted to realize the "magic bullet" concept, resulting in the development of various carrier systems designed for controlled drug delivery. These systems include hydrogels, liposomes, microspheres, nanoparticles, and many others. Each of these carriers offers unique advantages and can be tailored to deliver drugs to specific target sites in a controlled and efficient manner, minimizing side effects and enhancing therapeutic efficacy [58]. Recently, CNTs have also been tested for targeted delivery of drugs to the target sites in a controlled manner [59–61]. They have been tested for the controlled delivery of drugs as well as genetic materials like DNA, genes, and antibodies. These drugs can be loaded either inside the hollow cavity of the CNT through van der Walls forces or linked to the outer walls of the CNTs by hydrophobic interactions [62]. The hollow monolithic structure of CNTs, which have an outer and inner core and can be modified for functionalization with desired groups, makes them promising candidates for drug delivery applications [63]. Functionalization of CNTs performs manifold functions such as enhancing their biocompatibility with the body and increasing their solubility and encapsulation capacity for multimodal drug delivery [64, 65]. Targeted drug delivery at the cellular and nuclear levels has been achieved, since it has been found that

CNTs loaded with drugs easily pass into the cells and inside of the nucleus as well [66]. It has been reported by Pantoratto et al. [65, 67] that dimensions of the CNTs are a critical factor in deciding the fate of CNTs in the body. They examined the uptake of CNTs at the cellular and sub-cellular levels by using TEM. They observed that the functionalized CNTs showed better uptake with targeted actin by attaching surface functional groups on cellular surfaces [67]. It was reported by Yang and coworkers that ibuprofen, an anti-inflammatory drug, can be delivered in a controlled manner by using amine-functionalized mesoporous silica nanotubes (NH_2-MSNTs), modified with 3-aminopropyltriethoxysilane having additional functionalization with blue fluorescent CdS quantum dots [68]. It has also been observed that the drug was released in a controlled manner at a slower rate as compared to that from amine-free systems. Further, the presence of CdS QDs did not affect the release of drug. Zhang et al. [69] conducted a study in which they utilized nanohybrid hydrogels created through the self-assembly of carboxy-functionalized MWNT-COOH) and poly(methacrylic acid) networks. These nanogels exhibited pH sensitivity and swelled in water, allowing for the controlled release of drugs, specifically theophylline and aminophylline. By adjusting the pH values, they were able to regulate the release of these drugs. Functionalized CNTs that carry drugs can encounter challenges in releasing their contents at specific target sites. To address this issue, Kulamarva et al. performed encapsulation of CNTs as part of their research [70] using alginate-poly l lysine alginate to form a novel polymeric membrane microcapsule in which CNTs were either embedded or attached to the surface for targeted drug delivery. The polymeric membrane of the microcapsule protected the CNTs from some external harsh environmental factors, which resulted in a safe and effective release of drugs to the target site [26]. Chitosan-modified CNTs have also been used for targeted delivery of drugs. Studies have revealed that the CNT-chitosan complex shows better drug targeting than chitosan-free CNTs; hence, they have found myriad of applications for the targeted delivery of drugs for the treatment of lung cancers [71].

Targeted drug delivery with functionalized CNTs to target malignant cells is a growing area of research due to better uptake by target malignant cells without disturbing the surrounding tissues [72, 73]. They are used to deliver drugs for the treatment of malignant disorders such as breast cancer, testicular tumors, carcinoma of the cervix, and choriocarcinoma. Take the example of methotrexate, which is an anticancer drug used in cancer chemotherapy but, when administered, exhibits poor absorption from gastrointestinal tract and low cellular uptake [74]. However, when the drug is administered using CNTs functionalized with amino by employing 1,3-dipolar cycloaddition reaction of azomethine ylides, which introduced amino groups in the orthogonal positions of CNTs, it shows a better cellular uptake [47]. Besides, functionalized CNTs are also used to deliver proteins, nucleic acids, and other therapeutic agents to the site of action [75]. Antibody-mediated drug delivery was earlier considered to be the best method for the targeted delivery of drugs. However, when binding with drug molecules, the antibodies displayed a loss of specificity [76]. It has been observed that when functionalized with antibodies, CNTs do not change the specificity of such functional moieties, which can result in targeted delivery of drugs in an effective manner. Tumor targeting using anticancer drugs has been achieved by McDevit et al. [77] employing SWCNTs covalently functionalized with monoclonal antibodies. The properties of the drug molecules to be delivered must be similar to the CNTs as well as antibodies. There should be proper compatibility between the three so as to preserve the antigenicity of the antibodies on attachment drug to the CNTs and CNTs to the antibodies for delivery to the desired site. On attachment to the surface of a single antibody, CNTs can deliver several types of drugs (taxol) to the target tumor site. ZME-108, skin cancer antibody alone, has been shown to deliver more than 40 nanotube-loaded drugs to the target sites [78]. Selective targeting of CD20 cell surface receptor on B cells with slight nonspecific binding to negative T-cells and with Herceptin to recognize HER2/neu positive breast cancer cells was done using CNTs functionalized with PEG and conjugated with monoclonal antibody Rituxan [79]. Antibody-functionalized CNTs have thus proved to be ideal agents for the targeted delivery of drug molecules to cancer cells [26]. GO, with 1–2 nm thick comprising one to three layers, is an ideal nanocarrier for the efficient delivery of drugs and genes [80–82]. It is synthesized by vigorous oxidation of graphite using the Hummers method [83]. Excellent biocompatibility, and physiological solubility and stability, is due to its unique structural features, such as high specific surface area (2630 m^2/g), large planar sp^2 hybridized carbon domain, and enriched oxygen-containing groups. Drug and gene loading of GO occur via physisorption or chemical conjugation. Conjugation with various systems, such as biomolecule ligands [82], polymers [84], protein [85–87], DNA [88], Fe_3O_4 nanoparticles [89], quantum dots [90], and others [91] is facilitated by the COOH and OH groups on GO, which make it multifunctional and multimodal for various medical and biological applications. Nanoscale GO (NGO), as an innovative and efficient nanocarrier for the delivery of water-insoluble aromatic anticancer drugs into cells, was explored for the first time by Dai et al. [23] by taking inspiration from the idea of carbon-nanotube-based drug delivery [92–95]. NGO was first conjugated with an amine-terminated six-armed PEG molecule, which was followed by loading of a water-in-soluble anticancer drug, SN38, onto NGO surface via π-π stacking which is a simple noncovalent adsorption. SN38 drug loaded on PEG-functionalized NGO showed potent cytotoxicity against HCT-116 cells, 1000-fold more than CPT-11. Rituxan ($CD20^+$ antibody) conjugated NGO-PEG was used by the same group for the targeted delivery of chemical drugs into cells [82, 96]. The possibility of pH-controlled drug release was suggested as it was demonstrated that the release of drugs from the surface of GO is pH-dependent. The pH-sensitive GO-based drug delivery systems was also studied by many research groups [97, 98]. Not only pH-sensitive drug release, a thermo-responsive drug delivery cargo comprising poly(N-isopropylacrylamide) grafted graphene sheets was also developed by Pan et al. [99].

In order to overcome drug resistance of cancer cells, an extensively embraced clinical practice in cancer therapy is the combined use of multiple drugs [100, 101]. Due to various technical difficulties, few reports are available in the literature for controlled loading and delivery of multiple drugs based on nanomaterial-based drug delivery systems. In order to address this issue, the feasibility of GO as a nanocarrier for controlled loading and targeted delivery of mixed chemical drugs has been explored [98] using folic acid and SO_3H groups conjugated GO. This conjugated GO was co-laden with doxorubicin (DOX) and camptothecin (CPT) via π-π stacking in a controlled manner. This co-loading of drugs on a single GO conjugated with folic acid ligands showed a high level of toxicity to human breast cancer cells, MCF-7 cells, with folic acid receptors with specific targeting than the GO loaded with a single drug alone. A drug delivery cargo based on GO-Fe_3O_4 nanoparticle hybrid exhibiting magnetic and bio dual targeting was prepared by Yang et al. [102] using in vitro studies, which demonstrated the specific targeting of the multifunctional drug carriers by SK3 human breast cancer cells.

5.4.2 In Imaging Techniques

Nanotubes have been shown to play a significant role in diagnostic procedures. Besides imaging, they have been used in diagnosis to help identify the site of action of drugs [103]. CNTS is proven to have a superior ability to act as a contrast agent to identify cancer cells and in imaging of various tissues or organs [73]. Methotrexate (an anticancer drug) showed better visibility in the body when given with fluorescein probe functionalized nanotubes owing to the fluorescence of the drug-carrying probe on its surface [104]. CNTs functionalized with fluorescent substances produce images of the desired in vivo organs by acting as radio-opaque substances. Once these functionalized nanoformulations are administered into the body, they attach to the specific site, which can then be studied with some special imaging techniques like radiolabeling and scintigraphy [39].

Nanotubes encapsulating miniaturized video systems, when administered in pill form, have been reported to produce the nanoimaging of various body parts. Such a system can be swallowed easily and can image those affected areas of a tissue or an organ in a nanoscale segment where colonoscopy and endoscopy fail due to their macro-size imaging [75]. CNT film with optimized morphology has been used in high-performance CNT-based field-emission cathodes for the design of X-ray tubes [105]. These X-ray tubes, which are produced in both continuous and pulsed mode with finite repetition form, are constructed using purified SWNT bundles produced by laser ablation. This CNT-based X-ray technology can result in the design of transportable minuscule X-ray sources for specific medical applications. A dynamic radiography system with a CNT-based microfocus X-ray tube system has been developed, which can generate X-ray radiation in a pulsed mode with a temporal resolution as short as 50 ns [105]. This CNT-based dynamic radiography system has proven to be versatile for noninvasive imaging devices for biomedical research despite having some limitations to overcome [105, 106]. Exploring the possibility of GO in bioimaging started almost a decade earlier [80, 82, 89, 107]. PEG-modified GO laden with chemical drugs was first time examined by Dai et al. [82] using the intrinsic fluorescence of GO in the NIR region. Gelatin-grafted GO tagged with a fluorescence dye for cellular imaging and drug delivery was also studied later [108]. Researchers are now focusing their attention on preparing GO of smaller sizes ($\leq$10 nm), referred to as graphene quantum dots (GQDS). These GQDS can be produced either by the bottom-up approach [109] or by chemical oxidation of graphite [110, 111]. GQDS possesses intrinsic fluorescence, which makes it useful for bioimaging. GQDs with blue emission were synthesized by Pan et al. [110] using hydrothermal cutting of GO. However, synthesized GQDS showed weak fluorescence, much lower than II-VI type QDs or conventional organic dyes, which rendered them almost unfeasible for bioimaging applications. However, a significant improvement in the fluorescence of chemically produced GQDS was observed by Eda et al. [112] by exposing them to hydrazine vapor. Dramatic enhancement of fluorescence was also observed by surface functionalization of GQDS with alkylamines [111]. Various groups have explored the possibility of GQDS for cellular imaging [113, 114]. GQDs have been shown to possess excellent biocompatibility, low cytotoxicity, and physiological solubility, as compared to CdSe and other II-VI-type QDs (Figure 5.6). GQDs also exhibit unique optical properties like pH-dependent and upconversion fluorescence behaviors, which can be directly used for intracellular imaging without further surface functionalization [115]. GQDS can be excited in the NIR region due to upconversion fluorescence behavior of GQDS, which makes in vivo and in vitro imaging and biodetection safe and efficient without interference from autofluorescence from the organs, tissues, or cells organs for that region [79]. Likewise, dextran-coated Fe_3O_4 nanoparticles (NPs) and GO (Fe_3O_4-GO) composites have been used as T2-weighted contrast agents for efficient cellular MRI [114], an extensively used technique in clinical practice [116, 117]. These Fe_3O_4 nanoparticles (NPs) and GO (Fe_3O_4-GO) composites showed enhanced cellular MRI with good physiological stability and low cytotoxicity. This composite was able to detect cells at iron concentration of 20 μg/mL with a cell density of 1000 cells/mL and at an iron concentration of 5 μg/mL with a cell density of 2×10^5 cells/mL.

5.4.3 In Medical Nanorobotics

Nanorobotics is an ultramodern vision that deals with the interaction between atomic and molecular-sized objects and their controlled manipulation [118]. Nanorobots are the devices having dimensions of about 0.5–3 μm which in future will be used for maintaining the human body by protecting it against various pathogens. Nanorobots will have components in the dimensions of 1–100 nm and will be powered by either external energy or by internal energy derived from the metabolization of local glucose and oxygen. Nanorobots will help the immune system to

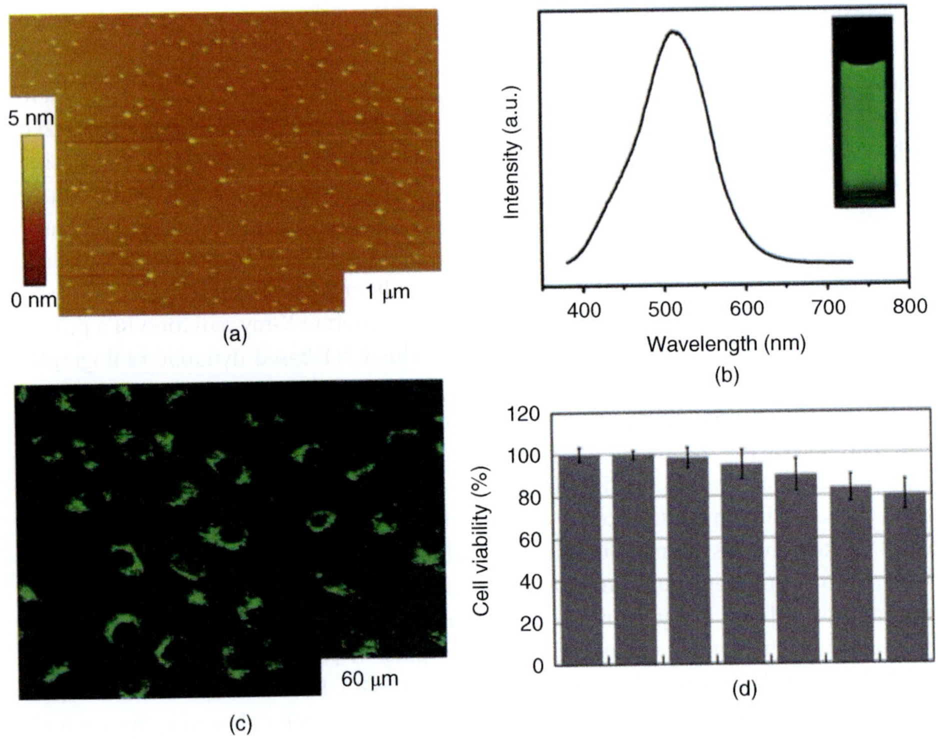

Figure 5.6 (a) Atomic force microscopy (AFM) image of the GQDs; (b) fluorescence spectra of aqueous solution of GQDs excited at 375 nm. Inset: aqueous solution of GQDs under UV light. (c) Cellular imaging of GQDs imaged under 405 nm. (d) Cytotoxicity of GQDs. Source: Reproduced with permission from Zhu et al. [114]/Royal Society of Chemistry.

ward off the danger by finding and disabling unwanted pathogens like bacteria and viruses. Once identified, these pathogenic invaders will be punctured with the help of nanorobots, which will result in spilling out its contents and thus ending its efficacy [106]. These nanorobots can also ingest these pathogens by using chemical or mechanical phagocytosis. They will have chemostatic sensors by which they will be able to distinguish between various cell types by checking their surface antigens. Nanorobots also work in the bloodstream, eat the arteriosclerotic deposits, and widen the blood vessels [106]. Nanorobots will allow doctors to perform direct in vivo surgery on individual human cells [118]. It is speculated that CNTs will offer promising support for the development of nanorobotic components [119]. Nano-manipulation refers to the 3D maneuvering of chemical components to build molecules, which then will be assembled further into larger devices. Nano-manipulation in living bodies involves nanoscale biological units such as DNA and proteins for scientific and clinical analysis. Nano-manipulation needs those materials with very specific properties for the fabrication of nanodevices. CNTs can be integrated into biological assemblies by manipulating CNTs at the atomic or molecular scale, making the integration of CNT-biological assembly more accessible [118]. CNTs and CNT composites have been employed in nanoelectromechanical systems (NEMS), as well as nano- and micro-robots. They escape an attack by the host's immune system due to their chemical inertness and high strength. New methods have been developed to construct the building blocks of CNTs at the nanometer scale to build 3D nanoassemblies that could be functionalized with biological specimens. These nano assemblies are based on controlled nanostructure interactions and complexity and, built with greater precision, can possess higher density, resulting in faster circuits with manifold functions.

They can become better electrical conductors with reduced power consumption [120]. Such sophisticated assemblies will be quite similar to biological moieties with nanoscale components, which will facilitate their easy integration into the living body. This integration will result in high-resolution monitoring of the phenomena due to the establishment of much more natural interactions with fewer secondary effects [118]. However, integration and manipulation of various components involves building high-level programming, which is in itself a big challenge [120].

5.4.4 In Drug and Gene Delivery

Nanocapsules, small nanosized particles, carry drug molecules and transport them to the target sites. Once they reach the target area, they release their contents in close proximity [115]. Drug delivery by the nanocapsule system has the main advantage of providing enough protection to the drug from the external environment. These nanocapsules are attached to special surface groups such as antibodies, functional groups, or ligands, which help them to make a safer entry into the cells with the least chance of rejection [10]. Nanocapsules are core-containing particles, whereas nanoparticles are hollow spherical particles [116]. CNTs bear some similarities with nanocapsules, such as the property of encapsulation and the property of not being rejected by the target cells. These and other properties make CNTs ideal candidates as carriers of drug delivery systems and for the treatment of such diseases where there are high chances of drug resistance due to the efflux of drug molecules from the cells by the action of P-glycoprotein such as cancer, bacterial or viral diseases, and tuberculosis [118]. One of the novel and promising techniques to treat various diseases related to genetic disorders like Parkinson's disease, cancer, and cystic fibrosis is gene therapy [49]. Gene therapy requires a gene vector that facilitates the cellular uptake of DNA by protecting DNA from nuclease degradation with high transmission efficiency [50]. However, the advancement of this approach faces some main challenges such as the lack of safe and efficient gene vectors [51]. GO has been studied for gene delivery. Polyethylenimine (PEI) conjugated GO (PEI-GO) has been shown to condense plasmid DNA onto the surface through electrostatic interaction owing to the cationic polymer [52]. The procedure is shown in Figure 5.7a. The transfection efficiency of GO-PEI-10K and GO-PEI-1.2K was compared with ungrafted polymers PEI-10K and PEI-1.2K, respectively (Figure 5.7b). The results show that the grafting of polymers with GO not only improved the transfection efficiency of the polymers but also significantly lowered the cytotoxicity of the polymer. Higher luciferase expression in HeLa cells (2.75×10^9 RLU/mg protein) was observed with PEI-GO/pGL-3 complexes than with PEI/pGL-3 complexes (2.07×10^9). PEI-GO shows high DNA transfection efficiency in HeLa cells even in the presence of 10% FBS [53]. These observations are somewhat surprising to the researchers owing to the fact that PEI, when attached to other systems, often leads to a reduction in transfection efficiency [53]. Although these research findings are preliminary, these findings provide a clue that PEI-GO can act as a favorable candidate for efficient gene delivery. PEI-GO nanocarrier has also been studied for the delivery of DOX, and Bcl-2 targeted siRNA to HeLa cells sequentially. The results of the cytotoxicity assay revealed that PEI-GO/Bcl-2 targeted siRNA complexes increase the cytotoxicity of the PEI-GO/DOX dramatically as compared to the PEI-GO/scrambled siRNA complexes, which was evident by the significant decline in IC_{50} of DOX due to strong synergistic anticancer effect of the PEI-GO/DOX and PEI-GO/Bcl-2 targeted siRNA. Likewise, a finding by a research group has reported the synthesis of chitosan-functionalized GO (GO-CS) with high drug payload capability for application in drug and gene delivery [58] CPT-loaded GO-CS showed better cancer cell killing capability than pure CPT. Further GO-CS showed better DNA transfection efficiency in HeLa cells at a certain nitrogen/phosphate ratio due to condensation of plasmid DNA into stable nanoparticles. However, further research is needed in this direction to gain further insights into GO-CS nanocarrier for chemo and gene therapy.

5.4.5 In Biosensing Applications

Graphene and its derivatives such as chemically reduced and doped graphene have been studied for biosensing and detection of various biomolecules such as amino acid [70], dopamine [71], oligonucleotide [70], thrombin [122], and ATP [68]. Various novel fluorescence resonance energy transfer (FRET)-based graphene biosensors have been developed by making use of the superefficient fluorescence quenching ability of graphene. Field-effect transistor (FET)-type biosensors have been developed which are based upon the unique electronic properties of graphene [73]. Highly ultrasensitive biosensors have also been developed based on controlled self-assembly of graphene-biomolecules for the detection of DNA and various other molecules [70, 76]. Graphene-based nano platform for matrix-assisted laser desorption/ionization time-of-flight mass spectrometry (MALDI-TOF MS) has been developed for the detection of various molecules [70, 77]. Due to excellent surface area, good payload capability, and good electrical conductivity, GO-based novel biosensors have been developed that bind to the biomolecules through chemical or physical interactions [71, 78]. There are various excellent research articles available on the GO-based biosensors [20, 21, 79].

5.4.6 As Scaffold for Cell Culture

GO has been shown to act as surface coating material for the implants. NIH-3T3 fibroblasts, once grown on supported film of GO have displayed almost no significant harmful effect on the mammalian cell line [95]. The work also suggested a high gene transfection efficiency of about 250% as compared to those cells grown on a cover glass substrate. Likewise, graphene/chitosan film has been investigated as scaffold material that does not hamper the proliferation of mesenchymal stem cells (hMSCs) [8]. Rather, this film promotes the controlled differentiation of these cells into bone cells by using osteogenic inducers and growth factors. This work suggested the use of GO-based films for their potential use in the proliferation and transplantation of stem cells and bone regeneration therapy by their differentiation into various muscles, bones, and cartilages (Figure 5.8), [123]. Similarly, a graphene substrate produced by CVD has been shown to be biocompatible for human osteoblasts and hMSCs with higher cell proliferation efficiency than the SiO_2 substrate [97]. Graphene has also been investigated for its potential application as a novel material for neural interfacing. CVD-grown graphene film has also been investigated as a substrate for neuritis, and the results suggest the significant enhancement of neurite numbers average length during 2–7 days after cell seeding on graphene film as compared to the tissue culture polystyrene (TCPS) substrates [98]. Graphene resulted in the greater enhancement of GAP-43 expression as compared to TCPS substrate, resulting in boosting of neurite outgrowth and sprouting.

5.5 Biomedical Applications of Graphene

Two-dimensional nanomaterials are becoming increasingly recognized as promising options for a broad spectrum of biomedical uses such as tissue engineering, biosensing, pathogen neutralization, wound healing, and gene and drug delivery. Among these materials, graphene stands out owing to its substantial surface area, photothermal characteristics, high loading capacity, and efficient cellular uptake, positioning it prominently in this diverse research domain. However,

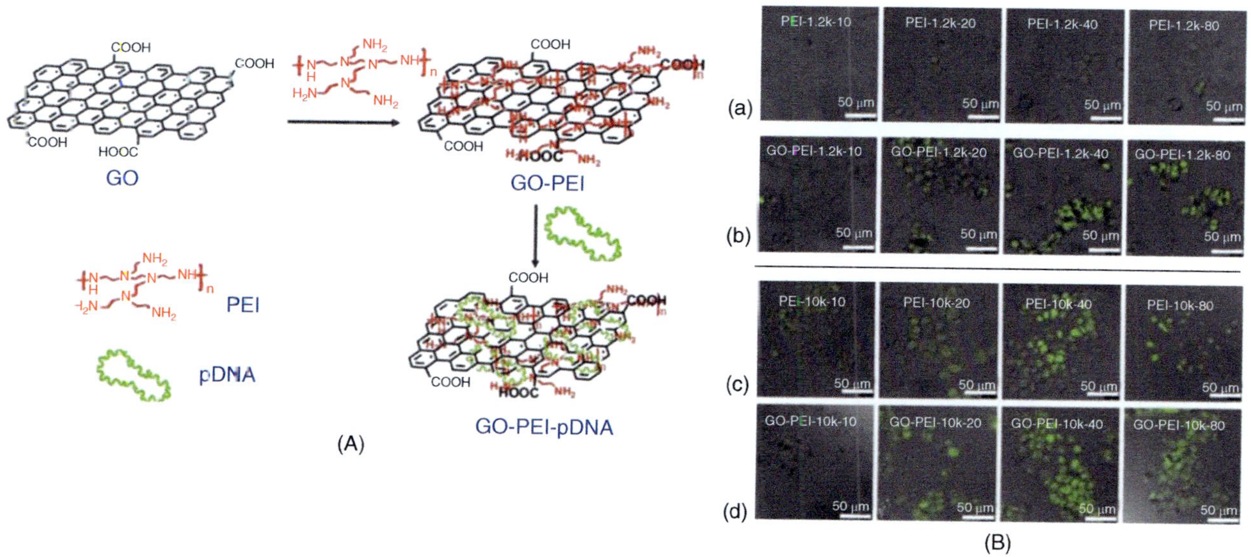

Figure 5.7 (A) Diagram illustrating the synthesis of GO-PEI-DNA complexes through electronic interactions. (B) Confocal fluorescence images displaying EGFP-transfected HeLa cells treated with PEI-1.2k (a), GO-PEI-1.2k (b), PEI-10k (c), GO-PEI-10k, (d) at varying N/P ratios from 10 to 80. Source: (A) Adapted from Feng et al. [121] and (B) Reproduced with permission from Feng et al. [121]/Royal Society of Chemistry.

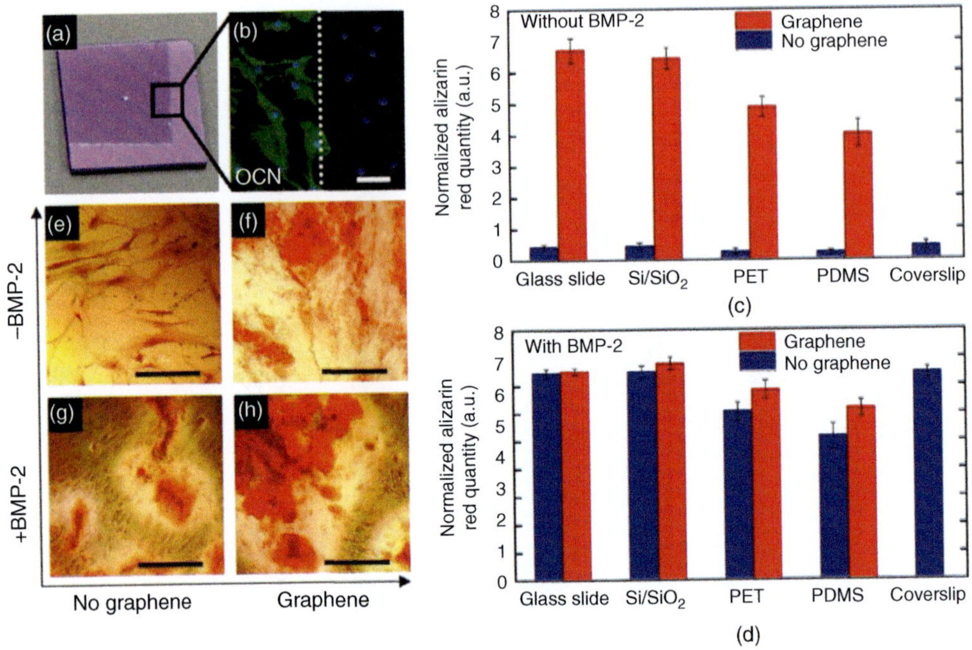

Figure 5.8 Graphene promotes osteogenic differentiation. (a) Visual depiction of a partially graphene-coated Si/SiO_2 chip measuring 1 × 1 cm, with the graphene boundary clearly visible. (b) The osteocalcin (OCN) marker indicates bone cell formation exclusively on the graphene-coated area of the chip; the boundary of the graphene coating is also evident. (c, d) Quantification of Alizarin Red staining from hMSCs cultured for 15 days on substrates with or without graphene. (c) Cells cultured without BMP-2. A control with coverslips is provided for reference. (d) Cells cultured with BMP-2. Conventional plain coverslips are used as a positive control. (e–h) Polyethylene terephthalate (PET) substrate stained with Alizarin Red to visualize calcium deposits resulting from osteogenesis. (e) PET without BMP-2 and graphene; (f) PET without BMP-2 but with graphene; (g) PET with BMP-2 but without graphene; (h) PET with both BMP-2 and graphene. Scale bars represent 100 μm. Source: Reproduced with permission from Nayak et al. [123]/American Chemical Society.

graphene's limited water dispersibility and functionality pose challenges to its integration into novel nanostructures for future nanomedicine applications. Functionalizing graphene, through either covalent or noncovalent methods, emerges as a crucial strategy to enhance its water dispersibility, augment functionality, and enable its processing into new materials and devices. This review explores recent advancements in the functionalization of graphene derivatives using various macromolecules tailored for potential biomedical uses, elucidating their mechanisms and implications. Specifically, emphasis is placed on hydrophilic functionalization of graphene and GO to enhance their water dispersibility and physicochemical properties. The focus is on the use of polyfunctional graphene sheets for anticancer drug delivery.

5.5.1 In Anticancer Drug Delivery Systems

Cancer stands as a significant global cause of mortality, and addressing this disease remains one of the foremost health challenges of our time. Chemotherapy has been a prevalent approach in the treatment of various cancers. Nevertheless, its effectiveness is hampered by several factors, encompassing multidrug resistance, systemic toxicity, immunogenicity, and limited bioavailability. Overcoming these restrictions necessitates the development of advanced drug delivery systems [124, 125]. Graphene and its derivatives have been found to be extensively used in a broad spectrum of biomedical applications due to their unique sheet-like structure, large specific surface area, and ease of functionalization [126]. Graphene and its derivatives offer a wide array of functionalization strategies, capitalizing on their multifunctionality. This versatility allows for the creation of drug-delivery systems with tailored biological properties. The functionalization of graphene derivatives with diverse polymers has the potential to enhance their reactivity, biocompatibility, in vivo circulation times, and anticancer efficacy. These graphene-polymer platforms, characterized by their high drug-loading capacity, have paved the way for innovative drug-delivery systems in various biomedical applications, addressing needs that conventional graphene or polymers alone cannot fulfill [127]. Functionalization of graphene and its derivatives with PEG have been extensively used for the construction of anticancer drug delivery systems. Liu and colleagues first introduced PEG-decorated GO as a nanocarrier for the efficient delivery of anticancer drugs. They harnessed noncovalent interactions to load hydrophobic anticancer drugs and observed that the synergy between PEG and graphene-based materials significantly enhances biocompatibility, cellular uptake, drug bioavailability, physiological stability, therapeutic effectiveness, and drug circulation time in the bloodstream [23]. Conversely, the PEGylation of graphene mitigates the aggregation of graphene in reticuloendothelial system (RES) organs, including the lungs, liver, spleen, and kidneys, along with reducing its retention and toxicity. In a separate study, a 6-armed PEG-functionalized graphene oxide-based nanocarrier was utilized for the delivery of anticancer drugs like oridonin and methotrexate (MTX) [128]. The synthesis of PEG-grafted GO and drug loading has been achieved through amidation and π-π stacking methods, respectively. The resulting complex, termed oridonin or MTX at GO-PEG, demonstrated enhanced cellular uptake and anticancer efficacy against CAL27 tumor cells compared to free drugs. Additionally, PEGylated graphene has been engineered as a dual-drug delivery system for DOX and platinum, exhibiting greater efficiency than single-drug systems. These studies demonstrated that the co-administration of drugs effectively circumvents cancer drug resistance and enhances their anticancer potency. Furthermore, this drug delivery system exhibited lower toxicity and reduced harm to normal cells in comparison to drugs like DOX and platinum-free compounds. The interactions between PEGylated graphene and drugs were modifiable based on the polymer chain length [129]. In a different investigation, PEGylation of GO was achieved using a PEG copolymer linked to multiple pyrene units, increasing the PEG coverage on the GO surface while concurrently reducing GO to reduced graphene oxide (rGO). In vitro assessments from this study unveiled that graphene with a PEG content exceeding 75% and longer PEG chains exhibited superior physicochemical characteristics compared to their counterparts with shorter PEG chains [130]. The incorporation of PEG with other polymers like hyaluronic acid, polyethyleneimine (PEI), and chitosan has been extensively explored to enhance the therapeutic efficacy of graphene-drug platforms [131–133]. PEI, a cationic polymer, is particularly notable for its surface modification of graphene-based materials [134, 135]. While bare PEI of high molecular weight can induce significant cytotoxicity due to increased cellular uptake and low biodegradability, its hybridization with carbon materials results in reduced cytotoxicity. PEI readily forms complexes with graphene derivatives through electrostatic interactions and amination methods, creating cationic platforms with advantages such as high hydrophilicity, biocompatibility, strong nucleic acid binding, effective cellular uptake, and thermal stability. Co-delivery of therapeutic genes and chemotherapy agents has been shown to down-regulate tumor cell resistance to anticancer drugs and overcome multidrug resistance (MDR) [136]. PEI-conjugated graphene has been engineered for the co-delivery of the DOX drug and the p53 tumor suppressor gene to inhibit HeLa cell growth effectively. The presence of the PEI cationic polymer and the synergistic effect of the drug and gene in the DOX/GO-PEI/p53 complex facilitate efficient internalization into HeLa cells via endocytosis, resulting in higher growth inhibition compared to single drug/gene delivery systems [137]. In 2019, Mallick et al. developed

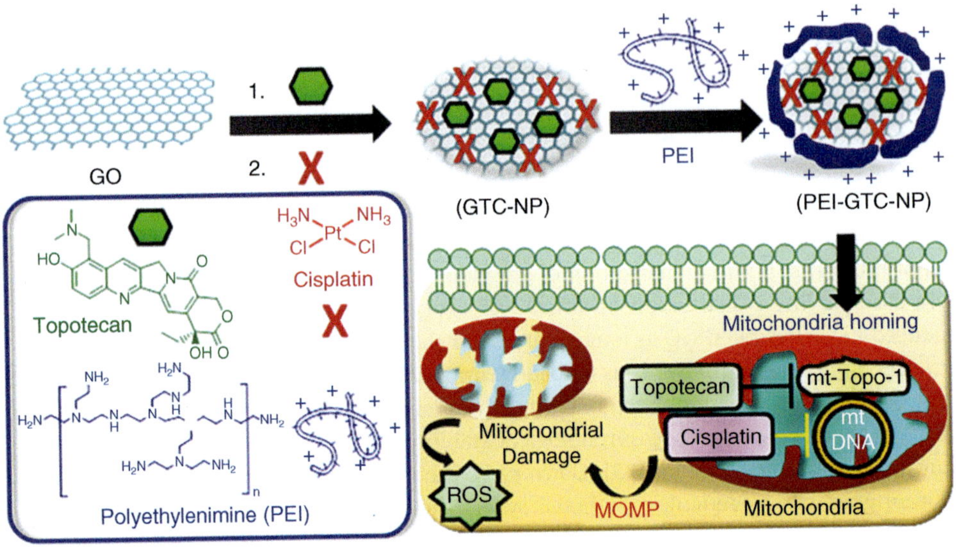

Figure 5.9 Utilizing GO with PEI coverage for co-delivery of cis-platin and topotecan to target mitochondria in cancer cells. Source: Reproduced with permission from Mallick et al. [137]/American Chemical Society.

a novel polymer-graphene platform based on self-assembled PEI-covered GO nanoparticles (PEI-GTC-NP) for the co-delivery of cisplatin and topotecan into mitochondria. This platform effectively targets drugs in the mitochondria of cancer cells, leading to the generation of reactive oxygen species (ROS) at specific sites within the mitochondria, ultimately inducing apoptosis in cancer cells (Figure 5.9) [137].

Topotecan was loaded onto GO through π-π stacking to enable visualization of the nanomaterial within cancer cell mitochondria via fluorescence emission. Cis-platin was utilized as a chemotherapy agent for the treatment of cervical cancer. The localization of PEI-GTC-NPs within cancer cells led to perforation of the mitochondrial membrane. MTT assay results indicated that the half-maximal inhibitory concentration (IC_{50}) for nanoparticles lacking PEI (GTC-NPs) was higher than PEI-GTC-NPs. These nanoparticles, characterized by a positive surface charge and appropriate size, demonstrated specific penetration into cancer cell mitochondria and exhibited promising potential for cancer treatment. Natural biopolymers, due to their low toxicity, biodegradability, renewability, environmental sensitivity, and biocompatibility, are also considered promising materials for enhancing the biological properties of graphene-based materials [138]. Chitosan, known for its anticancer properties and pH sensitivity, has been found to be extensively used in crafting polymer-graphene hybrids. These resultant multifunctional scaffolds have demonstrated exceptional loading capabilities and possess functional binding sites suitable for various drugs and targeting ligands [139]. Furthermore, graphene-chitosan platforms have been extensively investigated as pH-sensitive nanocarriers for targeted drug delivery to cancer cells. Dhanavel and colleagues developed graphene oxide-chitosan hybrids for the controlled release of 5-fluorouracil (5-FU) and curcumin (CUR). These hybrids were synthesized by linking chitosan to GO nanosheets using tripolyphosphate as a cross-linker. Chitosan/reduced graphene oxide (CS/rGO) has exhibited pH-dependent drug release characteristics. In 2018, Zhao et al. introduced a chitosan-graphene drug delivery system featuring a core–shell structure with graphene oxide nanoparticles (GONs) as the core and chitosan forming the surface charge-reversible shells, deposited on the graphene core through self-assembly (Figure 5.10) [140]. DOX adheres to the

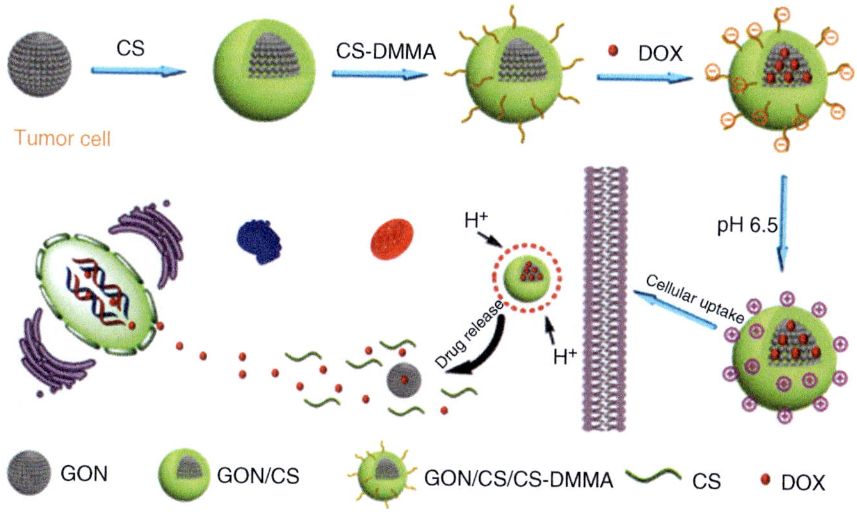

Figure 5.10 Schematic representation of the synthesis and cellular uptake of GO nanoparticle/chitosan hybrids as a drug-delivery system. Source: Reproduced with permission from Zhao et al. [140]/American Chemical Society.

extensive π-conjugated structure of GO through noncovalent interactions (π-π stacking) and is enclosed within the chitosan shells to prevent premature release into the surrounding medium. Upon encountering the acidic pH environment typical of cancerous tissue, the chitosan coating disintegrates, facilitating the rapid release of DOX. The charge of these nanomaterials shifts from negative at pH 7.4 in the bloodstream to positive at pH 6.5 in tumor tissue. The chitosan-graphene platform offers numerous advantages over conventional drug delivery systems, including high drug-loading capacity, exceptional encapsulation efficiency, prolonged circulation time in the bloodstream, enhanced cellular uptake, and customizable controlled release of anticancer agents. Despite its extensive applications in biomedical and environmental domains, chitosan faces challenges such as insolubility at neutral pH and slow degradation rates [141]. These problems have been overcome by combining CS with other biopolymers such as cellulose, dextrin, sodium alginate, and so on [142]. Multilayers of biopolymers can be disposed of on graphene through various techniques [143]. In a study, a layer-by-layer self-assembly technique has been explored for the deposition of chitosan and dextrin on the graphene surface [144]. Initially, GO is modified with chitosan (CS), and then dextrin (Dex) is conjugated to the CS-modified GO through electrostatic interaction with oppositely charged polyelectrolytes. The resulting GO CS/Dex conjugate has demonstrated superior drug-loading capacity, physiological stability, and cellular uptake compared to GO-CS alone. In a separate investigation, chitosan (CS) was modified with poly (itaconic acid-copolymerized acrylic acid) to enhance its hydrophilicity [145]. This new combination has been used to modify GO-amine (AGO) with an anticancer drug delivery application. CS-GO hybrid has been prepared via electrostatic interactions between chemically modified chitosan as an anionic polyelectrolyte and AGO which acts as a cationic polyelectrolyte [146]. Omidi et al. [147], Samadi et al. [148], and Lei et al. [149] have also verified the improvement of physicochemical and biological properties of graphene-chitosan composites through integration with other polymers. HA is recognized as another important polysaccharide for the improvement of the biological properties of graphene-based materials [150, 151]. Biomedical applications of HA and its usage for the production of anticancer drug delivery systems have been widely developed due to its excellent biocompatibility, biodegradability, strong binding affinity, and selective targeting of tumor markers, especially the cluster determinant (CD44) and hyaluronan receptor (HARE) [152]. Yin's group has prepared a redox-sensitive hyaluronic acid-graphene oxide (HSG) composite and studied its ability to target HA-receptor overexpressing tumors and the efficiency of redox-sensitive linkages to localize drugs at the target sites. This system was synthesized by conjugation of hyaluronic acid to GO nanosheet using disulfide linkages (HSG). HSG-DOX displayed accelerated release of DOX in acidic environments. The high activity of hyaluronidase (HAase) in acidic conditions resulted in the degradation of hyaluronic acid chains into small pieces and promoted the release of DOX. Owing to high drug-loading capacity, photothermally controlled effects, specific targeting of HA receptors overexpressing tumors, and redox-dependent response, HSG-DOX showed amplified chemotherapeutic outcomes [153]. Dendrimers and hyperbranched polymers, especially polyglycerol, are able to effectively improve biological properties and biofunctionality of graphene sheets, as demonstrated in our previous studies [154–157]. Our group implemented a new method for functionalization of graphene-based nanomaterials with hyperbranched polyglycerol [158]. Thermally reduced GO was functionalized by triazine and consequently conjugated with hyperbranched polyglycerol through triazine functional groups. The polyglycerol-covered nanographene with mitochondria-targeted ligands and pH-triggered surface charges showed a high loading capacity and triggered the release of DOX drug in the acidic environment (Figure 5.11). Owing to the photothermal properties of nanographene, this multifunctional drug delivery system exhibited accelerated drug release and enhanced chemotherapeutic effect after NIR irradiation. The authors concluded that the high accumulation of smart multifunctional drug delivery systems in mitochondria and nuclei, together with photothermal properties, were the reasons for the enhanced chemotherapeutic effects [158]. One of the cancer treatment methods that have received a great deal of attention is targeted delivery, in which anticancer drugs are transported to the target tissues specifically. The method is designed by attaching targeting agents to nanocarriers. Targeting drug delivery systems increases the localization of drugs at tumor sites and diminishes the side effects of chemotherapies. Accordingly, a drug-delivery system with targeting ligands and good emission properties in visible and infrared has been prepared by loading anticancer drugs onto NGO [82]. Zhang et al. have developed GO as a drug delivery system by loading DOX and CPT on its surface and conjugation of folic acid to transport cargo to the tumor sites. FA-conjugated NGO (FA–NGO) was able to target MCF-7 cells and human breast cancer cells by interaction with FA receptors. Some superparamagnetic nanocarrier based on attaching Fe_3O_4 as a magnetic nanocarrier on GO and aptamer (APT) as a targeting moiety for MCF-7 cancer cell is prepared, and then, paclitaxel (PAC) is loaded as an anticancer drug. This system has shown advantages including biocompatibility, pH-responsivity, thermal stability, and high drug-loading capacity over similar nanocarriers [159]. A rGO-based system with the ability to deliver DOX to human breast adenocarcinoma cancer cells (MDA-MB-231) is also reported. In this system, a thiol maleimide containing catechol (dopa-MAL) is assembled on the surface of rGO by π-π stacking. Then, a cancer cell targeting cyclic peptide is attached to dopa-MAL. This nanostructure has shown a high efficiency against MDA-MB-231 cells with an $IC_{50} = 58\,\mu g/mL$ for DOX-loaded rGO/dopa-MAL-c (RGDfC), corresponding to an $IC_{50} = 9.9\,\mu g/mL$ for DOX in the matrix (Figure 5.12) [160]. Hyaluronic acid (HA) is also used to target GO in cancer cells. GO with HA functionality (GO-HA) has been coloaded with DOX and paclitaxel (Ptx), and the obtained drug-delivery system (GO-HA-DOX/Ptx) has been used for the efficient incapacitation of cancer cells [160]. Another target drug delivery system is designed based on dopamine (DA) functionalized nGO nanocarrier (DA-nGO). DA improves the delivery of MTX) as an anticancer cargo into target cells [161]. Song et al. have produced a targeted drug delivery system by loading Fe_3O_4 nanoparticles on the surface of GO and conjugation of lactoferrin (Lf) as a targeting ligand to this platform. The results have shown that this system (Lf at GO at Fe_3O_4 at DOX) has the ability to deliver drugs into C6 glioma cells effectively [160]. Attaching polymers

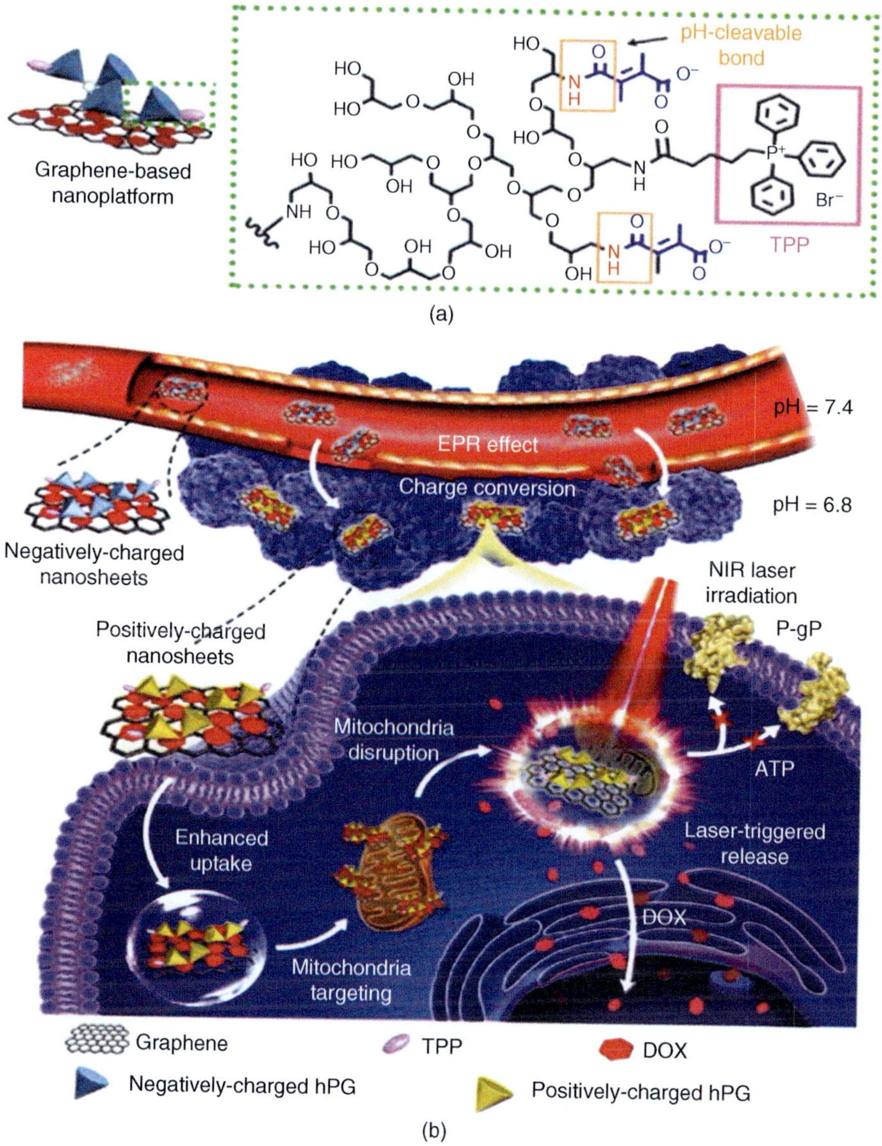

Figure 5.11 (a) The molecular composition of nanographene, coated with polyglycerol and featuring mitochondria-targeting ligands and charge conversional functional groups. (b) The multifunctional drug delivery system targets mitochondria using ligands and utilizes its photothermal properties under NIR laser irradiation to facilitate drug release, resulting in effective therapeutic outcomes. Source: Reproduced with permission from Tu et al. [158]/John Wiley and Sons.

and targeting ligands onto the surface of graphene derivatives enhances the therapeutic efficacy of these platforms. This process increases their loading capacity, reduces side effects, and enables the effective localization of drug delivery systems within targeted cells [162].

5.5.2 In Tissue Engineering

Tissue engineering is a scientific field that reproduces or regenerates damaged tissues or whole organs [163]. In pursuit of its objective, it employs a blend of cells, engineered scaffolds, and

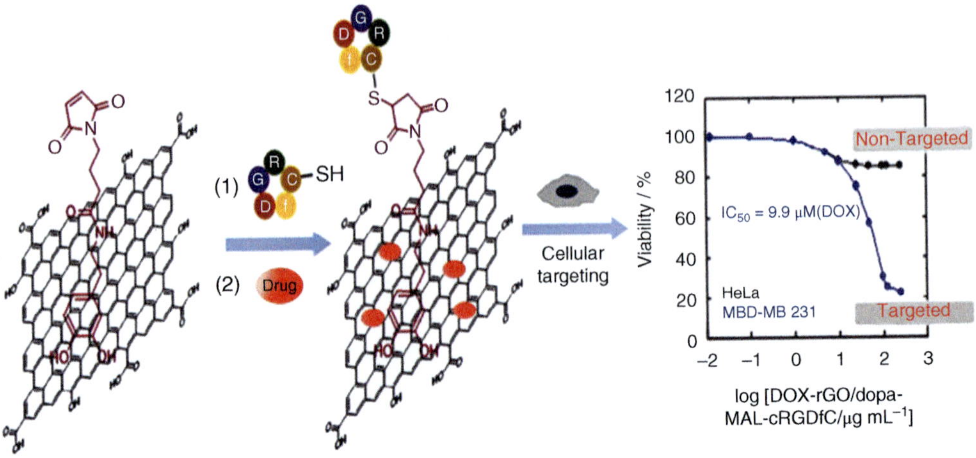

Figure 5.12 Thiol-maleimide-functionalized reduced graphene oxide (rGO) incorporating catechol (dopa-MAL) serves as a targeted drug delivery platform for doxorubicin (DOX), aimed at eradicating human breast adenocarcinoma cancer cells (MDA-MB-231). Source: Reproduced with permission from Oz et al. [160]/American Chemical Society.

biologically active compounds to fabricate functional constructs aimed at restoring, preserving, or enhancing tissue functionality [163]. Hence, the development of innovative scaffolds tailored for constructing intricate tissue constructs is pivotal in tissue engineering. An optimal scaffold should be capable of transporting active biomolecules, eliciting appropriate physiological signals, mimicking the mechanical properties of native tissue, and furnishing a substrate for interacting with living cells. This facilitates cell attachment, proliferation, and differentiation [164]. Graphene and GO, characterized by their expansive surface areas as two-dimensional materials, offer ample substrates for cellular interaction and can serve as carriers for various biomolecules like DNA, enzymes, proteins, or peptides. This transport can occur via either covalent bonds or noncovalent interactions, such as π-π stacking. Furthermore, their remarkable mechanical and electrical properties empower these materials to bolster the mechanical integrity of tissue substitutes and facilitate the application of electrical signals. Consequently, graphene and GO emerge as prime candidates for tissue engineering (Figure 5.13).

Graphene-based composites offer enhanced mechanical strength and are suitable for tissue engineering applications. Materials like bioceramics, hydrogels, films, fibers, and various tissue engineering scaffolds can significantly benefit from the reinforcement properties of graphene and GO. This enhancement in mechanical properties and stability enables their application across diverse tissue engineering domains. Particularly in bone tissue engineering, the integration of graphene or GO has led to substantial improvements in the mechanical strength of biocomposites. It is established that firm substrates can effectively support bone differentiation [166], and graphene-based composites can enhance osteogenic differentiation [167]. Consider, for instance, hydroxyapatite (HA), a primary inorganic component crucial for bone structure and regeneration. When combined with GO, composites like HA/silk fibroin (HA/SF) exhibit enhanced mechanical properties, displaying increased compressive strength and modulus. Moreover, GO-HA/SF composites have been shown to enhance the attachment and proliferation of mouse mesenchymal stem cells (MSCs). Additionally, they stimulate the expression of osteocalcin, an osteogenic gene, thereby fostering the differentiation of MSCs into bone tissue [168]. It has been documented that GO substantially enhances the tensile strength of poly(L-lactic-*co*-glycolic acid) (PLGA) and hydroxyapatite (HA) nanofibrous matrices. These matrices, composed of PLGA/GO/HA,

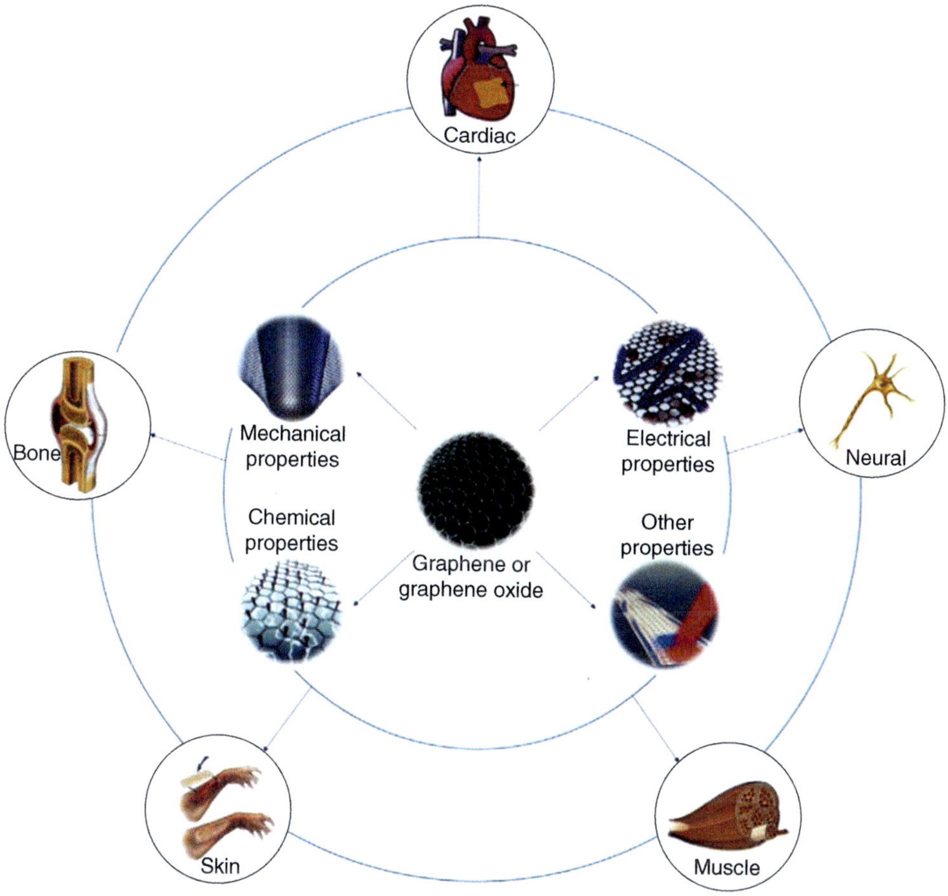

Figure 5.13 The collective tissue engineering applications of graphene and graphene oxide (GO) are determined by their respective properties. Source: Reproduced with permission from Shang et al. [165]/Elsevier.

demonstrate robust mechanical stability suitable for cell growth (Figure 5.14). Moreover, they effectively enhance alkaline phosphatase activity, express genes related to osteogenesis, and facilitate mineral deposition. Thus, they serve as exceptional and adaptable scaffolds for various applications in bone tissue engineering [169]. Other similar research works about GO-enhanced HA composites have also been reported for bone tissue engineering [170]. In addition to HA, graphene, and GO have been shown to improve the mechanical characteristics of hydrogels and biopolymers. For example, chemically converted graphene was combined with cellulose to create self-supporting graphene hydrogel films. These films were able to induce osteogenic differentiation of stem cells both in vitro and in vivo, without the need for additional inducers [171]. Furthermore, the application of a GO coating has been shown to enhance the biomedical characteristics of collagen scaffolds. This includes improvements in surface structure, compressive strength, and the ingrowth of osteoblastic cells, leading to enhanced biocompatibility and biodegradability [172]. Moreover, integrating GO notably enhanced the compressive modulus of SF scaffolds, with GO/SF demonstrating significantly improved osteogenic proliferation of osteoblastic cells compared to SF scaffolds lacking GO [173]. There are reports indicating that even at very low concentrations, GO can strengthen the biocompatible polymer polypropylene fumarate, resulting in significantly enhanced compression and flexural strength, making it suitable for applications in

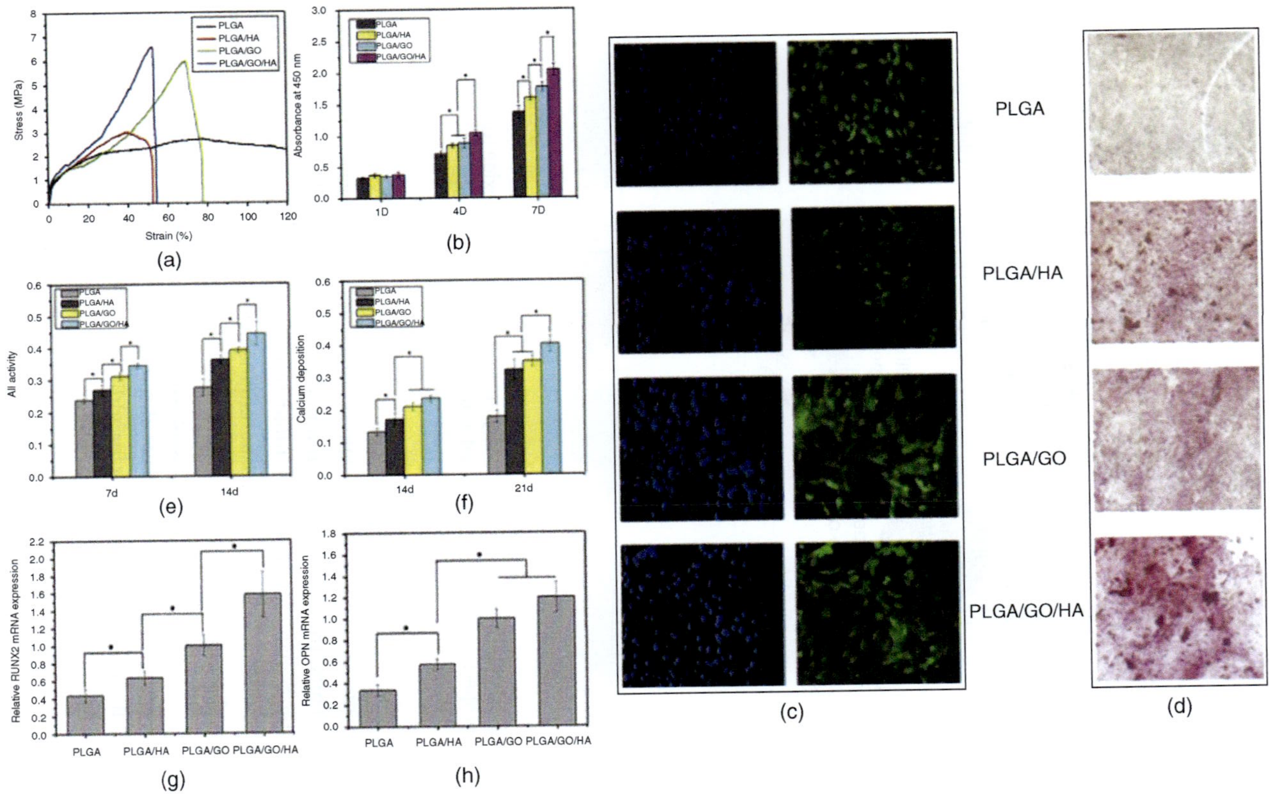

Figure 5.14 The PLGO/GO/HA matrices were employed as scaffolds to promote cell proliferation and osteogenic differentiation, with PLGA, PLGA/HA, and PLG/GO matrices serving as controls. (a) Stress–strain curves of various nanofibrous matrices were examined. (b) MC3T3-E1 cell proliferation on different nanofibrous matrices was assessed over 1–7 days in vitro. (c) Fluorescent staining observation of MC3T3-E1 cells cultured on various nanofibrous matrices for 4 days revealed cytoskeletal (fluorescein isothiocyanate [FITC], green) and nuclear (4′,6-diamidino-2-phenylindole [DAPI], blue) staining. (d) Alizarin Red staining of MC3T3-E1 cells on different nanofibrous matrices was conducted at 21 days. (e) Alkaline phosphatase (ALP) activities of MC3T3-E1 cells on various nanofibrous matrices during a 14-day in vitro culture were measured. (f) Calcium deposition after culturing on different nanofibrous matrices for 14 and 21 days was examined. Additionally, quantitative real-time PCR analysis of osteogenesis-related gene expression, including RUNX2 (g) and OPN (h), was performed after 7 days of MC3T3-E1 cell culture. ALP refers to alkaline phosphatase, PLGA denotes poly(L-lactic-*co*-glycolic acid), GO represents graphene oxide, and HA stands for hydroxyapatite. Source: Fu et al. [169]/PUBLIC LIBRARY OF SCIENCE (PLOS)/CC BY 4.0.

bone tissue engineering [174]. Furthermore, diverse organic and inorganic substrates coated with graphene have demonstrated the ability to regulate and expedite the osteogenic differentiation of human MSCs. This effect has been attributed to graphene's exceptionally high Young's modulus and its remarkable flexibility for out-of-plane deformation [123]. Besides augmenting mechanical strength, graphene can also enhance the toughness of hybrid materials intended for load-bearing implant applications aimed at bone tissue regeneration. In natural bone, a primary load-bearing component, microcracks often occur under routine physiological loads, triggering the bone remodeling process to halt the spread of microcracks and ultimately facilitate new bone formation in the affected area. This toughening mechanism can be replicated using composites incorporating graphene. The robust interaction between graphene or GO and the matrix materials can effectively bridge cracks and hinder crack propagation, thereby improving toughness [175]. For example, studies have shown that incorporating GO into poly(vinyl alcohol) (PVA) not only enhances its strength but also its toughness. The tensile strength and Young's modulus of PVA films containing GO increased significantly by 88% and 150%, respectively, compared to unfilled PVA. Additionally, the elongation at break was boosted by 22% in the presence of GO [176]. Therefore, the addition of graphene or GO can make brittle materials applicable for bone implants. Another significant application in tissue engineering involves enhancing the mechanical properties of graphene-based materials for skin and muscle regeneration. Creating biomaterials that mimic the mechanical microenvironment of the native extracellular matrix (ECM) is crucial for stimulating skin and skeletal muscle regeneration. However, many current biomaterials fall short of replicating the mechanical behavior of the ECM [177]. For instance, hydrogels have been extensively utilized as a type of ECM-mimicking material due to their hydrophilic polymeric network structure resembling the ECM and their adjustable physicochemical properties. However, enhancing the toughness of hydrogels while preserving their rigidity has proven to be quite challenging [[178]]. Moreover, electrospun nanofibers exhibit a nanoporous structure, providing a high specific surface area similar to the nanotopology of ECM. However, certain materials exhibit inherent brittleness, low toughness, and limited wear resistance, significantly restricting their use as scaffolds in tissue engineering [169]. Simultaneously, incorporating graphene or GO into biomaterials enables the adjustment of their mechanical properties to mimic the behavior of ECM. Du et al. demonstrated that the addition of rGO effectively modified the elastomeric behavior and mechanical strength of poly(citric acid octanediol-polyethylene glycol) (PCE) composites. The improvements in mechanical properties were ascribed to the robust interactions between rGO and the polymer matrix, resulting in PCE/GO nanocomposites with mechanical properties and elastomeric behavior closely resembling those of native skeletal muscle tissues. This led to significantly enhanced myogenic differentiation and regeneration [177]. Likewise, the integration of rGO into polyacrylamide (rGO/PAAm) hydrogels or PLGA/collagen hybrid matrices enhanced Young's moduli and elasticity of these materials owing to the robust molecular interactions between GO and polymer chains. By adjusting the concentrations of graphene materials, Young's moduli of these hydrogels or matrices could be tailored to resemble the mechanical properties of the native skeletal muscle ECM. This adjustment proved beneficial for promoting enhanced differentiation of myoblasts into myotubes [179]. Patel et al. presented a study on graphene-incorporated poly(ε-caprolactone) (PCL) nanocomposite, which promoted the formation of multinucleated myotubes. They illustrated that by adjusting the graphene concentration, the mechanical characteristics of the nanocomposite scaffolds could be finely adjusted, imparting them the ability to guide cell behavior and induce myoblast differentiation into multinucleated elongated myotubes, even in the absence of differentiation media. Furthermore, the pliability of graphene and GO, coupled with their distinct electromechanical properties, renders them suitable for constructing artificial muscle. Conventional rigid artificial devices often exhibit mechanical discrepancies with soft tissues, leading to diminished signal-to-noise ratios, mechanical fatigue, and scarring [180]. Hence, a major drawback of conventional artificial muscles is their limited long-term durability under continuous use, primarily attributed to electrolyte and hydrated cation leakage through cracks in the metallic electrodes [181]. In contrast, graphene enhances mechanical flexibility and stability during significant bending deformations of the electrode without requiring additional rigid metallic current collectors. Graphene-based materials have also demonstrated efficacy in enhancing the healing of wounds with poor regenerative capacity. Nyambat et al. created a composite sponge by cross-linking adipose stem cell-derived ECM with genipin and incorporating nanographene oxide (GO) for skin tissue engineering applications. Typically, ECM scaffolds derived from cells exhibit inferior mechanical properties compared to those derived from tissues. However, by employing GO and genipin cross-linking to enhance the mechanical strength of pure cell-derived ECM, they successfully engineered a biomimetic ECM sponge for use as a skin graft [182].

Collagen fibrin (CF) composite films integrated with GO were utilized as wound dressing materials for both in vitro and in vivo investigations. The inclusion of GO augmented the mechanical robustness of CF composite films, thereby promoting the process of wound healing [183]. Li et al. employed three-dimensional graphene foams (3D-GF) laden with bone marrow-derived MSCs to facilitate the healing of skin wounds with minimized scarring. Their findings revealed that the combination of bone marrow-derived MSCs with the biomechanical and biochemical attributes of 3D-GF worked synergistically to create an improved environment for wound healing, promoting early vascularization and diminishing scarring [184]. Hou et al. introduced a durable and flexible self-repairing film infused with graphene, envisaging its use in artificial skin applications. The film displayed significantly improved resilience and tensile strength, along with consistent sensitivity to even the slightest touch, maintaining its properties even when bent or stretched. Their study showcased how this thin film could emulate the mechanical self-repairing ability and pressure sensitivity characteristic of natural skin without requiring any external power source [185].

Furthermore, mechanically reinforced graphene-based materials have found applications in various other tissue engineering domains. For instance, in cardiology, myocardial infarction (MI) stands as a significant contributor to global morbidity and mortality. Numerous hydrogels with diverse mechanical properties have been developed to restore myocardial function. Interestingly, research indicates that hydrogels with higher stiffness are often more effective in restoring cardiac function [186]. For instance, incorporating pristine graphene into collagen boosted stiffness to align with native cardiac tissue, thereby enhancing the growth of human cardiac fibroblasts. Additionally, it hindered bacterial adhesion on the biohybrids by mechanically disrupting bacterial cells [187]. Incorporating GO nanoparticles into oligo(poly(ethylene glycol) fumarate) (OPF) hydrogels improved mechanical reinforcement and facilitated the formation of gap junctions. This enhancement promoted the development of cytoskeletal structures and the formation of intercalated discs crucial for cardiac integrity and function [188]. Likewise, the inclusion of rGO notably bolstered the mechanical characteristics of gelatin methacryloyl (GelMA) hydrogels. The scaffold comprising rGO-GelMA provided a microenvironment closer to natural conditions for cardiomyocytes, showing improved biological functions including cell viability, proliferation, and maturation [189]. Additionally, GO-infused methacryloyl-substituted recombinant human tropoelastin (MeTro) has been documented as a flexible biomaterial for cardiac tissue engineering. The introduction of GO enhanced both the elastic modulus and rupture strain. The MeTro/GO gel displayed remarkable resilience to rotational stress, exhibiting no signs of plastic deformation [190]. The inclusion of GO in multiarmed polyethylene glycol/melamine hydrogel via π-π conjugation stacking has resulted in a hydrogel that is softer but mechanically robust. This characteristic is beneficial for transmitting mechanical signals and synchronizing the contraction of myocardial tissue, thus greatly aiding in the restoration of cardiac function in MI cases [186]. In addition,

GO could also induce spontaneous cardiac differentiation in embryoid bodies (EBs) by changing the EBs' stiffness [191]. In the field of neural tissue engineering, while the electrical properties of graphene-based materials are a significant advantage, as we will explore later, their mechanical characteristics also facilitate their use in neural tissue engineering. Initially, graphene-based materials can serve as an appropriate scaffold for neural cell cultivation. Graphene nanofibers, produced through electrospinning techniques, exhibited Young's modulus akin to that of natural collagen fibers. This similarity offers optimal mechanical compatibility when utilized as a soft matrix for cell culture and regeneration of soft tissues, particularly in the context of brain and nerve tissue [192]. Surprisingly, the inclusion of rGO notably diminished the mechanical durability of PCL/rGO membranes during hydrolytic degradation. Nonetheless, the PCL/rGO membrane retained adequate mechanical characteristics, theoretically meeting the requirements for neural tissue regeneration specifications [193]. Furthermore, the flexibility of graphene enables the fabrication of graphene-based devices for neural imaging and optogenetic applications. These devices surpass the current indium tin oxide (ITO)-based electrode technology due to their significantly enhanced mechanical flexibility [194]. Hence, the robust interaction between graphene materials and biomaterials in tissue engineering scaffolds enhances their mechanical strength, aligning them with the mechanical properties of native tissues. This alignment promotes specific stem cell differentiation for various tissue engineering endeavors. Additionally, graphene-based materials enhance the toughness and flexibility of biohybrids, enabling their utilization in manufacturing biomedical devices for load-bearing implants in bone regeneration or stretchable devices in other tissue engineering domains.

5.5.3 In Bioimaging

Optical imaging is a method that does not require any invasive procedures. It harnesses visible light and the distinctive properties of photons to capture finely detailed images of organs, tissues, and even smaller entities such as cells and molecules [195]. Compared to other imaging techniques, optical imaging offers several benefits. It is relatively affordable, highly sensitive (detecting concentrations as low as 10^{-9}–10^{-12} mol/L), uses nonionizing radiation, provides real-time imaging, has a short acquisition time, and allows for multiplexing (simultaneous imaging of multiple targets). However, it does come with constraints, including restricted tissue penetration (0–2 cm) and considerable scattering of photons by tissues within the visible light spectrum (395–600 nm) [120, 121]. Moreover, there is notable background interference due to tissue autofluorescence and light absorption by proteins (257–280 nm), heme groups (absorbance maximum at 560 nm), and even water (above 900 nm) [196]. To tackle these challenges, researchers have delved into imaging techniques within the near-infrared window (NIR, 650–900 nm) [197] and the second NIR window (NIR-II, 1000–1700 nm) [198, 199]. These modalities offer advantages such as reduced autofluorescence, diminished tissue scattering, and enhanced depth of penetration for in vivo imaging. The exploration of graphene-based nanomaterials has been particularly active in the realm of optical imaging, encompassing fluorescence (FL) imaging, two-photon fluorescence imaging (TPFI), Raman imaging, and more. Various dyes, photosensitizers (PSs), quantum dots (QDs), graphene nanocomposites (GNCs), or upconversion nanoparticles (UCNPs) functionalized with graphene oxide/reduced GO/rGO) have been extensively investigated for fluorescence imaging [200]. GQDs, possessing inherent photoluminescence, can serve as probes for fluorescence (FL) imaging. Nitrogen-doped GQDs (N-GQDs) exhibit a robust two-photon absorption cross-section, reaching as high as 48,000 Göppert-Mayer units, showcasing their efficacy as two-photon fluorescent probes for TPFI [201].

Graphene and its derivatives have emerged as promising candidates for efficient quenchers in graphene-based nanosensors, capable of quenching various fluorescent substances like small molecule dyes, QDs, and conjugated polymers. This quenching mechanism is achieved through either FRET or charge transfer mechanisms [202]. Additionally, the intrinsic Raman signals of GOs, particularly the G line (E_2g mode of sp^2 carbon atoms) around $1600\,cm^{-1}$ and the D line (symmetry A1g mode) around $1350\,cm^{-1}$, can be intensified by metal nanoparticles such as Au or Ag. This enhancement finds applications in surface-enhanced Raman scattering (SERS) for Raman imaging [203].

Fluorescence imaging, a noninvasive technique relying on emitted photons from fluorescent probes [204, 205] has seen advancements in nanographene oxide (nGO) applications. The Dai group pioneered the use of B-cell-specific antibody Rituxan (anti-CD20) conjugated with PEG-modified GO (nGO-PEG-Rituxan) for targeted FL imaging of Raji B cells. Despite its first application, the low quantum yield of nGO-PEG FL limits in vivo imaging potential. Consequently, various groups turned to organic fluorescent dyes for GO/rGO functionalization, such as Liu's group employing Cy7-conjugated nGO-PEG (nGO-PEG-Cy7) for in vivo FL imaging of tumor xenografted mice. Similarly, the Chen group developed GO-IRDye800-VEGF for FL imaging of ischemic muscle tissues in a murine hindlimb ischemia model, demonstrating active targeting of hypoxic tissues. The importance of the "bridge" linker, such as PEG, between fluorescent dyes and GO/rGO to prevent FL quenching was emphasized. In addition to organic dyes, direct loading of dyes or PSs onto GO/rGO was explored for FL imaging. Unfortunately, the use of folic acid-conjugated GO and GO-PEG led to significant FL quenching of Ce6 and 2-(1-hexyloxyethyl)-2-devinyl pyropheophorbide-alpha (HPPH or Photochlor®), respectively. This quenching arises from strong π–π stacking and hydrophobic interactions between the single porphyrin rings of Ce6 and HPPH with the GO plane, hindering effective FL imaging. To address this limitation, a phototheranostic agent based on sinoporphyrin sodium (DVDMS) was loaded, aiming to overcome the challenges associated with FL quenching in photodiagnostics, particularly in FL imaging. A PEGylated graphene oxide (GO-PEG-DVDMS) was developed, demonstrating improved fluorescence (FL) properties for enhanced optical imaging-guided photodynamic therapy (PDT) (Figure 5.15a) [206]. The FL of the loaded DVDMS was significantly enhanced through intramolecular charge transfer, leading to shifted emission peaks at 644 nm and 670 nm in GO-PEG-DVDMS (Figure 5.15b). Notably, the FL intensities of GO-PEG-DVDMS were 3–8 times higher than those of DVDMS at various weight ratios of GO-PEG: DVDMS (0.1 : 1 ~ 2 : 1) (Figure 5.15c). In vivo, fluorescence signals were predominantly observed in the tumor area of mice intravenously injected with GO-PEG-DVDMS, indicating its potential for real-time FL visualization of in vivo DVDMS delivery and distribution, as well as imaging-guided PDT (Figure 5.15d). To address potential photobleaching issues associated with fluorescent dyes or PSs, inorganic nanomaterials like QDs, GNCs, or UCNPs have been integrated with GO/rGO for FL imaging. Chen et al. developed a multifunctional QD-tagged reduced graphene oxide (QD-rGO), which served as both an imaging agent in the visible region and a photothermal agent in the near-infrared (NIR) region. This design mitigated QD toxicity through surfactant coating and reduced FL quenching by introducing spacers [207]. Wang et al. anchored GNCs on rGO (GNC-rGO) to retain the NIR fluorescence property of GNCs for cellular imaging [208]. Zhang et al. created a multifunctional nanoplatform by grafting core-shell-structured UCNPs with nGO via PEG and loading phthalocyanine (ZnPc) on the nGO surface, demonstrating its use as an upconversion luminescence imaging probe with high contrast for cells and whole-body animals [209]. Photoluminescent GQDs, known for good photostability, nontoxicity, and easy functionalization, have been extensively explored for FL imaging. The FL mechanism of GQDs may stem from quantum size effects, electron-hole recombination, zigzag sites, and defect effects. GQDs, such as GQD-hyaluronic acid (GQD-HA), have shown efficient CD44-targeted delivery to tumor-bearing mice with bright FL in the tumor tissue. Further studies by Ge et al. in 2014 demonstrated that GQDs with a broad absorption spanning the UV and visible regions and a strong deep-red emission peak at 680 nm act as superior PDT agents and FL contrast agents for in vivo

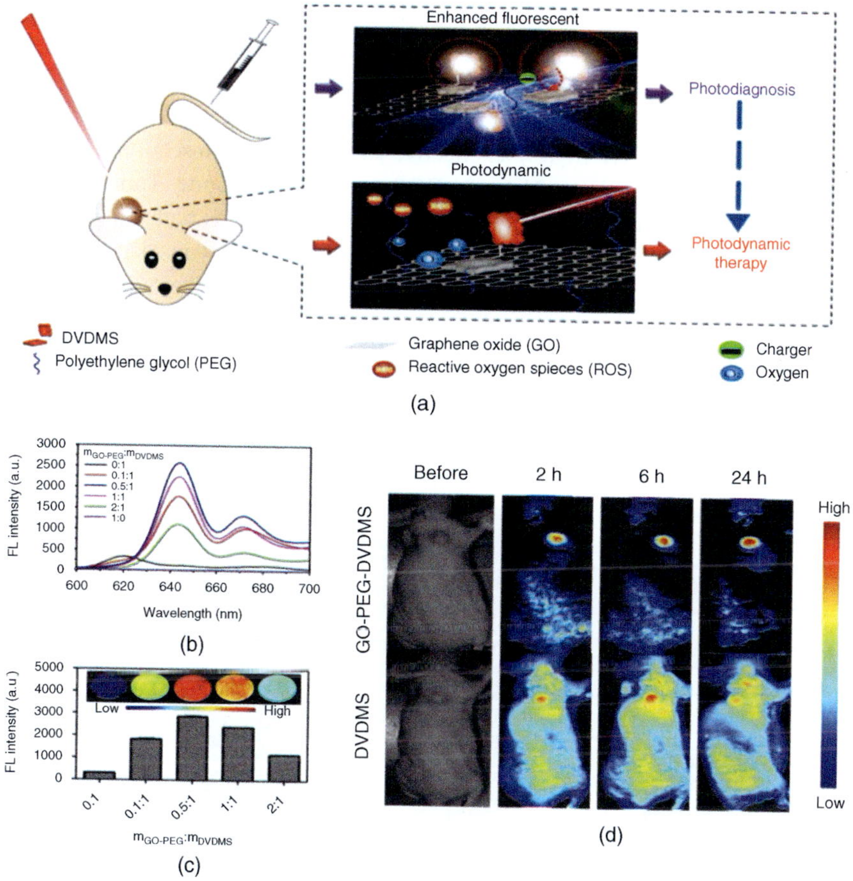

Figure 5.15 (a) Graphene oxide (GO) is used to enhance the fluorescence imaging-guided photodynamic therapy of sinoporphyrin sodium (DVDMS). (b) The fluorescence spectra of GO-PEG-DVDMS were measured at different weight ratios of GO-PEG to DVDMS. The DVDMS concentration was kept constant at 100 mg/mL, while the GO-PEG concentrations varied from 10 to 200 mg/mL. (c) The fluorescence intensity of DVDMS at 620 nm and different mixing ratios with GO-PEG-DVDMS at 640 nm were compared. (d) Mice with U87MG tumors were intravenously administered with GO-PEG-DVDMS or DVDMS (at a dosage of 2 mg/kg), and their in vivo fluorescence imaging was monitored at various time points. Source: Reproduced with permission from Yan et al. [206]/Elsevier.

FL imaging. While GQDs are promising candidates for FL imaging, efforts are needed to improve their quantum yield and optimize surface modifications to enhance their optical properties and tumor accumulation rates and reduce RES retention for broader bioimaging applications [206].

TPFI has garnered significant interest due to its potential uses in fundamental research and medical diagnostics. This interest stems from its advantages over single-photon fluorescence imaging, including a minimal autofluorescence background, increased imaging depth due to low Rayleigh scattering and NIR light's low tissue absorption, as well as reduced photobleaching and phototoxicity [200]. This method enables a more thorough examination of cellular and subcellular processes within complex biological samples. In contrast to single-photon excitation utilizing continuous-wave lasers, two-photon nonlinear excitation, commonly facilitated by a femtosecond laser, necessitates a significant input of excitation photons. This nonlinear excitation mode results in higher spatial resolution and decreased photobleaching. Additionally, the two-photon excitation wavelength, falling within the range of 700–1350 nm, is well-suited for the utilization of carbon-based nanomaterials such as carbon dots, GO, and GQDS, has recently garnered interest in the realm of TPFI for visualizing internal organs and tissues. Wang and Gu, among others, were the first to introduce transferrin-functionalized GO-PEG with robust two-photon luminescence capabilities as a long-lasting optical tool for three-dimensional TPFI and laser-assisted cancer microsurgery, employing an ultrafast pulsed laser for excitation. Subsequently, the Gong research team developed nitrogen-doped GQDs (N-GQDs) as effective two-photon fluorescent agents for imaging cellular structures and deep tissues. These N-GQDs, averaging around 3 nm in size, were synthesized via a solvothermal approach utilizing dimethylformamide as both the solvent and nitrogen source. The application of TPFI with N-GQDs as a fluorescent marker demonstrated significant efficacy in vivo for exploring biological structures within the 800–1500 μm range, exhibiting enhanced depth penetration capabilities compared to traditional one-photon fluorescence imaging techniques (Figure 5.16).

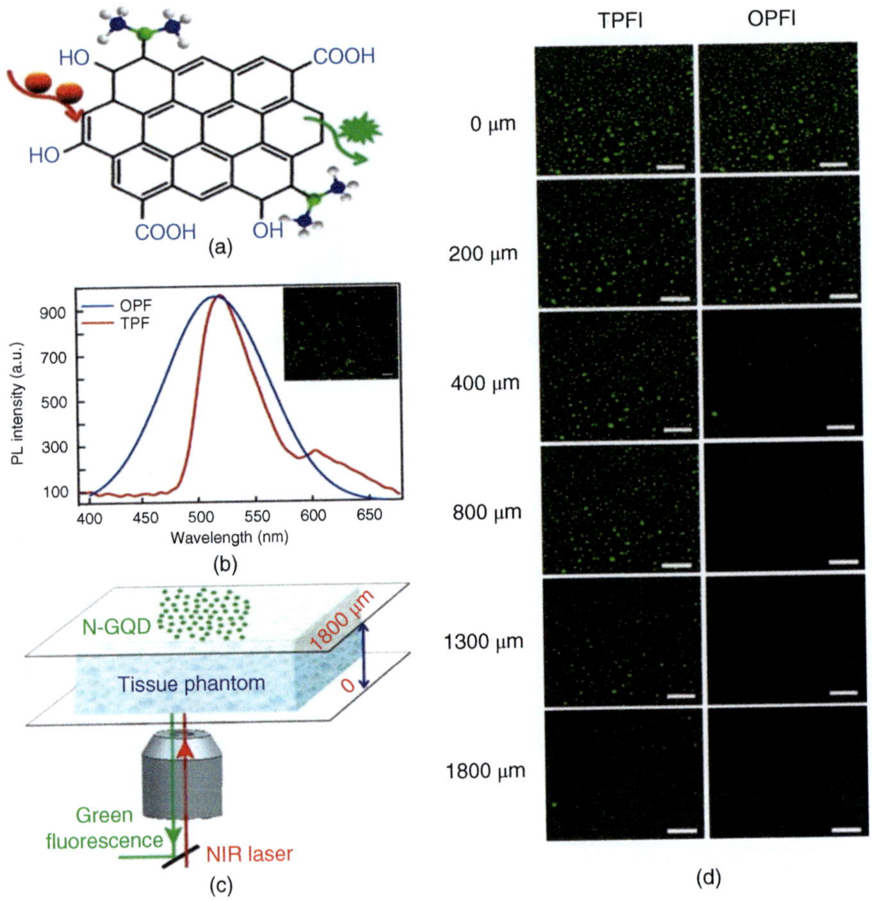

Figure 5.16 Effective two-photon fluorescent markers for imaging cellular structures and deep tissues utilizing nitrogen-doped graphene quantum dots (N-GQDs). (a) Illustration depicting the molecular structure of N-GQDs. (b) Fluorescence spectrum generated by two-photon excitation of a solution containing N-GQDs using an 800 nm femtosecond laser. Inset: Two-photon fluorescence visualization of solid N-GQD (scale bar: 10 μm). (c) Schematic representation of the setup for TPFI of N-GQDs within tissue phantoms of varying thickness. (d) Depth of penetration achieved by N-GQDs in TPFI (right panel) compared to one-photon fluorescence imaging (left panel) within tissue phantoms (all scale bar: 100 μm). Source: Reproduced with permission from Liu et al. [201]/American Chemical Society.

Raman imaging utilizes the inelastic scattering of phonons originating from molecular vibrational excitation modes and graphene oxides (GOs), which act as potent Raman markers with inherently strong D and G peaks. The incorporation of metal nanoparticles amplifies these peaks, particularly in Ag/GO hybrids, which exhibit impressive surface-enhanced Raman scattering properties. Folic acid-conjugated Ag/GO hybrids have been created for precise SERS imaging targeting folate receptor (FR)-positive cancer cells [200, 210, 211]. Similarly, in situ reduction of Au^+ on GO yields Au/GO structures that exhibit enhanced Raman signals [203]. The hybrids cultivated with HeLa 229 cells demonstrate notably enhanced Raman signals and clearer Raman images than those cultured with pristine graphene oxide (GO). Gold nanoparticles enclosed compactly within nanosized graphene oxides (NGOs) were presented by Ma et al. [212] as a dual-purpose Raman imaging probe and drug delivery system. The D and G peaks of Au at NGOs exhibited approximately one order of magnitude higher intensity than those of the NGO alone, enabling sensitive imaging of internalized Au at NGOs in HeLa cells. Liz-Marzán's group developed a seed-mediated growth method for reduced graphene oxide-gold nanostar (rGO-NS) nanocomposites, serving as active SERS materials for loading and releasing the anticancer drug DOX. These rGO-NS nanohybrids demonstrated tunable optical properties, enhanced stability compared to bare Au nanostars, and a sensitive SERS response to aromatic organic molecules. Successful SERS applications of rGO-NS were demonstrated for probing DOX loading and pH-dependent release. Additionally, the Zhang group synthesized Au/GO and Au/rGO composites via the noncovalent attachment of Au nanoparticles pre-modified with 2-mercaptopyridine to GO and rGO sheets, respectively. These composites exhibited improved SERS substrates compared to bare Au nanoparticles. The group also loaded 20 nm Au nanoparticles onto sub-200 nm GOs via amide coupling between PEG and dimercaptosuccinic acid (DMSA) ligands, utilizing the prepared Au/GO hybrids as Raman imaging probes to study cellular uptake mechanisms in the Ca Ski cell line. Meanwhile, the Tan group reported the facile synthesis of alkyne-PEG-functionalized AgCu at graphene (ACG) nanoparticles (NPs) using the CVD method to grow a few layers of graphene on the surface of AgCu NPs. These ACG NPs enhanced unique Raman signals from the graphitic shell, enabling their use for cell labeling, Raman imaging, and SERS detection. Alkyne-PEG was efficiently conjugated on the surface of ACGs through strong π-π stacking of the diphenylacetylene tail and graphene. High-resolution Raman images of MCF-7 cells treated with functionalized ACGs demonstrated good intracellular localization capability, utilizing all three modes (D, G, and alkyne) for the colocalization of ACGs, accurately pinpointing their location throughout the cell cytoplasm (Figure 5.17) [214].

5.5.4 As Biosensors

Because of its large surface area, electrical conductivity, high electron mobility, and capability to immobilize a variety of substances, graphene has been utilized in numerous biosensors utilizing diverse transduction modes. The conjugated structure of graphene facilitates electron flow between the bio-receptor and transducer, thereby boosting the signal sensitivity of electrochemical sensors. Additionally, graphene-based nanomaterials can serve as quenchers in the transducer, enabling the creation of fluorescent biosensors. Graphene based glucose sensors [215, 216] can be developed by immobilizing glucose oxidase on a graphene surface. These sensors are effectively utilized to monitor glucose levels in the human body. Furthermore, DNA biosensors based on graphene are employed to identify complementary DNA sequences through DNA-DNA hybridization. The fluorescence quenching ability of GO is harnessed by researchers to identify hybridization processes. When a dye-labeled single-stranded DNA (ssDNA) is noncovalently bound to GO [217], the fluorescence is quenched. Despite both GO and ssDNA being negatively charged, DNA can be adsorbed onto GO [218]. Recent advancements in graphene and its derivatives play a crucial

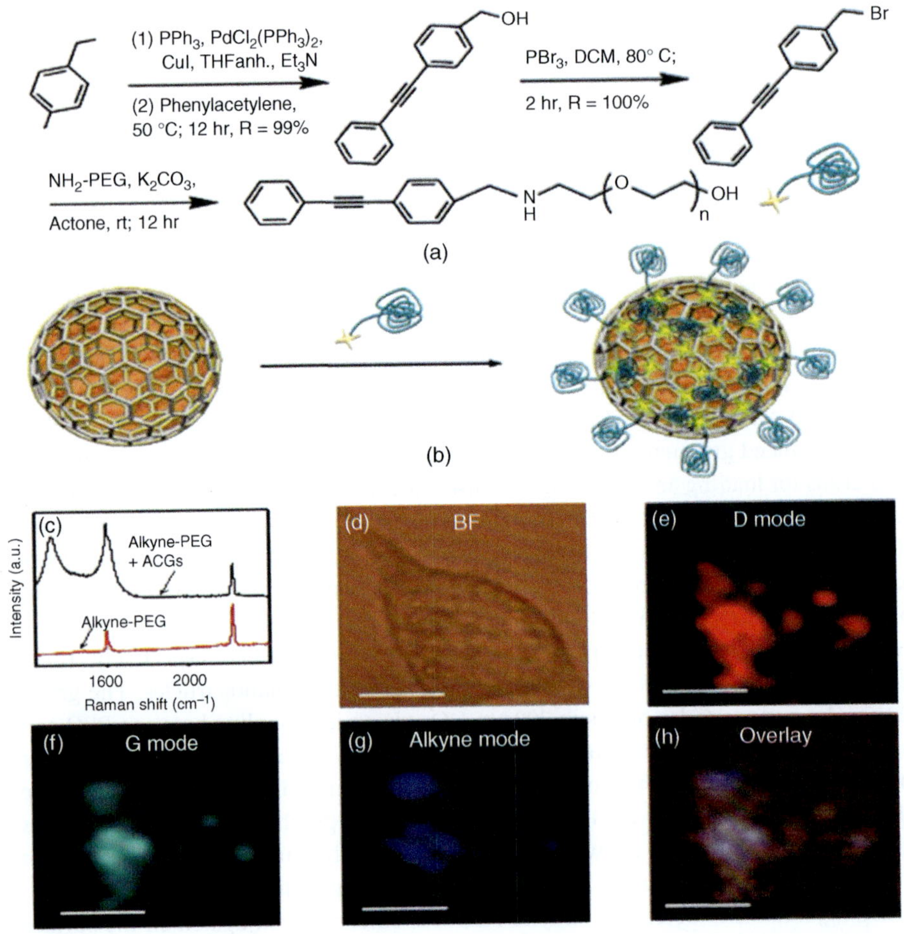

Figure 5.17 (a) Procedure for synthesizing alkyne-PEG. (b) Diagram depicting the alkyne-PEG functionalization process of ACGs. (c) Raman spectra comparison of alkyne-PEG with (black) and without (red) ACGs. (d–h) Raman imaging of MCF-7 cells treated with alkyne-PEG-modified ACGs using 1 s integration/pixel. BF represents bright field; scale bar indicates 10 μm. Source: (d)–(h) Reproduced with permission from Song et al. [213]/with permission of American Chemical Society and Reproduced with permission from Wang et al. [238]/Elsevier.

role in the early detection of life-threatening diseases such as diabetes, cancer, HIV, and more, as elaborated in Table 5.1 below. For the early identification of diseases, antibody biosensors utilizing graphene-based nanomaterials have been developed. These nanosensors have been designed to detect *Escherichia coli* [229, 230] and the *Zika virus* [231]. GO has been effectively utilized in detecting the dengue virus [232], rotavirus [233], and cardiovascular diseases [234].

The challenge in developing enzymatic biosensors lies in achieving enzyme functionalization and maintaining stability on a specific transducing layer. GO has been frequently employed to create enzymatic biosensors, and some noteworthy findings in this area are discussed in this section, with a particular emphasis on the detection mechanism. In a study by Zhou et al. [235], they highlighted the performance enhancement of an enzymatic glucose sensor based on zinc oxide (ZnO) nanorods. In this setup, reduced graphene oxide (rGO) was introduced between the ZnO nanorods and the ITO electrode, followed by stimulation under UV irradiation. Electrochemical characterization revealed that the presence of rGO not only facilitated electron transfer but also contributed to catalytic biochemical reactions. The characterization results indicated that reduced

Table 5.1 Examples of biosensing applications.

Type of sensors	Scaffold	Target	References
Electrochemistry	Graphene-poly (3-aminobenzoic acid)	PSA	[219]
Electrochemistry	rGO-Au	Folic acid	[220]
Electrochemistry	Graphene ink in roll-by-roll form with 25 wt% acetone	HIV1, HSV1	[221]
Electrochemistry	DNA-aptamer	Interferon gamma	[222]
SPR	GO-bacterial	Lysozyme	[223]
SPR	Graphene-coated POF	DNA (DXB) lattice, Protein streptavidin	[224]
SPR	Au/CdSQDs-rGO/Ab graphene sensor film, chitosan-GO-CdS QDs	Dengue, Co_{2+}	[225]
SERS	Gold/graphene-graphene	miRNA	[226]
SERS	GO-AuNPs-antibody magnetic bead	IgG	[227]
SERS	Ag–Cu_2O/reduced graphene oxide	H_2O_2, glucose	[228]

graphene oxide (rGO) played a dual role by facilitating electron transfer through the ZnO nanorods to the ITO electrode and inhibiting the rapid recombination of photogenerated electrons and holes. Under ultraviolet (UV) irradiation, holes were generated in the valence band of the ZnO nanorods, enhancing the catalytic activity of glucose oxidase (GOx) toward glucose [236, 237] (Figure 5.18). Incorporating rGO increased the sensitivity of the ZnO-nanorod-based glucose sensor by 1.6-fold and reduced the detection limit by 2.3-fold. The sensor demonstrated functionality with serum samples. Additionally, when combined with UV irradiation, the sensitivity further increased by 1.7-fold, albeit with a 2-fold reduction in the detection limit. In another study by Wang et al. [238], an electrochemical transistor for glucose sensing was developed using polypyrrole (PPy) nanowires and rGO (Figure 5.18). The composite matrix of PPy and rGO not only provided a platform for glucose oxidase but also facilitated electron transfer from the electrolyte to the collector. Incorporating rGO nanosheets facilitated the growth and proliferation of PPy nanowires within the sensor, resulting in enhanced electrical properties of fiber transistors, such as on/off ratio, switching speed, and cycling stability. Glucose sensors utilizing these transistors demonstrated remarkable sensitivities, rapid response times (approximately 0.5 seconds), an extended linear range (1 nM–5 μM), and a low limit of detection. Moreover, GO modified with chitosan, known for its biocompatibility, proved to be an efficient transduction material for glucose sensing [239, 240]. In the context of cholesterol sensing, an electrochemical biosensor was developed using the surface of an AuNP-modified screen-printed carbon electrode. Cholesterol present on the electrode surface underwent hydrolysis, generating H_2O_2. The produced H_2O_2 subsequently facilitated the reduction of silver ions within an $AgNO_3$ solution containing cholesterol, leading to the formation of metallic Ag. Anodic stripping voltammetry was employed as the electrochemical method, providing a reported limit of detection for cholesterol.

5.5.5 In Photothermal and Photodynamic Therapy

In recent times, graphene-based nanomaterials have attracted considerable interest in photothermal therapy (PTT) because of their distinctive physical and chemical characteristics. Graphene,

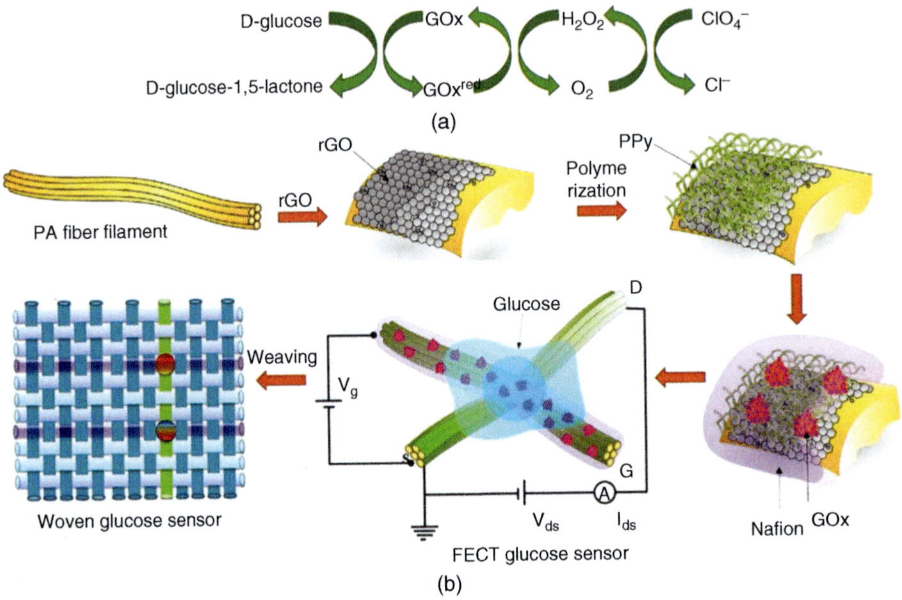

Figure 5.18 (a) The biochemical reaction illustrates glucose sensing utilizing GOx, PPy, and rGO (b) A transistor-based glucose sensor configuration is depicted. In this sensor, GOx was immobilized on PPy nanowires and the rGO surface, further coated with Nafion to enhance sensor selectivity. Source: Reproduced with permission from Wang et al. [238]/Elsevier.

CNTs, and their derivatives possess strong near-infrared absorbance, making them efficient photothermal agents. Moreover, their extensive surface areas enable functionalization with anticancer drugs, molecules enhancing biocompatibility, and biomolecules targeting specific cells, making them significant in the advancement of anticancer nanodrugs. Liu et al. pioneered the use of GO in combination with PTT to suppress tumors [241]. PEGylated GO, when exposed to near-infrared light, releases energy, generating excessive heat in cells and inducing apoptosis. Concurrently, the covalent binding of Cy7 allows for the tracking of GO metabolism. To enhance cancer treatment efficacy, PTT is being integrated with other strategies. A chemo-photothermal approach involves PEGylated GO loaded with DOX [98, 242, 243]. Wang et al. combined Fe_2O_3 at Au with rGO, loading the material with DOX for chemotherapy [98]. In this study, a magnetic field guided the transport of DOX, resulting in synergistic effects from combining chemotherapy and PTT. GO has also been utilized as a carrier for the controlled loading and release of multiple anticancer drugs, combining stability enhancement, specificity increase, and the use of drugs like camptothecin and DOX [98]. Loading PSs onto GO enables the combination of PDT and PTT methods [244–246]. Folic acid-conjugated GO loaded with Ce_6, reported by Cui and colleagues, demonstrated targeted delivery with controlled photo hyperthermia and synergistic singlet oxygen generation [244]. Other PSs like methylene blue [246] and hypocrellin B [247] have also been employed to eliminate cancer cells. To address the depth limitation of laser irradiation, ultrasound-activated PDT has been developed, enhancing treatment depth while reducing invasiveness [248]. Chen and colleagues introduced a novel approach involving a mesoporous silica layer grown on rGO nanosheets, combined with PEG-conjugated iron-oxide nanoparticles, for applications in sonodynamic therapy and ultrasound hyperthermia [249]. This nanocarrier, characterized by its composite structure of rGO, mesoporous silica, and superparamagnetic properties, demonstrates high photosensitivity, biocompatibility, and magnetic targeting capabilities. When loaded with Rose Bengal, the PEGylated nrGO/MSN-ION nanocarrier generates singlet oxygen under burst-mode ultrasound irradiation, enabling effective therapeutic outcomes. In addition to these applications, graphene-based nanomaterials have been utilized in gene therapy [250, 251], incorporating specific targeting molecules [252] and undergoing modifications aimed at extending circulation time in the bloodstream [253, 254].

5.6 Discussion on Toxicity of CNMs

Despite the numerous advantages of nanomaterials, the anticipation of toxicity effects is commonplace. These effects may arise from factors such as high-dose usage by consumers, non-standard practices by producers, environmental leakage, and occupational exposure for workers [255–257]. Globally, there is a concerted effort to mitigate these negative impacts through the establishment of regulations governing the production, handling, and utilization of nanomaterials and their end-products [258–260]. To address this concern, the initial step involves a comprehensive understanding of the potential toxic effects of each nanomaterial. This understanding is achieved through various experiments assessing the impact on the environment or living organisms, encompassing both in vitro and in vivo effects on organisms, tissues, or cells [261, 262]. Researchers, in their pursuit of comprehending nano–bio interactions, have dedicated significant efforts to conducting experiments, revealing new insights daily through both in vivo and in vitro studies [263]. Additionally, computer technology and modeling techniques have been employed to investigate interactions between nanomaterials and biomolecules [263]. Emerging research suggests the importance of considering factors such as biomolecular corona and cell parameters (type, sex, size, passage number, etc.) in understanding nano–bio interfaces [264–266]. Presently, a more nuanced understanding of nanotoxicity has been achieved based on knowledge derived from nano-bio interactions. This portion of the chapter revealed a comprehensive report on the toxicity associated with CNMs. The rise and swift advancement of nanotechnology have prompted extensive inquiries into the environmental and health impacts of nanomaterials [267]. In recent times, carbon-based nanomaterials (CNMs) have undergone extensive investigation for medical applications, including drug delivery [92], electrochemical sensors [268], imaging systems [269], regenerative medicine [270, 271], and cancer diagnosis and therapy [272–274]. Nanoparticles have been observed to accumulate in various organs, such as the heart, brain, kidneys, spleen, liver, and lungs, through inhalation, intravenous injection, and ingestion [275].

5.6.1 Neurotoxicity

Neurotoxicity denotes any detrimental impact on the structure, chemistry, or operation of the nervous system caused by either a chemical or physical stimulus [276]. While traditional pharmacological treatments encounter difficulties in accessing the central nervous system (CNS) due to the blood-brain barrier (BBB), specific nanoparticles possess the ability to penetrate this barrier. The BBB, a distinctive feature of the human body, regulates compound transportation and safeguards the brain from harmful substances, thereby maintaining brain homeostasis. Certain nanoparticles, characterized by their diminutive size (<1 μm) and extensive surface area, can breach the BBB and gather in different regions of the CNS. As a result, nanoparticles offer promise as carriers for pharmacological interventions in the diagnosis, monitoring, and treatment of CNS-related disorders [277]. Nonetheless, their utilization as nanocarriers necessitates thorough examination owing to the potential negative impacts on the CNS. Numerous in vitro and in vivo investigations have highlighted the neurotoxic repercussions of specific nanoparticles, culminating in neuro-inflammation and cognitive dysfunction [278]. Among various carbon nanomaterials, functionalized CNTs have demonstrated translocation into different subcellular compartments, including brain cells [279]. Visalli et al. explored the neurotoxic and neuroinflammatory effects caused by pristine and functionalized MWCNTs, specifically MWCNT-COOH, in differentiated SH-SY5 cells, which act as an in vitro model resembling neuroblastoma cells. Their results unveiled that MWCNTs triggered DNA damage and excessive generation of ROS within neuronal cells, leading to a proinflammatory reaction characterized by elevated levels of TNFα, IL-1β, IL-6, and cytokines [280]. In a separate investigation conducted by Bussy et al., the cytotoxic effects of four chemically functionalized MWCNTs, including pristine MWNTs, carboxylated MWNTs, amino-functionalized oxidized MWNTs, and amino-functionalized MWNTs, were assessed on glial and primary neuronal cells isolated from the fetal rat frontal cortex (FCO) and striatum (ST). The results from mLDH assays revealed a significant decline in the viability of mixed glial cells (non-neuronal cells) derived from the ST region after exposure to all types of MWCNTs (5–100 μg/mL), except for pristine MWCNTs at a concentration of 5 μg/mL, which exhibited no toxicity after 24 hours of incubation. Conversely, mixed glial cells obtained from the FCO brain region remained unaffected by MWCNTs at any concentration ranging from 5 to 100 μg/mL. Furthermore, exposure of neurons isolated from both FCO and ST regions to the aforementioned functionalized MWCNTs at doses up to 100 μg/mL did not result in any adverse effects. Microglial cells, responsible for maintaining brain homeostasis and protecting against injury, exhibited significantly higher uptake of CNTs, with the exception of pristine MWCNTs, compared to astrocytes. These findings suggested that the decreased viability of glial cells from the ST region compared to those from the FCO region was attributed to a higher proportion of microglial cells in the former. Recent investigations have also indicated that GO can penetrate animal cells through clathrin-mediated endocytosis, leading to cytotoxic effects [281]. Moreover, GO has been reported to cross the BBB due to its amphiphilic properties, resembling biological membranes [267]. In 2017, Hu et al. illustrated the neurotoxic impacts of GO nanosheets on zebrafish offspring following parental exposure. The parental zebrafish were subjected to an aqueous environment containing GO nanosheets at concentrations ranging from 0.01 to 1 μg/L. Although the parental zebrafish did not exhibit any signs of neurotoxicity in Hu et al.'s investigation, their offspring displayed various neurotoxic reactions to GO. These included the depletion of dopaminergic neurons, a reduction in acetylcholinesterase activity, disruptions to the endoplasmic reticulum, autophagy, down-regulation of ubiquitin, and heightened β-galactosidase activity (serving as an early indicator of cellular senescence). Furthermore, a direct correlation was observed between the impairment of carbohydrate and fatty acid metabolism and the loss of dopaminergic neurons at the metabolic level [267]. Onoda et al. examined the dose-dependent impacts of carbon black nanoparticles (CB-NP) administered to pregnant Institute of Cancer Research (ICR) mice on the brain astrocytes of their offspring. The researchers observed a dose-dependent rise in the expression of glial fibrillary acidic protein (GFAP) in the cerebral cortex and aquaporin-4 (Aqp-4) in the parenchyma region around blood vessels following maternal exposure to CB-NP. The changes in GFAP and Aqp-4 expression levels in the offspring after maternal CB-NP exposure mirrored those seen in older mice. Moreover, the study revealed extensive alterations in the expression levels of cortical mRNAs associated with proliferation, angiogenesis, chemotaxis, cell migration, and growth factor production in the offspring following maternal CB-NP exposure. These findings indicate that maternal exposure to CB-NP triggered astrocyte activation, leading to reactive astrogliosis and an increased susceptibility to age-related neurodegenerative diseases [282].

5.6.2 Hepatotoxicity

The hepatic sinusoid and Kupffer cells, crucial components for metabolism and detoxification, are vulnerable to drug-induced toxicity and nanoparticle accumulation [283]. About 75% of blood reaching the liver comes from the gastrointestinal viscera and spleen via the portal vein, carrying drugs absorbed by the intestines to the liver [281]. The liver accumulates 30–99% of administered nanoparticles from the bloodstream, leading to increased toxicity to hepatic cells. In a study conducted by Ahmadi et al. [284], they investigated the hepatocytotoxicity of PEG-labeled SWCNTs and Tween-functionalized SWCNT (Tween-SWCNT) in BALB/c mice after intravenous injection at doses of 75 and 150 μg/mouse. The evaluation of oxidative stress biomarkers and biochemical serum factors, such as malondialdehyde (MDA), glutathione (GSH), and liver enzymes alanine aminotransferase (ALT) and aspartate transaminase (AST), did not reveal any significant difference in the treated groups compared to the controls. However, histological injuries were observed in both groups, particularly in mice exposed to Tween-SWCNT. Proteomic and Western blot analyses indicated a notable increase in the expression of liver proteins with antioxidant and detoxifying activities following the intravenous administration of PEG- and Tween-SWCNT. These proteins, including Trap-1, Prx-6, TPx, Rgn, Gnmt, and INMT, contribute to cellular protection against oxidative stress and detoxification [284]. Adedara et al. demonstrated that repeated short-term intraperitoneal administration of purified carboxylated MWCNTs-induced hepatic oxidative damage in pubertal rats. This damage manifested as disruptions in antioxidant defense systems, elevated levels of cyclooxygenase-2 (COX-2) and inducible nitric oxide synthase (iNOS), and increased production of proinflammatory cytokines. Liver injury in pubertal rats exposed to MWCNTs was associated with significantly elevated serum activities of ALT, AST, gamma-glutamyl transferase, and alkaline phosphatase compared to the control group. Additionally, MWCNT exposure resulted in heightened activity of hepatic glutathione peroxidase (GPx) and increased levels of H_2O_2 and lipid peroxidation (LPO), suggesting inadequate GPx-mediated reduction of oxidative damage in the treated rats' livers. Lin et al. investigated the hepatotoxic effects of SWCNTs following intratracheal injection in male Wistar rats over a 15-day period. Biochemical analysis of plasma from the SWCNT-treated animals revealed significant alterations in alkaline phosphatase, total cholesterol, and total protein levels. Furthermore, 1H nuclear magnetic resonance (NMR) spectra of blood plasma and liver tissue extracts indicated changes in the concentrations of glutamate, creatine, lactate, glucose, trimethylamine oxide, choline, HDL, and VLDL, serving as potential indicators of hepatotoxicity. These findings collectively indicated liver damage and disturbances in energy, amino acid, and fat metabolism resulting from SWCNT exposure [285]. In 2009, Ji et al. documented that the intravenous administration of MWCNTs at dosages of 10 and 60 mg/kg in Kunming mice resulted in changes in the expression of cytochrome P450, a crucial enzyme in drug metabolism, leading to hepatotoxicity. It is important to note that various other types of CNMs, such as GO, CB, and fullerene, have exhibited hepatotoxic effects and the ability to modify gene and protein expression [286].

5.6.3 Nephrotoxicity

Nephrotoxicity refers to the negative impact of substances on renal function. The nephron, which constitutes the fundamental structural and functional unit of the kidney, is pivotal in crucial kidney functions such as waste filtration, regulating body fluid balance, maintaining blood pH levels, supporting bone health, producing hormones for red blood cell production, and regulating blood pressure. Any damage or loss of cells within the nephron can lead to kidney impairment and eventual renal failure [287]. Because of their inadequate dispersion in diverse solvents, CNTs provoke an immune response in the body, greatly constraining their potential applications [288]. Enhancing the solubility of CNTs can be achieved significantly through noncovalent functionalization with surfactants, peptides, polymers, nucleic acids, and oligomers, or by introducing moieties on the external surface of the tube [289]. In a study investigating the cytotoxicity of polyethylene glycol-modified CNTs on human embryonic kidney cells (HEK-293T), cell viability gradually decreased with increasing concentrations of PEG-CNTs. Exposure to PEG-CNTs at doses below 100 μg/mL did not induce toxicity in HEK-293T cells; however, concentrations ranging from 100 to 200 μg/mL exhibited mild toxicity. Previous research by Reddy et al. revealed that MWCNTs of two different sizes (60–80 nm and 90–150 nm)-induced kidney toxicity in a dose-dependent manner. HEK cells exposed to MWCNTs with sizes in the range of 60–80 nm displayed significantly higher cytotoxicity and oxidative stress compared to the 90–150 nm nanotubes. Another study assessed the impact of various types of MWCNTs, administered via intravenous injection to healthy mice, on the histology of different tissues, including the kidney. The findings suggested that a higher degree of ammonium modification on the nanotube surface led to reduced tissue accumulation. Histological examination of kidney slides taken 24 hours postadministration revealed no damage to glomerular physiology, and no accumulation was observed for all types of MWCNTs studied, including pristine, diethylenetriaminepentaacetic (DTPA), and ammonium-functionalized nanotubes. These findings raise concerns regarding the escalating human exposure to GO and its potential adverse effects and cytotoxicity [290]. Various studies have reported that GO can activate multiple signaling pathways related to autophagy and necrosis in different cell types [291, 292]. Nanotoxicity induced by epigenetic regulatory mechanisms, including histone modification, DNA methylation, chromosome remodeling, and small RNA regulation, has also been documented [290]. In a study conducted by Sun et al., the nanotoxic effects of GO on dynamic chromatin structural changes were evaluated in vitro using kidney cells (HEK-293T). The researchers demonstrated that GO treatment dynamically altered the chromatin architecture at the cox2 locus, which is involved in inflammation, via activation of the TNF-α–NF-κB signaling pathway. This process was associated with the activation of cox2 and induction of apoptosis through epigenetic mechanisms. Cox2, a key enzyme catalyzing prostaglandin production in response to various stimuli, regulates biological processes such as proliferation, apoptosis, cell death, and inflammation. The investigators suggested that the epigenetic effects of GO might occur through interactions between GO and cell membranes, leading to an increase in TNF-α secretion. Elevated TNF-α levels result in the formation of the p56–p300 complex, altering the higher-order chromatin structure at the cox2 locus by binding to its promoter and enhancer regions. Their findings indicated that both pristine GO and aminated GO (GO-NH_2) significantly increased the expression of cox2, whereas polyacrylic acid-modified GO (PAA-GO) had little effect on cox2 expression. Chromosome conformation capture (3C) assay results showed that PAA-GO did not significantly enhance dynamic chromatin interactions and, consequently, cox2 protein expression, leading to a weaker inflammatory response. These results are consistent with previous studies [293], suggesting that surface modification of GO can enhance its biocompatibility [290].

5.6.4 Immunotoxicity

The immune system, a finely regulated network comprising cells, organs, and molecules, serves a critical function in identifying and removing foreign particles [294]. The distinctive physicochemical characteristics of engineered nanomaterials (ENMs) render them susceptible to uptake by phagocytic cells and interaction with the immune system, giving rise to a range of undesired effects such as inflammation, immunomodulation, and hypersensitivity. This interaction is influenced by various chemical and physical attributes, including size and surface modifications. Studies have indicated that the size of ENMs can trigger activation of the complement system. The extensive diversity among ENMs in terms of size, composition, physicochemical properties, and surface modifications presents challenges in categorizing them as a homogeneous class of agents, leading to a wide spectrum of toxicological outcomes across different ENMs [295]. Mesoporous carbon nanomaterials (MCN), characterized by their carbonaceous and mesoporous structure, have garnered considerable attention in clinical and pharmaceutical domains owing to their distinctive features. These include a high surface area, substantial pore volume, capacity for drug loading, and customizable pore size [296]. In 2018, Li et al. examined the impact of unmodified and modified MCNs with polyvinylpyrrolidon (PVP) or PEG on dendritic cells (DCs), T-lymphocytes, and RAW264.7 macrophages in vitro. All MCNs studied in this research facilitated DC maturation and proliferation. However, MCN-PVP and MCN-PEG notably inhibited the secretion of TNF-α and IL-6 in these cells. Additionally, the modified MCNs induced apoptosis in T-lymphocytes [297]. The immunogenicity of PVP and PEG-modified MCNs was found to be less compared to pristine MCNs, and the immunotoxic effects of CNTs have raised concerns due to their extensive exploration as drug nanocarriers and diagnostic tools [298]. When administered intravenously, CNTs can directly engage with proteins and immune cells present in the bloodstream and tissues, triggering the activation of transcription factors such as activator protein 1 (AP-1), tumor necrosis factor-α (TNF-α), and nuclear factor kappa B (NF-κB) within macrophages. This activation prompts T cell activation, the release of proinflammatory cytokines, oxidative stress, necrosis, chromosomal abnormalities, recruitment of leukocytes, and the induction of protective and anti-apoptotic gene expression [298]. Examination under electron microscopy unveiled that MWCNTs disrupt the integrity of the plasma membrane in macrophages by binding to surface proteins like the macrophage receptor with collagenous structure (MACRO) [299]. Zhang et al. conducted a study on BALB/c mice to explore the immunotoxic effects of intravenously administered pristine MWCNTs compared to PEG-modified MWCNTs. They observed that pristine MWCNTs induced changes in lymphocyte populations, elevated levels of IgM and IgG, reduced immune response to sheep erythrocytes, and histological alterations. Conversely, PEG-modified MWCNTs showed reduced immunotoxicity and histological examination indicated improved solubility and dispersion within cells without affecting viability [298].

5.6.5 Cardiotoxicity

Cardiotoxicity, which can affect the heart either directly or indirectly through hemodynamic flow or thrombotic events, has been documented in association with nano-diamonds (ND). Although NDs find applications across diverse fields, such as biomedicine, biosensors, and imaging probes, both in vitro and in vivo investigations have revealed that inhalation exposure to NDs triggers cardiotoxic effects in animals. These effects are evidenced by elevated ROS production, lipid peroxidation, perturbation of mitochondrial membrane potential, fluctuations in glutathione levels, and diminished vascularization [300]. On the contrary, fullerene has exhibited protective effects on the heart against doxorubicin-induced cardiotoxicity in rats. Administration of Fullerenol $C_{60}(OH)_{24}$ prevented doxorubicin-induced reflex bradycardia and morphological alterations, indicating a protective function [301]. Graphene, composed of a single layer of carbon atoms, has garnered significant attention in various fields, such as drug delivery, medical imaging, tissue engineering, and solar cells. However, its release into aquatic environments raises concerns regarding potential health risks due to cytotoxic effects in aquatic animals. Exposure to graphene has been found to trigger cardiotoxicity, cardiovascular defects, and abnormal vascular growth in zebrafish embryos. Graphene derivatives, like ginsenoside Rh2-functionalized graphene oxide (GO-Rh2) and basic amino acid-functionalized graphene oxide (GO-Arg-Rh2, GO-Lys-Rh2), have been investigated for their ability to mitigate cardiac tissue damage compared to pristine GO [302].

5.6.6 Genotoxicity and Epigenetic Toxicity

Genotoxicity, which refers to processes altering DNA structure, information content, and segregation, has been linked to fibrous CNMs. These materials induce prolonged genotoxic stress either through direct interaction with DNA and mitotic apparatus or indirectly via inflammatory responses and oxidative stress. Cationic functionalized CNTs may cause DNA condensation, and studies on CNTs have reported both genotoxic and non-genotoxic effects, possibly influenced by factors such as concentration, route of exposure, duration, and dispersing agents. Recent research conducted by Demir et al. investigated the impacts of pristine MWCNTs, amide-functionalized MWCNTs, and graphene on gene expression and mutagenicity in a mouse lymphoma cell line. The study found no mutagenic effects or significant alterations in stress gene expression at doses ≤250 μg/mL [303]. Moreover, exposure to both MWCNTs and SWCNTs has been linked to abnormal epigenetic modifications, encompassing both hyper- and hypomethylation in the promoter region of certain genes observed in lung or bronchial cells and Allium cepa root cells [304]. Epigenetic modifications, which involve alterations in gene expression without changing the DNA sequence, have been associated with the development of various diseases. Oner et al. examined the effects of MWCNTs and SWCNTs on DNA methylation patterns. They found that MWCNTs caused hypomethylation of 2398 gene promoters, while SWCNTs induced both hyper- and hypomethylation at 589 CpG sites. These changes in DNA methylation led to diverse alterations in gene expression, impacting cellular pathways, the cell cycle, and DNA damage repair mechanisms. Notably, SWCNTs-induced hypermethylation in key genes such as shroom family member 2 (SHROOM2), glutathione S-transferase pi 1 (GTSP1), and SKI proto-oncogene (SKI), as well as both hypo- and hypermethylation of the neurofibromatosis type I (NF1) gene. Nuclear deposition studies revealed a higher accumulation of SWCNTs at both cellular and nuclear levels compared to MWCNTs, with both types of carbon nanotubes in direct contact with the nuclei. However, exposure to either type of carbon nanotube did not result in global methylation changes at 5-methylcytosine (5-mC) sites [301]. CB, categorized as a particle with low solubility and toxicity, has been linked to direct genotoxic effects, particularly at high concentrations. Research suggests that CB does not directly bind to DNA but can induce mutations in lung cells through mechanisms involving inflammation and oxidative stress [305]. Modrzynska et al. discovered that both intratracheal (IT) and intravenous (IV) exposure to carbon black (CB)-induced prolonged pulmonary inflammation, acute phase responses, and potential translocation of nanoparticles from the lung to the liver. Their study revealed that IV injection of CB resulted in increased levels of DNA strand breaks in the liver after 1, 28, and 180 days, while IT installation caused DNA strand breaks 28 and 180 days postexposure. Conversely, oral administration of CB did not lead to hepatic genotoxicity due to the absence of particle translocation, inflammation, and acute phase responses [306].

5.6.7 Dermatotoxicity

Dermatotoxicity, the skin's reaction to chemical substances resulting in damage, encompasses various responses such as allergic contact dermatitis, irritation, contact urticaria, photosensitization, and other cutaneous reactions, triggered by chemical penetration through different skin layers [307]. Understanding the potential risks and molecular mechanisms involved is crucial due to the skin being a probable route of exposure to nanomaterials [308]. When applied topically, CNMs need to permeate through the outer layer of the epidermis, known as the stratum corneum, which consists of keratinized dead cells embedded in a lipid matrix. Although this lipid barrier typically prevents the penetration of substances and pathogens, disruptions in the stratum corneum caused by conditions like disease or occupational factors may facilitate CNM penetration [309]. Limited research has delved into the cytotoxic impacts of CNMs on the skin. In a study conducted in 2005, Ding et al. explored the cytotoxic mechanisms of MWCNTs and multi-walled carbon nano-onions (MWCNOs) in human skin fibroblast cells. They observed that cytotoxic levels of MWCNTs and MWCNOs triggered cell cycle arrest, apoptosis, and necrosis and disrupted various cellular pathways linked to stress response, cell cycle regulation, cellular transport, and metabolism. Additionally, Liao et al. evaluated the cytotoxicity of graphene and GO in skin fibroblasts by assessing mitochondrial activities. Their findings revealed that densely packed graphene sheets induced more cytotoxicity in fibroblasts generated higher levels of ROS, and exhibited a strong association with fibroblast cell surfaces compared to less densely packed GO [310]. The toxic effects of graphene and GO were also evident in human HaCaT skin keratinocytes. Pelin et al. observed substantial mitochondrial depolarization following 72 hours of exposure to 100 μg/mL few-layer graphene and GO. This effect was likely mediated by ROS production induced by the activation of NADH dehydrogenase and xanthine oxidase, which are flavoprotein-based oxidative enzymes [311].

5.6.8 Carcinogenicity

The ability of certain types of MWCNTs to cause tumor development, known as carcinogenicity, has been associated with characteristics resembling asbestos fibers in animal studies. Factors such as length, rigidity, diameter, and surface modifications play a role in determining the carcinogenic potential of these materials. It is important to note that there is currently no available information on the carcinogenicity of SWCNTs [312]. In 2016, Kasai et al. conducted a study investigating the carcinogenic impacts of MWCNT-7 through whole-body inhalation in male and female rats. Male rats exposed to concentrations of 0.2 and 2 mg/L and female rats exposed to 2 mg/L exhibited a significant increase in bronchiolo-alveolar carcinoma, combined carcinomas, and adenomas. Interestingly, no cases of pleural mesothelioma were observed in the rats subjected to treatment [313]. Sakamoto et al. undertook a comparative investigation into the carcinogenic potential of seven types of MWCNTs in male Fischer rats. M-CNT, N-CNT, WL-CNT, and SD1-CNT, characterized by longer and thicker structures, resulted in mesotheliomas with a 100% absolute incidence. Conversely, curled and densely entangled MWCNTs such as WS-CNT, SD2-CNT, and T-CNT, exhibiting smaller length and thickness, did not induce mesothelioma, except for SD2-CNT and T-CNT, which had an incidence rate of 1 per 15 rats [314]. The results align with research by Nagai et al., indicating that MWCNTs with a diameter around 50 nm and high crystallinity exhibit increased potential for causing mesothelioma. In contrast, MWCNTs that are entangled (with diameters approximately ranging from 2 to 20 nm) and thicker (with diameters around 150 nm) show reduced levels of toxicity, carcinogenicity, and inflammation [315]. According to the International Agency for Research on Cancer (IARC), epidemiological studies within the rubber industry and CB production have indicated insufficient evidence of carcinogenicity. However, extensive evidence from rodent studies on the carcinogenicity of CB is available [316].

5.7 Conclusion and Future Outlook

CNTs have gained substantial attention in the field of biomedicine, finding applications as delivery vehicles and bioelectronic devices like biosensors. Their unique properties, especially their electronic characteristics and nanoscale size, make them ideal for various biomedical uses. CNTs' small diameter and considerable length enable them to interact with biological systems, including cells, without causing significant harm. They serve as effective carriers for delivering payloads such as DNA for therapeutic purposes. Despite remarkable progress, several challenges persist.

Processing CNTs must become more straightforward, and their integration with biomolecules in a biocompatible manner is another hurdle. The resulting bioelectronic devices should also be cost-effective and accessible in the commercial market. Toxicological concerns remain, with uncertainty surrounding the potential health risks of nanostructures. Although extensive toxicological studies have been conducted, standard testing protocols are needed to ensure comparability across experiments. Additionally, industrial hygiene practices must be adopted to address potential toxicological effects related to CNT inhalation during handling. Addressing these concerns is crucial given the growing interest in CNTs for advanced nano-biomaterials in biomedical applications. Likewise, significant advancements have been made in research regarding the biomedical applications of graphene. While these advancements are promising, challenges persist. Understanding the interactions between graphene and cells, tissues, or organs is critical, as is comprehending the mechanisms behind cellular uptake. To develop more efficient graphene or GO-based nanoplatforms for applications like biosensing and drug delivery, control over size, morphology, structural defects, size distribution, and oxygen-containing groups on GO is needed. Another challenge involves tuning the electronic properties of GO for biosensing applications based on the FRET principle. Techniques must be developed to integrate GO into practical devices that are cost-effective, highly sensitive, reproducible, and reliable. Enhancing the fluorescence quantum yield of GQDs is essential for their use in bioimaging. Methods to control the size, size distribution, surface defects, and functionalization of GQDs are urgently needed. Furthermore, evaluating the toxicity of graphene and GO at in vivo and in vitro levels is crucial. A systematic toxicity assessment is required to address safety concerns before practical biomedical applications of graphene.

References

1 Sahoo S K and Labhasetwar V 2003 Nanotech approaches to drug delivery and imaging *Drug Discov. Today* 8 1112–20.

2 Kroto H W, Heath J R, O'Brien S C, Curl R F and Smalley R E 1985 C60: buckminsterfullerene *Nature* 318 162–3.

3 Zheng L X, O'connell M J, Doorn S K, Liao X Z, Zhao Y H, Akhadov E A, Hoffbauer M A, Roop B J, Jia Q X and Dye R C 2004 Ultralong single-wall carbon nanotubes *Nat. Mater.* 3 673–6.

4 Iijima S 1991 Helical microtubules of graphitic carbon *Nature* 354 56–8.

5 Dresselhaus M, Dresselhaus G, Charlier J-C and Hernandez E 2004 Electronic, thermal and mechanical properties of carbon nanotubes *Philos. Trans. R. Soc. Lond. Ser. Math. Phys. Eng. Sci.* 362 2065–98.

6 Hilder T and Hill J 2008 Theoretical comparison of nanotube materials for drug delivery *Micro Nano Lett.* 3 18–24.

7 Boczkowski J and Lanone S 2007 Potential uses of carbon nanotubes in the medical field: how worried should patients be? *Nano Med.* 2 407–10.

8 Foldvari M and Bagonluri M 2008 Carbon nanotubes as functional excipients for nanomedicines: I. Pharmaceutical properties *Nanomedicine Nanotechnol. Biol. Med.* 4 173–82.

9 Novoselov K S, Geim A K, Morozov S V, Jiang D, Zhang Y, Dubonos S V, Grigorieva I V and Firsov A A 2004 Electric field effect in atomically thin carbon films *Science* 306 666–9.

10 Zhang K, Zhang L L, Zhao X and Wu J 2010 Graphene/polyaniline nanofiber composites as supercapacitor electrodes *Chem. Mater.* 22 1392–401.

11 Rao C N R, Sood A K, Subrahmanyam K S and Govindaraj A 2009 Graphene: the new two-dimensional nanomaterial *Angew. Chem. Int. Ed.* 48 7752–77.

12 Yang X, Tu Y, Li L, Shang S and Tao X 2010 Well-dispersed chitosan/graphene oxide nanocomposites *ACS Appl. Mater. Interfaces* 2 1707–13.

13 Bai H, Li C, Wang X and Shi G 2010 A pH-sensitive graphene oxide composite hydrogel *Chem. Commun.* 46 2376–8.

14 Fang M, Long J, Zhao W, Wang L and Chen G 2010 pH-responsive chitosan-mediated graphene dispersions *Langmuir* 26 16771–4.

15 Xuan Y, Wu Y, Shen T, Qi M, Capano M A, Cooper J A and Ye P 2008 Atomic-layer-deposited nanostructures for graphene-based nanoelectronics *Appl. Phys. Lett.* 92 013101.

16 Liu C, Alwarappan S, Chen Z, Kong X and Li C-Z 2010 Membraneless enzymatic biofuel cells based on graphene nanosheets *Biosens. Bioelectron.* 25 1829–33.

17 Wang L, Lee K, Sun Y-Y, Lucking M, Chen Z, Zhao J J and Zhang S B 2009 Graphene oxide as an ideal substrate for hydrogen storage *ACS Nano* 3 2995–3000.

18 Stoller M D, Park S, Zhu Y, An J and Ruoff R S 2008 Graphene-based ultracapacitors *Nano Lett.* 8 3498–502.

19 Feng L and Liu Z 2011 Graphene in biomedicine: opportunities and challenges *Nanomedicine* 6 317–24.

20 Wang Y, Li Z, Wang J, Li J and Lin Y 2011 Graphene and graphene oxide: biofunctionalization and applications in biotechnology *Trends Biotechnol.* 29 205–12.

21 Jiang H 2011 Chemical preparation of graphene-based nanomaterials and their applications in chemical and biological sensors *Small* 7 2413–27.

22 Guo S and Dong S 2011 Graphene nanosheet: synthesis, molecular engineering, thin film, hybrids, and energy and analytical applications *Chem. Soc. Rev.* 40 2644–72.

23 Liu Z, Robinson J T, Sun X and Dai H 2008 PEGylated nanographene oxide for delivery of water-insoluble cancer drugs *J. Am. Chem. Soc.* 130 10876–7.

24 Yaqoob A A, Mohd Noor N H B, Serrà A and Mohamad Ibrahim M N 2020 Advances and challenges in developing efficient graphene oxide-based ZnO photocatalysts for dye photo-oxidation *Nanomaterials* 10 932.

25 Joselevich E 2004 Electronic structure and chemical reactivity of carbon nanotubes: a chemist's view *ChemPhysChem* 5 619–24.

26 Beg S, Rizwan M, Sheikh A M, Hasnain M S, Anwer K and Kohli K 2011 Advancement in carbon nanotubes: basics, biomedical applications and toxicity *J. Pharm. Pharmacol.* 63 141–63.

27 Schouml C and Forró L 2000 Multiwall carbon nanotubes *Phys. World* 13 37.

28 Flahaut E, Bacsa R, Peigney A and Laurent C 2003 Gram-scale CCVD synthesis of double-walled carbon nanotubes *Chem. Commun.* 12 1442–3.

29 Danailov D, Keblinski P, Nayak S and Ajayan P 2002 Bending properties of carbon nanotubes encapsulating solid nanowires *J. Nanosci. Nanotechnol.* 2 503–7.

30 Rathinavel S, Priyadharshini K and Panda D 2021 A review on carbon nanotube: an overview of synthesis, properties, functionalization, characterization, and the application *Mater. Sci. Eng. B* 268 115095.

31 Iijima S, Yudasaka M, Yamada R, Bandow S, Suenaga K, Kokai F and Takahashi K 1999 Nano-aggregates of single walled graphitic carbon nano-horns *Chem. Phys. Lett.* 309 165–70.

32 Murakami T, Sawada H, Tamura G, Yudasaka M, Iijima S and Tsuchida K 2008 Water-dispersed single-wall carbon nanohorns as drug carriers for local cancer chemotherapy *Nanomedicine* 3 453–63.

33 Liu L, Guo G, Jayanthi C and Wu S 2002 Colossal paramagnetic moments in metallic carbon nanotori *Phys. Rev. Lett.* 88 217206.

34 Nasibulin A G, Pikhitsa P V, Jiang H, Brown D P, Krasheninnikov A V, Anisimov A S, Queipo P, Moisala A, Gonzalez D and Lientschnig G 2007 A novel hybrid carbon material *Nat. Nanotechnol.* 2 156.

35 Nguyen C V, Delzeit L, Cassell A M, Li J, Han J and Meyyappan M 2002 Preparation of nucleic acid functionalized carbon nanotube arrays *Nano Lett.* 2 1079–81.

36 Katz E and Willner I 2004 Biomolecule-functionalized carbon nanotubes: applications in nanobioelectronics *ChemPhysChem* 5 1084–104.

37 Bahr J L and Tour J M 2002 Covalent chemistry of single-wall carbon nanotubes *J. Mater. Chem.* 12 1952–8.

38 Mickelson E, Huffman C, Rinzler A, Smalley R, Hauge R and Margrave J 1998 Fluorination of single-wall carbon nanotubes *Chem. Phys. Lett.* 296 188–94.

39 Yang W, Thordarson P, Gooding J J, Ringer S P and Braet F 2007 Carbon nanotubes for biological and biomedical applications *Nanotechnology* 18 412001.

40 Niyogi S, Hamon M A, Hu H, Zhao B, Bhowmik P, Sen R, Itkis M E and Haddon R C 2002 Chemistry of single-walled carbon nanotubes *Acc. Chem. Res.* 35 1105–13.

41 Yu X, Munge B, Patel V, Jensen G, Bhirde A, Gong J D, Kim S N, Gillespie J, Gutkind J S, Papadimitrakopoulos F and Rusling J F 2006 Carbon nanotube amplification strategies for highly sensitive immunodetection of cancer biomarkers *J. Am. Chem. Soc.* 128 11199–205.

42 Yang W R, Hibbert D B, Zhang R, Willett G D and Gooding J J 2005 Stepwise synthesis of Gly–Gly–his on gold surfaces modified with mixed self-assembled monolayers *Langmuir* 21 260–5.

43 Lee C-S, Baker S E, Marcus M S, Yang W, Eriksson M A and Hamers R J 2004 Electrically addressable biomolecular functionalization of carbon nanotube and carbon nanofiber electrodes *Nano Lett.* 4 1713–6.

44 Wang J, Musameh M and Lin Y 2003 Solubilization of carbon nanotubes by Nafion toward the preparation of amperometric biosensors *J. Am. Chem. Soc.* 125 2408–9.

45 Moghaddam M J, Taylor S, Gao M, Huang S, Dai L and McCall M J 2004 Highly efficient binding of DNA on the sidewalls and tips of carbon nanotubes using photochemistry *Nano Lett.* 4 89–93.

46 Prato M, Li Q C, Wudl F and Lucchini V 1993 Addition of azides to fullerene C60: synthesis of azafulleroids *J. Am. Chem. Soc.* 115 1148–50.

47 Pantarotto D, Partidos C D, Graff R, Hoebeke J, Briand J-P, Prato M and Bianco A 2003 Synthesis, structural characterization, and immunological properties of carbon nanotubes functionalized with peptides *J. Am. Chem. Soc.* 125 6160–4.

48 Wang D, Wang Q, Javey A, Tu R, Dai H, Kim H, McIntyre P C, Krishnamohan T and Saraswat K C 2003 Germanium nanowire field-effect transistors with SiO_2 and high-κ HfO_2 gate dielectrics *Appl. Phys. Lett.* 83 2432–4.

49 Chen R J, Bangsaruntip S, Drouvalakis K A, Kam N W S, Shim M, Li Y, Kim W, Utz P J and Dai H 2003 Noncovalent functionalization of carbon nanotubes for highly specific electronic biosensors *Proc. Natl. Acad. Sci.* 100 4984–9.

50 Shim M, Shi Kam N W, Chen R J, Li Y and Dai H 2002 Functionalization of carbon nanotubes for biocompatibility and biomolecular recognition *Nano Lett.* 2 285–8.

51 Chen R J, Zhang Y, Wang D and Dai H 2001 Noncovalent sidewall functionalization of single-walled carbon nanotubes for protein immobilization *J. Am. Chem. Soc.* 123 3838–9.

52 Lynam C, Moulton S E and Wallace G G 2007 Carbon-nanotube biofibers *Adv. Mater.* 19 1244–8.

53 Hermanson G T 2013 *Bioconjugate Techniques* (London: Academic Press).

54 Thordarson P, Atkin R, Kalle W H, Warr G G and Braet F 2006 Developments in using scanning probe microscopy to study molecules on surfaces — from thin films and single-molecule conductivity to drug–living cell interactions *Aust. J. Chem.* 59 359–75.

55 Grady B P 2006 The use of solution viscosity to characterize single-walled carbon nanotube dispersions *Macromol. Chem. Phys.* 207 2167–9.

56 Mohajeri M, Behnam B and Sahebkar A 2019 Biomedical applications of carbon nanomaterials: drug and gene delivery potentials *J. Cell. Physiol.* 234 298–319.

57 Laurent C, Flahaut E and Peigney A 2010 The weight and density of carbon nanotubes versus the number of walls and diameter *Carbon* 48 2994–6.

58 Hui H-W, Robinson J and Lee V H 1987 Design and Fabrication of Oral Controlled Release Drug Delivery Systems *Controlled Drug Delivery Fundamentals and Applications* ed J Robinson and V H Lee (Boca Raton: CRC Press) pp 373–4.

59 Shiba K, Yudasaka M and Iijima S 2006 Carbon nanohorns as a novel drug carrier *Nihon Rinsho Jpn. J. Clin. Med.* 64 239–46.

60 Riggio C, Ciofani G, Raffa V, Cuschieri A and Micera S 2009 Combination of polymer technology and carbon nanotube array for the development of an effective drug delivery system at cellular level *Nanoscale Res. Lett.* 4 668–73.

61 Liu Z, Sun X, Nakayama-Ratchford N and Dai H 2007 Supramolecular chemistry on water-soluble carbon nanotubes for drug loading and delivery *ACS Nano* 1 50–6.

62 Gao H, Kong Y, Cui D and Ozkan C S 2003 Spontaneous insertion of DNA oligonucleotides into carbon nanotubes *Nano Lett.* 3 471–3.

63 Mitchell D T, Lee S B, Trofin L, Li N, Nevanen T K, Söderlund H and Martin C R 2002 Smart nanotubes for bioseparations and biocatalysis *J. Am. Chem. Soc.* 124 11864–5.

64 Pastorin G, Wu W, Wieckowski S, Briand J-P, Kostarelos K, Prato M and Bianco A 2006 Double functionalisation of carbon nanotubes for multimodal drug delivery *Chem. Commun.* 0 1182–4.

65 Pantarotto D, Singh R, McCarthy D, Erhardt M, Briand J-P, Prato M, Kostarelos K and Bianco A 2004 Functionalized carbon nanotubes for plasmid DNA gene delivery *Angew. Chem. Int. Ed.* 43 5242–6.

66 Martin C R and Kohli P 2003 The emerging field of nanotube biotechnology *Nat. Rev. Drug Discov.* 2 29.

67 Pantarotto D, Briand J-P, Prato M and Bianco A 2004 Translocation of bioactive peptides across cell membranes by carbon nanotubes *Chem. Commun.* 1 16–7.

68 Yang Y-J, Tao X, Hou Q and Chen J-F 2009 Fluorescent mesoporous silica nanotubes incorporating CdS quantum dots for controlled release of ibuprofen *Acta Biomater.* 5 3488–96.

69 Zhang C-H, Luo Y-L, Chen Y-S, Wei Q B and Fan L-H 2009 Preparation and theophylline delivery applications of novel PMAA/MWCNT-COOH nanohybrid hydrogels *J. Biomater. Sci. Polym. Ed.* 20 1119–35.

70 Kulamarva A, Raja P M V, Bhathena J, Chen H, Talapatra S, Ajayan P M, Nalamasu O and Prakash S 2008 Microcapsule carbon nanotube devices for therapeutic applications *Nanotechnology* 20 025612.

71 Mohapatra S S and Kumar A 2013 Method of drug delivery by carbon nanotube-chitosan nanocomplexes. USF Patents. 240. Patent Number; 8536324 https://digitalcommons.usf.edu/usf_patents/240.

72 Kam N W S, O'Connell M, Wisdom J A and Dai H 2005 Carbon nanotubes as multifunctional biological transporters and near-infrared agents for selective cancer cell destruction *Proc. Natl. Acad. Sci.* 102 11600–5.

73 Ferrari M 2005 Cancer nanotechnology: opportunities and challenges *Nat. Rev. Cancer* 5 161.

74 Quintana A, Raczka E, Piehler L, Lee I, Myc A, Majoros I, Patri A K, Thomas T, Mulé J and Baker J R 2002 Design and function of a dendrimer-based therapeutic nanodevice targeted to tumor cells through the folate receptor *Pharm. Res.* 19 1310–6.

75 Prakash S and Kulamarva A G 2007 Recent advances in drug delivery: potential and limitations of carbon nanotubes *Recent Pat. Drug Deliv. Formul.* 1 214–21.

76 Singh R and Lillard J W 2009 Nanoparticle-based targeted drug delivery *Exp. Mol. Pathol.* 86 215–23.

77 McDevitt M R, Chattopadhyay D, Kappel B J, Jaggi J S, Schiffman S R, Antczak C, Njardarson J T, Brentjens R and Scheinberg D A 2007 Tumor targeting with antibody-functionalized, radiolabeled carbon nanotubes *J. Nucl. Med.* 48 1180–9.

78 Ashcroft J M, Tsyboulski D A, Hartman K B, Zakharian T Y, Marks J W, Weisman R B, Rosenblum M G and Wilson L J 2006 Fullerene (C 60) immunoconjugates: interaction of water-soluble C 60 derivatives with the murine anti-gp240 melanoma antibody *Chem. Commun.* 28 3004–6.

79 Welsher K, Liu Z, Daranciang D and Dai H 2008 Selective probing and imaging of cells with single walled carbon nanotubes as near-infrared fluorescent molecules *Nano Lett.* 8 586–90.

80 Loh K P, Bao Q, Eda G and Chhowalla M 2010 Graphene oxide as a chemically tunable platform for optical applications *Nat. Chem.* 2 1015.

81 Kovtyukhova N I, Ollivier P J, Martin B R, Mallouk T E, Chizhik S A, Buzaneva E V and Gorchinskiy A D 1999 Layer-by-layer assembly of ultrathin composite films from micron-sized graphite oxide sheets and polycations *Chem. Mater.* 11 771–8.

82 Sun X, Liu Z, Welsher K, Robinson J T, Goodwin A, Zaric S and Dai H 2008 Nano-graphene oxide for cellular imaging and drug delivery *Nano Res.* 1 203–12.

83 Hummers W S Jr, and Offeman R E 1958 Preparation of graphitic oxide *J. Am. Chem. Soc.* 80 1339.

84 Shan C, Yang H, Han D, Zhang Q, Ivaska A and Niu L 2009 Water-soluble graphene covalently functionalized by biocompatible poly-L-lysine *Langmuir* 25 12030–3.

85 Zhang J, Zhang F, Yang H, Huang X, Liu H, Zhang J and Guo S 2010 Graphene oxide as a matrix for enzyme immobilization *Langmuir* 26 6083–5.

86 Zhang F, Zheng B, Zhang J, Huang X, Liu H, Guo S and Zhang J 2010 Horseradish peroxidase immobilized on graphene oxide: physical properties and applications in phenolic compound removal *J. Phys. Chem. C* 114 8469–73.

87 Lee D Y, Khatun Z, Lee J-H, Lee Y and In I 2011 Blood compatible graphene/heparin conjugate through noncovalent chemistry *Biomacromolecules* 12 336–41.

88 Lei H, Mi L, Zhou X, Chen J, Hu J, Guo S and Zhang Y 2011 Adsorption of double-stranded DNA to graphene oxide preventing enzymatic digestion *Nanoscale* 3 3888–92.

89 Chen W, Yi P, Zhang Y, Zhang L, Deng Z and Zhang Z 2011 Composites of aminodextran-coated Fe_3O_4 nanoparticles and graphene oxide for cellular magnetic resonance imaging *ACS Appl. Mater. Interfaces* 3 4085–91.

90 Dong H, Gao W, Yan F, Ji H and Ju H 2010 Fluorescence resonance energy transfer between quantum dots and graphene oxide for sensing biomolecules *Anal. Chem.* 82 5511–7.

91 Shen J, Shi M, Li N, Yan B, Ma H, Hu Y and Ye M 2010 Facile synthesis and application of Ag-chemically converted graphene nanocomposite *Nano Res.* 3 339–49.

92 Liu Z, Robinson J T, Tabakman S M, Yang K and Dai H 2011 Carbon materials for drug delivery & cancer therapy *Mater. Today* 14 316–23.

93 Dai H 2002 Carbon nanotubes: opportunities and challenges *Surf. Sci.* 500 218–41.

94 Bianco A, Kostarelos K, Partidos C D and Prato M 2005 Biomedical applications of functionalised carbon nanotubes *Chem. Commun.* 571–7.

95 Lu F, Gu L, Meziani M J, Wang X, Luo P G, Veca L M, Cao L and Sun Y 2009 Advances in bioapplications of carbon nanotubes *Adv. Mater.* 21 139–52.

96 Shen H, Zhang L, Liu M and Zhang Z 2012 Biomedical applications of graphene *Theranostics* 2 283.

97 Depan D, Shah J and Misra R 2011 Controlled release of drug from folate-decorated and graphene mediated drug delivery system: synthesis, loading efficiency, and drug release response *Mater. Sci. Eng. C* 31 1305–12.

98 Zhang L, Xia J, Zhao Q, Liu L and Zhang Z 2010 Functional graphene oxide as a nanocarrier for controlled loading and targeted delivery of mixed anticancer drugs *Small* 6 537–44.

99 Pan Y, Bao H, Sahoo N G, Wu T and Li L 2011 Water-soluble poly (N-isopropylacrylamide)–graphene sheets synthesized via click chemistry for drug delivery *Adv. Funct. Mater.* 21 2754–63.

100 Andersson M, Madsen E L, Overgaard M, Rose C, Dombernowsky P, Mouridsen H and Danish Breast Cancer Cooperative Group 1999 Doxorubicin versus methotrexate both combined with cyclophosphamide, 5-fluorouracil and tamoxifen in postmenopausal patients with advanced breast cancer — a randomised study with more than 10 years follow-up from the danish breast cancer cooperative group *Eur. J. Cancer* 35 39–46.

101 Gavrilov V, Steiner M and Shany S 2005 The combined treatment of 1, 25-dihydroxyvitamin D3 and a non-steroid anti-inflammatory drug is highly effective in suppressing prostate cancer cell line (LNCaP) growth *Anticancer Res.* 25 3425–9.

102 Yang X, Wang Y, Huang X, Ma Y, Huang Y, Yang R, Duan H and Chen Y 2011 Multi-functionalized graphene oxide based anticancer drug-carrier with dual-targeting function and pH-sensitivity *J. Mater. Chem.* 21 3448–54.

103 Pramanik M, Swierczewska M, Green D, Sitharaman B and Wang L V 2009 Single-walled carbon nanotubes as a multimodal-thermoacoustic and photoacoustic-contrast agent *J. Biomed. Opt.* 14 034018.

104 Prato M, Kostarelos K and Bianco A 2007 Functionalized carbon nanotubes in drug design and discovery *Acc. Chem. Res.* 41 60–8.

105 Cheng Y, Zhang J, Lee Y, Gao B, Dike S, Lin W, Lu J and Zhou O 2004 Dynamic radiography using a carbon-nanotube-based field-emission x-ray source *Rev. Sci. Instrum.* 75 3264–7.

106 Polizu S, Savadogo O, Poulin P and Yahia L 2006 Applications of carbon nanotubes-based biomaterials in biomedical nanotechnology *J. Nanosci. Nanotechnol.* 6 1883–904.

107 Peng C, Hu W, Zhou Y, Fan C and Huang Q 2010 Intracellular imaging with a graphene-based fluorescent probe *Small* 6 1686–92.

108 Liu K, Zhang J-J, Cheng F-F, Zheng T-T, Wang C and Zhu J-J 2011 Green and facile synthesis of highly biocompatible graphene nanosheets and its application for cellular imaging and drug delivery *J. Mater. Chem.* 21 12034–40.

109 Yan X, Cui X and Li L 2010 Synthesis of large, stable colloidal graphene quantum dots with tunable size *J. Am. Chem. Soc.* 132 5944–5.

110 Pan D, Zhang J, Li Z and Wu M 2010 Hydrothermal route for cutting graphene sheets into blue-luminescent graphene quantum dots *Adv. Mater.* 22 734–8.

111 Mei Q, Zhang K, Guan G, Liu B, Wang S and Zhang Z 2010 Highly efficient photoluminescent graphene oxide with tunable surface properties *Chem. Commun.* 46 7319–21.

112 Eda G, Lin Y, Mattevi C, Yamaguchi H, Chen H, Chen I, Chen C and Chhowalla M 2010 Blue photoluminescence from chemically derived graphene oxide *Adv. Mater.* 22 505–9.

113 Zhang L, Xing Y, He N, Zhang Y, Lu Z, Zhang J and Zhang Z 2012 Preparation of graphene quantum dots for bioimaging application *J. Nanosci. Nanotechnol.* 12 2924–8.

114 Zhu S, Zhang J, Qiao C, Tang S, Li Y, Yuan W, Li B, Tian L, Liu F and Hu R 2011 Strongly green-photoluminescent graphene quantum dots for bioimaging applications *Chem. Commun.* 47 6858–60.

115 Shen J, Zhu Y, Chen C, Yang X and Li C 2011 Facile preparation and upconversion luminescence of graphene quantum dots *Chem. Commun.* 47 2580–2.

116 Sipkins D A, Cheresh D A, Kazemi M R, Nevin L M, Bednarski M D and Li K C 1998 Detection of tumor angiogenesis in vivo by α v β 3-targeted magnetic resonance imaging *Nat. Med.* 4 623.

117 Bonnemain B 1998 Superparamagnetic agents in magnetic resonance imaging: physicochemical characteristics and clinical applications a review *J. Drug Target.* 6 167–74.

118 Freitas R A 2005 Current status of nanomedicine and medical nanorobotics *J. Comput. Theor. Nanosci.* 2 1–25.

119 Haberzettl C 2002 Nanomedicine: destination or journey? *Nanotechnology* 13 R9.

120 Requicha A A 2003 Nanorobots, NEMS, and nanoassembly *Proc. IEEE* 91 1922–33.

121 Feng L, Zhang S and Liu Z 2011 Graphene based gene transfection *Nanoscale* 3 1252–7.

122 Wang W, Su H, Wu Y, Zhou T and Li T 2019 Review — biosensing and biomedical applications of graphene: a review of current progress and future prospect *J. Electrochem. Soc.* 166 B505.

123 Nayak T R, Andersen H, Makam V S, Khaw C, Bae S, Xu X, Ee P-L R, Ahn J-H, Hong B H, Pastorin G and Özyilmaz B 2011 Graphene for controlled and accelerated osteogenic differentiation of human mesenchymal stem cells *ACS Nano* 5 4670–8.

124 Mu W, Chu Q, Liu Y and Zhang N 2020 A review on nano-based drug delivery system for cancer chemoimmunotherapy *Nano-Micro Lett.* 12 142.

125 Tu Z, Donskyi I S, Qiao H, Zhu Z, Unger W E S, Hackenberger C P R, Chen W, Adeli M and Haag R 2020 Graphene oxide-cyclic R10 peptide nuclear translocation nanoplatforms for the surmounting of multiple-drug resistance *Adv. Funct. Mater.* 30 2000933.

126 Liu J, Dong J, Zhang T and Peng Q 2018 Graphene-based nanomaterials and their potentials in advanced drug delivery and cancer therapy *J. Controlled Release* 286 64–73.

127 Fusco L, Gazzi A, Peng G, Shin Y, Vranic S, Bedognetti D, Vitale F, Yilmazer A, Feng X, Fadeel B, Casiraghi C and Delogu L G 2020 Graphene and other 2D materials: a multidisciplinary analysis to uncover the hidden potential as cancer theranostics *Theranostics* 10 5435–88.

128 Chai D, Hao B, Hu R, Zhang F, Yan J, Sun Y, Huang X, Zhang Q and Jiang H 2019 Delivery of oridonin and methotrexate via PEGylated graphene oxide *ACS Appl. Mater. Interfaces* 11 22915–24.

129 Zhang Y, Liu C, Shi W, Wang Z, Dai L and Zhang X 2007 Direct measurements of the interaction between pyrene and graphite in aqueous media by single molecule force spectroscopy: understanding the π–π interactions *Langmuir* 23 7911–5.

130 Demirel E, Karaca E and Yuksel Durmaz Y 2020 Effective PEGylation method to improve biocompatibility of graphene derivatives *Eur. Polym. J.* 124 109504.

131 Olabi A G, Abdelkareem M A, Wilberforce T and Sayed E T 2021 Application of graphene in energy storage device – a review *Renew. Sustain. Energy Rev.* 135 110026.

132 Wang B, Ruan T, Chen Y, Jin F, Peng L, Zhou Y, Wang D and Dou S 2020 Graphene-based composites for electrochemical energy storage *Energy Storage Mater.* 24 22–51.

133 Tehrani Z, Whelan S P, Mostert A B, Paulin J V, Ali M M, Ahmadi E D, Graeff C F O, Guy O J and Gethin D T 2020 Printable and flexible graphene pH sensors utilising thin film melanin for physiological applications *2D Mater.* 7 024008.

134 Fávaro W J, de Souza J G, Ferreira L A B, de Jesus M B, Durán M, Bockelmann P K, Bernardes J S and Durán N 2020 Hybrid graphene oxide as carrier of doxorubicin: cytotoxicity and preliminary in vivo assays against bladder cancer *Adv. Nat. Sci. Nanosci. Nanotechnol.* 11 025016.

135 Jiao J, Pan M, Liu X, Liu J, Li B and Chen Q 2020 An ultrasensitive non-enzymatic sensor for quantitation of anti-cancer substance chicoric acid based on bimetallic nanoalloy with polyetherimide-capped reduced graphene oxide *Nanomaterials* 10 499.

136 Soheili N, Eshghi M, Emadi-Baygi M, Mirzaei S A, Heidari R and Hosseinzadeh M 2020 Design and evaluation of biological gate circuits and their therapeutic applications in a model of multidrug resistant cancers *Biotechnol. Lett.* 42 1419–29.

137 Mallick A, Nandi A and Basu S 2019 Polyethylenimine coated graphene oxide nanoparticles for targeting mitochondria in cancer cells *ACS Appl. Bio Mater.* 2 14–9.

138 Kalkhoran A H Z, Naghib S M, Vahidi O and Rahmanian M 2018 Synthesis and characterization of graphene-grafted gelatin nanocomposite hydrogels as emerging drug delivery systems *Biomed. Phys. Eng. Express* 4 055017.

139 Khoshoei A, Ghasemy E, Poustchi F, Shahbazi M-A and Maleki R 2020 Engineering the pH-sensitivity of the graphene and carbon nanotube based nanomedicines in smart cancer therapy by grafting Trimetyl Chitosan *Pharm. Res.* 37 160.

140 Zhao X, Wei Z, Zhao Z, Miao Y, Qiu Y, Yang W, Jia X, Liu Z and Hou H 2018 Design and development of graphene oxide nanoparticle/chitosan hybrids showing pH-sensitive surface charge-reversible ability for efficient intracellular doxorubicin delivery *ACS Appl. Mater. Interfaces* 10 6608–17.

141 Gooneh-Farahani S, Naghib S M and Naimi-Jamal M R 2019 A critical comparison study on the pH-sensitive nanocomposites based on graphene-grafted chitosan for cancer theragnosis *Multidiscip. Cancer Investig.* 3 5–16.

142 Sattari S, Tehrani A D, Adeli M, Soleimani K and Rashidipour M 2020 Fabrication of new generation of co-delivery systems based on graphene-g-cyclodextrin/chitosan nanofiber *Int. J. Biol. Macromol.* 156 1126–34.

143 Xie M, Zhang F, Peng H, Zhang Y, Li Y, Xu Y and Xie J 2019 Layer-by-layer modification of magnetic graphene oxide by chitosan and sodium alginate with enhanced dispersibility for targeted drug delivery and photothermal therapy *Colloids Surf. B Biointerfaces* 176 462–70.

144 Xie M, Lei H, Zhang Y, Xu Y, Shen S, Ge Y, Li H and Xie J 2016 Non-covalent modification of graphene oxide nanocomposites with chitosan/dextran and its application in drug delivery *RSC Adv.* 6 9328–37.

145 Wan Y, Lin Z, Zhang Q, Gan D, Gama M, Tu J and Luo H 2020 Incorporating graphene oxide into biomimetic nano-microfibrous cellulose scaffolds for enhanced breast cancer cell behavior *Cellulose* 27 4471–85.

146 Anirudhan T S, Chithra Sekhar V and Athira V S 2020 Graphene oxide based functionalized chitosan polyelectrolyte nanocomposite for targeted and pH responsive drug delivery *Int. J. Biol. Macromol.* 150 468–79.

147 Omidi S, Pirhayati M and Kakanejadifard A 2020 Co-delivery of doxorubicin and curcumin by a pH-sensitive, injectable, and in situ hydrogel composed of chitosan, graphene, and cellulose nanowhisker *Carbohydr. Polym.* 231 115745.

148 Samadi S, Moradkhani M, Beheshti H, Irani M and Aliabadi M 2018 Fabrication of chitosan/poly(lactic acid)/graphene oxide/TiO_2 composite nanofibrous scaffolds for sustained delivery of doxorubicin and treatment of lung cancer *Int. J. Biol. Macromol.* 110 416–24.

149 Lei H, Xie M, Zhao Y, Zhang F, Xu Y and Xie J 2016 Chitosan/sodium alginate modificated graphene oxide-based nanocomposite as a carrier for drug delivery *Ceram. Int.* 42 17798–805.

150 Lima-Sousa R, de Melo-Diogo D, Alves C G, Costa E C, Ferreira P, Louro R O and Correia I J 2018 Hyaluronic acid functionalized green reduced graphene oxide for targeted cancer photothermal therapy *Carbohydr. Polym.* 200 93–9.

151 Guo Y, Xu H, Li Y, Wu F, Li Y, Bao Y, Yan X, Huang Z and Xu P 2017 Hyaluronic acid and Arg-Gly-Asp peptide modified Graphene oxide with dual receptor-targeting function for cancer therapy *J. Biomater. Appl.* 32 54–65.

152 Kim K, Choi H, Choi E S, Park M-H and Ryu J-H 2019 Hyaluronic acid-coated nanomedicine for targeted cancer therapy *Pharmaceutics* 11 301.

153 Yin T, Liu J, Zhao Z, Zhao Y, Dong L, Yang M, Zhou J and Huo M 2017 Redox sensitive hyaluronic acid-decorated graphene oxide for photothermally controlled tumor-cytoplasm-selective rapid drug delivery *Adv. Funct. Mater.* 27 1604620.

154 Tu Z, Guday G, Adeli M and Haag R 2018 Multivalent Interactions between 2D nanomaterials and biointerfaces *Adv. Mater.* 30 1706709.

155 Gheybi H, Sattari S, Soleimani K and Adeli M 2020 Graphene-dendritic polymer hybrids: synthesis, properties, and applications *J. Iran. Chem. Soc.* 17 735–64.

156 Tu Z, Wycisk V, Cheng C, Chen W, Adeli M and Haag R 2017 Functionalized graphene sheets for intracellular controlled release of therapeutic agents *Nanoscale* 9 18931–9.

157 Tan K H, Sattari S, Donskyi I S, Cuellar-Camacho J L, Cheng C, Schwibbert K, Lippitz A, Unger W E S, Gorbushina A, Adeli M and Haag R 2018 Functionalized 2D nanomaterials with switchable binding to investigate graphene–bacteria interactions *Nanoscale* 10 9525–37.

158 Tu Z, Qiao H, Yan Y, Guday G, Chen W, Adeli M and Haag R 2018 Directed graphene-based nanoplatforms for hyperthermia: overcoming multiple drug resistance *Angew. Chem.* 130 11368–72.

159 Hussien N A, Işıklan N and Türk M 2018 Aptamer-functionalized magnetic graphene oxide nanocarrier for targeted drug delivery of paclitaxel *Mater. Chem. Phys.* 211 479–88.

160 Oz Y, Barras A, Sanyal R, Boukherroub R, Szunerits S and Sanyal A 2017 Functionalization of reduced graphene oxide via thiol–maleimide “click” chemistry: facile fabrication of targeted drug delivery vehicles *ACS Appl. Mater. Interfaces* 9 34194–203.

161 Mauro N, Scialabba C, Agnello S, Cavallaro G and Giammona G 2020 Folic acid-functionalized graphene oxide nanosheets via plasma etching as a platform to combine NIR anticancer phototherapy and targeted drug delivery *Mater. Sci. Eng. C Mater. Biol. Appl.* 107 110201.

162 Teramoto A, Tsujimoto M, Inoue T, Tsukamoto T, Imaizumi K, Toyama H, Saito K and Fujita H 2019 Automated classification of pulmonary nodules through a retrospective analysis of conventional CT and two-phase PET images in patients undergoing biopsy *Asia Ocean. J. Nucl. Med. Biol.* 7 29–37.

163 Heineken F G and Skalak R 1991 Tissue engineering: a brief overview *J. Biomech. Eng.* 113 111–2.

164 Zhao H, Ding R, Zhao X, Li Y, Qu L, Pei H, Yildirimer L, Wu Z and Zhang W 2017 Graphene-based nanomaterials for drug and/or gene delivery, bioimaging, and tissue engineering *Drug Discov. Today* 22 1302–17.

165 Shang L, Qi Y, Lu H, Pei H, Li Y, Qu L, Wu Z and Zhang W 2019 7 – Graphene and Graphene Oxide for Tissue Engineering and Regeneration *Theranostic Bionanomaterials: Micro and Nano Technologies* ed W Cui and X Zhao (Amsterdam, Netherlands: Elsevier) pp 165–85.

166 Engler A J, Sen S, Sweeney H L and Discher D E 2006 Matrix elasticity directs stem cell lineage specification *Cell* 126 677–89.

167 Venkatesan J, Pallela R and Kim S-K 2014 Applications of carbon nanomaterials in bone tissue engineering *J. Biomed. Nanotechnol.* 10 3105–23.

168 Wang Q, Chu Y, He J, Shao W, Zhou Y, Qi K, Wang L and Cui S 2017 A graded graphene oxide-hydroxyapatite/silk fibroin biomimetic scaffold for bone tissue engineering *Mater. Sci. Eng. C* 80 232–42.

169 Fu C, Bai H, Zhu J, Niu Z, Wang Y, Li J, Yang X and Bai Y 2017 Enhanced cell proliferation and osteogenic differentiation in electrospun PLGA/hydroxyapatite nanofibre scaffolds incorporated with graphene oxide *PLOS ONE* 12 e0188352.

170 Rajesh R and Ravichandran Y D 2015 Development of new graphene oxide incorporated tricomponent scaffolds with polysaccharides and hydroxyapatite and study of their osteoconductivity on MG-63 cell line for bone tissue engineering *RSC Adv.* 5 41135–43.

171 Lu J, He Y-S, Cheng C, Wang Y, Qiu L, Li D and Zou D 2013 Self-supporting graphene hydrogel film as an experimental platform to evaluate the potential of graphene for bone regeneration *Adv. Funct. Mater.* 23 3494–502.

172 Nishida E, Miyaji H, Takita H, Kanayama I, Tsuji M, Akasaka T, Sugaya T, Sakagami R and Kawanami M 2014 Graphene oxide coating facilitates the bioactivity of scaffold material for tissue engineering *Jpn. J. Appl. Phys.* 53 06JD04.

173 Wang L, Lu C, Li Y, Wu F, Zhao B and Dong X 2015 Green fabrication of porous silk fibroin/graphene oxide hybrid scaffolds for bone tissue engineering *RSC Adv.* 5 78660–8.

174 Lalwani G, Henslee A M, Farshid B, Lin L, Kasper F K, Qin Y-X, Mikos A G and Sitharaman B 2013 Two-dimensional nanostructure-reinforced biodegradable polymeric nanocomposites for bone tissue engineering *Biomacromolecules* 14 900–9.

175 Gao C, Liu T, Shuai C and Peng S 2014 Enhancement mechanisms of graphene in nano-58S bioactive glass scaffold: mechanical and biological performance *Sci. Rep.* 4 4712.

176 Cheng H K F, Sahoo N G, Tan Y P, Pan Y, Bao H, Li L, Chan S H and Zhao J 2012 Poly(vinyl alcohol) Nanocomposites Filled with Poly(vinyl alcohol)-Grafted Graphene Oxide *ACS Appl. Mater. Interfaces* 4 2387–94.

177 Du Y, Ge J, Li Y, Ma P X and Lei B 2018 Biomimetic elastomeric, conductive and biodegradable polycitrate-based nanocomposites for guiding myogenic differentiation and skeletal muscle regeneration *Biomaterials* 157 40–50.

178 Cha C, Shin S R, Gao X, Annabi N, Dokmeci M R, Tang X S and Khademhosseini A 2014 Controlling mechanical properties of cell-laden hydrogels by covalent incorporation of graphene oxide *Small* 10 514–23.

179 Jo H, Sim M, Kim S, Yang S, Yoo Y, Park J-H, Yoon T H, Kim M-G and Lee J Y 2017 Electrically conductive graphene/polyacrylamide hydrogels produced by mild chemical reduction for enhanced myoblast growth and differentiation *Acta Biomater.* 48 100–9.

180 Kim S J, Cho K W, Cho H R, Wang L, Park S Y, Lee S E, Hyeon T, Lu N, Choi S H and Kim D-H 2016 Stretchable and transparent biointerface using cell-sheet–graphene hybrid for electrophysiology and therapy of skeletal muscle *Adv. Funct. Mater.* 26 3207–17.

181 Wang F, Kong Y, Shen F, Wang Y, Wang D and Li Q 2022 High-performance microfibrillated cellulose-based low voltage electroactive ionic artificial muscles in bioinspired applications *Compos. Part B Eng.* 228 109436.

182 Nyambat B, Chen C-H, Wong P-C, Chiang C-W, Satapathy M K and Chuang E-Y 2018 Genipin-crosslinked adipose stem cell derived extracellular matrix-nano graphene oxide composite sponge for skin tissue engineering *J. Mater. Chem. B* 6 979–90.

183 Deepachitra R, Ramnath V and Sastry T P 2014 Graphene oxide incorporated collagen–fibrin biofilm as a wound dressing material *RSC Adv.* 4 62717–27.

184 Li Z, Wang H, Yang B, Sun Y and Huo R 2015 Three-dimensional graphene foams loaded with bone marrow derived mesenchymal stem cells promote skin wound healing with reduced scarring *Mater. Sci. Eng. C* 57 181–8.

185 Hou C, Huang T, Wang H, Yu H, Zhang Q and Li Y 2013 A strong and stretchable self-healing film with self-activated pressure sensitivity for potential artificial skin applications *Sci. Rep.* 3 3138.

186 Bao R, Tan B, Liang S, Zhang N, Wang W and Liu W 2017 A π-π conjugation-containing soft and conductive injectable polymer hydrogel highly efficiently rebuilds cardiac function after myocardial infarction *Biomaterials* 122 63–71.

187 Ryan A J, Kearney C J, Shen N, Khan U, Kelly A G, Probst C, Brauchle E, Biccai S, Garciarena C D, Vega-Mayoral V, Loskill P, Kerrigan S W, Kelly D J, Schenke-Layland K, Coleman J N and O'Brien F J 2018 Electroconductive biohybrid collagen/pristine graphene composite biomaterials with enhanced biological activity *Adv. Mater.* 30 e1706442.

188 Zhou J, Yang X, Liu W, Wang C, Shen Y, Zhang F, Zhu H, Sun H, Chen J, Lam J, Mikos A G and Wang C 2018 Injectable OPF/graphene oxide hydrogels provide mechanical support and enhance cell electrical signaling after implantation into myocardial infarct *Theranostics* 8 3317–30.

189 Shin S R, Zihlmann C, Akbari M, Assawes P, Cheung L, Zhang K, Manoharan V, Zhang Y S, Yüksekkaya M, Wan K, Nikkhah M, Dokmeci M R, Tang X S and Khademhosseini A 2016 Reduced graphene oxide-GelMA hybrid hydrogels as scaffolds for cardiac tissue engineering *Small* 12 3677–89.

190 Annabi N, Shin S R, Tamayol A, Miscuglio M, Bakooshli M A, Assmann A, Mostafalu P, Sun J-Y, Mithieux S, Cheung L, Tang X S, Weiss A S and Khademhosseini A 2016 Highly elastic and conductive human-based protein hybrid hydrogels *Adv. Mater.* 28 40–9.

191 Ahadian S, Zhou Y, Yamada S, Estili M, Liang X, Nakajima K, Shiku H and Matsue T 2016 Graphene induces spontaneous cardiac differentiation in embryoid bodies *Nanoscale* 8 7075–84.

192 Feng Z-Q, Wang T, Zhao B, Li J and Jin L 2015 Soft graphene nanofibers designed for the acceleration of nerve growth and development *Adv. Mater.* 27 6462–8.

193 Sánchez-González S, Diban N and Urtiaga A 2018 Hydrolytic degradation and mechanical stability of poly(ε-Caprolactone)/reduced graphene oxide membranes as scaffolds for in vitro neural tissue regeneration *Membranes* 8 12.

194 Park D-W, Schendel A A, Mikael S, Brodnick S K, Richner T J, Ness J P, Hayat M R, Atry F, Frye S T, Pashaie R, Thongpang S, Ma Z and Williams J C 2014 Graphene-based carbon-layered electrode array technology for neural imaging and optogenetic applications *Nat. Commun.* 5 5258.

195 Janib S M, Moses A S and MacKay J A 2010 Imaging and drug delivery using theranostic nanoparticles *Adv. Drug Deliv. Rev.* 62 1052–63.

196 Park K, Lee S, Kang E, Kim K, Choi K and Kwon I C 2009 New generation of multifunctional nanoparticles for cancer imaging and therapy *Adv. Funct. Mater.* 19 1553–66.

197 Smith A M, Mancini M C and Nie S 2009 Second window for in vivo imaging *Nat. Nanotechnol.* 4 710–1.

198 Hong G, Zou Y, Antaris A L, Diao S, Wu D, Cheng K, Zhang X, Chen C, Liu B and He Y 2014 Ultrafast fluorescence imaging in vivo with conjugated polymer fluorophores in the second near-infrared window *Nat. Commun.* 5 4206.

199 Robinson J T, Hong G, Liang Y, Zhang B, Yaghi O K and Dai H 2012 In vivo fluorescence imaging in the second near-infrared window with long circulating carbon nanotubes capable of ultrahigh tumor uptake *J. Am. Chem. Soc.* 134 10664–9.

200 Yoo J M, Kang J H and Hong B H 2015 Graphene-based nanomaterials for versatile imaging studies *Chem. Soc. Rev.* 44 4835–52.

201 Liu Q, Guo B, Rao Z, Zhang B and Gong J R 2013 Strong two-photon-induced fluorescence from photostable, biocompatible nitrogen-doped graphene quantum dots for cellular and deep-tissue imaging *Nano Lett.* 13 2436–41.

202 Chung C, Kim Y-K, Shin D, Ryoo S-R, Hong B H and Min D-H 2013 Biomedical applications of graphene and graphene oxide *Acc. Chem. Res.* 46 2211–24.

203 Wang X, Huang P, Feng L, He M, Guo S, Shen G and Cui D 2012 Green controllable synthesis of silver nanomaterials on graphene oxide sheets via spontaneous reduction *Rsc Adv.* 2 3816–22.

204 Baker S N and Baker G A 2010 Luminescent carbon nanodots: emergent nanolights *Angew. Chem. Int. Ed.* 49 6726–44.

205 Zhu S, Meng Q, Wang L, Zhang J, Song Y, Jin H, Zhang K, Sun H, Wang H and Yang B 2013 Highly photoluminescent carbon dots for multicolor patterning, sensors, and bioimaging *Angew. Chem. Int. Ed.* 52 3953–7.

206 Yan X, Niu G, Lin J, Jin A J, Hu H, Tang Y, Zhang Y, Wu A, Lu J and Zhang S 2015 Enhanced fluorescence imaging guided photodynamic therapy of sinoporphyrin sodium loaded graphene oxide *Biomaterials* 42 94–102.

207 Hu S, Chen Y, Hung W, Chen I and Chen S 2012 Quantum-dot-tagged reduced graphene oxide nanocomposites for bright fluorescence bioimaging and photothermal therapy monitored in situ *Adv. Mater.* 24 1748–54.

208 Wang C, Li J, Amatore C, Chen Y, Jiang H and Wang X 2011 Gold nanoclusters and graphene nanocomposites for drug delivery and imaging of cancer cells *Angew. Chem.* 123 11848–52.

209 Wang Y, Wang H, Liu D, Song S, Wang X and Zhang H 2013 Graphene oxide covalently grafted upconversion nanoparticles for combined NIR mediated imaging and photothermal/photodynamic cancer therapy *Biomaterials* 34 7715–24.

210 Liu Z, Guo Z, Zhong H, Qin X, Wan M and Yang B 2013 Graphene oxide based surface-enhanced Raman scattering probes for cancer cell imaging *Phys. Chem. Chem. Phys.* 15 2961–6.

211 Kim Y-K, Han S W and Min D-H 2012 Graphene oxide sheath on ag nanoparticle/graphene hybrid films as an antioxidative coating and enhancer of surface-enhanced Raman scattering *ACS Appl. Mater. Interfaces* 4 6545–51.

212 Ma X, Qu Q, Zhao Y, Luo Z, Zhao Y, Ng K W and Zhao Y 2013 Graphene oxide wrapped gold nanoparticles for intracellular Raman imaging and drug delivery *J. Mater. Chem. B* 1 6495–500.

213 Song Z-L, Chen Z, Bian X, Zhou L-Y, Ding D, Liang H, Zou Y-X, Wang S-S, Chen L, Yang C, Zhang X-B and Tan W 2014 Alkyne-functionalized superstable graphitic silver nanoparticles for Raman imaging *J. Am. Chem. Soc.* 136 13558–61.

214 Wang Y, Polavarapu L and Liz-Marzán L M 2014 Reduced graphene oxide-supported gold nanostars for improved SERS sensing and drug delivery *ACS Appl. Mater. Interfaces* 6 21798–805.

215 Huang Y, Dong X, Shi Y, Li C M, Li L-J and Chen P 2010 Nanoelectronic biosensors based on CVD grown graphene *Nanoscale* 2 1485–8.

216 Szunerits S and Boukherroub R 2018 Graphene-based biosensors *Interface Focus* 8 20160132.

217 Wang K, Ruan J, Song H, Zhang J, Wo Y, Guo S and Cui D 2010 Biocompatibility of graphene oxide *Nanoscale Res. Lett.* 6 8.

218 Nguyen T V T, Le B H and Seo Y J 2017 Highly fluorescence quenching graphene oxide-based oligodeoxynucleotide hairpin systems for probing CNG DNA repeat sequences *Tetrahedron Lett.* 58 3301–5.

219 Pothipor C, Wiriyakun N, Putnin T, Ngamaroonchote A, Jakmunee J, Ounnunkad K, Laocharoensuk R and Aroonyadet N 2019 Highly sensitive biosensor based on graphene–poly (3-aminobenzoic acid) modified electrodes and porous-hollowed-silver-gold nanoparticle labelling for prostate cancer detection *Sens. Actuators B Chem.* 296 126657.

220 He L, Wang Q, Mandler D, Li M, Boukherroub R and Szunerits S 2016 Detection of folic acid protein in human serum using reduced graphene oxide electrodes modified by folic-acid *Biosens. Bioelectron.* 75 389–95.

221 Schultz A, Knoll T, Urban A, Schuck H, von Briesen H, Germann A and Velten T 2021 Novel cost-efficient graphene-based impedance biosensor for the analysis of viral cytopathogenicity and the effect of antiviral drugs *Front. Bioeng. Biotechnol.* 9 718889.

222 Farid S, Meshik X, Choi M, Mukherjee S, Lan Y, Parikh D, Poduri S, Baterdene U, Huang C-E and Wang Y Y 2015 Detection of Interferon gamma using graphene and aptamer based FET-like electrochemical biosensor *Biosens. Bioelectron.* 71 294–9.

223 Vasilescu A, Gáspár S, Gheorghiu M, David S, Dinca V, Peteu S, Wang Q, Li M, Boukherroub R and Szunerits S 2017 Surface plasmon resonance based sensing of lysozyme in serum on micrococcus lysodeikticus-modified graphene oxide surfaces *Biosens. Bioelectron.* 89 525–31.

224 Kim J A, Hwang T, Dugasani S R, Amin R, Kulkarni A, Park S H and Kim T 2013 Graphene based fiber optic surface plasmon resonance for bio-chemical sensor applications *Sens. Actuators B Chem.* 187 426–33.

225 Omar N A S, Fen Y W, Saleviter S, Daniyal W M E M M, Anas N A A, Ramdzan N S M and Roshidi M D A 2019 Development of a graphene-based surface plasmon resonance optical sensor chip for potential biomedical application *Materials* 12 1928.

226 Song Y, Xu T, Xu L-P and Zhang X 2019 Nanodendritic gold/graphene-based biosensor for tri-mode miRNA sensing *Chem. Commun.* 55 1742–5.

227 Ali A, Hwang E Y, Choo J and Lim D W 2018 PEGylated nanographene-mediated metallic nanoparticle clusters for surface enhanced Raman scattering-based biosensing *Analyst* 143 2604–15.

228 Guo Y, Wang H, Ma X, Jin J, Ji W, Wang X, Song W, Zhao B and He C 2017 Fabrication of Ag–Cu_2O/reduced graphene oxide nanocomposites as surface-enhanced Raman scattering substrates for in situ monitoring of peroxidase-like catalytic reaction and biosensing *ACS Appl. Mater. Interfaces* 9 19074–81.

229 Huang Y, Dong X, Liu Y, Li L-J and Chen P 2011 Graphene-based biosensors for detection of bacteria and their metabolic activities *J. Mater. Chem.* 21 12358–62.

230 Pandey A, Gurbuz Y, Ozguz V, Niazi J H and Qureshi A 2017 Graphene-interfaced electrical biosensor for label-free and sensitive detection of foodborne pathogenic *E. coli* O157: H7 *Biosens. Bioelectron.* 91 225–31.

231 Afsahi S, Lerner M B, Goldstein J M, Lee J, Tang X, Bagarozzi D A Jr, Pan D, Locascio L, Walker A and Barron F 2018 Novel graphene-based biosensor for early detection of Zika virus infection *Biosens. Bioelectron.* 100 85–8.

232 Navakul K, Warakulwit C, Yenchitsomanus P, Panya A, Lieberzeit P A and Sangma C 2017 A novel method for dengue virus detection and antibody screening using a graphene-polymer based electrochemical biosensor *Nanomedicine Nanotechnol. Biol. Med.* 13 549–57.

233 Liu F, Choi K S, Park T J, Lee S Y and Seo T S 2011 Graphene-based electrochemical biosensor for pathogenic virus detection *BioChip J.* 5 123–8.

234 Kailashiya J, Singh N, Singh S K, Agrawal V and Dash D 2015 Graphene oxide-based biosensor for detection of platelet-derived microparticles: a potential tool for thrombus risk identification *Biosens. Bioelectron.* 65 274–80.

235 Zhou F, Jing W, Xu Y, Chen Z, Jiang Z and Wei Z 2019 Performance enhancement of ZnO nanorod-based enzymatic glucose sensor via reduced graphene oxide deposition and UV irradiation *Sens. Actuators B Chem.* 284 377–85.

236 Devadoss A, Sudhagar P, Das S, Lee S Y, Terashima C, Nakata K, Fujishima A, Choi W, Kang Y S and Paik U 2014 Synergistic metal–metal oxide nanoparticles supported electrocatalytic graphene for improved photoelectrochemical glucose oxidation *ACS Appl. Mater. Interfaces* 6 4864–71.

237 Zheng M, Cui Y, Li X, Liu S and Tang Z 2011 Photoelectrochemical sensing of glucose based on quantum dot and enzyme nanocomposites *J. Electroanal. Chem.* 656 167–73.

238 Wang Y, Qing X, Zhou Q, Zhang Y, Liu Q, Liu K, Wang W, Li M, Lu Z, Chen Y and Wang D 2017 The woven fiber organic electrochemical transistors based on polypyrrole nanowires/reduced graphene oxide composites for glucose sensing *Biosens. Bioelectron.* 95 138–45.

239 Poletti F, Favaretto L, Kovtun A, Treossi E, Corticelli F, Gazzano M, Palermo V, Zanardi C and Melucci M 2020 Electrochemical sensing of glucose by chitosan modified graphene oxide *J. Phys. Mater.* 3 014011.

240 Kang X, Wang J, Wu H, Aksay I A, Liu J and Lin Y 2009 Glucose Oxidase–graphene–chitosan modified electrode for direct electrochemistry and glucose sensing *Biosens. Bioelectron.* 25 901–5.

241 Yang K, Zhang S, Zhang G, Sun X, Lee S-T and Liu Z 2010 Graphene in mice: ultrahigh in vivo tumor uptake and efficient photothermal therapy *Nano Lett.* 10 3318–23.

242 Zhang W, Guo Z, Huang D, Liu Z, Guo X and Zhong H 2011 Synergistic effect of chemo-photothermal therapy using PEGylated graphene oxide *Biomaterials* 32 8555–61.

243 Chen H, Liu F, Lei Z, Ma L and Wang Z 2015 Fe_2O_3@Au core@shell nanoparticle–graphene nanocomposites as theranostic agents for bioimaging and chemo-photothermal synergistic therapy *RSC Adv.* 5 84980–7.

244 Tian B, Wang C, Zhang S, Feng L and Liu Z 2011 Photothermally enhanced photodynamic therapy delivered by nano-graphene oxide *ACS Nano* 5 7000–9.

245 Huang P, Xu C, Lin J, Wang C, Wang X, Zhang C, Zhou X, Guo S and Cui D 2011 Folic acid-conjugated graphene oxide loaded with photosensitizers for targeting photodynamic therapy *Theranostics* 1 240–50.

246 Sahu A, Choi W I, Lee J H and Tae G 2013 Graphene oxide mediated delivery of methylene blue for combined photodynamic and photothermal therapy *Biomaterials* 34 6239–48.

247 Zhou L, Jiang H, Wei S, Ge X, Zhou J and Shen J 2012 High-efficiency loading of hypocrellin B on graphene oxide for photodynamic therapy *Carbon* 50 5594–604.

248 Bai W-K, Shen E and Hu B 2012 Induction of the apoptosis of cancer cell by sonodynamic therapy: a review *Chin. J. Cancer Res.* 24 368–73.

249 Chen Y-W, Liu T-Y, Chang P-H, Hsu P-H, Liu H-L, Lin H-C and Chen S-Y 2016 A theranostic nrGO@MSN-ION nanocarrier developed to enhance the combination effect of sonodynamic therapy and ultrasound hyperthermia for treating tumor *Nanoscale* 8 12648–57.

250 Feng L, Yang X, Shi X, Tan X, Peng R, Wang J and Liu Z 2013 Polyethylene glycol and polyethylenimine dual-functionalized nano-graphene oxide for photothermally enhanced gene delivery *Small* 9 1989–97.

251 Kim H and Kim W J 2014 Photothermally controlled gene delivery by reduced graphene oxide–polyethylenimine nanocomposite *Small* 10 117–26.

252 Wang Y, Li Z, Hu D, Lin C-T, Li J and Lin Y 2010 Aptamer/graphene oxide nanocomplex for in situ molecular probing in living cells *J. Am. Chem. Soc.* 132 9274–6.

253 Sasidharan A, Panchakarla L S, Sadanandan A R, Ashokan A, Chandran P, Girish C M, Menon D, Nair S V, Rao C N R and Koyakutty M 2012 Hemocompatibility and macrophage response of pristine and functionalized graphene *Small* 8 1251–63.

254 Singh S K, Singh M K, Kulkarni P P, Sonkar V K, Grácio J J A and Dash D 2012 Amine-modified graphene: thrombo-protective safer alternative to graphene oxide for biomedical applications *ACS Nano* 6 2731–40.

255 Roy S, Rahman I A, Santos J H and Ahmed M U 2016 Meat species identification using DNA-redox electrostatic interactions and non-specific adsorption on graphene biochips *Food Control* 61 70–8.

256 Dekkers S, Oomen A G, Bleeker E A J, Vandebriel R J, Micheletti C, Cabellos J, Janer G, Fuentes N, Vázquez-Campos S, Borges T, Silva M J, Prina-Mello A, Movia D, Nesslany F, Ribeiro A R, Leite P E, Groenewold M, Cassee F R, Sips A J A M, Dijkzeul A, van Teunenbroek T and Wijnhoven S W P 2016 Towards a nanospecific approach for risk assessment *Regul. Toxicol. Pharmacol.* 80 46–59.

257 Gubala V, Johnston L J, Liu Z, Krug H, Moore C J, Ober C K, Schwenk M and Vert M 2018 Engineered nanomaterials and human health: Part 1. Preparation, functionalization and characterization (IUPAC technical report) *Pure Appl. Chem.* 90 1283–324.

258 Kuhlbusch T A J, Wijnhoven S W P and Haase A 2018 Nanomaterial exposures for worker, consumer and the general public *NanoImpact* 10 11–25.

259 Salieri B, Turner D A, Nowack B and Hischier R 2018 Life cycle assessment of manufactured nanomaterials: where are we? *NanoImpact* 10 108–20.

260 Joris F, Manshian B B, Peynshaert K, Smedt S C D, Braeckmans K and Soenen S J 2013 Assessing nanoparticle toxicity in cell-based assays: influence of cell culture parameters and optimized models for bridging the in vitro–in vivo gap *Chem. Soc. Rev.* 42 8339–59.

261 Hischier R 2021 17 – Life Cycle Assessment of Engineered Nanomaterials *Health and Environmental Safety of Nanomaterials (Second Edition)* Woodhead Publishing Series in Composites Science and Engineering ed J Njuguna, K Pielichowski and H Zhu (Sawston, Cambridge: Woodhead Publishing) pp 443–58.

262 Donaldson K and Poland C A 2013 Nanotoxicity: challenging the myth of nano-specific toxicity *Curr. Opin. Biotechnol.* 24 724–34.

263 Ding H and Ma Y 2018 Computational approaches to cell–nanomaterial interactions: keeping balance between therapeutic efficiency and cytotoxicity *Nanoscale Horiz.* 3 6–27.

264 Mahmoudi M 2018 Debugging nano–bio interfaces: systematic strategies to accelerate clinical translation of nanotechnologies *Trends Biotechnol.* 36 755–69.

265 Mahmoudi M, Bertrand N, Zope H and Farokhzad O C 2016 Emerging understanding of the protein corona at the nano-bio interfaces *Nano Today* 11 817–32.

266 Gagner J E, Shrivastava S, Qian X, Dordick J S and Siegel R W 2012 Engineering nanomaterials for biomedical applications requires understanding the nano-bio interface: a perspective *J. Phys. Chem. Lett.* 3 3149–58.

267 Hu X, Wei Z and Mu L 2017 Graphene oxide nanosheets at trace concentrations elicit neurotoxicity in the offspring of zebrafish *Carbon* 117 182–91.

268 Qureshi A, Kang W P, Davidson J L and Gurbuz Y 2009 Review on carbon-derived, solid-state, micro and nano sensors for electrochemical sensing applications *Diam. Relat. Mater.* 18 1401–20.

269 Liu Z, Tabakman S, Welsher K and Dai H 2009 Carbon nanotubes in biology and medicine: in vitro and in vivo detection, imaging and drug delivery *Nano Res.* 2 85–120.

270 Tran P A, Zhang L and Webster T J 2009 Carbon nanofibers and carbon nanotubes in regenerative medicine *Adv. Drug Deliv. Rev.* 61 1097–114.

271 Murugan R and Ramakrishna S 2006 Nano-featured scaffolds for tissue engineering: a review of spinning methodologies *Tissue Eng.* 12 435–47.

272 Hola K, Zhang Y, Wang Y, Giannelis E P, Zboril R and Rogach A L 2014 Carbon dots — emerging light emitters for bioimaging, cancer therapy and optoelectronics *Nano Today* 9 590–603.

273 Ji S, Liu C, Zhang B, Yang F, Xu J, Long J, Jin C, Fu D, Ni Q and Yu X 2010 Carbon nanotubes in cancer diagnosis and therapy *Biochim. Biophys. Acta BBA – Rev. Cancer* 1806 29–35.

274 Gannon C J, Cherukuri P, Yakobson B I, Cognet L, Kanzius J S, Kittrell C, Weisman R B, Pasquali M, Schmidt H K, Smalley R E and Curley S A 2007 Carbon nanotube-enhanced thermal destruction of cancer cells in a noninvasive radiofrequency field *Cancer* 110 2654–65.

275 Reddy A R N, Reddy Y N, Krishna D R and Himabindu V 2010 Multi wall carbon nanotubes induce oxidative stress and cytotoxicity in human embryonic kidney (HEK293) cells *Toxicology* 272 11–6.

276 Costa L G, Pellacani C and Guizzetti M 2017 Chapter 14 - In Vitro and Alternative Approaches to Developmental Neurotoxicity *Reproductive and Developmental Toxicology (Second Edition)* ed R C Gupta (Cambridge, MA: Academic Press) pp 241–53.

277 Wei L, Shao L, Feng X, Chen A, Zhang Y and Wang J 2015 Central nervous system toxicity of metallic nanoparticles *Int. J. Nanomedicine* 4321.

278 Leite P E C, Pereira M R and Granjeiro J M 2015 Hazard effects of nanoparticles in central nervous system: searching for biocompatible nanomaterials for drug delivery *Toxicol. In Vitro* 29 1653–60.

279 Bussy C, Al-Jamal K T, Boczkowski J, Lanone S, Prato M, Bianco A and Kostarelos K 2015 Microglia determine brain region-specific neurotoxic responses to chemically functionalized carbon nanotubes *ACS Nano* 9 7815–30.

280 Visalli G, Currò M, Iannazzo D, Pistone A, Pruiti Ciarello M, Acri G, Testagrossa B, Bertuccio M P, Squeri R and Di Pietro A 2017 In vitro assessment of neurotoxicity and neuroinflammation of homemade MWCNTs *Environ. Toxicol. Pharmacol.* 56 121–8.

281 Jaeschke H, Gores G J, Cederbaum A I, Hinson J A, Pessayre D and Lemasters J J 2002 Mechanisms of hepatotoxicity *Toxicol. Sci.* 65 166–76.

282 Onoda A, Takeda K and Umezawa M 2017 Dose-dependent induction of astrocyte activation and reactive astrogliosis in mouse brain following maternal exposure to carbon black nanoparticle *Part. Fibre Toxicol.* 14 4.

283 Yang Y, Qin Z, Zeng W, Yang T, Cao Y, Mei C and Kuang Y 2017 Toxicity assessment of nanoparticles in various systems and organs *Nanotechnol. Rev.* 6 279–89.

284 Ahmadi H, Ramezani M, Yazdian-Robati R, Behnam B, Razavi Azarkhiavi K, Hashem Nia A, Mokhtarzadeh A, Matbou Riahi M, Razavi B M and Abnous K 2017 Acute toxicity of functionalized single wall carbon nanotubes: a biochemical, histopathologic and proteomics approach *Chem. Biol. Interact.* 275 196–209.

285 Lin B, Zhang H, Lin Z, Fang Y, Tian L, Yang H, Yan J, Liu H, Zhang W and Xi Z 2013 Studies of single-walled carbon nanotubes-induced hepatotoxicity by NMR-based metabonomics of rat blood plasma and liver extracts *Nanoscale Res. Lett.* 8 236.

286 Sumi N and Chitra K C 2017 Histopathological alterations in gill, liver and muscle tissues of the freshwater fish, Pseudetroplus maculatus exposed to fullerene C60 *Int. J. Fisher. Aquat. Stud.* 5 604–8.

287 Barnett L M A and Cummings B S 2018 Nephrotoxicity and renal pathophysiology: a contemporary perspective *Toxicol. Sci.* 164 379–90.

288 Yu S, Su X, Du J, Wang J, Gao Y, Zhang L, Chen L, Yang Y and Liu X 2018 The cytotoxicity of water-soluble carbon nanotubes on human embryonic kidney and liver cancer cells *New Carbon Mater.* 33 36–45.

289 Klumpp C, Kostarelos K, Prato M and Bianco A 2006 Functionalized carbon nanotubes as emerging nanovectors for the delivery of therapeutics *Biochim. Biophys. Acta BBA – Biomembr.* 1758 404–12.

290 Sun Y, Dai H, Chen S, Xu M, Wang X, Zhang Y, Xu S, Xu A, Weng J, Liu S and Wu L 2018 Graphene oxide regulates cox2 in human embryonic kidney 293T cells via epigenetic mechanisms: dynamic chromosomal interactions *Nanotoxicology* 12 117–37.

291 Chen G-Y, Yang H-J, Lu C-H, Chao Y-C, Hwang S-M, Chen C-L, Lo K-W, Sung L-Y, Luo W-Y, Tuan H-Y and Hu Y-C 2012 Simultaneous induction of autophagy and toll-like receptor signaling pathways by graphene oxide *Biomaterials* 33 6559–69.

292 Chen G-Y, Meng C-L, Lin K-C, Tuan H-Y, Yang H-J, Chen C-L, Li K-C, Chiang C-S and Hu Y-C 2015 Graphene oxide as a chemosensitizer: diverted autophagic flux, enhanced nuclear import, elevated necrosis and improved antitumor effects *Biomaterials* 40 12–22.

293 Zhi X, Fang H, Bao C, Shen G, Zhang J, Wang K, Guo S, Wan T and Cui D 2013 The immunotoxicity of graphene oxides and the effect of PVP-coating *Biomaterials* 34 5254–61.

294 Yu Y, Zhang Q, Mu Q, Zhang B and Yan B 2008 Exploring the immunotoxicity of carbon nanotubes *Nanoscale Res. Lett.* 3 271.

295 Smith M J, Brown J M, Zamboni W C and Walker N J 2014 From immunotoxicity to nanotherapy: the effects of nanomaterials on the immune system *Toxicol. Sci.* 138 249–55.

296 Li X, Wang X, Sha L, Wang D, Shi W, Zhao Q and Wang S 2018 Thermosensitive lipid bilayer-coated mesoporous carbon nanoparticles for synergistic thermochemotherapy of tumor *ACS Appl. Mater. Interfaces* 10 19386–97.

297 Li X, Wang L, She L, Sun L, Ma Z, Chen M, Hu P, Wang D and Yang F 2018 Immunotoxicity assessment of ordered mesoporous carbon nanoparticles modified with PVP/PEG *Colloids Surf. B Biointerfaces* 171 485–93.

298 Zhang T, Tang M, Zhang S, Hu Y, Li H, Zhang T, Xue Y and Pu Y 2017 Systemic and immunotoxicity of pristine and PEGylated multi-walled carbon nanotubes in an intravenous 28 days repeated dose toxicity study *Int. J. Nanomedicine* 12 1539–54.

299 Hirano S, Kanno S and Furuyama A 2008 Multi-walled carbon nanotubes injure the plasma membrane of macrophages *Toxicol. Appl. Pharmacol.* 232 244–51.

300 Khosravi Y, Salimi A, Pourahmad Naserzadeh P and Seydi E 2018 Inhalation exposure of nano diamond induced oxidative stress in lung, heart and brain *Xenobiotica* 48 860–6.

301 Öner D, Ghosh M, Bové H, Moisse M, Boeckx B, Duca R C, Poels K, Luyts K, Putzeys E, Van Landuydt K, Vanoirbeek J A, Ameloot M, Lambrechts D, Godderis L and Hoet P H 2018 Differences in MWCNT- and SWCNT-induced DNA methylation alterations in association with the nuclear deposition *Part. Fibre Toxicol.* 15 11.

302 Zare-Zardini H, Taheri-Kafrani A, Ordooei M, Amiri A and Karimi-Zarchi M 2018 Evaluation of toxicity of functionalized graphene oxide with ginsenoside Rh2, lysine and arginine on blood cancer cells (K562), red blood cells, blood coagulation and cardiovascular tissue: In vitro and in vivo studies *J. Taiwan Inst. Chem. Eng.* 93 70–8.

303 Demir E and Marcos R 2018 Toxic and genotoxic effects of graphene and multi-walled carbon nanotubes *J. Toxicol. Environ. Health A* 81 645–60.

304 Ghosh M, Öner D, Duca R C, Bekaert B, Vanoirbeek J A J, Godderis L and Hoet P H M 2018 Single-walled and multi-walled carbon nanotubes induce sequence-specific epigenetic alterations in 16 HBE cells *Oncotarget* 9 20351–65.

305 Chaudhuri I, Fruijtier-Pölloth C, Ngiewih Y and Levy L 2018 Evaluating the evidence on genotoxicity and reproductive toxicity of carbon black: a critical review *Crit. Rev. Toxicol.* 48 143–69.

306 Modrzynska J, Berthing T, Ravn-Haren G, Jacobsen N R, Weydahl I K, Loeschner K, Mortensen A, Saber A T and Vogel U 2018 Primary genotoxicity in the liver following pulmonary exposure to carbon black nanoparticles in mice *Part. Fibre Toxicol.* 15 2.

307 Ali N and Oehme F W 1992 A literature review of dermatotoxicity *Vet. Hum. Toxicol.* 34 428–37.

308 Ding L, Stilwell J, Zhang T, Elboudwarej O, Jiang H, Selegue J P, Cooke P A, Gray J W and Chen F F 2005 Molecular characterization of the cytotoxic mechanism of multiwall carbon nanotubes and nano-onions on human skin fibroblast *Nano Lett.* 5 2448–64.

309 Monteiro-Riviere N A and Inman A O 2006 Challenges for assessing carbon nanomaterial toxicity to the skin *Carbon* 44 1070–8.

310 Liao K-H, Lin Y-S, Macosko C W and Haynes C L 2011 Cytotoxicity of graphene oxide and graphene in human erythrocytes and skin fibroblasts *ACS Appl. Mater. Interfaces* 3 2607–15.

311 Pelin M, Fusco L, Martín C, Sosa S, Frontiñán-Rubio J, Miguel González-Domínguez J, Durán-Prado M, Vázquez E, Prato M and Tubaro A 2018 Graphene and graphene oxide induce ROS production in human HaCaT skin keratinocytes: the role of xanthine oxidase and NADH dehydrogenase *Nanoscale* 10 11820–30.

312 Tamokou J-D and Kuete V 2014 10 – Mutagenicity and Carcinogenicity of African Medicinal Plants *Toxicological Survey of African Medicinal Plants* ed V Kuete (Amsterdam, Netherlands: Elsevier) pp 277–322.

313 Kasai T, Umeda Y, Ohnishi M, Mine T, Kondo H, Takeuchi T, Matsumoto M and Fukushima S 2016 Lung carcinogenicity of inhaled multi-walled carbon nanotube in rats *Part. Fibre Toxicol.* 13 53.

314 Sakamoto Y, Hojo M, Kosugi Y, Watanabe K, Hirose A, Inomata A, Suzuki T and Nakae D 2018 Comparative study for carcinogenicity of 7 different multi-wall carbon nanotubes with different physicochemical characteristics by a single intraperitoneal injection in male Fischer 344 rats *J. Toxicol. Sci.* 43 587–600.

315 Nagai H, Okazaki Y, Chew S H, Misawa N, Yamashita Y, Akatsuka S, Ishihara T, Yamashita K, Yoshikawa Y, Yasui H, Jiang L, Ohara H, Takahashi T, Ichihara G, Kostarelos K, Miyata Y, Shinohara H and Toyokuni S 2011 Diameter and rigidity of multiwalled carbon nanotubes are critical factors in mesothelial injury and carcinogenesis *Proc. Natl. Acad. Sci.* 108 E1330–8.

316 Baan R A 2007 Carcinogenic hazards from inhaled carbon black, titanium dioxide, and talc not containing asbestos or asbestiform fibers: recent evaluations by an IARC monographs working group *Inhal. Toxicol.* 19 213–28.

6

Biomedical Applications of Silica-Based Nanomaterials and Polymeric Nanomaterials

Ajit Khosla[1], Irshad A. Wani[2], and Mohammad N. Lone[3]

[1] *School of Advanced Materials and Nanotechnology, Xidian University, Xian, Shaanxi, China*
[2] *Department of Chemistry, Govt. Degree College Anantnag, Anantnag, University of Kashmir, Jammu and Kashmir, India*
[3] *Department of Chemistry, Central University of Kashmir, Ganderbal, Jammu and Kashmir, India*

6.1 Biomedical Applications of Silica Nanoparticles

Silica nanoparticles have been extensively investigated in biomedical contexts, particularly in drug delivery and cancer therapy. These nanoparticles boast exceptional qualities such as high biocompatibility, chemical and thermal stability, and ease of modification. Additionally, silica is utilized to shield magnetic nanoparticles from acid corrosion and aggregation, thereby enhancing cytocompatibility. This book chapter highlights recent advancements in silica-based magnetic nanoparticles, with a focus on drug delivery, drug targeting systems, and their utility in magnetohyperthermia and magnetic resonance imaging. Furthermore, it explores their application in various other biomedical domains. Finally, this part outlines the challenges and prospects associated with the use of silica nanomaterials.

6.1.1 As Nanocarriers and Biomodulators

Given the limited impact of the silica framework's chemistry on therapeutic effectiveness, there was speculation that creating an organic–inorganic hybrid framework for silica nanoparticles (SNPs) would provide chemically enabled functionalities not achievable with inorganic SNPs alone. This concept was put to the test approximately a decade ago when we observed the increased drug loading capacity and sustained drug release profile of mesoporous organosilica nanoparticles (OSNPs). In comparison to mesoporous SNPs [1], a new type of multishelled dendritic OSNPs was fabricated in 2017, along with the development of the first nanovaccine based on organosilica [2]. This research demonstrated that organic–inorganic hybrid silica nanoparticles (OSNPs), characterized by their multishelled architecture, exhibited significantly heightened immunostimulatory effects compared to conventional silica nanoparticles (SNPs). The hydrophobic structure of OSNPs facilitated sustained release of antigens and triggered danger signals in immune cells, resulting in endosome rupture – an essential process for MHC I cross-presentation and activation of CD8 T cells. As a result, OSNPs-mediated superior tumor inhibition activity in a therapeutic vaccination model. To further enhance the immune modulation capability of OSNPs, disulfide bonds, which are highly reactive in biological environments, were introduced into the hybrid silica structure [3]. Through the redox reaction between —S—S— and glutathione (GSH), an intracellular antioxidant, the disulfide bond-integrated OSNPs effectively depleted the intracellular GSH of immune cells [4],

Metallic, Magnetic and Carbon-Based Nanomaterials: Synthesis and Biomedical Applications, First Edition.
Ajit Khosla, Irshad A. Wani, and Mohammad N. Lone.

leading to the accumulation of reactive oxygen species (ROS) a danger messenger that triggers the implementation of our strategy indeed resulted in enhanced maturation of antigen-presenting cells (APCs), thereby improving cytotoxic T lymphocyte-mediated antitumor immunity compared to ethyl bridged OSNPs. Additionally, the cleavage of —S—S— bonds led to the breakdown of nanoparticles into smaller debris, reducing their potential risk of long-term accumulation in organs and enhancing their safety profile. Recently, we have employed PEI-conjugated disulfide bond-incorporated OSNPs (referred to as DMONs-PEI) as vectors for mRNA transfection in challenging-to-transfect cells, such as macrophages and 4T1 cancer cells [5]. It was demonstrated that the high intracellular GSH level, associated with suppressed mRNA translation, was overcome by GSH depletion mediated by DMONs-PEI, resulting in superior mRNA transfection efficiency both *in vitro* and *in vivo* (Figure 6.1) [6]. Besides modulating immune cells, we also showcased the capacity to control the intracellular environment of cancer cells. Our approach involved creating yolk-shell structured mesoporous OSNPs with a framework incorporating disulfide bonds, with the aim of circumventing the oxidative defense mechanism of cancer cells [7]. These OSNPs feature an inner yolk with small pores and an outer shell featuring large pores and demonstrating a distinctive stepwise degradation response to GSH, and these yolk-shell structured mesoporous OSNPs offer unique structural and compositional characteristics. Leveraging these attributes, we constructed a codelivery system comprising glucose oxidase and hypoxia-activated prodrugs. Glucose oxidase facilitated glucose depletion, inducing cancer starvation and generating H_2O_2 to inflict oxidative damage on cancer cells. Furthermore, the enzymatic reaction of glucose oxidase consumed significant amounts of oxygen, leading to tumor hypoxia, which in turn activated a prodrug (AQ4N) to execute chemotherapy. Consequently, this nanosystem could concurrently modulate multiple signals within the tumor microenvironment including GSH, glucose, and hypoxia levels through cascade chemical reactions, resulting in exceptional antitumor efficacy. Building upon our previous success in individually modulating the intracellular microenvironments of immune and cancer cells, we have demonstrated the feasibility of concurrently modulating both via designed biomodulators. Hybrid nanoparticles were fabricated by coincorporating Cu^{2+} ions and disulfide bonds (designated as Cu-DMONs), which not only acted as potent immunoadjuvants to activate immune cells but also directed cell death from tolerogenic to immunogenic [8]. This approach synergistically improved intracellular ROS levels resulting from the depletion of intracellular GSH and the initiation of Fenton's reaction, leading to a significant escalation of oxidative stress within cancer cells. This amplified oxidative stress contributed to immunogenic cell death (ICD) triggered by doxorubicin (Dox). Through the concurrent amplification of ICD and activation of the immune system, this approach facilitated potent chemoimmunotherapy when combined with immune checkpoint inhibitors, effectively suppressing both treated primary tumors and untreated distant tumors (Figure 6.2). Importantly, this readily available nanosystem only necessitates a blend of Dox and nanoparticles without encapsulation, thereby greatly streamlining the nanomedicine manufacturing process and boosting translational feasibility. Beyond enhancing the potency of drugs through chemical means, there is potential to design SNP-based biomodulators to impart inherent therapeutic activities – an emerging research domain that has recently attracted considerable interest. A frequently employed approach entails modifying the silica framework or surface chemistry with cytotoxic agents like metal ions to heighten intracellular oxidative stress and trigger cell death [9]. Despite yielding promising results, the high cytotoxicity and limited selectivity of these biomodulators for cancer cells have raised concerns regarding their safety and translational applicability. In response, our laboratory has redirected therapeutic efforts from targeting cancer cells directly to other cell types as a means to indirectly inhibit cancer cell growth. To implement this approach, we have developed SNP-based biomodulators capable of modulating the immune and tumor vascular systems.

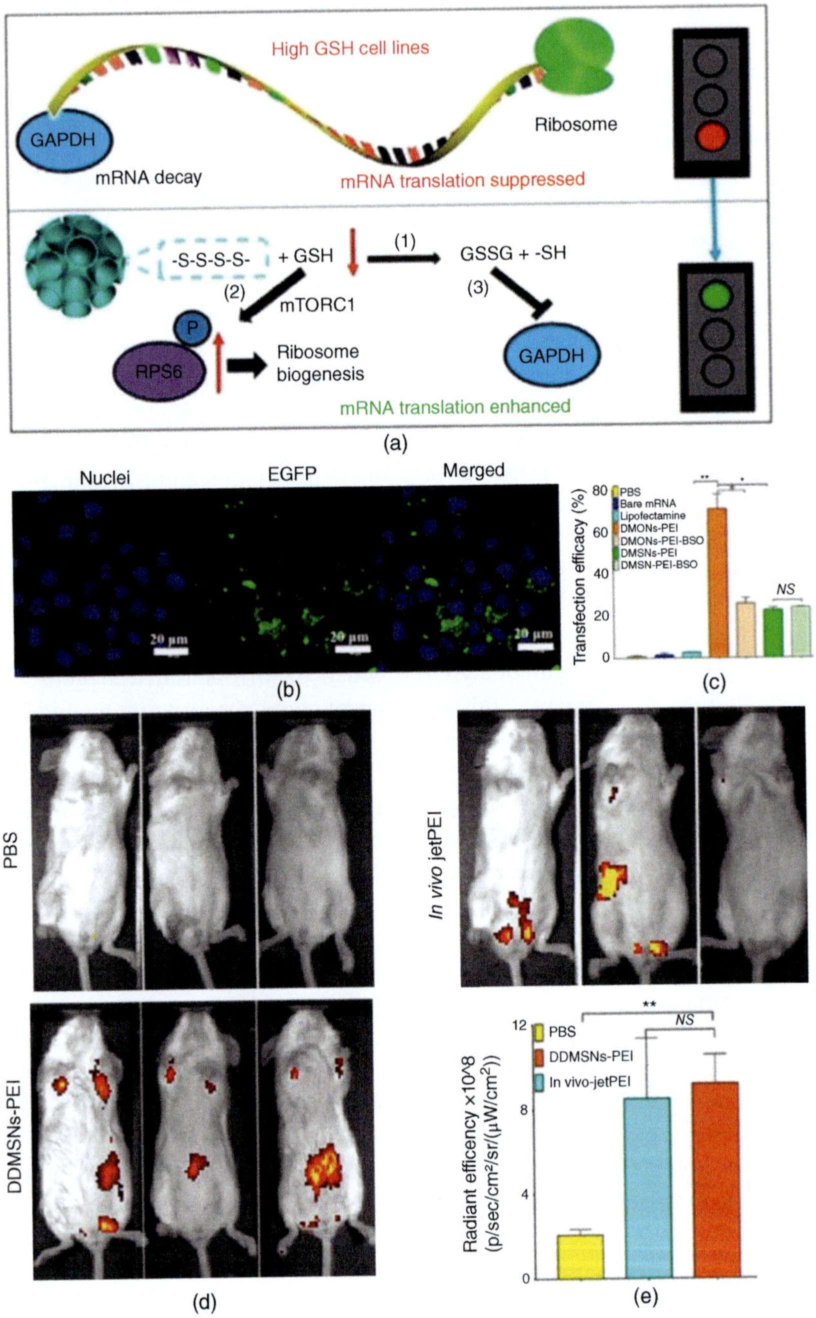

Figure 6.1 Disulfide bond-incorporated DMONs-PEI enhance mRNA transfection in vitro and in vivo in hard-to-transfect cells. (a) Illustration depicting nanoparticle chemistry facilitating upregulated mRNA translation. (b) Confocal images of RAW264.7 cells treated with DMONs-PEI/mRNA complexes for 48 hours. (c) Flow cytometry analysis showing the percentage of EGFP-positive cells among RAW264.7 cells treated with NP-PEI-mRNA complexes for 48 hours. (d) In vivo fluorescence imaging of mCherry mRNA delivery in BALB/c mice, accompanied by quantitative fluorescence results (e). Source: Reproduced with permission from Wang et al. [6]/John Wiley & Sons.

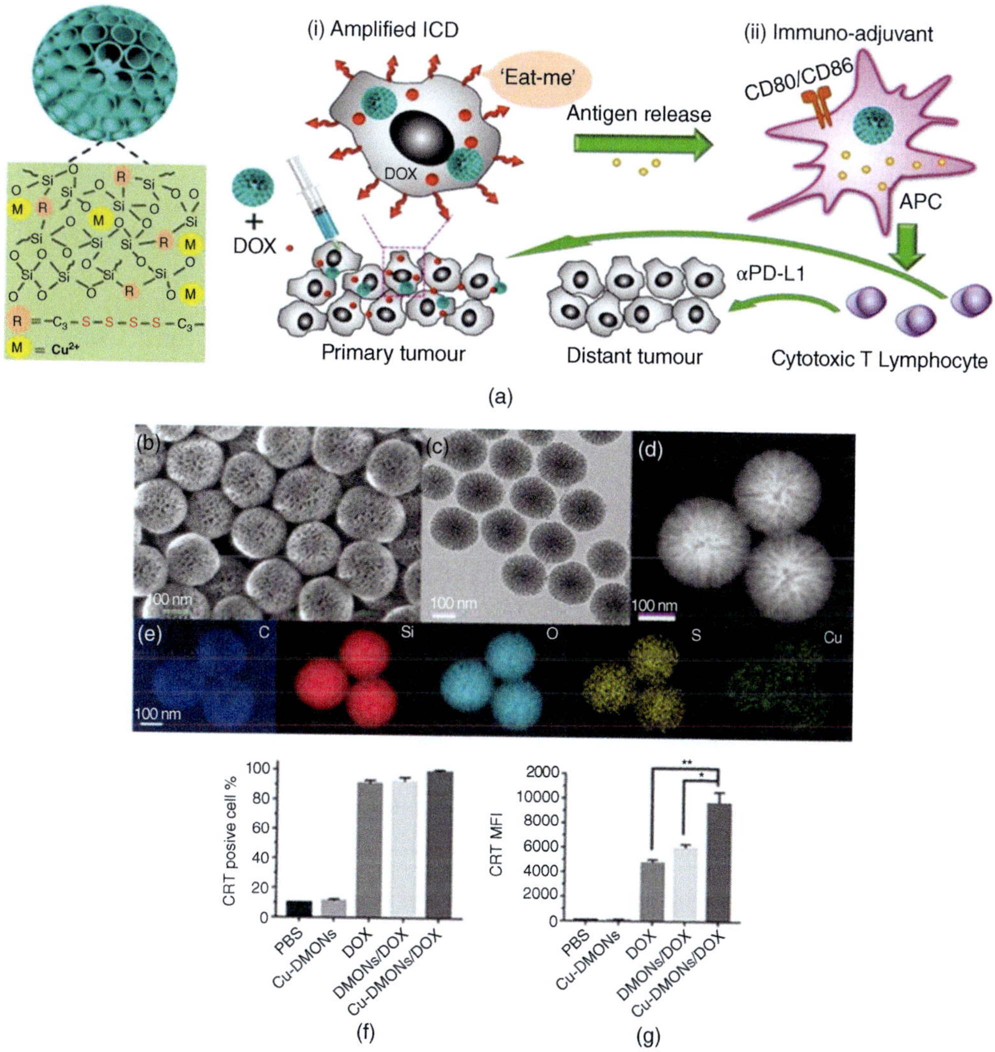

Figure 6.2 Enhancing cancer chemo-immunotherapy synergistically with immune checkpoint blockades through simultaneous modulation of cancer cells and immune cells using hybrid organosilica-based nanoreactors. (a) Illustration schematically depicting how nanoreactors improve cancer chemo-immunotherapy. (b) SEM, (c) TEM, (d) dark-field scanning TEM, and (e) energy dispersive X-ray spectroscopy elemental mapping images of the nanoreactors. (f, g) Modulation of immunogenic cell death (ICD) by nanoreactors, demonstrated by the surface exposure of calreticulin proteins. Source: Reproduced with permission from Yang et al. [8]/John Wiley & Sons.

Tumor-associated macrophages, which are abundant in solid tumors, exhibit high plasticity and play a crucial role in tumor progression. Recent studies have shown that iron-based nanoparticles can shift tumor-associated macrophages from the tumor-promoting M2 phenotype to the tumor-suppressing M1 phenotype. Building upon this discovery, our research group has engineered composite materials consisting of magnetite or hematite iron oxide nanoparticles (IONPs) embedded in mesoporous silica to regulate the M1 phenotype and elucidate the underlying biological mechanisms. Our findings have demonstrated that magnetite nanoparticles are more effective than hematite in promoting M1 polarization and suppressing tumors. Additionally, we have elucidated the molecular mechanism driving this effect, which involves signaling through regulatory factor 5 for M1 polarization, favoring cytotoxic T lymphocyte activation in antitumor activity and downregulation of M2-associated arginase. This mechanism offers new possibilities for developing macrophage-assisted anticancer therapies [10]. Interfering with the formation of tumor blood vessels (antiangiogenesis) or compromising the function of established vessels has demonstrated effectiveness in both preclinical and clinical tumor suppression. However, depending solely on a single mechanism for inhibiting tumors can be unpredictable. Recently, our team has pioneered the development of nanochelators aimed at modulating the tumor vascular system. These nanochelators are designed to simultaneously inhibit the formation of new vessels and disrupt the function of existing ones (Figure 6.3) [11]. These nanochelators, measuring less than 6 nm and functionalized with imidazole, are composed of organosilica nanodots. They can remain dispersed in circulation, targeting copper ions to impede angiogenesis selectively. In the tumor microenvironment, they transition to a larger aggregated state in response to elevated levels of copper and phosphate, thereby obstructing tumor blood vessels. By simultaneously targeting both newly formed and established tumor blood vessels, these nanochelators demonstrate significantly improved efficacy in inhibiting tumor growth compared to tetrathiomolybdate, a copper chelator used in clinical trials for antiangiogenesis. The clinical translation of SNP-based nanomedicine has been impeded by concerns regarding their biosafety, primarily stemming from two factors. Firstly, nanomedicine containing highly cytotoxic agents, such as chemodrugs, may lead to acute damage to normal organs due to off-target effects. Secondly, SNPs themselves may accumulate in normal organs over the long term, potentially causing chronic damage. To address these concerns, current design principles prioritize site-specific targeting, on-demand drug release profiles, and the high biodegradability of SNPs. Exploring the use of renal-clearable SNPs for indirectly controlling tumor growth, without carrying highly cytotoxic agents, holds promise as a viable research direction. Preliminary *in vivo* evaluations of these nanochelators indicate no weight loss or organ damage, suggesting excellent biosafety [11]. Applications of SNPs as nanocarriers and biomodulators are summarized in Table 6.1.

6.1.2 In Drug Delivery

While conventional treatments effectively manage and cure many diseases, pharmacological therapies still suffer from various limitations including nonspecific action, numerous side effects, toxicity, and susceptibility to changes in pH and light [20, 21]. Nanostructured systems have found extensive applications in enhancing the biocompatibility and systemic circulation of drugs (Figure 6.4), while also reducing immunogenicity [23]. Some investigations indicated that silica nanoparticles could enhance drug delivery and improve treatment response. In this context, Pinna et al. [24] carried out a study utilizing MIL-88A (Fe) MOF crystals nucleated and grown around a polymer core containing superparamagnetic nanoparticles, with a focus on magnetophoretic drug release of dopamine. The safety of MOF was assessed using PC12 cells as the neuronal cell model. The results suggested that dopamine could be delivered into the intracellular compartment using the PMP at MIL-88A carrier while minimizing side effects. Moreover, Balasamy et al. [25] explored the potential of nanoformulations utilizing biocompatible metal–organic frameworks (MOFs) and magnetic nanocarriers for drug delivery and tumor imaging, aiming to advance cancer therapy. A nanocomposite, Fe/SBA-16 and ZIF-8 (Fe/S-16/ZIF-8), was developed through ultrasonic irradiation. Drug delivery of cis-platin was analyzed using an automated diffusion cell system equipped with a flow-type Franz cell. The results indicated that the nanocarrier exhibited promising characteristics for cis-platin delivery, facilitating slow and sustained release. Follmann et al. [26] synthesized innovative hybrid organic–inorganic aerogel materials with one-dimensional aligned pores to create an extended drug delivery system for a hydrophobic drug.

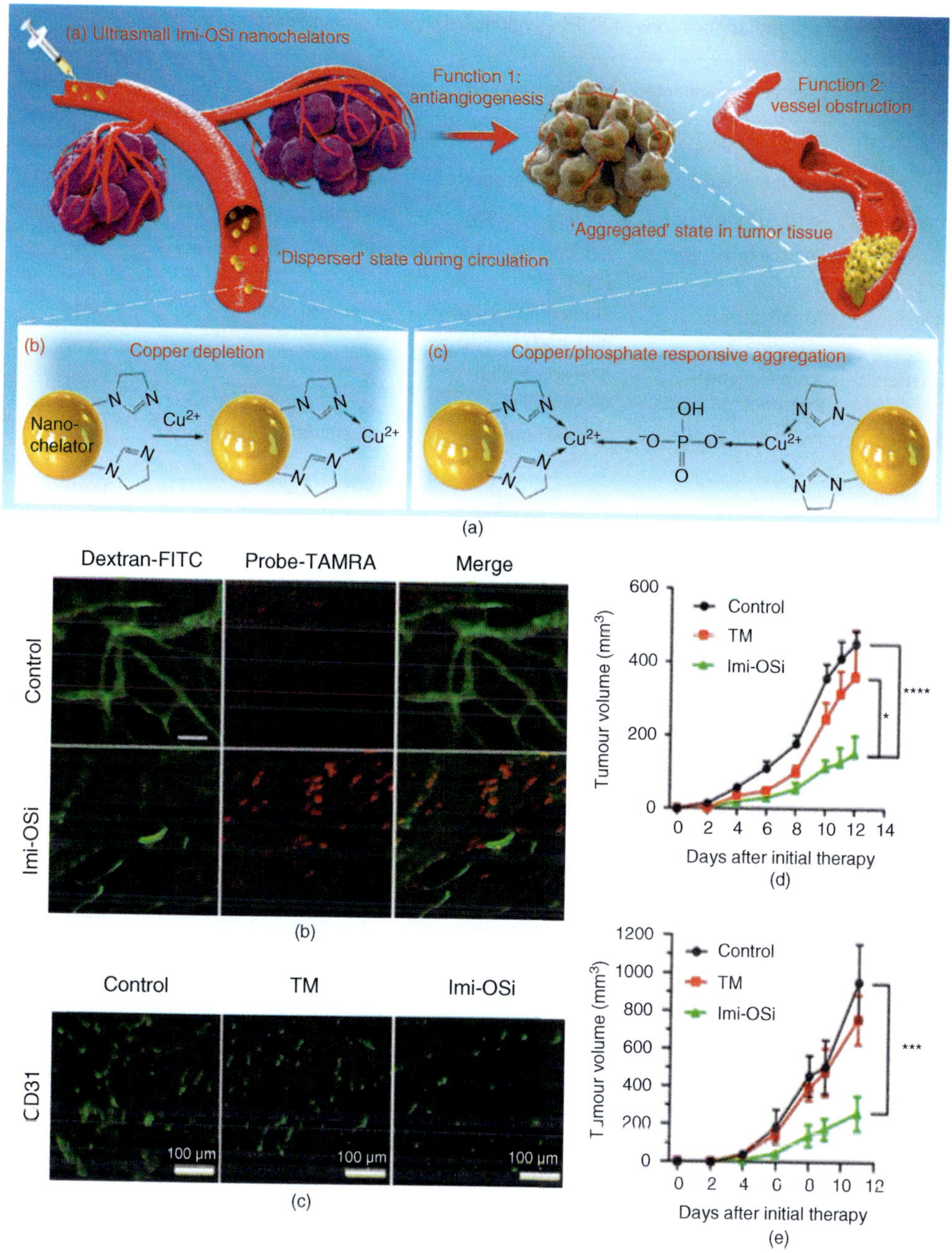

Figure 6.3 Nanochelators induce both antiangiogenesis and obstruction of tumor blood vessels simultaneously for tumor inhibition. (a) Illustration depicting how nanochelators induce concurrent antiangiogenesis and tumor vessel obstruction. (b) Real-time imaging of tumor blood vessel obstruction. (c) Visualization of antiangiogenic activity of nanochelators using CD31 antibody staining. (d) Antitumor efficacy of nanochelators in 4T1 and (e) CT26 xenograft tumor models. Source: Reproduced with permission from Yang et al. [11]/American Chemical Society.

Table 6.1 Summary of the applications of SNPs as nanocarriers and biomodulators.

Role of SNPs	Engineered parameter	Materials property	Application	References
Nanocarriers	Surface topology	Cellular uptake	siRNA delivery	[12]
Nanocarriers	Surface topology	Adhesive property	Lysozyme delivery	[13]
Nanocarriers	Surface topology	Hydrophobicity	Antibiotics delivery	[14]
Nanocarriers	Surface topology	DNA binding and protection	Plasmid DNA delivery	[15]
Nanocarriers	Architectural asymmetry	Hemocompatibility and adjuvanticity	Nanovaccine	[16]
Nanocarriers	Architectural asymmetry	Transepithelial capability	—	[17]
Nanocarriers	Pore geometry	Encapsulating large biomolecule	β-galactosides delivery	[18]
Nanocarriers	Pore geometry	High loading, biodegradability	Hyaluronan synthase type 2 delivery	[19]
Nanocarriers, biomodulators	Ethyl groups incorporation	Immunostimulatory activity	Cancer vaccine	[2]
Nanocarriers, biomodulators	Disulfide incorporation	Biodegradability, GSH depletion	Cancer vaccine	[5]
Nanocarriers, biomodulators	Disulfide incorporation	GSH depletion	mRNA delivery	[6]
Nanocarriers, biomodulators	Disulfide incorporation	GSH depletion	Prodrug/enzyme co-delivery	[7]
Biomodulators	Cu^{2+}/disulfide incorporation	Fenton reaction, GSH depletion	Cancer chemo-immunotherapy	[8]
Biomodulators	Iron oxide encapsulation	Macrophage polarization	Cancer immunotherapy	[10]
Biomodulators	Imidazole incorporation	Copper chelation and responsive aggregation	Vasculature targeted therapy	[11]

Due to their hydrophobic nature, these compounds displayed a remarkable capacity for drug loading and demonstrated an excellent release profile for dexamethasone, serving as a model drug. The findings revealed sustained drug delivery for over 50 days. Chitosan-modified magnetic silica nanoparticles and *N*-isopropylacrylamide (MagSi at Chi-g-NIPAAm) was used as a doxorubicin (Dox) carrier [27]. The drug release investigation was conducted under different pH conditions (4.0 and 7.4) to replicate physiological and tumor environments. Additionally, two temperatures were employed to simulate normal body temperature and hyperthermia conditions (37 and 45 °C, respectively). *In vitro* findings demonstrated that the drug release exhibited a significant pH and temperature responsiveness, with the highest release rate observed in acidic conditions at 45 °C. This behavior was attributed to polymer swelling at pH 4.0 (facilitated protonation of amine groups) and increased nanoparticle rupture. The magnetic nanocarrier loaded with cis-platin exhibited increased drug release at pH 5.5 compared to pH 7.4, likely due to the protonation of NH_2 groups in an acidic environment. Silica-based magnetic nanoparticles were utilized for the targeted delivery and cellular recognition of doxorubicin and miRNA-21.

The study examined the impact of varying amounts of doxorubicin on loading efficiency (%) and assessed how surface modification with DNA and miRNA-21 affected it. *In vitro* findings

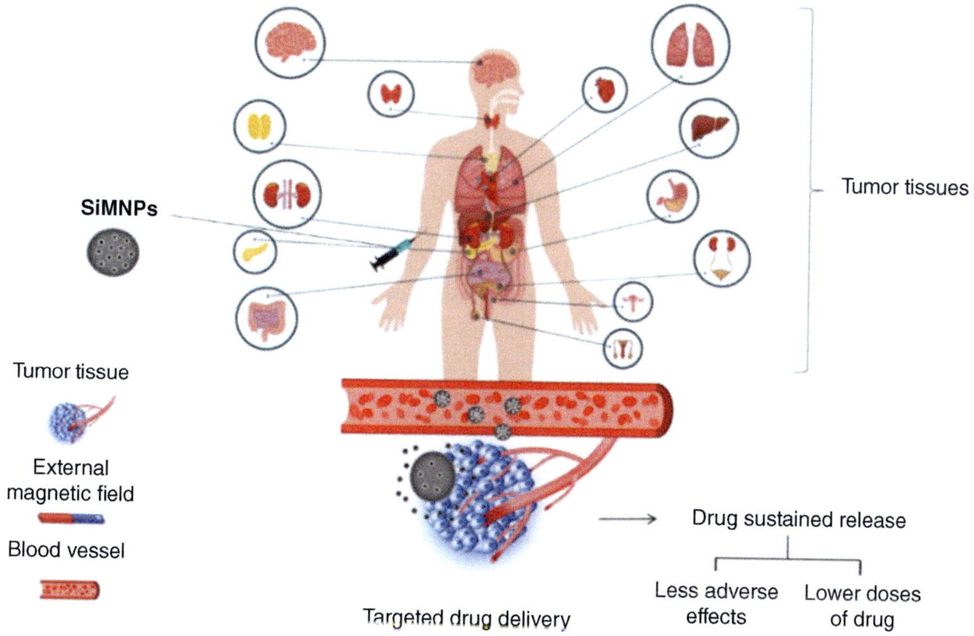

Figure 6.4 Drug delivery systems employing magnetic nanoparticles based on silica. Source: da Bruckmann et al. [22]/MDPI/CC BY 4.0.

demonstrated that increasing the amount of Dox progressively enhanced drug loading efficiency. Additionally, the inclusion of DNA and miRNA-21 in the nanocarrier structure influenced drug release. In a study reported by Espinoza et al. [28]. MNPs-SiO_2-F127–2% and MNPs-SiO2-F127–4% nanocomposites were utilized as carriers for doxorubicin. Experiments were conducted at two distinct pH levels, 5.4 and 7.4, representing physiological and tumoral conditions, respectively. Analysis of drug release data indicated a substantial pH-dependent effect, with higher release observed under acidic conditions compared to pH 7.4, possibly due to the cleavage of amide bonds. Moreover, the content of Pluronic F127 influenced the rate of drug release. Along with this, Wu and co-workers [29] used a nanocomposite (Fe_3O_4 at $mSiO_2$ and CS/Fe_3O_4 at $mSiO_2$) to deliver the same drug (Dox) and noted that under acidic conditions (pH 4.0), a higher drug release percentage (%) was reached. Both nanocarriers exhibit a time-dependent behavior; initially, a high concentration was released to the medium, followed by a slower release. The nanocarrier containing chitosan (CS) showed the highest drug release rate due to breaking the "gatekeeper" faster than Fe_3O_4 at $mSiO_2$. Moorthy et al. [30] investigated the effectiveness of a fucoidan-coated core–shell magnetic mesoporous silica drug carrier system (FeNP at SiOH at Fuc NPs) in the human breast adenocarcinoma cell line (MDA-MB-231). This system comprises a magnetic iron oxide (FeNP) core, a mesoporous silica shell (SiOH), and fucoidan, a marine biopolymer, aimed at delivering Dox while considering pH and hyperthermia conditions. Similarly, a novel Dox nanocarrier was developed using nanoparticles loaded into multiwalled carbon nanotubes with mesoporous silica ($mSiO_2$) (MWCNTs at $CoFe_2O_4$ at $mSiO_2$). The findings demonstrated a notable pH-responsive drug delivery capability within 48 hours, offering the potential for enhanced cancer therapy [31]. Wang et al. [32] produced Fe_2B at SiO_2 nanoparticles covalently grafted from graphene oxide containing 90% entrapped epirubicin (EPI). The drug delivery was pH-dependent, with higher EPI release achieved at pH 5.7. Wang et al. [33] examined the combined anticancer impact of the hydrogel-based nanocomposite (MMSN-RBITCs) for delivering Dox using the

MTT assay. Free Dox exhibited minimal cytotoxicity toward the 4T1 cell line during a 24-hour exposure. Conversely, the Dox-loaded hydrogel led to a reduction in cell viability. Furthermore, the cytotoxicity notably escalated when the hydrogel underwent magnetic hyperthermia therapy, suggesting that heat enhances the cytotoxic effects of nanoparticles. In a study reported by Solak et al. [34], demonstrated the loading of disulfiram onto mesoporous silica magnetic nanoparticles modified with acid-linked polyethylenimine (mMDPF) and investigated *in vitro* drug release. Under slightly acidic conditions (pH 6.0), the drug exhibited a faster and more controlled release compared to physiological conditions (pH 7.4). These findings suggest a potential application of the nanoparticles as a platform for disulfiram release, particularly due to the accelerated release in an acidic environment, which helps evade drug inactivation by P-glycoprotein. Laranjeira et al. [35] synthesized a nanodevice to achieve the slow release of exemestane for intravenous administration. The encapsulation efficiency and drug loading capacity were 90% and 37.7%, respectively. Exemestane was released over 72 hours, indicating its adsorption primarily on the inner pore walls of the nanoparticles. These findings suggest the potential application of silica-coated mesoporous magnetic nanoparticles to enhance drug release. Additionally, a novel dehydropeptide hydrogelator, Npx-L-Met-Z-ΔPhe-OH, was developed and coupled with two different plasmonic/magnetic nanoparticle architectures: core/shell manganese ferrite/gold nanoparticles and gold-decorated manganese ferrite nanoparticles, aimed at cancer therapy. This composite was evaluated using curcumin, a model antitumor drug, demonstrating promising outcomes for photothermal and phototriggered drug delivery [36].

The mechanism of uptake and release of drugs by mesoporous NPs relies on retaining the drug within the nanoparticle pores, often facilitated by adjunct molecules like gold nanoparticles [37], as caps to close the pores. This concept is called gatekeeping [38]. Lai et al. [39] reported the synthesis of mesoporous silicon nanoparticles (SiNPs) with chemically removable CdS nanoparticle (NP) caps as carrier systems for drug delivery was reported. Figure 6.5 illustrates the uptake and release mechanism, where CdSNPs act as caps to retain the drug within the pores. Covalent interaction occurs between the CdSNPs and the pore surface after drug loading to ensure retention within the pores. To release the drug, a disulfide bond-reducing molecule, such as dithiothreitol (DTT) or mercaptoethanol (ME), is required to break the covalent bond and open the pores. These systems offer efficient stimuli-controlled release, require no additional modifications of the drug for loading in the pores, and allow for the transfer of a large number of drugs. Functionalized silica nanoparticles (FSN) are employed for gene delivery applications, where they must exhibit two different states under different conditions. During gene loading, good affinity between the genes and nanoparticles (NPs) is crucial. This process relies on electrostatic interactions, as DNA molecules are negatively charged. Therefore, researchers aim to prepare positively charged NPs, often incorporating amino groups [40]. In the second condition, gene release is facilitated by preferring weak interactions between the gene and FSN, aiding the release of genes from the nanoparticles into the target. Bharali et al. [41] reported the *in vivo* utilization of amino-surface functionalized SiNPs for gene delivery. SiNPs capable of binding with plasmid DNA were injected into mouse brains alongside free plasmids. Notably, only the plasmid DNA-binding SiNPs demonstrated substantial gene expression, indicating effective protection and transfer of the gene into the cell nuclei.

6.1.3 In Bioimaging

Silica nanoparticles (SiNPs) find utility in medical imaging, where they serve as adjuncts by encapsulating contrast agent particles like gold [42], silver [43], iron oxide [44], organic dyes [45], and quantum dots [46]. SiNPs are commonly employed in medical imaging due to their attributes, including enhanced biocompatibility, tunable size, and size distribution, protection of contrast

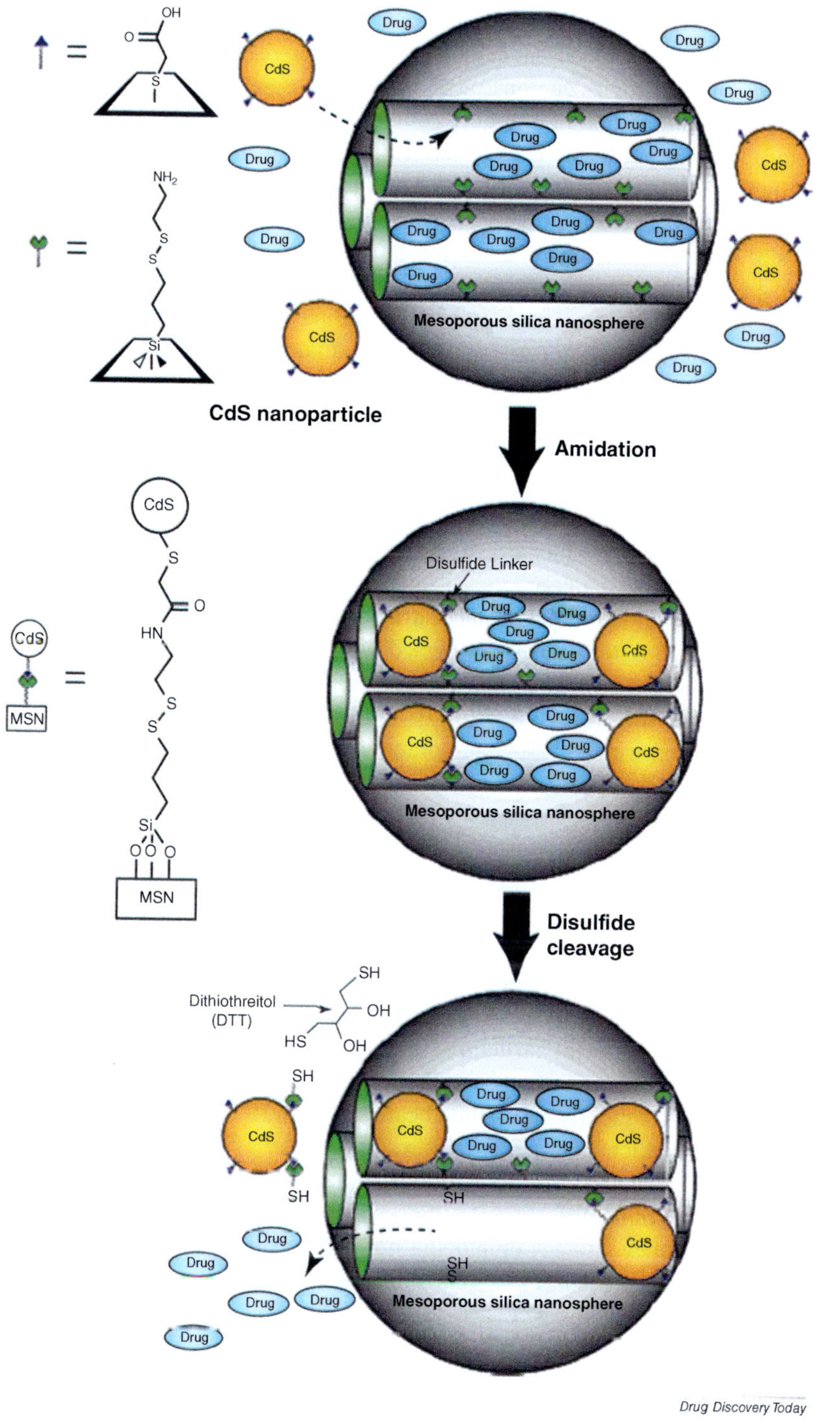

Figure 6.5 A drug delivery system based on mesoporous silica nanoparticles (MSN) capped with CdS nanoparticles is described. The CdS nanoparticles (2 nm) are linked to the MSN (200 nm) via a disulfide bond, which can be chemically reduced to release the drug from the pores of the MSN (2.3 nm). Source: Reproduced with permission from Ref. [39] Copyright (2003) American Chemical Society.

agents, and a wide range of potential surface functionalities. Ow et al. [47] reported fluorophore nanoparticles coated with silica were synthesized, exhibiting high fluorescence and sizes ranging between 20 and 30 nm. These nanoparticles demonstrated remarkable photostability and were 20 times brighter compared to their individual fluorophore components. The silica shell outer layer offers an extra avenue for targeting particular cells or tissues through surface functionalization. Bakalova et al. [48] highlighted the intracellular positioning of silica-coated quantum dot micelles functionalized with amino groups. These nanoparticles showed effective intracellular transport owing to their compact size and positively charged surface. Moreover, their minimal cytotoxicity allows for their versatile application as contrast agents and carriers for drugs, genes, or proteins. Iron oxide-doped silica nanoparticles are employed for cell labeling in MRI applications [49]. Silica-coated dye nanoparticles have undergone grafting to create dual imaging probes capable of fluorescence and MRI functions. The utilization of multicomponent nanostructures in imaging techniques showcases the potential of multifunctional particles. The technique of a core and multilayer shell was applied to synthesize gold-coated and/or silica-coated iron oxide NPs [50] as bifunctional NPs suitable for both MRI and photothermal therapy. Lee et al. [51] carried out the work where mesoporous silica-coated dye nanoparticles were combined with iron oxide nanoparticles (Fe_3O_4-MSN). Here, iron oxide served as a contrast agent for MRI, the mesoporous silica shell facilitated drug delivery, and the dye nanoparticles enabled fluorescence imaging. These trifunctional nanoparticles were loaded with the anticancer drug doxorubicin (DOX) and evaluated in two scenarios: (i) *in vitro*, where B16-F10 cells confirmed NP uptake through both MRI and fluorescence imaging and (ii) *in vivo*, employing mice to assess the ability of Fe_3O_4-MSN to accumulate at the tumor site and deliver the drug. As a result, Fe_3O_4-MSN nanoparticles were found to accumulate at the tumor site, as confirmed by MRI and orange RITC fluorescence. Furthermore, antitumor activity was observed at the tumor site, validating the drug delivery capability of the investigated nanoparticles.

6.1.4 As Bioadhesives

Tissue bioadhesives are essential for managing hemostasis and wound care. A range of both natural and synthetic materials, including fibrin glue, proteins, and synthetic adhesives, have been created for traditional hemostatic and wound closure applications [52–54]. Among these, polymer-based tissue adhesives offer adjustable chemical structures and controllable material properties, including mechanical strength, degradation rate, and adhesion properties [55–57]. Nevertheless, conventional polymeric adhesives frequently necessitate heating, chemical reactions, and specific pH conditions during synthesis, which may potentially result in toxic effects. Hence, there is a demand for developing functional bioadhesive materials that are biocompatible, nontoxic, sterilizable, cost-effective, and easy to manufacture. To improve bioadhesive effectiveness, Leibler et al. recently proposed an innovative strategy employing nonporous silica nanoparticles (NPs) to bind swollen polymer networks together (Figure 6.6). This method capitalizes on the robust affinity between NPs and the polymeric network chains [58]. Spreading TM-50 silica solution onto two slices of calf liver and pressing them together for about 30 seconds resulted in immediate tissue adhesion. The robust bonding at the tissue interface is credited to the optimal match between the size of NPs and the mesh size of the polymeric network on the tissue surface, with NPs acting as "connectors" between polymer chains. When subjected to external stress, the polymer chains bound to NPs can rearrange, allowing for substantial deformation and energy dissipation, thereby forming a sturdy bond without compromising the assembly's stiffness. Moreover, the same research group introduced a nanobridging technique for wound closure using silica NPs. After applying a silica NP solution to the surface of the wounded tissue and gently pressing the wound

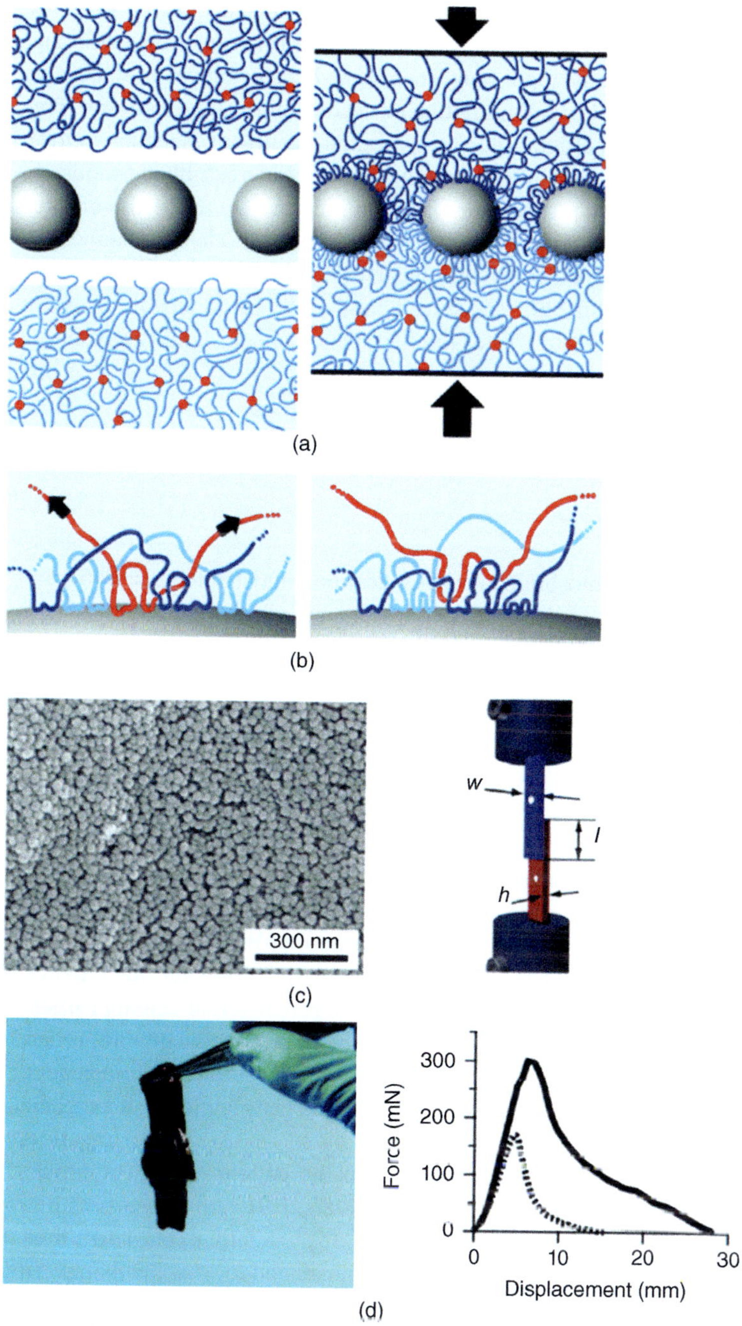

Figure 6.6 (a) The concept of bonding swollen polymer networks together using silica nanoparticles (NPs). (b) The irreversible adsorption onto the surface of the NPs. (c) Scanning electron microscopy (SEM) characterization of TM-50 silica NPs adhered to an S0.1 gel surface after multiple soaking and washing cycles in water (left); lap-joint geometry. Video extensometers marked by white dots were utilized to measure displacement (right). (d) Two slices of calf liver adhered together by applying TM-50 silica NP solution to the interface and pressing (left); corresponding normalized force–displacement curves of the liver slice joint (right). Source: Reproduced with permission from Rose et al. [58]/Springer Nature.

edges together, numerous connectors formed by silica NPs effectively linked the wound edges. This approach demonstrated effective healing of skin injuries on both thin abdominal skin and thick dorsal skin of Wistar rats. Additionally, rapid hemostasis post-hepatectomy was achievable with silica NPs, without inducing pathological inflammation or necrosis at the treatment site. This straightforward and swift nanobridging method holds the potential for urgent bleeding control, tissue repair, and other clinical applications. While silica nanoparticles have exhibited promise as tissue bioadhesive materials, current techniques primarily rely on the inherent properties of silica nanoparticles. Conversely, several strategies exist to enhance silica nanoparticle-based adhesives by utilizing biocompatible polymers such as polyethylene glycol (PEG) or other synthetic polymers for silica nanoparticle functionalization. Furthermore, silica nanoparticle-based bioadhesives can be loaded with prodrugs or biological factors to bolster the therapeutic effect, significantly aiding tissue regeneration during wound healing [59, 60].

6.2 Biomedical Applications of Polymeric Nanoparticles

6.2.1 In Antibacterial Activity

The integration of antibacterial agents into biomedical applications and related scientific and technological domains has become essential. Annually, a considerable number of fatalities result from infections caused by pathogens [61]. The pharmaceutical industry is particularly concerned about bacteria among various types of pathogens, which also include viruses, fungi, algae, and others. This concern primarily stems from the emergence of multidrug-resistant bacteria (MDR). According to the World Health Organization, antibiotic resistance is recognized as one of the greatest threats to global health and food security, affecting individuals of all ages and from any part of the world. Common bacteria of concern include Acinetobacter, Pseudomonas, and certain enterobacteria such as *Klebsiella*, *Escherichia coli*, *Serratia*, and *Proteus*. These bacteria can cause severe infections such as bloodstream infections, pneumonia, and, in some cases, fatalities. Infections associated with implants and medical devices are commonly known as device-related infections (DRI) [62]. The growth and proliferation of bacteria on the surfaces of biomedical devices and implants are considered the primary causes of DRI [63]. Recent investigations into the effects of bacterial colonization leading to infections on medical devices have revealed that urinary catheter infections and central venous infections are the most widespread, with orthopedic implant-related infections following closely. Among these, catheter-associated urinary tract infection stands out as particularly prevalent in hospital environments [64]. Bacteria or pathogens adhere to the catheter surface, forming biofilms, which heighten the chances of secondary bloodstream infections. Microbes or bacteria can infiltrate via the catheter hub or from the patient's skin, resulting in contamination and eventual biofilm development. Besides catheters, DRI can impact prosthetic cardiac valves, pacemakers, vascular grafts, orthopedic implants or prosthetic joints, and contact lenses, among other medical devices [65]. Post-operative infections related to orthopedic implants are also frequently observed [66]. Methicillin-resistant *Staphylococcus aureus* and *Escherichia coli* are the predominant pathogens in DRI [67]. A prevalent approach to counteracting these bacteria involves employing controlled-release drugs or antibiotics. This method presents various benefits, such as impeding biofilm formation and lowering infection risks, ultimately diminishing patient mortality rates. Nevertheless, the utilization of broad-spectrum antibiotics could potentially foster the emergence of MDR [68] and systemic toxicity [69]. Figure 6.7 depicts a schematic diagram of microbial contamination on a catheter. Microorganisms or bacteria can infiltrate through the catheter hub or originate from the patient's skin (**top portion of** Figure 6.7), resulting in contamination and consequent biofilm development. Once bacteria firmly adhere, microcolony formation

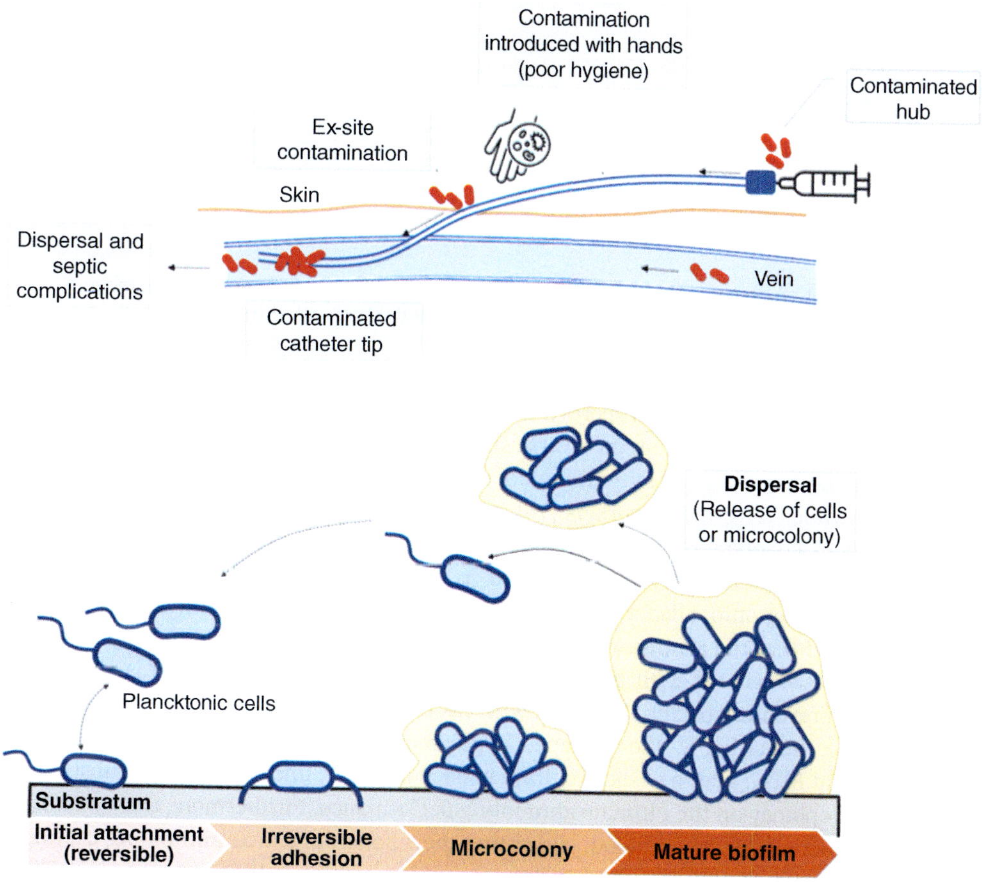

Figure 6.7 (Top) Illustration depicting the process of biofilm formation in various sections of a venous catheter. Microbial contamination may originate from the catheter hub, the patient's skin due to inadequate skin antisepsis, or from the bloodstream. (Bottom): Progression of biofilm formation and the dispersion of bacteria or microcolonies. Source: Olmos et al. [70]/MDPI/CC BY 4.0.

and mature biofilm establishment ensue (**bottom portion of** Figure 6.7). Multiple approaches have been utilized to tackle DRI. Alongside antibiotic therapy, a prevalent tactic involves employing antifouling or antimicrobial coatings. According to Francolini and colleagues [71], antifouling coatings can be crafted from PEG, enzyme based coatings, zwitterionic coatings, and super hydrophobic coatings. Surface functionalization with these coatings can efficiently deter bacterial attachment and biofilm formation, thereby mitigating the likelihood of DRI. PEG or oligo(ethylene glycol) is recognized for its ability to resist protein adhesion and other biological agents [72]. Steric hindrance, resulting from the physically adsorbed or chemically covalently bonded PEG chains, typically reduces protein adsorption, thereby limiting the attachment of proteins or other biomolecules [72]. In a recent study, Hendrick, Yang, and their colleagues [73] devised a coating method to create antibacterial and antifouling surfaces on catheters in a single step. This technique involved the formation of brush-like structures comprising polycarbonates with pendant adhesive dopamine, PEG, and antibacterial cations, which exhibited prolonged efficacy against *S. aureus* and *E. coli*. However, the development of antifouling materials is not solely reliant on chemical modifications; surface micro- or nanostructures and architecture also play pivotal roles. These aspects are illustrated in Figure 6.8. Several studies have explored the impact of surface architecture

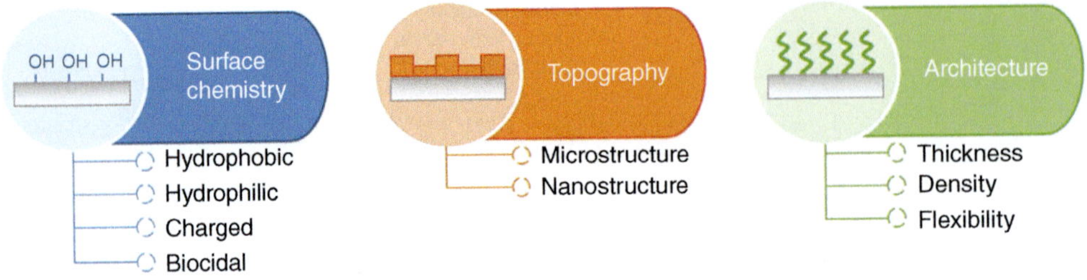

Figure 6.8 The illustration presents three methods for imparting antifouling properties to a surface: (1) altering surface chemistry, (2) modifying surface topography, and (3) adjusting the coating architecture. Source: Maan et al. [74]/John Wiley & Sons/CC BY 4.0.

on the antifouling and antibacterial properties of triblock copolymers. Zhou et al. [75] reviewed various strategies for brush fabrication on diverse surfaces, achieved through physical adsorption or chemical coupling reactions, or by grafting copolymers using surface-initiated radical polymerization. Meanwhile, Yang and coworkers [76] explored the antibacterial properties relative to the structure of block copolymers. These copolymers consisted of three segments: an antifouling PEG, an antibacterial cationic polycarbonate, and a maleimide-functionalized polycarbonate, engineered for anchoring onto silicone rubber surfaces like those found in catheter tubing. The investigation revealed that the arrangement of the blocks affected the antifouling characteristics of the coatings. Surfaces coated with the tethering block, referred to as V-shaped copolymers, were ineffective in preventing bacterial fouling, whereas those coated with S-shaped copolymers, showed no fouling. Remarkably, the S-shaped copolymers demonstrated effective resistance against *S. aureus* and *E. coli* after a week of culture in each instance, underscoring the substantial influence of block topology on the ultimate antifouling performance. Furthermore, these coatings exhibited no detrimental effects on blood, suggesting their potential use as antifouling coatings for intravenous catheters. Another emerging domain in biomedicine is tissue engineering (TE), spurred by the growing demand for organs and the pursuit of tissue and organ reconstruction. As defined by Langer and Vacanti [77], TE is an interdisciplinary domain that merges principles from engineering and life sciences to develop biological substitutes with the aim of restoring, maintaining, or enhancing the function of biological tissues or entire organs. Unlike traditional organ transplants, TE seeks to seamlessly integrate laboratory-fabricated tissues into the body's architecture, aiding in the recovery of damaged organs or tissues. Within the field of Materials Science and Engineering, the development of functional materials suitable for use as scaffolds in TE is of paramount importance. Infections associated with tissues encompass a wide array of chronic conditions, including otitis, sinusitis, endocarditis, lung and biliary tract infections, urinary tract infections, osteomyelitis, and chronic wounds. A diverse range of polymers, both natural and synthetic, or their combinations, can be employed in TE endeavors [78]. In addition to inherent biocompatibility, antibacterial or antimicrobial properties are desirable characteristics in scaffolds to prevent biofilm formation during tissue growth. While polymers generally lack intrinsic antibacterial properties, exceptions like CS, which exhibits antibacterial properties, exist [79]. To confer antibacterial properties onto polymer materials, antimicrobial or antibacterial additives are introduced into the polymer matrix using various methods. Among the commonly employed systems, those based on biodegradable and biocompatible polymers filled with antibacterial particles are highly favored. For example, polyhydroxyalkanoates such as polylactic acid (PLA), polyglycolic acid (PGA), and polyhydroxybutyrate (PHB) are frequently utilized in such systems. TE applications often require multifunctional materials [80]. For instance, a polymeric periodontal membrane with antibacterial and osteogenic properties was successfully developed [81].

The membrane consisted of films made from poly(caprolactone) (PCL), which were osteoconductive, antibacterial, and flexible, incorporating zinc oxide (ZnO) nanoparticles and prepared through electrospinning. Porous films with precisely controlled mechanical characteristics were produced to address periodontitis through guided tissue regeneration (GTR). In a rat model defect, the PCL/ZnO membrane is positioned between the gum flap and the alveolar. In a study by Wang and coworkers [82], a novel polymer-based hydrogel approach for GTR aimed at treating periodontitis, a bacterial infection leading to tissue loss supporting teeth, was introduced. Their method involved incorporating cuprous oxide (Cu_2O) and polydopamine-coated titanium dioxide nanoparticles into a hydrogel to facilitate the regeneration of new tissue, as illustrated in Figure 6.9. Thermosensitive hydrogels, incorporating CS modified with beta-glycerophosphate (β-GP), were formulated for bone defect treatment [83]. These materials are garnering increased interest in the

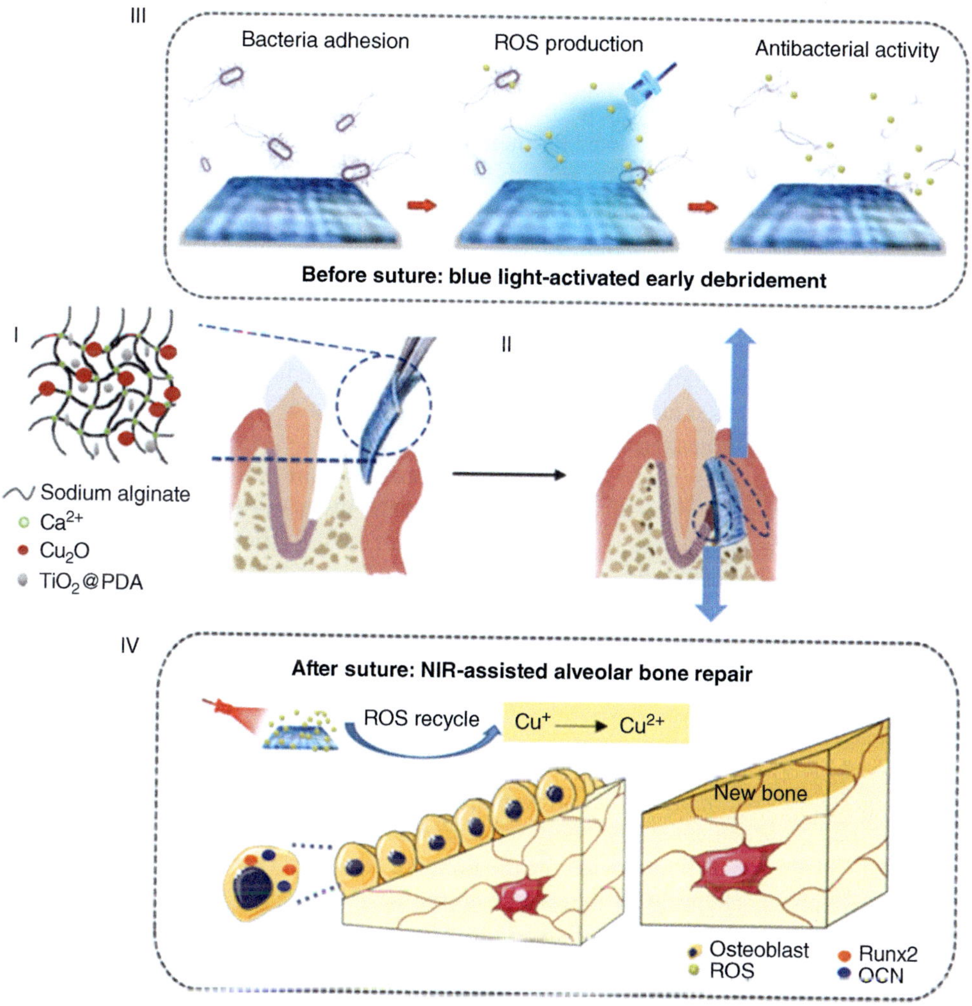

Figure 6.9 This figure illustrates injectable hydrogel composite materials inspired by jelly for guided tissue regeneration strategies. Prior to suturing, when exposed to blue light, the sodium alginate hydrogel composite (CTP-SA) was activated to generate ROS, effectively eliminating bacteria and facilitating early debridement. Following suturing under near-infrared (NIR) irradiation, the photothermal properties of CTP-SA, coupled with ROS recycling, promote osteogenesis. Source: Reproduced with permission from Xu et al. [82]/American Chemical Society.

field of bone tissue engineering (BTE) due to their capacity to regenerate bone tissue effectively without necessitating surgical intervention, thus reducing treatment invasiveness [83]. Another intriguing material type is the system developed by Tamayol et al. [84], comprising poly(glycerol sebacate) (PGS) and poly(ε-caprolactone) (PCL) filled with calcium peroxide, enabling oxygen release. Oxygen deprivation can impede tissue growth by inducing a hypoxic environment for cells. Scaffold development with oxygen-releasing capabilities enhances cellular metabolic activity, facilitating new tissue growth. Additionally, silk fibroin scaffolds have shown promise in BTE [85], given silk fibroin's natural protein composition with favorable mechanical properties and tunable degradation rates. BTE has received significant research attention due to the scarcity of suitable autografts or allografts. Nevertheless, extensive investigations in TE encompass a wide array of tissues, including skin, heart, and kidney. Skin TE, motivated by the need to address wounds resulting from burns, is another area of focus. Skin TE studies prominently feature natural polymers such as CS, hyaluronic acid (HA), collagen, or gelatin, as well as synthetic polymers like PEG or polylactic-co-glycolic acid (PLGA) [86]. CS materials loaded with selenium nanoparticles have demonstrated efficacy against infections in burn wounds [87]. Moreover, scaffolds composed of silk fibroin-collagen loaded with titania nanoparticles have shown promise in skin tissue regeneration [88]. Polyurethanes, renowned for their diverse mechanical properties ranging from flexibility to rigidity, find wide application in both industrial and biomedical fields [89]. Hybrid polyurethane-polylactide porous scaffolds, modified with ciprofloxacin, have been developed as antibacterial scaffolds for skin regeneration applications [90]. With the emergence of 3D printing or additive manufacturing, TE has witnessed a surge in the fabrication of various scaffold types. For instance, Li and colleagues engineered a 3D printed structure using PLA filled with nanohydroxyapatite or modified with polydopamine for BTE applications [89], although numerous other instances can be found in the literature. In addition to their role as scaffolds, 3D printed materials are also being utilized for external prostheses or in the production of medical devices such as surgical instruments with antibacterial or antimicrobial properties at a reduced cost [91]. In a succinct report, J.M. Zuniga presented a practical example involving the fabrication of a prosthetic material using a commercial antibacterial PLA filament, PLACTIVE™. The antibacterial attributes were achieved through the incorporation of small quantities of copper particles (1–3%). Figure 6.10 showcases some illustrations of 3D printed medical devices and prostheses developed by.

6.2.2 As Vaccine Adjuvants

Polymer nanoparticles present a promising adjuvant platform for various reasons (Figure 6.11). One advantage of nanoparticle-based vaccines lies in the ability to manipulate their biophysical and biochemical characteristics to enhance antigen uptake, processing, and presentation. Furthermore, the careful selection of polymer chemistry, whether natural or synthetic, can improve antigen stability, regulate release kinetics, and modulate the immune response. This section will delineate the comprehensive physicochemical attributes, encompassing size, shape, charge, and hydrophobicity, of nanoparticle platforms. Additionally, specific instances showcasing the distinct benefits of particle platforms, such as controlled release, co-encapsulation, and the induction of cell-mediated immunity (CMI), will be outlined below.

Adjuvants for vaccine formulations now incorporate hybrid systems comprising diverse material classes, such as natural/synthetic polymers and lipids. One prevalent method involves employing PEG as a surfactant to bolster stability and confer stealth properties upon the systems. For nasal administration aimed at enhancing the mucoadhesive and mucopenetrating properties of vaccines, researchers have modified the surface of chitosan-based nanoemulsions with PEG, as

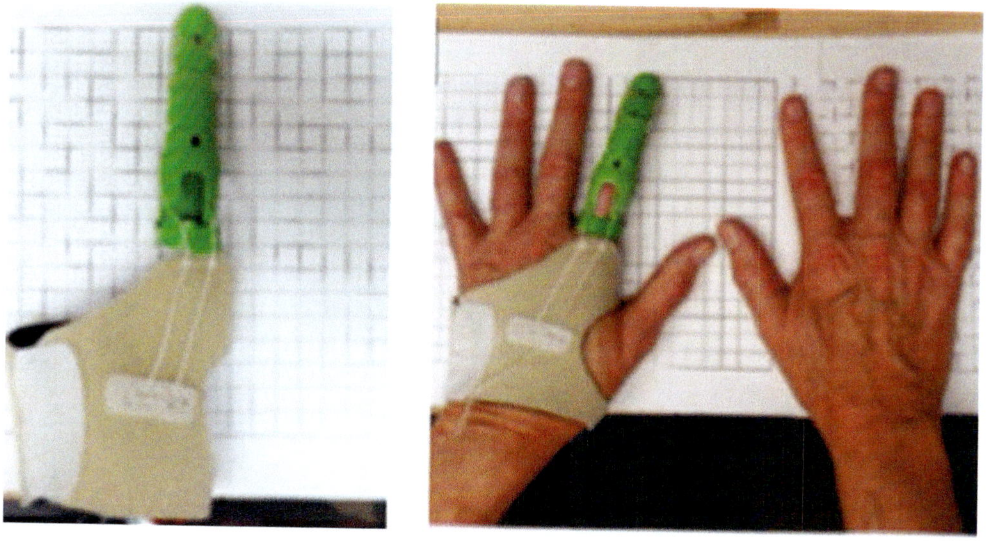

Figure 6.10 Instances of antibacterial prosthetic materials produced through 3D printing utilizing a commercial filament containing PLA infused with copper nanoparticles, specifically PLACTIVE. Source: Zuniga et al. [91]/MDPI/CC BY 3.0.

demonstrated in a study by Di Cola et al. in 2019 [93]. Another common strategy involves synergistically utilizing carbohydrates to coat synthetic polymer surfaces, thereby augmenting cellular uptake. Mannosylated poly(beta-amino esters) (PBAEs), devoid of inflammatory co-adjuvants, exhibited enhanced gene delivery, leading to improved activation of APCs and immune response, as demonstrated by Jones et al. in 2015 [94]. In another instance, alginate was utilized to create cationic nanogels cross-linked with polyethylenimine (PEI) for the delivery of OVA. It was observed that the OVA-containing nanogels significantly enhanced the generation of tumor-specific CD^{8+} T cells and the production of antigen-specific antibodies, as reported by Li et al. in 2013 [95]. The cationic properties of CS and its derivatives have been harnessed to coat synthetic polymer particles, thereby increasing antigen loading efficiency and/or adjuvanticity. For instance, Wusiman et al. in 2019 [96] demonstrated that PLGA nanoparticles loaded with Hepatitis B surface antigen (HBsAg) and coated with trimethyl CS (TMC) exhibited efficient absorption and increased antibody titers when administered nasally. Similarly, chitosan-coated PLA particles enhanced antigen adsorption capacity and macrophage uptake, resulting in the upregulation of MHC I and MHC II and the secretion of pro-inflammatory cytokines, as shown by Chen et al. in 2014 [97]. In comparison to a commercially available vaccine, Another study showed that applying a coating of CS onto poly-ε-caprolactone (PCL) led to increased IgG responses against HBsAg, while preventing the induction of IgE. Furthermore, the nanoparticle formulation stimulated CMI against HBsAg, resulting in the secretion of IL-17 and IFN-γ. Similarly, peptides containing B and T cell epitopes were conjugated with PGA and combined with TMC to create self-adjuvanting nanovaccines, as demonstrated by Nevagi et al. in 2019 [98]. Hybrid polymer systems can also integrate different polymer systems to elicit synergistic immune responses against a wide range of pathogens and delivery routes. In a recent study, Bartlett et al. in 2020 [99] achieved an increased humoral immune response upon oral delivery by developing a hybrid delivery system wherein self-assembled lipopeptide nanoparticles were coated with layers of alginate and TMC via electrostatic interactions. Similarly, a hybrid nanoparticle system comprising cationic liposomes and natural HA exhibited enhanced stability and antigen release characteristics, leading to a significantly improved serum IgG response to *Yersinia pestis* F1—V upon intranasal administration, as reported by Fan

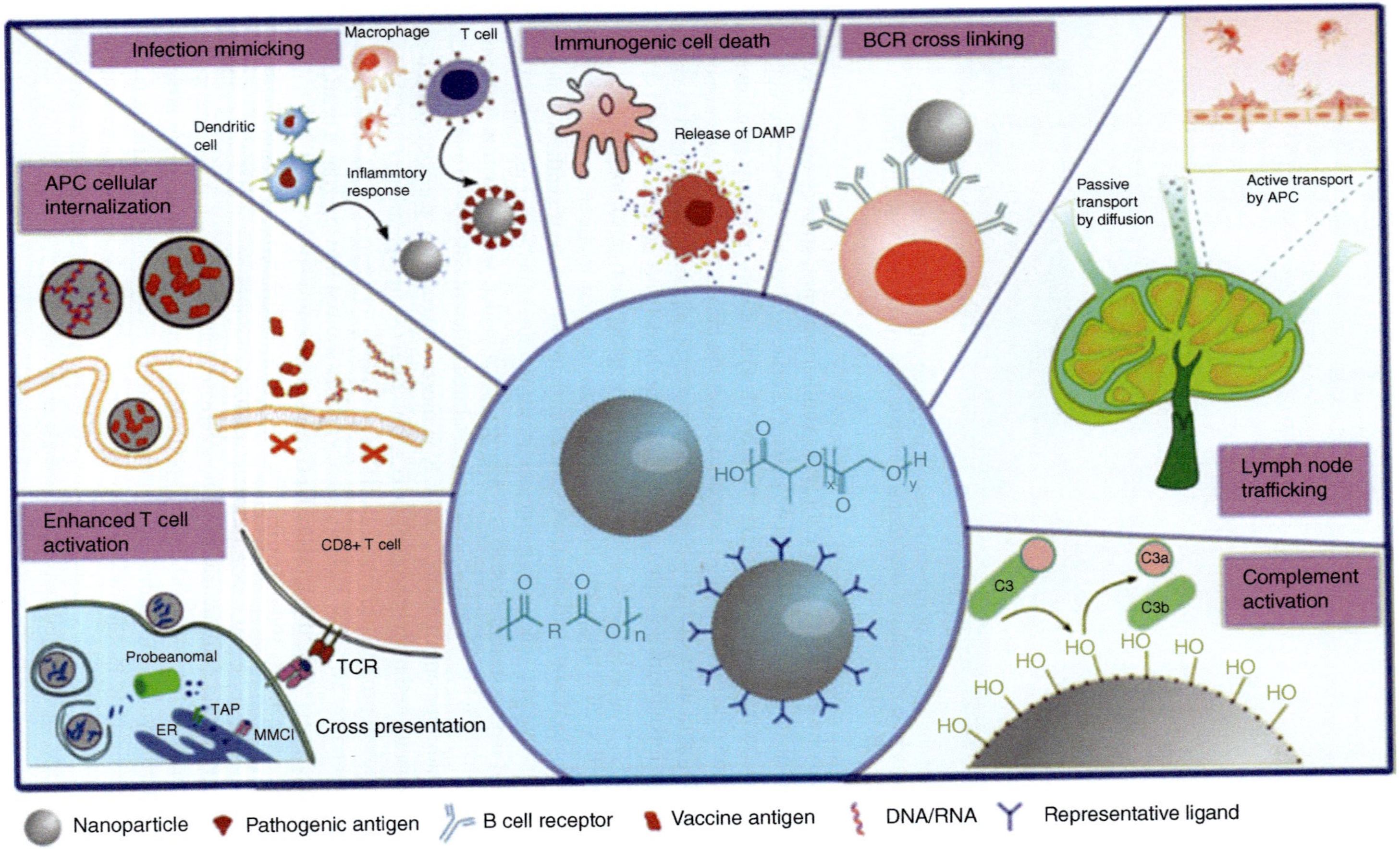

Figure 6.11 A schematic depicting the advantages provided by polymeric nanoparticle adjuvant systems in vaccination. Progressing clockwise from left to right: heightened T cell activation facilitated by nanoparticles, illustrating antigen cross-presentation; antigen-presenting cell (APC) internalization of vaccine payload conveyed by nanoparticles compared to free diffusion; induction of immune recognition and pro-inflammatory responses via nanoparticle-facilitated infection mimicking; immune activation initiated by immunogenic cell death; cross-linking of B cell receptors (BCRs) facilitated by nanoparticles; nanoparticle migration to lymph nodes; and complement activation prompted by nanoadjuvants. Source: Reproduced with permission from Grego et al. [92]/Springer Nature.

et al. in 2015 [100]. In a separate investigation, an adjuvant based on a liposome-PEG-PEI complex exhibited improved uptake, surface marker expression, pro-inflammatory cytokine induction, and antigen presentation, as demonstrated by Chen et al. in 2012 [101]. Moreover, intranasal immunization utilizing lipid-polymer hybrid nanoparticles coated with CS augmented mucosal immune responses by eliciting both humoral and CMI, as evidenced by Rose et al. in 2018 [102]. Moreover, liposomes engineered with pH-sensitive polymers could function as delivery platforms to trigger the generation of antigen-specific CD^{8+} T cells. For instance, incorporating cationic lipids into polymer-modified liposomes enhanced the expression of costimulatory molecules and the release of IL-12 and TNF-α from dendritic cells (DCs), along with antigen-specific secretion of IFN-γ from splenocytes, as demonstrated by Yoshizaki et al. in 2017 [103].

6.2.3 As Delivery Carries for Vaccines

In recent times, researchers have shifted their attention toward developing thermostable vaccines using particle adjuvants and delivery systems. While liquid formulations have been traditionally favored for their ease of manufacturing and packaging, there is growing interest in dry formulations achieved through spray drying to improve vaccine stability. For instance, Saboo et al. demonstrated that a spray-dried virus-like particle vaccine remained stable and immunogenic even after storage at 37 °C for up to 14 months [104]. Similarly, the adsorption or encapsulation of vaccines within polymer matrices has shown promise in enhancing vaccine stability. A recent study found that a vaccine for the Ebola virus coated onto PLGA particles maintained stability for six weeks at 37 °C [105]. Moreover, PLGA particles have been shown to preserve the stability of encapsulated inactivated poliovirus [106]. Additionally, polyanhydride nanoadjuvants have exhibited the capability to maintain the stability of various subunit vaccine antigens [107]. In contrast to alum, Petersen et al. [108] demonstrated that encapsulation into polyanhydride nanoparticles preserved PA from *Bacillus anthracis* over a wide temperature range (−20 to 40 °C) for up to four months nanoadjuvants are a promising platform to enhance the thermostability of vaccines and eliminate the vaccine cold chain. Polymeric nanovaccines offer a compelling alternative to existing influenza vaccines, serving as both adjuvants and delivery systems for various payloads, including influenza virus antigens such as hemagglutinin, nucleoprotein, neuraminidase, and additional co-adjuvants. These nanovaccines enable the delivery of multiple vaccine formulations tailored to specific age groups or individual patient needs. They benefit from efficient and large-scale production, providing a thermostable product that is cost-effective to manufacture, stockpile, and distribute. Polymers such as polysaccharides, polyesters, and polyanhydrides have been utilized to formulate nanoparticles for delivering influenza virus antigens [109]. These versatile nanovaccines have been successfully administered intranasally, intramuscularly, and subcutaneously. Intranasal immunizations with nanovaccines for respiratory pathogens like the influenza virus have been shown to induce local tissue-resident T cell immunity, providing protection against heterologous challenges. This represents a significant advantage over current influenza vaccines [110], paving the way for the development of a "universal flu vaccine." Intranasal vaccinations mimic both the natural site and route of infection, harnessing the natural immune response to safeguard the body against future infections [107].

6.3 Toxicity Studies

Silica (SiO_2) can be manufactured at a low cost, is generally considered relatively safe, and finds application in various fields. Its utilization in biomedical contexts is particularly noteworthy.

The cytotoxicity of amorphous, nonporous silica nanoparticles has been studied, and a concise overview of notable findings is provided here for comparison with porous silica nanomaterials, zeolites, and mesoporous silica. Hoet and coworkers examined the cytotoxicity of amorphous silica nanoparticles (ranging in size from 14 to 335 nm in diameter) in endothelial cells (EAHY926) [111]. Their amorphous silica samples existed as stable monodisperse dispersions resistant to aggregation. Particle size distributions were determined via dynamic light scattering. They found that smaller silica nanoparticles exhibited greater toxicity when particle morphologies were comparable and doses were expressed in mass concentration. Specifically, silica nanoparticles with diameters of 14 and 16 nm demonstrated ED_{50} values of 33 and 47 mg/cm^2, respectively, in contrast to particles sized 104 and 335 nm, which showed ED_{50} values of 1095 and 1087 mg/cm^2, respectively. Surface area also played a significant role in toxicity, with toxicity being similar for silica particles of different sizes when compared in terms of surface area (cm^2/cm^2). The mechanism of cell death was investigated, revealing that necrosis was observed within a few hours for smaller particles. In a subsequent investigation, Martens and colleagues explored the cytotoxicity of amorphous silica nanoparticles (ranging in size from 2 to 335 nm in diameter) in endothelial cells (EAHY926) and mouse macrophages (J774) [112]. The correlation between particle size and ED_{50} depicted in Figure 6.12 revealed an increase in cytotoxicity with decreasing particle size. Additionally, the silica's capacity to produce hydroxyl radicals from hydrogen peroxide was investigated using electron paramagnetic resonance (EPR) spectroscopy with the spin trap, 5,5-dimethyl-1-pyrroline-N-oxide (DMPO), to detect the hydroxyl adduct, 2,2-dimethyl-5-hydroxy-1-pyrrolidinyloxyl (DMPO-OH). On a per-mass basis, the activity of radical generation decreased as particle size increased, suggesting that radical formation, being a surface property, correlates with cytotoxicity. This study marked the first observation of radical formation on amorphous silica nanoparticles in an aqueous solution [112]. Ma and colleagues assessed the cytotoxicity of SiO_2 nanoparticles (15 and 46 nm) in human bronchoalveolar carcinoma-derived cells, utilizing crystalline silica as a positive control [113].

In vitro toxicological studies reveal that silica-amorphous silica nanoparticles (SASNs) provoke harmful effects across different cell types, inducing pro-inflammatory responses, DNA damage, autophagy, cell-cycle arrest, and cell death. Nevertheless, understanding the structure–activity relationships dictating SASN toxicity remains limited. To clarify these relationships, it is imperative to explore the interaction between SASNs and the cell surface. Hemolytic potential serves as a common metric for evaluating nanoparticle toxicity. Research indicates that hemolysis is contingent upon the concentration of silanols [114]. At physiological pH, individual silanols establish hydrogen bonds with phosphate ester groups of membrane phospholipids and secondary amide groups of proteins. Anionic deprotonated geminal and vicinal silanols interact electrostatically with positively charged tetraalkylammonium-containing phospholipids. These noncovalent interactions disrupt the cellular membrane and contribute to hemolysis. Colloidal particles that are fully hydroxylated (such as crystalline quartz), predominantly featuring hydrogen-bonded silanols, exhibit low hemolytic activity. The porosity of mesoporous silica nanoparticles (MSNs) reduces the nanoparticle area in contact with the membrane, thus reducing the effective silanol concentration and decreasing hemolysis compared to solid, colloidal (Stöber) SASNs. Silsesquioxane-based SASNs (e.g. $SiO_{1.5}$-C_2H_4-$SiO_{1.5}$) further mitigate this effect. Plasma membranes are susceptible to oxidative damage, leading to membrane lipid peroxidation and lysis. Three-membered rings act as reservoirs for generating ROS, while other surface radicals may coexist on the amorphous silica surface, leading to radical-transfer reactions in the presence of water and hydrogen peroxide, resulting in hydroxyl radical generation. The ranking of hemolytic activity follows fumed silica (2–3 OH per nm^2, highly EPR active) $\approx$ Min-U-Sil (crystalline silica nanoparticles used as a positive control) $\gg$ Stöber silica (5 OH per nm^2, weakly EPR active) $>$ MSN (4 OH per nm^2, weakly EPR active) $>$ silsesquioxane-based SASNs ($\sim$1–2 OH per nm^2, weakly EPR active). As the surface

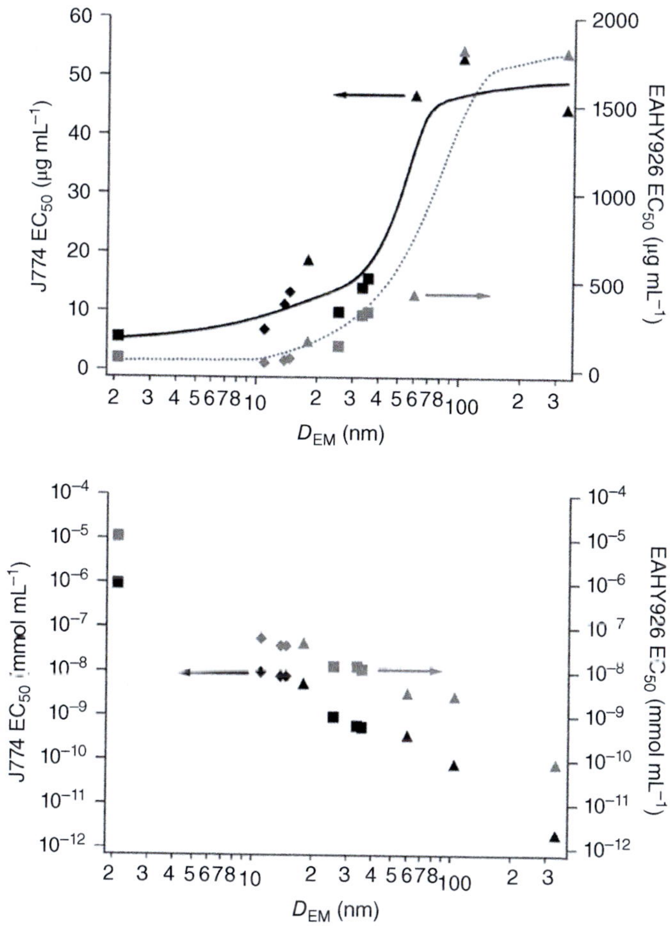

Figure 6.12 The cytotoxicity, represented by the half-maximum effective dose (ED50 value), is shown in both g/mL (upper panel) and mmol/mL (lower panel) of silica nanoparticles (SNP) in human endothelial (EAHY926, gray) and mouse monocyte macrophage (J774, black) cell lines, as a function of the silica nanoparticle diameter. Various symbols are employed to differentiate between different SNP synthesis procedures: squares denote Lys-Sil, triangles represent Stober, and diamonds indicate Ludox. Source: Reproduced with permission from Thomassen et al. [112]/American Chemical Society.

area-to-volume (mass) ratio increases with decreasing size, the hemolysis potential of MSNs also increases with decreasing size. Heat treatments, which reduce the overall silanol concentration, or modification of SASNs with carboxylic acid, amines, PEG, or lipids that screen silanols, largely eliminate the hemolytic potential.

Upon nanoparticle exposure, most cells internalize nanoparticles via energy-dependent internalization pathways. Specifically, SASNs are enclosed in membrane-bound compartments (e.g. endosomes, phagosomes, or macropinosomes) that are transported within the cell via microtubules to late endosomes, lysosomes, or multivesicular bodies and ultimately excreted by exocytotic pathways. The same noncovalent interactions that underlie hemolysis, along with receptor binding, stimulate and direct these internalization pathways. Additionally, in biological fluids, protein corona formation mediates attractive interactions, reduces the internalization rate, and alters the trafficking pathway. Although systematic studies on the effects of size, shape, and charge on internalization efficiency and intracellular trafficking are limited, a comparison of positively and negatively charged MSN uptake in macrophages and epithelial cells revealed a greater dose-dependent uptake of larger, positively charged nanoparticles compared to smaller or negatively charged nanoparticles. In a separate study, rod-shaped cationic MSNs induced shape-dependent microscopic absorption, resulting in 40 times more internalization than spherical MSNs. Fumed silica aggregates perturb the cell membrane but predominantly remain associated with it in a process akin to frustrated phagocytosis, similar to interactions with crystalline asbestos [115]. Targeting modifies internalization pathways; for instance, motor neurons internalize cholera-toxin-B-modified MSNs through a combination of macropinocytosis and lipid-raft-mediated endocytosis. While endocytosed SASNs are generally observed within membrane-bound vesicles, free nanoparticles have been detected, attributed to passive diffusion through damaged cellular membranes or escape from membrane-bound vesicles. Transmission electron microscopy investigations reveal that SASN internalization by multiple routes leads to intracellular damage, including damaged vacuoles of mitochondria, swelling and disappearance of mitochondrial cristae, degranulation of the endoplasmic reticulum, damage of lysosomal vesicles, and formation of autophagosomes. Finally, SASNs can undergo partial or complete degradation in the cell or can be expelled from cells within membrane-bound compartments through various exocytotic processes, including lysosome secretion, vesicle-related secretion, and tunneling nanotubes. Membrane perturbation and intracellular damage from SASN association and internalization lead to several dose-dependent toxicity pathways. For instance, the NLRP3-inflammasome-activation pathway, implicated in diseases like asbestosis and silicosis, is initiated by exposure to fumed silica. ROS generation by fumed silica serves as Signal 2, alongside its membrane association, activating the inflammasome. In contrast to asbestos and mineral-derived quartz, which often contain Fe impurities that generate ROS through Fenton-like reactivity, fumed silica's ROS generation stems from its strained 3-membered rings. Stöber colloidal silica, with a greater population of hydrogen-bonded silanols and no strained 3-membered rings, exhibits limited potential for NLRP3-inflammasome activation. Other identified toxicity pathways for SASNs include autophagy, lysosomal damage leading to apoptosis, mitochondrial dysfunction, and pro-inflammatory responses, all implicated in ROS and/or membrane damage. Silica toxicity arises from its ability to generate ROS or damage membranes, facilitated by attractive hydrogen bonding and electrostatic interactions between silanol species and membrane-bound cellular constituents. The physicochemical properties of the silica surface, particularly the abundance, type, and extent of hydrogen bonding of surface silanols, dictate its hemolytic potential and cytotoxicity, influenced by processing pathways and environmental exposure. Processing conditions resulting in fully hydrogen-bonded silanols or low overall silanol concentrations reduce membrane interactions and increase biocompatibility. High-temperature processing leads to strained rings as sources of ROS generation, enhancing toxicity. Surface modification, such as shielding silanols with carboxylic acids, amines, or PEG, or inclusion of R groups or metal dopants, decreases toxicity. For example, the addition of silsesquioxanes-PEG onto MSNs suppresses hemolytic activity, and hollow phenylene-bridged silsesquioxane nanoparticles exhibit no hemolysis compared with hollow MSNs. Additionally, 6% TiO_2 doping of fumed silica arrests NLRP3-inflammasome activation and pro-inflammatory IL-1β generation. The apparent contradictory reports on silica toxicity are largely attributable to the considerable variability in physicochemical properties of the silica surface. Unfortunately, few studies conduct the necessary characterization steps to establish structure–activity relationships and unravel the causes of toxicity. Furthermore, studies inconsistently use bare nanoparticles or pretreated nanoparticles with proteins or conduct *in vitro* studies with or without serum, failing to account for opsonization, which mediates cellular interactions and can affect the mechanisms of nanoparticle uptake by cells. While it is expected that reduced silanol concentration and electrostatic interactions with cell membranes should decrease the toxicity of MSNs, *in vivo* toxicity studies have yielded variable results, with mounting evidence of *in vitro* toxicity. For instance, one study of 120 nm-diameter MSNs found the maximum tolerated dose (MTD) to be 30 mg/kg when injected intravenously. Conversely, other studies of similar-sized MSNs reported no major effect at doses of 100–200 mg/kg and an MTD of 500 mg/kg. The variability in toxicity results is partly attributed to toxic surfactant residual amounts, which depend on processing. Using FTIR to confirm surfactant removal prior to testing MSNs resulted in MTDs of 1000 mg/kg by intravenous (IV) injection, highlighting the nontoxic nature of the MSN silica framework itself. Size also plays a crucial role in nanomaterial toxicity, as a decrease in size leads to an increase in surface area per gram. In the case of Stöber nanoparticles, increased toxicity was observed with decreasing size. For instance, while 300 nm nanoparticles were nontoxic at 200 mg/kg when delivered intravenously, 70 nm nanoparticles were lethal at 45 mg/kg and 30 nm nanoparticles were lethal at 12 mg/kg in mice. Examination of organs of animals injected with 115 nm Stöber nanoparticles at 450 mg/kg showed major damage to the heart, lung, and spleen, associated with mechanical obstruction of small vessels rather than specific cellular interactions with silica. However, at lower than lethal doses, Stöber nanoparticle toxicity, if present, primarily affects the liver and is size-dependent and dose-dependent, demonstrating a safe dose threshold. The route of administration also influences SASN toxicity. For example, while the MTD for oral exposure to Stöber nanoparticles exceeds 2000 mg/kg and that for MSNs exceeds 5000 mg/kg, the MTD is reduced to 600 mg/kg for MSNs after intraperitoneal (IP) injection. Moreover, the type of injection also affects toxicity; the MTD of 100 nm MSNs varies depending on the injection type. Additionally, specific toxicities such as immunotoxicity, genotoxicity, and fetal toxicity are areas of concern. Genotoxicity studies after oral exposure to 20 and 100 nm Stöber nanoparticles showed no genotoxicity in the bone marrow in mice or rats. However, oral exposure may not result in the systemic distribution required for bone marrow exposure. Another study of 20 nm Stöber nanoparticles via IV delivery found no genotoxicity in rat. More research, including chronic exposure studies and assessments of specific toxicities, is necessary to fully understand the potential risks of SASN exposure [116–121].

6.4 Conclusions and Future Scope

Silicon nanoparticles (SiNPs) play an ongoing pivotal role in biomedical nanotechnology, facilitating the design of advanced tools and systems for both *in vitro* and *in vivo* applications. Utilizing the interaction between silica surfaces and proteins enables specific protein detection, while the reverse interaction aids in separating proteins from biological media. Functionalized silica NPs exhibit high specificity in targeting proteins, benefiting from the diverse options for surface functionalization provided by silica. Moreover, SiNPs find extensive utility in gene detection and purification due to their specific affinity for nucleic acids. Employing SiNPs in this context offers numerous advantages, including the reduction of organic solvent usage, expedited processes, and cost-effectiveness. Given the importance of DNA detection and extraction in diagnostics and gene therapy, the utilization of SiNPs holds promise for the development of novel and rapid techniques in these fields, positioning them as revolutionary tools for nucleic acid treatments. Additionally, SiNPs serve as crucial auxiliary materials. Silica shells demonstrate high efficiency in protecting imaging agents, thereby enhancing their efficacy. Furthermore, SiNPs have an application in pharmaceutical production as additives, enhancing the mechanical properties of powders. Silica-based nanoparticles stand out as a prominent category of nanomaterials with distinct advantages in biomedical domains, encompassing drug delivery, bioimaging, and bioadhesives. In comparison to many other inorganic counterparts, silica-based NPs demonstrate reduced toxicity and heightened biocompatibility. Moreover, their degradation byproducts can be effectively eliminated through renal excretion. Nonporous silica NPs primarily serve as carriers for theranostic agents via encapsulation or covalent conjugation, ensuring stability and minimizing nonspecific toxicity to healthy tissues. These nonporous silica NPs offer adjustable size control and versatile functionalization strategies, rendering them highly advantageous for theranostic applications. Utilizing the Stöber method enables precise size adjustment of these NPs down to 20 nm with exceptional monodispersity and morphological control, rendering them suitable for various biomedical applications. Incorporating functional ligands within or onto the NP surface enhances their interaction with biological systems, facilitating disease-specific cellular internalization, stimuli-responsive cargo release, and improving the stability, biocompatibility, and degradation rates of the carrier. In conclusion, it is noteworthy that due to their distinctive attributes, silica-based NPs rank among the most crucial materials for advancing biomedical applications, with the potential to significantly impact future healthcare practices.

Polymeric nanoparticles present significant potential as versatile carriers for diverse therapeutics across biomedical domains. Their advancement and efficacy hinge on the manageability of parameters like size, shape, geometry, charge, structure, and composition. Nonetheless, challenges in their development include the complexity of both the nanoparticles themselves and the available evaluation methodologies, covering physical, chemical, and biological assessments. Concerns also emerge regarding potential toxicity, immunological responses, scalability challenges, batch-to-batch variability, and the need for personalized evaluation. For intravenous administration, nanoparticles of sizes conducive to kidney clearance or capable of dissociating into individual polymer chains with molecular weights below the renal cutoff size are appealing options to mitigate toxicity and immunogenicity concerns. Alternatively, hepatobiliary clearance presents another viable route. Nevertheless, alternative administration and clearance pathways are also plausible. Biodegradable nanoparticles offer a safer option across all administration routes, with biocompatibility being a crucial consideration, necessitating thorough *in vitro* and *in vivo* evaluations. While these constraints are stringent for systemic nanoparticle administration, they are less pronounced for nanoparticles intended for topical applications. Polymer chemistry and supramolecular assembly can be manipulated to tailor polymeric nanoparticles for various applications, including imaging, pH sensing, cancer therapy, and treatment of infections in pulmonary and urinary tracts. Moreover, copolymer chemistry can be regulated to engineer stimuli-responsive and self-communicating nanoparticles with diverse structures and morphologies, such as spherical and cylindrical nanoparticles. Robust formulations, like cross-linked nanoparticles, further bolster their utility and stability. Enhancing the *in vitro* and *in vivo* durability, extending circulation duration, and increasing accumulation at targeted sites are primary objectives. However, it is vital to also address the elimination of these entities and the impediment of cargo release. By controlling polymer chemistry and supramolecular assembly, and leveraging novel targets and cell-recognition components, the stability and effectiveness of polymeric nanoparticles can be significantly bolstered.

References

1 Lin C X C, Qiao S Z, Yu C Z, Ismadji S and Lu G Q M 2009 Periodic mesoporous silica and organosilica with controlled morphologies as carriers for drug release *Microporous Mesoporous Mater.* 117 213–9.

2 Yang Y, Lu Y, Abbaraju P L, Zhang J, Zhang M, Xiang G and Yu C 2017 Multi-shelled dendritic mesoporous organosilica hollow spheres: roles of composition and architecture in cancer immunotherapy *Angew. Chem.* 129 8566–70.

3 Yang Y, Wan J, Niu Y, Gu Z, Zhang J, Yu M and Yu C 2016 Structure-dependent and glutathione-responsive biodegradable dendritic mesoporous organosilica nanoparticles for safe protein delivery *Chem. Mater.* 28 9008–16.

4 Yan W, Chen W and Huang L 2008 Reactive oxygen species play a central role in the activity of cationic liposome based cancer vaccine *J. Control. Release* 130 22–8.

5 Lu Y, Yang Y, Gu Z, Zhang J, Song H, Xiang G and Yu C 2018 Glutathione-depletion mesoporous organosilica nanoparticles as a self-adjuvant and co-delivery platform for enhanced cancer immunotherapy *Biomaterials* 175 82–92.

6 Wang Y, Tang J, Yang Y, Song H, Fu J, Gu Z and Yu C 2020 Functional nanoparticles with a reducible tetrasulfide motif to upregulate mRNA translation and enhance transfection in hard-to-transfect cells *Angew. Chem.* 132 2717–21.

7 Yang Y, Lu Y, Abbaraju P L, Azimi I, Lei C, Tang J, Jambhrunkar M, Fu J, Zhang M, Liu Y, Liu C and Yu C 2018 Stepwise degradable nanocarriers enabled cascade delivery for synergistic cancer therapy *Adv. Funct. Mater.* 28 1800706.

8 Yang Y, Tang J, Abbaraju P L, Jambhrunkar M, Song H, Zhang M, Lei C, Fu J, Gu Z, Liu Y and Yu C 2018 Hybrid nanoreactors: enabling an off-the-shelf strategy for concurrently enhanced chemo-immunotherapy *Angew. Chem. Int. Ed.* 57 11764–9.

9 Lin L, Song J, Song L, Ke K, Liu Y, Zhou Z, Shen Z, Li J, Yang Z, Tang W, Niu G, Yang H and Chen X 2018 Simultaneous fenton-like ion delivery and glutathione depletion by MnO_2 based nanoagent to enhance chemodynamic therapy *Angew. Chem.* 130 4996–5000.

10 Gu Z, Liu T, Tang J, Yang Y, Song H, Tuong Z K, Fu J and Yu C 2019 Mechanism of iron oxide-induced macrophage activation: the impact of composition and the underlying signaling pathway *J. Am. Chem. Soc.* 141 6122–6.

11 Yang Y, Tang J, Zhang M, Gu Z, Song H, Yang Y and Yu C 2019 Responsively aggregatable sub-6 nm nanochelators induce simultaneous antiangiogenesis and vascular obstruction for enhanced tumor vasculature targeted therapy *Nano Lett.* 19 7750–9.

12 Niu Y, Yu M, Hartono S B, Yang J, Xu H, Zhang H, Zhang J, Zou J, Dexter A, Gu W and Yu C 2013 Nanoparticles mimicking viral surface topography for enhanced cellular delivery *Adv. Mater.* 25 6233–7.

13 Fu J, Jiao J, Song H, Gu Z, Liu Y, Geng J, Jack K S, Du A, Tang J and Yu C 2020 Fractal-in-a-sphere: confined self-assembly of fractal silica nanoparticles *Chem. Mater.* 32 341–7.

14 Ahmad Nor Y, Niu Y, Karmakar S, Zhou L, Xu C, Zhang J, Zhang H, Yu M, Mahony D, Mitter N, Cooper M A and Yu C 2015 Shaping nanoparticles with hydrophilic compositions and hydrophobic properties as nanocarriers for antibiotic delivery *ACS Cent. Sci.* 1 328–34.

15 Song H, Yu M, Lu Y, Gu Z, Yang Y, Zhang M, Fu J and Yu C 2017 Plasmid DNA delivery: nanotopography matters *J. Am. Chem. Soc.* 139 18247–54.

16 Abbaraju P L, Jambhrunkar M, Yang Y, Liu Y, Lu Y and Yu C 2018 Asymmetric mesoporous silica nanoparticles as potent and safe immunoadjuvants provoke high immune responses *Chem. Commun.* 54 2020–3.

17 Fu J, Gu Z, Liu Y, Zhang J, Song H, Yang Y, Yang Y, Noonan O, Tang J and Yu C 2019 Bottom-up self-assembly of heterotrimeric nanoparticles and their secondary janus generations *Chem. Sci.* 10 10388–94.

18 Meka A K, Abbaraju P L, Song H, Xu C, Zhang J, Zhang H, Yu M and Yu C 2016 A vesicle supra-assembly approach to synthesize amine-functionalized hollow dendritic mesoporous silica nanospheres for protein delivery *Small* 12 5169–77.

19 Li H, Guo H, Lei C, Liu L, Xu L, Feng Y, Ke J, Fang W, Song H, Xu C, Yu C and Long X 2019 Nanotherapy in joints: increasing endogenous hyaluronan production by delivering hyaluronan synthase 2 *Adv. Mater.* 31 1904535.

20 González-Sarrías A, Tomé-Carneiro J, Bellesia A, Tomás-Barberán F A and Espín J C 2015 The ellagic acid-derived gut microbiota metabolite, urolithin A, potentiates the anticancer effects of 5-fluorouracil chemotherapy on human colon cancer cells *Food Funct.* 6 1460–9.

21 Jiang A, Song B, Ji X, Peng F, Wang H, Su Y and He Y 2018 Doxorubicin-loaded silicon nanoparticles impregnated into red blood cells featuring bright fluorescence, strong photostability, and lengthened blood residency *Nano Res.* 11 2285–94.

22 da Bruckmann F, S, Nunes F B, da Salles T, R, Franco C, Cadoná F C and Bohn Rhoden C R 2022 Biological applications of silica-based nanoparticles *Magnetochemistry* 8 131.

23 Suk J S, Xu Q, Kim N, Hanes J and Ensign L M 2016 PEGylation as a strategy for improving nanoparticle-based drug and gene delivery *Adv. Drug Deliv. Rev.* 99 28–51.

24 Pinna A, Migheli R, Rocchitta G, Serra P A, Falcaro P, Malfatti L and Innocenzi P 2018 A MOF-based carrier for in situ dopamine delivery *RSC Adv.* 8 25664–72.

25 Balasamy R J, Ravinayagam V, Alomari M, Ansari M A, Almofty S A, Rehman S, Dafalla H, Marimuthu P R, Akhtar S and Al Hamad M 2019 Cisplatin delivery, anticancer and antibacterial properties of Fe/SBA-16/ZIF-8 nanocomposite *RSC Adv.* 9 42395–408.

26 Follmann H D M, Oliveira O N, Lazarin-Bidóia D, Nakamura C V, Huang X, Asefa T and Silva R 2018 Multifunctional hybrid aerogels: hyperbranched polymer-trapped mesoporous silica nanoparticles for sustained and prolonged drug release *Nanoscale* 10 1704–15.

27 Pon-On W, Tithito T, Maneeprakorn W, Phenrat T and Tang I-M 2019 Investigation of magnetic silica with thermoresponsive chitosan coating for drug controlled release and magnetic hyperthermia application *Mater. Sci. Eng. C* 97 23–30.

28 Carrera Espinoza M J, Lin K-S, Weng M-T, Kunene S C and Wang S S-S 2021 *In vitro* studies of Pluronic F127 coated magnetic silica nanocarriers for drug delivery system targeting liver cancer *Eur. Polym. J.* 153 110504.

29 Wu J, Jiang W, Shen Y, Jiang W and Tian R 2017 Synthesis and characterization of mesoporous magnetic nanocomposites wrapped with chitosan gatekeepers for pH-sensitive controlled release of doxorubicin *Mater. Sci. Eng. C* 70 132–40.

30 Moorthy M S, Subramanian B, Panchanathan M, Mondal S, Kim H, Lee K D and Oh J 2017 Fucoidan-coated core–shell magnetic mesoporous silica nanoparticles for chemotherapy and magnetic hyperthermia-based thermal therapy applications *New J. Chem.* 41 15334–46.

31 Fan H, Xing X, Yang Y, Li B, Wang C and Qiu D 2017 Triple function nanocomposites of porous silica-$CoFe_2$ O_4-MWCNTs as a carrier for pH-sensitive anti-cancer drug controlled delivery *Dalton Trans.* 46 14831–8.

32 Wang X, Yin H, Guan Y, Yang Y, Huang Y, Yuan H, Liu X and Ni C 2020 Graphene oxide covalently grafted Fe_2B@ SiO_2 nanoparticles for Epirubicin loading and releasing *J. Nanosci. Nanotechnol.* 20 2104–13.

33 Wang C, Zhao N, Huang Y, He R, Xu S and Yuan W 2020 Coordination of injectable self-healing hydrogel with Mn-Zn ferrite@ mesoporous silica nanospheres for tumor MR imaging and efficient synergistic magnetothermal-chemo-chemodynamic therapy *Chem. Eng. J.* 401 126100.

34 Solak K, Mavi A and Yılmaz B 2021 Disulfiram-loaded functionalized magnetic nanoparticles combined with copper and sodium nitroprusside in breast cancer cells *Mater. Sci. Eng. C* 119 111452.

35 Laranjeira M S, Ribeiro T P, Magalhães A I, Silva P C, Santos J A and Monteiro F J 2022 Magnetic mesoporous silica nanoparticles as a theranostic approach for breast cancer: loading and release of the poorly soluble drug exemestane *Int. J. Pharm.* 619 121711.

36 Veloso S R, Martins J A, Hilliou L, Amorim C O, Amaral V S, Almeida B G, Jervis P J, Moreira R, Pereira D M and Coutinho P J 2020 Dehydropeptide-based plasmonic magnetogels: a supramolecular composite nanosystem for multimodal cancer therapy *J. Mater. Chem. B* 8 45–64.

37 Torney F, Trewyn B G, Lin V S-Y and Wang K 2007 Mesoporous silica nanoparticles deliver DNA and chemicals into plants *Nat. Nanotechnol.* 2 295–300.

38 Slowing I I, Vivero-Escoto J L, Wu C-W and Lin V S-Y 2008 Mesoporous silica nanoparticles as controlled release drug delivery and gene transfection carriers *Adv. Drug Deliv. Rev.* 60 1278–88.

39 Lai C-Y, Trewyn B G, Jeftinija D M, Jeftinija K, Xu S, Jeftinija S and Lin V S-Y 2003 A mesoporous silica nanosphere-based carrier system with chemically removable CdS nanoparticle caps for stimuli-responsive controlled release of neurotransmitters and drug molecules *J. Am. Chem. Soc.* 125 4451–9.

40 Kneuer C, Sameti M, Haltner E G, Schiestel T, Schirra H, Schmidt H and Lehr C-M 2000 Silica nanoparticles modified with aminosilanes as carriers for plasmid DNA *Int. J. Pharm.* 196 257–61.

41 Bharali D J, Klejbor I, Stachowiak E K, Dutta P, Roy I, Kaur N, Bergey E J, Prasad P N and Stachowiak M K 2005 Organically modified silica nanoparticles: a nonviral vector for *in vivo* gene delivery and expression in the brain *Proc. Natl. Acad. Sci.* 102 11539–44.

42 Viarbitskaya S, Ryderfors L, Mikaelsson T, Mukhtar E and Johansson L B-Å 2011 Luminescence enhancement from silica-coated gold nanoparticle agglomerates following multi-photon excitation *J. Fluoresc.* 21 257–64.

43 Gong J-L, Liang Y, Huang Y, Chen J-W, Jiang J-H, Shen G-L and Yu R-Q 2007 Ag/SiO_2 core-shell nanoparticle-based surface-enhanced Raman probes for immunoassay of cancer marker using silica-coated magnetic nanoparticles as separation tools *Biosens. Bioelectron.* 22 1501–7.

44 Lee K, Moon H-Y, Park C, Kim O R, Ahn E, Lee S Y, Park H E, Ihm S-H, Seung K-B and Chang K 2009 Magnetic resonance imaging of macrophage activity in atherosclerotic plaques of apolipoprotein E-deficient mice by using polyethylene glycolated magnetic fluorescent silica-coated nanoparticles *Curr. Appl. Phys.* 9 S15–8.

45 Yuan J, Zhou S, Gu G and Wu L 2005 Encapsulation of organic pigment particles with silica via sol-gel process *J. Sol-Gel Sci. Technol.* 36 265–74.

46 Hagura N, Ogi T, Shirahama T, Iskandar F and Okuyama K 2011 Highly luminescent silica-coated ZnO nanoparticles dispersed in an aqueous medium *J. Lumin.* 131 921–5.

47 Ow H, Larson D R, Srivastava M, Baird B A, Webb W W and Wiesner U 2005 Bright and stable core–shell fluorescent silica nanoparticles *Nano Lett.* 5 113–7.

48 Bakalova R, Zhelev Z, Aoki I, Ohba H, Imai Y and Kanno I 2006 Silica-shelled single quantum dot micelles as imaging probes with dual or multimodality *Anal. Chem.* 78 5925–32.

49 Zhang C, Wangler B, Morgenstern B, Zentgraf H, Eisenhut M, Untenecker H, Krüger R, Huss R, Seliger C, Semmler W and Kiessling F 2007 Silica- and alkoxysilane-coated ultrasmall superparamagnetic iron oxide particles: a promising tool to label cells for magnetic resonance imaging *Langmuir* 23 1427–34.

50 Ji X, Shao R, Elliott A M, Stafford R J, Esparza-Coss E, Bankson J A, Liang G, Luo Z-P, Park K, Markert J T and Li C 2007 Bifunctional gold nanoshells with a superparamagnetic iron oxide–silica core suitable for both MR imaging and photothermal therapy *J. Phys. Chem. C* 111 6245–51.

51 Lee J E, Lee N, Kim H, Kim J, Choi S H, Kim J H, Kim T, Song I C, Park S P, Moon W K and Hyeon T 2010 Uniform mesoporous dye-doped silica nanoparticles decorated with multiple magnetite nanocrystals for simultaneous enhanced magnetic resonance imaging, fluorescence imaging, and drug delivery *J. Am. Chem. Soc.* 132 552–7.

52 Lee H, Lee B P and Messersmith P B 2007 A reversible wet/dry adhesive inspired by mussels and geckos *Nature* 448 338–41.

53 Wheat J C and Wolf J S 2009 Advances in bioadhesives, tissue sealants, and hemostatic agents *Urol. Clin.* 36 265–75.

54 Spotnitz W D and Burks S 2008 Hemostats, sealants, and adhesives: components of the surgical toolbox *Transfusion (Paris)* 48 1502–16.

55 Bharti B, Meissner J and Findenegg G H 2011 Aggregation of silica nanoparticles directed by adsorption of lysozyme *Langmuir* 27 9823–33.

56 Mehdizadeh M and Yang J 2013 Design strategies and applications of tissue bioadhesives *Macromol. Biosci.* 13 271–88.

57 Xu Z, Ma X, Gao Y-E, Hou M, Xue P, Li C M and Kang Y 2017 Multifunctional silica nanoparticles as a promising theranostic platform for biomedical applications *Mater. Chem. Front.* 1 1257–72.

58 Rose S, Prevoteau A, Elzière P, Hourdet D, Marcellan A and Leibler L 2014 Nanoparticle solutions as adhesives for gels and biological tissues *Nature* 505 382–5.

59 Meddahi-Pellé A, Legrand A, Marcellan A, Louedec L, Letourneur D and Leibler L 2014 Organ repair, hemostasis, and in vivo bonding of medical devices by aqueous solutions of nanoparticles *Angew. Chem. Int. Ed.* 53 6369–73.

60 Pedrazzoli D and Pegoretti A 2013 Silica nanoparticles as coupling agents for polypropylene/glass composites *Compos. Sci. Technol.* 76 77–83.

61 Kamaruzzaman N F, Tan L P, Hamdan R H, Choong S S, Wong W K, Gibson A J, Chivu A and de Pina M, F 2019 Antimicrobial polymers: the potential replacement of existing antibiotics? *Int. J. Mol. Sci.* 20 2747.

62 Sun D, Babar Shahzad M, Li M, Wang G and Xu D 2015 Antimicrobial materials with medical applications *Mater. Technol.* 30 B90–5.

63 Beloin C, Fernández-Hidalgo N and Lebeaux D 2017 Understanding biofilm formation in intravascular device-related infections *Intensive Care Med.* 43 443–6.

64 Andersen M J and Flores-Mireles A L 2019 Urinary catheter coating modifications: the race against catheter-associated infections *Coatings* 10 23.

65 Lebeaux D, Ghigo J-M and Beloin C 2014 Biofilm-related infections: bridging the gap between clinical management and fundamental aspects of recalcitrance toward antibiotics *Microbiol. Mol. Biol. Rev.* 78 510–43.

66 Macha I J, Karacan I, Ben-Nissan B, Cazalbou S and Müller W H 2019 Development of antimicrobial composite coatings for drug release in dental, orthopaedic and neural prostheses applications *SN Appl. Sci.* 1 68.

67 Olivares E, Badel-Berchoux S, Provot C, Prévost G, Bernardi T and Jehl F 2020 Clinical impact of antibiotics for the treatment of *Pseudomonas aeruginosa* biofilm infections *Front. Microbiol.* 10 2894.

68 Ventola C L 2015 The antibiotic resistance crisis: part 1: causes and threats *Pharm. Ther.* 40 277.

69 Golestaneh L and Mokrzycki M H 2018 Prevention of hemodialysis catheter infections: ointments, dressings, locks, and catheter hub devices *Hemodial. Int.* 22(S2) S75–82.

70 Olmos D and González-Benito J 2021 Polymeric materials with antibacterial activity: a review *Polymers* 13 613.

71 Francolini I, Vuotto C, Piozzi A and Donelli G 2017 Antifouling and antimicrobial biomaterials: an overview *APMIS* 125 392–417.

72 Banerjee I, Pangule R C and Kane R S 2011 Antifouling coatings: recent developments in the design of surfaces that prevent fouling by proteins, bacteria, and marine organisms *Adv. Mater.* 23 690–718.

73 Yang C, Ding X, Ono R J, Lee H, Hsu L Y, Tong Y W, Hedrick J and Yang Y Y 2014 Brush-like polycarbonates containing dopamine, cations, and PEG providing a broad-spectrum, antibacterial, and antifouling surface via one-step coating *Adv. Mater.* 26 7346–51.

74 Maan A M C, Hofman A H, de Vos W M and Kamperman M 2020 Recent developments and practical feasibility of polymer-based antifouling coatings *Adv. Funct. Mater.* 30 2000936.

75 Ma S, Zhang X, Yu B and Zhou F 2019 Brushing up functional materials *NPG Asia Mater.* 11 24.

76 Voo Z X, Khan M, Narayanan K, Seah D, Hedrick J L and Yang Y Y 2015 Antimicrobial/antifouling polycarbonate coatings: role of block copolymer architecture *Macromolecules* 48 1055–64.

77 Vacanti J P and Langer R 1999 Tissue engineering: the design and fabrication of living replacement devices for surgical reconstruction and transplantation *Lancet* 354 S32–4.

78 Simões D, Miguel S P, Ribeiro M P, Coutinho P, Mendonça A G and Correia I J 2018 Recent advances on antimicrobial wound dressing: a review *Eur. J. Pharm. Biopharm.* 127 130–41.

79 Mouriño V and Boccaccini A R 2010 Bone tissue engineering therapeutics: controlled drug delivery in three-dimensional scaffolds *J. R. Soc. Interface* 7 209–27.

80 Li G, Qin S, Zhang D and Liu X 2019 Preparation of antibacterial degummed silk fiber/nano-hydroxyapatite/polylactic acid composite scaffold by degummed silk fiber loaded silver nanoparticles *Nanotechnology* 30 295101.

81 Nasajpour A, Ansari S, Rinoldi C, Rad A S, Aghaloo T, Shin S R, Mishra Y K, Adelung R, Swieszkowski W, Annabi N, Khademhosseini A, Moshaverinia A and Tamayol A 2018 A multifunctional polymeric periodontal membrane with osteogenic and antibacterial characteristics *Adv. Funct. Mater.* 28 1703437.

82 Xu Y, Zhao S, Weng Z, Zhang W, Wan X, Cui T, Ye J, Liao L and Wang X 2020 Jelly-inspired injectable guided tissue regeneration strategy with shape auto-matched and dual-light-defined antibacterial/osteogenic pattern switch properties *ACS Appl. Mater. Interfaces* 12 54497–506.

83 Saravanan S, Vimalraj S, Thanikaivelan P, Banudevi S and Manivasagam G 2019 A review on injectable chitosan/beta glycerophosphate hydrogels for bone tissue regeneration *Int. J. Biol. Macromol.* 121 38–54.

84 Abdullah T, Gauthaman K, Hammad A H, Joshi Navare K, Alshahrie A A, Bencherif S A, Tamayol A and Memic A 2020 Oxygen-releasing antibacterial nanofibrous scaffolds for tissue engineering applications *Polymers* 12 1233.

85 Melke J, Midha S, Ghosh S, Ito K and Hofmann S 2016 Silk fibroin as biomaterial for bone tissue engineering *Acta Biomater.* 31 1–16.

86 Przekora A 2020 A concise review on tissue engineered artificial skin grafts for chronic wound treatment: can we reconstruct functional skin tissue in vitro? *Cells* 9 1622.

87 Dorazilová J, Muchová J, Šmerková K, Kočiová S, Diviš P, Kopel P, Veselý R, Pavliňáková V, Adam V and Vojtová L 2020 Synergistic effect of chitosan and selenium nanoparticles on biodegradation and antibacterial properties of collagenous scaffolds designed for infected burn wounds *Nanomaterials* 10 1971.

88 Khalid H, Iqbal H, Zeeshan R, Nasir M, Sharif F, Akram M, Irfan M, Khan F A, Chaudhry A A and Khan A F 2021 Silk fibroin/collagen 3D scaffolds loaded with TiO_2 nanoparticles for skin tissue regeneration *Polym. Bull.* 78 7199–218.

89 Li X, Wang Y, Guo M, Wang Z, Shao N, Zhang P, Chen X and Huang Y 2018 Degradable three dimensional-printed polylactic acid scaffold with long-term antibacterial activity *ACS Sustain. Chem. Eng.* 6 2047–54.

90 Iga C, Agata T, Marcin Ł, Natalia F and Justyna K-L 2020 Ciprofloxacin-modified degradable hybrid polyurethane-polylactide porous scaffolds developed for potential use as an antibacterial scaffold for regeneration of skin *Polymers* 12 171.

91 Zuniga J M 2018 3D printed antibacterial prostheses *Appl. Sci.* 8 1651.

92 Grego E A, Siddoway A C, Uz M, Liu L, Christiansen J C, Ross K A, Kelly S M, Mallapragada S K, Wannemuehler M J and Narasimhan B 2021 Polymeric nanoparticle-based vaccine adjuvants and delivery vehicles *Curr. Top. Microbiol. Immunol.* 433 29–76.

93 Di Cola E, Cantu' L, Brocca P, Rondelli V, Fadda G C, Canelli E, Martelli P, Clementino A, Sonvico F, Bettini R and Del Favero E 2019 Novel O/W nanoemulsions for nasal administration: structural hints in the selection of performing vehicles with enhanced mucopenetration *Colloids Surf. B Biointerfaces* 183 110439.

94 Jones C H, Chen M, Ravikrishnan A, Reddinger R, Zhang G, Hakansson A P and Pfeifer B A 2015 Mannosylated poly(beta-amino esters) for targeted antigen presenting cell immune modulation *Biomaterials* 37 333–44.

95 Li P, Luo Z, Liu P, Gao N, Zhang Y, Pan H, Liu L, Wang C, Cai L and Ma Y 2013 Bioreducible alginate-poly(ethylenimine) nanogels as an antigen-delivery system robustly enhance vaccine-elicited humoral and cellular immune responses *J. Control. Release* 168 271–9.

96 Wusiman A, Gu P, Liu Z, Xu S, Zhang Y, Hu Y, Liu J, Wang D and Huang X 2019 Cationic polymer modified PLGA nanoparticles encapsulating Alhagi honey polysaccharides as a vaccine delivery system for ovalbumin to improve immune responses *Int. J. Nanomedicine* 14 3221–34.

97 Chen X, Liu Y, Wang L, Liu Y, Zhang W, Fan B, Ma X, Yuan Q, Ma G and Su Z 2014 Enhanced humoral and cell-mediated immune responses generated by cationic polymer-coated PLA microspheres with adsorbed HBsAg *Mol. Pharm.* 11 1772–84.

98 Nevagi R J, Skwarczynski M and Toth I 2019 Polymers for subunit vaccine delivery *Eur. Polym. J.* 114 397–410.

99 Bartlett S, Eichenberger R M, Nevagi R J, Ghaffar K A, Marasini N, Dai Y, Loukas A, Toth I and Skwarczynski M 2020 Lipopeptide-based oral vaccine against hookworm infection *J. Infect. Dis.* 221 934–42.

100 Fan Y, Sahdev P, Ochyl L J, Akerberg J and Moon J J 2015 Cationic liposome–hyaluronic acid hybrid nanoparticles for intranasal vaccination with subunit antigens *J. Control. Release* 208 121–9.

101 Chen C-H, Lin Y-L, Liu Y-K, He P-J, Lin C-M, Chiu Y-H, Wu C-J, Cheng T-L, Liu S-J and Liao K-W 2012 Liposome-based polymer complex as a novel adjuvant: enhancement of specific antibody production and isotype switch *Int. J. Nanomedicine* 7 607–21.

102 Rose F, Wern J E, Gavins F, Andersen P, Follmann F and Foged C 2018 A strong adjuvant based on glycol-chitosan-coated lipid-polymer hybrid nanoparticles potentiates mucosal immune responses against the recombinant *Chlamydia trachomatis* fusion antigen CTH522 *J. Control. Release* 271 88–97.

103 Yoshizaki Y, Yuba E, Sakaguchi N, Koiwai K, Harada A and Kono K 2017 pH-sensitive polymer-modified liposome-based immunity-inducing system: effects of inclusion of cationic lipid and CpG-DNA *Biomaterials* 141 272–83.

104 Saboo S, Tumban E, Peabody J, Wafula D, Peabody D S, Chackerian B and Muttil P 2016 Optimized formulation of a thermostable spray-dried virus-like particle vaccine against human papillomavirus *Mol. Pharm.* 13 1646–55.

105 Yang H-W, Ye L, Guo X D, Yang C, Compans R W and Prausnitz M R 2017 Ebola vaccination using a DNA vaccine coated on PLGA-PLL/γPGA nanoparticles administered using a microneedle patch *Adv. Healthc. Mater.* 6 1600750.

106 Tzeng S Y, Guarecuco R, McHugh K J, Rose S, Rosenberg E M, Zeng Y, Langer R and Jaklenec A 2016 Thermostabilization of inactivated polio vaccine in PLGA-based microspheres for pulsatile release *J. Control. Release* 233 101–13.

107 Ross K A, Loyd H, Wu W, Huntimer L, Ahmed S, Sambol A, Broderick S, Flickinger Z, Rajan K, Bronich T, Mallapragada S, Wannemuehler M J, Carpenter S and Narasimhan B 2015 Hemagglutinin-based polyanhydride nanovaccines against H5N1 influenza elicit protective virus neutralizing titers and cell-mediated immunity *Int. J. Nanomedicine* 10 229–43.

108 Petersen L K, Phanse Y, Ramer-Tait A E, Wannemuehler M J and Narasimhan B 2012 Amphiphilic polyanhydride nanoparticles stabilize *Bacillus anthracis* protective antigen *Mol. Pharm.* 9 874–82.

109 Ross K, Senapati S, Alley J, Darling R, Goodman J, Jefferson M, Uz M, Guo B, Yoon K-J, Verhoeven D, Kohut M, Mallapragada S, Wannemuehler M and Narasimhan B 2019 Single dose combination nanovaccine provides protection against influenza a virus in young and aged mice *Biomater. Sci.* 7 809–21.

110 Lau Y-F, Wright A R and Subbarao K 2012 The contribution of systemic and pulmonary immune effectors to vaccine-induced protection from H5N1 influenza virus infection *J. Virol.* 86 5089–98.

111 Napierska D, Thomassen L C J, Rabolli V, Lison D, Gonzalez L, Kirsch-Volders M, Martens J A and Hoet P H 2009 Size-dependent cytotoxicity of monodisperse silica nanoparticles in human endothelial cells *Small* 5 846–53.

112 Thomassen L C J, Aerts A, Rabolli V, Lison D, Gonzalez L, Kirsch-Volders M, Napierska D, Hoet P H, Kirschhock C E A and Martens J A 2010 Synthesis and characterization of stable monodisperse silica nanoparticle sols for in vitro cytotoxicity testing *Langmuir* 26 328–35.

113 Lin W, Huang Y, Zhou X-D and Ma Y 2006 In vitro toxicity of silica nanoparticles in human lung cancer cells *Toxicol. Appl. Pharmacol.* 217 252–9.

114 Spyrogianni A, Herrmann I K, Keevend K, Pratsinis S E and Wegner K 2017 The silanol content and *in vitro* cytolytic activity of flame-made silica *J. Colloid Interface Sci.* 507 95–106.

115 Zhao Y, Sun X, Zhang G, Trewyn B G, Slowing I I and Lin V S-Y 2011 Interaction of mesoporous silica nanoparticles with human red blood cell membranes: size and surface effects *ACS Nano* 5 1366–75.

116 Francia V, Yang K, Deville S, Reker Smit C, Nelissen I and Salvati A 2019 Corona composition can affect the mechanisms cells use to internalize nanoparticles *ACS Nano* 13 11107–21.

117 Gonzalez Porras M A, Durfee P, Giambini S, Sieck G C, Brinker C J and Mantilla C B 2018 Uptake and intracellular fate of cholera toxin subunit b-modified mesoporous silica nanoparticle-supported lipid bilayers (aka protocells) in motoneurons *Nanomed. Nanotechnol. Biol. Med.* 14 661–72.

118 Oh N and Park J-H 2014 Endocytosis and exocytosis of nanoparticles in mammalian cells *Int. J. Nanomedicine* 9 51–63.

119 Fu C, Liu T, Li L, Liu H, Chen D and Tang F 2013 The absorption, distribution, excretion and toxicity of mesoporous silica nanoparticles in mice following different exposure routes *Biomaterials* 34 2565–75.

120 He L, Lai H and Chen T 2015 Dual-function nanosystem for synergetic cancer chemo–/radiotherapy through ROS-mediated signaling pathways *Biomaterials* 51 30–42.

121 Park E-J and Park K 2009 Oxidative stress and pro-inflammatory responses induced by silica nanoparticles in vivo and in vitro *Toxicol. Lett.* 184 18–25.

7

Nanotechnology-Based Biomedical Products, Devices, and Applications

Ajit Khosla[1], Irshad A. Wani[2], and Mohammad N. Lone[3]

[1] *School of Advanced Materials and Nanotechnology, Xidian University, Xian, Shaanxi, China*
[2] *Department of Chemistry, Govt. Degree College Anantnag, Anantnag, University of Kashmir, Jammu and Kashmir, India*
[3] *Department of Chemistry, Central University of Kashmir, Ganderbal, Jammu and Kashmir, India*

7.1 Nanotechnology-Based Treatments for Heart Disease

Nanotechnology has opened up promising avenues in the field of cardiovascular medicine, offering the potential to revolutionize the treatment of heart disease [1]. The unique properties and capabilities of nanomaterials present exciting opportunities for targeted and efficient interventions in managing heart conditions. As cardiovascular disease continues to be a significant contributor to illness and death on a global scale, these innovative approaches hold the promise of significantly improving patient outcomes and overall cardiac health.

A key application of nanotechnology in heart disease lies in targeted drug delivery. By engineering nanoparticles (NPs) to carry therapeutic agents directly to affect cardiac tissues, precise and localized drug release becomes possible [2]. This approach not only maximizes drug effectiveness but also minimizes off-target effects, minimizing the necessary dosage and alleviating potential side effects. Moreover, nanomaterials facilitate the creation of advanced diagnostic instruments for the early identification and monitoring of heart disease [3]. Nanosensors and imaging agents provide real-time information on structural and functional changes in the heart, facilitating timely interventions and personalized treatment strategies.

In the realm of regenerative medicine, nanotechnology plays a pivotal role in heart repair and tissue engineering. Nanoscale scaffolds and biomaterials support facilitating the growth and differentiation of cardiac stem cells, promoting tissue regeneration and repair after myocardial injury [4]. Additionally, nanotechnology offers opportunities to augment the electrical and mechanical properties of cardiac tissue constructs, further advancing cardiac tissue engineering.

In this section, we will explore the latest developments and research in nanotechnology-based treatments for heart disease. Understanding the possibilities and limitations of these novel approaches will open avenues for revolutionary progress in handling and treating cardiovascular conditions, benefiting millions of patients worldwide.

Metallic, Magnetic and Carbon-Based Nanomaterials: Synthesis and Biomedical Applications, First Edition.
Ajit Khosla, Irshad A. Wani, and Mohammad N. Lone.

7.1.1 Nanomaterial-Based Biosensors for Early Detection of Coronary Artery Diseases (CADs)

Early detection of coronary artery diseases (CADs) plays a critical role in improving treatment success, potential cure, and overall patient prognosis prolonged survival periods. Biomarkers associated with CAD, such as myoglobin (Myo), cardiac troponins (cTns), C-reactive protein (CRP), creatinine kinase MB (CK-MB), and various microRNAs (miRNAs), are discharged into the bloodstream in response to heart damage or stress [5]. Creating accurate, targeted, uncomplicated, durable, and swift blood analyses for these biomolecules presents a hopeful strategy for early CAD diagnosis. Although mass spectrometry is commonly employed to identify potential CAD biomarkers, it faces challenges in sensitivity and specificity due to the low concentrations of biomarkers in human plasma. Integrating nanotechnologies with biosensors provides a potential resolution for the early detection of CADs [6]. Biosensors comprise two primary components: a biological sensing element for target recognition and a transducer for converting data into electrical signals. Nanomaterials with exceptional optoelectronic properties notably amplify detection sensitivity. For example, $ZnSnO_3$ perovskite nanomaterial-decorated glassy carbon electrodes (GCEs) exhibited subfemtomolar detection sensitivity for troponin T (TnT) due to their ferroelectric property [7]. An additional illustration involves the utilization of a gold triangular nanoprism (AuTNP)-based localized surface plasmon resonance biosensor, which attains an atto-molar (a.m.) limit of detection (LOD) for cardiac troponin T (cTnT). This makes it at least 50-fold more sensitive than alternative label-free techniques [8]. The morphology of nanomaterials, exemplified by 3D silver anisotropic nano-pinetree array-modified indium tin oxide (Ag NPT/ITO) substrates, also influences sensitive detection, as it provides numerous hotspots that enhance performance. Dual-hybrid or multihybrid nanostructure materials can incorporate the benefits of multiple structures, effectively achieving signal amplification [9, 10]. Ren's group introduced an ultrasensitive sandwich-type electrochemical (EC) immunosensor utilizing TiO_2-PPy-Au as the substrate material and TB-Au-CO-Fs as signal labels for the quantitative detection of cTnI [11]. Utilizing hybrid nanomaterials to enhance electron transfer, the mentioned sensor exhibited a linear range from 0.5 to 10.0 ng/mL and a low LOD of 0.17 pg/mL. Additionally, CdTe at $IRMOF^{-3}$ at CdTe nanocomposites also amplified cTnI detection signals [12] (Figure 7.1A). The research reveals the formation of nanomaterials $IRMOF^{-3}$ and CdTe at $IRMOF^{-3}$. $IRMOF^{-3}$ showed a diameter of approximately 500 ± 50 nm with similar spherical structure. The XPS analysis confirmed the presence of C_{1s}, O_{1s}, N_{1s}, and Zn_{2p} in $IRMOF^{-3}$. CdTe at $IRMOF^{-3}$ exhibited a clearly defined spherical structure with a particle size of approximately 1 μm, and the transmission electron microscopy (TEM) images revealed the successful loading of CdTe onto the surface of CdTe at $IRMOF^{-3}$ with a size mainly around 5 nm. The electrochemiluminescence (ECL) measurements indicated that $IRMOF^{-3}$ enhanced the ECL signal of the $CdTe/S_2O_8{}^{2-}$ system, acting as a co-reaction accelerator for $S_2O_8{}^{2-}$ rather than directly interacting with CdTe. The cyclic voltammetry (CV) responses also supported the role of $IRMOF^3$ in promoting the oxygenation between CdTe and $S_2O_8{}^{2-}$. The findings suggest that $IRMOF^{-3}$ can significantly influence the ECL signal in the $CdTe/S_2O_8{}^{2-}$ system. An immunosensor was later constructed and its performance was evaluated for detecting cardiac troponin I (cTnI), a crucial biomarker for myocardial injury. The construction of the immunosensor was assessed through CV and EC impedance spectroscopy (EIS) (Figure 7.1B). Initially, the bare GCE exhibited symmetric redox peaks for the $[Fe(CN)_6]^{3-/4-}$ redox indicator. Subsequently, the GCE was modified with Au at TiO_2, and the peak currents declined as a result of the limited electron-transporting capability of Au at TiO_2. The electrode was further functionalized with Ab_1, leading to a reduction in peak current owing to Ab1's electron inert feature. The nonelectroactive protein BSA was then used to hinder nonspecific sites, resulting in a subsequent reduction in peak current. Upon incubating cTnI on the BSA/Ab_1/Au at TiO_2/GCE, the peak current exhibited a notable decrease, primarily attributed to the formation of a sizable immune complex that impeded electron transfer. The immunosensor's EIS results mirrored the CV findings, showing an increase in electron transfer resistance (Ret) with each modification step. A clear GCE yielded a small semicircle in the EIS, while the modification with Au at TiO_2 increased Ret. Ab_1 bonding to Au at TiO_2 via Au-S or Au-N bonds. The formation of Au-S or Au-N bonds resulted in a further rise in Ret, highlighting the electron-inert nature of Ab_1. Blocking with BSA increased Ret further, ascribed to the insulating nature of protein molecules. Upon incubating cTnI on the surface, Ret increased significantly, indicating that the formation of the immunocomplex hindered the transport of electrons from the bulk of the solution to the electrode interface. The signal probe CdTe at $IRMOF^3$ at CdTe exhibited enhanced ECL intensity. Comparative studies with various Ab_2 signal probes confirmed that the CdTe at $IRMOF^{-3}$ at CdTe labeled Ab_2 probe displayed the highest ECL emission, owing to the co-reaction accelerator role of $IRMOF^{-3}$ in the CdTe/$S_2O_8^{2-}$ ECL system (Figure 7.1C). This improved sensitivity made it the optimal choice for detecting cTnI. The immunosensor's detection ability for cTnI was evaluated with varying concentrations of cTnI. The ECL signal elevated with an increase in cTnI concentration, showing an excellent linear relationship on a logarithmic scale. The estimated LOD was impressively low at 0.46 fg m/L, highlighting the sensor's high sensitivity. Compared to previous reports, the immunosensor exhibited superior performance. The stability and selectivity of the immunosensor were also assessed. The immunosensor demonstrated excellent stability over 15 consecutive cycles, exhibiting relative standard deviations (RSD) of ECL peaks at only 0.92%. When tested with nontarget proteins as interfering substances, the immunosensor exhibited nearly identical responses to the blank solution, demonstrating its good selectivity and specificity for detecting cTnI. To evaluate the immunosensor's potential application in clinical settings, a recovery experiment was conducted using human serum samples containing predetermined concentrations of cTnI. The immunosensor achieved impressive recovery rates between 93.6% and 109.7%, showcasing its promise for clinical determination. Overall, the immunosensor's step-by-step assembly and thorough characterization demonstrate its excellent sensitivity, stability, selectivity, and potential for clinical applications in detecting cTnI, which is essential for diagnosing myocardial injury. Due to their high surface area-to-volume ratio, nanomaterials, particularly those with porous structures, can accommodate significant quantities of recognition elements (such as antibodies or aptamers) and transducer elements, leading to signal cascade amplification [13, 14]. In accordance with this, Zhang et al. [15] synthesized nanodiamond hybrid hydrogen-substituted graphdiyne (HsGDY at NDs) to construct EC aptasensors for the detection of Myo and cTnI. The HsGDY at NDs exhibited nanospheric morphology with sizes ranging from 200 to 500 nm. Embedded within these nanospheres, nanodiamonds (NDs) were incorporated into the HsGDY network. This nanostructure integrated the favorable characteristics of HsGDY's chemical stability and three-dimensional porous networks with NDs' biocompatibility and EC activity. This allowed for the immobilization of diverse aptamer strands and the recognition of target biomarkers. In comparison to HsGDY and NDs-based aptasensors, HsGDY at NDs-based aptasensors exhibited superior sensing capabilities for Myo and cTnI, attaining low detection limits of 6.29 and 9.04 fg m/L, respectively. Moreover, the HsGDY at NDs-based aptasensors demonstrated high selectivity, reproducibility, stability, and applicability in real human serum. Overall, the development of the HsGDY at NDs-based aptasensor holds promise for expanding the utilization of porous organic frameworks in sensing applications and providing a potential strategy for the early recognition of disease biomarkers. Similar research was documented by Zhao et al. [16], where they synthesized Pd at Au nanocube-doped three-dimensional porous graphene to develop an ultrasensitive EC immunosensor for detection of cTnI. In this research, a novel composite, CDs-3D-PG-Pd at Au NCs, was fabricated through a simple method, where β-cyclodextrins (CDs) played a key role in enhancing the dispersibility of three-dimensional porous graphene (3D-PG) and effectively capturing secondary antibodies (Ab_2). The incorporation of Pd at Au NCs onto CDs-3D-PG further enhanced the EC signals. These CDs-3D-PG-Pd at Au NCs were employed as signal amplification labels. Additionally, sensor platforms were prepared by decorating AuNPs and thionine (Th) onto microporous carbon spheres functionalized with amino groups (AuNPs-FMCS-Th). This platform not only efficiently immobilized primary antibodies (Ab1) through Au-NH_2 interactions but also facilitated electron transfer on the electrode surface through the facilitative effect of Th, leading to enhanced signal responses. The nanomaterials were thoroughly characterized using scanning electron microscope (SEM), TEM UV-vis spectroscopy, and Raman spectroscopy. The electrocatalytic reduction of H_2O_2 by CDs-3D-PG-Pd at Au NCs was investigated using CV and amperometric (i-t) methods with mediation of Th for electron transfer. The proposed immunosensor exhibited high selectivity, satisfactory stability, and good reproducibility for detecting cTnI, achieving a low detection limit of 33.3 fg m/L under optimal conditions. Promising results were achieved when analyzing real serum samples, indicating that this designed method could serve as an effective strategy in clinical research. In another study, a highly sensitive and accurate EC biosensor was developed for detecting cTnI, a critical biomarker for myocardial infarction (MI) diagnosis [16]. The biosensor utilized a dual-aptamer approach, employing DNA nanotetrahedron (NTH) capture probes and multifunctional hybrid nanoprobes. Initially, Tro_4 aptamer probes, based on NTH, were immobilized onto a screen-printed gold electrode (SPGE) through Au-S bonding, thereby offering enhanced spatial dimension and accessibility for capturing cTnI. The hybrid nanoprobes were then created by utilizing magnetic Fe_3O_4 NPs as carriers to load Tro_6 aptamer, natural horseradish peroxidase (HRP), HRP-mimicking Au at Pt nanozymes, and G-quadruplex/hemin DNAzyme. These signaling nanoprobes specifically recognized cTnI through the Tro_6 aptamer and amplified the signals via enzymatic processes, significantly improving detection sensitivity. The heightened EC signal was ascribed to the co-catalysis effect of hybrid nanozymes, HRP, and DNAzyme. The target cTnI was sandwiched between the two types of aptamers (Tro_4 and Tro_6) at the electrode interface. The EC aptasensing platform demonstrated outstanding analytical performance with a broad dynamic range of 0.01–100 ng/mL and an impressively low detection limit of 7.5 pg/mL for cTnI. The platform showcased high selectivity, reliability, and presenting substantial potential for clinical disease diagnostics. In recent times, nanotechnology-driven multimodal or multitarget point-of-care testing (POCT) assays have surfaced for the detection of CAD biomarkers. These advancements aid in a more accurate, rapid, and convenient assessment of the disease process [17–20]. For instance, Huang et al. [17] have made significant progress by introducing paper-based fluorogenic immunodevices capable of simultaneously detecting three cardiac biomarkers, viz., human heart-type fatty acid-binding protein (FABP), cTnI, and myoglobin. Enhanced detection is achieved through the strategic use of zinc oxide nanowires (ZnO NWs) significantly amplify fluorescence signals, approximately fivefold higher than that observed on pure paper. Remarkably, the immunodevices demonstrated outstanding sensitivity and selectivity for FABP, cTnI, and myoglobin, with LOD of 1.36 ng/mL, 1.00 ng/mL, and 2.38 ng/mL, respectively. Furthermore, the paper-based immunoassay is rapid, completing the test in approximately 5 minutes, and provides portability through a simple homemade chamber equipped with a smartphone and an ultraviolet lamp. By integrating ZnO NWs into these devices, sensitive and simultaneous detection of multiple cardiac biomarkers has been achieved at point-of-care settings in a quantitative manner. This innovative approach holds great promise for monitoring AMI diseases and has potential applications in other medical diagnostics and environmental assessments. The application of ratiometric ECL has garnered significant attention in the field of biological analysis due to its capability to mitigate environmental interference and facilitate precise measurements. In a study conducted by Yang et al. [21], they developed a dual-wavelength ratiometric ECL biosensor designed for the detection of cTnI. This biosensor design incorporated the use of (4,4′-dicarboxylic acid-2,2′-bipyridyl) ruthenium(II) ($Ru(dcbpy)_3^{2+}$) and Au nanoparticle-loaded graphene oxide/polyethylenimine (GPRu-Au) nanomaterial as an acceptor, coupled with Au nanoparticle-modified graphitic phase carbon nitride nanosheet composite (Au-CNN) serving as a donor. The Au-CNN exhibited a stable and intense ECL signal at 455 nm, aligning effectively with the adsorption of GPRu-Au. This alignment resulted in the establishment of a highly effective electrochemiluminescent resonance energy transfer (ECL-RET) sensing platform. The presence of AuNPs facilitated the immobilization of antibodies on the surface of nanomaterials through Au-N bonds. The graphene oxide/polyethylenimine's large surface area allowed for a substantial loading of $Ru(dcbpy)3^{2+}$, resulting in significant ECL signal amplification. The developed biosensor exhibited excellent analytical performance in detecting cTnI, achieving an impressive LOD of 3.94 fg/mL (S/N = 3). The ratiometric ECL biosensor demonstrated high reliability, selectivity, and sensitivity, making it a versatile platform for bioanalysis.

Early detection of acute myocardial infarction (AMI) at the onset of chest pain symptoms is crucial for patient survival. However, this proves challenging, particularly in cases without a persistent elevation of the ST-segment in electrocardiograms or blood tests. Although many POCT devices can accurately and swiftly diagnose AMI at intermediate and later stages using specific cardiac biomarkers (e.g., troponin I or TnT, creatine kinase-MB, and myoglobin), early diagnosis during myocardial ischemia remains unattainable. In an effort to overcome this limitation, a research team led by Lim et al. [22] devised a rapid and sensitive microfluidic paper-based device (μPAD) for the simultaneous detection of multiple cardiac biomarkers, aiming to facilitate early and late AMI diagnosis. The μPAD platform allowed detection of glycogen phosphorylase isoenzyme BB (GPBB) during early ischemic myocardial injury (within the first four hours). Moreover, the device enabled the identification of sustained elevation in cTnT and CK-MB levels, which typically manifest six hours after the onset of chest pain in human serum. The sandwich immunoassay conducted on the μPAD demonstrated commendable reproducibility (RSD, RSD approximately 10%) and intra- as well as inter-day precision (Coefficient of Variation, CV 10–20%, 99th percentile). The results remained consistent over 28 days, showing a robust correlation ($r^2 > 0.962$) with the standard Siemens Centaur XPT Immunoassay system. The results underscore the μPAD platform's potential as a point-of-care device for early diagnosis and prognosis of AMI. This technology's capability to identify cardiac biomarkers in the early stages of myocardial ischemia suggests the potential to enhance patient outcomes and survival rates.

Currently, several POCT methods have been developed to identify biomarkers related to CAD [23]. These methods utilize paper based or microfluidic portable devices in combination with nanomaterials that offer signal amplification capabilities, such as ZnO nanowires, Au at AgPtNPs, core–shell upconversion NPs, and TiO_2 nanofibers [24]. As a result, the time taken from collecting blood samples to obtaining detection results can be as short as 20 minutes. However, it is unfortunate that the current POCT methods still have a limitation in their LOD, which remains in the ng/mL range. Further improvements in LOD are necessary to enhance the sensitivity of these methods [25]. Currently, a range of POCT methods has been developed for the detection of CAD biomarkers utilizing paper-based or microfluidic portable devices, leveraging the signal amplification capabilities of nanomaterials such as ZnO nanowires [17], Au at AgPtNPs [26], and TiO_2 nanofibers [27]. It has been observed that the duration from blood sample collection to obtaining detection results can be accomplished in less than 20 minutes. Unfortunately, current POCT methods still possess limits of detection (LODs) in the ng/mL range, indicating a need for further improvement.

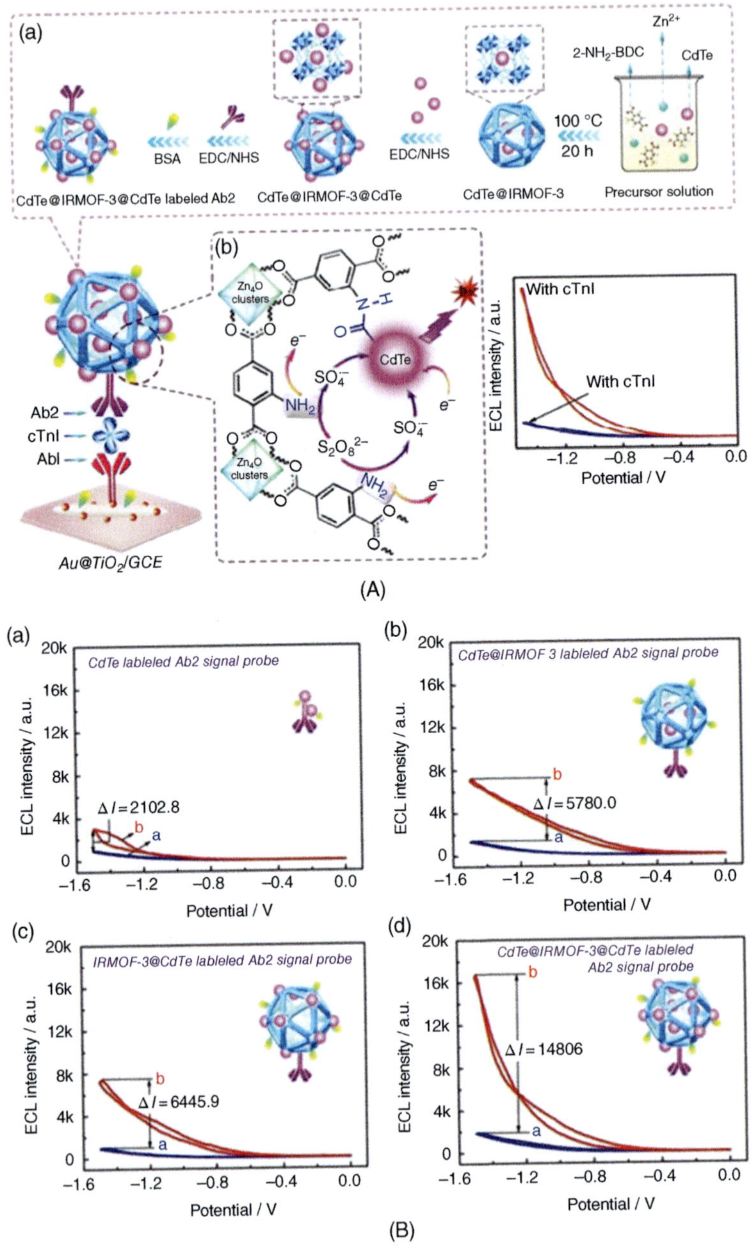

Figure 7.1 (A) Schematic illustrations of the electrochemiluminescence (ECL) immunosensor. (a) Synthesis of the signal probe labeled with CdTe at IRMOF-3 at CdTe. (b) Exploring the potential mechanism by which IRMOF-3 acts as an accelerator in the detection of cardiac troponin I (cTnI) within the CdTe/$S_2O_8{}^{2-}$ system. (B) Electrochemiluminescence (ECL) profiles of various sandwich-format immunosensors in the presence (b, red line) and absence (a, blue line) of cTnI, employing different signal probes (Ab2): (a) Signal probe labeled with CdTe, (b) signal probe labeled with CdTe at IRMOF-3, (c) signal probe labeled with IRMOF-3 at CdTe, and (d) the suggested probe, labeled with CdTe at IRMOF-3 at CdTe. The concentration of cTnI used was 11 ng m/L. (C) Electrochemical evaluation of sequentially modified electrodes, including cyclic voltammetry (CV) (a) and electrochemical impedance spectroscopy (EIS) (b), conducted on glassy carbon electrode (GCE) (curve a), Au at TiO_2-coated GCE (curve b), Ab1/Au at TiO_2/GCE (curve c), BSA/Ab1/Au at TiO_2/GCE (curve d), and cTnI/BSA/Ab1/Au at TiO_2/GCE (curve e). Measurements were taken in 0.1 M phosphate-buffered solution (PBS) at pH 7.4, comprising 5.0 mM $[Fe(CN)6]^{3-/4-}$ as the redox probe and 0.1 M KCl, with a scan rate of 100 mV/s. Source: Reproduced with permission from Yang et al. [12]/American Chemical Society.

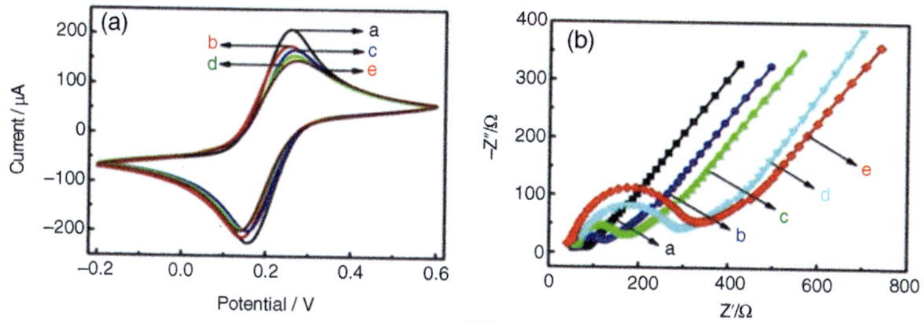

(C)

Figure 7.1 (*Continued*)

In summary nanotechnology is transforming the field of cardiovascular medicine, holding the potential to revolutionize heart disease treatment. This section explores the diverse applications of nanotechnology, such as targeted drug delivery, advanced diagnostic tools for early detection, and tissue engineering in regenerative medicine. The focus is on enhancing sensitivity, precision, and personalized treatment approaches for heart disease. Notably, nanomaterial-based biosensors for early diagnosis of CADs are discussed in detail, showcasing the potential of nanotechnology to improve disease diagnostics, with a focus on detection limits and the need for further sensitivity enhancements in POCT.

7.1.2 Nanomaterials: Advancing Cardiac Regeneration Through Targeted Therapies

In recent years, the field of nanotechnology has opened exciting possibilities for revolutionizing regenerative medicine, particularly in the context of cardiac regeneration. The heart's limited regenerative capacity following injury, such as myocardial infarction, has spurred intense research efforts to develop innovative therapeutic approaches. NPs, with their unique properties and versatile functionalities, have emerged as a promising platform for targeted drug delivery and tissue regeneration in the cardiovascular system. These minute engineered particles, varying in size from tens to hundreds of nanometers, offer the potential to enhance the precision, efficacy, and safety of cardiac regenerative therapies. In this rapidly evolving field, researchers are focused on designing NPs that can deliver therapeutic agents to the site of cardiac injury, stimulate tissue repair and regeneration, and provide a supportive microenvironment for healing. This section explores the design principles, characteristics, and potential applications of NPs in cardiac regeneration, highlighting the exciting advancements that hold promise for improving the treatment and prognosis of cardiovascular diseases.

Nanomedicine is a swiftly advancing field that applies nanotechnology to medicine, encompassing both diagnosis and therapy. The term "nanomedicine" encompasses various innovative therapeutic approaches employing precisely bioengineered nanostructured materials, typically with at least one dimension within the range of 1–100 nm [28]. However, recent advancements have expanded the definition to include structures above 100 nm, including sub-micrometer and nanostructured microparticles, classified as nanomaterials and employed in various nanomedical applications. The primary goal of nanomedicine is to reduce the side effects of therapeutic drugs and augment their selective accumulation, thereby enhancing treatment efficacy in clinical settings. Conventional therapies often suffer from significant side effects due to drug toxicity, broad activity spectrum, and poor control over drug delivery [29]. NPs present a promising solution to address these limitations, owing to their tunable properties. Various forms of NPs have been successfully loaded with therapeutic agents like microRNAs (miRNAs) and drugs, facilitating their delivery through different routes of administration. These nanocarriers demonstrate distinct advantages over standard therapies, presenting a potential paradigm shift in the treatment of diseases. One significant challenge with therapeutic miRNAs is their rapid degradation and clearance in the bloodstream and cellular cytoplasm, resulting in a short half-life. Additionally, miRNAs struggle to efficiently penetrate cell membranes [30, 31]. Nature-inspired engineered miRNA nanocarriers, such as exosomes, have appeared as a promising solution to address these limitations, facilitating effective intercellular communication and improved delivery of therapeutic miRNAs [32]. The integration of nanomedicine with nanoparticle-based systems offers exciting possibilities for advancing the field of therapeutic medicine and addressing various medical challenges.

In addition to facilitating miRNA delivery, the utilization of bioengineered nanocarriers offers various advantages in improving the circulation time, bioavailability, and biodistribution of various drugs and proteins [33]. Many existing drugs, being lipophilic, face challenges in systemic administration due to their restricted aqueous solubility, leading to suboptimal delivery and therapeutic efficiency [34]. To overcome this, encapsulating these drugs within amphiphilic systems can improve their efficacy and enable sustained release at the targeted site. Protein therapy, known for its higher specificity and reduced toxicity compared to conventional drugs, faces hurdles such as enzymatic degradation or inactivation, short circulation half-lives, and inadequate membrane permeability [35]. NPs present a promising solution to safeguard therapeutic proteins against proteolysis, improve delivery efficiency, and maintain sustained release at the desired location.

Considering these advantages, nanotechnology has emerged as a promising tool for the delivery of cardioprotective drugs and for facilitating the prolonged release of growth factors to restore compromised heart function [36, 37]. Different administration routes, customized to the physicochemical characteristics of the drug/nanoparticle and the intended outcome, have been investigated to achieve optimal delivery for treating cardiomyopathies [38, 39]. Common routes include intravascular (intra-cardiac and intravenous injection) and extravascular (inhalation) methods, each offering unique advantages and drawbacks for addressing various pathological conditions, including compromised vascularization, fibrosis, and inflammation, aiming to ultimately enhance cardiac functionality.

In this section, we will present a comprehensive overview of the current advancements in nanomedicine concerning cardiac regeneration.

Polymeric NPs have attracted considerable interest owing to their outstanding characteristics, versatility, and tunable properties, making them highly attractive tools for controlled drug encapsulation and release. These NPs offer finely tunable physicochemical features such as surface charge, surface functionalities, and hydrophobicity, which enable efficient accommodation of nucleic acids, drugs, and proteins for effective intracellular release [40, 41]. This varied category of NP-based systems encompasses amphiphilic micelles, vesicles, dendrimers, and polymerases. Each of them exhibits distinct structures and properties, which can be effectively tailored during synthesis to accommodate various cargo types [42]. Numerous designed polymeric NPs integrate novel synthetic copolymers, amalgamating diverse functionalities such as targeting and selective cargo delivery systems [43]. Considering the growing utilization of miRNAs in cardiac regeneration, it is unsurprising that several studies have formulated polymeric NPs as carriers for miRNAs. These NPs are developed either independently or in conjunction with targeting elements or therapeutic drugs. In a notable study conducted by Yang and collaborators, polymeric NPs were employed, comprising a blend of poly(9,9-dioctylfluorene-alt-benzothiadiazole) (PBFT) and 1,2-distearoylphosphatidyl-ethanolamine-PEG-amino (DSPE-PEG-NH2) or DSPE-PEG-maleimide. These NPs were designed for the attachment of transactivator of transcription (TAT) peptide, facilitating the targeted delivery of therapeutic miRNA to the infarcted myocardium [44]. The miRNA NPs (miNPs) created a protective polymeric matrix for miR-199a, shielding it from enzymatic degradation, and allowed functionalization with TAT for enhanced cell uptake. In vitro experiments conducted on human embryonic stem cell-derived cardiac cells demonstrated that miNPs exhibited reduced cytotoxicity in comparison to commercially available lipocomplexes. Additionally, these miNPs showed comparable transfection efficiency under both normoxia and hypoxia conditions. Significantly, miNPs selectively induced the proliferation of hESC-derived cardiomyocytes and endothelial cells, while sparing human cardiac fibroblasts (hCFs). In vivo studies illustrated that the injection of miNPs into mice with beating hearts, in conjunction with an injectable hydrogel, yielded promising results which resulted in reduced scar size, enhanced muscle and vessel restoration, leading to long-term therapeutic effects, and restored cardiac function in the neo-vascularized myocardium for over three months. In a noteworthy investigation, Xue et al. [45] focused on the targeted delivery of NPs carrying a miRNA inhibitor, in contrast to a therapeutic miRNA. In this investigation, a novel dendrimer nanovector termed AT1-PEG-DGL, derived from PEGylated dendrigraft poly-L-lysine (PEG-DGL) and anchored with the AT1 targeting peptide, was created. The structure of AT1-PEG-DGL was confirmed through 1HNMR characterizations. The objective of this thoughtfully designed nanovector was to attain targeted delivery of the microRNA-1 inhibitor (AMO-1) to ischemic cardiac tissues, effectively exerting its antiapoptotic cardioprotective effects (Figure 7.2A). DGL, a derivative of naturally occurring lysine, is commonly employed for gene delivery owing to its distinct structure. To mitigate potential cytotoxicity and enhance stability, PEGylation was utilized, which prevented recognition by the reticuloendothelial system (RES) and prolonged in vivo circulation time. Prior studies have demonstrated that PEGylated DGL effectively suppresses VEGF expression in tumor mice using siRNA-VEGF. In this investigation, AT1 peptide was linked to PEGylated DGL to facilitate the specific targeting of hypoxic cardiac cells overexpressing AT1R during the initial phase of myocardial infarction (MI). AT1-PEG-DGL displayed remarkable gene compactness and formed 200 nm NPs (AT1-PEG-DGL at AMO-1 NPs) with a minimal positive charge, rendering them suitable for endocytic intracellular uptake due to effective electrostatic interactions with negatively charged cell membranes. Furthermore, AT1-PEG-DGL at AMO-1 remained stable in physiological environments containing serum for up to 96 hours, shielded by the PEG layer. At concentrations up to 200 μg/mL, AT1-PEG-DGL showed no significant cytotoxic effects and displayed high biosafety in gene vector delivery to cardiomyocytes compared to the commonly used bPEI, which has been associated with pronounced cytotoxicity. Furthermore, in vitro cellular uptake studies demonstrated that both PEG-DGL and AT1-PEG-DGL successfully delivered AMO-1 to the cytoplasm, where miR-1 becomes mature and functional (Figure 7.2B). However, the cellular uptake of AMO-1 was considerably higher in the AT1-PEG-DGL group in comparison to the PEG-DGL group, revealing the enhanced efficiency of AT1-PEG-DGL in targeting hypoxia cardiac cells that overexpress AT1 receptors. The results of the in vivo imaging further confirmed that AT1-PEG-DGL effectively targeted the infarcted heart in MI mice after intravenous injection, with fast accumulation within 30 minutes and peak accumulation at one-hour post-injection (Figure 7.2C). In contrast, PEG-DGL showed minimal accumulation until 3 hours post-injection. Therapeutic in vivo experiments were conducted to evaluate the antiapoptotic and cardioprotective effects of AT1-PEG-DGL at AMO-1 in MI mice. Following a 24-hour administration, the miR-1 level in the left ventricular tissue significantly decreased, while the levels of the antiapoptotic proteins Bcl-2 and PKCε increased in the AT1-PEG-DGL at AMO-1 group compared to the MI, AT1-PEG-DGL at scramble AMO-1, and PEG-DGL at AMO-1 groups. PKCε and Bcl-2 are recognized to inhibit apoptotic signaling pathways crucial for protecting the heart against MI and are targets of miR-1. Moreover, AT1-PEG-DGL at AMO-1 notably diminished cardiomyocyte apoptosis in the infarct border zone, resulting in a smaller MI size and more viable tissue within the infarct region. This is indicative of cardiac remodeling post-MI. Furthermore, AT1-PEG-DGL at AMO-1 demonstrated effective promotion of cardiac functional recovery compared to other treatments. The findings demonstrate that AT1-PEG-DGL nanovectors hold great potential for targeted delivery of AMO-1 to ischemic myocardial tissue, providing timely and emergency treatment of MI and eliciting cardioprotective effects against MI. They created a pegylated dendrigraft poly-L-lysine (PEG-DGL) dendrimer functionalized with an early myocardium targeting peptide (AT1) and an antisense oligonucleotide capable of inhibiting miR-1 (AMO-1) termed AT1-PEG-DGL-AMO-1. In vivo experiments revealed that the nanovector effectively targeted the infarcted heart of mouse within 30–60 minutes following a single intravenous injection, leading to a significant reduction in the infarcted area. The inhibition of miR-1 successfully attenuated cardiomyocyte apoptosis, leading to reduced cell death and enhanced cardiac repair. Yokoyama et al. [46] employed an innovative approach by combining nanomaterials with stem cell therapy. They utilized adipose-derived stem cells (ADSCs), known for their ability to promote neovascularization (NV) and prevent cell death after MI. To achieve this, the researchers loaded simvastatin-conjugated PLGA NPs onto ADSCs in vitro. This resulted in a natural recovery of the infarcted myocardium and an augmentation in vascularity. The ADSCs, laden with statin-PLGA NPs, were effectively directed to the ischemic myocardium, facilitating local release of the therapeutic payload. The combination of nanomaterials and stem cell therapy significantly enhanced cardiac regeneration, even with a small number of cells (10,000 cells per mouse), thanks to the gradual release of statin.

Vascular endothelial growth factor (VEGF), a well-studied proangiogenic cytokine, is recognized for its role in fostering NV in ischemic hearts. However, VEGF's therapeutic application has been limited by its potential adverse effects, which can vary depending on the dosage and delivery method. To overcome these constraints, Oduk et al. [47] investigated the potential of attaining prolonged exposure to a low dose of VEGF by encapsulating it in polylactic coglycolic acid NPs. The aim was to investigate whether treatment with VEGF-containing NPs could enhance cardiac function and protection against left ventricular remodeling in mice with experimentally generated MI. To accomplish this, polylactic coglycolic acid NPs of an average diameter of approximately 113 nm using double emulsion were produced, and they were loaded with VEGF. The encapsulation efficacy of VEGF in the NPs was found to be $53.5 \pm 1.7\%$ (equivalent to 107.1 ± 3.3 ng VEGF per milligram of NPs). In culture, these VEGF NPs demonstrated continuous release of VEGF for a minimum of 31 days. In a mouse model of MI, the administration of VEGF NPs resulted in significantly higher vascular thickness in the peri-infarct region, reduced infarct size, and improved left ventricular contractile function four weeks after the treatment. In the field of cell reprogramming, scientists developed NPs with spermine modification on acetylated dextran (AcDXSp) to encapsulate poorly water-soluble drugs – SB431542 (a transforming growth factor b (TGFb) inhibitor) and CHIR99021 (a glycogen synthase kinase-3 (GSK3) inhibitor) – for the purpose of reprogramming fibroblasts [48]. These NPs were additionally endowed with a targeting peptide (atrial natriuretic peptide, ANP) designed specifically for cardiac fibroblasts. The combined targeting and therapeutic impact of these NPs demonstrate potential in addressing the constraints of local injection. However, in vivo experiments to substantiate this hypothesis are still pending.

Liposomes, extensively investigated as carriers for drug delivery and transfection, consist of artificial vesicles composed of phospholipids and cholesterol that can encapsulate drugs, proteins/peptides, and DNA [49]. Although there may be concerns about potential stability issues during blood circulation, liposomes remain attractive due to their cost-effectiveness and scalability in synthesis. These characteristics have played a role in their clinical development, as demonstrated by Doxil, the first FDA-approved nanodrug. In the context of treating cardiomyopathy, liposomes have been employed to deliver therapeutic agents [50]. For instance, in a rat model of MI, the effects of anti-P-selectin-conjugated liposomes containing VEGF were investigated [51]. The targeted delivery of VEGF to the post-MI tissue resulted in noteworthy improvements in

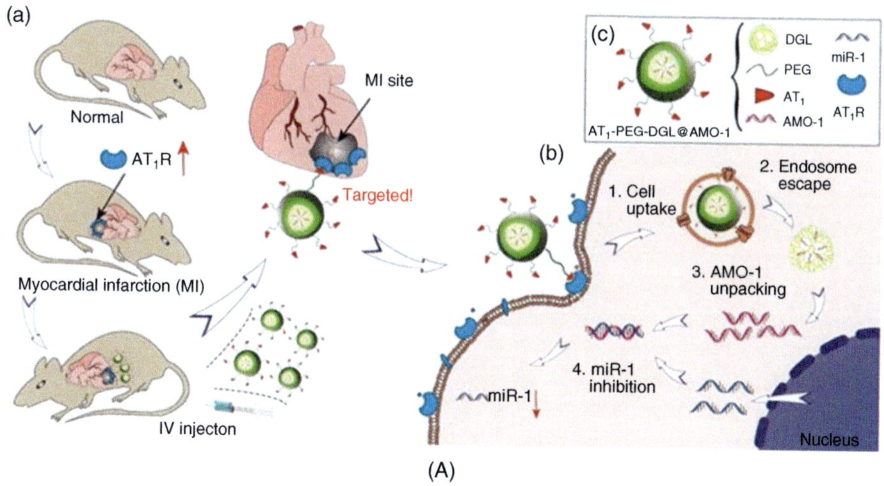

Figure 7.2 (A) Diagrams depicting the formation and delivery of the AT1-PEG-DGL at AMO-1 complex. (a) Upon intravenous injection, the AT1-PEG-DGL at AMO-1 complexes swiftly accumulate near the cardiac cells affected by ischemia, facilitated by the specific binding of the myocardium-targeted peptide AT1 to the AT1 receptor (AT1R). (b) Ischemic cardiac cells subsequently internalize the AT1-PEG-DGL at AMO-1 complexes through endocytosis, leading to the downregulation of miR-1 expression. (c) The AT1-PEG-DGL at AMO-1 complexes consist of an inner dendrimer nanovector for loading the miR-1 inhibitor and surface-bound AT1 for targeted delivery. (B) In vitro cellular uptake of AMO-1 when complexed with different agents (PEG-DGL, AT1-PEG-DGL, and bPEI) in H9C2 cells and live/dead staining analysis. (a) Representative confocal microscopy images of H9C2 cells treated with various complexes, showing the localization of FAM-labeled AMO-1 (in green) after a four-hour incubation. The cell nuclei were stained with DAPI (in blue). Scale bar: 25 μm. (b) Flow cytometry-based quantitative analysis of the cellular uptake intensity of different NPs in H9C2 cells after a four-hour treatment (left), with statistical analysis of the mean fluorescence intensity of these NPs (right). (c) Live/dead staining of H9C2 cells following a four-hour incubation with different complexes at the same concentration used in the cellular uptake experiments, visualized by microscopy. Live cells are shown in green and dead cells in red. Scale bar: 100 μm. (d) Statistical analysis of the percentage of viable cells. The data are presented as the mean ± standard deviation (S.D.) for a group of $n = 3$. Significance levels: $^{**}P < 0.01$, $^{***}P < 0.001$. (C) Overexpression of AT1 receptor in ischemic heart tissue 1–24 hours and 3–7 days following myocardial infarction, and myocardium-targeted visualization of AT1-PEG-DGL in mice with myocardial infarction. (a) Microscopy images depicting the expression of AT1 receptors in the ischemic myocardium at two different time points post-myocardial infarction (1–24 hours and 3–7 days). The control group consists of myocardium from Sham-operated mice, while the MI group includes myocardial infarction-induced heart tissue at the specified time intervals. AT1 receptors are highlighted in red (Alexa 647), and the cell nuclei are stained in blue (DAPI). Scale bar: 50 μm. (b) In vivo imaging of experimental mice, with Sham mice receiving saline as a control. Myocardial infarction (MI) mice were administered BODIPY-labeled AT1-PEG-DGL or PEG-DGL. (c) Measurement of fluorescence intensity in isolated hearts one hour after administration. MI mice injected with AT1-PEG-DGL exhibited significantly stronger fluorescence intensity in the isolated heart compared to the PEG-DGL group. Source: Reproduced with permission from Xue et al. [45]/Elsevier.

cardiac function and vasculature. Specifically, animals that received treatment exhibited a notable rise in fractional shortening and improved systolic function. These functional improvements were accompanied by a noteworthy 21% rise in the count of anatomical vessels and a substantial 74% increase in the count of perfused vessels within the MI region. Conversely, animals subjected to systemic VEGF treatment did not show significant improvements in cardiac function, nontargeted liposome treatment, or blank immunoliposomes, indicating the importance of targeted delivery in achieving the desired therapeutic effects. Another investigation utilized PEGylated liposomes that carried a targeting ligand against the angiotensin II type 1 (AT1) receptor, a receptor widely expressed in the infarcted heart [52]. In experiments with isolated cardiac cells, the AT1 NPs

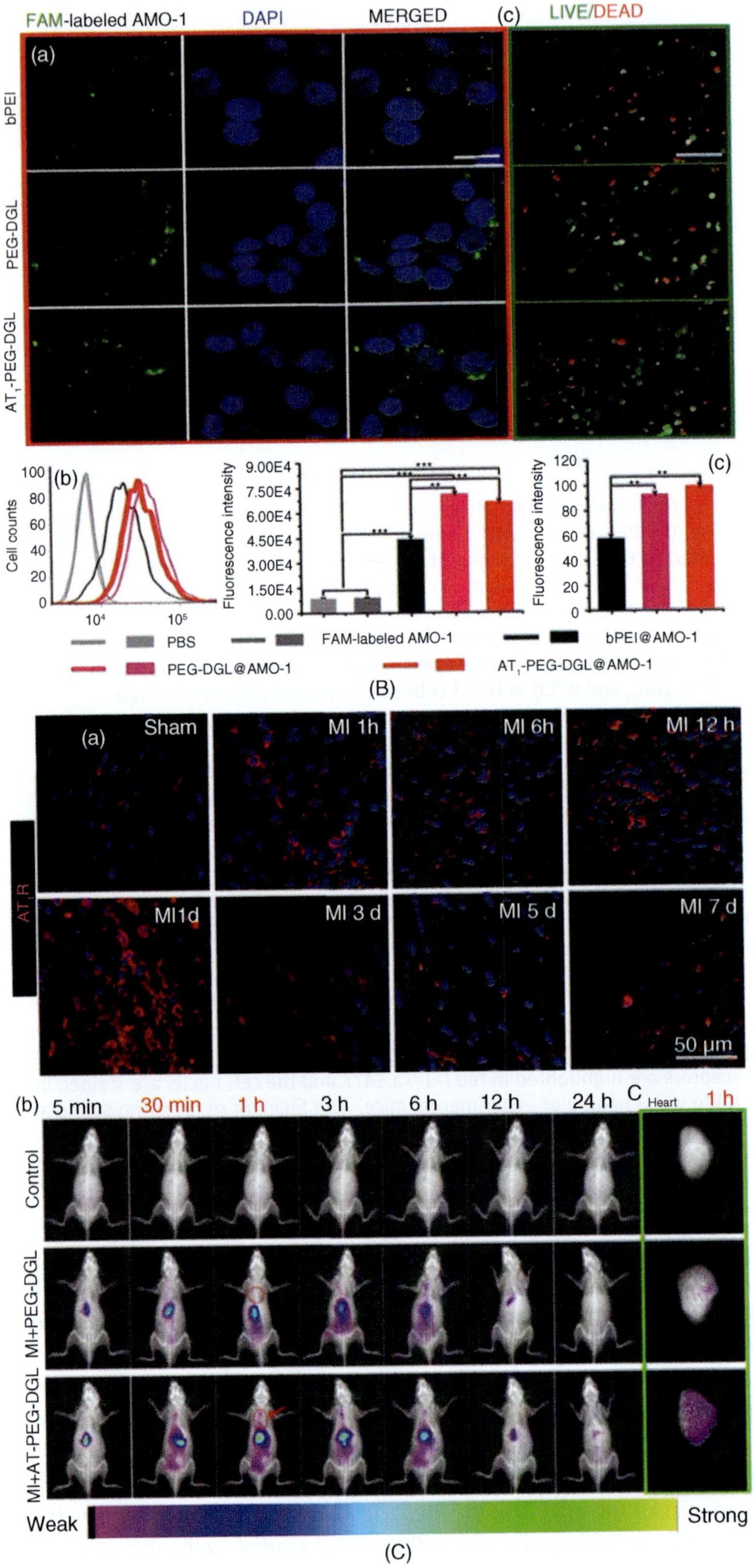

Figure 7.2 (*Continued*)

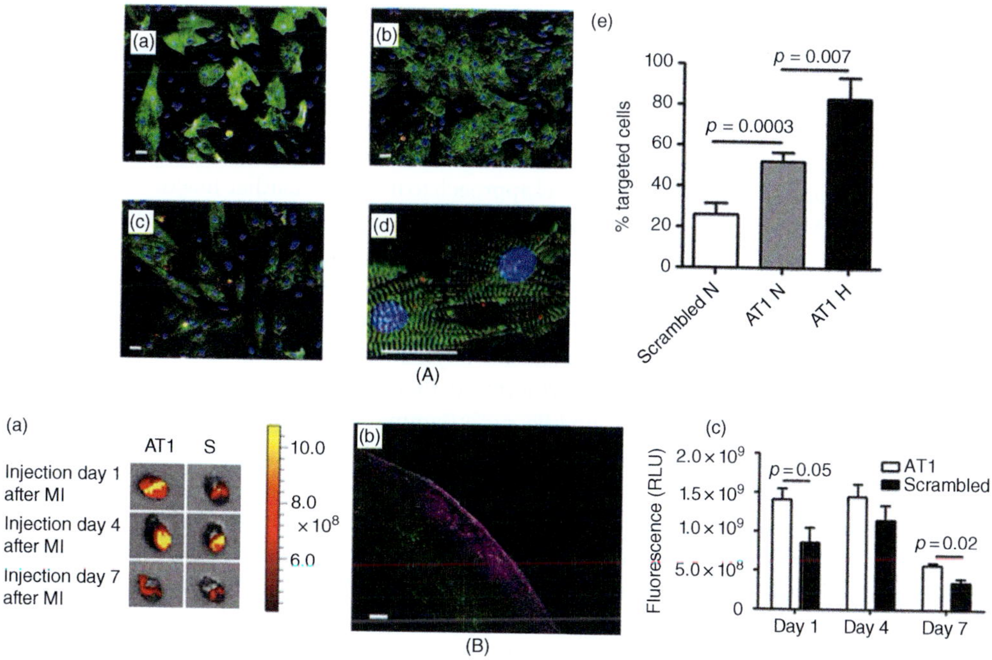

Figure 7.3 (A) In vitro targeting of cardiac cells. (a) Targeting of cardiac cells using nanoparticles that are linked with a non-specific scrambled peptide as a targeting agent. (b) Targeting using AT1 nanoparticles. (c) Cardiac cells cultured in a hypoxic environment and targeted by AT1 nanoparticles. (d) A closer view of panel C, highlighting multiple particles binding to each individual cell. Sarcomeric actinin is shown in green, cell nuclei in blue, and nanoparticles in red. The scale bar is set at 20 μm for all panels. (e) Quantitative analysis of the percentage of targeted cells ($n \geq 6$). The data are presented as means (±standard deviation). (B) Targeted delivery to the infarcted heart. (a) Imaging using IVIS of hearts that received injections of AT1 or scrambled (S) nanoparticles at 1, 4, and 7 days following myocardial infarction. The hearts were excised and imaged 24 hours after injection. The left ventricle (LV) is located in the lower right section of each image. (b) Accumulation of AT1 nanoparticles injected one day after myocardial infarction in the left ventricle of the infarcted heart. Nanoparticles are represented in red, while tissue autofluorescence is shown in green. The scale bar is set at 200 μm. (c) Quantitative assessment of fluorescence in isolated hearts collected one day after injection. Source: Reproduced with permission from Dvir et al. [52]/American Chemical Society.

demonstrated a significantly greater ability to target cardiac cells in contrast to NPs holding a targeter with scrambled amino acids. After 30 minutes, 52% of the cells incubated with AT1 NPs were targeted, compared to only 27% with scrambled particles (Figure 7.3A). This targeting ability was further enhanced to over 83% after 48 hours of hypoxia. In vivo experiments in mice with induced MIs showed that fluorescently labeled AT1 NPs accumulated predominantly in the left ventricle. Quantifying fluorescence levels verified a markedly higher accumulation of AT1 NPs in comparison to scrambled particles on days 1 and 7 post-infarction (Figure 7.3B). Additionally, negligible binding of NPs was observed in the hearts of sham-operated mice without coronary occlusion. The quantities of AT1 NPs gathered in the infarcted heart diminished over time (48%, 39%, and 27% on days 1, 4, and 7, respectively). This decline is probably attributed to alterations in inflammation, perfusion, and vascularization following the infarction. In summary, the findings underscore the potential of AT1 NPs as a targeted drug delivery system for the infarcted heart, offering valuable insights for potential therapeutic applications in cardiac regeneration following MI.

Investigations have been carried out into the utilization of liposomes to target immune system cells that regulate the inflammatory response in the infarcted heart. Directing nanoparticle (NP)-based therapies to immune cells at the inflammation site presents a promising and effective approach. This can be ascribed to the inherent inclination of the body's components to naturally recognize and internalize external materials, including NPs, when introduced into the body [53]. Cohen S. and colleagues have undertaken a novel approach to modulate cardiac macrophages into a reparative state after MI with the aim of promoting inflammation resolution and facilitating infarct repair [54]. Their strategy involves intravenous injections of liposomes presenting phosphatidylserine (PS), which imitates the anti-inflammatory effects of apoptotic cells. When these PS-liposomes were taken up by macrophages in both laboratory settings and living organisms, the cells significantly increased their exudation of anti-inflammatory cytokines, including transforming growth factor β (TGFβ) and interleukin 10 (IL-10). Furthermore, there was an upregulation in the expression of the mannose receptor-CD206; meanwhile, proinflammatory markers such as tumor necrosis factor α (TNFα) and the surface marker CD86 experienced downregulation. In an AMI rat model, researchers utilized magnetic resonance imaging (MRI) to showcase the successful targeting of PS-presenting liposomes to infarct macrophages following administration via the femoral vein. The treatment with these liposomes led to notable outcomes, including angiogenesis promotion, conservation of small scars, and prevention of ventricular dilatation and remodeling. In a more recent study by Cheng and colleagues [55], novel approach was introduced using biomimicking platelet-like proteoliposomes with their ability of interaction with monocytes. This method hinges on the interaction between immune system cells and platelet membrane proteins (PMPs), facilitating the enhanced concentration of liposomes at the injured site of the heart, where monocytes are recruited. The proteoliposomes were stocked with cobalt protoporphyrin IX (CoPP), a therapeutic compound recognized for its ability to suppress the inflammatory action of macrophages. These proteoliposomes were effectively "pulled" by the monocytes to the site of infarction. In vivo experiments exhibited promising outcomes in comparison to the systemic administration of free CoPP. Despite the potential benefits of this system, there are some concerns regarding its practicality. The unpredictability in the surface coating could create difficulties for reproducing the synthetic technique. As PMPs originate from diverse samples or batches, this can result in discrepancies in the composition and biophysical characteristics of the eventual particles. Additionally, the incomplete comprehension of the interaction with monocytes may constrain effective control over this mechanism, which in turn hinders the likelihood of clinical translation for proteoliposomes as a viable therapeutic option.

Recently, Schiffelers and his team introduced a liposomal carrier to deliver berberine, a natural compound recognized for its anti-inflammatory, antioxidative, and cardioprotective attributes [56]. Berberine, in its unbound state, faces limitations due to its low solubility in aqueous solutions and its brief presence in the bloodstream. These liposomes outperformed both control liposomes and free berberine in their effects. By administering the liposomes intravenously, they successfully targeted the inflammatory site to release the drug due to uptake by macrophages, resulting in a decrease in both inflammatory damage and systemic adversative effects. The berberine-loaded liposomes were carefully characterized, revealing a diameter of 0.11 μm and containing 0.3 mg/mL of the drug. In vitro experiments revealed that while free berberine significantly inhibited interleukin-6 (IL-6) secretion in RAW 264.7 macrophages (IC50 = 10.4 μM), the encapsulated berberine failed to replicate the same effect, as it remained trapped within the formulation throughout the study period. However, in vivo experiments showcased the efficacy of berberine-loaded liposomes, effectively preserving cardiac ejection fraction by 64% at day 28 post-MI, surpassing the effects of both control liposomes and free berberine. The study shows great promise because of the consistent and reproducible synthesis of the liposome formulation, its effective targeting capabilities, and its long-term stability. This outcome suggests that the delivery of berberine through liposomes substantially enhances its therapeutic availability, making it a promising treatment option for preserving cardiac function and mitigating adverse ventricular remodeling after MI. Lati et al. [57] concentrated on exploring the immunomodulatory function of a liposomal azithromycin (AZM) formulation (L-AZM) in a clinically pertinent model, with the objective of augmenting its therapeutic effectiveness and mitigating off-target effects. L-AZM was intravenously administered promptly after MI and assessed in comparison to free AZM (F-AZM). The findings demonstrated that L-AZM markedly decreased cardiac toxicity and related mortality by 50% in mice. Moreover, it prompted a substantial shift in favor of reparatory/anti-inflammatory macrophages, resulting in a noteworthy reduction in cardiac inflammatory neutrophils and the penetration of inflammatory monocytes. This modulation of immune cells was associated with the downregulation of pro-inflammatory genes and the upregulation of anti-inflammatory genes. The immunomodulatory effects of L-AZM were correlated with a decrease in cardiac cell death and scar size, along with an improvement in angiogenesis. Overall, the use of L-AZM proved to enhance the recovery and survival of cardiac cells after MI, while also protecting from the cardiac off-target effects associated with F-AZM. This study illustrated that the liposomal formulation of AZM not only enhances the drug's effectiveness but also guarantees its safety in an animal model of acute myocardial injury. It is noteworthy that this study is the first to establish the immunomodulatory properties of liposomal AZM formulations. These encouraging findings strongly advocate for the initiation of clinical trials to assess L-AZM as a novel and clinically relevant therapeutic strategy for improving cardiac recovery and mitigating heart failure post-(MI in human patients. In summary, liposomes offer a promising avenue for drug delivery in cardiac regeneration therapy, holding the potential to improve treatment effectiveness and target specific areas within the damaged heart following MI.

Inorganic NPs possess favorable physical properties that can be harnessed for both diagnosis and therapy simultaneously, a concept known as theranostics, applicable to a range of medical conditions [57]. Typically, these NPs are composed of an inorganic core enveloped by an organic or inorganic shell, augmenting their biocompatibility and interactions with the biological environment [58]. Cells (hMSCs) affect vital molecular processes and promote cell adhesion both in laboratory settings and when injected into infarcted rat hearts [59]. The internalization of SiO_2-NPs also facilitates communication between transplanted cells and the host a critical factor for achieving successful engraftment and tissue regeneration. The study revealed that when SiO_2-NPs are administered to cells, they form stable agglomerates of approximately 1 μm in size, with a protein corona that alters the ζ-potential of the NPs (Figure 7.4A). During incubation, the SiO_2-NPs were retained inside the cells and initiated an endocytic process, leading to invaginations of the cell plasma membrane. Moreover, SiO_2-NP internalization was found to impact cellular focal adhesion complexes (FACs), critical for cell adhesion, and migration. This internalization resulted in elongated and larger FACs, indicating possible FAC stabilization, leading to enhanced cell adhesion capability in comparison to untreated cells. The adhesive strength of SiO_2-NP-treated cells was significantly greater than that of untreated cells as demonstrated by trypsin/EDTA-induced detachment and adhesion assays (Figure 7.4B).

Furthermore, SiO_2 NP internalization also influenced the surface expression of Connexin 43 (Cx_{43}), an essential gap junction protein in the heart. SiO_2-NP treatment elevated Cx_{43} surface expression in hMSCs, indicating enhanced communication between cells. This was substantiated by dye transfer assays, demonstrating heightened dye exchange between SiO_2-NP-treated hMSCs and cardiomyoblasts, signifying improved intercellular communication. (Figure 7.5A). The study also observed that the internalized SiO_2-NPs accumulated in lysosomes, temporarily impairing lysosomal degradation activity. This impairment was linked to the transient increase

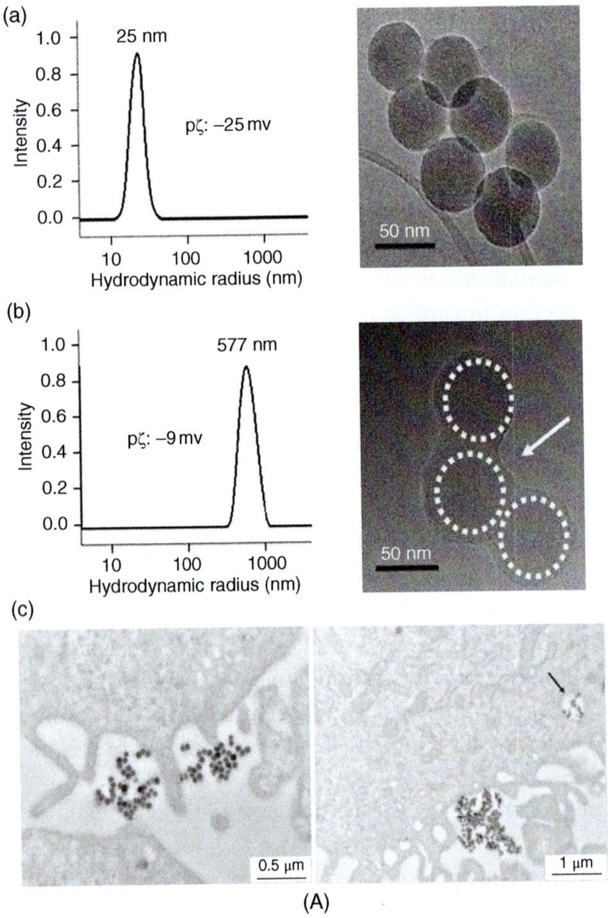

Figure 7.4 (A) Silica nanoparticle characterization and interaction with human mesenchymal stem cells (a) The first section involves examining the hydrodynamic radius through dynamic light scattering and ζ-potential (seen in the left panel) of silica NPs (SiO_2-NP) dispersed in water. Additionally, a transmission electron microscopy (TEM) image (shown in the right panel) illustrates the NPs in this aqueous dispersion. (b) The second part investigates the changes in hydrodynamic radius (determined via dynamic light scattering) and ζ-potential (highlighted in the left panel) of SiO_2-NP after they were suspended for 1 hour in a complete medium containing 1% fetal bovine serum. The TEM image (depicted in the right panel) reveals the NPs with a visible corona formed by serum proteins adsorbed onto the NPs, and their shape is outlined by a dotted circle. (c) The final section includes transmission electron microscopy images depicting human mesenchymal stem cells after 1 hour of incubation with SiO_2-NP in a complete medium supplemented with 1% fetal bovine serum. The magnification level used for these images is 50,000×. (B) Impact of silica NPs on adhesion of human mesenchymal stem cells (a) evaluation of hMSC adhesion using a trypsin-based cell detachment assay. Both control (CTR) and silica nanoparticle-human mesenchymal stem cells (SiO_2-NP-hMSCs) were treated with 0.025% trypsin-0.02% EDTA for specified time intervals. The study assessed the ability of hMSCs to remain attached to the substrate. (b) Adhesion examination. CTR and SiO_2-NP-hMSCs were allowed to adhere to fibronectin-coated plates for the indicated time points. In both assays (A and B), adherent cells were stained with 0.1% crystal violet, and their relative absorbance was determined spectrophotometrically at 540 nm. The data are presented as the mean ± standard deviation. (c) Visualization of CMFDA-labeled hMSCs within infarcted hearts. (d) Visualization of hMSCs labeled with red fluorescent cyanine dye-doped SiO_2-NPs within infarcted hearts. For both images (C and D), a 10 μm transverse slice reconstruction illustrates the distribution of labeled hMSCs in infarcted hearts, with green/red points highlighting hMSC clusters. Magnification: 10×; scale bar: 200 μm. (e and f) Superimposed images display sarcomeric α-actinin staining in red (E) and green (F), alongside labeled hMSCs, which appear green (CMFDA in E) and red (SiO_2-NPs in F). Magnification: 20×. Significance levels: $^{**}p < 0.01$; $^{***}p < 0.005$. CMFDA stands for 5-chloromethylfluorescein diacetate. Source: Reproduced with permission from Popara et al. [59]/FUTURE MEDICINE LTD.

Figure 7.4 (*Continued*)

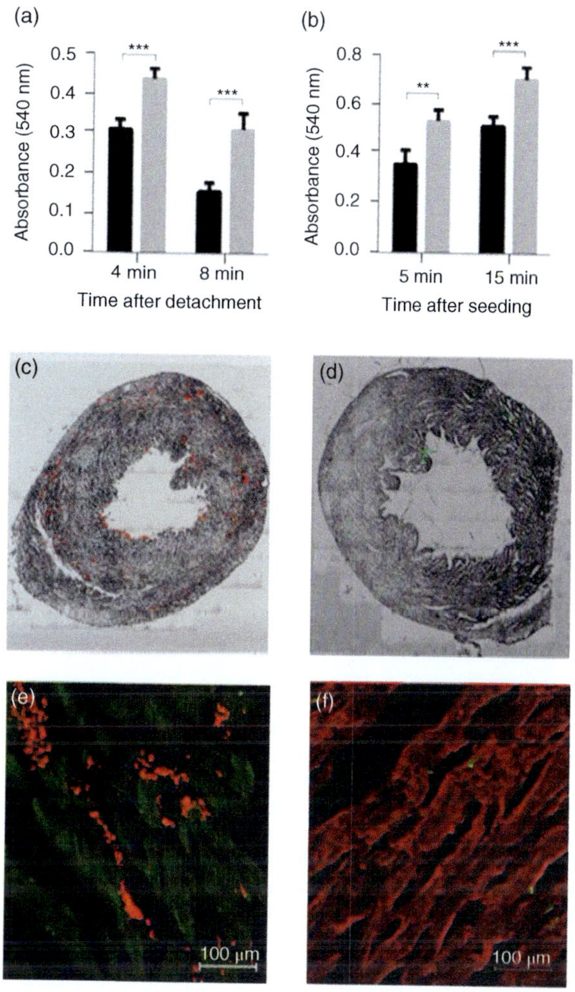

(B)

in cell adhesion, Cx43 expression, and accumulation of autophagosomes in SiO_2-NP-treated cells (Figure 7.5B). Altered autophagic flux was evident from increased levels of LC3-II and p62 proteins. Overall, these findings suggest that SiO_2-NP internalization enhances the adhesion ability of hMSCs, improves their communication with surrounding cells, and influences cellular lysosomal function. These findings carry implications for regenerative medicine, especially in the realm of cardiac regenerative therapies. Nevertheless, it is crucial to address potential long-term toxicity and side effects linked to the prolonged intracellular accumulation of SiO_2-NPs.

The potential of biodegradable calcium phosphate NPs (CaPs) for targeted delivery of therapeutic peptides to the heart through inhalation was evaluated by Miragoli et al. [60]. In vitro evaluation procedures were conducted before in vivo administration to assess the stability of CaPs in aqueous suspension and their interaction with proteins. It was observed that the surface charge of CaPs remained stable over time, while the particle size decreased slightly and then remained stable. The interaction of CaPs with proteins was additionally confirmed, signifying that the NPs were dispersed in the presence of serum proteins. Next, near-infrared fluorophore-loaded CaPs were generated and characterized to assess their stability and encapsulation efficiency (EE). It was discovered that the fluorophore was efficiently encapsulated within the CaPs, and the NPs did not

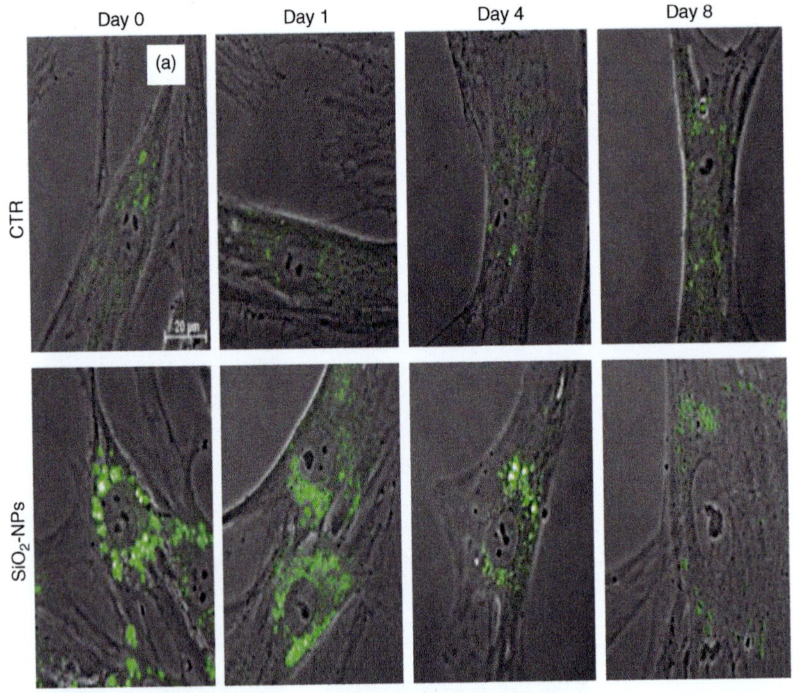

Figure 7.5 Assessment of the temporal impact, metabolism, and cellular characteristics in response to silica nanoparticles on human mesenchymal stem cells (A) Human mesenchymal stem cells (hMSCs) subjected to control conditions (CTR) and treatment with silica nanoparticles (SiO_2-NP) were labeled with LysoTracker (in green) at various time points ranging from day 0 to day 8. Representative images were captured using confocal microscopy, employing a 63× magnification. (b) Analysis of CellTiter-Blue fluorescence was performed at day 0, day 1, day 4, and day 8, with fluorescence values being normalized against control cells and expressed as the mean ± standard deviation (SD). (c) After undergoing recovery in complete medium for 4, 8, and 14 days, CTR and SiO_2-NP-treated hMSCs participated in an adhesion assay, with adherent cells being stained with 0.1% crystal violet. The relative absorbance was quantified spectrophotometrically at 540 nm, and the data are presented as the mean ± SD; $^*p < 0.05$. (d) Following recovery in complete medium for 1, 4, and 8 days, the expression of Cx43 was assessed through confocal microscopy in both CTR and SiO_2-NP-treated hMSCs. The box plot illustrates the quantification of Cx43 area in these cells, with an analysis involving at least 120 cells. An unpaired t-test was performed for statistical analysis; $^{**}p < 0.01$. (B) Suppression of lysosomal degradation functionality (a) adhesion test: Human mesenchymal stem cells (hMSCs) were exposed to either control medium or Bafilomycin A1 at a concentration of 100 nM for a duration of 2 hours. Following treatment, the cells were permitted to adhere to fibronectin-coated plates for specific time intervals. Subsequently, the adherent cells were stained with 0.1% crystal violet, and their relative absorbance was measured spectrophotometrically at 540 nm. The data are presented as the mean ± standard deviation (SD); $^{***}p < 0.001$, $^{****}p < 0.0001$. (b) Cx43 expression analysis: hMSCs were subjected to a two-hour treatment with either control medium or Bafilomycin A1 (100 nM), after which the expression of Cx43 was assessed using confocal microscopy. An unpaired t-test was employed for statistical analysis, and the data are displayed as the mean ± SD; $^{***}p < 0.001$. (c) Western Blotting: Control (CTR) and silica nanoparticle (SiO_2-NP) treated hMSCs were analyzed via western blotting to evaluate the levels of p62 and the LC3-II/LC3-I ratio. The quantification was performed and normalized to untreated control cells, and the data are illustrated as the mean ± SD; $^*p < 0.05$, $^{***}p < 0.001$. (d) Confocal imaging: Left panel exhibits representative confocal images of CTR or SiO_2-NP treated hMSCs under the aforementioned conditions. Post-fixation, the cells were stained for LC3. In the right panel, quantification of LC3 puncta per cell is shown. Data are presented as the mean ± SD; $^{***}p < 0.001$, $^{****}p < 0.0001$. (e) Left panel displays representative confocal images of SiO_2-NP treated hMSCs stained with LC3 and/or LysoTracker green. The right panel reveals Pearson's correlation analysis, which indicates minimal colocalization between LC3 and LysoTracker (0.223 ± 0.172) and strong colocalization between SiO_2-NPs and LysoTracker (0.909 ± 0.255). The data are represented as the mean ± SEM, with $n = 3$ independent experiments. Source: Reproduced with permission from Popara et al. [59]/FUTURE MEDICINE LTD.

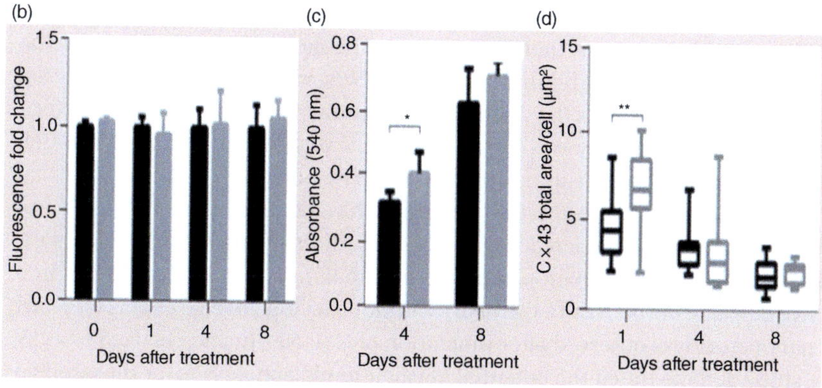

(A)

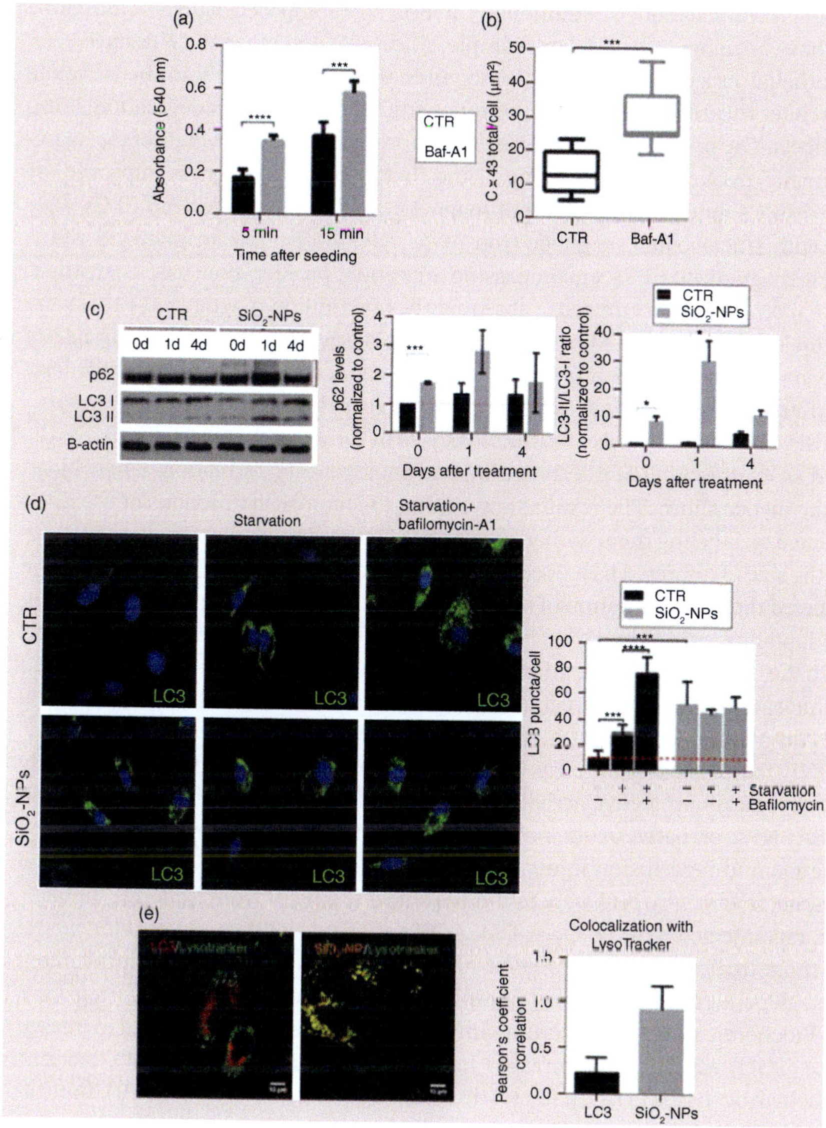

(B)

Figure 7.5 *(Continued)*

release the fluorophore under various conditions. Administering CaPs to the heart through various routes of administration in healthy mice was then investigated. The most efficient and rapid delivery to the myocardium was achieved through inhalation of CaP NPs, while enteral administration did not effectively target the heart. To assess the therapeutic potential of CaPs for peptide delivery, a mouse model of diabetic cardiomyopathy was employed, and it underwent treatment through the inhalation of CaPs laden with a therapeutic peptide (MP). The outcomes demonstrated that employing this approach resulted in the restoration of cardiac function and the amelioration of molecular and cellular abnormalities associated with diabetic cardiomyopathy. Apart from mice, the viability of cardiac targeting through nanoparticle-based inhalation was also evaluated in a porcine model. Effective delivery of the NPs to the heart without causing adverse effects on heart function or systemic parameters was observed after inhalation of CaP NPs in anesthetized healthy pigs. In summary, the study demonstrated the potential of calcium phosphate NPs for the targeted delivery of therapeutic peptides to the heart through inhalation, presenting a promising approach for cardiac treatment in preclinical models. Additionally, hybrid NPs composed of diverse inorganic core–shell structures have been investigated. For example, silicate-coated magnetic NPs have been employed to tag endothelial progenitor cells (EPCs) to guide them magnetically to the ischemic heart [61]. EPCs are frequently utilized in cell therapy following MI due to their mobilization, homing, and angiogenic effects. The primary aim of this investigation was to evaluate whether the aggregation of EPCs in ischemic areas could be augmented, and therapeutic effectiveness improved, by employing magnetized EPCs guided by an external magnetic field. Male rat-derived EPCs were isolated and labeled with silica-coated magnetic iron oxide NPs to generate magnetized EPCs. Diverse aspects of these magnetized EPCs, encompassing migration, proliferation, vascularization, and cytophenotypic markers, were scrutinized. Subsequently, one million magnetized EPCs were transplanted into a female rat model with MI through the tail vein, seven days after the initiation of MI. Some rats underwent transplantation aided by an external magnet positioned above the infarct area, while others did not. Following a four-week period, the researchers assessed cardiac function, myocardial fibrosis, and the apoptosis of cardiomyocytes in the treated rats. Additionally, the investigation included an examination of EPC retention and angiogenesis (formation of new blood vessels) in the ischemic myocardium. The results showed that the biological functions of the EPCs were minimally impacted by labeling them with magnetic NPs. Moreover, cardiac function was significantly improved, the size of infarcted tissue reduced, and myocardial apoptosis decreased in the rats with MI who received the transplantation of magnetized EPCs directed by an external magnet. Moreover, this transplantation guided by an external magnet resulted in augmented aggregations of magnetized EPCs in the border zone of the infarcted area. This was accompanied by a substantial rise in the density of microvessels and an upregulation of proangiogenic factors (factors promoting blood vessel growth) compared to the rats without external magnet guidance.

Iron oxide NPs (IONPs) have been explored as potential cardioprotective agents [62]. The study investigated the potential cardioprotective activity of 2,3-dimercaptosuccinic acid stabilized iron oxide (Fe_2O_3 at DMSA) NPs, prepared using a co-precipitation method, in both animal and cell models of heart ischemia and reperfusion injury. The Fe_2O_3 at DMSA NPs possessed a spherical core of an average diameter of 9.8 nm and were found to protect the heart from ischemic damage. In the animal model, rats injected with Fe_2O_3 at DMSA NPs showed significantly smaller infarct regions compared to those treated with normal saline after 30 minutes of injury. The improvement in infarct size was dose-dependent, with the most significant enhancement observed at 0.5 mg/kg of Fe_2O_3 at DMSA NPs. Biochemical analysis of serum samples from the treated rats showed increased superoxide dismutase (SOD) activity and decreased levels of malondialdehyde (MDA), creatine kinase (CK), lactate dehydrogenase (LDH), and creatine kinase isoenzyme-MB (CK-MB), indicating reduced oxidative stress and damage to heart tissue. To further confirm the cardioprotective activity, the study used a Guinea pig Langendorff heart model. Fe_2O_3 at DMSA nanoparticle-treated hearts showed considerable recovery in left ventricular developed pressure (LVDP) compared to untreated hearts during reperfusion after ischemia. The treated hearts also appeared normal in color, whereas the untreated hearts looked pale. In neonatal rat cardiomyocytes, Fe_2O_3 at DMSA NPs were internalized by the cells in a dose-dependent manner. At nontoxic doses, these NPs elevated cell viability and SOD activity, while concurrently diminishing the content of MDA and reactive oxygen species (ROS). The NPs also impacted cell energy metabolism and increased the generation of nitric oxide synthase (NOS), nitric oxide (NO), and S-nitrosothiols, further indicating their protective effects. The study explored the role of surface charges, surface molecules, and iron ions in the cardioprotective activity of Fe_2O_3 at DMSA NPs. Surface charges and surface molecules did not exert a substantial influence, and intact NPs were essential for optimal activity. Interestingly, the study found that Fe_2O_3 at DMSA NPs with a smaller core size (9.8 nm) were more effective in heart recovery than those with a larger core size (35.2 nm). Comparison with other cardiovascular drugs, Verapamil (synthetic drug) and Salvia miltiorrhiza extract (natural product), showed that Fe_2O_3 at DMSA NPs were more potent in improving heart recovery, and they did not have the same toxic effects observed with Verapamil. In summary, the study demonstrated that Fe_2O_3 at DMSA NPs had a strong cardioprotective activity in both animal and cell models of heart ischemia and reperfusion injury. These NPs demonstrate promising potential as a novel therapeutic strategy for shielding the heart from ischemic damage. However, additional research is necessary to comprehensively comprehend the mechanisms involved and evaluate their long-term safety and efficacy in clinical applications.

Inorganic NPs have also been devised for diagnostic applications. Gold NPs, in particular, exhibit promise as contrast agents for computed tomography (CT), an X-ray-based imaging technique that exploits variations in tissue absorption to generate images of bodily structures. Kee and Danila [63] used Au NPs coated with CNA35, collagen-binding adhesion protein 35, for molecular-level CT imaging of infarcted hearts. A clinically promising contrast agent for X-ray molecular imaging has been developed using gold NPs (AuNPs) functionalized with CNA35. This novel contrast agent offers two key capabilities: persistent circulation enhancement in blood pool for vascular imaging of coronary arteries and targeted binding to collagen within myocardial scar tissue. At a concentration of approximately 45 mg Au/mL, the CNA35-AuNPs demonstrated stable and enhanced blood pool circulation at 142–160 HU within one hour of administration intravenously. Furthermore, at the six-hour mark, specific signal enhancement was observed in the myocardial scar tissue of rats inoculated with CNA35-AuNPs, while control AuNPs or noninjected control animals did not exhibit such enhancement. In conclusion, CNA35-AuNPs exhibit significant potential as a contrast agent for CT, serving well in vascular imaging for detecting CAD and molecular imaging to identify myocardial scar tissue within the heart. This study underscores the advantages of using functionalized Au NPs for targeted and improved detection of infarcted hearts compared to traditional iodinated agents. Numerous clinical trials are currently exploring the potential of inorganic NPs for addressing cardiovascular diseases [64]. Yet, concerning MI, the emphasis has predominantly centered on MRI applications employing Ferumoxytol, an ultrasmall superparamagnetic iron oxide nanoparticle (USPION) preparation [65]. Unfortunately, these MRI applications using USPION NPs have not yielded the expected outcomes, leading to a rather disappointing performance. Consequently, the overall use of USPION NPs for MRI has fallen short of meeting the desired standards, and presently, Ferumoxytol finds predominant utilization in iron replacement therapy for adults grappling with chronic kidney diseases (CKDs) [66]. Despite the extensive history of employing inorganic NPs in diverse nanomedical applications, their application in cardiac regeneration remains relatively limited, with only a few successful examples reported. Consequently, the effective translation of these findings into successful clinical interventions for cardiac regeneration still remains a significant challenge.

To summarize, nanomedicine is transforming cardiac regeneration by addressing long-standing challenges in drug delivery and cardiac repair. Bioengineered nanocarriers, such as polymeric NPs, provide precise control over drug release and have been used to deliver therapeutic miRNAs, facilitating cardiac healing while sparing non-cardiac tissues. The combination of nanomaterials and stem cell therapy enhances cardiac regeneration by enabling the targeted delivery of therapeutic agents at the specific site. Notably, VEGF-loaded NPs have displayed potential in improving vascularization and cardiac function in ischemic hearts. Additionally, NPs designed for cell reprogramming offer opportunities to overcome treatment barriers through their targeted and therapeutic effects. These advances underscore the potential of nanotechnology in advancing cardiac regeneration and providing more effective, targeted therapies for heart-related conditions. Liposomes, as extensively studied drug delivery and transfection carriers, offer a cost-effective and scalable method for encapsulating therapeutic agents like proteins, peptides, and DNA. Liposomes have demonstrated promise in cardiomyopathy treatment, particularly when used for targeted delivery. For instance, liposomes containing VEGF have shown significant enhancements in cardiac function and vasculature in models of MI, highlighting the significance of targeted delivery for therapeutic effectiveness. Additionally, liposomes have been employed to convert cardiac macrophages into a reparative state post-MI, promoting inflammation resolution and infarct repair. Moreover, liposomes have been utilized for delivering natural compounds like berberine and therapeutic peptides such as azithromycin, leading to successful outcomes in preserving cardiac function and mitigating adverse ventricular remodeling. Inorganic NPs, including silicon oxide and calcium phosphate NPs, hold the potential for improving cell adhesion, intercellular communication, and therapeutic peptide delivery, offering valuable contributions to cardiac regenerative therapies. These NPs have also shown promise in inhalation-based cardiac targeting in preclinical models. In summary, both liposomes and inorganic NPs present innovative solutions for drug delivery and cardiac regeneration, offering the potential to address specific targets within the damaged heart after MI. However, further research is needed to ensure clinical translation and long-term safety while refining these approaches.

7.2 Nanotechnology for the Treatments for Diabetes

Nanotechnology has great potential to revolutionize the landscape of diabetes treatment and management. Researchers are leveraging the ability to manipulate materials at the nanoscale, one billionth of a meter, to unlock innovative methods for enhancing insulin delivery, enabling continuous glucose monitoring, and advancing regenerative medicine strategies. Nanotechnology-driven interventions aspire to notably enhance the quality of life for individuals facing the challenges of diabetes by providing more efficient, targeted, and personalized therapies. As ongoing research continues to reveal the full potential of nanotechnology in diabetes treatments, there is a strong sense of hope for safer, more effective, and transformative advancements in managing this prevalent chronic condition. This section delves into cutting-edge advancements and ongoing research exploring nanotechnology's application in diabetes, with a focus on improving patient outcomes, enhancing quality of life, and shaping a brighter and more effective future for diabetes management.

7.2.1 Nanotechnology for Diabetic Retinopathy Treatment

The hyperglycemic condition associated with diabetes significantly contributes to the onset of diabetic retinopathy (DR) [67]. Elevated blood glucose levels within retinal vessels induce heightened vascular permeability, resulting in the accumulation of fluid in and beneath the retina [67]. This process leads to edema and necrosis of nerve cells. Additionally, hyperglycemia causes harm to the blood-retinal barrier (BRB), causing blood leakage from different retinal vessels [68]. This, in turn, initiates ischemic changes in the adjacent retina. The ensuing ischemic condition prompts the activation of the body's compensatory mechanisms, fostering the formation and proliferation of new blood vessels. However, this process can result in complications such as retinal detachment and vision loss due to stretching [68, 69]. Nanotechnology has surfaced as a focal point in addressing the pathogenesis of DR, presenting potential solutions [69]. DR is a chronic complication with the potential to cause blindness, particularly in severe cases, and is more prevalent in individuals with diabetes for over a decade [70]. This condition can be categorized into two types with a total of six stages: nonproliferative diabetic retinopathy (NPDR) and proliferative diabetic retinopathy (PDR). Current treatments include laser therapy for timely intervention and drugs to relieve macular edema. Vitrectomy is widely recognized as the most efficient approach for advanced DR due to its ability to restore the anatomical integrity of the retina. However, conventional drug treatments face certain limitations in addressing the complexities of the condition. Nanotechnology has tackled these challenges by providing benefits such as reduced particle size, enhanced penetration through the BRB, favorable biocompatibility, and prolonged drug release, all contributing to effective nerve tissue regeneration. An increasing array of nanomaterials has been applied in clinical DR treatment, providing potential solutions and advantages. Nevertheless, the safety and feasibility of these nanomaterials remain subjects of ongoing research, underscoring the importance of thorough investigation before their application in DR treatment. This subsection is dedicated to examining the application and advancement of nanotechnology in the treatment of DR. It seeks to explore the role of nanotechnology in preventing and treating DR, as well as its potential contributions to the regeneration of nerve tissue damaged by the condition.

In order to prevent DR, blood pressure, sugar, and intense physical activity management are not the only measures to take, regular fundus examinations play a crucial role in early identification. Since the 1960s, fluorescein angiography – which includes injecting sodium fluorescein, a fluorescent dye, into the antebrachial vein – has been regarded as the gold standard for DR fundus investigation. Since the 1960s, fluorescein angiography – which includes injecting sodium fluorescein, a fluorescent dye, into the antebrachial vein – has been regarded as the gold standard for DR fundus investigation. Angiography is a technique used to depict dynamic blood circulation within the retina by taking pictures of the bright green light released as the dye passes through retinal arteries, showing microstructural and microcirculatory changes in fundus vessels. Researchers like Vedarethinam et al. [71] suggested employing vanadium core–shell nanorods to identify DR metabolic alterations which allowed for early detection and diagnosis. Laser desorption ionization mass spectrometry and vanadium core–shell nanorods were used to identify and stage diabetic retinopathy (LDI MS). Core–shell nanorods with a vanadium outer layer were fabricated using silica as the core material, employing a layer-by-layer assembly approach and modification with resorcinol formaldehyde (RF) to facilitate vanadium anchoring. The resulting nanorods exhibited increased roughness and nanocrevices, which were considered crucial for their application as matrices in LDI MS. Optimization was carried out on the core–shell structures and surface roughness of the nanorods. SiO_2 at V_2O_5-2 nanorods exhibited superior performance for laser desorption/ionization mass spectrometry (LDI MS). Using the optimized SiO_2 at V_2O_5-2 nanorods as matrices, the detection of small metabolites in bio-fluids, including serum, urine, cerebrospinal fluid, and plasma, was successfully achieved. A noninvasive and rapid approach with several advantages over traditional mass spectrometry techniques was demonstrated, including high analysis throughput, minimal sample pretreatment, and simple spectra. The application of the vanadium core–shell nanorods and LDI MS for the detection and staging of DR was carried out through untargeted metabolomic profiling of plasma samples from DR and non-DR control patients. The resulting plasma metabolite fingerprints, referred to as metabolic features (MFs), were used to create classification models using the least absolute shrinkage, and selector operator (LASSO) technique was used to create classification models using the LASSO technique. High sensitivity, specificity, and area under the curve (AUC) values for distinguishing DR patients from non-DR controls in both the discovery and validation cohorts were achieved by the classification models based on the MFs. Notably, the AUC values obtained for the LDI MS-based approach were superior to those achieved using traditional clinical measurements such as fasting plasma glucose (FPG) and hemoglobin A1c (HbA1c), which are now advised for DR screening. Furthermore, a panel of 20 metabolic biomarkers was identified from the LASSO-generated models, which were strongly predictive of DR progression. The metabolite profiling throughout various stages of DR, ranging from mild to severe, revealed substantial variations in the metabolic panel. These findings provide valuable insights into the pathogenic pathways associated with the development of DR. The clinical implications of this study are noteworthy. The use of Laser Desorption/Ionization Mass Spectrometry (LDI MS) facilitated a quantitative, rapid, and straightforward staging of DR. This technique, enhanced by vanadium core–shell nanorods, accelerates the monitoring of DR progression in a more efficient and personalized manner. The early identification and intervention enabled by this approach hold the potential to mitigate the risk of vision loss in diabetic patients. Additionally, the metabolic signature panel identified in this study offers promising biomarkers for future clinical applications related to DR. Beyond routine monitoring and supplementary examinations, attention to hypoxia emerges as another critical factor in preventing the onset of DR. Hypoxia serves as the precursor stage for retinal NV, emphasizing the critical importance of pinpointing hypoxic regions within retinal tissue [72]. Addressing hypoxia at an early stage becomes imperative in the effort to impede the progression of DR. Recognizing and mitigating hypoxia can play a pivotal role in delaying the onset of retinal NV and consequently preventing further complications associated with DR [73]. This underscores the need for proactive strategies that target hypoxia as a key factor in the development and progression of DR. Research findings suggest that Hypoxia-inducible factor-1 (HIF-1) plays a crucial role in mitigating the expression of inflammatory factors, such as IL-6 and TNF-α, in the context of DR [74]. Furthermore, HIF-1 is implicated in the regulation of microRNA expression, specifically influencing miR-125b-5p and miR-146a-5p within the retina. These regulatory mechanisms involving HIF-1 showcase its potential as a key player in modulating the inflammatory response and microRNA expression patterns associated with DR. Understanding these molecular interactions provides valuable insights into potential therapeutic targets for managing inflammation and associated complications in DR [75]. In addition to its role in regulating inflammatory factors and microRNA expression, Hypoxia-inducible factor-1 (HIF-1) also enhances the expression of pivotal growth factors, including VEGF and platelet-derived growth factor (PDGF) [76]. The upregulation of these growth factors is associated with angiogenesis and vascular permeability, processes implicated in the pathogenesis of DR. Targeting HIF-1 antagonists to hypoxic tissues emerges as a promising strategy to effectively prevent the progression of DR.

Notably, among these antagonists, daunorubicin, and doxorubicin (DXR) have been identified for their ability to inhibit HIF-1's transcriptional activity by obstructing its DNA binding. This inhibition of HIF-1's function holds potential in mitigating the overexpression of growth factors linked to angiogenesis, offering a targeted therapeutic approach for managing DR. The exploration of HIF-1 antagonists, particularly daunorubicin and doxorubicin, represents a novel avenue for developing interventions aimed at modulating the molecular pathways involved in DR [77]. The challenge of limited water solubility leading to precipitation on the retinal surface after intraocular injection has prompted researchers to explore innovative delivery systems. In a study conducted by Iwase et al. [78] using diabetic rabbits, they investigated the potential of polysebacic acid (PSA)-polyethylene glycol (PEG) NPs for achieving sustained delivery of daunorubicin to the retina. While the intraocular injections of both daunorubicin (DNR) and DXR efficiently suppressed choroidal and retinal NV, a notable observation was the disruption of retinal function. This impact on retinal function was assessed through electroretinograms (ERGs). Despite the successful suppression of NV, the potential adverse effects on retinal function raise important considerations for the clinical translation of these intraocular injection strategies. Further investigations are warranted to optimize delivery systems and dosages to minimize any detrimental effects on retinal function while maintaining the therapeutic efficacy against NV in DR. To address this concern, they combined DXR with innovative copolymers of branched polyethylene glycol and poly (sebacic acid) to develop DXR–PSA–PEG_3 NPs, which exhibited a gradual release of small DXR-conjugates when placed in an aqueous buffer. Upon intraocular injection of DXR–PSA–PEG_3 NPs (1–10 μg DXR content), the researchers observed a reduction in HIF-1-responsive gene products, a robust suppression of choroidal and retinal NV, and no signs of retinal toxicity. Furthermore, in a study involving transgenic mice expressing VEGF in photoreceptors, the administration of DXR–PSA–PEG_3 NPs through intraocular injection (with 10 μg DXR content) demonstrated significant efficacy in suppressing NV, sustaining this effect for a minimum of 35 days. Likewise, in rabbit models, the intraocular injection of DXR–PSA–PEG_3 NPs (containing 2.7 mg DXR) facilitated a prolonged release of DXR-conjugates, maintaining detectable levels in both the aqueous humor and vitreous for a remarkable duration of at least 105 days. This comprehensive study underscores the significance of a novel HIF-1-inhibitor–polymer conjugate formulated into controlled-release particles. This formulation not only maximizes efficacy and the duration of activity but also minimizes potential toxicity. The findings introduce a promising new chemical entity with considerable potential for treating ocular NV, presenting a novel avenue for addressing DR and related ocular conditions.

Nanotechnology presents distinctive advantages in the treatment of DR [79]. Drug-carrying nanomolecules, typically ranging in diameter from 1 to 100 nm, are widely utilized. Their small size facilitates easy passage through the BRB [80]. Furthermore, these nanomolecules exhibit an exceptional ability to bind to specific receptors in epithelial cells of retinal pigment. This targeted delivery mechanism enhances their effectiveness in achieving the desired outcomes in DR treatment [80]. Badr et al. [81] conducted experimental research that confirmed a reduced expression of glucose transporter 1 (GLUT1) in isolated retinal capillaries and the neural retina. This finding contributes valuable insights into the molecular mechanisms associated with altered glucose transport in the context of DR. In their study, Badr et al. [81] performed a comparative analysis of GLUT1 and glucose transporter 3 (GLUT3) expression levels in the cerebrum, retina, and their respective micro vessels. This analysis was carried out using Western blot techniques, providing a comprehensive examination of the differential expression of these glucose transporters in different tissues and microvascular environments. In normal nondiabetic animals, the retina and its microvessels exhibited significantly higher levels of GLUT1 protein compared to the gray matter of the cerebral cortex and its microvessels. Upon inducing diabetes with Streptozotocin for either two weeks or two months, there was an approximate 50% reduction in GLUT1 expression in the retina and its microvasculature, but no similar decrease was observed in the cerebrum or its microvessels. It is interesting to note that in diabetic rats, the neural retina had higher levels of GLUT1 than the cerebrum, even though the condition decreased GLUT1 expression in the retina. Furthermore, diabetes had no effect on the expression of GLUT1, which was found in greater quantities in the retinal pigment epithelium (RPE) than in the neural retina or its micro vessels. However, GLUT3 expression in both the retina and the cerebral gray matter was not significantly affected by diabetes, and GLUT3 levels were higher in the former than in the latter. Different regulatory mechanisms for glucose transport in these similar embryological organs are suggested by the different effects of diabetes on GLUT1 expression in the retina and cerebral cortex. Notably, the repression of GLUT1 in retinal microvessels, while leaving RPE unaffected, suggests that a larger fraction of glucose passing into the retina during diabetes may occur across the RPE rather than the retinal vasculature. These findings shed light on the complexities of glucose transport regulation in the context of DR. Ulas et al. later noted that diabetes induced by Streptozotocin reduced GLUT1 expression by approximately 50% in the retina and its microvasculature but not in the cerebrum or its microvessels. They concluded that the accumulation of glucose in the diabetic retina, leading to a high glucose microenvironment and the occurrence of DR, was due to elevated expression levels of GLUT1 and GLUT3. In a study by Mahaling et al. [82], a nanoparticle-based delivery system with a core–shell structure was engineered. This setup comprised a polycaprolactone core that is hydrophobic and a Pluronic® F68 shell that is hydrophilic. Triamcinolone acetonide was incorporated into these NPs, and their effectiveness was evaluated in a DR rat model. The NPs loaded with the drug were applied as eye drops, resulting in significant enhancements in the structural integrity of the retina, including retinal thickness and vascular health. Additionally, functional improvements were observed in terms of rod and cone activity when compared to DR controls that received the drug alone or placebo NPs. Furthermore, retinal inflammation was effectively reduced by the drug-loaded NPs, as evidenced by the decline in NF-κB, ICAM-1, and TNFα expression after a 20-day treatment period. Moreover, the NPs loaded with the drug demonstrated an effective reduction in retinal inflammation, as demonstrated by the decreased expression of NF-κB, ICAM-1, and TNFα following a 20-day treatment duration. Furthermore, a decrease in glial cell hyperplasia, indicated by reduced GFAP expression, and a reduction in microvascular complications, characterized by diminished VEGF secretion and decreased formation of microvascular tufts, were observed in rat retinas after 40 days of treatment. The simultaneous decrease in retinal inflammation and vascular abnormalities, both well-known hallmarks of DR, underscores the considerable potential of the nanoparticulate delivery system as a topical formulation for treating DR. Triamcinolone acetonide acetate (TAA) was encapsulated within nanostructured lipid carriers (NLC) for the purpose of studying its in vitro transdermal absorption and establishing a rat model of DR [83]. The objective was to evaluate the influence of nanoparticle capsules loaded with triamcinolone acetonide on the structural enhancement of both the retina and pancreas. Blood glucose levels were monitored, and an examination of pathological alterations in the retina and pancreas was conducted to assess the efficacy of the triamcinolone acetonide-loaded nanoparticle capsules. The study involved four groups for comparison: the normal group, positive control group, model control group, and triamcinolone acetonide-loaded nanoparticle group. Following the administration of triamcinolone acetonide-loaded nanoparticle capsules for two months, notable decreases in blood glucose and glycosylated hemoglobin levels were evident in the triamcinolone acetonide-loaded nanoparticle group ($P < 0.01$). Moreover, in the triamcinolone acetonide-loaded nanoparticle group, the retina displayed an organized arrangement of the photoreceptor cell layers, diminished intracellular and extracellular edema in each layer, unobstructed capillary lumens, and slight edema in peripheral cells. Additionally, this group showed less severe pathological changes compared to the model control group. The results suggest that the triamcinolone acetonide-loaded nanoparticle capsules efficiently reduced blood glucose and glycosylated hemoglobin levels, concurrently improving the structural integrity of both the retina and pancreas in rats with DR. Laddha and Kshirsagar [84] designed surface-modified NPs using poly (D, L-lactide-co-glycolide), specifically PLGA 75 : 25 and PLGA (50 : 50). These NPs were developed for the delivery of pioglitazone, a peroxisome proliferator-activated receptor-gamma agonist, to the rear segment of the eye through topical administration. Polysorbate 80 was employed for surface modification. The NPs were crafted utilizing the method of single-emulsion solvent evaporation, and the process was fine-tuned through optimization using a 3-factor 3-level Box–Behnken statistical design. In the optimized batches, PLGA 75 : 25 exhibited a mean particle size of 163.23 nm, a PDI of 0.286, and an entrapment efficiency of 91%. Meanwhile, PLGA 50 : 50 showed values of 171.7 nm, 0.280, and 93% for mean particle size, PDI, and entrapment efficiency, respectively. Differential scanning calorimetry (DSC) verified the appropriate molecular dispersion of the drug within the polymer. In vitro release studies demonstrated a biphasic drug release pattern, with cumulative drug release of $58.48 \pm 1.38\%$ for PLGA 75 : 25 NPs and $74.17 \pm 1.38\%$ for PLGA 50 : 50 NPs at the conclusion of 10 hours. The pioglitazone release from the NPs followed the Higuchi model. Additionally, an in vivo study on rats revealed a dose-dependent decrease in concentration of VEGF in the vitreous fluid. The findings from this study emphasize the noteworthy contribution of peroxisome proliferator-activated receptor-gamma in the management of DR. A formulation was devised involving curcumin, recognized as a peroxisome proliferator-activated receptor-gamma agonist, encapsulated within surface-modified PLGA (50,50) NPs. The purpose was to target DR through topical administration [85]. The optimization process employed a 32-factorial design with the PLGA concentration and the number of high-pressure homogenization (HPH) cycles as independent variables, and particle size, polydispersity index (PDI), and percentage of entrapment efficiency as dependent variables. The optimized batch, designated as F6, exhibited a particle size of 194.7 nm, a PDI of 0.231, and % age entrapment efficiency of 90%. Following optimization, the selected batch underwent spray drying and was subjected to evaluation for several parameters, including remaining solvent analysis using gas chromatography. The in vitro drug release study demonstrated a biphasic pattern, with a cumulative drug release of $70.17 \pm 1.38\%$ achieved at the end of 10 hours. The release pattern of curcumin from the NPs demonstrated a close correspondence with the Higuchi model, affirming the controlled release behavior. Moreover, the X-ray diffraction (XRD) study provided additional confirmation of the appropriate molecular dispersion of the drug within the NPs. Moreover, the formulation successfully passed sterility testing, histology examination, and isotonicity testing, affirming its safety and appropriateness for ocular administration. In vivo studies conducted on rats demonstrated a dose-dependent decrease in concentration of VEGF concentration the vitreous fluid, substantiating the formulation's potential in the management of DR. Based on the available evidence, the prepared formulation is deemed to hold significant promise as an effective approach for managing DR.

The anti-inflammatory and antiangiogenic properties of PEDF34-NP in cultured cells under hypoxic conditions were investigated by Zhang et al. [86]. To achieve this, PEDF34mer peptide was encapsulated in PLGA NPs (PLGA34-NP) and used to treat ARPE19 and BRCEC cells. It was observed that hypoxia-induced VEGF protein expression in both cell types was effectively reduced by PEDF34-NP (Figure 7.6a,b). Additionally, a concentration-dependent decrease in hypoxia-induced VEGF secretion and Vegf mRNA expression was demonstrated by PEDF34-NP (Figure 7.6c,d). Moreover, the MTT assay confirmed the specific inhibition of BRCEC growth under hypoxia by PEDF34-NP, with no effect on ARPE19 cell viability (Figure 7.6e,f), indicating its endothelial cell-specific inhibitory effect. In the oxygen-induced retinopathy (OIR) rat model, a notable decrease in neovascular tufts was noted in the retinas of OIR rats treated with PEDF34-NP in comparison to both the PLGA-NP control and untreated OIR rats (Figures 7.7a–i) following intravitreal injection at postnatal day 12. Additionally, the retinas of OIR rats treated with PEDF34-NP exhibited a considerably smaller central nonperfusion area (Figure 7.7j) and a decreased number of pre-retinal nuclei (Figure 7.7k) compared to the control groups. Furthermore, in the OIR rats, PEDF34-NP treatment resulted in decreased retinal VEGF and ICAM-1 expression, both known contributors to BRB breakdown and NV. Vascular permeability assessments using the Evans blue-albumin tracer validated the decrease in vascular leakage following PEDF34-NP injection. In streptozotocin-induced diabetic rats, PEDF34-NP injection led to a significant reduction in retinal vascular leakage as compared to the control-NP treatment. Western blot analysis demonstrated a decrease in the expression of VEGF, ICAM-1, TNF-α, and nonphosphorylated β-catenin in the retinas of diabetic rats treated with PEDF34-NP. Furthermore, PEDF34-NP led to a downregulation of mRNA levels related to inflammatory and fibrogenic factors (Vegf, Tnf-α, Icam-1, Mcp-1, and Ctgf) in the retina. Importantly, PEDF34-NP treatment did not induce any detectable ocular toxicities in both non-diabetic and streptozotocin-induced diabetic rats, as confirmed by electroretinogram (ERG) recording and histological examination. These results suggest that PEDF34-NP has promising therapeutic potential without causing adverse effects on visual function or retinal structure.

A groundbreaking investigation was conducted by Amadio et al. [87] to devise a promising therapeutic approach for DR by selectively targeting human antigen R (HuR), a key constituent of the embryonic lethal abnormal vision (ELAV) protein family. To achieve this, nanosystems loaded with siRNA were harnessed to effectively silence HuR expression, and they were cleverly formulated into lipoplexes comprising solid lipid NPs (SLN) and liposomes (SUV). A relevant diabetic rat model was established by inducing diabetes in the test subjects using streptozotocin (STZ), and then the HuR-silencing siRNA-loaded lipoplexes were precisely administered into the rats' ocular regions. The expression levels of HuR and VEGF genes in the retinas of both experimental and control groups were assessed with great precision, along with thorough histological examinations. The insightful findings showcased a significant upsurge in HuR and VEGF levels within the STZ-induced diabetic rats. However, this elevation was remarkably curtailed by the administration of HuR-silencing siRNA-loaded lipoplexes, underscoring the efficacy of the treatment. Impressively, the nanocarriers employed for siRNA delivery exhibited a feeble positive surface charge and a well-balanced 4 : 1 N/P ratio (nitrogen from cationic lipid to siRNA phosphate), which correlated with enhanced transfection efficiency and an effective reduction in HuR and VEGF expression levels in the retina. In sum, the remarkable potential of lipid-based nanocarriers, specifically lipoplexes, in facilitating the efficient delivery of siRNA into the rat retina to silence HuR expression was effectively demonstrated by this pioneering study. The promise shown by these lipoplexes positions them as strong contenders for the management and treatment of retinal ailments, particularly DR. Separate release of Dex (dexamethasone) and Avastin (bevacizumab) from a nanocarrier that is a supramolecular hydrogel was successfully achieved by Zamani et al. [88]. The nanocarrier, comprising methoxy polyethylene glycol/polycaprolactone (mPEG-PCL) micelles and α-cyclodextrin (α-CD), harnessed the "host–guest" interaction facilitated by MPEG and α-CD. The study's objective was to investigate inflammatory corneal NV in a model of rats. According to the trial findings, 70% of Avastin was released after five days and 70% of Dex was released after one day. This observation suggests that the combination of DEX and Avastin might have the potential to contribute to the inhibition of retinal NV. Moreover, the PEG-PCL hydrogel, functioning as a drug carrier, exhibited efficient and sustained drug release capabilities, thereby further elevating the therapeutic potential of this system. In 2020, a novel and potentially effective method for treating diabetic retinopathy (DR) was devised by Amato et al. [89]. Their methodology entailed combining the somatostatin equivalent octreotide (OCT), known for its neuroprotective and antiangiogenic properties, with MNPs. Subsequently, this combination was injected into rat eyes to observe its effects. The findings showed that MNP-OCT displayed a significant therapeutic effect at lower doses in comparison to free OCT. Specifically, the bioactivity of the drug was enhanced 100-fold in retinal endothelial cells (HRECs) and 10-fold in retinal explants of humans upon binding to MNPs. Notably, MNPs exhibited excellent biocompatibility and elicited no toxicity in the rats. Furthermore, the attachment of OCT to MNPs facilitated the sustained release of the drug in the rat retina for few months without compromising its efficacy. The utilization of

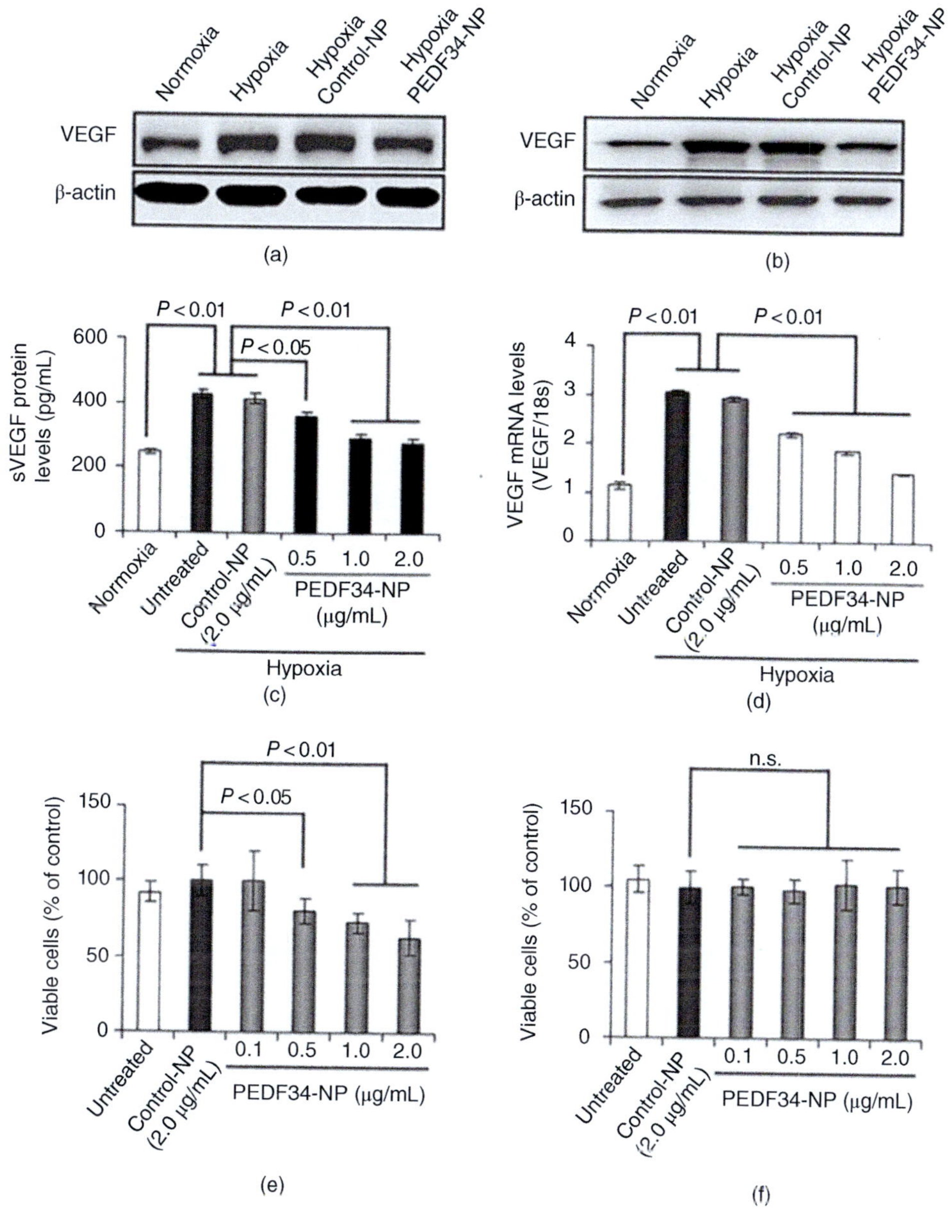

Figure 7.6 Suppression of hypoxia-induced VEGF expression by PEDF34-NP in ARPE19 cells and BRCEC (a) ARPE19 cells were cultivated to approximately 70% confluence in a culture medium containing 2% fetal bovine serum (FBS) and subsequently exposed to PEDF34-NP at a concentration of 1 μg/mL for 16 hours under hypoxic conditions (1% oxygen). Equal amounts of total cellular proteins (30 μg) were utilized for Western blot analysis employing an antibody specific to VEGF. (b) BRCEC cells were subjected to hypoxia along with treatment involving PEDF34-NP for 16 hours. The levels of VEGF within the cells were assessed using Western blot analysis. (c) and (d) ARPE19 cells were treated with control-NP and varying concentrations of PEDF34-NP for 24 hours under hypoxic conditions. The VEGF released into the cell culture media (sVEGF) was quantified using an ELISA assay (c). Simultaneously, VEGF mRNA levels were measured through real-time reverse transcription-polymerase chain reaction (RT-PCR) and subsequently normalized using 18S ribosomal RNA (rRNA) levels (d). (e) and (f) BRCEC (e) and ARPE19 cells (f) were exposed to PEDF34-NP at concentrations as indicated for a 24-hour duration. The quantification of viable cells was conducted using the MTT assay. All presented values represent the mean ± standard deviation ($n = 3$). Source: Qu et al. [86]/Frontiers Media S.A./CC BY 4.0.

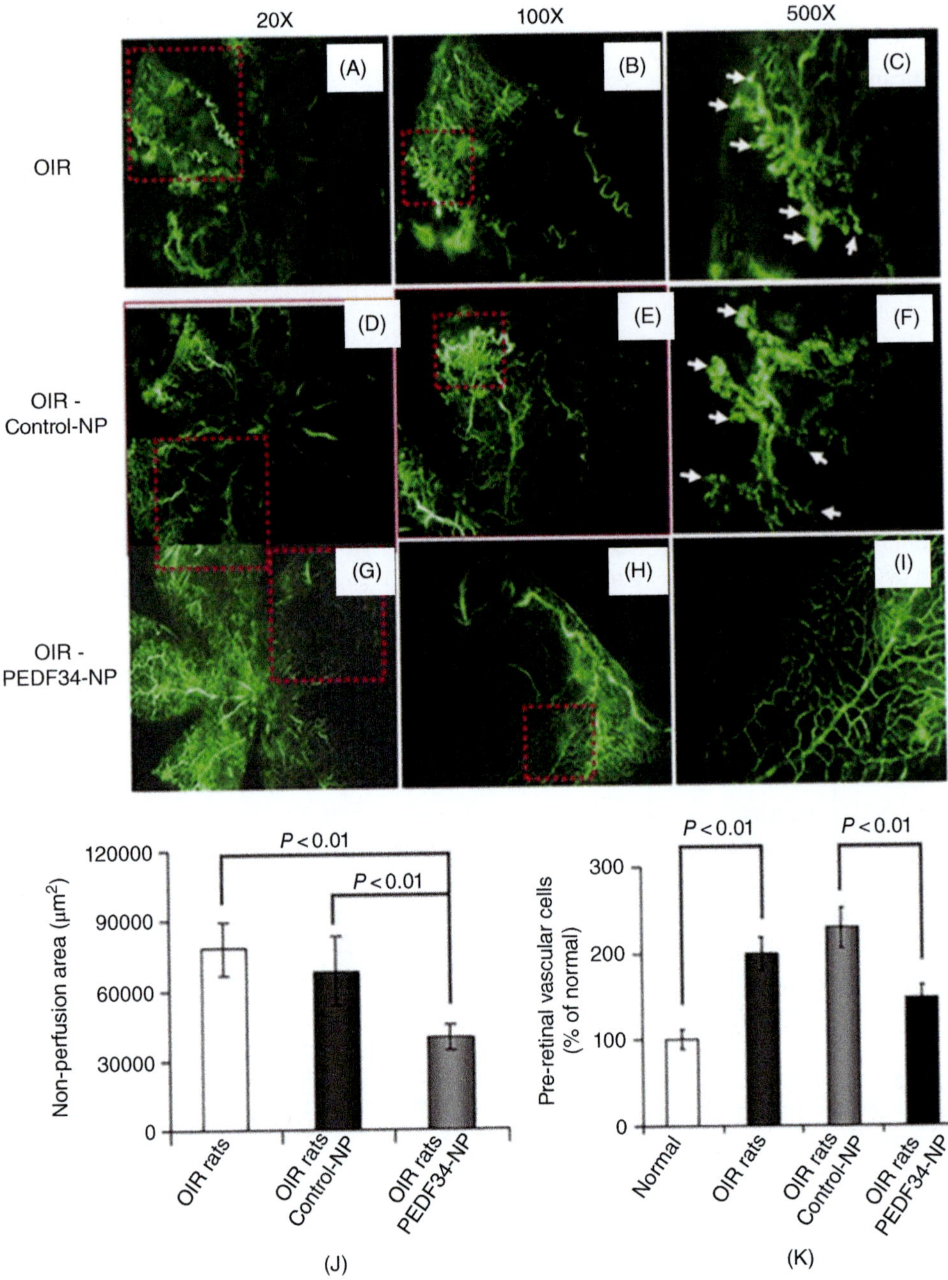

Figure 7.7 Suppression of ischemia-induced retinal neovascularization by PEDF34-NP. During this experiment, rats were exposed to an oxygen concentration of 75% ± 2% from postnatal day 7 (P7) to P12. On P18, the condition of the retinal vasculature was assessed through fluorescein angiography, as detailed in the Methods section. (A–C) Depict rats subjected to oxygen-induced retinopathy (OIR) without any treatment. (D–F) Represent OIR rats that were administered control-NP (5 μg/eye). (G–I) Show OIR rats that received intravitreal injections of PEDF34-NP (5 μg/eye) on P12. (J) Quantifies the nonperfusion area as a percentage of the total retinal area, with results presented as mean ± standard deviation (SD) for a sample size of 6. (K) At P18, the eyes were fixed, sectioned, and stained with hematoxylin and eosin (H&E). Preretinal vascular cells were counted in eight noncontinuous sections per eye and averaged, as outlined in the Methods section, and the results are reported as mean ± SD for a sample size of 6. Source: Qu et al. [86]/Frontiers Media S.A./CC BY 4.0.

MNPs as a drug carrier offers the advantage of reducing the frequency of anti-VEGF antibody injections, thereby achieving a sustained and effective drug concentration. This advancement holds significant promise in enabling more manageable and efficient treatment strategies for DR.

The advantages of nanotechnology in DR treatment go beyond its nano-scale size, enabling drugs to cross through the BRB smoothly. NPs possess unique properties that offer the ease of monitoring the therapeutic drugs. In 2019, Shoval et al. [90] conducted a study where they applied carbon nanodot (C-dots) coatings to two anti-VEGF medications, bevacizumab and aflibercept. These drug–nanoparticle combinations were separately injected into diabetic rats. The researchers carried out noninvasive intraocular medication concentration monitoring by utilizing the inherent fluorescence of C-dots, eliminating the need for invasive procedures that could damage the eyes. The results indicated that incorporating C-dots as the outer envelope for bevacizumab and aflibercept successfully prevented drug degradation prior to reaching the retina. As a result, the effective concentration of drug in the retinal tissue increased, leading to more efficient inhibition of NV in choroidal blood vessels induced by VEGF. Through the utilization of this approach, the transition from nonproliferative to proliferative DR can be averted, effectively safeguarding the retinal tissue. This signifies a promising development in DR treatment utilizing nanotechnology.

Nanotechnology is thus playing a critical role in advancing the treatment of DR, a condition exacerbated by elevated blood sugar levels in diabetic patients. DR is a severe and potentially blinding complication of diabetes that necessitates early detection and intervention. The condition is characterized by microstructural and microcirculatory changes in retinal blood vessels and can be effectively monitored through fundus examinations. Traditional methods, such as fluorescein angiography, are being complemented by innovative approaches using nanomaterials like vanadium core–shell nanorods for the early detection and diagnosis of DR. These nanorods facilitate fast and noninvasive staging of DR, lowering the chances of vision loss in diabetic patients. Additionally, addressing hypoxia, a precursor to retinal NV in DR, is critical for prevention. Hypoxia-inducible factor-1 (HIF-1) and HIF-1 antagonists, such as daunorubicin and doxorubicin, hold promise in managing DR. NPs like polysebacic acid-polyethylene glycol NPs offer sustained delivery of HIF-1 antagonists, reducing NV while minimizing retinal toxicity. New nanomaterials are continually being developed for DR treatment, including nanoparticle-based systems and lipid-based nanocarriers that enable the efficient delivery of drugs, siRNA, and peptides. These novel systems show promise in addressing DR's underlying mechanisms and reducing vascular leakage, inflammation, and NV while maintaining ocular safety. Furthermore, nanotechnology allows for controlled and sustained drug release, minimizing the need for frequent injections, which is particularly beneficial for anti-VEGF therapy. Strategies like coating anti-VEGF drugs with carbon nanodots enhance drug stability and efficacy, reducing the risk of progression from nonproliferative to proliferative DR.

Overall, nanotechnology is revolutionizing the management of DR by offering innovative tools for early diagnosis, targeted treatment, and monitoring of the disease, ultimately improving the quality of life for individuals with diabetes.

7.2.2 Nanotechnology in Retinal Tissue Regeneration

PDR impacts the retina, comprising seven key cell types: amacrine, cone, rod, bipolar, horizontal, retinal ganglion, and RPE cells. RPE stands out as the retina's primary dietary source, prompting research to focus on retinal regeneration. Presently, retinal regeneration relies on two primary types of nanomaterials: electro spun and self-assembled nanofibers. Natural polymers like fibrin, collagen, laminin, hyaluronic acid, and gelatin/chitosan are used in electrospun nanofibers, but synthetic polymers like PCL, poly(L-lactic-*co*-ε-caprolactone) (PLCL), poly (lactic-*co*-glycolic acid) (PLGA), and poly (L-lactic acid) (PLA) are frequently used. Notably, natural polymers exhibit better cell attachment and biological activity owing to their distinct chemical properties compared to synthetic polymers. However, synthetic polymers offer advantages in terms of transport characteristics, mechanical strength, and biodegradation, making them suitable for designing and controlling an extracellular analog matrix that supports cell division and growth. Despite natural polymers' cell attraction, synthetic polymers outperform them in terms of mechanical strength and durability.

A research study by Stanzel et al. [91] demonstrated that nanofibers with approximately 200 nm diameter closely imitate the nerve fibers in terms of their distribution, resulting in improved retinal regeneration compared to those with a diameter of about 1000 nm. The primary goal of this research was to develop a supporting scaffold with a favorable topography that could assist in maintaining the functioning of RPE monolayer while being compatible with the retina. To achieve this, films and electrospun substrates with fiber diameters ranging from 200 to 1000 nm were created using poly(l-lactide-co-ε-caprolactone) or polyethylene terephthalate. These substrates were then subjected to in vitro testing employing human fetal RPE cells, which were subsequently subretinally injected into rabbits. The results showed that the largest cell densities were seen in RPE cells grown on the 200 nm fiber versions, which formed adherent monolayers with more consistent hexagonal tight junctions and deeper pigmentation. By electrospinning flat 200 nm fiber membranes onto a porous rigid-elastic carrier, the researchers were able to perform simple subretinal implantation. Using spectral-domain optical coherence tomography, the retina was observed to reconnect and thinned marginally throughout the course of the two-week observation period. Histology revealed that reactive gliosis and degradation of some photoreceptors were seen along with the variable migration of native RPE cells onto the nanofibers. Finally, scaffolds with a fiber topography of 200 nm showed improved RPE culture along with subretinal biocompatibility, which hold potential use in cell-based treatments for retinal disorders that cause blindness. It becomes imperative to take action to promote the regeneration of vulnerable neurons in response to the loss of nerve function of retinal neurons on the verge of cell death. To tackle this issue, a novel self-assembling peptide multichannel nanofiber scaffold has been created, integrating nerve growth factor (NGF) and brain-derived growth factor (BDGF) [92]. This biodegradable multichannel scaffold is characterized by aligned electro spun nanofibers and a neurotrophic gradient (MC/AN/NG). The gelatin-based conduits of the scaffold closely mimic the architecture of natural nerve extracellular matrix (ECM), ensuring effective guidance for axon outgrowth. The microbial transglutaminase-crosslinked MC scaffolds provide long-term mechanical stability. Furthermore, a temporal approach is utilized to produce the regulated release of both NGF and BDNF, neurotrophic factors. Neural stem cells successfully expanded their neurites along the allied scaffold nanofibers, as shown by in vitro investigations. Additionally, in the treated group, myelination was markedly accelerated by BDNF, and cell density rose in gradient membrane regions with high NGF concentrations. The MC/AN/NG scaffold was shown to be superior to autograft in a rat model involving in vivo rabbit sciatic nerve transection, which resulted in improved nerve repair and decreased muscle atrophy. The MC/AN/NG scaffolds functioned as efficient nerve-guiding conduits and presented a promising alternative to autologous nerve grafts by integrating many techniques to support peripheral nerve regeneration. In a similar vein [93], the scaffold was enhanced with single-electron layer graphite NPs and conductive polypyrrole, resulting in a remarkable 137% rise in retinal ganglion cells (RGCs) growth and development when subjected to electrical stimulation. To achieve this, a novel method known as polymerization-enhanced ball milling was utilized to successfully synthesize polypyrrole functionalized graphene (PPy-G). This method introduced nitrogen electron–acceptor to enhance capacitance while maintaining a conductive platform through the π–π conjugated carbon plane for effective charge transportation. The ES treatment resulted in significantly improved viability, antiaging capabilities and neurite outgrowth of RGCs, hinting at the potential for optic nerve regeneration using ES on suitable nanoelectrodes. Exciting possibilities for advancing retinal nerve regeneration through the implementation of these functionalized nanofibers are opened by this groundbreaking research.

Soleimani et al. [94] conducted a study to evaluate the appropriateness of fibrin gel as a framework for the differentiation of conjunctival mesenchymal stem cells (CJMSCs) into photoreceptor cells. The surface of poly-L-lactic acid (PLLA) nanofibrous scaffolds were seeded CJMSCs and prompted to differentiate into lineage of photoreceptor cells. The results revealed that fibrin scaffold was favorable due to its suitable porous structure, enabling effective combination of CJMSCs with the gel, and demonstrating excellent compatibility among the cells and the fibrin gel. The physical properties of the fibrin gel could also be easily adjusted according to the environment. The differentiated cells and their expression of genes specific to photoreceptors were examined by quantitative real-time RT-PCR (qPCR), SEM imaging, and immunocytochemistry. The results demonstrated that CJMSCs could successfully differentiate into photoreceptor cells on the surface of nanofibrous scaffolds, indicating a possible use for these cells in retinal regeneration. Moreover, SEM imaging indicated a well-oriented spindle-shaped morphology of CJMSCs on the nanofiber scaffolds. Remarkably, the expression of genes specific to rod photoreceptors was notably higher in CJMSCs that underwent differentiation on randomly oriented nanofibers in comparison to those on aligned nanofibers. In conclusion, the nanofibrous PLLA scaffold demonstrated in this study holds promise in retinal tissue engineering as a potential cell carrier. The integration of electrospun nanofiber scaffolds with conjunctiva stromal cells derived from mesenchymal stem cells (MSCs) holds promise for potential applications in retinal regenerative therapy. The research provides optimistic insights into the formulation of effective strategies for retinal regeneration and therapeutic interventions. In related studies, the creation of fibrin hydrogel scaffolds has been explored, serving dual functions of safeguarding stem cells during delivery and triggering endogenous regeneration through interactions between the transplanted stem cells and retinal tissue [95]. This investigation specifically employed fibrin gel as a hydrogel scaffold for cell immobilization. Following the isolation of CJMSCs, these cells underwent differentiation into photoreceptor-like cells by exposure to taurine for 14 days, with comparisons made between a tissue culture plate (TCP) group and the fibrin hydrogel (3D) group. Cell attachment was assessed through scanning electron microscopy (SEM) and 3-(4,5-dimethylthiazol-2-yl)-2,5-diphenyltetrazolium bromide (MTT) assays. Moreover, the assessment of specific markers for photoreceptor-like cells, such as rhodopsin, CRX, PKC, peripherin, recoverin, nestin and RPE65, was conducted through immunocytochemistry and quantitative real-time PCR (RT-PCR) in both the TCP and 3D groups. The SEM analysis revealed that CJMSCs attached well within the fibrin gels, indicating good compatibility between the cells and the scaffold. The gel's elastic modulus and constant degradation contributed to the growth and proliferation of cells, and importantly, there was no toxic effect of the fibrin hydrogel on the cells. The viability of cultured cells was higher in the 3D fibrin gels compared to the TCP groups. After two weeks, the expression levels of rhodopsin, PKC, CRX, peripherin, recoverin, nestin, and RPE65, all markers for photoreceptor cells, were detected using real-time PCR and immunofluorescence. Notably, these expressions were higher in the 3D groups than in the TCP groups. The findings of this study demonstrate the potential of readily available sources of adult stem cells, such as human conjunctiva stem cells encapsulated in fibrin gel, as an intriguing strategy to enhance photoreceptor progenitor cell numbers for the repair and regeneration of retina diseases, including photoreceptor injury. Another study by Ai et al. [96] explored the differentiation capability of human endometrial stromal cells (hEnSCs) when grown on a nanofibrous scaffold made of polylactic acid and chitosan (PLA/CS). The induction of the differentiation process occurred through the utilization of conditioned medium derived from BE(2)-C human neuroblastoma cells, coupled with the existence of growth factors. The study aimed to employ the PLA/CS framework as a three-dimensional (3D) platform conducive to the differentiation of neuroglial cells. To commence the differentiation process, hEnSCs were grown on the PLA/CS scaffold following three passages and subjected to neuroblastoma-conditioned medium supplemented with FGF2/PDGF-AA. Following an 18-day period, the analyzed cells, which had undergone differentiation, were examined for the presence of neuroglial markers using qRT-PCR and immunofluorescence techniques. The results demonstrated that hEnSCs successfully adhered, grew, and differentiated on the nanofibrous PLA/CS scaffolding. Additionally, the investigation demonstrated the mRNA expression of markers associated with neural and glial lineages, including NF-L, Nestin, PDGFRa, MAP2, Olig2, CNP, MBP, and GFAP. At the protein level, the differentiation period of 18 days revealed the presence of Tuj-1, MAP2, and NF-L. The results present persuasive proof that hEnSCs, when cultivated on the PLA/CS nanofibrous scaffold, possess the capability to undergo differentiation into both neuronal and glial cells when exposed to conditioned medium of neuroblastoma. Deb et al. [97] investigated the use of 180 nm nanofibers scaffolds of polyamide to facilitate the regeneration of fetal and adult RPE cells. They discovered that the scaffold significantly enhanced the regeneration of both fetal and adult RPE cells in comparison to the traditional glass surface. This finding highlights the promising potential of polyamide nanomaterials as efficient and viable scaffolds for retinal tissue regeneration. Researcher team led by Chen et al. [98] successfully produced induced pluripotent stem cells (iPSCs) from mononuclear cells of peripheral blood, using nano printing and these iPSCs exhibited characteristics of embryonic stem cells with high nucleus-to-cytoplasm ratios and pluripotency marker expression. The iPSCs were then differentiated into RGCs using a modified method, resulting in the formation of ocular vesicles (OVs) following an 18-day differentiation period. Day 26 expression of neuron marker TUJ1 and RGC progenitor marker Math5 confirmed the successful differentiation of iPSCs into RGCs. To enhance neurite outgrowth from the OVs, the researchers used a copolymer composed of poly(ethylene-co-vinyl acetate) (EVA) as a scaffold, creating a straightforward groove pattern using nano-imprinting lithography. When the OVs were cultured on this scaffold, the RGCs exhibited lengthy and noticeable neurites, indicating the scaffold's ability to guide RGC axon growth. Furthermore, functional analysis revealed that the scaffold supported RGCs demonstrated mature physiological function, including firing of action potentials. The effectiveness of the EVA scaffold was also tested with RGCs derived from patients with RGC degenerative disease. The patient-derived RGCs showed enhanced neurite outgrowth and fired repetitive action potentials when grown on the EVA scaffold, further confirming the scaffold's ability to promote RGC growth. Overall, the results demonstrated that EVA scaffold with the straight groove pattern was successful in guiding RGC neurite outgrowth, both in iPSC-derived RGCs and patient-derived RGCs from individuals with RGC degenerative disease. This innovative scaffold shows promising potential for enhancing RGC regeneration and could have significant implications for retinal tissue engineering and regenerative therapies for optic nerve-related diseases. The studies mentioned highlight the utilization of different nanofiber scaffolds for retinal tissue regeneration, with promising outcomes in iPSC regeneration, cell growth, and specific biological processes critical for tissue engineering. These advancements pave the way for potential applications of regenerative medicine in this field.

In summary, nanotechnology has shown promise as an innovative strategy for addressing DR, encompassing prevention, treatment, and regeneration of damaged nerve tissue. The hyperglycemic state in diabetes significantly contributes to DR development, resulting in increased vascular permeability, ischemic changes, and the growth of new blood vessels. Although current DR treatments have limitations, nanotechnology offers unique advantages, including targeted drug delivery, sustained drug release, and improved penetration through the BRB. Various studies have proved the effectiveness of nanomaterial-based therapies in preclinical models, effectively reducing inflammation, NV, and vascular leakage associated with DR. The versatility of nanotechnology enables personalized medicine, noninvasive monitoring, and the potential possibility of combination therapies to improve treatment outcomes. Despite these promising advancements, the safety and feasibility of nanomaterials in DR treatment necessitate further investigation, including rigorous clinical trials, to validate their effectiveness in human patients. Additionally, considerations of scalability, manufacturing processes, and cost-effectiveness are essential for making these treatments widely accessible and commercially viable. Overall, nanotechnology holds tremendous potential in transforming the management of DR, resulting in improved patient outcomes and the prevention of vision loss in individuals with diabetes. Ongoing research and progress in this domain will have a crucial impact in fully harnessing the potential of nanomaterial-based therapies for the benefit of those affected by DR.

7.3 Nanotechnology for Kidney Disease Treatments

Nanotechnology, an ever-advancing field that intersects science and engineering, has brought about significant transformations across diverse medical domains, providing innovative solutions to complex diseases. Notably, kidney diseases have become a focal point of attention due to their substantial impact on global health, affecting millions of individuals worldwide and frequently leading to chronic kidney failure, necessitating dialysis or kidney transplantation. Conventional treatments for kidney diseases have encountered limitations, including inadequate delivery of drugs to the affected kidney tissues, undesirable side effects, and difficulties in targeting specific cellular pathways. However, the integration of nanotechnology into kidney disease treatments introduces a new paradigm, offering unprecedented opportunities to enhance drug delivery, augment therapeutic effectiveness, and facilitate tissue repair and regeneration. In this section, the role of nanotechnology in the context of kidney disease treatments is explored, with light being shed on its potential to address crucial challenges encountered in current therapeutic approaches. Researchers are actively investigating novel methods utilizing nanomaterials, NPs, and nanoscale drug delivery systems to target kidney diseases at the cellular and molecular levels. Whether addressing CKDs or acute kidney injuries, nanotechnology holds the promise of revolutionizing the management of kidney diseases, ultimately leading to improved patient outcomes and an elevated quality of life for those grappling with these debilitating conditions. Through a comprehensive exploration of recent advancements and ongoing research, we delve into the potential advantages and challenges of employing nanotechnology-based treatments for kidney diseases. Kidney diseases can be categorized into CKD and acute kidney injury (AKI) based on the progression of kidney functioning. AKI is characterized by elevated serum creatinine levels within a short period, often caused by factors like hypovolemia or urinary obstruction. On the other hand, CKD is characterized by the sustained impairment of kidney function or structure, and it is commonly associated with conditions like diabetes, hypertension, and primary glomerular disease. CKD represents a worldwide health issue, impacting over 10% of the population, with a heightened prevalence observed among the elderly. It leads to multiple complications and adverse outcomes, placing a significant financial burden on individuals and society. The timely identification and prevention of the progression of CKD are becoming crucial priorities in public health.

Studies have indicated that NPs exhibit potential in both the diagnosis and treatment of CKD. NPs provide precise and accurate approaches for assessing kidney function and morphology, facilitating targeted drug delivery to specific tissues, thereby enhancing renal targeting, localization, and retention. Despite their potential, systematic studies on the application of NPs in CKD are limited. This section focuses on the use of NPs for detecting kidney injury biomarkers and advanced imaging techniques. Furthermore, a new treatment strategy for CKD, along with renal replacement therapy for individuals with end-stage renal disease (ESRD), is outlined. Challenges associated with NP technology and future perspectives are also discussed.

7.3.1 Nanotechnology in Detection of Kidney Disease Biomarkers

In the context of CKD diagnosis, early and specific markers play a crucial role in predicting the disease's onset and progression. Applying effective intervention therapies can help prevent CKD and minimize associated complications. However, traditional diagnostic methods have limitations in terms of sensitivity and convenience. NPs offer a potential solution for early CKD detection, providing considerable sensitivity and accuracy. Albuminuria serves as a predictive factor in both the onset and advancement of CKD [99, 100]. Conventionally, positive outcomes from the urinalysis dipstick were observed exclusively when the urinary albumin concentration exceeded 30 mg/dL. For the detection of microalbuminuria, specific methods such as urinary albumin dipsticks or various antibody-based techniques have been employed [101]. Nevertheless, these approaches have drawbacks in terms of sensitivity and convenience. In contrast, surface-enhanced Raman scattering (SERS) utilizes surfaces with copper, silver, or gold NPs, providing a technology that involves the inelastic scattering of laser energy [102]. SERS offers several benefits, including high sensitivity, uncomplicated sample processing, swift analysis, and the availability of commercially accessible portable and compact Raman spectrometers [103]. For instance, the use of silver nanoparticle surfaces enhances the Raman signal by absorbing the analyte significantly [104]. A significant unmet medical requirement is posed by the absence of a precise point-of-care system of detection for microalbuminuria, contributing to adverse health outcomes and mortality in individuals with kidney diseases. This proof-of-concept study, SERS spectroscopy was utilized for the identification of urinary albumin with the concentrations within the normal to slightly elevated albuminuria range. This approach proves valuable for the early detection of renal injury linked to uncontrolled hypertension, diabetes, or cardiovascular disease. The examination comprised 27 urine samples, employing iodide-modified Ag NPs for the SERS analysis. Effective discrimination between groups with low and high albumin concentrations was accomplished through linear discriminant analysis using cutoff values with the concentration of 3, 6, and 10 μg/mL for albumin. An overall accuracy of 89%, 93%, and 89% was attained, respectively. The achieved detection limit for human serum albumin was 3 μg/mL, relying on the 1002 cm SERS band associated with the ring breathing vibration of phenylalanine. The attained detection limit closely aligns with that of immunoturbidimetric assays and is roughly one order of magnitude lower than the detection limit of commonly used urinary dipsticks for the detection of microalbuminuria. A spectral model was developed to quantify albumin using principal least squares regression, and when applying an independent prediction set, an R^2 value of 0.982 and a root mean squared error of prediction of 2.82 were achieved, showcasing the accuracy of the predictions compared to reference values of human serum albumin concentrations. The findings demonstrate that the requisite sensitivity to detect clinically significant concentrations of urinary albumin is exhibited by direct SERS spectroscopy. This breakthrough offers a promising strategy for potential utilization in future point-of-care examination of microalbuminuria, allowing for early diagnosis and intervention in individuals at risk of kidney impairment due to various underlying conditions such as hypertension, cardiovascular disease, or diabetes. The design of the immunosensor involved utilizing screen-printed carbon microelectrodes on a flexible plastic substrate. EIS was employed for direct and label-free immunosensing by analyzing changes in the interfacial properties on the surface of the electrode [105]. Through electrodeposition utilizing linear sweep voltammetry, Au NPs are added to the screen-printed electrodes to enhance their conductivity and biocompatibility. Novel PS/Ag/ab-HSA nanoprobes, which consist of a polystyrene nanoparticle core with Ag nanoshells covalently attached to HSA antibodies, are created for the efficient immobilization of HSA antibodies. Dielectrophoresis was then used to trap these nanoprobes on the electrode surface (Figure 7.8A). Two sensing sites, designated as the sensing and control sites, have been integrated into the immunosensor, enabling the conduct of a differential analysis and enhancing specificity. The feasibility of the proposed immunosensor for detecting microalbuminuria has been confirmed through preliminary tests, wherein PBS spiked with different concentrations of HSA was utilized. A linear response to HSA concentration within the therapeutically relevant range and successful immobilization of nanoprobes are evidenced by impedance values at various stages of electrode modification and following immunosensing. The proposed immunosensor demonstrates stability and repeatability, making it suitable for quantitative screening of microalbuminuria. A low-cost prototype impedance readout module has been developed to facilitate POCT. This module utilizes bluetooth low energy (BLE) for the wireless transmission of data to a mobile device, such as a tablet or smartphone (Figure 7.8B). The readout module shows high accuracy compared to conventional laboratory LCR meters, making it a favorable platform for point-of-care immunosensing applications. The disposable chip and portable readout module included in the point-of-care immunosensing technology offer the possibility for early kidney disease identification and enhanced treatment outcomes. The study by Park et al. [106] developed a colorimetric method for the creatinine detection using gold NPs (GNPs) and three various peptides (APC01C, APC02C, and APC03C). The peptides were tested for their sensitivity in detecting creatinine at various concentrations (1–100 ppm). For APC01C and APC02C, the addition of creatinine resulted in the aggregation of peptide-GNPs, turning the solution purple blue. In contrast, APC03C caused the GNPs to disperse and turn red upon interaction with creatinine. The detection time for each peptide-GNP system was optimized, and a linear correlation was observed between the absorbance ratios and the logarithm of creatinine concentration (1–100 ppm) for all three peptides. The detection limits were found to be 0.025 ppm, 0.346 ppm, and 0.9796 ppm for APC01C, APC02C, and APC03C, respectively. GNPs/peptide specificity toward creatinine was confirmed by the absence of distinguishable changes in absorbance ratios for other anions and metals present in biological fluids. The method was successfully evaluated for the determination of creatinine concentration in urine mimic samples and showed excellent performance in measuring creatinine levels accurately. Copper nanoclusters (Cu NCs) with red fluorescence properties were successfully synthesized by employing 2-mercaptopyrimidine (2-MPY) through a straightforward one-step process [107]. These 2-MPY-Cu NCs exhibit remarkable promise as a sensing tool for the discerning and highly sensitive identification of creatinine, a vital biomarker in biological fluids. Upon excitation at 390 nm, an intense fluorescence peak at 728 nm is emitted by 2-MPY-Cu NCs. The fluorescence recognition of creatinine is achieved through various interactions, including intramolecular H-bonding and π–π interactions, between the 2-MPY-Cu NCs and creatinine. A significant fluorescence quenching of approximately 92% is observed, enabling efficient fluorescent sensing of creatinine. An impressive quantitative fluorescence response to varying concentrations of creatinine, ranging from 0.0625×10^{-6} to 5×10^{-6} mol dm^3(–|-), with a low LOD of 9.91×10^{-9} mol dm^3(–|-), is exhibited by the 2-MPY-Cu NCs. The successful application of this analytical approach in assessing creatinine levels in biological fluids has yielded encouraging outcomes.

A sensor based on barium hexaferrite, exhibiting high sensitivity, selectivity, and stability, was successfully created for the detection of minute quantities of ammonia vapor in exhaled human breath. This served as a biomarker for renal diseases [108]. Barium hexaferrite NPs used for the fabrication of the sensor were produced by a solid-state reaction route. Various characterization techniques were employed to analyze the nano powder and the sensor film, including XRD, TEM, field emission scanning electron microscopy (FESEM), X-ray photoelectron spectroscopy (XPS), energy-dispersive X-ray spectroscopy (EDX), Brunauer–Emmett–Teller (BET), and I–V

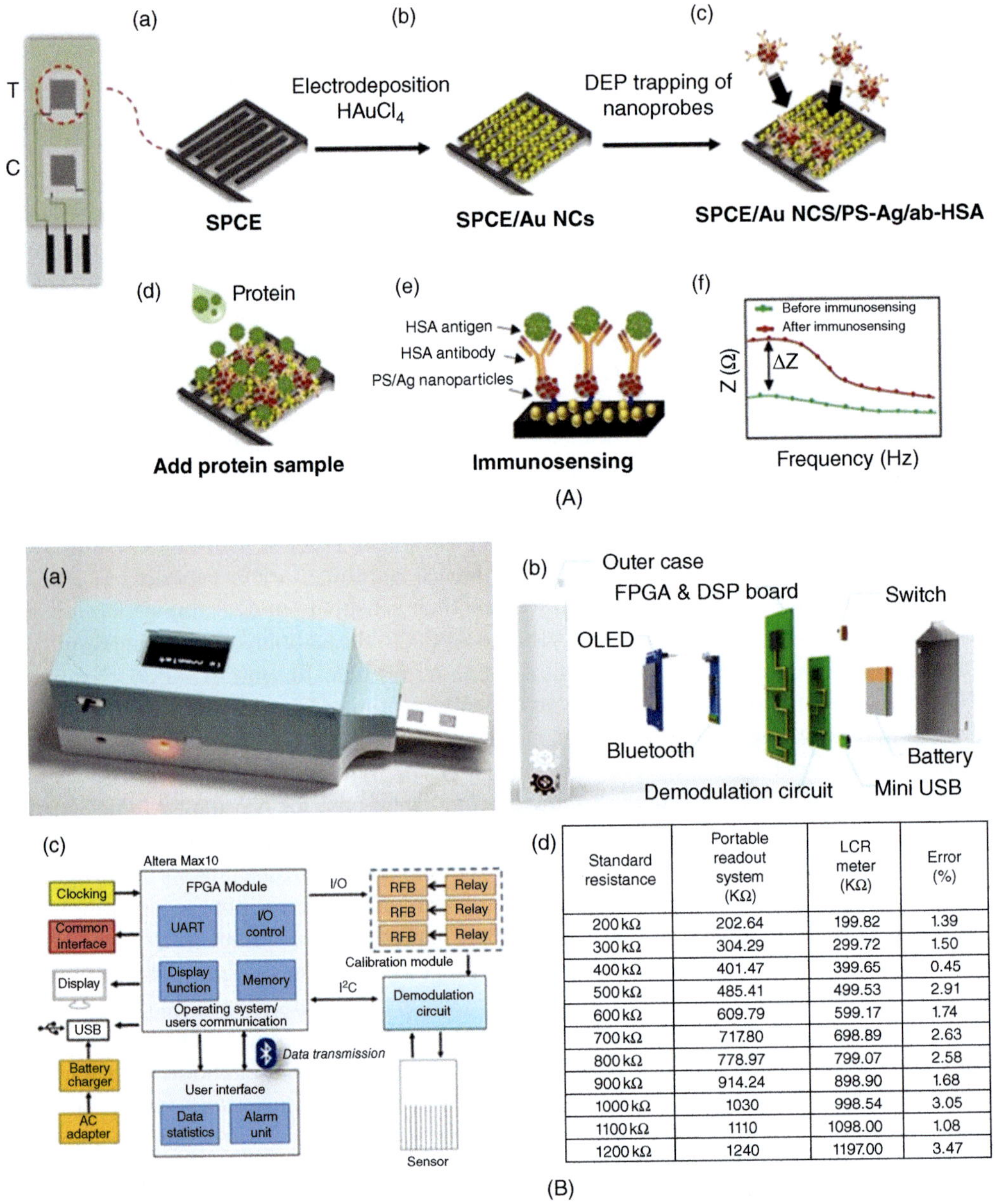

Standard resistance	Portable readout system (KΩ)	LCR meter (KΩ)	Error (%)
200 kΩ	202.64	199.82	1.39
300 kΩ	304.29	299.72	1.50
400 kΩ	401.47	399.65	0.45
500 kΩ	485.41	499.53	2.91
600 kΩ	609.79	599.17	1.74
700 kΩ	717.80	698.89	2.63
800 kΩ	778.97	799.07	2.58
900 kΩ	914.24	898.90	1.68
1000 kΩ	1030	998.54	3.05
1100 kΩ	1110	1098.00	1.08
1200 kΩ	1240	1197.00	3.47

Figure 7.8 (A) Illustration outlining the systematic procedure for the construction and functioning of the immunosensor. (B); (a) Picture, (b) diagram depicting internal elements, and (c) flowchart of the prototype impedance measurement module. (d) Chart illustrating a contrast between the impedance readings acquired with the measurement module and a traditional laboratory LCR meter. Source: Reproduced with permission from Shaikh et al. [105]/Elsevier.

measurements. Exhibiting *p*-type behavior, the sensor showcased the ability to detect ammonia at concentrations as low as 0.2 ppm, displaying a high response of approximately 1.46-folds. It also exhibited a remarkable response of approximately 2.34-folds for 1 ppm of ammonia. Importantly, the sensor showed no significant response to additional significant interfering volatile compounds in breath, such as ethanol, acetone, and saturated moisture, indicating its specificity for ammonia detection. The sensor showed a fast response time of about 2.88 s and a recovery time of approximately 39.4 s, enabling breath analysis in real time. Moreover, the sensor's extended-term stability, persisting for over three months, renders it appropriate for commercial applications. The operational temperature of the sensor was fine-tuned within the range of 260–400 °C, reaching a response saturation point at approximately 50 ppm of ammonia. The sensor demonstrated a substantial response to ammonia at the concentration of 0.2 ppm, attributed to its significant specific surface area, providing numerous active sites for ammonia oxidation. Ammonia molecules displayed weak interaction on the sensor surface and physisorption based on Freundlich absorption isotherm analysis, leading to complete and rapid recovery of the sensor. The sensor exhibited outstanding sensitivity when detecting minimal concentrations of ammonia, establishing its applicability for identifying kidney malfunction or renal diseases. It also displayed excellent selectivity toward ammonia over other interfering gases and volatiles found in human breath. The sensor's stability over a period of 90 days confirmed its commercial significance, and its nearly spherical/ovoid morphology with large particle size contributed to its long-term stability by slowing down defect annihilation kinetics. In general, this research presents a promising method for the real-time, selective, and sensitive identification of ammonia vapor in human breath, suggesting potential applications in the realm of biomedical analysis. The study by Lopes et al. [109] aimed to create a specific and sensitive immunosensor for detecting Cystatin C (CysC), a biomarker for CKD. Utilizing a sandwich-type immunoassay with matched monoclonal antibodies, the process involved immobilizing a capture antibody on screen-printed carbon electrodes (SPCEs) modified with Au NPs, followed by blocking to reduce nonspecific binding. Cystatin C (CysC) and a biotinylated anti-CysC detection antibody were introduced, and the formation of immunocomplexes was identified through an enzymatic label, streptavidin-alkaline phosphatase (S-AP). EC signals resulted from enzymatically generated metallic silver oxidation and reduction, proportional to CysC concentration. Optimization steps included investigating the impact of gold NPs (AuNPs) on signal enhancement, optimizing antibody concentrations, enzymatic label concentration, and incubation times. The immunosensor exhibited a linear relationship between the peak current intensity and CysC concentration within the range of 10–100 ng/mL. It achieved a detection limit of 6.0 ng/mL and demonstrated a sensitivity of $6.4 \pm 0.3\,\mu A\,ng/mL/cm^2$. Reproducibility tests indicated an RSD of less than 6.2% for six sensors detecting Cys C at 50 ng/mL. The sensor's performance was validated with real CKD serum samples, aligning with the gold standard method. Overall, this immunosensor, incorporating a sandwich immunoassay format, AuNP-modified SPCE, and enzymatic label, holds promise for precise and sensitive CysC detection. Further its simplicity and cost-effectiveness makes it suitable for point-of-care CKD diagnostics with potential for integration into portable devices. An EC immunoassay devoid of enzymatic involvement was developed for the detection of neutrophil gelatinase-associated lipocalin (NGAL), a kidney disease biomarker [110]. The procedure involves the deposition of Prussian Blue (PB) NPs with inherent redox activity onto graphitic C_3N_4 nanosheets (g-C_3N_4) through in situ reduction. Through modification of SPCE with NGAL-specific antibodies, and utilization of PB-g-C_3N_4 nanohybrids as a signal-generating label for the subsequent NGAL-specific secondary antibody, a sandwich configuration is constructed on the SPCE. When target NGAL and the secondary antibody are introduced, a distinctive voltametric peak is discernible under applied potential of 0.13 V, attributable to the PB attached to the secondary antibody. This peak current displays a linear augmentation within the NGAL concentration range of 0.01–10 ng/mL, with the LOD of $2.8\,pg/mL^{1}$ under ideal conditions. Remarkably, the precision across different batches is consistently below 12%. Notably, the presence of other disease-related biomarkers does not cause interference. To establish the reliability and cross-laboratory validity of this technique, it was evaluated against a commercial ELISA for detecting NGAL in spiked human serum samples. The outcomes generated by the two methodologies exhibit notable agreement.

This investigation conducted by Yuan et al. [111] reveals a notable dual enhancement phenomenon displayed by poly[9,9-dioctylfluorenyl-2,7-diyl] (PFO) dots within the ECL system incorporating N-(aminobutyl)-N-(ethylisoluminol), ABEI /hydrogen peroxide, H_2O_2. This phenomenon not only increases the ECL intensity of ABEI but also acts as a catalyst for the decomposition of H_2O_2, consequently enhancing the ECL signal emitted by ABEI even further. Because PFO dots are electronegative, electropositive ABEI-PEI, an ECL reagent, could be adsorbed onto their surface. This created a new type of light emitter known as ABEI-PEI-PFO dots. Electrostatic attraction allows this emitter to have an exceptional ECL efficiency. The amine-rich properties of ABEI-PEI also led to improvements in the emitter's aqueous solubility and stability, which boosts the luminous efficacy of ECL processes in aqueous conditions. For the improvement of electron transfer efficacy, a one-pot synthetic method is utilized to affix Pt NPs (PtNPs) onto reduced graphene oxide nanosheets. The intricate structure is employed as an immobilization platform for the ECL emitter, specifically ABEI-PEI-PFO dots. Consequently, this dual-amplified ABEI-PEI-PFO dots-RGOs/PtNPs architecture is employed as a promising nanocarrier for capturing the detection antibody (Ab2). An incredibly sensitive ECL immunosensor was created using a sandwiched immunoreaction mechanism, enabling the detection of kidney injury molecule-1 (KIM-1) at concentrations between 50 fg/mL and 1 ng/mL. The achieved LOD is notably low at 16.7 fg/mL. The innovative ECL emitter, with its dual enhancement attributes for signal augmentation, presents a novel avenue with significant potential to improve sensitive bioanalysis and clinical applications.

Thus, the coming together of nanotechnology, advanced sensing techniques, and diagnostic methodologies holds a lot of potential for the future of early detection and management of CKDs. The ongoing progress in creating sensors that are highly sensitive and specific, as well as immunosensors and point-of-care devices, has the potential to bring about a transformation in the field of nephrology. These advancements could lead to timely interventions, improved patient outcomes, and a reduction in the burden that CKD places on healthcare systems. Moreover, as our understanding of CKD biomarkers continues to evolve, the incorporation of emerging technologies such as nanomaterials, SERS, and sophisticated data analysis methods shows significant potential for pushing the boundaries of diagnostic sensitivity, accuracy, and efficiency. Collaborative efforts among researchers, clinicians, and engineers will be instrumental in translating these innovative ideas into practical clinical tools that can be easily implemented within healthcare environments. In summation, the integration of nanotechnology and advanced sensing approaches brings a sense of optimism for the early detection and improved management of CKDs. Through sustained research and interdisciplinary teamwork, these technologies have the capability to reshape the landscape of nephrology, ultimately enhancing the quality of life for individuals dealing with kidney-related issues and contributing to more impactful healthcare strategies.

7.3.2 Kidney Disease Treatment Using Nanotechnology-Based Targeted Drug Delivery

In recent times, nanotechnology has surfaced as a revolutionary approach in the realm of drug delivery, presenting significant potential for the management of kidney diseases [112]. Leveraging the unique properties of NPs (NPs), this innovative strategy offers the potential to reform the way drugs are delivered to address various renal conditions [113]. By harnessing the advantages of nanoscale materials and tailored surface modifications, researchers are exploring novel avenues to enhance drug efficiency, controlled release, and targeted delivery specifically for kidney diseases [114]. In this section, we delve into the remarkable potential and possibilities of nanotechnology in transforming the landscape of drug delivery for kidney disorders, shedding light on its key mechanisms, applications, and the transformative impact it could have on patient outcomes [115, 116].

PEG linked with PLGA, (poly(lactic-*co*-glycolic acid) stands as a frequently employed polymer combination for the production of polymeric NPs (NPs), as highlighted in an investigation by Danhier et al. [117]. The safety profile of PLGA-PEG has been extensively examined, receiving FDA endorsement for human applications. An exemplar of this is the mesoscale nanoparticle (MNP) about 400 nm across in diameter, originating from the fusion of PLGA-PEG components, as demonstrated in research by Williams et al. [118]. The study involved the creation and assessment of anionic MSNPs and cationic-MNPs, each around 400 nm in diameter, derived from PLGA functionalized with PEG. A fluorescent dye was loaded into these NPs for biodistribution studies. The process included the conjugation of PLGA terminated with carboxylic acid ends with heterobifunctional PEG with amine and carboxylic acid linkages confirmed by 1H NMR. Nanoprecipitation was utilized to fabricate A-MNPs having a diameter of 386.7 nm and a ζ-potential of −19.5 mV, while C-MNPs were generated with a diameter of 402.8 nm and a ζ-potential of 18.3 mV. Both particle types encapsulated 2.2 μg of fluorescent dye per loading per 1 mg of NPs, and their fluorescence emission was comparable. Stability analyses involved aggregation, ζ-potential assessments and dye release indicating stable behavior in PBS over 3 days and serum up to 48 hours, beyond which aggregation began. The ζ-potential of both types of NPs converged upon the introduction of fetal bovine serum (FBS). Dye release kinetics were comparable in PBS and FBS. Upon intravenous administration to mice, the NPs exhibited preferential accumulation in the kidneys. Biodistribution analysis indicated the kidney as the primary site of accumulation, confirmed by fluorescence imaging over several days. Further investigation using immunofluorescence and immunohistochemistry techniques revealed the NPs' localization in the basolateral domain of epithelia cells of the proximal tubules. The study's results suggest that effective kidney targeting involves a combination of nanoparticle size, surface functionalization, and opsonization resistance. The mechanism proposed involves endocytosis by endothelial cells of peritubular capillaries, followed by potential transcytosis and endocytosis by renal epithelial cells. This targeting approach holds potential for diseases affecting proximal tubules and could contribute to advancements in renal physiology research. Utilizing extended PEG chains can diminish opsonization, consequently preventing uptake by the mononuclear phagocyte system. Administration of MNPs to mice resulted in predominant localization within the tubules, with no apparent adverse consequences [119].

The clinical utilization of triptolide (TP) in treating renal conditions is still hindered by its limited water solubility and potential toxicity. In a study by Deng et al. [120], precisely defined, evenly distributed, and uniformly sized mesoscale NPs (TP-MNPs) encapsulating TP were developed using a nanoprecipitation technique. The distinct kidney-targeting capabilities, gradual release characteristics, and high efficacy in addressing renal ischemia-reperfusion injury (IRI) were exhibited by these NPs. Favorable compatibility with various TP concentrations (0–500 ng/mL) and exposure durations (6–24 hours) was demonstrated by TP-MNPs. Exceptional kidney-targeting ability with prolonged retention duration (7 days) was indicated by pharmacokinetic (PK) analysis and ex vivo organ fluorescence imaging for TP MNPs. Notably, significant kidney protection against IRI was provided by even a minute dose of TP (0.01 mg/kg) within TP-MNPs, whereas the same dosage of TP alone was ineffectual. Following TP-MNP treatment, substantially reduced levels of serum creatinine, blood urea nitrogen, C_3 complement expression, and extracellular signal-regulated kinase with phosphorylation were exhibited by the IRI mice, compared to TP-treated mice, with respective reductions of 5.9-, 2.0-, 5.4-, and 2.8-fold. Additionally, minimal reproductive toxicity, hepatotoxicity, and immunotoxicity were demonstrated by TP-MNPs compared to TP alone. These effects included reduced levels of alanine aminotransferase and aspartate aminotransferase showing 0.5 fold and 0.2 fold decrease, respectively, as well as an elevated CD^{4+}/CD^{8+} ratio (2.2-fold). The potential of uniformly sized TP-MNPs with specialized kidney-targeting and sustained-release capabilities is highlighted by this study, offering a promising avenue for novel therapeutic strategies in renal diseases. In the study by Tan et al. [121], a drug delivery system was established by linking KTP to PLGA-PEG NPs formed by PLGA-PEG via a covalent linkage. The SEM image depicted that the KTP-NPs obtained displayed a spherical morphology, characterized by a consistently uniform size of 200 nm. Dynamic light scattering (DLS) measurements verified a hydrodynamic diameter of approximately 200 nm for the KTP-NPs, aligning with the observations from the SEM analysis (Figure 7.9a). The zeta potential of the NPs was found to be −10.6 mV, which decreased to −34.4 mV after KTP conjugation, possibly due to the presence of the attached peptide, indicating the incorporation of KTP on the nanoparticle surface. Through the SRB assay assessing cell viability, the KTP-NPs system displayed a time-dependent enhancement in the responsiveness of HK2 cells toward AA (Figure 7.9b–d). The IC50 values of KTP-NPs-AA were notably reduced after 12 hours, 24 hours, and 48 hours, to 31.63 μM, 26.4 μM, and 28.17 μM, respectively, than those of free AA (52.14 μM, 47.1 μM, and 43.35 μM, respectively). The research also explored how KTP influences specificity of PLGA-PEG NPs to kidneys. Upon administration to Sprague–Dawley rats, the utilization of FITC-labeled NPs (NPs) and FITC-labeled KTP-NPs revealed a substantial increase in the accumulation of NPs in kidney tissue facilitated by KTP (Figure 7.9e), while NPs without KTP primarily accumulated in the liver (Figure 7.9f). Quantitative analysis indicated a 30-fold increase in NP accumulation in the kidney due to KTP, with a noticeable enhancement also observed in heart distribution. Over time, KTP-NPs exhibited sustained kidney retention and reduced liver accumulation (Figure 7.9g). Renal slices were imaged using fluorescence microscopy to assess the intrarenal distribution of KTP-NPs. Proximal tubules were the primary location of the KTP-NPs with strong fluorescence signals observed in these tubules compared to glomeruli or distal renal tubules (Figure 7.10a). Staining of immunohistochemistry with an anti-megalin antibody confirmed the concentration of megalin in the brush border of proximal tubules. Relative to megalin, FITC fluorescence indicated that the basolateral side of the proximal tubules was the site of KTP-NP concentration. To investigate KTP-NP uptake and cell-type binding, in vitro tests were carried out. Flow cytometric analysis revealed that among the cell types tested, HK2 exhibited the highest binding capacity for KTP-NPs, while other cell types showed lower binding abilities or none at all (Figure 7.10b). Confocal fluorescence images further supported these findings, showing maximal uptake of KTP-NPs by HK2 cells (Figure 7.10c). The target-specific therapeutic outcomes of asiatic acid (AA) using the KTP-NPs system were evaluated in UUO rats. KTP-NPs-AA treatment resulted in significant improvements in left kidneys, along with decreased biochemical parameters, indicating reduced renal dysfunction compared to untreated UUO rats. Free drug AA showed minimal therapeutic effects, and NPs-AA exhibited moderate improvements in renal morphology. Moreover, the study explored the effects of targeted AA on biomarkers associated with fibrosis and oxidative stress. Collagen III, TGF-β1, FN, α-SMA, collagen I, and PAI-1 were evaluated, with targeted AA effectively mitigating the UUO-induced increase in these markers. Protein expression in the TGF-β/Smad signaling pathway was also quantified, indicating that AA-targeted treatment resulted in a decrease in the upregulation of TGF-β1, Smad2, and Smad3, which are associated with renal fibrosis. Additionally, AA demonstrated antioxidative stress activity, with targeted AA effectively decreasing the expression of proteins associated with oxidative stress like NOX4 and Rac1, as well as 12-LOX. Mesangial proliferative glomerulonephritis (MsPGN), a frequently encountered pathological form of glomerulonephritis, often progressed to end-stage renal damage over an extended period. However, the lack of specificity and the induction of severe adverse effects by the current approach to MsPGN treatment highlight the pressing need for novel therapeutic options and precise targeting strategies. Through the synergistic integration of the benefits of PEG-PLGA NPs and the size-based mechanism of filtration of renal glomeruli, an innovative PEG-PLGA NP conduit system proficient in transporting A-DEX (dexamethasone acetate) to the glomerular mesangium has been conceptualized and formulated by Zhang et al. [122]. The glomerular size-selection mechanism was used to determine the ideal NP size of 90 nm for encasing A-DEX. Following intravenous injection in rats, higher accumulation of 90 nm DiD-loaded NPs within the kidney and kidney cortex was observed in comparison to the free DiD solution. Notably, the 90 nm A-DEX NPs demonstrated improved stability at room temperature and displayed a sustained release profile. In vitro studies conducted with rat glomerular mesangial cells (HBZY-1) indicated that the uptake of 90 nm A-DEX NPs depended on both temperature and energy, predominantly facilitated through clathrin-dependent endocytosis pathways. Tain et al. fabricated a novel nanoplatform for drug release that responds to stimuli [123]. Nano assemblies are formed through the coordination interactions between glutathione (GSH)-modified gold NPs (GLAuNPs) and Co^{2+}, resulting in GLAuNPs-Co. The assembly occurs due to coordination interactions between the empty orbitals of Co^{2+} and the lone pairs present on GSH. GLAuNPs displayed an impressive ability to load drugs as a drug carrier. With Co^{2+} assembly triggering pH-dependent drug release. Upon interaction with Co^{2+}, the GLAuNPs exhibit a reaction to the acidic lysosomal environment found in renal fibrosis tissues, resulting in the disassembly of GLAuNPs-Co and the subsequent release of Co^{2+}. Additionally, encapsulating Co^{2+} ions within the GLAuNPs significantly lowered Co^{2+}'s cytotoxicity in cells of the kidney tubules. Fluorescence imaging of tissues demonstrated the specific accumulation of GLAuNPs-Co within the kidneys, particularly in the proximal renal tubules. Following intraperitoneal injection of GLAuNPs-Co into mice with ureter obstruction, a notable reduction in interstitial fibrosis was observed. The beneficial outcomes are primarily attributed to the restoration of expression of miR-29c facilitated by activation of HIF-α. These observations underscore the pH-dependent drug release and renal targeting abilities of GLAuNPs-Co, positioning them as a potential platform for the delivery of drugs for addressing kidney disease. Zhang et al. demonstrated the use of HSA for the fabrication of albumin NPs, enabling selective transport of the drug CLT to mesangial cells [124]. The objective was to determine the most effective particle size for targeting mesangial cells, encapsulating CLT within albumin NPs (ANs), and evaluating the therapeutic effectiveness and safety in a rat model of anti-Thy1.1 nephritis. The researchers initially examined how the size of NPs influenced their localization on mesangial cells. ANs with a size of around 95 nm (AN-95) demonstrated the highest accumulation in mesangial cells, facilitated through mechanisms such as macropinocytosis, clathrin-mediated endocytosis, and caveolae-mediated endocytosis. Subsequently, CLT was loaded into ANs of comparable size, and the mesangial cell-specific delivery of CLT-AN was confirmed in vivo. In terms of therapeutic evaluation, superior efficacy in comparison to free CLT in terms of proteinuria reduction, managing glomerular lesions both in acute and progressive stages in rat models was demonstrated by CLT-AN. The attributed treatment effects were linked to the CLT-ANs antiproliferative, anti-inflammatory, and antifibrotic mechanisms. Additionally, a decrease in systemic toxicity was observed in CLT-AN as opposed to free CLT, as supported by limited effects on kidney, liver, heart, and brain tissues. Investigation into the biodistribution of CLT-AN revealed a greater propensity for mesangial cell targeting, leading to decreased drug concentrations in nontargeted organs and thus a reduction in systemic toxicity. The labeling of CLT-AN with FITC confirmed that the predominant accumulation of these NPs occurred within Kupffer cells of the liver, potentially accounting for the observed decrease in hepatotoxicity. Overall, the study showcased the effective mesangial cell targeting of CLT-AN with a size of approximately 95 nm. This targeting strategy enhanced the therapeutic efficiency of CLT and lowered its systemic toxicity. The approach holds promise as a strategy for augmenting drug delivery precision and safety in conditions such as mesangial proliferative glomerulonephritis.

Chitosan, renowned for its remarkable biocompatibility and biodegradability, stands as another widely utilized material for NPs [125]. Chitosan-based NPs carrying siRNA have demonstrated the capability to direct siRNA toward tubular cells in mice. The therapeutic potential of RNA interference (RNAi)-based approaches in addressing diverse human conditions, particularly kidney disorders, is substantial. Chimeric mice, featuring a conditional knockout of the megalin gene, received administration of chitosan/siRNA NPs [126]. Intriguingly, these NPs displayed a nearly exclusive distribution in cells expressing megalin, suggesting that the NPs employing the systemic intravenous administration of chitosan carrying water channel aquaporin 1 (AQP1) siRNA in rats led to a significant knockdown of up to 50% in the AQP1 within PTECs. In summary, the targeted delivery of chitosan NPs to PTECs holds promise as a valuable approach for gene knockdown within these cells. This avenue presents a potential therapeutic strategy for addressing a spectrum of kidney diseases. The study by Wei et al. [127] aimed to create Rhein (RH) conjugates with improved solubility and bioavailability by utilizing low-molecular-weight chitosan (LMWC) and alanine, an amino acid linker. The use of LMWC–drug conjugates offers a promising avenue to enhance RH's water solubility, facilitate oral absorption, and ultimately improve its availability in biological systems. The conjugates were created using a carbodiimide reaction and their formation was verified through FTIR, UV-vis, and 1H NMR spectroscopy. The solubility in water and in vitro release characteristics were evaluated. In Sprague Dawley rats, PK studies involved the oral administration of free RH and RH-LMWC conjugates, both at a consistent dose of 35 mg/kg of RH. Successfully synthesized conjugates, incorporating 9.65% RH, demonstrated advantageous water solubility at 9.73 mg/mL. These conjugates demonstrated a prolonged release profile lasting 72 hours, with potential enzymatic involvement in the degradation process that increased Rhein release. The administration of RH-LMWC conjugates orally resulted in a notable sevenfold augmentation in half-life (T1/2) and a 3.1-fold rise in the AUC in comparison to RH suspension. This study showcased the sustained release capabilities of RH-LMWC conjugates, coupled with significant enhancements in oral bioavailability when compared to administering RH alone. These findings suggest that RH-LMWC conjugates could serve as a promising basis for the development of an advanced platform for oral delivery of Rhein. Renal fibrosis stands as a prevailing progressive kidney ailment, yet effective treatment options are limited. A targeted nanocomplex designed for the kidneys was developed through the orchestrated assembly of low molecular weight chitosan derived from catechol (HCA-Chi), metal ions, and active medicinal compounds [128]. Diverse metal-ligand coordination activities were probed alongside assessments of immunogenicity, cytotoxicity, and biodistribution pertaining to HCA-Chi. By employing autofluorescent (DOX), the construction of ternary nanocomplex of HCA-Chi-Cu-DOX was achieved, allowing for an inquiry into cellular uptake dynamics, transmembrane behavior, and targeting attributes. This nanodevice displayed commendable stability under typical physiological settings and unveiled a pH-triggered drug liberation in acidic environments. The HCA-Chi-Cu-DOX uptake by HK-2 cells hinged on factors such as concentration, exposure duration and temperature, with megalin receptor blockers impeding the process. Analysis of tissue distribution unveiled the kidney-specific accumulation of the HCA-Chi-Cu-DOX nanocomplex, exhibiting a rate of renal uptake (re) of 25.6. Through the integration of the active antifibrosis agent emodin into HCA-Chi-Zn-emodin, and subsequent intravenous administration to mice with ureter obstruction, a distinct mitigation of fibrotic advancement was evident. In conclusion, the coordination-induced nanocomplex of HCA-Chi displayed specialized renal targeting capabilities, holding promise for the developing the drug delivery systems tailored to address renal fibrosis. A kidney-specific nanocomplex has been formulated through the via means of the coordination-driven assembly of Sal B, calcium ions,

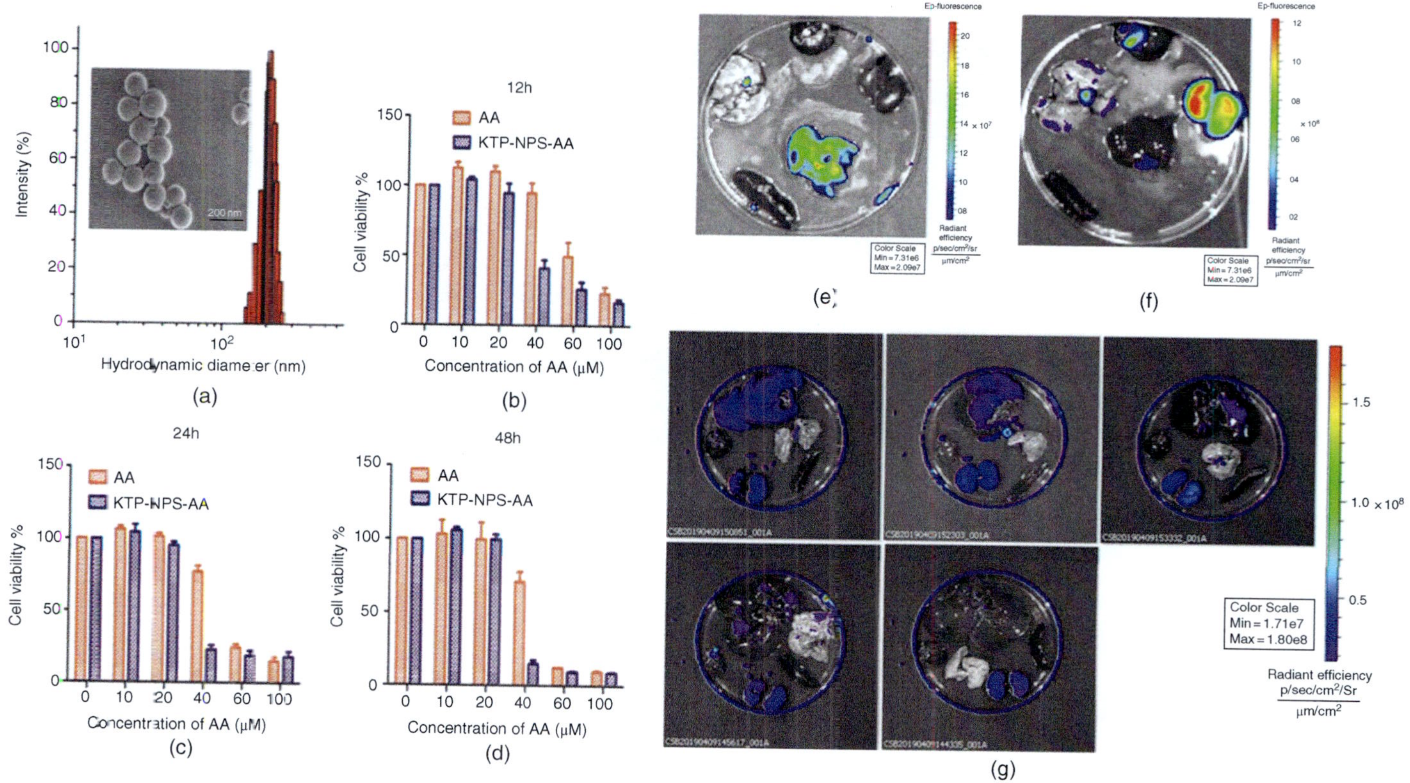

Figure 7.9 Analysis of KTP-NPs properties. (a) Examination of KTP-NPs dimensions using DLS and a SEM portrayal of KTP-NPs. Scale indicator: 200 nm. (b–d) Appraisal of in vitro cell harmlessness for HK2 cells following 12 hours, 24 hours, and 48 hours of exposure to AA and KTP-NPs-AA. Assessment of the tissue-targeting effectiveness of the KTP-NPs drug delivery system. (e) Fluorescently-labeled PLGA-PEG NPs were constructed and introduced, and tissues were harvested two hours later for ex vivo fluorescence imaging. (f) Fluorescence imaging of rat tissues two hours after the injection of FITC-labeled KTP-NPs. (g) Rats were administered fluorescently labeled KTP-NPs, and the distribution in organs was determined through ex vivo fluorescence imaging at 15 minutes, 30 minutes, 1 hour, 2 hours, and 4 hours post-injection. Quantitative analysis demonstrated a significant enhancement in kidney specificity of the NPs enabled by KTP. Source: Reproduced with permission from He et al. [121]/Elsevier.

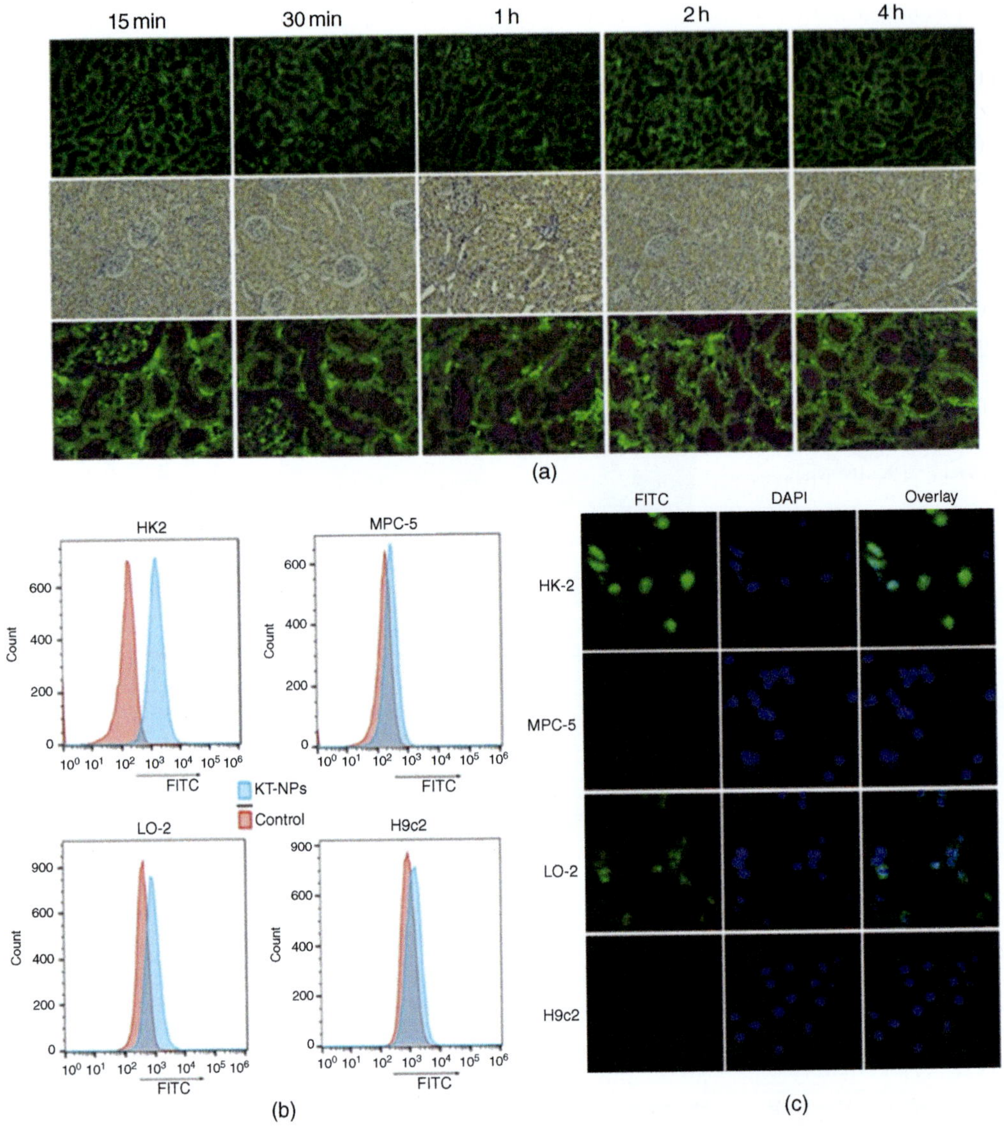

Figure 7.10 KTP-NPs' intrarenal location and cell-type specificity. (a). KTP-NPs that were FITC-labeled were primarily found in proximal tubules. Megalin protein immunohistochemistry suggested brush-bordered tubules, and the magnified images of the merged images demonstrated that KTP-NPs were time-dependently concentrated at the basolateral side of proximal tubules. 200 times magnification. (b). HK2 cells, MPC-5 cells, LO-2 cells, and H9c2 cells were subjected to flow cytometric tests after being exposed to KTP-NPs for four hours (c). HK2 cells, MPC-5 cells, LO-2 cells, and H9c2 cells in confocal fluorescent pictures. 20 m scale bar. Source: Reproduced with permission from He et al. [121]/Elsevier.

and low molecular weight modified with catechol (HCA-Chi) [129]. The resulting nanocomplex, denoted as HChi-Ca-Sal B, has exhibited the ability to reverse epithelial–mesenchymal transition (EMT) in HK-2 cells triggered by TGF-β1. The biodistribution of this nanocomplex was found to be localized specifically within the kidneys through in vivo imaging. An animal model of unilateral ureteral obstruction (UUO) was used to evaluate HChi-Ca-Sal B's antifibrotic properties. Significant amelioration of morphological lesions and decreased levels of ECM proteins in the tubulointerstitium were observed in mice subjected to HChi-Ca-Sal B treatment. This suggests that the nanocomplex exhibits enhanced efficacy in combating fibrosis when compared to the administration of free Sal B. The findings further suggest that the nanocomplex exhibited precise renal targeting capabilities and holds promise for augmenting Sal B delivery for renal fibrosis treatment. Utilizing a combination of miRi (microRNA-21 inhibitor), PLGA (polylactide-co-glycoside) NPs (PLNPs), and low-molecular-weight chitosan (LMWC), specialized kidney-targeting, and highly efficient antifibrosis treatment is achieved through the creation of small-sized cationic miRi-PCNPs (miRi-encapsulated PCNPs) [130]. These miRi-PCNPs effectively shield miRi from nuclease degradation by encapsulating it within the PCNPs. Extensive assessments encompassing cytotoxicity, immunotoxicity, systemic toxicity, together with fluorescence imaging in vitro and ex vivo, demonstrate the remarkable biocompatibility, increased cellular absorption efficiency, and specific kidney-targeting capacity of PCNPs. together with fluorescence imaging in vitro and ex vivo, demonstrated remarkable biocompatibility, increased cellular absorption efficiency, and specific kidney-targeting capacity of PCNPs. By employing Western blotting, real-time polymerase chain reaction analyses, and pathological staining, it is evidenced that the therapeutic efficacy of miRi-PCNPs in mitigating kidney fibrosis exceeds that of miRi administered alone. This outcome can be predominantly ascribed to the inhibition of the expression of microRNA-21, leading to the suppression of the TGF-β1/drosophila mothers against decapentaplegic protein 3 (TGF-β1/Smad3) and extracellular signaling pathways linked to kinases/mitogen-activated protein kinase. As an illustration in the miRi-PCNPs group, the tubule damage index and the tubulointerstitial fibrosis area exhibit a reduction of approximately 2.5 times compared to the corresponding values in both the saline and bare miRi groups. The innovative miRi-PCNPs, boasting specialized kidney-targeting and high-efficiency antifibrosis capabilities, hold significant potential as a promising avenue for conceiving and creating therapeutic interventions for kidney fibrosis. In another investigation, the utilization of glycol chitosan is introduced as a vehicle for the targeted conveyance of hydrophobic drugs, such as tacrolimus (TAC), to the kidney [131]. A polymeric nanomicelle, created by hydrophobic modification of glycol chitosan (HGC), was synthesized. Subsequently, TAC was integrated, leading to the formation of TAC-loaded HGC nanomicelles (HGC-TAC). A spherical structure was exhibited by these HGC-TAC nanomicelles, featuring enhanced drug loading capacity and encapsulation efficacy. In both in vitro and in vivo evaluations, it was confirmed that the HGC-TAC nanomicelles exhibit no toxicity and show a preference for delivering TAC to the kidneys, concurrently lowering plasma concentrations. The therapeutic impacts of HGC-TAC and unencapsulated TAC on renal function were demonstrated, suggesting comparable efficacy to that accomplished by intraperitoneal TAC infusions administered daily over a 14-day period, but without inducing any systemic adverse effects, was accomplished through a singular intravenous administration of HGC-TAC. In light of these findings, the nanomicelles of HGC-TAC emerge as a viable strategy for the directed transport of TAC to the kidney, underscoring their ability as a secure approach for targeted renal drug delivery. The carrier based on l-serine-modified chitosan (SC), as designed by Liu et al. [132], exhibited the capability to accumulate efficiently in the kidneys of AKI mice, while being effectively eliminated by healthy mice. The design of candidates for targeted delivery of drug to AKI kidneys involved the creation of three chitosan-based carriers: oligolysine polyethylene glycol-modified chitosan (PEG-C), L-serine-modified chitosan (PEG-C), and L-serine-modified PEG C (S-PEG-C). The synthetic paths for these carriers were illustrated, and their chemical structures were analyzed through various spectroscopic techniques. SC synthesis involved attaching boc-l-Ser(tBu)-OH groups to chitosan, followed by deprotection. Successful SC synthesis was confirmed using 1H NMR spectrum and gel permeation chromatography (GPC), revealing a M. Wt. of 2.84 kDa and a serine substitution ratio of approximately 4/1 (serine/C, mol/mol). PEG-C and S-PEG-C were synthesized through amidation reactions, and their successful synthesis was confirmed by 1H

NMR. Renal targeting efficacy of the carriers was evaluated through in vivo distribution experiments using fluorescent labeling. Mice were administered Cy5-labeled chitosan (Cy5-C), Cy5-SC, Cy5-PEG-C, and Cy5-S-PEG-C. The fluorescent pictures were taken at various moments in time. Cy5-SC exhibited the highest kidney accumulation in both normal and AKI mice. Increased kidney fluorescence was observed in AKI mice for all carriers, with Cy5-SC showing the most substantial increase. These findings indicated SC's superior renal targeting ability. Cellular uptake studies on HK-2 cells treated with or without H_2O_2 indicated that Cy5-SC demonstrated greater cellular internalization in H_2O_2-stimulated cells compared to normal cells. This was attributed to the specific interaction between SC and the upregulated transmembrane protein, Kim-1, in renal cells with sustained injuries. Confocal microscopy confirmed colocalization of SC and Kim-1, supporting this specific binding. SC-TK-SS31, ROS-sensitive prodrug was synthesized, and its ROS-triggered cleavage and mitochondria-targeted capability were explored. In the presence of H_2O_2, SC-TK-SS31 exhibited gradual SS31 release, confirmed by HPLC-MS. Cellular colocalization studies using Cy5-labeled prodrugs showed that SC-TK-SS31 released SS31 in H_2O_2-stimulated cells while remaining stable in normal cells. Antioxidative stress and antiapoptotic effects of the prodrugs were evaluated through in vitro assays. SC-TK-SS31 demonstrated superior antioxidative stress and antiapoptotic effects compared to SS31, SC-NO-SS31, and C-TK-SS31. The therapeutic efficacy of SC-TK-SS31 was assessed in an AKI mouse model induced by ischemia reperfusion (IR). SC-TK-SS31 treatment led to reduced blood urea nitrogen and serum creatinine levels compared to other treatments, indicating enhanced renal function restoration. Histological analyses demonstrated that SC-TK-SS31 treatment preserved renal architecture and reduced mitochondrial damage, oxidative stress, inflammation, and apoptosis in the kidney. Concentration analysis of SS31 in various organs indicated that SC-TK-SS31 achieved higher drug accumulation in the kidney compared to SS31 and C-TK-SS31, further underscoring its superior AKI kidney-targeted ability. These results imply that SC-TK-SS31 has potential as a therapeutic approach for AKI treatment.

Polyamidoamine (PAMAM) dendrimer-based NPs have been extensively investigated for drug delivery. Dendrimers are densely branched macromolecules that possess compact structures, enabling the binding of targeting ligands [133]. Numerous studies have showcased the efficacy of dendrimers comprised of L-serine-modified PAMAM (G3) as targeted drug carriers for the renal proximal tubules [134, 135]. These dendrimers preferentially accumulate in the kidney while exhibiting negligible levels in other organs. They hold potential for preventing kidney injury caused by ischemia/reperfusion by acting as scavengers of ROS. The dendrimers undergo filtration through the membrane and are then reabsorbed by tubular cells. The reabsorption process encompasses diverse mechanisms, including micropinocytosis, caveolae-mediated, and megalin-mediated endocytosis. Effective kidney-specific delivery necessitates a high level of SER modification ($\geq$80%) [136]. Utilizing dendrimers, captopril can be efficiently delivered to the kidney, thereby mitigating its side effects [134]. In a similar vein, a macromolecular conjugate of a multitargeted sunitinib analog, based on PAMAM dendrimers, exhibited targeted delivery to proximal tubular cells [137]. This conjugate entailed connecting the sunitinib equivalent to NH_2-PAMAM-G3 dendrimers through a platinum-based linker, resulting in notable deposition in rodent kidneys [138]. The same group also investigated conjugating the sunitinib analog to lysozyme, a kidney-specific carrier, demonstrating antifibrotic benefits that suggested lysozymes' potential as targeting ligands. Therapeutic efficacy in kidney diseases has been exhibited by *Tripterygium wilfordii*, an herbal remedy. Considerable toxicity is displayed by one of its active components, celastrol (CLT). Given the unique structure and pathological characteristics of the glomerulus, a targeted drug delivery strategy called PC-PLNs (peptides coupled CLT-phospholipid lipid NPs) has been created [139]. In this system, the efficient transportation of CLT to impaired podocytes and endothelial cells within the glomerulus is aimed for, with a focus on CKD treatment and exploration. Inflammation was effectively curbed, endothelial damage minimized, CKD severity ameliorated, and CLT's toxicity mitigated by PC-PLNs. Mechanistic exploration into CLT's nephropathy treatment showcased its potential for elevating nitric oxide (NO) levels through enhanced eNOS expression while suppressing VCAM-1 expression, resulting in an anti-inflammatory effect. In summary, not only was a promising CKD drug formulation introduced for future development through our investigation, but novel insights into CKD's medical understanding were also extended. The management of AKI predominantly involves targeting renal tubular epithelial cells (RTECs), crucial cells in this context. Several approaches have been investigated to target RTECs by employing ligand-modified NPs after their systemic administration. Nevertheless, the efficacy of these approaches is compromised by the nonspecific interactions between NPs and plasma proteins, resulting in reduced targeting efficacy and NP stability. The study by Fu et al. [140] presented an approach where CLT is encapsulated within D-α-tocopherol PEG 1000 succinate (TPGS), creating a nanocomplex core packed with CLT (referred to as CT). This core exhibited a remarkable loading capacity of approximately 50%. Subsequently, BSA is adsorbed on CT nanocomplex surface, forming a complete albumin corona while maintaining BSA's structural integrity (referred to as CTB). The CTB formulation displayed consistent size distribution and demonstrated substantial stability both in vitro and in vivo. In addition to the routes involving clathrin-mediated and macropinocytosis, the CTB nanocomplex underwent active internalization through megalin receptor-facilitated endocytosis within HK-2 cells. Biodistribution investigations indicate that CTB demonstrates superior specificity in distribution for renal tubules with enhanced targetability in mice in comparison to CT with no albumin corona. Moreover, in vivo pharmacodynamic investigations provide further evidence supporting the efficacy of CTB in mitigating injuries induced by ischemia-reperfusion without causing significant systemic side effects in mouse models of AKI. The CTB nanocomplex's ability to actively target and efficiently deliver CLT to impaired renal tubular epithelial cells makes it a promising candidate for the AKI treatment and offers insights for future therapeutic development in this area.

In conclusion, nanotechnology has ushered in a new era of drug delivery, particularly for kidney diseases. NPs, often constructed from materials like PEG, PLGA, chitosan, and PAMAM dendrimers, have demonstrated remarkable potential in revolutionizing drug delivery strategies for kidney disorders. Through tailored surface modifications and size-based filtration mechanisms, researchers have harnessed the advantages of NPs to enhance drug efficiency, controlled release, and targeted delivery. Studies have shown the effectiveness of NPs in targeting specific renal cells and addressing various kidney conditions, from renal IRI to glomerulonephritis and fibrosis. These NPs exhibit the ability to accumulate selectively in the kidneys, delivering therapeutic agents precisely to the affected regions. The integration of active compounds, such as triptolide (TP), Rhein (RH), and CLT, into NPs has improved their solubility, bioavailability, and therapeutic efficacy.

Furthermore, the development of kidney-targeted nanocomplexes, such as those formed with l-serine-modified chitosan, glycol chitosan, and D-α-tocopherol PEG 1000 succinate (TPGS), has enabled the precise delivery of therapeutic agents to renal cells, reducing systemic toxicity and enhancing treatment outcomes. The incorporation of ligands and albumin coronas has further improved nanoparticle stability, cellular uptake, and specific targeting, resulting in superior therapeutic effects in animal models.

These advancements hold immense promise for the future of kidney disease treatment. Nanotechnology-based drug delivery strategies offer the potential to mitigate kidney injuries, reduce inflammation, suppress fibrosis, and enhance overall renal function. The continued exploration of nanoscale materials, surface modifications, and targeting mechanisms could lead to transformative breakthroughs in renal physiology research and the development of novel therapeutic interventions for kidney disorders.

7.4 Nanotechnology-Based Wound Dressings for Accelerated Wound Healing

In recent times, the fusion of nanotechnology has inaugurated a revolutionary phase in wound healing and antibacterial methodologies. Positioned at the intersection of these scientific realms, nanotechnology presents itself as a transformative influence, providing unparalleled prospects for augmenting wound-healing processes and simultaneously redefining strategies to counter bacterial infections. Through harnessing the distinctive attributes of nanoscale materials and capitalizing on their intricate interactions within biological systems, researchers have unveiled novel pathways for expediting tissue regeneration and formulating precise antibacterial interventions. This section delves into the notable implications of nanotechnology in reshaping wound healing and antibacterial approaches, illuminating inventive remedies that have surfaced and the auspicious prospects that lie on the horizon.

Recent advancements in tissue regeneration have led to a surge in innovative nano therapies undergoing clinical trials. A number of nanoscale methodologies, including NPs, have been under scrutiny for their efficacy in addressing specific phases of the wound-healing process [141]. Two categories of nanomaterials are harnessed within the healing context: first, nanomaterials with inherent attributes that hold promise in wound treatment, and second, NPs employed as carriers for pharmaceutical substances [142, 143].

In the modern era, there has been an increasing interest in enhancing the effectiveness of common antibacterial wound dressings by incorporating bioactive agents. This strategy holds promise as a potent therapeutic strategy for the treatment of chronic wounds. The integration of bioactive compounds into these dressings allows for their controlled release, primarily facilitated by the hydrolytic action of enzymes within wound exudes [144]. Furthermore, various mechanisms, including hydration, diffusion, and swelling, are employed to facilitate the release of bioactive molecules from the dressings to the injury site [145]. Nonetheless, the effectiveness of these bioactive compounds in the dressings may be compromised due to their instant absorption by wound exudations [146, 147]. NPs offer a means to externally deliver substances produced at the injury site, including the short-lived molecule nitric oxide (NO). In conditions like diabetic wounds/ulcers, where endogenous NO generation is compromised, providing exogenous NO at the injury site holds great promise. However, practical implementation has been hindered by a lack of effective delivery methods [148]. In a study conducted in 2012, Blecher and colleagues established that the topical use of diazeniumdiolate (NO-releasing NPs, NO-NP) expedited wound healing in diabetic mice, promoting augmented blood vessel formation, increased fibroblasts, and enhanced organized collagen levels [149]. The study utilized amperometric analysis to monitor the release of nitric oxide (NO) using NO-NPs (NO-releasing NPs). The results showed an immediate release of NO when the NO-NPs were introduced into the solution. Specifically, a 1 mg/mL solution of these particles exhibited a gradual increase in the rate of NO release over time. After 5 hours, a consistent equilibrium level of approximately 245 nM/min was accomplished, which was maintained for approximately 24 hours. The practical application of NO for wound healing was investigated in NOD-SCID mice. The effect of different treatments, including DETA NONOate, NO-NPs, control NPs, and no treatment, on the healing of wounds was assessed. Remarkably, the application of NO-NPs topically to wound sites led to a considerable decrease in eschar size compared to the other groups receiving treatment. For example, by day 5, the wound area in mice healed with NO-NPs had reduced by 29.4% compared to the initial measurement on day 0. In contrast, control mice showed an increase of 12.5% in wound area. Compared to day 0, mice cured with NO-NPs showed an astounding 56.7% wound closure by day 7, while control groups exhibited only approximately 15% closure. The positive effects of NO-NPs continued, with mice treated with NO-NPs achieving 84% wound closure by day 9. In histological analysis, it was observed that NO-NPs enhanced fibroplasia, collagen deposition, and NV in the granulation tissue of treated mice compared to control and other treatment groups. Collagen staining intensity significantly increased in the NO-NP-treated group, according to Masson's Trichrome staining. This staining revealed that collagen fibers in the NO-NP group were well organized, with both immature and mature fibers present. In contrast, control and NP-treated groups displayed less organized collagen fibers. Moreover, the CD34 staining indicated elevated NV in mice subjected to NO-NP treatment in comparison to both the control and NP groups. In summary, the study's results demonstrated the substantial wound-healing potential of NO-NPs in NOD-SCID mice. These effects were evident clinically through accelerated wound closure and supported by histological findings, including enhanced collagen deposition and NV. The study highlights the promise of NO-NPs as a novel therapeutic approach for chronic wounds, particularly in cases where wound healing is compromised due to immunocompromised conditions. Another study in BALB/c mice confirmed NO-NP's role in enhancing wound closure, anti-inflammatory cytokines, growth factors, fibroblast activity, and collagen expression [150]. The research produced several significant findings that provided insights into the mechanisms by which NO-NP can enhance the process of wound healing. One of the key observations was the accelerated closure of wounds in mice treated with NO-NPs. In contrast to mice that were untreated or received only NP treatment, wounds in the NO-NP-treated group exhibited rapid progress toward complete closure, achieving full healing within 12 days after surgery. This outcome suggests that the utilization of NO-NP significantly hastened the overall wound closure process, reducing the time needed for tissue repair by at least three days. Histological analysis of the wounds provided further insights. Untreated wounds displayed intense localized inflammatory infiltrates in the epidermis, which can impede the healing process and lead to complications. On the contrary, wounds that underwent treatment with NO-NPs exhibited significantly less inflammation and preserved the structural integrity of the tissue. This histological improvement suggests that NO-NP treatment promotes a more favorable wound microenvironment, reducing excessive inflammation that can hinder healing. The research also illustrated that the treatment with NO-NPs improved the migration and proliferation of fibroblasts, along with enhanced collagen deposition. Fibroblasts treated with NO-NP exhibited faster migration toward the wounded area, and the collagen content in wounds subjected to NO-NP treatment was significantly elevated. Moreover, the expression of collagen types I and III, important for tissue regeneration, was significantly increased in fibroblasts of human dermal tissue exposed to NO-NPs. Angiogenesis is another pivotal process in wound healing as it ensures adequate blood supply to the injured area. NO-NP treatment resulted in increased angiogenesis, as evidenced by higher levels of CD34 expression in the wounded skin. This suggests that NO-np not only enhances cellular components of wound healing but also promotes the formation of a robust vascular network, facilitating nutrient and oxygen delivery to the wound site. The study also explored the levels of nitric oxide (NO) in the wounds. The utilization of NO-NP resulted in substantially increased levels of NO in wounds, surpassing those in untreated or NP-only treated mice. This additional supply of exogenous NO appears to be instrumental in accelerating wound healing. Furthermore, inhibiting NO production using a specific inhibitor (AG) resulted in decreased NO levels in wounds, stressing the importance of NO in the recovery process. Finally, the research explored the influence of NO-NP on the infiltration of leukocytes into the wound area. Neutrophils and macrophages are key immune cells involved during the inflammatory stage of healing a wound. NO-np treatment reduced neutrophil infiltration, which can be beneficial as excessive neutrophil activity can potentially damage healthy tissue. Conversely, NO-np increased macrophagelike cell infiltration, suggesting a role in modulating the type and extent of leukocyte recruitment during wound healing. Ma et al. [151] have designed a novel nanosystem consisting of hollow CeO_2 NPs with a textured surface and L-arginine encapsulated within (referred to as $AhCeO_2$ NPs). This nano system is designed to effectively facilitate the sequential stages of hemostasis followed by inflammation and proliferation in wound healing. The textured surface of $AhCeO_2$ NPs acted as a bridge, swiftly closing wounds and accelerating the hemostasis phase. Meanwhile, the hollow structure of $AhCeO_2$ NPs enables efficient light harvesting and increases electron–hole pairs within the particles. Additionally, the porous shell of $AhCeO_2$ NPs aided in separating electron–hole pairs and generating ROS, thus inhibiting wound infections, and assisting wound healing throughout the inflammation phase. Moreover, $AhCeO_2$ NPs exhibit enzyme-mimicking properties that mitigate oxidative damage within the wound. Additionally, the liberated L-arginine had the potential to transform into nitric oxide (NO) through inducible NO synthase, thereby further enhancing the proliferation phase of wound healing.

A novel formulation was created by modifying the surface of polyethyleneimine 25k functionalized graphene oxide (PEI25k-GO) NPs with polyethyleneimine 1.8k (PEI1.8k) [152]. This nano formulation denoted as GO-PEI25k/NO-PEI1.8k NPs has been designed to improve their antibacterial properties and enhance the healing of infected wounds by binding them to bacterial surfaces. The antibacterial activity of nano formulation has been carried out in vitro and assessed their effectiveness in promoting wound healing using an infected wound model. The introduction of PEI1.8k surface modification has proven effective in enhancing the NPs' affinity for bacteria. The NPs demonstrated sustained NO release over 48 hours and exhibited remarkable bactericidal activity, achieving a 99.99% reduction in the viability of methicillin-resistant *Staphylococcus aureus* (MRSA) and multidrug-resistant *Pseudomonas aeruginosa* (MRPA) while ensuring non-toxicity toward L929 mouse fibroblast cells at a concentration of 0.1 mg/mL. In a wound model infected with MRPA, GO-PEI25k/NO-PEI1.8k NPs exhibited a significant 87% reduction in wound size, whereas GO-PEI25k/NO NPs achieved only a 23% reduction by the 9th day post-injury. Histological examinations, including Masson trichrome, eosin, and hematoxylin staining, indicated that GO-PEI25k/NO-PEI1.8k NPs promoted re-epithelialization and collagen deposition, resembling the characteristics observed in skin tissue of the healthy mouse. These results imply that GO-PEI25k/NO-PEI1.8k NPs show significant potential as effective agents for both antibacterial treatment and wound-healing applications. Several studies have confirmed the accelerated wound-healing ability of NP-based materials owing to their ability to induce prolonged and enhanced nitric oxide releasing ability [153–155].

Nanotechnology has emerged as a tool for delivering bioactive plant compounds with antimicrobial and wound-healing properties. Curcumin, a prominent component of turmeric, is renowned for its anti-inflammatory, antimicrobial, and antioxidant properties. Studies in diabetic rats have consistently shown that topically applied curcumin enhances reepithelialization, increases proliferation of fibroblasts, boosts deposition of collagen, and stimulates several growth factors [156, 157]. However, the practical utility of curcumin is limited due to its short lifespan and high solubility in aqueous media. This challenge can be addressed through the use of curcumin-loaded nanohybrids. For instance, a novel strategy utilizing collagen cryostructurates loaded with curcumin has been developed for applications related to wound healing [158]. Curcumin, acknowledged for its powerful antioxidant, antiseptic, and anti-inflammatory characteristics, and collagen, renowned for its capacity to enhance cell adhesion, cell migration, and differentiation, were employed in this investigation. However, a challenge was encountered when attempting to load curcumin directly into collagen hydrogels, leading to the creation of substantial molecular aggregates and the obstruction of matrix pores. To address this issue, a double-encapsulation strategy was devised, wherein curcumin was first loaded into lipid NPs (LNPs), and these NPs were subsequently implanted within collagen scaffolds. The cryostructurates formed by collagen/LNP combination displayed an ideal fibrous arrangement, featuring an average pore size of around 100 μm, which supported prolonged cell migration. The research results indicated that the integrity of the collagen structure was maintained, and the NPs were evenly dispersed among the collagen fibers. Upon immersion in a saline buffer, about 20–30% of the hydrogels' nanoparticle content was released within the initial 24 hours, reaching complete release after 25 days. Additionally, upon introducing NIH 3T3 fibroblasts to these hydrogels, they functioned as an effective scaffold for cell interaction within just 4 hours of seeding, and no cytotoxic effects were observed. The encouraging attributes of collagen/lipid cryostructurates position them as a prospective material worthy of further investigation in the realm of wound healing. Zhang developed a nano-micelle based on curcumin-alginate (C-A-NM) to overcome curcumin's bioavailability limitations and improve its delivery to the colon [159]. The release kinetics of curcumin in response to the gastrointestinal conditions were assessed after fabricating C-A-NM, characterized by a size of 55.5 nm. Furthermore, the MTT method was employed to assess the impact of C-A-NM on HCT-8 cell viability at 24 and 48 hours, as well as to evaluate its antibacterial properties. Colon tissue regeneration on the 7th and 14th days, coupled with a colonoscopy investigation on the 14th day, was performed using the hematoxylin and eosin method to elucidate its wound-healing effects in rats. Additionally, measurements of protein/collagen concentration and the assessment of expression of TGF-1/NF-B gene were conducted to appraise colon wound healing. Toxicological investigations revealed that the administration of C-A-NM at a dose of 7.5 mg did not have any adverse effects on cell viability. Significantly, when exposed to C-A-NM, the lowest IC values of 153, 245, and 319 μg/mL indicated diminished activity against *S. aureus*, *P. aeruginosa*, and *Escherichia coli*. Unlike NF-B, the use of C-A-NM increased TGF-1 expression and boosted protein and collagen levels in the affected regions. These findings, in conjunction with colonoscopy and histological observations, underscored C-A-NM's potential in expediting gastrointestinal wound healing. In conclusion, the use of C-A-NM, attributed to its capacity to enhance collagen production and reduce bacterial activity, holds substantial promise for accelerating the healing of gastrointestinal wounds. Sandhu et al. [160] developed a method to create curcumin-loaded solid lipid NPs (CSLNs) with the objective of enhancing the healing process. Initially, the Taguchi design was employed, and subsequently, the central composite design was used to optimize the production of CSLN through hot and HPH. These optimized CSLNs exhibited a particle size below 200 nm and remarkable assay/drug content (0.6% w/w), and solubility (6105-fold increase), along with EE of 75%. Moreover, these CSLNs displayed controlled drug release characteristics, demonstrated safety both in vitro and in vivo experiments, exhibited photostability, could withstand autoclaving, and maintained stability for a period of one year at 30 °C. In a significant comparison with free curcumin dispersion, CSLNs showcased remarkable inhibition of *S. aureus* 9144 growth, with minimum inhibitory concentrations (MIC) of 64 μg/mL against planktonic cells, 512 μg/mL for inhibiting biofilm formation, and 2 mg/mL for eradicating mature biofilm. The main objective of this research was to investigate the capability of CSLNs in disrupting mature biofilms. Additionally, the in vivo assessment of the wound healing capacity of CSLNs, when integrated into a hydrogel, and the in vitro evaluation of their cell proliferation potential was conducted. In a full-thickness excision wound model and a nitrogen mustard gas model, CSLNs demonstrated significantly enhanced wound closure, improved histological and immunohistochemical healing, decreased levels of oxidative stress (LPO) and inflammation (TNF), and heightened angiogenesis (VEGF). Moreover, elevated levels of antioxidant enzymes like catalase and GSH were observed. To sum up, CSLNs hold substantial promise as an innovative wound therapy, especially for infected wounds, due to their ability to target the disintegration of mature biofilms. Sonamuthu et al. [161] conducted research on the wound-healing potential of a silk protein-based hydrogel loaded with L-carnosine and curcumin, denoted as (L-car at cur/SF). The antibacterial effectiveness of the developed L-car at cur/SF hydrogel matrix was assessed using a biofilm assay for bacterial pathogens. In this assay, the distinction between live and dead bacterial cells was determined by the presence or absence of green fluorescence, observed through confocal laser scanning microscopy (CLSM). The findings indicated that the L-car at cur/SF hydrogel demonstrated noteworthy suppression of bacterial biofilm growth in comparison to the control. Confocal microscopy confirmed the reduction in biofilm formation in the presence of hydrogel. Quantitative analysis supported the conclusion that the L-car at cur/SF hydrogel had a pronounced antibacterial effect. To assess the compatibility of the hydrogel with cell culture of human fibroblast, cell growth, and viability were investigated. Hydrogel cultures were utilized for the cultivation of fibroblast cells, a pivotal contributor to tissue repair processes. The results demonstrated excellent cell viability and growth on the L-car at cur/SF hydrogel. Microscopic examinations showed that cells treated with the L-car at cur/SF hydrogel exhibited strong migration and growth in contrast to the cur/SF and control samples. Additionally, the L-car at cur/SF hydrogel demonstrated anti-inflammatory properties by decreasing the production of nitric oxide (NO) in macrophages. This reduction in NO production was observed when macrophages were activated by lipopolysaccharide (LPS), a potent inflammatory stimulant. The combined effects of curcumin and L-carnosine within the hydrogel contributed to the attenuation of inflammation. Confocal microscopy showed spindle-shaped macrophages, indicating a reduction in LPS activation in the presence of the hydrogel. The hydrogel exhibited potent radical scavenging capability, as evaluated through DPPH assay. This capacity to scavenge ROS highlighted the antioxidant potential of the hydrogel. Furthermore, the L-car at cur/SF hydrogel effectively inhibited collagenase activity and demonstrated the ability to bind zinc ions (Zn^{2+}). These properties are crucial for wound healing, as they inhibit collagen degradation and the activity of matrix metalloproteinases (MMPs), which are involved in tissue remodeling. The in vivo diabetic wound-healing model in BALB/c mice was utilized to assess the L-car at cur/SF hydrogel. Photographic images revealed the progress of wound closure over time. Quantitative analysis revealed that the L-car at cur/SF hydrogel markedly hastened the closure of wounds in comparison to both the cur/SF hydrogel and the control. The wound closure rate with L-car at cur/SF reached 55% after 6 days, surpassing the rates of the other groups. Histological analysis of wound tissues revealed enhanced tissue regeneration, collagen deposition, and NV in wounds subjected to the L-car at cur/SF hydrogel treatment. Microscopic observations showed that the hydrogel promoted the formation of a dense epidermis layer and facilitated re-epithelialization, minimizing scar formation. Li et al. [162] conducted an investigation in which they examined the wound-healing capabilities of a film loaded with a nanoformulation of curcumin. The film was constructed using methoxy poly(ethylene glycol)-graft-chitosan through a solvent/casting evaporation technique. The study results demonstrated notable enhancements in re-epithelialization and collagen synthesis at the wound site. Furthermore, in a rat model, this composite film exhibited outstanding wound dressing properties and improved the delivery of curcumin to facilitate wound healing. Nanofibrous scaffolds imitate the structural features of the extracellular matrix in the skin, allowing them to bind to cells and support processes such as differentiation, migration, growth, proliferation, and angiogenesis [162]. Nanofibrous membranes have garnered considerable recognition in the realm of tissue engineering [163]. In a study conducted by Mahmud et al. [164] curcumin-loaded nanofibers were fabricated using PVA through the electrospinning technique. The nanofiber mats displayed remarkable antibacterial characteristics effective against both gram-positive and gram-negative bacteria. Meanwhile, Mohammadi et al. [165] combined polycaprolactone and polyethylene glycol to produce chrysin-curcumin nanofibers via electrospinning. Different concentrations of chrysin (5%, 10%, and 15% w/w relation to the PCL-PEG-PCL ratio) were added to the solvent to produce nanofibers loaded with chrysin. The created nanofibers exhibited notable anti-inflammatory effects throughout various stages of the wound-healing process. This study effectively addressed the solubility challenge associated with chrysin and curcumin-based materials by introducing nanofibers as a novel solution to expedite wound healing. Moreover, incorporating curcumin into an electrospun nanofiber system using chitosan as the polymer has shown synergistic effects with silver, contributing to the management of wound infections and the facilitation of wound healing [166].

In short, the integration of nanotechnology into the realms of wound healing and antibacterial strategies represents a transformative leap forward in medical science. By harnessing the unique properties of nanomaterials and their intricate interactions within biological systems, researchers have unveiled innovative pathways to accelerate tissue regeneration and combat bacterial infections effectively. Recent advancements in nanotherapies, including NPs and nanofibers, have shown encouraging results in various stages of wound healing. These nanoscale interventions hold the potential to revolutionize wound management by providing controlled drug release, promoting angiogenesis, and enhancing cellular components of healing. Moreover, the utilization of nanotechnology for delivering bioactive compounds, such as curcumin, has opened new horizons in wound treatment. Curcumin-loaded nanohybrids and NPs address challenges related to solubility and bioavailability, offering efficient wound-healing solutions. These innovations present exciting prospects for future wound care and tissue regeneration. Furthermore, the application of NO-NPs has demonstrated remarkable wound-healing potential, particularly in conditions like diabetic wounds. NO-NPs offer a means to overcome deficiencies in endogenous nitric oxide production, accelerating wound closure and enhancing tissue regeneration. Overall, the convergence of nanotechnology with wound healing and antibacterial strategies brings us closer to realizing advanced, targeted, and efficient treatments for wounds and infections. As ongoing research continues to unveil new possibilities, the future holds great promise for the integration of nanotechnology into clinical practice, improving the lives of patients and advancing the field of medicine.

7.5 Nanotechnology in Cancer Treatment

The advancements in cancer treatment and diagnosis in the preclinical domain have been substantial, particularly in the context of nanotechnology implementation. This field continues to evolve, awaiting the full realization of its potential impact. Originally, there was great promise for oncology, driven by the prospect of enhancing drug delivery to tumors with reduced side effects and the early identification of cancer using in vitro diagnostic tools or improved in vivo imaging precision. However, addressing these challenges has proven to be a complex task due to the heterogeneous nature of cancer [167].

Until now, the translation of nanotechnology into clinical application has advanced at a corresponding pace with the complexity of the endeavor. The tangible benefits for cancer patients and their care have seen only modest improvements. Significantly, the initial clinical achievement in this field took place in 1995 when Doxil, a liposomal formulation of doxorubicin, received approval [168]. This accomplishment was supported by a comprehensive comprehension of the in vivo properties of liposomes, a research area dating back to the 1960s. While nanomedicines receiving clinical approval have been introduced in the following years, it is only in recent times that approvals for advanced platforms like targeted liposomes and combination therapies have started to emerge [169]. Apart from these advancements, there are presently over 80 ongoing clinical trials focused on cancer applications of nanotechnology. FDA-approved nanopharmaceuticals such as Abraxane, Doxil, Onyvide, Marqibo, Vyxeos, and others have effectively alleviated the severe toxicities associated with the active pharmaceutical ingredient (API) [170]. Hence, they have played a role in bringing about moderate enhancements in the overall survival rates of patients. However, even with these significant advancements, the application of therapeutic and imaging agents based on NPs faces restrictions due to immunological, biological, and translational hurdles. Consequently, they have not fulfilled the heightened expectations set by preclinical results.

The clinical effectiveness of NPs depends on various crucial factors, such as (i) stability and duration in the bloodstream, (ii) capability to overcome physiological barriers and reach specific anatomical locations, (iii) bioavailability at the disease site, and (iv) safety profile. It is essential to enhance these characteristics to improve the efficacy of nanomedicines, aiding their incorporation into established cancer treatment protocols [171]. Progress in particle design can be tracked concurrently with alterations to the tumor microenvironment, which can significantly impact particle accumulation. Simultaneously, addressing the efficacy challenge involves identifying more potent APIs in combination with specific and specialized interventions for cancer. In this regard, nanoparticle-based delivery offers significant advantages [171]. Ultimately, a promising approach involves carrying out well-structured and rigorous clinical trials that include the possibility of preselecting patients. By combining these strategies, we can work toward achieving a substantial enhancement in the efficiency of nanomedicines.

7.5.1 Cancer Immunotherapy and the Role of Nanomedicine

Cancer immunotherapy has surfaced as a powerful, enduring approach in the field of cancer treatment [172]. The field of cancer immunotherapy, after facing multiple disappointing efforts and unequivocal clinical failures, recently experienced a significant resurgence. The resurgence in interest was chiefly triggered by the authorization of sipuleucel-T, an autologous cellular immunotherapy approved for prostate cancer treatment in 2010 [173], and the subsequent authorizations of ipilimumab, an antibody targeting cytotoxic T lymphocyte-associated protein 4 (CTLA-4), and antibodies against programmed cell death protein 1 (PD1) for melanoma treatment in 2011 and 2014, respectively [174]. These achievements injected new life into the field and underscored the potential of immunotherapeutic approaches.

Antibodies targeting negative immune regulatory pathways, such as CTLA-4 and PD-1 receptors, have improved survival in challenging cancer treatment like bladder, melanoma, renal, and lung cancers like durvalumab, pembrolizumab, and nivolumab [172, 175, 176]. These treatments, referred to as checkpoint inhibitors, are supported by thorough preclinical research dating back to the discovery in 1987 by Sharma and Allison [177]. In addition to checkpoint blockade agents, therapies involving dendritic cells and chimeric antigen receptor (CAR) T-cell have also demonstrated clinical success [178, 179]. Similar to checkpoint inhibitors, there are more than 230 clinical trials dedicated to CAR-T cell therapies. Recent data also suggest that certain cancer vaccines and immunotherapies, when used in combination with other treatment modalities, have the potential to enhance patient survival. These achievements have spurred interest in enhancing cancer immunotherapy.

Checkpoint inhibitors face the challenge of low rates of durable responses, with many patients ultimately developing resistance and progressive disease. Engineering and sustaining CAR-T cells are intricate processes, carrying a notable risk of toxicity if the normal cells carry the targeted antigens. Therefore, CAR-T cell therapy is limited to specific hematological cancers that can be controlled after therapy. Dendritic cell treatment, on the other hand, exhibits low potency, with therapeutic benefits realized several years after treatment. Key distinctions among these approaches include patient response rates, aptness for solid tumors contrasted with hematologic cancers, robustness of response, and posttreatment toxicity profiles. This implies there is space for alternative approaches and therapeutics to improve cancer immunotherapy [167].

Stimulating the immune response against tumor cells is a multifaceted procedure, and nanomedicine presents numerous benefits for enhancing cancer immunotherapy (Figure 7.11). Because of their size resembling that of viruses, NPs easily trigger the immune system when administered in vivo. Unlike previous efforts to minimize immune activation, this immune response is beneficial for cancer immunotherapy and can be harnessed for therapeutic development. NPs are employed to transport tumor antigens to antigen-presenting cells (APCs), thereby augmenting immune responses. Additionally, antigens bound to NPs elicit more robust immune responses compared to free antigens [180]. NPs can also function as immune adjuvants, amplifying responses when used in conjunction with cancer vaccines. Moreover, NPs can be designed to transport tumor antigens along with proinflammatory and pro-immune molecules, thereby intensifying immune responses [181]. In addition, nanotechnology plays a role in the manipulation of tumor-targeting T-cells and the development of CAR-T cell treatments for solid tumors, providing potential solutions for these intricate therapies [182]. Targeting and controlling subpopulations of regulatory T-cell (Treg) using nanoparticle delivery systems is another promising avenue, as Treg cells play a role in immune evasion in cancer [183]. NPs can target cancer-specific Treg cells, silence specific genes, reprogram these cells, or enable diagnostic imaging of tumor burden. Reprogramming tumor-associated macrophages (TAMs) using nanomaterials is also being explored as a therapeutic strategy [184]. Combining nanomedicine with checkpoint blockade inhibitors shows promise, and future research should focus on identifying synergistic agents and regimens. Additionally, local immune responses induced by nanotherapeutics can translate into systemic immune responses, making NPs suitable for local applications and vaccines. NPs can target lymphatic systems and enhance systemic responses [185]. Finally, nanoscale platforms play a crucial role in evaluating patient responses to immunotherapy, detecting side effects, and distinguishing responders from nonresponders, ultimately improving cancer management.

NPs can be tailored to simultaneously deliver proinflammatory and pro-immune molecules with tumor antigens, thereby boosting immune responses [186]. This approach is likely to activate APCs, leading to vigorous immune reactions. Another encouraging use of NPs in immunotherapy centers around the advancement of tumor-targeting T-cells, especially in the treatment of solid tumors, and in the development of CAR-T cell therapies [187]. In particular, NPs hold the potential to enhance the modification of CAR-T cells and other varieties of tumor-targeting T-cells in vivo by transporting molecular agents such as CRISPR-CAS [188, 189]. Moreover, utilizing nanotechnology to separate subpopulations of tumor-infiltrating lymphocytes in vivo has the potential to enhance the effectiveness of adoptive cell transfer strategies for treating solid tumors. Despite initial success with metastatic melanoma, these approaches have faced limitations and are in need of solutions that can be facilitated through nanotechnology [190–192].

Another possible application involves targeting and regulating Treg subpopulations, which contribute to maintaining immune response homeostasis [193, 194]. Nevertheless, these subpopulations can also facilitate tumor progression by facilitating immune evasion. Treg cells regulate the targeting and activation of APCs by effector T-cells and induce cytokine starvation. They are drawn to tumor sites and amass both in peripheral locations such as peripheral blood and spleen and within the tumor tissue [193]. The extent of Treg upregulation frequently corresponds to heightened tumor burden and adverse patient outcomes. This interaction constitutes a primary immune evasion mechanism in cancer, alongside other regulatory pathways that diminish the effectiveness of checkpoint inhibitors [195]. Research has been conducted to elucidate T-cell subpopulations and activation mechanisms. Nanotechnology offers the potential to target cancer-derived Treg cell subpopulations by leveraging their commonly upregulated surface markers. This can involve silencing-specific genes, labeling or reprogramming, for visualizing and assessing the tumor load through diagnostic imaging. Altering the phenotypic characteristics of TAMs is also an intriguing

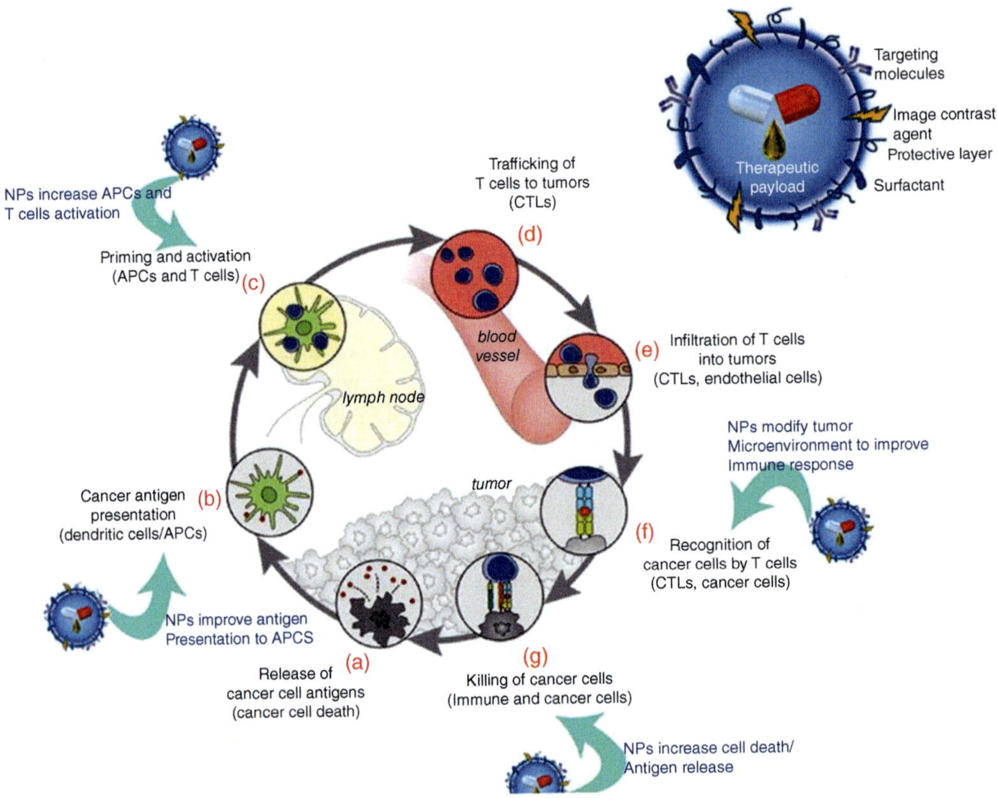

Figure 7.11 Illustration of the complicated immunotherapy pathway for cancer. At several stages along this pathway, nanoparticle delivery vehicles may be involved. Antigens are released upon cancer cell demise, allowing for antigen presentation (a). In the end, this enables lymph node-based priming and activation of T-cells that are then transported back to the tumor as cytotoxic T-cells (CTL). Following tumor invasion (e), CTL detection of cancer cell antigens (f), and CTL death of particular cancer cells (g), occur. The immune response to cancer is an ideal scenario that, because of the tumor's constantly developing immune evasion systems, does not happen frequently. Source: Reproduced with permission from Hartshorn et al. [167]/American Chemical Society.

therapeutic avenue currently under exploration [196]. Several types of nanomaterials naturally gather within TAMs in the tumor microenvironment [197]. This passive strategy leverages the existing PK (PK)/pharmacodynamic (PD) properties of current nanomaterials to selectively transport immunotherapeutic or induce phenotype changes, such as through colony-stimulating factor 1 receptor (CSF1R), among other mechanisms [198]. Given the encouraging clinical results of checkpoint blockade inhibitors, the combination of nanomedicine and checkpoint blockade inhibitors is poised to make an immediate clinical difference. Subsequent research efforts should concentrate on identifying complementary checkpoint blockade agents and schedules alongside NPs. Additionally, there is a need for the incorporation of nanoparticle-based agents into checkpoint blockade treatments and their combinations with various therapeutic modalities [199]. The objective of this approach is to attain durable response rates that extend beyond the primary tumor to metastatic sites. Different alternative pathways for checkpoint inhibitors, beyond CTLA-4 and PD-1, might demonstrate success in clinical trials, thereby creating opportunities for nano-enabled delivery vehicles [200]. Additionally, given that local immune responses can lead to systemic immune reactions, nanotherapeutics have the potential to be locally applied in clinical applications, either as vaccines or devices [201]. Nanoscale systems can also target lymphatics and trigger the complement system to promote an enhanced systemic response. Therefore, the ongoing advancement of nanoscale systems that replicate biological processes or engage with other cells will have a noteworthy impact on cancer immunotherapy [202].

In summary, cancer immunotherapy has surfaced as a powerful and enduring strategy in the realm of cancer treatment, marked by significant advancements and clinical successes in recent years. The approval of therapies such as sipuleucel-T, ipilimumab, and anti-PD1 antibodies has reinvigorated the field, emphasizing the potential of immunotherapeutic approaches.

Checkpoint inhibitors, dendritic cell therapy, and CAR-T cell therapies have shown clinical promise, yet they face challenges such as low rates of durable responses and potential toxicity risks. These limitations suggest the need for alternative strategies to enhance cancer immunotherapy.

Nanomedicine presents a promising avenue for improving cancer immunotherapy. NPs can activate the immune system, deliver tumor antigens, serve as immune adjuvants, and co-deliver pro-immune molecules, enhancing immune responses. Nanotechnology also aids in engineering tumor-targeting T-cells, controlling regulatory T-cell subpopulations, and reprogramming TAMs. Combining nanomedicine with checkpoint blockade inhibitors holds great potential and should be further explored in research. Additionally, nanotherapeutics can induce local immune responses that translate into systemic effects, making them suitable for local applications and vaccines. NPs can also target lymphatics and enhance systemic immune responses.

Continued development in nanoscale systems that mimic biological processes and interact with various immune cells will undoubtedly play a significant role in advancing cancer immunotherapy and improving patient outcomes. Cancer immunotherapy has thus, made substantial progress, marked by approvals of immunotherapies like sipuleucel-T, ipilimumab, and anti-PD1 antibodies, reinvigorating the field and emphasizing its potential. However, challenges persist, including low rates of durable responses and potential toxicity issues. This suggests the need for alternative strategies to enhance cancer immunotherapy. Nanomedicine offers a promising avenue for improving cancer immunotherapy. NPs can activate the immune system, deliver tumor antigens, serve as immune adjuvants, and co-deliver pro-immune molecules, enhancing immune responses. Nanotechnology aids in engineering tumor-targeting T-cells, controlling regulatory T-cell subpopulations, and reprogramming TAMs. Combining nanomedicine with checkpoint blockade inhibitors holds potential and should be explored in research.

Moreover, nanotherapeutics induce local immune responses that translate into systemic effects, making them suitable for local applications and vaccines. NPs can also target lymphatics and activate complement to stimulate increased systemic immune responses. Continued development in nanoscale systems that mimic biological processes and interact with various immune cells. It is poised to play a crucial role in advancing cancer immunotherapy and enhancing patient outcomes.

7.5.2 Nanotechnology in Intraoperative Imaging for Enhanced Cancer Surgery

Identifying, categorizing, and addressing cancer in its initial phases play a crucial role in minimizing mortality associated with cancer [203]. Although surgery is a fundamental component of cancer treatment, its applicability is primarily limited to early stage tumors. Unfortunately, when surgery becomes impractical or is no longer a feasible choice patient outcomes often deteriorate significantly [204]. The primary objective of cancer surgery is to accurately differentiate cancerous tissue from normal tissue, enabling effective surgical intervention while simultaneously preserving aesthetics, minimizing the risk of soft tissue damage, and improving surgical efficiency [205]. However, achieving this objective remains a considerable technical challenge.

Current surgical techniques rely on tactile examination and visual examination to identify subtle variations in texture rather than providing insights into the molecular characteristics specific to each disease stage [206]. This limitation severely restricts the surgeon's ability to precisely delineate the degree of malignancy, identify microscopic tumor remnants, and assess the presence of residual tissue, especially at surgical margins. As a result, the likelihood of incomplete tumor removal is elevated. Given that these factors jointly influence treatment results, prognosis, and therapeutic management, there is a critical need for targeted optical visualization tools and sensors in the intraoperative setting. These instruments are crucial for directing minimally invasive procedures with precision at the cellular level, improving surgical visibility, and offering immediate evaluations of both healthy and afflicted tissues through real-time structural and functional imaging. It is anticipated that such tools will significantly impact the field of oncology by playing a crucial role in aiding surgical diagnosis and treatment, potentially revolutionizing the practice of surgery [167].

Although considerable advancements have been achieved in the creation of extensive imaging apparatus like PET-CT and MRI, their application in intraoperative settings has been hindered by limitations in attainable spatial precision, sensitivity, and pragmatic obstacles [207]. Moreover, conventional optical imaging techniques have encountered limitations in surgical scenarios. These include a restricted array of targeted near-infrared agents, shallow penetration into tissues, limited contrast resolution, quick photobleaching, and the incapability to concurrently identify multiple cancer and normal tissue targets governing various biological processes [208]. Nevertheless, significant endeavors have been focused on surmounting these constraints through the creation of diverse optically active and targeted imaging agents. This includes the development of particle-based probes with the aim of enhancing sensitivity, specificity, and penetration depth into tissues. These probes hold the potential to substantially improve the immediate identification of molecular signals and therapeutic interventions of early-stage disease when used in conjunction with advanced multichannel fluorescence camera systems that meet clinical standards. In contrast to alternative imaging methods, this combination allows for the simultaneous evaluation of biological processes in real-time and the concurrent detection of one or more biomarkers specific to tissues, facilitating precise identification of cancerous areas, accurate staging, and personalized treatment plans (Figure 7.12). The credibility of these markers can be additionally supported through the conduct of clinical trials. Considerable research has been carried out using inherent tissue characteristics or nonspecific optical agents like an FDA-approved NIR fluorescent dye such as indocyanine green (ICG), with specific clinical applications. However, their efficacy in directly identifying cancer cells is limited due to the absence of precise targeting. To introduce targeting precision, various optically active imaging agents, including conjugates of antibodies and peptides with dyes, have been employed as visualization tools to enhance the evaluation of surgical margins and metastatic lymph nodes [209]. These agents bind to cancer receptors that are overexpressed in these areas. While optical imaging has not fully unleashed its ability in surgical environments, initial signs of its advantages have been noticed in clinical studies using specific molecular probes. In certain cases, the attachment to hydrophobic dyes has constrained the bioactivity, penetration depth, and tissue contrast of these probes.

Besides their therapeutic applications, NPs play a pivotal role in the realm of medical imaging [210]. An illustrative case is ferumoxytol (FMX), an iron oxide NP that gained FDA approval for treating chronic renal failure patients by serving as an iron replacement therapy. This FDA endorsement has expanded the off-label utilization of FMX as a contrast agent for MRI, drawing on earlier advancements involving SPIONs for assessing lymph nodes [211]. The potential of nanomaterials as imaging tools is extensive, owing to their versatile and finely adjustable properties that can be customized to suit particular imaging needs [212]. Significantly, it is feasible to incorporate several contrast agents within a singular NP, each situated in distinct compartments within the nanoparticle structure [213]. These versatile multimodal nanoimaging agents can harness the advantages of various imaging techniques when combined. A practical illustration of

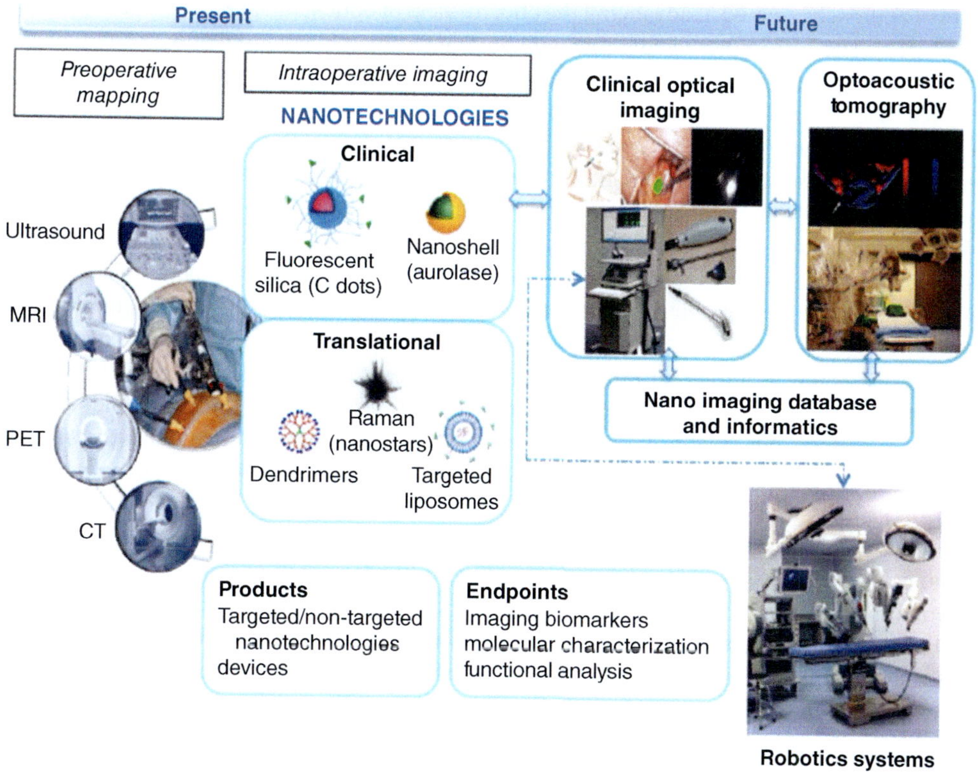

Figure 7.12 Image-guided surgical suites for nano-oncology: present and future. Optical-driven minimally invasive and open surgical techniques are guided by preoperative conventional imaging technologies that are used to screen for disease. Real-time monitoring of clinically applicable particle platforms is performed with portable multichannel camera systems. Additionally exhibited are representative translational probes and tools for potential clinical application. In the future, the operating surgeon will choose the best probe-device combinations for particular indications and receive information on tissue state at the structural, functional, and/or molecular levels for subsequent therapeutic management. Source: Reproduced with permission from Hartshorn et al. [167]/American Chemical Society.

this concept involves the integration of a contrast agent suitable for imaging of the entire body with another characterized by high resolution and sensitivity, thus capitalizing on their complementary attributes. A pioneering instance of multimodal imaging entailed combining an ultrasmall SPION with fluorochromes, yielding an imaging agent capable of both MRI and fluorescence, a concept that has since seen further refinements with similar NPs [214].

A more recent strategy in integrating whole-body imaging with localized intraoperative imaging revolves around a technology referred to as C-dots or Cornell dots [215]. These C-dots are tiny silica NPs, measuring less than 8 nm, and are excreted through the kidneys. These dots have near infrared fluorochromes at their core and are enveloped by a PET tracer molecule. This combination serves a dual function: the PET tracer aids in preoperative mapping, while the fluorochromes allow for real-time visualization during surgery [216]. Clinical trials are presently underway for these nanoscale constructs. In addition to C-dots, a variety of nanoparticle structures have been developed to enhance the signal intensity in optical imaging techniques. Specifically, formulations of polymeric NPs derived from hyaluronic acid have shown significantly improved tumor contrast when contrasted with the use of fluorescent dyes alone. These innovations represent a broad spectrum of approaches aimed at advancing the effectiveness of optical imaging in various applications [217, 218]. These examples vividly illustrate the transformative potential of nanotechnology in improving intraoperative imaging, a field previously reliant on conventional fluorescent dyes, which had their limitations. An alternative high-resolution optical imaging technique that addresses the constraints associated with fluorescent-based methods is surface-enhanced (resonance) Raman spectroscopy (SE(R)RS) imaging. SE(R)RS leverages the surface plasmon resonance effects occurring on noble metal NP cores, resulting in a significant augmentation in signal intensity when compared to intrinsic Raman spectroscopy. The resonant version surface-enhanced Raman spectroscopy (SERRS) of these NPs has demonstrated potential in intraoperative scenarios [219]. These agents for nanoimaging circumvent issues like autofluorescence and photobleaching, possess unique spectral fingerprints for unequivocal in vivo identification, and exhibit detection sensitivity levels as low as attomolar range. Due to these exceptional characteristics, SERRS NPs hold potential as universal cancer imaging agents, enabling the visualization of microscopic spread, including locoregional micrometastases, in mouse models [220]. Alterations to the core–shell architecture of gold-silica SE(R)RS NPs also allow detection using whole-body imaging methods like MRI or PET, as well as techniques that facilitate deep-tissue, three-dimensional imaging during surgery, such as photoacoustic tomography [221, 222]. Continuous progress in nanofabrication and NP synthesis techniques presents opportunities for creating modular NP designs. The potential integration of therapeutic and imaging capabilities is on the horizon, holding the promise of developing sophisticated and adaptable theranostic nano agents in the future.

Nevertheless, despite the accumulation of substantial data endorsing the optically active particle probes' possible use in intraoperative cancer therapy, the translation of these agents into clinical practice poses a significant challenge, particularly those designed for molecular-specific applications, which has encountered challenges. Significantly, the FDA has granted approvals for Investigational New Drug (IND) exemption for targeted particle-based drug technologies and Investigational Device Exemption (IDE) for device technologies [223]. These approvals mark significant progress and set the stage for the translation of additional targeted optical technologies into clinical settings, particularly for image-guided surgeries [224].

As a result, surgeons will increasingly have access to an expanding array of specialized intraoperative tools, rather than depending solely on visual and tactile cues for guiding disease assessment and therapeutic decision-making. It is expected that surgical visualization enhanced by fluorescence will have a substantial impact and likely bring about a transformation in conventional surgical practices within the field of oncology over the next 5–15 years [225]. This transformation will be characterized by heightened sensitivity and precision in surgical procedures focused on the mapping of surgical margins, identification of loco-regional cancer-affected nodes, and the detection of micro metastases with single-cell accuracy [226].

During intraoperative procedures, real-time decision-making for treatment management will involve the acquisition of structural imaging information alongside functional optical assessments. This comprehensive approach considers factors such as three-dimensional (3D) tissue function mapping (e.g., oxygenation status, tissue viability, perfusion), assessments of deep tissue imaging, and volumetric imaging of in situ lesions. These elements collectively facilitate the navigation of surgical procedures [227]. Prior to the implementation of high-precision perioperative therapies, the copious structural and functional data can be seamlessly integrated with pre-treatment planning volumes and augmented reality presentations. This integration enhances the identification of normal or pathologic tissue structures [228].

Nanoparticle surfaces' adaptability is anticipated to be crucial in the fields of pathology and precision surgery, given their distinctive physicochemical and biological characteristics [229]. This opens up exciting prospects for investigating crucial cancer targets and identifying potent imaging biomarkers that can reflect a wide spectrum of biological processes. Importantly, these findings can undergo rigorous validation through clinical trials.

Various particle-based therapeutic approaches are under development, though they are still in their initial stages, aiming to treat intraoperative disorders effectively alongside standard-of-care medications. A noteworthy strategy includes employing targeted particle radiotherapies and integrating nanomaterials with elements of high atomic numbers (Z) known for their radio-sensitizing properties. These advancements have the potential to substantially increase the radiation dose administered to the tumor tissue, thereby enhancing therapeutic efficiency while minimizing the impact on adjacent normal tissues [230]. Thus, early detection and precise cancer treatment are vital for improving patient outcomes. Surgery plays a pivotal role in the treatment of cancer, especially for early-stage tumors. However, conventional surgical techniques relying on tactile examination and visual inspection have limitations in accurately identifying the extent of malignancy and surgical margins.

The emergence of targeted optical visualization tools and sensors holds significant potential in transforming cancer surgery. These instruments promise cellular-level precision, improved surgical visualization, and immediate structural and functional imaging of both normal and diseased tissues. While large-scale imaging machines like PET-CT and MRI have limitations in intraoperative settings, real-time molecular detection and individualized treatment planning could be significantly improved by using innovative agents such as nanoparticle-based probes.

NPs, such as IONPs like ferumoxytol, serve as contrast agents for enhancing imaging, including MRI. Their versatility and ability to carry multiple contrast agents within a single particle open up opportunities for multimodal imaging. Other nanoparticle constructs, such as C-dots and polymeric formulations, enhance optical imaging and overcome the limitations of traditional, fluorescent-based methods. Techniques like SERRS using NPs offer high-resolution imaging with unique spectral fingerprints.

The integration of these advanced imaging techniques into surgery is expected to bring a transformation in conventional practices over the next 5–15 years. Surgeons will have access to a range of dedicated intraoperative tools, enhancing sensitivity and precision in procedures, from mapping surgical margins to detecting micrometastases. Nanoparticle surfaces, with their versatile characteristics and unique physicochemical and biological properties, are poised to play a crucial role in the fields of precision surgery and pathology. The ongoing development of particle-based therapeutic strategies, such as targeted particle radiotherapies, holds the promise of augmenting radiation dose delivery to tumors while minimizing impact on normal tissues.

In conclusion, progress in intraoperative imaging holds the promise of enhancing the precision and efficiency of cancer surgery. This opens up new avenues for cancer detection and treatment, ultimately contributing to improved patient outcomes.

7.5.3 Nanotherapeutics: Enhancing Cancer Treatment Through Multifaceted Strategies

The pursuit of improving the effectiveness of nanotherapeutics is poised to advance along several parallel avenues, encompassing the following key elements: (i) the formulation of strategies to heighten the accumulation and penetration of NPs within tumors; (ii) the expansion of the range of molecules deliverable via NPs, thereby enabling therapies beyond the reach of traditional drug delivery methods; (iii) the identification of specific niche areas in cancer treatment uniquely poised to gain advantages from nanotechnology; and, lastly, (iv) the enhancement of the design and execution of clinical trials involving nanotherapeutics [171]. Notably, clinically authorized and readily applicable drug delivery systems have been developed to surmount a multitude of technical obstacles that have historically curtailed the therapeutic capabilities of initial versions of cancer nanomedicines. Due to the fact that most of these nanoformulated therapeutic platforms have sizes surpassing ~10 nm, their successful nontargeted delivery has traditionally hinged on leveraging the enhanced permeability and retention (EPR) effect. Significant obstacles impeding translation have encompassed inefficiencies in transporting therapeutics to specific sites, due to lack of consistent intratumorally distributions, dose-limiting toxicity, hostile PK profiles, and narrow therapeutic margins [231]. Moreover, challenges like opsonization and uptake by the RES's own phagocytic cells (including the spleen, liver, blood, and bone marrow,) continue to pose significant hurdles in the thoughtful design of NPs, considering factors like size and shape. It is worth mentioning that only a restricted subset of existing drug delivery vehicles, which have been modified with surface-targeting ligands to improve specificity, is advancing to clinical trials. This is likely attributed to the complexities associated with chemistry and PKs. Consequently, even though numerous NP-based imaging probes have been created in research laboratories, only a small fraction exhibits the necessary characteristics for practical clinical translation.

Besides tackling the previously mentioned challenges, it is clear that, given the complex nature of cancer, compounded by factors like tumor heterogeneity and drug resistance, relying on single-agent therapeutic platforms is inadequate. To augment the efficacy of cancer treatments, there has been a progressive shift in delivery strategies toward multifaceted combination approaches. This endeavor has entailed significant efforts in the development of (i) Nanocarriers tailored with a combination of drugs that operate through different pathways and mechanisms; (ii) treatments that incorporate biological therapeutics, such as antibodies, genes, or siRNAs, alongside other drugs; and (iii) intelligent nanocarriers, commonly referred to as stimuli-responsive platforms [232, 233]. These stimuli-sensitive platforms can be employed individually or in combination to actively target cancer, providing advanced spatiotemporal regulation over the release of cargo, such as drugs, in response to internal or external cues within the microenvironment of tumor. Internal cues that trigger release include low pH levels, overexpressed enzymes, and increased redox potential. External stimuli applied to enable on-demand drug release include the use of ultrasound, NIR light, or magnetic fields. Upon encountering these triggers, responsive systems undergo modifications in composition or conformation, leading to changes in physicochemical properties. Several research teams have employed these stimuli-sensitive approaches to overcome size limitations and the associated challenges of larger formulations. This reduction in particle dimensions has facilitated the development of renally clearable platforms, with applications in both diagnostics and therapeutics. This involves creating enzymatically cleavable silica nanoparticle drug conjugates (NDCs) with sizes below 10 nm, targeting genetic mutations expressed in various cancer types. Furthermore, there is progress in creating enzymatically triggered nanoaggregate constructs designed to monitor treatment response. This is accomplished through the self-assembly of endogenic biochemical species and biological pathways [234, 235].

Target-enabled in situ ligand assembly (TESLA) is a new precision medicine strategy that has made it easier to monitor the progression of tumors and more precisely, to evaluate the response of tumors to therapies in a clinical setting [236]. The difficulties encountered in cancer imaging using nanotechnology to monitor treatment response may be addressed by TESLA probes, given the great prognostic value of early assessment of drug-induced tumor cell death, which permits the prompt selection of the most effective course of action and precise delivery of therapeutics [237]. This strategy is fundamentally different from the traditional techniques used by nanoscale platforms since the particles are assembled on the spot at the locations of tumors. The assembly originates from low molar mass building blocks that are internalized by target cells. The smaller size and precisely controlled chemical structures of these probes ease their passage through the vasculature, enabling them to reach tumors and diffuse into the interstitial space. These probes activate upon coming into contact with the tumor-specific target, which could be an enzyme or other endogenous biomolecules; they then create NPs at the tumor site. Longer retention and better imaging contrast result from this phenomenon [238]. This method enables the imaging of diverse tumor-specific biomarkers, allowing the exploration of cancer biology in vivo. Examples include assessing redox potential in cancer cells and detecting caspase-3/7 activity in tumor apoptosis [239]. A notable agent, [^{18}F]-C-SNAT, is an [^{18}F]-labeled, caspase 3/7-sensitive nano aggregation PET probe. It has been verified for direct PET imaging of caspase-3 activity, demonstrated in a HeLa tumor xenograft-induced tumor apoptosis model in nude mice using doxorubicin [240]. A direct comparison of its efficacy for the noninvasive detection of cell death with other widely used radiotracers, such as ^{18}F-FDG, ^{99m}Tc-Annexin V, and ^{18}F-ML-10, has demonstrated positive image contrast. Additionally, in the lymphoma model, it has demonstrated an improved ability to precisely and sensitively identify tumor cell death [241].

A recent example of a clinically applicable "smart" particle-based system, demonstrating anticancer properties with internally activated signals, is an exceptionally small NDC. Derived from a clinically translatable fluorescent sub-10 nm core–shell silica NPs known as C dots, this NDC showcases anticancer properties with internally activated signals [224]. The NDCs are engineered to include tyrosine kinase inhibitors, attached to the particle's surface through a protease-sensitive self-immolative linker. This linker is responsive to lysosomal cysteine proteases, such as cathepsin B. Due to this architecture, highly focused tumor therapy is possible at doses that are almost a thousand times lower than those of the free drug. Furthermore, only a few number of "smart" platforms that react to outside stimuli – such as magnetic IONPs (NanoTherm AS1, MagForce Nanotechnologies) and thermosensitive liposomes (ThermoDox) have lately progressed to clinical trial stages and/or been approved for the treatment of cancer [234].

Moreover, nanopreparations responsive to multiple stimuli, loaded with diverse drug classes, hold promise as a flexible platform for enhancing the targeted delivery of numerous anticancer drugs with low solubility. This can be achieved by responding to locally heightened enzymatic activities within different tumors and individual tumor cells. Continuing investigations are delving into additional internal stimuli, including GSH and MMPs [242]. Elevated expression of MMPs in particular types of tumors serves as a trigger for the precise delivery of contrast agents or therapeutics. Several other potential internal stimuli show potential for the self-assembly and successive biodegradation of nanoscale systems for delivering diagnostic and therapeutic agents. A shared characteristic among all these systems is their ability to directly observe and track real-time biological procedures that are specific to the changing tumor and its surrounding microenvironment. These nanotechnology platforms offer advantages by overcoming common limitations encountered in delivering both small molecule and nanoscale materials. These challenges include issues related to low bioavailability, limited effectiveness in cellular uptake, and poor solubility. Moreover, their translation into practical applications is anticipated to be simpler, given that they are administered as small to medium-sized molecules or molecular assemblies. This simplification reduces both manufacturing complexities and regulatory hurdles.

In brief, the field of nanotherapeutics is rapidly evolving to overcome the limitations associated with traditional cancer treatment. Current efforts focus on several key avenues, including strategies to improve the accumulation and infiltration of NPs within tumors, expanding the range of molecules deliverable via NPs, identifying niche areas in cancer treatment benefiting from nanotechnology, and enhancing the design and execution of clinical trials. Clinically approved nanomedicines have addressed multiple technical hurdles, yet significant challenges persist, such as inefficient delivery, uneven distributions within tumors, suboptimal PKs, toxicity restricted by dosage, and opsonization.

To address the intricacies of cancer, characterized by tumor diversity and resistance to drugs, there is a shift toward combination therapies. This includes the development of nanocarriers delivering multiple drugs with distinct mechanisms of action, the integration of biological therapeutics like genes and antibodies, and "smart" nanocarriers responsive to internal and external cues for precise drug release. These stimuli-sensitive systems hold promise for improved drug delivery, overcoming size limitations and reducing manufacturing and regulatory complexities. One pioneering approach, TESLA, offers real-time monitoring of tumor advancement and response to treatment. By assembling NPs at the site of tumor in reaction to endogenous cues, TESLA provides a means to address the challenges of cancer imaging and treatment response, offering enhanced imaging contrast and specificity.

These advancements reflect a promising era in nanotherapeutics, providing innovative solutions for effective cancer treatment and diagnosis, while overcoming historical barriers in drug delivery and translational complexities.

7.6 Conclusion and Future Perspectives

In an era marked by rapid advancements in nanotechnology, this chapter has explored its profound impact on diverse biomedical applications, ranging from cardiac regeneration to DR, kidney disease treatments, wound healing, and cancer therapy. As we conclude this exploration, it becomes evident that nanotechnology is not merely a scientific endeavor but a transformative force that holds the promise to reform the landscape of modern medicine.

The journey through nanotechnology-based treatments for heart disease revealed the promise of precision drug delivery and cardiac repair using bioengineered nanocarriers. However, the chapter acknowledges the need for further research to ensure the clinical translation of these innovations and to establish long-term safety parameters. The combination of nanomaterials with stem cell therapy and the use of liposomes and inorganic NPs open avenues for targeted therapies, emphasizing the necessity of refining these approaches for practical application.

In the realm of diabetes treatment, particularly DR, nanotechnology has offered innovative tools for early diagnosis, targeted treatment, and monitoring of the disease. The chapter has highlighted the potential of nanomaterials, NPs, and lipid-based nanocarriers in addressing the underlying mechanisms of DR. The extended research recognizes the ongoing efforts to validate the safety and feasibility of nanomaterials in clinical trials and emphasizes the importance of scalability and cost-effectiveness for widespread accessibility.

The application of nanotechnology in kidney disease treatments has showcased the promise of NPs in both detection and targeted drug delivery. Although there are limited systematic studies on the application of NPs in CKD, the research emphasizes the potential of nanotechnology to bring about a transformation in nephrology. The future perspectives involve overcoming challenges associated with NP technology, expanding research efforts, and enhancing interdisciplinary collaborations.

Moving to the domain of wound healing, the integration of nanotechnology has presented innovative pathways for tissue regeneration and antibacterial strategies. The discussion in this chapter acknowledges the exciting prospects brought about by recent advancements in nano therapies, including nitric oxide-releasing NPs. It highlights ongoing research as essential in unveiling new possibilities for future wound care and tissue regeneration, ultimately improving patient lives.

Finally, the chapter concludes with a comprehensive exploration of nanotechnology in cancer treatment, covering aspects such as cancer immunotherapy, intraoperative imaging, and nanotherapeutics. The chapter recognizes the current challenges associated with nanotherapeutics, including issues of delivery efficiency, toxicity, and nonuniform intratumoral distributions. Future perspectives involve a shift toward combination therapies, responsive nanocarriers, and real-time monitoring approaches, reflecting a promising era in nanotherapeutics with the potential to transform cancer treatment. Further, while nanotechnology has unlocked remarkable possibilities in various biomedical applications, the journey is ongoing. Future endeavors must focus on refining these technologies, ensuring their safety and efficacy, and fostering collaborative efforts to bridge the gap between innovative ideas and practical clinical implementation. The transformative potential of nanotechnology in medicine continues to unfold, promising a future where these advancements become integral components of healthcare, contributing to better patient outcomes and advancements in medical science as a whole.

References

1. Lao Y-H, Phua K K L and Leong K W 2015 Aptamer nanomedicine for cancer therapeutics: barriers and potential for translation *ACS Nano* 9 2235–54 https://doi.org/10.1021/nn507494p.
2. Suri S S, Fenniri H and Singh B 2007 Nanotechnology-based drug delivery systems *J. Occup. Med. Toxicol.* 2 16 https://doi.org/10.1186/1745-6673-2-16.
3. Liu Y, Ashton J R, Moding E J, Yuan H, Register J K, Fales A M, Choi J, Whitley M J, Zhao X, Qi Y, Ma Y, Vaidyanathan G, Zalutsky M R, Kirsch D G, Badea C T and Vo-Dinh T 2015 A plasmonic gold nanostar theranostic probe for in vivo tumor imaging and photothermal therapy *Theranostics* 5 946–60 https://doi.org/10.7150/thno.11974.
4. Cassani M, Fernandes S, Vrbsky J, Ergir E, Cavalieri F and Forte G 2020 Combining nanomaterials and developmental pathways to design new treatments for cardiac regeneration: the pulsing heart of advanced therapies *Front. Bioeng. Biotechnol.* 8 323 https://www.frontiersin.org/articles/10.3389/fbioe.2020.00323 (accessed July 28, 2023).
5. Schulte C, Barwari T, Joshi A, Theofilatos K, Zampetaki A, Barallobre-Barreiro J, Singh B, Sörensen N A, Neumann J T, Zeller T, Westermann D, Blankenberg S, Marber M, Liebetrau C and Mayr M 2019 Comparative analysis of circulating noncoding RNAs versus protein biomarkers in the detection of myocardial injury *Circ. Res.* 125 328–40 https://doi.org/10.1161/CIRCRESAHA.119.314937.
6. Yeh H-W and Ai H-W 2019 Development and applications of bioluminescent and chemiluminescent reporters and biosensors *Annu. Rev. Anal. Chem.* 12 129–50 https://doi.org/10.1146/annurev-anchem-061318-115027.
7. Supraja P, Sudarshan V, Tripathy S, Agrawal A and Singh S G 2019 Label free electrochemical detection of cardiac biomarker troponin T using $ZnSnO_3$ perovskite nanomaterials *Anal. Methods* 11 744–51 https://doi.org/10.1039/C8AY02617C.
8. Liyanage T, Sangha A and Sardar R 2017 Achieving biosensing at attomolar concentrations of cardiac troponin T in human biofluids by developing a label-free nanoplasmonic analytical assay *Analyst* 142 2442–50 https://doi.org/10.1039/C7AN00430C.
9. Gupta A, Kumar Sharma S, Pachauri V, Ingebrandt S, Singh S, Sharma A L and Deep A 2021 Sensitive impedimetric detection of troponin I with metal organic framework composite electrode *RSC Adv.* 11 2167–74 https://doi.org/10.1039/D0RA06665F.
10. Adeel M, Rahman M M and Lee J-J 2019 Label-free aptasensor for the detection of cardiac biomarker myoglobin based on gold nanoparticles decorated boron nitride nanosheets *Biosens. Bioelectron.* 126 143–50 https://doi.org/10.1016/j.bios.2018.10.060.

11 Zhang T, Ma N, Ali A, Wei Q, Wu D and Ren X 2018 Electrochemical ultrasensitive detection of cardiac troponin I using covalent organic frameworks for signal amplification *Biosens. Bioelectron.* 119 176–81 https://doi.org/10.1016/j.bios.2018.08.020.

12 Yang X, Yu Y-Q, Peng L-Z, Lei Y-M, Chai Y-Q, Yuan R and Zhuo Y 2018 Strong electrochemiluminescence from MOF accelerator enriched quantum dots for enhanced sensing of trace cTnI *Anal. Chem.* 90 3995–4002 https://doi.org/10.1021/acs.analchem.7b05137.

13 Sun D, Luo Z, Lu J, Zhang S, Che T, Chen Z and Zhang L 2019 Electrochemical dual-aptamer-based biosensor for nonenzymatic detection of cardiac troponin I by nanohybrid electrocatalysts labeling combined with DNA nanotetrahedron structure *Biosens. Bioelectron.* 134 49–56 https://doi.org/10.1016/j.bios.2019.03.049.

14 Hong D, Jo E-J, Kim K, Song M-B and Kim M-G 2020 Ru(bpy)32+-loaded mesoporous silica nanoparticles as electrochemiluminescent probes of a lateral flow immunosensor for highly sensitive and quantitative detection of troponin I *Small* 16 2004535 https://doi.org/10.1002/smll.202004535.

15 Wang C, Li J, Kang M, Huang X, Liu Y, Zhou N and Zhang Z 2021 Nanodiamonds and hydrogen-substituted graphdiyne heteronanostructure for the sensitive impedimetric aptasensing of myocardial infarction and cardiac troponin I *Anal. Chim. Acta* 1141 110–9 https://doi.org/10.1016/j.aca.2020.10.044.

16 Zhang X, Lv H, Li Y, Zhang C, Wang P, Liu Q, Ai B, Xu Z and Zhao Z 2019 Ultrasensitive sandwich-type immunosensor for cardiac troponin I based on enhanced electrocatalytic reduction of H_2O_2 using β-cyclodextrins functionalized 3D porous graphene-supported Pd@Au nanocubes *J. Mater. Chem. B* 7(9) 1460–8.

17 Guo X, Zong L, Jiao Y, Han Y, Zhang X, Xu J, Li L, Zhang C, Liu Z, Ju Q, Liu J, Xu Z, Yu H-D and Huang W 2019 Signal-enhanced detection of multiplexed cardiac biomarkers by a paper-based fluorogenic immunodevice integrated with zinc oxide nanowires *Anal. Chem.* 91 9300–7 https://doi.org/10.1021/acs.analchem.9b02557.

18 Chen X and Liang Y 2021 Colorimetric sensing strategy for multiplexed detection of proteins based on three DNA-gold nanoparticle conjugates sensors *Sens. Actuators B Chem.* 329 129202 https://doi.org/10.1016/j.snb.2020.129202.

19 Xin Y, Yang R, Qu Y, Liu H, Feng Y, Li L, Shi W and Liu Q 2021 Novel, highly sensitive, and specific assay to monitor acute myocardial infarction (AMI) by the determination of cardiac troponin I (cTnI) and heart-type fatty acid binding protein (H-FABP) by a colloidal gold-based immunochromatographic test strip *Anal. Lett.* 54 1329–50 https://doi.org/10.1080/00032719.2020.1802594.

20 Ji J, Lu W, Zhu Y, Jin H, Yao Y, Zhang H and Zhao Y 2019 Porous hydrogel-encapsulated photonic barcodes for multiplex detection of cardiovascular biomarkers *ACS Sens.* 4 1384–90 https://doi.org/10.1021/acssensors.9b00352.

21 Ye J, Zhu L, Yan M, Zhu Q, Lu Q, Huang J, Cui H and Yang X 2019 Dual-wavelength ratiometric electrochemiluminescence immunosensor for cardiac troponin I detection *Anal. Chem.* 91 1524–31 https://doi.org/10.1021/acs.analchem.8b04640.

22 Lim W Y, Thevarajah T M, Goh B T and Khor S M 2019 Paper microfluidic device for early diagnosis and prognosis of acute myocardial infarction via quantitative multiplex cardiac biomarker detection *Biosens. Bioelectron.* 128 176–85 https://doi.org/10.1016/j.bios.2018.12.049.

23 Ouyang M, Tu D, Tong L, Sarwar M, Bhimaraj A, Li C, Coté G L and Di Carlo D 2021 A review of biosensor technologies for blood biomarkers toward monitoring cardiovascular diseases at the point-of-care *Biosens. Bioelectron.* 171 112621 https://doi.org/10.1016/j.bios.2020.112621.

24 Hu Q, Fang Z, Ge J and Li H 2022 Nanotechnology for cardiovascular diseases *The Innovation* 3 100214 https://doi.org/10.1016/j.xinn.2022.100214.

25 Karimi M, Zare H, Bakhshian Nik A, Yazdani N, Hamrang M, Mohamed E, Sahandi Zangabad P, Moosavi Basri S M, Bakhtiari L and Hamblin M R 2016 Nanotechnology in diagnosis and treatment of coronary artery disease *Nanomed.* 11 513–30 https://doi.org/10.2217/nnm.16.3.

26 Huang D, Lin B, Song Y, Guan Z, Cheng J, Zhu Z and Yang C 2019 Staining traditional colloidal gold test strips with pt nanoshell enables quantitative point-of-care testing with simple and portable pressure meter readout *ACS Appl. Mater. Interfaces* 11 1800–6 https://doi.org/10.1021/acsami.8b15562.

27 Seok Lee W, Sunkara V, Han J-R, Park Y-S and Cho Y-K 2015 Electrospun TiO_2 nanofiber integrated lab-on-a-disc for ultrasensitive protein detection from whole blood *Lab. Chip* 15 478–85 https://doi.org/10.1039/C4LC00900B.

28 Zhang L, Gu F, Chan J, Wang A, Langer R and Farokhzad O 2008 Nanoparticles in medicine: therapeutic applications and developments *Clin. Pharmacol. Ther.* 83 761–9 https://doi.org/10.1038/sj.clpt.6100400.

29 Jabir N R, Tabrez S, Ashraf G M, Shakil S, Damanhouri G A and Kamal M A 2012 Nanotechnology-based approaches in anticancer research *Int. J. Nanomedicine* 7 4391–408 https://doi.org/10.2147/IJN.S33838.

30 Zhang S, Zhao B, Jiang H, Wang B and Ma B 2007 Cationic lipids and polymers mediated vectors for delivery of siRNA *J. Controlled Release* 123 1–10 https://doi.org/10.1016/j.jconrel.2007.07.016.

31 Boon R A and Vickers K C 2013 Intercellular transport of MicroRNAs *Arterioscler. Thromb. Vasc. Biol.* 33 186–92 https://doi.org/10.1161/ATVBAHA.112.300139.

32 Valadi H, Ekström K, Bossios A, Sjöstrand M, Lee J J and Lötvall J O 2007 Exosome-mediated transfer of mRNAs and microRNAs is a novel mechanism of genetic exchange between cells *Nat. Cell Biol.* 9 654–9 https://doi.org/10.1038/ncb1596.

33 Patra J K, Das G, Fraceto L F, Campos E V R, Rodriguez-Torres M D, Acosta-Torres L S, Diaz-Torres L A, Grillo R, Swamy M K, Sharma S, Habtemariam S and Shin H-S 2018 Nano based drug delivery systems: recent developments and future prospects *J. Nanobiotechnology* 16 71 https://doi.org/10.1186/s12951-018-0392-8.

34 Kalepu S and Nekkanti V 2015 Insoluble drug delivery strategies: review of recent advances and business prospects *Acta Pharm. Sin. B* 5 442–53 https://doi.org/10.1016/j.apsb.2015.07.003.

35 Yu D, Khan O F, Suvà M L, Dong B, Panek W K, Xiao T, Wu M, Han Y, Ahmed A U, Balyasnikova I V, Zhang H F, Sun C, Langer R, Anderson D G and Lesniak M S 2017 Multiplexed RNAi therapy against brain tumor initiating cells via lipopolymeric nanoparticle infusion delays glioblastoma progression *Proc. Natl. Acad. Sci.* 114 E6147–56 https://doi.org/10.1073/pnas.1701911114.

36 Durán-Lobato M, López-Estévez A M, Cordeiro A S, Dacoba T G, Crecente-Campo J, Torres D and Alonso M J 2021 Nanotechnologies for the delivery of biologicals: historical perspective and current landscape *Adv. Drug Deliv. Rev.* 176 113899 https://doi.org/10.1016/j.addr.2021.113899.

37 Ud Din F, Aman W, Ullah I, Qureshi O S, Mustapha O, Shafique S and Zeb A 2017 Effective use of nanocarriers as drug delivery systems for the treatment of selected tumors *Int. J. Nanomedicine* 12 7291–309 https://doi.org/10.2147/IJN.S146315.

38 Chenthamara D, Subramaniam S, Ramakrishnan S G, Krishnaswamy S, Essa M M, Lin F-H and Qoronfleh M W 2019 Therapeutic efficacy of nanoparticles and routes of administration *Biomater. Res.* 23 20 https://doi.org/10.1186/s40824-019-0166-x.

39 Prabhakar U, Maeda H, Jain R K, Sevick-Muraca E M, Zamboni W, Farokhzad O C, Barry S T, Gabizon A, Grodzinski P and Blakey D C 2013 Challenges and key considerations of the enhanced permeability and retention effect for nanomedicine drug delivery in oncology *Cancer Res.* 73 2412–7 https://doi.org/10.1158/0008-5472.CAN-12-4561.

40 Patil Y and Panyam J 2009 Polymeric nanoparticles for siRNA delivery and gene silencing *Int. J. Pharm.* 367 195–203 https://doi.org/10.1016/j.ijpharm.2008.09.039.

41 Fortuni B, Inose T, Ricci M, Fujita Y, Van Zundert I, Masuhara A, Fron E, Mizuno H, Latterini L, Rocha S and Uji-i H 2019 Polymeric engineering of nanoparticles for highly efficient multifunctional drug delivery systems *Sci. Rep.* 9 2666 https://doi.org/10.1038/s41598-019-39107-3.

42 Chandarana M, Curtis A and Hoskins C 2018 The use of nanotechnology in cardiovascular disease *Appl. Nanosci.* 8 1607–19 https://doi.org/10.1007/s13204-018-0856-z.

43 El-Say K M and El-Sawy H S 2017 Polymeric nanoparticles: promising platform for drug delivery *Int. J. Pharm.* 528 675–91 https://doi.org/10.1016/j.ijpharm.2017.06.052.

44 Yang H, Qin X, Wang H, Zhao X, Liu Y, Wo H-T, Liu C, Nishiga M, Chen H, Ge J, Sayed N, Abilez O J, Ding D, Heilshorn S C and Li K 2019 An in vivo miRNA delivery system for restoring infarcted myocardium *ACS Nano* 13 9880–94 https://doi.org/10.1021/acsnano.9b03343.

45 Xue X, Shi X, Dong H, You S, Cao H, Wang K, Wen Y, Shi D, He B and Li Y 2018 Delivery of microRNA-1 inhibitor by dendrimer-based nanovector: an early targeting therapy for myocardial infarction in mice *Nanomedicine Nanotechnol. Biol. Med.* 14 619–31 https://doi.org/10.1016/j.nano.2017.12.004.

46 Yokoyama R, Ii M, Masuda M, Tabata Y, Hoshiga M, Ishizaka N and Asahi M 2019 Cardiac regeneration by statin-polymer nanoparticle-loaded adipose-derived stem cell therapy in myocardial infarction *Stem Cells Transl. Med.* 8 1055–67 https://doi.org/10.1002/sctm.18-0244.

47 Oduk Y, Zhu W, Kannappan R, Zhao M, Borovjagin A V, Oparil S and Zhang J 2018 VEGF nanoparticles repair the heart after myocardial infarction *Am. J. Physiol.-Heart Circ. Physiol.* 314 H278–84 https://doi.org/10.1152/ajpheart.00471.2017.

48 Ferreira M P A, Talman V, Torrieri G, Liu D, Marques G, Moslova K, Liu Z, Pinto J F, Hirvonen J, Ruskoaho H and Santos H A 2018 Dual-drug delivery using dextran-functionalized nanoparticles targeting cardiac fibroblasts for cellular reprogramming *Adv. Funct. Mater.* 28 1705134 https://doi.org/10.1002/adfm.201705134.

49 Akbarzadeh A, Rezaei-Sadabady R, Davaran S, Joo S W, Zarghami N, Hanifehpour Y, Samiei M, Kouhi M and Nejati-Koshki K 2013 Liposome: classification, preparation, and applications *Nanoscale Res. Lett.* 8 102 https://doi.org/10.1186/1556-276X-8-102.

50 Sercombe L, Veerati T, Moheimani F, Wu S Y, Sood A K and Hua S 2015 Advances and challenges of liposome assisted drug delivery *Front. Pharmacol.* 6 286 https://www.frontiersin.org/articles/10.3389/fphar.2015.00286 (accessed August 2, 2023).

51 Scott R C, Rosano J M, Ivanov Z, Wang B, Chong P L-G, Issekutz A C, Crabbe D L and Kiani M F 2009 Targeting VEGF-encapsulated immunoliposomes to MI heart improves vascularity and cardiac function *FASEB J.* 23 3361–7 https://doi.org/10.1096/fj.08-127373.

52 Dvir T, Bauer M, Schroeder A, Tsui J H, Anderson D G, Langer R, Liao R and Kohane D S 2011 Nanoparticles targeting the infarcted heart *Nano Lett.* 11 4411–4 https://doi.org/10.1021/nl2025882.

53 Fadeel B 2019 Hide and seek: nanomaterial interactions with the immune system *Front. Immunol.* 10 https://www.frontiersin.org/articles/10.3389/fimmu.2019.00133 (accessed August 3, 2023).

54 Harel-Adar T, Mordechai T B, Amsalem Y, Feinberg M S, Leor J and Cohen S 2011 Modulation of cardiac macrophages by phosphatidylserine-presenting liposomes improves infarct repair *Proc. Natl. Acad. Sci.* 108 1827–32 https://doi.org/10.1073/pnas.1015623108.

55 Cheng B, Toh E K W, Chen K-H, Chang Y-C, Hu C-M J, Wu H-C, Chau L-Y, Chen P and Hsieh P C H 2016 Biomimicking platelet–monocyte interactions as a novel targeting strategy for heart healing *Adv. Healthc. Mater.* 5 2686–97 https://doi.org/10.1002/adhm.201600724.

56 Allijn I E, Czarny B M S, Wang X, Chong S Y, Weiler M, da Silva A E, Metselaar J M, Lam C S P, Pastorin G, de Kleijn D P V, Storm G, Wang J-W and Schiffelers R M 2017 Liposome encapsulated berberine treatment attenuates cardiac dysfunction after myocardial infarction *J. Controlled Release* 247 127–33 https://doi.org/10.1016/j.jconrel.2016.12.042.

57 Giner-Casares J J, Henriksen-Lacey M, Coronado-Puchau M and Liz-Marzán L M 2016 Inorganic nanoparticles for biomedicine: where materials scientists meet medical research *Mater. Today* 19 19–28 https://doi.org/10.1016/j.mattod.2015.07.004.

58 Conde J, Dias J T, Grazú V, Moros M, Baptista P V and de la Fuente J M 2014 Revisiting 30 years of biofunctionalization and surface chemistry of inorganic nanoparticles for nanomedicine *Front. Chem.* 2 48 https://www.frontiersin.org/articles/10.3389/fchem.2014.00048 (accessed August 3, 2023).

59 Popara J, Accomasso L, Vitale E, Gallina C, Roggio D, Iannuzzi A, Raimondo S, Rastaldo R, Alberto G, Catalano F, Martra G, Turinetto V, Pagliaro P and Giachino C 2018 Silica nanoparticles actively engage with mesenchymal stem cells in improving acute functional cardiac integration *Nanomed.* 13 1121–38 https://doi.org/10.2217/nnm-2017-0309.

60 Miragoli M, Ceriotti P, Iafisco M, Vacchiano M, Salvarani N, Alogna A, Carullo P, Ramirez-Rodríguez G B, Patrício T, Esposti L D, Rossi F, Ravanetti F, Pinelli S, Alinovi R, Erreni M, Rossi S, Condorelli G, Post H, Tampieri A and Catalucci D 2018 Inhalation of peptide-loaded nanoparticles improves heart failure *Sci. Transl. Med.* 10 eaan6205 https://doi.org/10.1126/scitranslmed.aan6205.

61 Zhang B, Jiang H, Chen J, Hu Q, Yang S and Liu X 2019 Silica-coated magnetic nanoparticles labeled endothelial progenitor cells alleviate ischemic myocardial injury and improve long-term cardiac function with magnetic field guidance in rats with myocardial infarction *J. Cell. Physiol.* 234 18544–59 https://doi.org/10.1002/jcp.28492.

62 Xiong F, Wang H, Feng Y, Li Y, Hua X, Pang X, Zhang S, Song L, Zhang Y and Gu N 2015 Cardioprotective activity of iron oxide nanoparticles *Sci. Rep.* 5 8579 https://doi.org/10.1038/srep08579.

63 Kee P H and Danila D 2018 CT imaging of myocardial scar burden with CNA35-conjugated gold nanoparticles *Nanomedicine Nanotechnol. Biol. Med.* 14 1941–7 https://doi.org/10.1016/j.nano.2018.06.003.

64 Yilmaz A, Rösch S, Klingel K, Kandolf R, Helluy X, Hiller K-H, Jakob P M and Sechtem U 2013 Magnetic resonance imaging (MRI) of inflamed myocardium using iron oxide nanoparticles in patients with acute myocardial infarction – preliminary results *Int. J. Cardiol.* 163 175–82 https://doi.org/10.1016/j.ijcard.2011.06.004.

65 Stirrat C G, Alam S R, MacGillivray T J, Gray C D, Dweck M R, Raftis J, Jenkins W S, Wallace W A, Pessotto R, Lim K H, Mirsadraee S, Henriksen P A, Semple S I and Newby D E 2017 Ferumoxytol-enhanced magnetic resonance imaging assessing inflammation after myocardial infarction *Heart* 103 1528–35 https://doi.org/10.1136/heartjnl-2016-311018.

66 Bobo D, Robinson K J, Islam J, Thurecht K J and Corrie S R 2016 Nanoparticle-based medicines: a review of FDA-approved materials and clinical trials to date *Pharm. Res.* 33 2373–87 https://doi.org/10.1007/s11095-016-1958-5.

67 Stitt A W, Curtis T M, Chen M, Medina R J, McKay G J, Jenkins A, Gardiner T A, Lyons T J, Hammes H-P, Simó R and Lois N 2016 The progress in understanding and treatment of diabetic retinopathy *Prog. Retin. Eye Res.* 51 156–86 https://doi.org/10.1016/j.preteyeres.2015.08.001.

68 Duh E J, Sun J K and Stitt A W 2017 Diabetic retinopathy: current understanding, mechanisms, and treatment strategies *JCI Insight* 2 e93751 https://doi.org/10.1172/jci.insight.93751.

69 Mansour S E, Browning D J, Wong K, Flynn H W Jr, and Bhavsar A R 2020 The evolving treatment of diabetic retinopathy *Clin. Ophthalmol.* 14 653–78 https://doi.org/10.2147/OPTH.S236637.

70 Fangueiro J F, Silva A M, Garcia M L and Souto E B 2015 Current nanotechnology approaches for the treatment and management of diabetic retinopathy *Eur. J. Pharm. Biopharm.* 95 307–22 https://doi.org/10.1016/j.ejpb.2014.12.023.

71 Vedarethinam V, Huang L, Zhang M, Su H, Hu H, Xia H, Liu Y, Wu B, Wan X, Shen J, Xu L, Liu W, Ma J and Qian K 2020 Vanadium core–shell nanorods inspect metabolic changes of diabetic retinopathy *Adv. Funct. Mater.* 30 2002791 https://doi.org/10.1002/adfm.202002791.

72 Akwii R G and Mikelis C M 2021 Targeting the angiopoietin/tie pathway: prospects for treatment of retinal and respiratory disorders *Drugs* 81 1731–49 https://doi.org/10.1007/s40265-021-01605-y.

73 Forrester J V, Kuffova L and Delibegovic M 2020 The role of inflammation in diabetic retinopathy *Front. Immunol.* 11 583687 https://www.frontiersin.org/articles/10.3389/fimmu.2020.583687 (accessed January 7, 2024).

74 Li B-Y, Tan W, Zou J-L, He Y, Yoshida S, Jiang B and Zhou Y-D 2021 Role of interferons in diabetic retinopathy *World J. Diabetes* 12 939–53 https://doi.org/10.4239/wjd.v12.i7.939.

75 Gao X, Li Y, Wang H, Li C and Ding J 2017 Inhibition of HIF-1α decreases expression of pro-inflammatory IL-6 and TNF-α in diabetic retinopathy *Acta Ophthalmol.* 95 e746–50 https://doi.org/10.1111/aos.13096.

76 Ramakrishnan S, Anand V and Roy S 2014 Vascular endothelial growth factor signaling in hypoxia and inflammation *J. Neuroimmune Pharmacol.* 9 142–60 https://doi.org/10.1007/s11481-014-9531-7.

77 Gong Q, Xie J, Li Y, Liu Y and Su G 2019 Enhanced ROBO4 is mediated by up-regulation of HIF-1α/SP1 or reduction in miR-125b-5p/miR-146a-5p in diabetic retinopathy *J. Cell. Mol. Med.* 23 4723–37 https://doi.org/10.1111/jcmm.14369.

78 Iwase T, Fu J, Yoshida T, Muramatsu D, Miki A, Hashida N, Lu L, Oveson B, Silva R L, Seidel C, Yang M, Connelly S, Shen J, Han B, Wu M, Semenza G L, Hanes J and Campochiaro P A 2013 Sustained delivery of a HIF-1 antagonist for ocular neovascularization *J. Controlled Release* 172 625–33 https://doi.org/10.1016/j.jconrel.2013.10.008.

79 Li S, Chen L and Fu Y 2023 Nanotechnology-based ocular drug delivery systems: recent advances and future prospects *J. Nanobiotechnology* 21 232 https://doi.org/10.1186/s12951-023-01992-2.

80 Wu K Y, Joly-Chevrier M, Akbar D and Tran S D 2023 Overcoming treatment challenges in posterior segment diseases with biodegradable nano-based drug delivery systems *Pharmaceutics* 15 1094 https://doi.org/10.3390/pharmaceutics15041094.

81 Badr G A, Tang J, Ismail-Beigi F and Kern T S 2000 Diabetes downregulates GLUT1 expression in the retina and its microvessels but not in the cerebral cortex or its microvessels *Diabetes* 49 1016–21 https://doi.org/10.2337/diabetes.49.6.1016.

82 Mahaling B, Srinivasarao D A, Raghu G, Kasam R K, Reddy G B and Katti D S 2018 A non-invasive nanoparticle mediated delivery of triamcinolone acetonide ameliorates diabetic retinopathy in rats *Nanoscale* 10 16485–98 https://doi.org/10.1039/C8NR00058A.

83 Li C, Zhao G, Lu L and Du X 2020 Effect of triamcinolone acetonide-loaded nanoparticles on diabetic retinopathy in rats *Mater. Express* 10 1511–7 https://doi.org/10.1166/mex.2020.1784.

84 Laddha U D and Kshirsagar S J 2020 Formulation of PPAR-gamma agonist as surface modified PLGA nanoparticles for non-invasive treatment of diabetic retinopathy: in vitro and in vivo evidences *Heliyon* 6 e04589 https://doi.org/10.1016/j.heliyon.2020.e04589.

85 Laddha U D, Kshirsagar S J, Sayyad L S, Ahmed M T, Gaikwad S S, Udavant P B and Kulkarni A D 2022 Development of surface modified nanoparticles of curcumin for topical treatment of diabetic retinopathy: in vitro, ex vivo and in vivo investigation *J. Drug Deliv. Sci. Technol.* 76 103835 https://doi.org/10.1016/j.jddst.2022.103835.

86 Qu Q, Park K, Zhou K, Wassel D, Farjo R, Criswell T, Ma J and Zhang Y 2022 Sustained therapeutic effect of an anti-inflammatory peptide encapsulated in nanoparticles on ocular vascular leakage in diabetic retinopathy *Front. Cell Dev. Biol.* 10 1049678 https://www.frontiersin.org/articles/10.3389/fcell.2022.1049678 (accessed August 5, 2023).

87 Amadio M, Pascale A, Cupri S, Pignatello R, Osera C, Agata V D, Amico A G D, Leggio G M, Ruozi B, Govoni S, Drago F and Bucolo C 2016 Nanosystems based on siRNA silencing HuR expression counteract diabetic retinopathy in rat *Pharmacol. Res.* 111 713–20 https://doi.org/10.1016/j.phrs.2016.07.042.

88 Zamani M, Shirinzadeh A, Aghajanzadeh M, Andalib S and Danafar H 2019 In vivo study of mPEG–PCL as a nanocarriers for anti-inflammatory drug delivery of simvastatin *Pharm. Dev. Technol.* 24 663–70 https://doi.org/10.1080/10837450.2018.1556689.

89 Amato R, Giannaccini M, Dal Monte M, Cammalleri M, Pini A, Raffa V, Lulli M and Casini G 2020 Association of the somatostatin analog octreotide with magnetic nanoparticles for intraocular delivery: a possible approach for the treatment of diabetic retinopathy *Front. Bioeng. Biotechnol.* 8 00144 https://www.frontiersin.org/articles/10.3389/fbioe.2020.00144 (accessed August 5, 2023).

90 Shoval A, Markus A, Zhou Z, Liu X, Cazelles R, Willner I and Mandel Y 2019 Anti-VEGF-aptamer modified C-Dots – a hybrid nanocomposite for topical treatment of ocular vascular disorders *Small* 15 1902776 https://doi.org/10.1002/smll.201902776.

91 Liu Z, Yu N, Holz F G, Yang F and Stanzel B V 2014 Enhancement of retinal pigment epithelial culture characteristics and subretinal space tolerance of scaffolds with 200 nm fiber topography *Biomaterials* 35 2837–50 https://doi.org/10.1016/j.biomaterials.2013.12.069.

92 Chang Y-C, Chen M-H, Liao S-Y, Wu H-C, Kuan C-H, Sun J-S and Wang T-W 2017 Multichanneled nerve guidance conduit with spatial gradients of neurotrophic factors and oriented nanotopography for repairing the peripheral nervous system *ACS Appl. Mater. Interfaces* 9 37623–36 https://doi.org/10.1021/acsami.7b12567.

93 Yan L, Zhao B, Liu X, Li X, Zeng C, Shi H, Xu X, Lin T, Dai L and Liu Y 2016 Aligned nanofibers from polypyrrole/graphene as electrodes for regeneration of optic nerve via electrical stimulation *ACS Appl. Mater. Interfaces* 8 6834–40 https://doi.org/10.1021/acsami.5b12843.

94 Nadri S, Kazemi B, Eeslaminejad M B, Yazdani S and Soleimani M 2013 High yield of cells committed to the photoreceptor-like cells from conjunctiva mesenchymal stem cells on nanofibrous scaffolds *Mol. Biol. Rep.* 40 3883–90 https://doi.org/10.1007/s11033-012-2360-y.

95 Soleimannejad M, Ebrahimi-Barough S, Soleimani M, Nadri S, Tavangar S M, Roohipoor R, Yazdankhah M, Bayat N, Riazi-Esfahani M and Ai J 2018 Fibrin gel as a scaffold for photoreceptor cells differentiation from conjunctiva mesenchymal stem cells in retina tissue engineering *Artif. Cells Nanomedicine Biotechnol.* 46 805–14 https://doi.org/10.1080/21691401.2017.1345922.

96 Ebrahimi-Barough S, Hoveizi E, Norouzi Javidan A and Ai J 2015 Investigating the neuroglial differentiation effect of neuroblastoma conditioned medium in human endometrial stem cells cultured on 3D nanofibrous scaffold *J. Biomed. Mater. Res. A* 103 2621–7 https://doi.org/10.1002/jbm.a.35397.

97 Deb P, Deoghare A B, Borah A, Barua E and Das Lala S 2018 Scaffold development using biomaterials: a review *Mater. Today Proc.* 5 12909–19 https://doi.org/10.1016/j.matpr.2018.02.276.

98 Yang T-C, Chuang J-H, Buddhakosai W, Wu W-J, Lee C-J, Chen W-S, Yang Y-P, Li M-C, Peng C-H and Chen S-J 2017 Elongation of axon extension for human iPSC-derived retinal ganglion cells by a nano-imprinted scaffold *Int. J. Mol. Sci.* 18 2013 https://doi.org/10.3390/ijms18092013.

99 Reichel H, Zee J, Tu C, Young E, Pisoni R L, Stengel B, Duttlinger J, Lonnemann G, Robinson B M, Pecoits-Filho R and Fliser D 2020 Chronic kidney disease progression and mortality risk profiles in Germany: results from the Chronic Kidney Disease Outcomes and Practice Patterns Study *Nephrol. Dial. Transplant.* 35 803–10 https://doi.org/10.1093/ndt/gfz260.

100 Nelson R G, Grams M E, Ballew S H, Sang Y, Azizi F, Chadban S J, Chaker L, Dunning S C, Fox C, Hirakawa Y, Iseki K, Ix J, Jafar T H, Köttgen A, Naimark D M J, Ohkubo T, Prescott G J, Rebholz C M, Sabanayagam C, Sairenchi T, Schöttker B, Shibagaki Y, Tonelli M, Zhang L, Gansevoort R T, Matsushita K, Woodward M, Coresh J, Shalev V and for the CKD Prognosis Consortium 2019 Development of risk prediction equations for incident chronic kidney disease *JAMA* 322 2104–14 https://doi.org/10.1001/jama.2019.17379.

101 Keane W F and Eknoyan G 1999 Proteinuria, albuminuria, risk, assessment, detection, elimination (PARADE): a position paper of the National Kidney Foundation *Am. J. Kidney Dis.* 33 1004–10 https://doi.org/10.1016/S0272-6386(99)70442-7.

102 Mosier-Boss P A 2017 Review of SERS substrates for chemical sensing *Nanomaterials* 7 142 https://doi.org/10.3390/nano7060142.

103 Huang Z, Zhang A, Zhang Q and Cui D 2019 Nanomaterial-based SERS sensing technology for biomedical application *J. Mater. Chem. B* 7 3755–74 https://doi.org/10.1039/C9TB00666D.

104 Phuong N T T, Quang Dang V, Hieu L V, Ngoc Bach T, Xuan Khuyen B, Ta H K T, Ju H, Thang Phan B and Tran N H T 2022 Functionalized silver nanoparticles for SERS amplification with enhanced reproducibility and for ultrasensitive optical fiber sensing in environmental and biochemical assays *RSC Adv.* 12 31352–62 https://doi.org/10.1039/D2RA06074D.

105 Shaikh M O, Zhu P-Y, Wang C-C, Du Y-C and Chuang C-H 2019 Electrochemical immunosensor utilizing electrodeposited Au nanocrystals and dielectrophoretically trapped PS/Ag/ab-HSA nanoprobes for detection of microalbuminuria at point of care *Biosens. Bioelectron.* 126 572–80 https://doi.org/10.1016/j.bios.2018.11.035.

106 Feng S, Shi R, Xu P, Bhamore J R, Bal J, Baek S H, Park C Y, Park J P and Park T J 2020 Colorimetric detection of creatinine using its specific binding peptides and gold nanoparticles *New J. Chem.* 44 15828–35 https://doi.org/10.1039/D0NJ03860A.

107 Borse S, Murthy Z V P and Kailasa S K 2022 Synthesis of red emissive copper nanoclusters with 2-mercaptopyrimidine for promoting selective and sensitive fluorescent sensing of creatinine as a kidney disease biomarker in biofluids *J. Mol. Liq.* 368 120705 https://doi.org/10.1016/j.molliq.2022.120705.

108 Das T, Das S, Karmakar M, Chakraborty S, Saha D and Pal M 2020 Novel barium hexaferrite based highly selective and stable trace ammonia sensor for detection of renal disease by exhaled breath analysis *Sens. Actuators B Chem.* 325 128765 https://doi.org/10.1016/j.snb.2020.128765.

109 Lopes P, Costa-Rama E, Beirão I, Nouws H P A, Santos-Silva A and Delerue-Matos C 2019 Disposable electrochemical immunosensor for analysis of cystatin C, a CKD biomarker *Talanta* 201 211–6 https://doi.org/10.1016/j.talanta.2019.04.006.

110 Zhang F, Zhong H, Lin Y, Chen M, Wang Q, Lin Y and Huang J 2018 A nanohybrid composed of Prussian blue and graphitic C_3N_4 nanosheets as the signal-generating tag in an enzyme-free electrochemical immunoassay for the neutrophil gelatinase-associated lipocalin *Microchim. Acta* 185 327 https://doi.org/10.1007/s00604-018-2865-8.

111 Yang H, Wang H, Xiong C, Chai Y and Yuan R 2018 Highly sensitive electrochemiluminescence immunosensor based on ABEI/H_2O_2 system with PFO dots as enhancer for detection of kidney injury molecule-1 *Biosens. Bioelectron.* 116 16–22 https://doi.org/10.1016/j.bios.2018.05.032.

112 Huang X, Ma Y, Li Y, Han F and Lin W 2021 Targeted drug delivery systems for kidney diseases *Front. Bioeng. Biotechnol.* 9 https://www.frontiersin.org/articles/10.3389/fbioe.2021.683247 (accessed August 11, 2023).

113 Guo J, Gao X, Su L, Xia H, Gu G, Pang Z, Jiang X, Yao L, Chen J and Chen H 2011 Aptamer-functionalized PEG–PLGA nanoparticles for enhanced anti-glioma drug delivery *Biomaterials* 32 8010–20 https://doi.org/10.1016/j.biomaterials.2011.07.004.

114 Williams R M, Jaimes E A and Heller D A 2016 Nanomedicines for kidney diseases *Kidney Int.* 90 740–5 https://doi.org/10.1016/j.kint.2016.03.041.

115 Kim S-H, Moon J-H, Jeong S-U, Jung H-H, Park C-S, Hwang B Y and Lee C-K 2019 Induction of antigen-specific immune tolerance using biodegradable nanoparticles containing antigen and dexamethasone *Int. J. Nanomedicine* 14 5229–42 https://doi.org/10.2147/IJN.S210546.

116 Nieto K, Pei P, Wang D, Mallery S R and Schwendeman S P 2018 In vivo controlled release of fenretinide from long-acting release depots for chemoprevention of oral squamous cell carcinoma recurrence *Int. J. Pharm.* 538 48–56 https://doi.org/10.1016/j.ijpharm.2017.11.037.

117 Danhier F, Ansorena E, Silva J M, Coco R, Le Breton A and Préat V 2012 PLGA-based nanoparticles: an overview of biomedical applications *J. Controlled Release* 161 505–22 https://doi.org/10.1016/j.jconrel.2012.01.043.

118 Williams R M, Shah J, Ng B D, Minton D R, Gudas L J, Park C Y and Heller D A 2015 Mesoscale nanoparticles selectively target the renal proximal tubule epithelium *Nano Lett.* 15 2358–64 https://doi.org/10.1021/nl504610d.

119 Williams R M, Shah J, Tian H S, Chen X, Geissmann F, Jaimes E A and Heller D A 2018 Selective nanoparticle targeting of the renal tubules *Hypertension* 71 87–94 https://doi.org/10.1161/HYPERTENSIONAHA.117.09843.

120 Deng X, Zeng T, Li J, Huang C, Yu M, Wang X, Tan L, Zhang M, Li A and Hu J 2019 Kidney-targeted triptolide-encapsulated mesoscale nanoparticles for high-efficiency treatment of kidney injury *Biomater. Sci.* 7 5312–23 https://doi.org/10.1039/C9BM01290G.

121 He J, Chen H, Zhou W, Chen M, Yao Y, Zhang Z and Tan N 2020 Kidney targeted delivery of asiatic acid using a FITC labeled renal tubular-targeting peptide modified PLGA-PEG system *Int. J. Pharm.* 584 119455 https://doi.org/10.1016/j.ijpharm.2020.119455.

122 Li S, Zeng Y, Peng K, Liu C, Zhang Z and Zhang L 2019 Design and evaluation of glomerulus mesangium-targeted PEG-PLGA nanoparticles loaded with dexamethasone acetate *Acta Pharmacol. Sin.* 40 143–50 https://doi.org/10.1038/s41401-018-0052-4.

123 Tan L, Lai X, Zhang M, Zeng T, Liu Y, Deng X, Qiu M, Li J, Zhou G, Yu M, Geng X, Hu J and Li A 2019 A stimuli-responsive drug release nanoplatform for kidney-specific anti-fibrosis treatment *Biomater. Sci.* 7 1554–64 https://doi.org/10.1039/C8BM01297K.

124 Guo L, Luo S, Du Z, Zhou M, Li P, Fu Y, Sun X, Huang Y and Zhang Z 2017 Targeted delivery of celastrol to mesangial cells is effective against mesangioproliferative glomerulonephritis *Nat. Commun.* 8 878 https://doi.org/10.1038/s41467-017-00834-8.

125 Kravanja G, Primožič M, Knez Ž and Leitgeb M 2019 Chitosan-based (nano) materials for novel biomedical applications *Molecules* 24 1960 https://doi.org/10.3390/molecules24101960.

126 Gao S, Hein S, Dagnæs-Hansen F, Weyer K, Yang C, Nielsen R, Christensen E I, Fenton R A and Kjems J 2014 Megalin-mediated specific uptake of chitosan/siRNA nanoparticles in mouse kidney proximal tubule epithelial cells enables AQP1 gene silencing *Theranostics* 4 1039–51 https://doi.org/10.7150/thno.7866.

127 Luo J, Sun J, Luo X, Wei Y, Zheng H, Mu C and Yao W 2019 Low molecular weight chitosan-based conjugates for efficient Rhein oral delivery: synthesis, characterization, and pharmacokinetics *Drug Dev. Ind. Pharm.* 45 96–104 https://doi.org/10.1080/03639045.2018.1522326.

128 Qiao H, Sun M, Su Z, Xie Y, Chen M, Zong L, Gao Y, Li H, Qi J, Zhao Q, Gu X and Ping Q 2014 Kidney-specific drug delivery system for renal fibrosis based on coordination-driven assembly of catechol-derived chitosan *Biomaterials* 35 7157–71 https://doi.org/10.1016/j.biomaterials.2014.04.106.

129 Li J, Zhang C, He W, Qiao H, Chen J, Wang K, Oupický D and Sun M 2018 Coordination-driven assembly of catechol-modified chitosan for the kidney-specific delivery of salvianolic acid B to treat renal fibrosis *Biomater. Sci.* 6 179–88 https://doi.org/10.1039/C7BM00811B.

130 Geng X, Zhang M, Lai X, Tan L, Liu J, Yu M, Deng X, Hu J and Li A 2018 Small-Sized Cationic miRi-PCNPs Selectively Target the Kidneys for High-Efficiency Antifibrosis Treatment *Adv. Healthc. Mater.* 7 1800558 https://doi.org/10.1002/adhm.201800558.

131 Kim C S, Mathew A P, Uthaman S, Moon M J, Bae E H, Kim S W and Park I-K 2020 Glycol chitosan-based renal docking biopolymeric nanomicelles for site-specific delivery of the immunosuppressant *Carbohydr. Polym.* 241 116255 https://doi.org/10.1016/j.carbpol.2020.116255.

132 Liu D, Shu G, Jin F, Qi J, Xu X, Du Y, Yu H, Wang J, Sun M, You Y, Zhu M, Chen M, Zhu L, Shen Q, Ying X, Lou X, Jiang S and Du Y 2020 ROS-responsive chitosan-SS31 prodrug for AKI therapy via rapid distribution in the kidney and long-term retention in the renal tubule *Sci. Adv.* 6 eabb7422 https://doi.org/10.1126/sciadv.abb7422.

133 Huang D and Wu D 2018 Biodegradable dendrimers for drug delivery *Mater. Sci. Eng. C* 90 713–27 https://doi.org/10.1016/j.msec.2018.03.002.

134 Matsuura S, Katsumi H, Suzuki H, Hirai N, Hayashi H, Koshino K, Higuchi T, Yagi Y, Kimura H, Sakane T and Yamamoto A 2018 l-Serine – modified polyamidoamine dendrimer as a highly potent renal targeting drug carrier *Proc. Natl. Acad. Sci.* 115 10511–6 https://doi.org/10.1073/pnas.1808168115.

135 Katsumi H, Takashima R, Suzuki H, Hirai N, Matsuura S, Kimura H, Morishita M and Yamamoto A 2020 S-nitrosylated l-serine-modified dendrimer as a kidney-targeting nitric oxide donor for prevention of renal ischaemia/reperfusion injury *Free Radic. Res.* 54 841–7 https://doi.org/10.1080/10715762.2019.1697437.

136 Matsuura S, Katsumi H, Suzuki H, Hirai N, Takashima R, Morishita M, Sakane T and Yamamoto A 2018 l-cysteine and l-serine modified dendrimer with multiple reduced thiols as a kidney-targeting reactive oxygen species scavenger to prevent renal ischemia/reperfusion injury *Pharmaceutics* 10 251 https://doi.org/10.3390/pharmaceutics10040251.

137 Dolman M E, van Dorenmalen K M A, Pieters E H E, Sparidans R W, Lacombe M, Szokol B, Őrfi L, Kéri G, Bovenschen N, Storm G, Hennink W E and Kok R J 2012 Dendrimer-based macromolecular conjugate for the kidney-directed delivery of a multitargeted sunitinib analogue *Macromol. Biosci.* 12 93–103 https://doi.org/10.1002/mabi.201100277.

138 Dolman M M, Harmsen S, Pieters E H, Sparidans R W, Lacombe M, Szokol B, Őrfi L, Kéri G, Storm G, Hennink W E and Kok R J 2012 Targeting of a platinum-bound sunitinib analog to renal proximal tubular cells *Int. J. Nanomedicine* 7 417–33 https://doi.org/10.2147/IJN.S26485.

139 Wu Q, Wang J, Wang Y, Xiang L, Tan Y, Feng J, Zhang Z and Zhang L 2022 Targeted delivery of celastrol to glomerular endothelium and podocytes for chronic kidney disease treatment *Nano Res.* 15 3556–68 https://doi.org/10.1007/s12274-021-3894-x.

140 Qin S, Wu B, Gong T, Zhang Z-R and Fu Y 2022 Targeted delivery via albumin corona nanocomplex to renal tubules to alleviate acute kidney injury *J. Controlled Release* 349 401–12 https://doi.org/10.1016/j.jconrel.2022.07.013.

141 Pormohammad A, Monych N K, Ghosh S, Turner D L and Turner R J 2021 Nanomaterials in Wound Healing and Infection Control *Antibiotics* 10 473 https://doi.org/10.3390/antibiotics10050473.

142 Wang L, Hu C and Shao L 2017 The antimicrobial activity of nanoparticles: present situation and prospects for the future *Int. J. Nanomedicine* 12 1227–49 https://doi.org/10.2147/IJN.S121956.

143 Khezerlou A, Alizadeh-Sani M, Azizi-Lalabadi M and Ehsani A 2018 Nanoparticles and their antimicrobial properties against pathogens including bacteria, fungi, parasites and viruses *Microb. Pathog.* 123 505–26 https://doi.org/10.1016/j.micpath.2018.08.008.

144 Boateng J and Catanzano O 2015 Advanced therapeutic dressings for effective wound healing – a review *J. Pharm. Sci.* 104 3653–80 https://doi.org/10.1002/jps.24610.

145 Dias A M A, Braga M E M, Seabra I J, Ferreira P, Gil M H and de Sousa H C 2011 Development of natural-based wound dressings impregnated with bioactive compounds and using supercritical carbon dioxide *Int. J. Pharm.* 408 9–19 https://doi.org/10.1016/j.ijpharm.2011.01.063.

146 Azuma K, Izumi R, Osaki T, Ifuku S, Morimoto M, Saimoto H, Minami S and Okamoto Y 2015 Chitin, chitosan, and its derivatives for wound healing: old and new materials *J. Funct. Biomater.* 6 104–42 https://doi.org/10.3390/jfb6010104.

147 Das S and Baker A B 2016 Biomaterials and nanotherapeutics for enhancing skin wound healing *Front. Bioeng. Biotechnol.* 4 82 https://www.frontiersin.org/articles/10.3389/fbioe.2016.00082 (accessed September 29, 2023).

148 Baltzis D, Eleftheriadou I and Veves A 2014 Pathogenesis and treatment of impaired wound healing in diabetes mellitus: new insights *Adv. Ther.* 31 817–36 https://doi.org/10.1007/s12325-014-0140-x.

149 Blecher K, Martinez L R, Tuckman-Vernon C, Nacharaju P, Schairer D, Chouake J, Friedman J M, Alfieri A, Guha C, Nosanchuk J D and Friedman A J 2012 Nitric oxide-releasing nanoparticles accelerate wound healing in NOD-SCID mice *Nanomedicine Nanotechnol. Biol. Med.* 8 1364–71 https://doi.org/10.1016/j.nano.2012.02.014.

150 Han G, Nguyen L N, Macherla C, Chi Y, Friedman J M, Nosanchuk J D and Martinez L R 2012 Nitric oxide–releasing nanoparticles accelerate wound healing by promoting fibroblast migration and collagen deposition *Am. J. Pathol.* 180 1465–73 https://doi.org/10.1016/j.ajpath.2011.12.013.

151 Ma X, Cheng Y, Jian H, Feng Y, Chang Y, Zheng R, Wu X, Wang L, Li X and Zhang H 2019 Hollow, rough, and nitric oxide-releasing cerium oxide nanoparticles for promoting multiple stages of wound healing *Adv. Healthc. Mater.* 8 1900256 https://doi.org/10.1002/adhm.201900256.

152 Cao J, Hlaing S P, Lee J, Kim J, Lee E H, Kang S H, Hong S W, Yoon I-S, Yun H, Jung Y and Yoo J-W 2022 Bacteria-adhesive nitric oxide-releasing graphene oxide nanoparticles for mrpa-infected wound healing therapy *ACS Appl. Mater. Interfaces* 14 50507–19 https://doi.org/10.1021/acsami.2c13317.

153 Nurhasni H, Cao J, Choi M, Kim I, Lee B L, Jung Y and Yoo J-W 2015 Nitric oxide-releasing poly(lactic-co-glycolic acid)-polyethylenimine nanoparticles for prolonged nitric oxide release, antibacterial efficacy, and in vivo wound healing activity *Int. J. Nanomedicine* 10 3065–80 https://doi.org/10.2147/IJN.S82199.

154 Mihu M R, Sandkovsky U, Han G, Friedman J M, Nosanchuk J D and Martinez L R 2010 The use of nitric oxide releasing nanoparticles as a treatment against Acinetobacter baumannii in wound infections *Virulence* 1 62–7 https://doi.org/10.4161/viru.1.2.10038.

155 Liang Z, Liu W, Wang Z, Zheng P, Liu W, Zhao J, Zhong Y, Zhang Y, Lin J, Xue W and Yu S 2022 Near-infrared laser-controlled nitric oxide-releasing gold nanostar/hollow polydopamine Janus nanoparticles for synergistic elimination of methicillin-resistant Staphylococcus aureus and wound healing *Acta Biomater.* 143 428–44 https://doi.org/10.1016/j.actbio.2022.02.029.

156 Chereddy K K, Coco R, Memvanga P B, Ucakar B, des Rieux A, Vandermeulen G and Préat V 2013 Combined effect of PLGA and curcumin on wound healing activity *J. Controlled Release* 171 208–15 https://doi.org/10.1016/j.jconrel.2013.07.015.

157 Kant V, Gopal A, Pathak N N, Kumar P, Tandan S K and Kumar D 2014 Antioxidant and anti-inflammatory potential of curcumin accelerated the cutaneous wound healing in streptozotocin-induced diabetic rats *Int. Immunopharmacol.* 20 322–30 https://doi.org/10.1016/j.intimp.2014.03.009.

158 Laghezza Masci V, Taddei A-R, Courant T, Tezgel O, Navarro F, Giorgi F, Mariolle D, Fausto A-M and Texier I 2019 Characterization of collagen/lipid nanoparticle–curcumin cryostructurates for wound healing applications *Macromol. Biosci.* 19 1800446 https://doi.org/10.1002/mabi.201800446.

159 Zhang Z and Zhang X 2021 Curcumin loading on alginate nano-micelle for anti-infection and colonic wound healing *J. Biomed. Nanotechnol.* 17 1160–9 https://doi.org/10.1166/jbn.2021.3089.

160 Sandhu S K, Kumar S, Raut J, Singh M, Kaur S, Sharma G, Roldan T L, Trehan S, Holloway J, Wahler G, Laskin J D, Sinko P J, Berthiaume F, Michniak-Kohn B, Rishi P, Ganesh N and Kaur I P 2021 Systematic development and characterization of novel, high drug-loaded, photostable, curcumin solid lipid nanoparticle hydrogel for wound healing *Antioxidants* 10 725 https://doi.org/10.3390/antiox10050725.

161 Sonamuthu J, Cai Y, Liu H, Kasim M S M, Vasanthakumar V R, Pandi B, Wang H and Yao J 2020 MMP-9 responsive dipeptide-tempted natural protein hydrogel-based wound dressings for accelerated healing action of infected diabetic wound *Int. J. Biol. Macromol.* 153 1058–69 https://doi.org/10.1016/j.ijbiomac.2019.10.236.

162 Li F, Shi Y, Liang J and Zhao L 2019 Curcumin-loaded chitosan nanoparticles promote diabetic wound healing via attenuating inflammation in a diabetic rat model *J. Biomater. Appl.* 34(4) 476–86.

163 Guo H, Tan S, Gao J and Wang L 2020 Sequential release of drugs form a dual-delivery system based on pH-responsive nanofibrous mats towards wound care *J. Mater. Chem. B* 8 1759–70 https://doi.org/10.1039/C9TB02522G.

164 Mahmud M M, Zaman S, Perveen A, Jahan R A, Islam M F and Arafat M T 2020 Controlled release of curcumin from electrospun fiber mats with antibacterial activity *J. Drug Deliv. Sci. Technol.* 55 101386 https://doi.org/10.1016/j.jddst.2019.101386.

165 Mohammadi Z, Sharif Zak M, Majdi H, Mostafavi E, Barati M, Lotfimehr H, Ghaseminasab K, Pazoki-Toroudi H, Webster T J and Akbarzadeh A 2019 The effect of chrysin–curcumin-loaded nanofibres on the wound-healing process in male rats *Artif. Cells Nanomedicine Biotechnol.* 47 1642–52 https://doi.org/10.1080/21691401.2019.1594855.

166 Liu C, Zhu Y, Lun X, Sheng H and Yan A 2022 Effects of wound dressing based on the combination of silver at curcumin nanoparticles and electrospun chitosan nanofibers on wound healing *Bioengineered* 13 4328–39 https://doi.org/10.1080/21655979.2022.2031415.

167 Hartshorn C M, Bradbury M S, Lanza G M, Nel A E, Rao J, Wang A Z, Wiesner U B, Yang L and Grodzinski P 2018 Nanotechnology strategies to advance outcomes in clinical cancer care *ACS Nano* 12 24–43 https://doi.org/10.1021/acsnano.7b05108.

168 Peer D, Karp J M, Hong S, Farokhzad O C, Margalit R and Langer R 2020 Nanocarriers as an emerging platform for cancer therapy *Nano-Enabled Medical Applications* ed L P Balogh (New York: Jenny Stanford Publishing) pp 61–91.

169 Anselmo A C and Mitragotri S 2016 Nanoparticles in the clinic *Bioeng. Transl. Med.* 1 10–29 https://doi.org/10.1002/btm2.10003.

170 van der Meel R, Lammers T and Hennink W E 2017 Cancer nanomedicines: oversold or underappreciated? *Expert Opin. Drug Deliv.* 14 1–5 https://doi.org/10.1080/17425247.2017.1262346.

171 Grodzinski P, Kircher M, Goldberg M and Gabizon A 2019 Integrating nanotechnology into cancer care *ACS Nano* 13 7370–6 https://doi.org/10.1021/acsnano.9b04266.

172 Farkona S, Diamandis E P and Blasutig I M 2016 Cancer immunotherapy: the beginning of the end of cancer? *BMC Med.* 14 73 https://doi.org/10.1186/s12916-016-0623-5.

173 Topalian S L, Weiner G J and Pardoll D M 2011 Cancer immunotherapy comes of age *J. Clin. Oncol.* 29 4828–36 https://doi.org/10.1200/JCO.2011.38.0899.

174 Sharma P and Allison J P 2015 Immune checkpoint targeting in cancer therapy: toward combination strategies with curative potential *Cell* 161 205–14 https://doi.org/10.1016/j.cell.2015.03.030.

175 Khalil D N, Smith E L, Brentjens R J and Wolchok J D 2016 The future of cancer treatment: immunomodulation, CARs and combination immunotherapy *Nat. Rev. Clin. Oncol.* 13 273–90 https://doi.org/10.1038/nrclinonc.2016.25.

176 Hahn A W, Gill D M, Pal S K and Agarwal N 2017 The future of immune checkpoint cancer therapy after PD-1 and CTLA-4 *Immunotherapy* 9 681–92 https://doi.org/10.2217/imt-2017-0024.

177 Sharma P and Allison J P 2015 The future of immune checkpoint therapy *Science* 348 56–61 https://doi.org/10.1126/science.aaa8172.

178 Chan J D, von Scheidt B, Zeng B, Oliver A J, Davey A S, Ali A I, Thomas R, Trapani J A, Darcy P K, Kershaw M H, Dolcetti R and Slaney C Y 2020 Enhancing chimeric antigen receptor T-cell immunotherapy against cancer using a nanoemulsion-based vaccine targeting cross-presenting dendritic cells *Clin. Transl. Immunol.* 9 e1157 https://doi.org/10.1002/cti2.1157.

179 Sadeghzadeh M, Bornehdeli S, Mohahammadrezakhani H, Abolghasemi M, Poursaei E, Asadi M, Zafari V, Aghebati-Maleki L and Shanehbandi D 2020 Dendritic cell therapy in cancer treatment; the state-of-the-art *Life Sci.* 254 117580 https://doi.org/10.1016/j.lfs.2020.117580.

180 Shao K, Singha S, Clemente-Casares X, Tsai S, Yang Y and Santamaria P 2015 Nanoparticle-based immunotherapy for cancer *ACS Nano* 9 16–30 https://doi.org/10.1021/nn5062029.

181 Stephan M T, Moon J J, Um S H, Bershteyn A and Irvine D J 2010 Therapeutic cell engineering with surface-conjugated synthetic nanoparticles *Nat. Med.* 16 1035–41 https://doi.org/10.1038/nm.2198.

182 Zhang F, Stephan S B, Ene C I, Smith T T, Holland E C and Stephan M T 2018 Nanoparticles that reshape the tumor milieu create a therapeutic window for effective T-cell therapy in solid malignancies *Cancer Res.* 78 3718–30 https://doi.org/10.1158/0008-5472.CAN-18-0306.

183 Abdalla A M E, Xiao L, Miao Y, Huang L, Fadlallah G M, Gauthier M, Ouyang C and Yang G 2020 Nanotechnology promotes genetic and functional modifications of therapeutic T cells against cancer *Adv. Sci.* 7 1903164 https://doi.org/10.1002/advs.201903164.

184 Zhang Y, Chen Y, Li J, Zhu X, Liu Y, Wang X, Wang H, Yao Y, Gao Y and Chen Z 2021 Development of toll-like receptor agonist-loaded nanoparticles as precision immunotherapy for reprogramming tumor-associated macrophages *ACS Appl. Mater. Interfaces* 13 24442–52 https://doi.org/10.1021/acsami.1c01453.

185 Peng X, Wang J, Zhou F, Liu Q and Zhang Z 2021 Nanoparticle-based approaches to target the lymphatic system for antitumor treatment *Cell. Mol. Life Sci.* 78 5139–61 https://doi.org/10.1007/s00018-021-03842-6.

186 Milling L, Zhang Y and Irvine D J 2017 Delivering safer immunotherapies for cancer *Adv. Drug Deliv. Rev.* 114 79–101 https://doi.org/10.1016/j.addr.2017.05.011.

187 Travis L B 2002 Therapy-associated solid tumors *Acta Oncol.* 41 323–33 https://doi.org/10.1080/028418602760169361.

188 Wang X, Meng F, Yen Y-T, Li R and Liu B 2021 Nanotechnology-based CAR-T strategies for improving efficacy and safety of tumor immunotherapy *Adv. Funct. Mater.* 31 2004713 https://doi.org/10.1002/adfm.202004713.

189 Manguso R T, Pope H W, Zimmer M D, Brown F D, Yates K B, Miller B C, Collins N B, Bi K, LaFleur M W, Juneja V R, Weiss S A, Lo J, Fisher D E, Miao D, Van Allen E, Root D E, Sharpe A H, Doench J G and Haining W N 2017 In vivo CRISPR screening identifies Ptpn2 as a cancer immunotherapy target *Nature* 547 413–8 https://doi.org/10.1038/nature23270.

190 Rosenberg S A, Restifo N P, Yang J C, Morgan R A and Dudley M E 2008 Adoptive cell transfer: a clinical path to effective cancer immunotherapy *Nat. Rev. Cancer* 8 299–308 https://doi.org/10.1038/nrc2355.

191 Asghari F, Khademi R, Ranjbar F E, Malekshahi Z V and Majidi R F 2019 Application of nanotechnology in targeting of cancer stem cells: a review *Int. J. Stem Cells* 12 227–39 10.15283/ijsc19006.

192 Yue M, Guo T, Nie D-Y, Zhu Y-X and Lin M 2023 Advances of nanotechnology applied to cancer stem cells *World J. Stem Cells* 15 514–29 https://doi.org/10.4252/wjsc.v15.i6.514.

193 Nishikawa H and Sakaguchi S 2014 Regulatory T cells in cancer immunotherapy *Curr. Opin. Immunol.* 27 1–7 https://doi.org/10.1016/j.coi.2013.12.005.

194 Sharma A and Rudra D 2018 Emerging functions of regulatory T cells in tissue homeostasis *Front. Immunol.* 9 883 https://www.frontiersin.org/articles/10.3389/fimmu.2018.00883 (accessed October 8, 2023).

195 Shevach E M 2018 Foxp3+ T regulatory cells: still many unanswered questions – a perspective after 20 years of study *Front. Immunol.* 9 1048 https://www.frontiersin.org/articles/10.3389/fimmu.2018.01048 (accessed October 8, 2023).

196 Mantovani A, Marchesi F, Malesci A, Laghi L and Allavena P 2017 Tumour-associated macrophages as treatment targets in oncology *Nat. Rev. Clin. Oncol.* 14 399–416 https://doi.org/10.1038/nrclinonc.2016.217.

197 Chen D S and Mellman I 2017 Elements of cancer immunity and the cancer–immune set point *Nature* 541 321–30 https://doi.org/10.1038/nature21349.

198 Miller M A, Chandra R, Cuccarese M F, Pfirschke C, Engblom C, Stapleton S, Adhikary U, Kohler R H, Mohan J F, Pittet M J and Weissleder R 2017 Radiation therapy primes tumors for nanotherapeutic delivery via macrophage-mediated vascular bursts *Sci. Transl. Med.* 9 eaal0225 https://doi.org/10.1126/scitranslmed.aal0225.

199 He C, Duan X, Guo N, Chan C, Poon C, Weichselbaum R R and Lin W 2016 Core-shell nanoscale coordination polymers combine chemotherapy and photodynamic therapy to potentiate checkpoint blockade cancer immunotherapy *Nat. Commun.* 7 12499 https://doi.org/10.1038/ncomms12499.

200 Le Q-V, Choi J and Oh Y-K 2018 Nano delivery systems and cancer immunotherapy *J. Pharm. Investig.* 48 527–39 https://doi.org/10.1007/s40005-018-0399-z.

201 van der Meel R, Sulheim E, Shi Y, Kiessling F, Mulder W J M and Lammers T 2019 Smart cancer nanomedicine *Nat. Nanotechnol.* 14 1007–17 https://doi.org/10.1038/s41565-019-0567-y.

202 Min Y, Roche K C, Tian S, Eblan M J, McKinnon K P, Caster J M, Chai S, Herring L E, Zhang L, Zhang T, DeSimone J M, Tepper J E, Vincent B G, Serody J S and Wang A Z 2017 Antigen-capturing nanoparticles improve the abscopal effect and cancer immunotherapy *Nat. Nanotechnol.* 12 877–82 https://doi.org/10.1038/nnano.2017.113.

203 Jemal A, Siegel R, Xu J and Ward E 2010 Cancer statistics, 2010 *CA: Cancer J. Clin.* 60 277–300 https://doi.org/10.3322/caac.20073.

204 Bray F, Ferlay J, Soerjomataram I, Siegel R L, Torre L A and Jemal A Global cancer statistics 2018: GLOBOCAN estimates of incidence and mortality worldwide for 36 cancers in 185 countries *CA: Cancer J. Clin.* 68(2018) 394–424 https://doi.org/10.3322/caac.21492.

205 Mao X, Mei R, Yu S, Shou L, Zhang W, Li K, Qiu Z, Xie T and Sui X 2022 Emerging technologies for the detection of cancer micrometastasis *Technol. Cancer Res. Treat.* 21 15330338221100355 https://doi.org/10.1177/15330338221100355.

206 Vahrmeijer A L, Hutteman M, van der Vorst J R, van de Velde C J H and Frangioni J V 2013 Image-guided cancer surgery using near-infrared fluorescence *Nat. Rev. Clin. Oncol.* 10 507–18 https://doi.org/10.1038/nrclinonc.2013.123.

207 Ehman E C, Johnson G B, Villanueva-Meyer J E, Cha S, Leynes A P, Larson P E Z and Hope T A 2017 PET/MRI: where might it replace PET/CT? *J. Magn. Reson. Imaging* 46 1247–62 https://doi.org/10.1002/jmri.25711.

208 Ben-Haim S and Ell P 2009 18F-FDG PET and PET/CT in the evaluation of cancer treatment response *J. Nucl. Med.* 50 88–99 https://doi.org/10.2967/jnumed.108.054205.

209 van Dam G M, Themelis G, Crane L M A, Harlaar N J, Pleijhuis R G, Kelder W, Sarantopoulos A, de Jong J S, Arts H J G, van der Zee A G J, Bart J, Low P S and Ntziachristos V 2011 Intraoperative tumor-specific fluorescence imaging in ovarian cancer by folate receptor-α targeting: first in-human results *Nat. Med.* 17 1315–9 https://doi.org/10.1038/nm.2472.

210 Ehlerding E B, Grodzinski P, Cai W and Liu C H 2018 Big potential from small agents: nanoparticles for imaging-based companion diagnostics *ACS Nano* 12 2106–21 https://doi.org/10.1021/acsnano.7b07252.

211 Harisinghani M G, Barentsz J, Hahn P F, Deserno W M, Tabatabaei S, van de Kaa C H, de la Rosette J and Weissleder R 2003 Noninvasive detection of clinically occult lymph-node metastases in prostate cancer *N. Engl. J. Med.* 348 2491–9 https://doi.org/10.1056/NEJMoa022749.

212 Jokerst J V and Gambhir S S 2011 Molecular imaging with theranostic nanoparticles *Acc. Chem. Res.* 44 1050–60 https://doi.org/10.1021/ar200106e.

213 Wang Y, Zhou K, Huang G, Hensley C, Huang X, Ma X, Zhao T, Sumer B D, DeBerardinis R J and Gao J 2014 A nanoparticle-based strategy for the imaging of a broad range of tumours by nonlinear amplification of microenvironment signals *Nat. Mater.* 13 204–12 https://doi.org/10.1038/nmat3819.

214 Kircher M F, Mahmood U, King R S, Weissleder R and Josephson L 2003 A multimodal nanoparticle for preoperative magnetic resonance imaging and intraoperative optical brain tumor delineation *Cancer Res.* 63 8122–5.

215 Bradbury M S, Phillips E, Montero P H, Cheal S M, Stambuk H, Durack J C, Sofocleous C T, Meester R J C, Wiesner U and Patel S 2013 Clinically-translated silica nanoparticles as dual-modality cancer-targeted probes for image-guided surgery and interventions *Integr. Biol.* 5 74–86 https://doi.org/10.1039/c2ib20174g.

216 Bradbury M S, Pauliah M, Zanzonico P, Wiesner U and Patel S 2016 Intraoperative mapping of sentinel lymph node metastases using a clinically translated ultrasmall silica nanoparticle, WIREs Nanomedicine *Nanobiotechnology* 8 535–53 https://doi.org/10.1002/wnan.1380.

217 Hill T K, Kelkar S S, Wojtynek N E, Souchek J J, Payne W M, Stumpf K, Marini F C and Mohs A M 2016 Near infrared fluorescent nanoparticles derived from hyaluronic acid improve tumor contrast for image-guided surgery *Theranostics* 6 2314 https://doi.org/10.7150/thno.16514.

218 Souchek J J, Wojtynek N E, Payne W M, Holmes M B, Dutta S, Qi B, Datta K, LaGrange C A and Mohs A M 2018 Hyaluronic acid formulation of near infrared fluorophores optimizes surgical imaging in a prostate tumor xenograft *Acta Biomater.* 75 323–33 https://doi.org/10.1016/j.actbio.2018.06.016.

219 Harmsen S, Bedics M A, Wall M A, Huang R, Detty M R and Kircher M F 2015 Rational design of a chalcogenopyrylium-based surface-enhanced resonance Raman scattering nanoprobe with attomolar sensitivity *Nat. Commun.* 6 6570 https://doi.org/10.1038/ncomms7570.

220 Harmsen S, Huang R, Wall M A, Karabeber H, Samii J M, Spaliviero M, White J R, Monette S, O'Connor R, Pitter K L, Sastra S A, Saborowski M, Holland E C, Singer S, Olive K P, Lowe S W, Blasberg R G and Kircher M F 2015 Surface-enhanced resonance Raman scattering nanostars for high-precision cancer imaging *Sci. Transl. Med.* 7 271ra7 https://doi.org/10.1126/scitranslmed.3010633.

221 Kircher M F, de la Zerda A, Jokerst J V, Zavaleta C L, Kempen P J, Mittra E, Pitter K, Huang R, Campos C, Habte F, Sinclair R, Brennan C W, Mellinghoff I K, Holland E C and Gambhir S S 2012 A brain tumor molecular imaging strategy using a new triple-modality MRI-photoacoustic-Raman nanoparticle *Nat. Med.* 18 829–34 https://doi.org/10.1038/nm.2721.

222 Wall M A, Shaffer T M, Harmsen S, Tschaharganeh D-F, Huang C-H, Lowe S W, Drain C M and Kircher M F 2017 Chelator-free radiolabeling of SERRS nanoparticles for whole-body PET and intraoperative Raman imaging *Theranostics* 7 3068–77 https://doi.org/10.7150/thno.18019.

223 Chapman S, Dobrovolskaia M, Farahani K, Goodwin A, Joshi A, Lee H, Meade T, Pomper M, Ptak K, Rao J, Singh R, Sridhar S, Stern S, Wang A, Weaver J B, Woloschak G and Yang L 2013 Nanoparticles for cancer imaging: the good, the bad, and the promise *Nano Today* 8 454–60 https://doi.org/10.1016/j.nantod.2013.06.001.

224 Phillips E, Penate-Medina O, Zanzonico P B, Carvajal R D, Mohan P, Ye Y, Humm J, Gönen M, Kalaigian H, Schöder H, Strauss H W, Larson S M, Wiesner U and Bradbury M S 2014 Clinical translation of an ultrasmall inorganic optical-PET imaging nanoparticle probe *Sci. Transl. Med.* 6 260ra149 https://doi.org/10.1126/scitranslmed.3009524.

225 de Boer E, Harlaar N J, Taruttis A, Nagengast W B, Rosenthal E L, Ntziachristos V and van Dam G M 2015 Optical innovations in surgery *Br. J. Surg.* 102 e56–72 https://doi.org/10.1002/bjs.9713.

226 Lal S, Clare S E and Halas N J 2008 Nanoshell-enabled photothermal cancer therapy: impending clinical impact *Acc. Chem. Res.* 41 1842–51 https://doi.org/10.1021/ar800150g.

227 Stummer W, Pichlmeier U, Meinel T, Wiestler O D, Zanella F and Reulen H-J 2006 Fluorescence-guided surgery with 5-aminolevulinic acid for resection of malignant glioma: a randomised controlled multicentre phase III trial *Lancet Oncol.* 7 392–401 https://doi.org/10.1016/S1470-2045(06)70665-9.

228 Belykh E, Martirosyan N L, Yagmurlu K, Miller E J, Eschbacher J M, Izadyyazdanabadi M, Bardonova L A, Byvaltsev V A, Nakaji P and Preul M C 2016 Intraoperative fluorescence imaging for personalized brain tumor resection: current state and future directions *Front. Surg.* 3 55 https://www.frontiersin.org/articles/10.3389/fsurg.2016.00055 (accessed October 9, 2023).

229 Nel A E, Mädler L, Velegol D, Xia T, Hoek E M V, Somasundaran P, Klaessig F, Castranova V and Thompson M 2009 Understanding biophysicochemical interactions at the nano–bio interface *Nat. Mater.* 8 543–57 https://doi.org/10.1038/nmat2442.

230 Retif P, Pinel S, Toussaint M, Frochot C, Chouikrat R, Bastogne T and Barberi-Heyob M 2015 Nanoparticles for radiation therapy enhancement: the key parameters *Theranostics* 5 1030–44 https://doi.org/10.7150/thno.11642.

231 Gabizon A, Bradbury M, Prabhakar U, Zamboni W, Libutti S and Grodzinski P 2014 Cancer nanomedicines: closing the translational gap *Lancet Lond. Engl.* 384 2175–6 https://doi.org/10.1016/S0140-6736(14)61457-4.

232 Jhaveri A, Deshpande P and Torchilin V 2014 Stimuli-sensitive nanopreparations for combination cancer therapy *J. Controlled Release* 190 352–70 https://doi.org/10.1016/j.jconrel.2014.05.002.

233 Kemp J A, Shim M S, Heo C Y and Kwon Y J 2016 "Combo" nanomedicine: co-delivery of multi-modal therapeutics for efficient, targeted, and safe cancer therapy *Adv. Drug Deliv. Rev.* 98 3–18 https://doi.org/10.1016/j.addr.2015.10.019.

234 Yoo B, Ma K, Zhang L, Burns A, Sequeira S, Mellinghoff I, Brennan C, Wiesner U and Bradbury M S 2015 Ultrasmall dual-modality silica nanoparticle drug conjugates: design, synthesis, and characterization *Bioorg. Med. Chem.* 23 7119–30 https://doi.org/10.1016/j.bmc.2015.09.050.

235 Park S, Aalipour A, Vermesh O, Yu J H and Gambhir S S 2017 Towards clinically translatable in vivo nanodiagnostics *Nat. Rev. Mater.* 2 1–20 https://doi.org/10.1038/natrevmats.2017.14.

236 Chen Z, Chen M, Zhou K and Rao J 2020 Pre-targeted imaging of protease activity through in situ assembly of nanoparticles *Angew. Chem.* 132 7938–44 https://doi.org/10.1002/ange.201916352.

237 Ye D, Shuhendler A J, Cui L, Tong L, Tee S S, Tikhomirov G, Felsher D W and Rao J 2014 Bioorthogonal cyclization-mediated in situ self-assembly of small-molecule probes for imaging caspase activity in vivo *Nat. Chem.* 6 519–26 https://doi.org/10.1038/nchem.1920.

238 Liang G, Ren H and Rao J 2010 A biocompatible condensation reaction for controlled assembly of nanostructures in living cells *Nat. Chem.* 2 54–60 https://doi.org/10.1038/nchem.480.

239 Palner M, Shen B, Jeon J, Lin J, Chin F T and Rao J 2015 Preclinical kinetic analysis of the caspase-3/7 PET tracer 18F-C-SNAT: quantifying the changes in blood flow and tumor retention after chemotherapy *J. Nucl. Med.* 56 1415–21 https://doi.org/10.2967/jnumed.115.155259.

240 Witney T H, Hoehne A, Reeves R E, Ilovich O, Namavari M, Shen B, Chin F T, Rao J and Gambhir S S 2015 A systematic comparison of 18F-C-SNAT to established radiotracer imaging agents for the detection of tumor response to treatment *Clin. Cancer Res.* 21 3896–905 https://doi.org/10.1158/1078-0432.CCR-14-3176.

241 Shuhendler A J, Ye D, Brewer K D, Bazalova-Carter M, Lee K-H, Kempen P, Dane Wittrup K, Graves E E, Rutt B and Rao J 2015 Molecular magnetic resonance imaging of tumor response to therapy *Sci. Rep.* 5 14759 https://doi.org/10.1038/srep14759.

242 Chien M-P, Carlini A S, Hu D, Barback C V, Rush A M, Hall D J, Orr G and Gianneschi N C 2013 Enzyme-directed assembly of nanoparticles in tumors monitored by in vivo whole animal imaging and ex vivo super-resolution fluorescence imaging *J. Am. Chem. Soc.* 135 18710–3 https://doi.org/10.1021/ja408182p.

Index

a

Metallic, Magnetic and Carbon-Based Nanomaterials: Synthesis and Biomedical Applications, First Edition.
Ajit Khosla, Irshad A. Wani, and Mohammad N. Lone.

b

c

o

p

q

r

S

W

Z

The Electrochemical Society Series

The Electron Microprobe
Edited by T. D. McKinley, K. F. J. Heinrich, and D. B. Wittry

Chemical Physics of Ionic Solutions
Edited by B. E. Conway and R. G. Barradas

High-Temperature Materials and Technology
Edited by Ivor E. Campbell and Edwin M. Sherwood

Alkaline Storage Batteries
S. Uno Falk and Alvin J. Salkind

The Primary Battery (in Two Volumes)
Volume I
Edited by George W. Heise and N. Corey Cahoon

Volume II
Edited by N. Corey Cahoon and George W. Heise

Zinc-Silver Oxide Batteries
Edited by Arthur Fleischer and J. J. Lander

Lead-Acid Batteries
Hans Bode
Translated by R. J. Brodd and Karl V. Kordesch

Thin Films-Interdiffusion and Reactions
Edited by J. M. Poate, M. N. Tu, and J. W. Mayer

Lithium Battery Technology
Edited by H. V. Venkatasetty

Quality and Reliability Methods for Primary Batteries
P. Bro and S. C. Levy

Techniques for Characterization of Electrodes and Electrochemical Processes
Edited by Ravi Varma and J. R. Selman

Electrochemical Oxygen Technology
Kim Kinoshita

Synthetic Diamond: Emerging CVD Science and Technology
Edited by Karl E. Spear and John P. Dismukes

Corrosion of Stainless Steels, Second Edition
A. John Sedriks

Semiconductor Wafer Bonding: Science and Technology
Q.-Y. Tong and U. Göscle

Fundamentals of Electrochemistry, Second Edition
V. S. Bagotsky

Fundamentals of Electrochemical Deposition, Second Edition
Milan Paunovic and Mordechay Schlesinger

Uhlig's Corrosion Handbook, Third Edition
Edited by R. Winston Revie

Fuel Cells: Problems and Solutions
Vladimir S. Bagotsky

Lithium Batteries: Advanced Technologies and Applications
Edited by B. Scrosati, K. M. Abraham, W. A. van Schalkwijk, and J. Hassoun

Modern Electroplating, Fifth Edition
Edited by Mordechay Schlesinger and Milan Paunovic

Electrochemical Power Sources: Batteries, Fuel Cells, and Supercapacitors
By V. S. Bagotsky, A. M. Skundin, and Y. M. Volfkovic

Molecular Modeling of Corrosion Processes: Scientific Development and Engineering Applications
Edited by C. D. Taylor and P. Marcus

Atmospheric Corrosion, Second Edition
Christofer Leygraf, Inger Odnevall Wallinder, Johan Tidblad, and Thomas Graedel

Electrochemical Impedance Spectroscopy, Second Edition
Mark E. Orazem and Bernard Tribollet

Electrochemical Systems, Fourth Edition
John Newman and Nitash P. Balsara

LaQue's Handbook of Marine Corrosion
David A. Shifler *LaQue's Handbook of Marine Corrosion*, Second Edition. Edited by David A. Shifler.

Metallic, Magnetic and Carbon-Based Nanomaterials: Synthesis and Biomedical Applications
Ajit Khosla, Irshad A. Wani and Mohammad N. Lone

Printed and bound by CPI Group (UK) Ltd, Croydon, CR0 4YY
17/09/2026
14964258-0003